Art W.

artweonsult (

FLUID DYNAMICS
AND HEAT TRANSFER
OF TURBOMACHINERY

FLUID DYNAMICS AND HEAT TRANSFER OF TURBOMACHINERY

Budugur Lakshminarayana

Department of Aerospace Engineering
The Pennsylvania State University
University Park, Pennsylvania

A Wiley-Interscience Publication

John Wiley & Sons, Inc.

New York • Chichester • Brisbane • Toronto • Singapore

Library of Congress Cataloguing-in-Publication Data:
Lakshminarayana, B.
    Fluid dynamics and heat transfer of turbomachinery / Budugur
Lakshminarayana.
        p.   cm.
    Includes index.
    ISBN 0-471-85546-4 (cloth)
    1. Turbomachines—Fluid dynamics.   2. Heat—Transmission.
I. Title.
TJ267.L35   1996
621.406—dc20   94-41844

Printed in the United States of America

10 9 8 7 6 5 4 3 2 1

I dedicate this book to
my parents, **Seethamma** and **Sreenivasachar,**
wife, **Saroja,**
children, **Anita** and **Arvind**.

PREFACE

It has been nearly three decades since the last book appeared that provided a unified treatment of turbomachinery fluid dynamics. Since then, considerable advances have been made in analyses, computation, experimental measurements, development, and design. The need for a modern book for use at universities, as well as by professional engineers, has been widely recognized. However, few have made the attempt to embark on the arduous task of compiling modern approaches and advances in turbomachinery fluid dynamics and heat transfer into one book, including the great body of material available in the open turbomachinery literature. This book represents an attempt to achieve this goal.

The objective of this book is to cover the aerodynamics, hydrodynamics, and heat transfer of (a) turbomachinery used for propulsion and power generation in aircraft, spacecraft, and marine and land vehicles and (b) turbomachinery used for land-based power plants, processing equipment, and other industrial applications. Emphasis is placed on basic analysis, computation, experimental measurement, and data correlations that have lasting value. An attempt has also been made to include more recent developments in the numerical computation of the flow field in turbomachinery. The reader should have a foundation in the fundamentals of fluid mechanics and thermodynamics and should also know some basic concepts of applied mathematics and computation. The material presented is intended for undergraduate seniors and graduate students at universities, as well as for professional engineers in industry engaged in analysis, design, and development of turbomachinery. In order to keep the book to a reasonable length, no attempt has been made to address structural or other mechanical aspects of turbomachinery and unsteady flow.

Turbomachinery is widely used in aircraft, automotive, marine, space, and industrial applications. The field is growing fast, and accordingly, turbomachinery represents a major industry worldwide. Fluid dynamics and heat transfer of turbomachinery have reached a level of sophistication very difficult to follow by self-study. There is then a need for the required information to be provided in a coherent, consistent, and unified manner. An effort has been made here to provide a unified approach that would be useful to an audience in academia, industry, government, laboratories, and defense establishments.

Coverage begins with the fundamental concepts and classifications of turbomachinery. The reader is then taken through increasingly sophisticated material in succeeding chapters. Chapter 2 deals with fundamental principles, analysis, and performance. It also includes sections on transonic and supersonic turbomachinery which, the author believes, will achieve prominence during the 1990s. This is followed by two chapters on cascade inviscid flow and three-dimensional inviscid (and quasi-viscous) flows, respectively. Chapter 5 deals with computational methods for the turbomachinery flow field. This is followed by a chapter on two- and three-dimensional viscous flows. The nature of three-dimensional viscous flows and correlations for estimating losses are included in Chapter 6. A final chapter on turbine blade cooling and heat transfer, which plays a fundamental role in turbine flow and heat transfer analysis, is included. Basic computational techniques are described in Appendix B.

This book could be used in a one-semester senior or graduate level course on fluid dynamics of turbomachinery covering material in Chapters 1, 2, 3, and 4, and it could be used in a second-semester course covering the material on three-dimensional viscid flows, computation, and heat transfer (Chapters 5, 6, and 7). The material in this book has been used by the author in a senior undergraduate course and in two advanced graduate courses taught at The Pennsylvania State University.

The author has benefited greatly from the association with his former and present students, as well as from the research projects carried out under the sponsorship of the U.S. National Aeronautics and Space Administration, U.S. Navy, U.S. National Science Foundation, U.S. Army Research Office, Pratt & Whitney Aircraft Division, Allison Engine Co., General Motors, Rolls Royce Inc., Allied Signal Inc. (Garrett), and Sociétè Européene De Propulsion (France). The author has also benefited greatly from his association with John H. Horlock (*Axial Flow Compressors*, 1958; *Axial Flow Turbines*, 1965, Butterworth), whose contribution to turbomachinery is very well known and the late George F. Wislicenus (*Fluid Mechanics of Turbomachinery*, 1947, McGraw-Hill), who was one of the earliest authors to provide a unified approach in presenting the theory of turbomachinery. My understanding of turbomachinery has greatly improved through my association with these two eminent men.

I have sought help from my professional colleagues and former students to ensure accuracy and selection of materials: R. Kunz for Chapters 1 and 5; J. H. Horlock for Chapters 2 and 4; P. Gostelow and H. Starken for Chapter 3; F. Leboeuf and N. Suryavamshi for Chapter 4; J. Schwab for Chapter 5; H. Weingold, A. Sehra, and J. Prato for Chapter 6; C. Camci, A. Arts, W. Heiser, and J. Luo for Chapter 7, and G. Dulikravich for Appendix B. I would like to thank these individuals and the anonymous reviewers for invaluable comments and criticisms. I take responsibility for remaining deficiencies and I request that readers send me corrections and comments for subsequent editions.

My parents instilled in me many of the qualities needed for such an undertaking. I owe a debt of gratitude to my family, Saroja, Arvind, Anita, and Andrew Silva for untold numbers of hours stolen from them and for their indulgence, patience and love, and cheerful support, without which this book would not have been completed. I would also like to acknowledge the excellent typing of Billie Hackney and Kathy Barry and assistance from Pat Ellenberger. Their patience in transcribing my handwriting into an acceptable manuscript is greatly appreciated. Clifford Warner Jr., who has provided drafting assistance during most of my professional career, transcribed my sketches into excellent illustrations. The cheerful and helpful staff at John Wiley & Sons, made the final stages of the book a pleasant experience.

B. LAKSHMINARAYANA

University Park, Pennsylvania
August 1995

CONTENTS

3 Cascade Inviscid Flows   189

NOMENCLATURE, ACRONYMS, AND TERMINOLOGY

COMMON NOMENCLATURE

In a book of this magnitude, it is impossible to avoid using the same symbols to represent different quantities. The notations commonly used (and employed throughout this book) are defined here. Various additional symbols used infrequently are introduced throughout the book, both in the text and in the figures.

a	Speed of sound; acceleration; radius of leakage vortex
A	Area; aspect ratio (length of the blade/chord); discretization coefficient; instantaneous flow variable
A^*	Throat area ($M = 1$)
AS	Aspect ratio (length/chord ratio)
AVR	Axial velocity ratio (V_{z_2}/V_{z_1})
b	Effective blade spacing ($1 - t/S$)
B	Body force; number of blades; blockage; stream-tube thickness
BHP	Brake horsepower
c_p	Specific heat at constant pressure
c_v	Specific heat at constant volume
C	Chord length; blade chord
C_D	Drag coefficient ($D/\frac{1}{2}\rho W_m^2 C$)
C_f	Skin friction coefficient $2\tau_w/\rho V_e^2$
C_F	Thrust coefficient ($F/\rho N^2 D^5$)
C_H	Head coefficient ($g\,\Delta H/N^2 D^2$)
C_L	Lift coefficient ($L/\frac{1}{2}\rho_m W_m^2 C$) based on mean velocity (V_m or W_m)
C_p	Pressure drop or rise coefficient ($p_{local} - p_1)/(P_{o1} - p_1$); coefficient of pressure; power coefficient of propeller (Eq. 2.52); blade pressure coefficient; power coefficient $P_{shaft}/\rho N^3 D^5$
C_Q	Flow rate coefficient (Q/ND^3)

C_{pt}	$(P_o - p_1)/\frac{1}{2}\rho V_1^2$
C_{pR}	Pressure rise or drop or recovery $\pm(p_2 - p_1)/(P_{o1} - p_1)$, ($+$ for compressor, $-$ for turbine)
$C_s, C_1, C_2, C_3, C_{\epsilon 1}, C_{\epsilon 2}$	
$C_{\epsilon 3}, C_\epsilon, C_S, C_K, C_\mu$	Constants in Reynolds stress and $k-\epsilon$ equations
C_T	Torque coefficient $(T/\rho N^2 D^5)$
dA	Infinitesimal area
D	Drag force; diameter of coolant jet or hole or pipe; diffusion factor (Eq. 2.47); dissipation term; diameter of turbomachine
DR	Density ratio (ρ_f/ρ_e) of coolant
e	Energy; stored energy per unit mass
E	Dilution ratio ($\dot{m}_e/\dot{m}_f$; see Fig. 7.25)
Ec	Eckert number $V_\infty^2/c_p \Delta T_o$
E, F, G, T, P, Q	Flux vectors (Eqs. 1.18–1.23)
e_o	$e + V^2/2$
f	Maximum camber height; force
F	Force exerted by the blade (or vice versa); force; complex potential; thrust force
F_μ	Damping function for low-Reynolds-number flows
Fr	Froude number
g	Acceleration due to gravity; constant
g_{ik}	Metric tensor
G_i	Apparent stresses due to spatial variation of flow properties (Eq. 4.74)
Gr	Grashof number (Eq. 7.27)
h	Static enthalpy; width of duct; height of blade; heat transfer coefficient (Eq. 7.14)
h_f	Heat transfer coefficients with film cooling (Eq. 7.59)
h_g	Gas heat transfer coefficient
h_o	Stagnation enthalpy $(h + V^2/2)$
h_{oR}	Stagnation enthalpy of relative flow $(h + W^2/2)$
h_v	Vapor head; p_v/ρ
H	Shape factor (δ^*/θ); total head
HP	High pressure; horsepower
H_{sv}	Net positive suction head (difference between inlet and vapor pressure)
i	Incidence angle (difference between flow angle and camber angle at inlet)
i, j, k	Unit vectors in the x–y–z direction; nodal points for numerical analysis
I	Identity matrix; rothalpy $(h + (W^2/2) - (\Omega^2 r^2/2))$; momentum flux ratio ($\rho_f u_f^2/\rho_e u_e^2$)
J	Rotational Reynolds number (Eq. 7.30); advance ratio for propeller; Jacobian

k	Thermal conductivity; turbulent kinetic energy $\frac{1}{2}((u')^2 + (\overline{v'})^2 + (\overline{w'})^2)$
k_{ij}	Kernel function in panel method
K	Pressure gradient parameter (Eq. 7.55)/slip factor (Eq. 2.58)/fraction of circulation or lift retained at the tip (Eq. 4.140)
L	Lift force; characteristic length; enthalpy of evaporation at the saturation condition; length scale; lower diagonal matrix; turbulent length scale
LE	Leading edge
LKE	Kinetic energy in leakage flow/vortex
LP	Low pressure
m	Mass flow; meridional direction
$\dot{m}$	Mass flow rate
m^*	Coolant mass flow parameter $= \dot{m}_c c_{pc}/h_g \text{Sh}$
$\dot{m}_p$	Rate of mass flow of primary air (through one blade passage)
m_R	Blowing or mass flow parameter ($\rho_f u_f / \rho_e u_e$) (see Fig. 7.25)
M	Mach number; mass
M^*	Mach number based on the sonic speed (speed of sound at $M = 1$)
M_{cr}	Critical Mach number
MR	Coolant to primary flow mass ratio ($\dot{m}_f/\dot{m}_p$)
n	Extensive property per unit mass; index; iteration number
$\mathbf{n}$	Unit normal vector
$\overline{N}$	Specific speed $(N\sqrt{Q})/(g\,\Delta H)^{3/4}$
$\overline{N}_p$	Power specific speed $N\sqrt{\text{BHP}} / \sqrt{\rho}(g\,\Delta H)^{5/4}$
N	Extensive property; speed of rotation; distance normal to the blade normalized by chord length; number of blades
N_S	Specific speed
Nu	Nusselt's number (Eq. 7.15)
Nu_x	Nusselt's number based on local distance (Eq. 7.19)
Nu_m	Mean Nusselt's number (based on mean heat transfer coefficient; k, characteristic length C or D)
p	Static pressure
p_m, p_c	Global pressure correction and pressure correction in transverse plane
ps, ss	Pressure and suction surfaces
$\bar{p}, p'$	Mean pressure, fluctuating pressure
P	Stagnation pressure; production of turbulent kinetic energy (Eq. 1.89)
P_{ij}	Production term in Reynolds stress equation (Eq. 1.89)

P_o	Stagnation pressure
P_{oR}	Stagnation pressure of relative flow
PQA	Power coefficient
PR	Prandtl number ($\mu c_p / k$)
PR_t	Turbulent Prandtl number (μ_t / ϵ_t)
P_{shaft}, P_{shear}	Rate of work done by shaft power and shear forces respectively
q_w	Heat flux at the wall
q	Strength of source or sink sheet; heat transfer per unit area; local velocity in s direction; variable vector $(\rho, \rho u, \rho v, \rho w, e_o, \rho k, \rho \epsilon)^T$
Q	Propeller torque; primary transport variable ($\rho, \rho u, \rho v, \rho w)^T$; volume flow rate; source strength; total velocity; heat transfer rate
r	Radial distance from the axis; total pressure drop or rise; radius of annulus or pipe; recover factor (Eq. 7.33)
r_m	Meridional streamline radius
r, θ, z	Radial, tangential, and axial directions, respectively
R	Radial distance normalized by the tip radius; radius of curvature; degree of reaction; relaxation factor; universal gas constant; radius of streamline
R_c	Radius of curvature of the camber line
Ra	Rayleigh number (Eq. 7.28)
Re	Reynolds number $\rho_\infty V_\infty L / \mu_\infty$ (e.g., cascade $L = C$, $V_\infty = V_1$)
R_{ht}	Hub/tip radius ratio
Ri	Gradient Richardson number (Eq. 7.31)
Ro	Rotation number ($\Omega L / V$, L is characteristic length, V characteristic velocity) (e.g., see Section 5.5.2 and Eq. 7.29)
R_θ	Reynolds number based on momentum thickness and free-stream velocity
RRa	Rotational Rayleigh number (Eq. 7.50)
$R_{ic} = \epsilon_{ipj} \Omega^p / u_{ij}$	Generalized gradient Richardson number
R_T	Turbulent Reynolds number ($k^2 / \nu \epsilon$)
s	Entropy
s, n, b	Intrinsic coordinate system (s is streamwise, n is principal normal, b is bi-normal directions; Fig. 4.19)
s', n', b'	Intrinsic coordinate in a rotating system (streamwise, principal normal and binormal directions)
s, n, r	Streamwise, normal, and radial directions (orthogonal to each other); n is the distance normal to a surface
S	Blade pitch; distance along the blade surface measured from the leading edge normalized by chord; source term vector; vane perimeter; spacing of the

	coolant jets; specific speed $(N\sqrt{Q}/(\Delta H)^{3/4})$
S_{ij}	Strain rate tensor of mean flow $\frac{1}{2}(u_{i,j} + u_{j,i})$
S'_{ij}	Fluctuating strain rate tensor $\frac{1}{2}(u'_{i,j} + u'_{j,i})$
SKE	Kinetic energy in secondary flow (Eq. 6.22)
SS	Suction specific speed $N\sqrt{Q}/(H_{sv})^{3/4}$; suction surface
St	Stanton number (Eq. 7.18)
t	Blade thickness; time; blade spacing
t, s	Quasi-orthogonal coordinate system
T	Static temperature; propeller torque; time; temperature; surface forces; period for the unsteady wave/torque
T'	Static temperature fluctuation
TE	Trailing edge
T_o	Stagnation temperature
T_{oR}	Stagnation temperature of relative flow
T'_{oR}	Stagnation temperature fluctuation
T_r	Recovery temperature (Eq. 7.34)
T_s	Saturation temperature
Tu, T_u, T_i	Turbulent intensity $\sqrt{\overline{(v'_i)^2}}/V_i$
u_e	Local free-stream (edge) velocity
u'_i	Turbulent velocity fluctuations; fluctuating velocity
u_i	Mean velocity components; velocity components in tensor form
$\overline{u'_i u'_j}$	Reynolds stress tensor; turbulent stresses
u, v, w	Velocity components in Cartesian coordinate (x, y, z) and s, n, b system; perturbations in velocity along x, y, z
u^*	Friction velocity $(\sqrt{\tau_w/\rho})$
U	Blade speed; upper diagonal matrix
U_t	Blade tip speed
U, V	Components of cascade mean velocity/contravariant velocity (Eq. A.3)
$\mathbf{V}$	Absolute velocity vector (stationary frame)
$\mathbf{V}_m$	Mean velocity vector $((\mathbf{V}_1 + \mathbf{V}_2)/2)$; meridional velocity $(\sqrt{V_z^2 + V_r^2})$
V_o	Flight speed
V_r, V_θ, V_z	Absolute velocity components along r, θ, z directions
w'	Fluctuating relative velocity
W	Total relative flow velocity; weight flow; complex velocity
$\mathbf{W}$	Relative velocity vector (rotating frame)
$\mathbf{W}_m$	Mean velocity vector $(\mathbf{W}_1 + \mathbf{W}_2)/2$; meridional velocity $(\sqrt{W_z^2 + W_r^2})$

W_r, W_θ, W_z	Relative velocity along r, θ, z directions
W_C	Maximum defect in total velocity in the wake
W_S	Work done by surface forces; streamwise velocity in intrinsic s, n, r coordinate system
x, y, z	Cartesian coordinate system, x along axial direction, y and z are transverse coordinates, y is usually normal to the surface or pitchwise direction, z is usually along the spanwise direction
X	Axial distance from leading edge normalized by blade chord; distance along chord
y_c	Camber function
y_t	Thickness function
y^+	$(\rho y_w/\mu)\sqrt{\tau_w/\rho}$
Y	Tangential distance (blade-to-blade) normalized by S (in many instances)
z	Complex coordinate $(x + iy)$
Z	Axial distance normalized by blade chord ($Z = 0$ at the leading edge); distance normal to end wall

Greek Letters

α	Absolute flow angle measured from the axial direction (angle of incidence for an isolated airfoil); streamwise angle; injection angle (Fig. 7.23)
α'	Blade angle measured from axial direction
α_i	Angle of attack (angle between chordwise direction and inlet flow angle; Fig. 2.11)
β	Relative flow angle measured from the axial direction; volumetric coefficient of thermal expansion $(\partial\rho/\partial T)$; injection angle (Fig. 7.23); pseudo compressibility parameter
β_m	Mean flow angle (the angle of $\mathbf{W}_m$ with respect to the axial direction) [For incompressible flow, $\tan\beta_m = (\tan\beta_1 + \tan\beta_2)/2$]
β_1, β_2	Relative inlet and outlet flow angle measured from the axial direction
γ	Vortex strength per unit length or angle; ratio of specific heats of a gas; thermal diffusivity
Γ	Vortex strength; circulation; diffusion coefficient (Eq. 5.56)
δ	Boundary layer thickness; deviation angle; infinitesimal angular extent; pressure ratio $(p_1/p_a, P_{o1}/p_a)$
δ^*	Displacement thickness
δ_{ij}	Kronecker tensor
δ_t	Thermal boundary layer thickness (Eq. 7.64)

$(\Delta P_o)_{\text{loss}}$	Stagnation pressure loss (relative for rotor)
$\Delta s, \Delta P_o, \Delta \alpha, \Delta h_o$	Change in entropy, stagnation pressure, flow angle, stagnation enthalpy, respectively
ΔV	Infinitesimal volume
$\Delta t, \Delta x, \Delta y$	Time and spatial steps
ϵ	Flow turning angle; lean angle (Fig. 4.10); turbulent energy dissipation rate $(2\nu \overline{S'_{ij}S'_{ij}})$; artificial dissipation coefficient; specific heat ratio correction factor $\{\gamma_o(2/(\gamma_o + 1))^{\gamma_o/(\gamma_o-1)}/\gamma(2/(\gamma+1))^{\gamma(\gamma-1)}\}$
ϵ_t	Eddy diffusivity (Eq. 7.20)
ϵ_T	Dissipation rate of $(T')^2$ (Eq. 7.105)
ϵ_w	Limiting streamline angle (see Fig. 5.8)
ϵ_{ipj}	Permutation tensor
ζ	Loss coefficient (loss in stagnation pressure between inlet and exit divided by upstream dynamic head for absolute flow; for axial rotor the loss is based on difference in relative stagnation pressure normalized by $0.5\rho U^2$); vorticity in relative flow $(\nabla \times \mathbf{W})$; mapping function; complex plane $(\xi + i\eta)$
ζ_{KE}	Loss in kinetic energy with film cooling (Eq. 7.90)
η	Isentropic efficiency (total-to-total unless otherwise stated (see Section 2.3 for definition); nondimensional radius; boundary layer coordinate $(n/\delta$ or $y/\delta)$
η_f	Film cooling effectiveness (Eq. 7.61)
θ	Camber angle; angular coordinate; momentum thickness; temperature ratio $(T_{o1}/T_a, T_1/T_a)$; Blade turning angle/nondimensional temperature $\{(T - T_W)/(T_e - T_w)\}$
λ	Stagger angle measured from axial direction; friction loss coefficient; characteristic vector
μ	Molecular viscosity; strength of the doublet
μ_e	Mach angle
μ_t	Turbulent or eddy viscosity
μ_{eff}, μ_e	$\mu + \mu_t$
ν	Volume; kinematic viscosity; Prandtl-Meyer function
ν_{art}	Artificial viscosity
ν_t	μ_t/ρ
ξ, η, ζ	Vorticity in the r, θ, and z directions, respectively; body-fitted coordinate system (nonorthogonal, ξ coincides with the blade surface at the body)
π	Buckingham nondimensionalized group
ρ	Fluid density
$\overline{\rho u'_i u'_j}, \overline{\rho w'_i w'_j}$ $\rho(u')^2, \rho(w'_\theta)^2$, etc.	Turbulent shear and normal stresses
$\overline{\rho u'_i T'}$	Turbulent heat flux

σ	Solidity (chord/spacing ratio); net positive suction head$/0.5V^2$; thickness
$\sigma_k, \sigma_\epsilon, \sigma_w$	Turbulence modeling constants
τ	Shear stress; turbulence intensity; Thoma cavitation parameter $2(P_{o1} - p_v)/\rho V_1^2$; transformation parameter; tip clearance height; sweep angle (Fig. 4.11)
τ_{ij}	Shear stress (first subscript indicates the axis to which the face is perpendicular, the second indicates the direction to which it is parallel)
ϕ	Flow coefficient (ratio of axial velocity/blade speed); dissipation function; potential function; streamline angle (Fig. 4.11); dependent variable; pitchangle (angle of streamline with reference to axis of rotation in rz plane)
Φ	Dissipation terms (Eq. 1.18); potential function
ψ	Stream function; loading coefficient $(\pm c_p(T_{o2} - T_{o1})/U^2)$ or $[\pm g(H_2 - H_1)]/U^2$ (+ for compressors and pumps, − for turbines)
ψ_p	Pressure rise or drop coefficient $\pm(P_{o2} - P_{o1})/\frac{1}{2}\rho U_t^2$
ψ_{loss}	Stagnation pressure loss coefficient (based on relative stagnation pressure and relative inlet velocity for rotor flows), $(P_{o1} - P_{o2})/\frac{1}{2}\rho V_1^2$ for stator and cascade
$\boldsymbol{\omega}$	Absolute vorticity $(\nabla \times \mathbf{V})$; frequency
Ω, Ω^P	Angular speed of the rotor

Subscripts

A	Absolute
a	Ambient conditions; adiabatic
b	Blade; blade-to-blade direction
c	Compressor; compressibility effect; coolant; correction; velocity defect at the center line of wake
d	Doublet; diffuser
e	Edge conditions; cascade entrance region; properties for mainstream flow with film cooling
ew	Endwall
f	Values refer to coolant in the film cooling (Chapter 7)
g	Gas (main)
h	Hub
i	Induced; incompressible; inviscid; ideal
i, j, k, l, m, n	Indices; grid points along $x(i)$, $y(j)$, and $z(k)$ directions
is	Isentropic
iso	Isolated blade
j	Coolant jet number
L, l	Liquid; local; perturbation and losses due to leakage flow; laminar; leading edge

loc	Local
LE	Leading edge
m	Mean; meridional; mixed gas
max	Maximum value
min	Minimum value
n	Nozzle
o	Stagnation conditions; inlet value; standard conditions; free-stream values
oR	Stagnation properties in relative frame of reference
p	Primary; main flow; polytropic; pressure side; power; propeller; profile; pump
q	Due to sources or sink
r	Radial direction
rel, R	Quantities in relative frame of reference
r, θ, z	Components in r, θ, z directions, respectively
s	Suction surface; isentropic; source; perturbations and losses due to secondary flow; shock
sh	Due to shock
ss	Isentropic condition in a stage
$s, n, b/s, n, r$	Streamwise, principal normal (secondary component) and binormal directions (see Figs. 4.19 and 5.8)
S	Separation
t	Tip; turbine; total to total; turbulent; thermal; transition
th	Thermal (efficiency)
ts	Total to static
T	Transition; total
TE	Trailing edge
v	Due to vortex; vapor
w	Values at the wall
x, y, z	Components in x, y, and z directions, respectively
γ	Refers to vortex
∞, o	Values away from the wall (in the free stream); far upstream and downstream values
μ	Doublet
ν	Vortex; viscous
0	Values without extra strain; edge; free stream; stagnation condition
1, 2	Inlet and outlet/exit of a blade row, respectively
1, 3	Inlet and exit to a stage

Superscripts

$'$	Fluctuating quantities; blade angle; fluctuation in density average (Eq. 4.53); quantities in rotating system; correction

$^{-}$	Algebraic average; time (Eq. 1.66)/space-averaged value; mass-averaged quantities; passage-averaged (e.g., Eq. 6.5, 4.54)
$''$	Fluctuation in algebraic average (Eq. 4.52)
$\wedge$	Density-averaged; transformed flux vectors (Appendix A)
$*$	Sonic conditions; initial guess

COMMON ACRONYMS

AIAA	American Institute of Aeronautics and Astronautics
ARC	(British) Aeronautical Research Council
ARSM	Algebraic Reynolds stress model
ASME	American Society of Mechanical Engineers
AEVM	Algebraic eddy viscosity model
BL	Boundary layer
CFD	Computational fluid dynamics
CFL	Courant, Friedrichs, and Lewy Condition
CPU	Central processor unit
DFVLR/DLR/DFL	Deutsche Forschungsanstalt Für Luft- und Raumfahrt
IGV	Inlet guide vane
$k-\epsilon$	(Turbulent) Kinetic energy/dissipation model
LDV	Laser doppler velocimeter
MIT	Massachusetts Institute of Technology
NASA	National Aeronautics and Space Administration
NS	Navier–Stokes equation
PDE	Partial differential equation
PNS	Parabolized Navier–Stokes equation
PSU	Pennsylvania State University
RNS	Reynolds-averaged Navier–Stokes equation
RSM	Reynolds stress model
SAE	Society of Automotive Engineers
TLNS	Thin-layer Navier–Stokes equation
VKI	Von Karman Institute
2D, 3D	Two-dimensional; Three-dimensional

COMMON TERMINOLOGY

The common terminology used in turbomachinery for denoting the blade geometry and flow variables are given below and are illustrated (some of them) in Fig. 2.11. The definitions are standard and are adopted from Glassman (1972).

Angle of attack (α_i). Angle between camber line and the chordwise direction.

Aspect ratio (A). The ratio of the blade height to the chord.

Axial chord (C_x). Chord length of the blade in the axial direction.

Blade exit angle (α'_2, β'_2). The angle between the tangent to the camber line at the trailing edge and the axial direction.

Blade inlet angle (α'_1, β'_1). The angle between the tangent to the camber line at the leading edge and the axial direction.

Camber angle (θ). The angle formed by the intersection of the tangents to the camber line at the leading and trailing edges.

Camber line. The mean line (between pressure and suction surfaces) of the blade profile.

Chord (C). The length of the perpendicular projection of the blade profile onto the chord line. It is approximately equal to the linear distance between the leading edge and the trailing edge.

Deflection and turning angle $(\epsilon)[(\alpha_1 - \alpha_2), (\beta_1 - \beta_2)]$. The total turning angle of the fluid. It is equal to the difference between the flow inlet angle and the flow exit angle.

Deviation angle (δ). The flow exit angle minus the blade exit angle.

Flow exit angle (α_2, β_2). The angle between the fluid flow direction at the blade exit and the axial direction.

Flow inlet angle (α_1, β_1). The angle between the fluid flow direction at the blade inlet and the axial direction.

Hub (h). The innermost section of the blade.

Hub-to-tip radius ratio, Hub-tip ratio (r_h / r_t). The ratio of the hub radius to the tip radius.

Incidence angle (i). The flow inlet angle minus the blade inlet angle.

Leading edge (LE). The front, or nose, of the blade.

Mean section (m). The blade section halfway between the hub and the tip.

Meridional velocity (V_m, W_m). The resultant of axial velocity and radial velocity in the meridional (rz) plane.

Pitch (S). The distance in the direction of rotation between corresponding points on adjacent blades.

Pressure surface (PS). The concave surface of the blade. Along this surface, pressures are highest.

Radius ratio. Same as hub-to-tip radius ratio.

Root. Same as hub.

Solidity (σ). The ratio of the chord to the spacing.

Spacing (S). Same as pitch.

Stagger angle (λ). The angle between the chord line and the axial direction.

Stator or nozzle blade. A stationary blade.

Suction surface (SS). The convex surface of the blade. Along this surface, pressures are lowest.

Tip (t). The outermost section of the blade.

Trailing edge (TE). The rear, or tail, of the blade.

1

CLASSIFICATION AND BASIC CONCEPTS OF FLUID MECHANICS

1.1 INTRODUCTION AND CLASSIFICATION

Turbomachines are devices in which energy is transferred either to or from a continuously flowing fluid by the dynamic action of one or more moving blade rows. The word *turbo* is of Latin origin, meaning "that which spins." The rotor changes the stagnation enthalpy, kinetic energy, and stagnation pressure of the fluid. In a compressor or pump, the energy is imparted to the fluid by a rotor. In a turbine, the energy is extracted from the fluid. Turbomachinery is a major component in (a) aircraft, marine, space (liquid rockets), and land propulsion systems, (b) hydraulic, gas, and steam turbines, (c) industrial pipeline and processing equipment such as gas, petroleum, and water pumping plants, and (d) a wide variety of other applications (e.g., heart-assist pumps, industrial compressors, and refrigeration plants).

If the turbomachinery is without a shroud or annulus wall near the tip, the machine is termed *extended*. Examples of this are aircraft and ship propellers, wind turbines, and so on. On the other hand, *enclosed* machines are accommodated in a casing so that a finite quantity of fluid passes through the machine per unit of time. Examples of this are jet engine compressors, turbines, and pumps.

The turbomachine is classified according to the type of flow path, as shown in Fig. 1.1. In axial flow turbomachinery, the meridional flow path is axial. In radial or centrifugal turbomachinery, the flow path is predominantly radial. If the flow path is partially axial and partially radial, the device is called *mixed-flow* turbomachinery. For example, the enclosed compressor of the axial type is called an *axial flow compressor*. If a liquid is used, the same machine would be called an *axial flow pump*. The turbine designation also depends on the flow path and the fluid used. For example, an axial flow gas

1

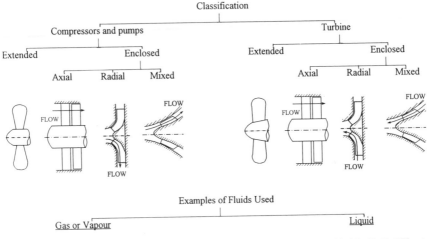

Figure 1.1   Classification of turbomachinery.

turbine has its flow path in the axial direction, and utilizes a gas. Typical examples of the fluids used are also shown in Fig. 1.1.

The major emphasis of the book is on enclosed turbomachinery. Performance analysis of prop fans and propellers are included. The analysis and computation described are of a general nature, and they can be easily adapted to other cases with minor modifications. There are several special-

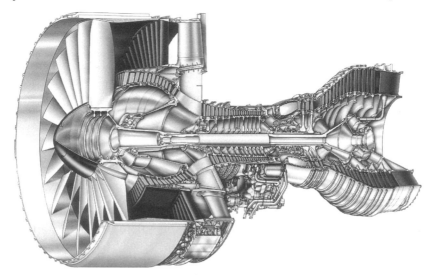

Figure 1.2   Pratt & Whitney PW 4084 turbo fan engine (for Boeing 777 and other aircraft). Length 4.87 m, diameter (fan) 2.84 m, bypass ratio 6.4, fan pressure ratio 1.70, overall pressure ratio 32, sea-level thrust 84,600 lbs. (Photograph courtesy of Pratt & Whitney.)

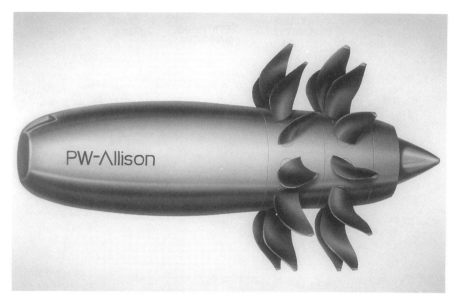

Figure 1.3   Pratt \Whitney–Allison prop fan. (Photograph courtesy of Pratt & Whitney.)

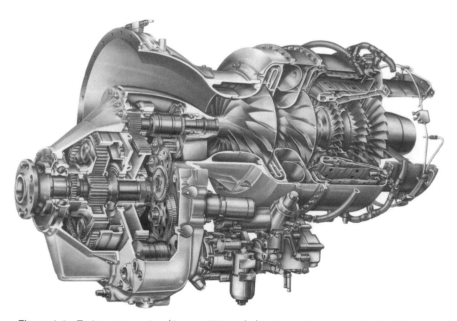

Figure 1.4   Turbo prop engine (Garrett TPE 331). (Photograph courtesy of Allied Signal, Inc.)

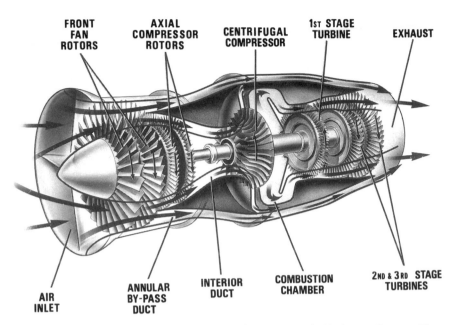

Figure 1.5 World's smallest turbo fan engine (Williams F107). Maximum diameter 30 cm, length 91 cm, sea-level thrust 600 lb$_f$, comp and fan pressure ratio 12 : 1. (Photograph courtesy of Williams International, Inc.)

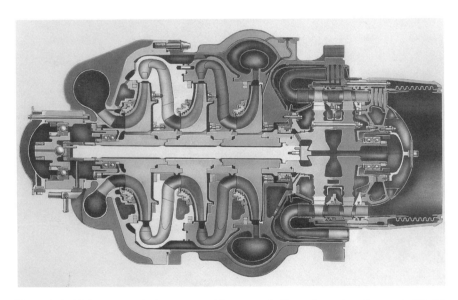

Figure 1.6 High-pressure fuel pump used in space shuttle main engine. Pressure ratio 27, turbine HP 61,000. (Photograph courtesy of Rockwell International Corporation's Rocketdyne Division.)

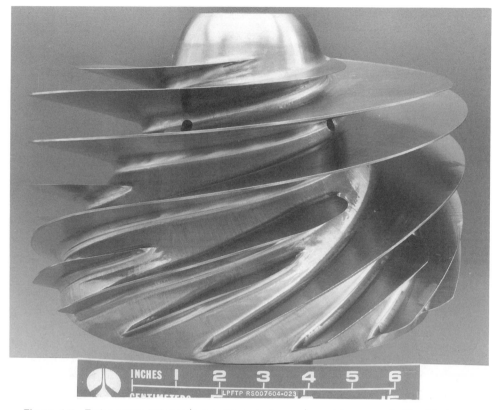

Figure 1.7 Turbo pump inducer (low-pressure fuel pump) used in space shuttle. (Photograph courtesy of Rockwell International Corporation's Rocketdyne Division.)

ized and general books in the field of turbomachinery. Some of these are listed in the reference section at the end of this book. The books by Dixon, Traupel, Csanady, Vavra, and Wislicenus treat the subject in a unified manner, which is the approach followed in this book.

Some typical examples of turbomachinery used in various applications are listed below. Photographs of some of this turbomachinery are given in Figs. 1.2 through 1.13.

1. *Aerospace Vehicle Application.* Compressors and turbines are used in gas turbines for power and propulsion of aircraft, helicopters, unmanned aerospace vehicles, V/STOL aircraft, missiles, and so on. Turbines and pumps are used in liquid rocket engines utilized for the propulsion of space vehicles. Typical examples shown are a large turbo fan (Fig. 1.2), a prop fan (Fig. 1.3), a turbo prop (Fig. 1.4), the smallest turbo fan (Fig. 1.5), a high-pressure rocket engine fuel pump (Fig. 1.6), and a rocket engine pump inducer (Fig. 1.7).

Figure 1.8 Ship propeller. (Photograph courtesy of Maritime Research Institute, Netherlands.)

2. *Marine Vehicle Application.* Turbomachinery components are used in (a) power plants for submarines, hydrofoil boats, Naval surface ships, hovercraft, and so on and (b) propeller and propulsion plants used in ships, underwater vehicles, hydrofoil boats, and so on. A typical example is ship propeller, as shown in Fig. 1.8.

3. *Land Vehicle Application.* Turbomachinery is an important component in the gas turbines used in trucks, cars, and high-speed train systems. An automotive gas turbine which utilizes a centrifugal compressor and a radial turbine is shown in Fig. 1.9.

4. *Energy Application.* Steam turbines are used in steam, nuclear, and coal power plants; hydraulic turbines are used in hydropower plants; gas turbines are used in gas turbine power plants; wind turbines also belong to this class. A large Francis turbine used in hydropower applications is shown in Fig. 1.10.

5. *Industrial Applications.* Compressors and pumping machinery are used in gas and petroleum transmissions and other industrial and processing applications; pumping machinery is used in fire fighting, water purification, and pumping plants; high-speed miniature turbo expanders are used in refrigeration equipment; compressors are used in refrigeration plants (industrial and other uses). A large industrial axial compressor is shown in Fig. 1.11, a centrifugal compressor is shown in Fig. 1.12,

Figure 1.9 Automotive gas turbine (AGT101). (Photograph courtesy of Allied Signal, Inc.)

and the smallest turbomachine used in refrigeration plants is shown in Fig. 1.13.

6. *Miscellaneous.* Pumps are used in heart-assist devices, automotive torque converters, swimming pools, and hydraulic brakes. One of the interesting applications of turbomachinery in the medical field, the artificial heart pump, is shown in Fig. 1.14.

It is thus clear that the range of applications where turbomachinery is used is enormous. Because the turbomachinery industry is a global one, any small gain in turbomachinery efficiency and performance translates into a major economic impact worldwide.

The flow field encountered in a turbomachine is one of the most complicated in the field of fluid dynamics practice. The flow may be incompressible,

Figure 1.10   Francis reversible pump-turbine, turbine mode: 258 MW, diameter (bottom axial exit) 5.7 m, 225 rpm, head 230 m, H_2O; $\eta_t = 90\%$. Pump mode: Axial inlet and radial exit. (Photograph courtesy of Voith Inc.)

subsonic, transonic, or supersonic. Many aircraft engines have mixed flows, in which all of these regimes are present in one single blade row. The flow may be single-phase or two-phase, as in pumps (liquid–gas) or turbines (gas laden with solid particles). The flow is viscous, with laminar, transitional, or turbulent regions present. Schematics of the types of flows encountered in turbomachinery are shown in Figs. 1.15 and 1.16. The presence of a rotor blade and wall boundary layers, as well as vortices at several locations, makes analysis complicated. The flow in a turbomachine is usually three-dimensional, and the velocities in the absolute frame are always unsteady. Additional effects, such as (a) noise generation and propagation and (b) fluid–blade interactions (forced and free vibrations), are also important. Not all of these effects are present in any given turbomachine, but most sources are present in these devices.

Tubomachinery research, analysis, design, computation, and development involve the interaction of various fields, as shown in Fig. 1.17. A large turbomachinery company (e.g., an aircraft jet engine manufacturer) will have experts and groups in most of the areas indicated in this figure. It is

Figure 1.11 Industrial gas turbine. (Photograph courtesy of FIAT.)

Figure 1.12   Centrifugal compressor. (8 : 1 pressure ratio, 75,000 rpm.) (Photograph courtesy of Creare, Inc.)

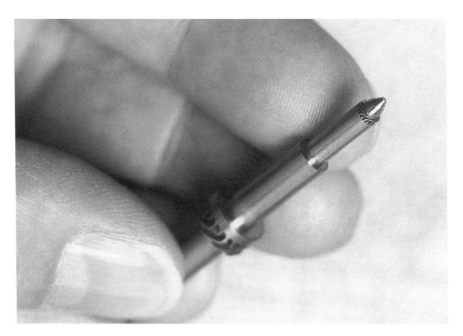

Figure 1.13   Smallest turbomachine (turbo expander used in medium-sized liquefiers and refrigerators. The photograph shows a turboexpander shaft incorporating a 3.18 mm diameter radial inflow turbine at the end and an 8 mm brake rotor near the middle, 11 full blades and 11 splitter blades, N = 510,000 rpm.) (Photograph courtesy of Creare, Inc.)

Figure 1.14 Artificial implantable axial flow heart pump. Left Ventricular Assist System (LVAS): 14 mm dia, 3–8 L / min, pressure range 0–180 mm Hg. power speed 10,000–15,000 rpm. Power range 7–20 watts. Courtesy Nimbus, Inc. and University of Pittsburgh Medical Center, further details in Butler, 1992 and Antaki, 1993.

impossible to cover all of these aspects of turbomachinery in a single text. The major emphasis in this text is on fluid dynamics, thermodynamics, and heat transfer.

1.2   SOME BASIC CONCEPTS IN FLUID MECHANICS

Before we introduce the performance and analysis of turbomachinery, it would be useful to review some basic concepts and equations in fluid mechanics that are useful in the analysis and design of turbomachinery. The reader is referred to some excellent books on the subject of fluid mechanics

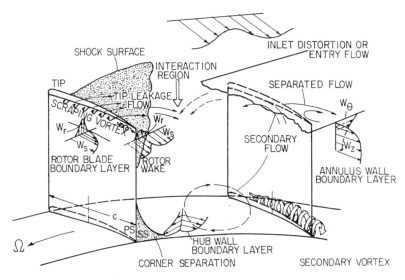

Figure 1.15   Nature of flow in an axial flow compressor rotor passage.

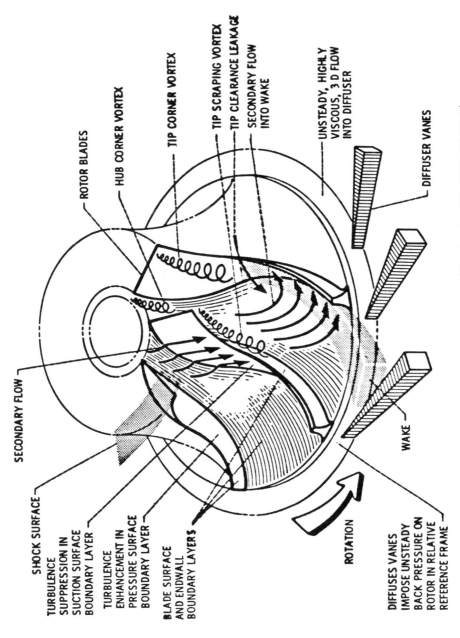

Figure 1.16   Centrifugal compressor flow phenomena (Wood et al., 1983, NASA TM 83398).

12

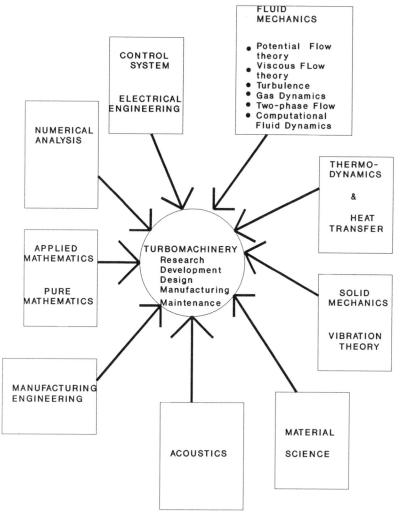

Figure 1.17   Interdisciplinary nature of turbomachinery field.

and gas dynamics (Fox and McDonald, 1985; Karamachetti, 1966; Bertin and Smith, 1989; Anderson, 1982; Schlichting, 1979; White, 1991; Sherman, 1990).

1.2.1 Control Volume Concept

One of the most useful concepts in the performance evaluation of turbomachinery is the control volume approach as applied to the basic laws of conservation of mass, momentum, and energy.

Consider a turbomachine enclosed in a control volume, fixed in space, such as the one shown in Fig. 1.18. The generalized equation relating the laws of mechanics (system approach) to the fluid which is instantaneously contained within the control volume is given by

$$\frac{dN}{dt} = \iint_{c.s.} n\rho \mathbf{V} \cdot d\mathbf{A} + \frac{\partial}{\partial t} \iiint_{c.v.} n\rho \, dv \tag{1.1}$$

| Rate of change of property N for the system (same as control volume at $t = 0$) | Flux of property N through the control surface | Rate of change of property N inside the control volume |

where N is any extensive property such as mass, linear momentum, angular momentum, or stored energy (internal plus kinetic energy), and n is the extensive property per unit mass.

The equations for continuity, linear momentum, angular momentum, and energy based on these equations are given by substituting $n = N/M = 1$, $\mathbf{V}, \mathbf{r} \times \mathbf{V}, e$, respectively, in Eq. 1.1:

$$0 = \iint_{c.s.} \rho \mathbf{V} \cdot d\mathbf{A} + \frac{\partial}{\partial t} \iiint_{c.v.} \rho \, dv \tag{1.2}$$

$$\iint_{c.s.} \mathbf{F} \, dA + \iiint_{c.v.} \mathbf{B}\rho \, dv = \iint_{c.s.} \mathbf{V}(\rho \mathbf{V} \cdot d\mathbf{A}) + \frac{\partial}{\partial t} \iiint_{c.v.} \mathbf{V}(\rho \, dv) \tag{1.3}$$

$$\iint_{c.s.} \mathbf{r} \times \mathbf{F} \, dA + \iiint_{c.v.} \mathbf{r} \times \mathbf{B}\rho \, dv = \iint_{c.s.} (\mathbf{r} \times \mathbf{V})(\rho \mathbf{V} \cdot d\mathbf{A})$$

$$+ \frac{\partial}{\partial t} \iiint_{c.v.} (\mathbf{r} \times \mathbf{V})\rho \, dv \tag{1.4}$$

$$\frac{dq}{dt} - \frac{dW_s}{dt} - \iiint_{c.v.} \mathbf{B} \cdot \mathbf{V}\rho \, dv = \iint_{c.s.} h_o \rho \mathbf{V} \cdot d\mathbf{A} + \frac{\partial}{\partial t} \iiint_{c.v.} e_0 \rho \, dv \tag{1.5}$$

Here $\mathbf{F}$ is the surface (normal and tangential) force per unit area acting on the control surface; $\mathbf{B}$ is the body force per unit mass, such as gravity, acting inside the control volume; $\mathbf{r}$ is the distance from the origin of the coordinate system (Fig. 1.18); q is the heat transferred to the control volume; W_s is the work done on the control volume by the rotor and by shear forces; and h_o is the stagnation enthalpy.

The laws of mechanics and thermodynamics are used to obtain expressions for the term dN/dt appearing on the left-hand side in Eqs. 1.1–1.5. For turbomachinery *rotor* flows, we are dealing with forces on a rotating blade; and, as will be seen, the angular momentum (or moment of momentum) equation (Eq. 1.4) is more useful than the linear momentum equation (Eq. 1.3).

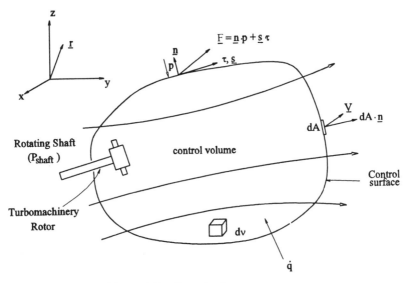

Figure 1.18 Control volume approach.

The rate of work (dW_s/dt) transferred across the control surface is in several forms. One of the dominant and most important rate of work transfer in turbomachinery is through a rotating shaft. This can be denoted by $P_{shaft} = T_{shaft}\,\omega$. The work can also be transferred by normal stresses, and this is included in the last term $(h_o = e_0 + p/\rho)$. The work transfer by tangential stresses is small and this is represented by P_{shear}. For steady-state flows, considering the more useful angular momentum equation in lieu of the linear momentum equation, the three integral conservation equations become:

Continuity

$$\iint_{c.s.} \rho \mathbf{V} \cdot d\mathbf{A} = 0 \tag{1.6}$$

Angular momentum

$$\iint_{c.s.} \mathbf{r} \times \mathbf{F}\, dA + \iiint_{c.v.} (\mathbf{r} \times \mathbf{B}) \rho\, dv = \iint_{c.s.} (\mathbf{r} \times \mathbf{V})(\rho \mathbf{V} \cdot d\mathbf{A}) \tag{1.7}$$

Energy

$$\dot{q} = P_{shaft} + P_{shear} + \iiint_{c.v.} \mathbf{B} \cdot \mathbf{V}\rho\, dv + \iint_{c.s.} h_o\, \rho \mathbf{V} \cdot d\mathbf{A} \tag{1.8}$$

The terms in the steady-state angular momentum equation represent, respectively, the moment of all surface forces (shear and normal forces) acting on the control surface, the moment of all body forces acting inside the control volume, and the flux of angular momentum across the control surfaces. In turbomachinery applications, the normal forces correspond to pressures acting normal to the control volume (Fig. 1.18), and the tangential forces correspond to viscous and turbulent stresses acting on the surfaces of the control volume and the shear forces due to a rotating shaft protruding from the control surface. In most instances, torque with respect to the axis of rotation only is important. For such a case the shaft torque (Fig. 1.18) is important. The net torque associated with the normal and tangential forces exerted on the control surface is small. Hence, the term $\int\int \mathbf{r} \times \mathbf{F} \, dA$ can be replaced by T_{shaft}, which is the shaft torque.

The terms in the steady-state energy equation (Eq. 1.8) represent, respectively, the heat transfer to the surroundings, the rate of work done by the shaft (such as a turbine or compressor shaft piercing through the control volume), the rate of work done by the shear stresses on the surface (e.g., $\tau \cdot \mathbf{V}$, where $\mathbf{V}$ is the local velocity), the rate of work done by body forces inside the control volume, and the flux of stagnation enthalpy through the control surface.

1.2.2 Equations of Motion for Compressible Viscous Flows

The differential form of the continuity and linear momentum conservation equations can be obtained from the integral relations 1.2 and 1.3 by application of the divergence theorem. If the working fluid is Newtonian and gravity is the only body force acting, expressions for $\mathbf{F}$ and $\mathbf{B}$ can be obtained (Schlichting, 1979; White, 1991) and the equations governing the flow field can be written as

$$\frac{\partial \rho}{\partial t} + \nabla \cdot (\rho \mathbf{V}) = 0 \tag{1.9}$$

$$\underset{(1)}{\rho \frac{D\mathbf{V}}{Dt}} = \rho \left[\underset{(2)}{(\mathbf{V} \cdot \nabla)\mathbf{V}} + \underset{(3)}{\frac{\partial \mathbf{V}}{\partial t}} \right] = \underset{(4)}{-\nabla p} + \underset{(5)}{\rho \mathbf{g}} + \underset{(6)}{\frac{\partial}{\partial x_j} \left[\mu \left\{ \frac{\partial u_i}{\partial x_j} + \frac{\partial u_j}{\partial x_i} \right\} + \underset{(7)}{\delta_{ij} \lambda \, \text{div} \, \mathbf{V}} \right]}$$

$$\tag{1.10}$$

Terms 1, 2, and 3 in Eq. 1.10 represent the total, convective, and local acceleration, respectively; term 4 represents the pressure force. Term 5 represents the gravitational force; terms 6 and 7 represent the viscous stresses, where normally the bulk viscosity, $\mu' \equiv \lambda + \frac{2}{3}\mu$, is taken as equal to zero (i.e., $\lambda = -\frac{2}{3}\mu$, Stokes' hypothesis). A detailed discussion of the viscous terms, including a discussion of the bulk viscosity, can be found in White (1991). For incompressible, laminar flow, the density and viscosity are taken as constant in Eqs. 1.9 and 1.10; therefore, these equations represent a complete set of four equations in the four unknowns, namely, pressure and three velocity components. For inviscid flows, the last two terms in Eq. 1.10 are zero. In most situations, the fifth term in Eq. 1.10, ρg, is neglected.

The energy equation for the laminar mean flow (based on the thermodynamics equation $\partial q/\partial t = dE/dt + dW/dt$, where the terms represent, respectively, rate of heat transfer, energy transport, and rate of work done) is given by

$$\rho \frac{Dh}{Dt} = \frac{Dp}{Dt} + \nabla(k\nabla T) + \phi \tag{1.11}$$

where ϕ is the dissipation function, representing the heat equivalent of mechanical energy due to viscous dissipation. Specifically, the rate of work done by shear stresses can be expressed as

$$\phi = \frac{\partial u_i}{\partial x_j} \tau_{ij}$$

$$= \mu \left[2\left(\frac{\partial u}{\partial x}\right)^2 + 2\left(\frac{\partial v}{\partial y}\right)^2 + 2\left(\frac{\partial w}{\partial z}\right)^2 + \left(\frac{\partial v}{\partial x} + \frac{\partial u}{\partial y}\right)^2 \right.$$

$$\left. + \left(\frac{\partial w}{\partial y} + \frac{\partial v}{\partial z}\right)^2 + \left(\frac{\partial u}{\partial z} + \frac{\partial w}{\partial x}\right)^2 \right]$$

$$+ \lambda \left(\frac{\partial u}{\partial x} + \frac{\partial v}{\partial y} + \frac{\partial w}{\partial z}\right)^2 \tag{1.12}$$

The first term on the left side of Eq. 1.11 is the total rate (local, as well as convective) of change of static enthalpy; Dp/Dt and ϕ are the rate of work done by pressure and shear stresses, respectively. $\nabla(k\nabla T)$ is heat transfer by steady conduction, and k is the coefficient of thermal conductivity [see Schlichting (1979) for the derivation of this equation].

For compressible flows, energy transport and generation is coupled to the dynamics of the fluid motion, and therefore the energy equation (Eq. 1.11) must be solved in conjunction with the continuity and momentum equations. Additionally, a constitutive equation relating density to pressure and temperature is required. For an ideal gas an equation of state is given as

$$p = \rho RT \tag{1.13}$$

Equations 1.9–1.13 provide six equations (in three-dimensional flow) for six unknowns: V, ρ, p, T (or h).

1.2.2.1 Equations in Conservative Form

In many situations (such as the numerical solution of the Navier–Stokes equations), the equations expressed in terms of "conservative" variables are most useful. Variables such as ρ, ρu, ρv, ρh_o, ρe, which include mass density (through the variable ρ), are called *conservative variables*. Variables which are expressed per unit mass flow (e.g., u, v, w, e, h_o) are termed *primitive variables*. The partial differential equations, Eqs. 1.9–1.11, are all based on the conservation laws of fluid mechanics. Equations 1.10 and 1.11 represents the conservation of ρu, ρv, ρw, and ρh though the primitive variables appear in the convective transport term (D/Dt).

When "conservative" variables are used in a finite difference scheme, the discretized governing equations more accurately conserve mass, momentum, and energy. This can be an advantage in shocked flows, because the equations in conservative form satisfy the Rankine–Hugoniot relations and will produce the correct jump conditions across shocks. Another advantage is that the finite difference form of these equations can be interpreted as integral laws over a computational cell control volume, as discussed (for example) in Hirsch (1990).

The momentum equations in conservative form can be derived by combining Eqs. 1.9 and 1.10 and by manipulating the resulting equation to give, for example, the following x momentum equation (assuming Stokes' hypothesis):

$$\frac{\partial \rho u}{\partial t} + \frac{\partial}{\partial x}(\rho u^2 + p) + \frac{\partial}{\partial y}(\rho uv) + \frac{\partial}{\partial z}(\rho uw)$$

$$= \rho g_x + \frac{\partial}{\partial x}\left[2\mu\frac{\partial u}{\partial x} - \frac{2}{3}\mu\,\mathrm{div}\,\mathbf{V}\right]$$

$$+ \frac{\partial}{\partial y}\left[\mu\left(\frac{\partial u}{\partial y} + \frac{\partial v}{\partial x}\right)\right] + \frac{\partial}{\partial z}\left[\mu\left(\frac{\partial u}{\partial z} + \frac{\partial w}{\partial x}\right)\right] \quad (1.14)$$

The energy equation (Eq. 1.11) can also be expressed in terms of stagnation enthalpy and conservative variables. Such a form is extremely useful in fluid dynamics and turbomachinery in situations where stagnation enthalpy changes are small (e.g., stator, relative enthalpy in a rotor along a streamline in an adiabatic flow). Let us dot the momentum equation with the velocity

$$\rho\frac{D\mathbf{V}}{Dt}\cdot\mathbf{V} = -\nabla p\cdot\mathbf{V} + \rho\mathbf{g}\cdot\mathbf{V} + (\nabla\cdot\tau_{ij})\cdot\mathbf{V} \quad (1.15)$$

Combine this equation with Eqs. 1.11 and 1.12 to give

$$\rho\frac{D(h_o)}{Dt} = \frac{\partial p}{\partial t} + \rho\mathbf{g}\cdot\mathbf{V} + \nabla(k\nabla T) + \nabla\cdot(\tau_{ij}\cdot\mathbf{V}) \quad (1.16)$$

where $h_o = h + u_i u_i/2 = h + (u^2 + v^2 + w^2)/2$. This equation can be combined with the continuity equation, yielding

$$\frac{\partial \rho h_o}{\partial t} + \nabla\cdot\rho\mathbf{V}h_o = \frac{\partial p}{\partial t} + \rho\mathbf{g}\cdot\mathbf{V} + \nabla(k\nabla T) + \nabla\cdot(\tau_{ij}\cdot\mathbf{V}) \quad (1.17)$$

Hence the equations in conservative form can be written (e.g., Eqs. 1.14 and 1.17) as follows [see Peyeret and Taylor (1983) for a detailed derivation of these equations];

$$\frac{\partial q}{\partial t} + \frac{\partial E}{\partial x} + \frac{\partial F}{\partial y} + \frac{\partial G}{\partial z} = \frac{1}{\mathrm{Re}}\left[\frac{\partial T}{\partial x} + \frac{\partial P}{\partial y} + \frac{\partial Q}{\partial z}\right] + S \quad (1.18)$$

where q, E, F, G, T, P, Q, and S are given by

$$q = \begin{bmatrix} \rho \\ \rho u \\ \rho v \\ \rho w \\ \rho e_o \end{bmatrix}, \quad E = \begin{bmatrix} \rho u \\ \rho u^2 + p \\ \rho u v \\ \rho u w \\ \rho h_o u \end{bmatrix}, \quad F = \begin{bmatrix} \rho v \\ \rho u v \\ \rho v^2 + p \\ \rho v w \\ \rho h_o v \end{bmatrix}, \quad G = \begin{bmatrix} \rho w \\ \rho u w \\ \rho v w \\ \rho w^2 + p \\ \rho h_o w \end{bmatrix}$$

$$(1.19)$$

$$T = \begin{bmatrix} 0 \\ 2\mu\dfrac{\partial u}{\partial x} - \dfrac{2}{3}\mu\left(\dfrac{\partial u}{\partial x} + \dfrac{\partial v}{\partial y} + \dfrac{\partial w}{\partial z}\right) \\ \mu\left(\dfrac{\partial v}{\partial x} + \dfrac{\partial u}{\partial y}\right) \\ \mu\left(\dfrac{\partial w}{\partial x} + \dfrac{\partial u}{\partial z}\right) \\ -Q_x + \Phi_1 \end{bmatrix} \quad (1.20)$$

$$P = \begin{bmatrix} 0 \\ \mu\left(\dfrac{\partial u}{\partial y} + \dfrac{\partial v}{\partial x}\right) \\ 2\mu\dfrac{\partial v}{\partial y} - \dfrac{2}{3}\mu\left(\dfrac{\partial u}{\partial x} + \dfrac{\partial v}{\partial y} + \dfrac{\partial w}{\partial z}\right) \\ \mu\left(\dfrac{\partial v}{\partial z} + \dfrac{\partial w}{\partial y}\right) \\ -Q_y + \Phi_2 \end{bmatrix} \quad (1.21)$$

$$Q = \begin{bmatrix} 0 \\ \mu\left(\dfrac{\partial w}{\partial x} + \dfrac{\partial u}{\partial z}\right) \\ \mu\left(\dfrac{\partial v}{\partial z} + \dfrac{\partial w}{\partial y}\right) \\ 2\mu\dfrac{\partial w}{\partial z} - \dfrac{2}{3}\mu\left(\dfrac{\partial u}{\partial x} + \dfrac{\partial v}{\partial y} + \dfrac{\partial w}{\partial z}\right) \\ -Q_z + \Phi_3 \end{bmatrix} \quad (1.22)$$

$$S = \left(0,\ \rho g_x,\ \rho g_y,\ \rho g_z,\ \rho g_x u + \rho g_y v + \rho g_z w\right)^T \quad (1.23)$$

where Q_x, Q_y, and Q_z are heat transfer rates and Φ_1, Φ_2, and Φ_3 are the viscous dissipation terms (in the energy equation)

$$\Phi_1 = 2\mu u \frac{\partial u}{\partial x} - \frac{2}{3}\mu u \left(\frac{\partial u}{\partial x} + \frac{\partial v}{\partial y} + \frac{\partial w}{\partial z} \right)$$

$$+ \mu v \left(\frac{\partial u}{\partial y} + \frac{\partial v}{\partial x} \right) + \mu w \left(\frac{\partial u}{\partial z} + \frac{\partial w}{\partial x} \right)$$

$$\Phi_2 = 2\mu v \frac{\partial v}{\partial y} - \frac{2}{3}\mu v \left(\frac{\partial u}{\partial x} + \frac{\partial v}{\partial y} + \frac{\partial w}{\partial z} \right)$$

$$+ \mu u \left(\frac{\partial u}{\partial y} + \frac{\partial v}{\partial x} \right) + \mu w \left(\frac{\partial v}{\partial z} + \frac{\partial w}{\partial y} \right) \qquad (1.24)$$

$$\Phi_3 = 2\mu w \frac{\partial w}{\partial z} - \frac{2}{3}\mu w \left(\frac{\partial u}{\partial x} + \frac{\partial v}{\partial y} + \frac{\partial w}{\partial z} \right)$$

$$+ \mu u \left(\frac{\partial u}{\partial z} + \frac{\partial w}{\partial x} \right) + \mu v \left(\frac{\partial v}{\partial z} + \frac{\partial w}{\partial y} \right)$$

$$\rho e_o = \rho h_o - p = \rho h + \rho \frac{V^2}{2} - p \qquad (1.25)$$

The viscous terms arise from the term $\nabla \cdot (\tau_{ij} \cdot \mathbf{V})$. The components of τ_{ij} in the Cartesian system are given by

$$\tau_{xx} = 2\mu \frac{\partial u}{\partial x} - \frac{2}{3}\mu \left(\frac{\partial u}{\partial x} + \frac{\partial v}{\partial y} + \frac{\partial w}{\partial z} \right)$$

$$\tau_{yy} = 2\mu \frac{\partial v}{\partial y} - \frac{2}{3}\mu \left(\frac{\partial u}{\partial x} + \frac{\partial v}{\partial y} + \frac{\partial w}{\partial z} \right)$$

$$\tau_{zz} = 2\mu \frac{\partial w}{\partial z} - \frac{2}{3}\mu \left(\frac{\partial u}{\partial x} + \frac{\partial v}{\partial y} + \frac{\partial w}{\partial z} \right) \qquad (1.26)$$

$$\tau_{yx} = \tau_{xy} = \mu \left(\frac{\partial u}{\partial y} + \frac{\partial v}{\partial x} \right), \qquad \tau_{xz} = \tau_{zx} = \mu \left(\frac{\partial w}{\partial x} + \frac{\partial u}{\partial z} \right),$$

$$\tau_{yz} = \tau_{zy} = \mu \left(\frac{\partial v}{\partial z} + \frac{\partial w}{\partial y} \right)$$

The variables in Eqs. 1.18–1.23 are all nondimensional quantities. The normalizing factor for length (x, y, z) is a characteristic length L, the velocities (u, v, w) are normalized by a characteristic velocity V_∞, where ∞ refers to upstream or unperturbed values of the velocity. Likewise, ρ, μ, and T are normalized by their upstream values; p and g by $\rho_\infty V_\infty^2$; e, h, and h_o by V_∞^2; and t by L/V_∞. Nondimensionalization of heat transfer terms in Eq. 1.18

needs further explanation. The nondimensional heat transfer rate is given by

$$q_x = -k \frac{\partial (T/T_\infty)}{\partial (x/L)} \frac{T_\infty}{L} \frac{1}{\rho_\infty V_\infty^3} = -k \frac{\partial \overline{T}}{\partial \overline{x}} \frac{T_\infty}{L} \frac{1}{\rho_\infty V_\infty V_\infty^2} \frac{\mu_\infty c_p}{\mu_\infty c_p} \tag{1.27}$$

$$= -\frac{1}{PR} \frac{1}{Re} \frac{\partial \overline{T}}{\partial \overline{x}} \frac{c_p T_\infty}{V_\infty^2} \tag{1.28}$$

For simplicity, the superscripts for nondimensional quantities are dropped. Therefore,

$$q_x = -\frac{1}{PR} \frac{1}{Re} \frac{dT}{dx} \frac{c_p T_\infty}{V_\infty^2} = \frac{1}{Re} Q_x \tag{1.29}$$

$$q_y = -\frac{1}{PR} \frac{1}{Re} \frac{dT}{dy} \frac{c_p T_\infty}{V_\infty^2} = \frac{1}{Re} Q_y \tag{1.30}$$

$$q_z = -\frac{1}{PR} \frac{1}{Re} \frac{dT}{dz} \frac{c_p T_\infty}{V_\infty^2} = \frac{1}{Re} Q_z \tag{1.31}$$

For compressible flow of a perfect gas, many authors (Schlichting, 1979) prefer to express the term

$$\frac{c_p T_\infty}{V_\infty^2} = \frac{c_p T_\infty}{M_\infty^2 \gamma R T_\infty} = \frac{1}{(\gamma - 1) M_\infty^2} \tag{1.32}$$

Several nondimensional groups appear in these equations:

$$Re = \frac{\rho_\infty V_\infty L}{\mu_\infty}, \qquad M = \frac{V_\infty}{a_\infty} \sim \frac{V_\infty}{\sqrt{P_\infty / \rho_\infty}}, \qquad PR = \frac{\mu_\infty c_p}{k} \tag{1.33}$$

The Reynolds number represents the ratio of inertial to viscous forces. The Mach number represents the compressibility effect, the ratio of inertial to pressure forces. The Prandtl number is a property of the medium. For air, it is approximately 0.73.

In many situations, such as heat transfer calculations, the variables of interest are temperature and velocity. Equation 1.11 can be nondimensionalized with respect to a quantity which characterizes the heat transfer, $T_w - T_\infty$, which is the difference between the wall temperature and the temperature at infinity. Let us denote this quantity by ΔT_o. Then, using the other nondimensionalizing factors employed before, assuming k and c_p are constants, the energy equation in Cartesian coordinates reduces to

$$\frac{\partial}{\partial t}(\rho T) + \frac{\partial}{\partial x}(\rho u T) + \frac{\partial}{\partial y}(\rho v T) + \frac{\partial}{\partial z}(\rho w T)$$

$$= \frac{1}{PR\,Re}\left(\frac{\partial^2 T}{\partial x^2} + \frac{\partial^2 T}{\partial y^2} + \frac{\partial^2 T}{\partial z^2}\right)$$

$$+ Ec\left(\frac{\partial p}{\partial t} + u\frac{\partial p}{\partial x} + v\frac{\partial p}{\partial y} + w\frac{\partial p}{\partial z}\right) + \frac{Ec}{Re}\phi \qquad (1.34)$$

The terms on the right side of Eq. 1.34 are, respectively, heat transfer, pressure work, and dissipation by viscous shearing. All quantities in the above equations are nondimensional. For example, T is the local temperature normalized by ΔT_o and all other nondimensional parameters are as before. The Eckert number (Ec) appearing in Eq. 1.34 is related to the adiabatic compression given by

$$Ec = \frac{V_\infty^2}{c_p \Delta T_o} = \frac{((\gamma - 1)M_\infty^2)T_\infty}{\Delta T_o} \qquad (1.35)$$

A detailed discussion of all nondimensional parameters, Re, Pr, Ec, and so on, can be found in Schlichting (1979).

The unknown variables are ρ, u, v, w, T, and p. The equation sets (1.9, 1.10, 1.13, and 1.34) or (1.18 and 1.13) are to be solved simultaneously to predict these quantities.

In many aerodynamic problems involving flow over bodies, it is convenient and more accurate to transform the transport equations into a body-fitted coordinate system. Equations 1.18–1.25 can be transformed into an arbitrary curvilinear system using the transformation

$$\xi = \xi(x, y, z), \qquad \eta = \eta(x, y, z), \qquad \zeta = \zeta(x, y, z)$$

The equations can be transformed using well known techniques. The transformed equations are given in Appendix A.

1.2.2.2 Vorticity Equations In many situations, the vorticity transport equations are of general interest. The vorticity transport equations for compressible viscous flow are obtained by taking the curl of the momentum equations (Eq. 1.10).

$$\frac{\partial \omega}{\partial t} + (\mathbf{V} \cdot \nabla)\omega = (\omega \cdot \nabla)\mathbf{V} - \omega(\nabla \cdot \mathbf{V}) - \nabla \times (\nabla p/\rho)$$

$$\text{(1)} \qquad \text{(2)} \qquad \qquad \text{(3)} \qquad \qquad \text{(4)} \qquad \qquad \text{(5)}$$

$$- \nabla \times \left[\frac{\mu}{\rho}\nabla \times \omega - \frac{4}{3}\frac{\mu}{\rho}\nabla(\nabla \cdot \mathbf{V})\right] + \nabla \times \left(\frac{\mathbf{B}}{\rho}\right) \qquad (1.36)$$

$$\text{(6)} \qquad \qquad \qquad \text{(7)}$$

The first two terms represent local and convective acceleration terms (inviscid transport of vorticity). Term 3 represents the rate of vorticity amplification due to the normal straining of vortex lines (vortex stretching). Terms 4 and 5 correspond, respectively, to vorticity production arising from volumetric expansion and the rotational motion induced by pressure forces acting on a fluid of nonuniform density. Both are zero in a constant density flow. Term 6 represents viscous diffusion of vorticity, and the last term represents the production of vorticity arising from the action of a nonconservative body force.

For an incompressible, homogeneous (constant ν), and steady flow, this equation reduces to

$$(\mathbf{V} \cdot \nabla)\boldsymbol{\omega} = (\boldsymbol{\omega} \cdot \nabla)\mathbf{V} + \nu\nabla^2\boldsymbol{\omega} \tag{1.37}$$

One advantage of using the vorticity transport equation is the absence of the pressure term in the incompressible flow case. For a three-dimensional compressible flow, the vorticity equation (Eq. 1.36) can be combined with the continuity equation (Eq. 1.9), the equation of state (Eq. 1.13), the energy equation (Eq. 1.11), and the definition of vorticity

$$\boldsymbol{\omega} = \nabla \times \mathbf{V} \tag{1.38}$$

to solve for three components of velocity and vorticity, pressure, and density.

For two-dimensional incompressible flow, a stream function ψ can be defined such that

$$\frac{\partial\psi}{\partial x} = -v, \qquad \frac{\partial\psi}{\partial y} = u, \qquad \omega = \omega_z \tag{1.39}$$

and the vorticity and continuity equations reduce to

$$\frac{D\omega}{Dt} = \nu\nabla^2\omega \tag{1.40}$$

$$\nabla^2\psi = -\omega \tag{1.41}$$

Equations 1.40 and 1.41 can be solved simultaneously to predict ω and ψ.

1.2.3 Simplified Equations of Motion

The governing equations can often be simplified to provide forms suitable for analytical or numerical treatment. Some of these simplifications are listed in Table 1.1.

In many situations, the flow is incompressible (density is constant) and viscosity gradients are small. In this case, the viscous terms in the momentum equations are simplified and the energy equation is not needed because

Table 1.1 Governing equations for special cases

Formulation	Equation in Primitive Variables	Equation in Conservative Variables	Vorticity Transport Equations
Generalized equation numbers	Eqs. 1.9–1.13	Eqs. 1.18–1.31 and 1.13	Eqs. 1.9, 1.11, 1.13 1.36, and 1.38
Thin-layer Navier–Stokes equation (TLNS)	In a body-fitted coordinate system streamwise viscous and turbulent diffusion terms are neglected. The pressure terms are retained in their entirety.		
Parabolized Navier–Stokes equations (PNS)	Same as above; if the pressure gradient in the streamwise direction $(\partial p/\partial x_1)$ is prescribed or guessed initially, the system of equations is parabolic in the x_1 direction.		
Two-dimensional viscous flow (incompressible) ψ–ω formulation	$\psi_y = u,\quad \psi_x = -v.$ Solve Eqs. 1.40 and 1.41		
Boundary layer equations (compressible flow)	The pressure gradient normal to the wall is neglected, normal momentum equation is not needed. The external pressure gradient is specified from an inviscid solution. Viscous diffusion in the streamwise and crossflow directions are neglected. Equations 1.9–1.13 are solved for u, v, w, T, and ρ.	Same as that for primitive variables. (all except ρv conservation equation are solved for ρ, ρu, ρw, and T.)	Approximations similar to those for primitive variables. Further approximations can be made in vorticity components.

Boundary layer equations (incompressible flow)	In addition to above assumptions, ρ = const. Equations 1.44–1.46 are solved.	Streamwise and crosswise diffusion terms are neglected in Eq. 1.40
Inviscid flow 3D compressible (Euler equation)	$\mu = 0$ in Eq. 1.10.　　　　$Re = \infty$ in Eq. 1.18.	$\mu = 0$ in Eq. 1.36
Inviscid flow 3D incompressible (homogeneous)	Equations 1.42 and 1.43 with $\mu = 0$, ρ = const.	Terms 4, 5, and 6 are zero in Eq. 1.36
Inviscid irrotational flow (compressible)	$\phi_x = u, \phi_y = v, \phi_z = w$; Equation 1.52.	$\omega = 0$
Inviscid irrotational incompressible flow	$\phi_x = u \ \ \phi_y = v \ \ \phi_z = w$; Equation 1.55.	$\omega = 0$
Two-dimensional potential flow (compressible) stream function	$u = (\rho_o/\rho)\psi_y; v = -(\rho_0/\rho)\psi_x, w = 0$. Equations 1.13, 1.57–1.60 and are solved simultaneously for ψ, u, υ, ρ, p, and T.	$\omega = 0$
Two-dimensional potential flow (incompressible) stream function equation	$u = \psi_y, v = -\psi_x, w = 0$　Equation 1.63 is solved.	$\omega = 0$.

density is treated as constant. The continuity equation in this case is given by

$$\nabla \cdot \mathbf{V} = 0 \tag{1.42}$$

and the momentum equation, neglecting gravitational effects, is given by

$$\rho \frac{D\mathbf{V}}{Dt} = -\nabla p + \mu \nabla^2 \mathbf{V} \tag{1.43}$$

$\rho \dfrac{D\mathbf{V}}{Dt}$	$-\nabla p$	$\mu \nabla^2 \mathbf{V}$
Local and convective acceleration	Pressure gradient	Viscous forces

In many practical situations, the thickness of the boundary layer developing on a surface is small compared to a characteristic length of the body. In such a situation, the viscous diffusion in the direction normal to the body (y) is locally dominant compared to viscous diffusion in the two other directions:

$$\mu \frac{\partial^2 \mathbf{V}}{\partial y^2} \gg \mu \left(\frac{\partial^2 \mathbf{V}}{\partial x^2} + \frac{\partial \mathbf{V}}{\partial z^2} \right)$$

If this assumption is incorporated into Eq. 1.18 or 1.43, the equation is generally called a *thin-layer Navier–Stokes equation* (TLNS) and is used by many computational fluid dynamicists.

If, in addition to the above assumptions, the streamwise pressure gradient $[-(1/\rho)(\partial p/\partial x_1), x_1$ is along the body surface] is known or prescribed, the governing equations are parabolic in nature. This assumption provides a major simplification to the governing equations, especially in numerical analysis, because the solution is independent of downstream conditions and the computation can be marched in the dominant flow direction. Such equation sets are called *parabolized Navier–Stokes equations* (PNS). There is one more level of approximation, called the *partially parabolized Navier–Stokes equations* (PPNS), where the elliptic effects of pressure are incorporated through a separate equation.

The next order of approximation is the "boundary layer approximation" introduced by Prandtl. In high-Reynolds-number flows where the flow remains attached to the surface, the pressure gradient normal to the wall ($\partial p/\partial y$) can be shown to be small, and the pressure gradient in the streamwise and cross-flow ($\partial p/\partial x, \partial p/\partial z$) directions are assumed to be known or prescribed. In most situations, these pressures are derived from an inviscid flow solution. This pressure is assumed to be "impressed" or constant in the shear layer. This simplification eliminates the momentum equation in the normal direction, and the analysis and computation are further simplified substantially. For example, the governing equations for a three-dimensional boundary layer in an incompressible steady flow (in the Cartesian system) are given by

$$\frac{\partial u}{\partial x} + \frac{\partial v}{\partial y} + \frac{\partial w}{\partial z} = 0 \tag{1.44}$$

$$u\frac{\partial u}{\partial x} + v\frac{\partial u}{\partial y} + w\frac{\partial u}{\partial z} = -\frac{1}{\rho}\frac{\partial p}{\partial x} + \nu\frac{\partial^2 u}{\partial y^2} \tag{1.45}$$

$$u\frac{\partial w}{\partial x} + v\frac{\partial w}{\partial y} + w\frac{\partial w}{\partial z} = -\frac{1}{\rho}\frac{\partial p}{\partial z} + \nu\frac{\partial^2 w}{\partial y^2} \tag{1.46}$$

where

$$-\frac{1}{\rho}\frac{\partial p}{\partial x} = U_e\frac{\partial U_e}{\partial x} + W_e\frac{\partial U_e}{\partial z} \tag{1.47}$$

$$-\frac{1}{\rho}\frac{\partial p}{\partial z} = U_e\frac{\partial W_e}{\partial x} + W_e\frac{\partial W_e}{\partial z} \tag{1.48}$$

Here the edge velocities $U_e(x, z)$ and $W_e(x, z)$ are known or prescribed. There are three unknown velocity components (u, v, w) and three governing equations. In view of these assumptions, the techniques used for the computation/analysis of the boundary layer equations are much simpler than those employed for the exact solution of the TLNS, PNS, and PPNS.

The next level of approximation neglects viscosity terms in the equation of motion. The compressible Euler equations are given by Equations 1.10 (with $\mu = 0$). There are five unknowns (u, v, w, p, e_0) which are closed using continuity, and appropriate (inviscid) energy equation, and an equation of state. The incompressible Euler equations are given by Eq. 1.43 (with $\mu = 0$). Here there are four unknowns (u, v, w, p), and closure is obtained by introducing continuity and the Bernoulli equation

$$p + \tfrac{1}{2}(u^2 + v^2 + w^2) = \text{const} \tag{1.49}$$

and Eq. 1.49 is applied along streamlines.

1.2.4 Potential Flow Equations

The next level of approximation is the assumption of irrotationality, $\boldsymbol{\omega} = \nabla \times \mathbf{V} = 0$. For any vector whose curl is zero, $\nabla \times \mathbf{A} = 0$, $\mathbf{A}$ can be expressed by $\nabla\zeta$, where ζ is the scalar. This arises directly from the vector relation $\nabla \times \nabla\zeta \equiv 0$. Applying this to irrotational flow, we can define a scalar velocity potential, ϕ, by

$$\mathbf{V} = \nabla\phi \tag{1.50}$$

Substitution of this equation in the continuity equation and elimination of ρ from the momentum equation

$$dp = -\rho d\left[(u^2 + v^2 + w^2)/2\right] \tag{1.51}$$

leads to the following equation in the Cartesian system (Anderson, 1982):

$$\left(1 - M_x^2\right)\phi_{xx} + \left(1 - M_y^2\right)\phi_{yy} + \left(1 - M_z^2\right)\phi_{zz} - 2M_x M_y \phi_{xy}$$
$$- 2M_x M_z \phi_{xz} - 2M_y M_z \phi_{yz} = 0 \tag{1.52}$$

where

$$M_x = u/a, \qquad M_y = v/a, \qquad M_z = w/a$$
$$\phi_x = u, \qquad \phi_y = v, \qquad \phi_z = w$$

and a is the speed of sound. This equation is hyperbolic for supersonic flows and elliptic for subsonic flows.

Many further simplifications can be made to Eq. 1.52, as follows:

1. If x is the dominant flow direction (M_y, $M_z \ll M_x$), Eq. 1.52 reduces to

$$\left(1 - M_x^2\right)\phi_{xx} + \phi_{yy} + \phi_{zz} = 0 \tag{1.53}$$

2. If the perturbations due to the body are small, the potential can be expressed in terms of a perturbation potential

$$\phi = \Phi + \phi'$$

where Φ is the potential of the unperturbed flow and ϕ' is the perturbation potential. Equation 1.53 can be simplified (Anderson, 1982) to

$$\left(1 - M_\infty^2\right)\phi'_{xx} + \phi'_{yy} + \phi'_{zz} = 0 \tag{1.54}$$

where M_∞ is the upstream Mach number of the unperturbed flow, which is parallel to the x axis. Equation 1.54 is a *linear equation* and, therefore, is easier to solve. Equations 1.52 and 1.53 are called *full potential equations*, and Eq. 1.54 is called a *perturbation potential equation*.

3. If the flow is incompressible, Eq. 1.52 reduces to

$$\phi_{xx} + \phi_{yy} + \phi_{zz} = 0 \tag{1.55}$$

In any given situation, if the flow is irrotational, one of the Eqs. 1.52–1.55 can be solved and the velocity field can be derived utilizing Eq. 1.50. In steady, inviscid flow without heat transfer, the energy equation (Eq. 1.16) can be written

$$\nabla \cdot \rho \mathbf{V} h_o = 0 \tag{1.56}$$

If, in addition, the flow is irrotational and the working fluid is a perfect gas, Eqs. 1.52 and 1.56 can be combined to yield a closure relation:

$$a^2 = a_0^2 - \frac{\gamma - 1}{2}(u^2 + v^2 + w^2) \tag{1.57}$$

from which pressure and density can be obtained. If the flow is incompressible, the Bernoulli equation (Eq. 1.49) is used instead of Eq. 1.57.

1.2.5 Stream Function Equations

In a *two-dimensional* steady potential flow, the Euler equation can be simplified by defining a new function related to the velocities. This is a point function and is related to the mass flow. The stream function for steady, two-dimensional flow can be defined as

$$\frac{\rho_o}{\rho} \frac{\partial \psi}{\partial y} = \frac{\rho_o}{\rho} \psi_y = u \tag{1.58}$$

$$-\frac{\rho_o}{\rho} \psi_x = v \tag{1.59}$$

Adopting these equations automatically enforces continuity equation, Eq. 1.9. Eliminating the density from the Euler equations (Eq. 1.10 with $\mu = 0$) results in the following expression (in Cartesian coordinates) for two-dimensional potential flow:

$$\left(1 - M_x^2\right)\psi_{xx} + \left(1 - M_y^2\right)\psi_{yy} - 2M_x M_y \psi_{xy} = 0 \tag{1.60}$$

Equations 1.60, 1.58, 1.59, 1.57 ($w = 0$), and 1.13 are solved simultaneously to derive values of u, v, ρ, p, and T, respectively.

Equation 1.52 (with $\phi_z = M_z = 0$) is similar to Eq. 1.60. The techniques for the solution of Eqs. 1.52 and 1.60 are also similar. Specifically, Eq. 1.60 can be simplified in many situations, as described below.

1. If the dominant flow direction is along x ($M_y \ll M_x$), Eq. 1.60 reduces to

$$\left(1 - M_x^2\right)\psi_{xx} + \psi_{yy} = 0 \tag{1.61}$$

2. If the perturbations due to the body are small, the stream function can be expressed in terms of the perturbation stream function

$$\psi = \Psi + \psi'$$

where Ψ is the stream function of the unperturbed flow. Neglecting second-order terms, Eq. 1.61 can be simplified to

$$\left(1 - M_\infty^2\right)\psi'_{xx} + \psi'_{yy} = 0 \tag{1.62}$$

where ψ' is the stream function of the perturbed flow, M_∞ is the freestream Mach number. This is a linear equation and is easier to solve analytically. If the flow is *incompressible*, Eq. 1.60 can be simplified to give

$$\psi_{xx} + \psi_{yy} = 0 \tag{1.63}$$

which is a Laplace equation (linear) and can be solved by various standard techniques.

It is advantageous to solve one partial differential equation (PDE; Eq. 1.60) for ψ as compared to two PDEs for u and v in two-dimensional flow. This approach is, therefore, widely used for two-dimensional flows.

The stream function approach is also used in the analysis of viscous flows. For example, the two-dimensional form of the incompressible boundary layer equation ($w = W_e = 0$) in Eqs. 1.44 and 1.45 can be expressed as

$$\psi_y \psi_{xy} - \psi_x \psi_{yy} = U_e \frac{dU_e}{dx} + \nu \frac{\partial^3 \psi}{\partial y^3} \tag{1.64}$$

where U_e is prescribed. Instead of two PDEs (Eqs. 1.44 and 1.45, with $w = 0$), only one PDE (Eq. 1.64) is solved for the solution of the flow field. The stream function equations are widely used in the following situations:

1. Two-dimensional inviscid, incompressible flows
2. Two-dimensional inviscid, irrotational compressible flows ($\omega = 0$)
3. Two-dimensional boundary layer flows

1.2.6 Basic Concepts of Turbulent Flows

Most turbomachinery flows are turbulent, with laminar and transitional regions occurring near the leading edge of the blades. Turbulence influences the aerodynamic and thermodynamic performance of turbomachines, and therefore its consideration is critical to turbomachinery analysis. A brief introductory treatment of the subject is provided in this section.

Turbulence is characterized by irregular or random fluctuations. It often originates as instability of high-Reynolds-number laminar flow which results in a transition to turbulence. Turbulence is a very complex phenomenon, from both the measurement and analysis point of view. Because of its random variation in space and time, much analysis relies on statistical approaches. The equations governing the statistical properties of a turbulent flow involve more unknowns than the equations available; additional terms which appear in these equations must be modeled, and therefore the problem of including turbulence in the formulation is often referred to as the *closure problem*. Statistical approaches vary from very crude (often purely empirical) models to highly sophisticated (often impractical) models.

Turbulence, in addition to being random and irregular, is three-dimensional. Velocity fluctuations exist in all directions, even if the mean (time average) flow is one- or two-dimensional. Turbulence is diffusive and dissipative. The former phenomenon gives rise to rapid mixing and increased rates of momentum, heat, and mass transfer. Dissipation is characterized by deformation work associated with velocity fluctuations, and it increases the internal energy of the fluid at the expense of the kinetic energy in the mean flow and the turbulence.

There are several books and reviews on the subject of turbulent flow. Tennekes and Lumley (1972) and Bradshaw (1970) provide introduction to the subject, and Hinze (1975), Lumley (1970), and Batchelor (1956), provide more advanced treatment. The Stanford Conference Proceedings (Kline et al., 1982) is a compilation of experimental data and provides (a) a review of various aspects of turbulence, and (b) a critique of various "closure models" used in analysis and computation. Schlichting (1979) and White (1991) also provide a good introductory treatment of the physics, modeling, and analysis of turbulent flows. This section summarizes basic concepts in turbulence and turbulence modeling. Advanced concepts and turbulence modeling for complex flows are briefly presented in Chapter 5.

A detailed classification of various turbulent flows is given in Kline et al. (1982). An attempt is made below to classify the turbulent flows encountered in turbomachinery:

1. Homogeneous or nonhomogeneous

2. Isotropic or nonisotropic

3. Steady or unsteady mean flow

4. Incompressible ($M < 0.3$) or compressible ($M > 0.3$); subsonic ($M < 1$), transonic ($M = 1$), supersonic ($M > 1$)

5. Two-dimensional mean flow (cascade) and three-dimensional mean flow (rotor, stator)

6. Internal or external flows (most turbomachinery flows are internal flows)

7. Simple strain or extra strain with curvature, rotation, buoyancy effects

8. Free shear flows (cascade wake, rotor wake, tip leakage jet)

9. Wall bounded shear flows (cascade and rotor blade boundary layer, annulus wall and hub wall boundary layers, etc.)

The technique used in solving the turbulent flow equations and the model or equations employed for representing turbulence depends on the type of flow listed. Most classifications are obvious, but some new definitions were introduced at the Stanford Conference. In a simple strain shear layer, there is only one significant rate-of-strain component, say $\partial u / \partial y$, the dominant velocity gradient in a two-dimensional boundary layer or wake. An extra strain is one in which additional effects such as rotation, curvature, and buoyancy give rise to turbulence structure modifications. The influence that curvature and rotation have on the turbulence field in turbomachinery can be significant.

Wall bounded and free turbulent shear flows are of importance in turbomachinery. Turbulent boundary layers develop on blade surfaces and endwalls. Cascade and rotor wakes, tip clearance flow, and turbine cooling jets are examples of turbulent free shear flows encountered in turbomachinery.

Further distinctions can be drawn between shear layers in the near and far fields. Specifically, the transition from a wall layer to a wake at the blade trailing edge is termed the *near field*. This region is characterized by large streamwise gradients and rapidly varying flow properties. This region may extend from the trailing edge to a small fraction of the blade chord length from the trailing edge. Similar effects are associated with the leakage jet emanating from the blade tip. The far field (e.g., wake about one chord downstream of the trailing edge) is where the rate of change of flow properties with distance is small and the turbulence has attained near-equilibrium conditions.

The basic assumption in the statistical representation of turbulence is that the influence of turbulence on average properties of the flow are of interest, and thus a time average for each of the instantaneous flow equations is considered. The instantaneous velocity $u_i(t)$ can be decomposed as follows:

$$u_i(t) = \bar{u}_i + u_i' \tag{1.65}$$

where u_i is the instantaneous velocity ($i = 1, 2, 3$), u' is the fluctuating component, and the overbar indicates time-averaged or Reynolds-averaged values, defined as

$$\bar{\phi} = \frac{1}{T} \int_0^T \phi \, dt \tag{1.66}$$

Here, T is large compared to the time scale of fluctuations. The instantaneous value of pressure can likewise be decomposed as

$$p = \bar{p} + p' \tag{1.67}$$

When Eqs. 1.65 and 1.67 are substituted into Eq. 1.10 and the entire equation is time-averaged, a new term appears:

$$\frac{\partial \bar{u}_i}{\partial t} + \bar{u}_j \frac{\partial \bar{u}_i}{\partial x_j} = -\frac{1}{\rho}\frac{\partial \bar{p}}{\partial x_i} + \frac{\partial}{\partial x_j}\left(\nu \frac{\partial \bar{u}_i}{\partial x_j} - \overline{u'_i u'_j}\right) \tag{1.68}$$

In the derivation of Eq. 1.68, the flow has been assumed incompressible. Time averaging of the entire equation gave rise to the elimination of several terms:

$$\frac{\partial \overline{\bar{u}_j u'_i}}{\partial x_j} = \frac{\partial \overline{u'_j \bar{u}_i}}{\partial x_j} = -\frac{1}{\rho}\frac{\overline{\partial p'}}{\partial x_i} = \frac{\partial}{\partial x_j}\left(\nu \frac{\overline{\partial u'_i}}{\partial x_j}\right) = 0$$

Further details of this Reynolds averaging procedure are available in White (1991). The new term which arises, $(\partial/\partial x_j)(-\overline{u'_i u'_j})$, distinguishes the form of Eq. 1.68 from the instantaneous incompressible momentum equation. By rewriting the last term on the right-hand side of Eq. 1.68 as

$$\frac{1}{\rho}\frac{\partial}{\partial x_j}(\tau_{ij}), \qquad \tau_{ij} = \mu \frac{\partial \bar{u}_i}{\partial x_j} - \rho\overline{u'_i u'_j} \tag{1.69}$$

we can identify $-\rho\overline{u'_i u'_j}$ with the molecular stress tensor $\tau_{ij} = \mu(\partial \bar{u}_i/\partial x_j)$. Accordingly, the term $-\rho\overline{u'_i u'_j}$ is termed the *Reynolds stress*. The Reynolds stresses account for the enhanced mixing observed in turbulent flows. The introduction of the Reynolds stress tensor has added six additional unknowns to the momentum equations.

Application of the Reynolds averaging procedure to the incompressible continuity equation yields

$$\frac{\partial \bar{u}_i}{\partial x_i} = 0 \tag{1.70}$$

which is identical in form to the steady instantaneous continuity equation. The six Reynolds stresses remain to be evaluated in order that the time-averaged continuity (Eq. 1.70) and momentum (Eq. 1.68) equations be closed.

Following a procedure analogous to that laid out above, the energy equation for turbulent mean flow can be derived. Limiting ourselves again to incompressible flow, we decompose temperature and velocity components as before:

$$T = \bar{T} + T', \qquad u_i = \bar{u}_i + u'_i \tag{1.71}$$

Substituting Eq. 1.71, into Eq. 1.11 or 1.18 or 1.34, and time averaging the resulting equation yields the mean energy equation in turbulent flow.

For example, the mean energy equation for incompressible flow, based on Eq. 1.34, is given by

$$\rho\frac{\partial \overline{T}}{\partial t} + \rho\frac{\partial}{\partial x_j}\left(\overline{u}_j\overline{T}\right) = \frac{1}{PR\,Re}\frac{\partial^2 \overline{T}}{\partial x_j\,\partial x_j} + Ec\left(\overline{u}_j\frac{\partial \overline{p}}{\partial x_j} + \overline{u'_j\frac{\partial p'}{\partial x_j}} + \frac{\partial \overline{p}}{\partial t}\right)$$

$$+ \frac{Ec}{Re}(\phi_l + \phi_t) - \rho\frac{\partial}{\partial x_j}\left(\overline{u'_j T'}\right) \qquad (1.72)$$

The total heat flux q_i is given by

$$q_i = q_t + q_l = -\frac{1}{PR\,Re}\frac{\partial^2 \overline{T}}{\partial x_j\,\partial x_j} + \rho\frac{\partial}{\partial x_j}\left(\overline{u'_j T'}\right) \qquad (1.73)$$

This is a nondimensional equation (similar to Eq. 1.34), $\overline{u'_j T'}$ is normalized by $V_\infty \Delta T_o$. The dissipation (ϕ_l) due to molecular viscosity is given by Eq. 1.12, and the turbulent dissipation is given by

$$\phi_t = \left(\frac{\partial \overline{u}_i}{\partial x_j}\right)\left(-\rho\overline{u'_i u'_j}\right) \qquad (1.74)$$

Equations 1.72–1.74 represent the Reynolds averaged energy equation for an incompressible flow with constant properties. As with the Reynolds stress in the momentum equations, the Reynolds heat flux, $\rho\overline{u'_j T'}$, remains to be modeled, and this is covered in Section 7.8.1. The fluctuating pressure gradient-velocity correlation is one of the most difficult closure problems in turbulence, and no satisfactory model is available for this term. Further discussion of the energy equation for incompressible and compressible turbulent flows with heat transfer can be found in Schlichting (1979) and White (1991).

In the remainder of the book, the superscript ($-$) on flow variables will be dropped for clarity. Hence quantities ($u, v, w, p, \rho, h_o, h, s, e$) would represent ensemble averaged (to remove random fluctuations) and the variables with superscript ($'$) represent the fluctuating quantities.

1.2.7 Turbulence Modeling

Turbulence closure requires modeling of the fluctuation terms in Eqs. 1.68 and 1.72. Modeling strategies for the velocity correlation, $\overline{u'_i u'_j}$, will be introduced in this section and considered again in Chapter 5, while the modeling of $\overline{u'_j T'}$ will be considered in Chapter 7. Several papers have appeared recently reviewing the state of the art of turbulence modeling and computation (Kline et al., 1982; Marvin, 1983; and Lakshminarayana, 1986).

A recent entry to this field is a book by Wilcox (1993). The closure techniques can be classified as follows:

1. *Zero-Equation or Algebraic Eddy Viscosity Models.* These models employ an algebraic form for the turbulent stresses $(\overline{u_i'u_j'})$.
2. *One-Equation Models.* These models employ an additional PDE for a turbulence velocity scale.
3. *Two-Equation Models.* These models employ one PDE for a turbulence length scale and one PDE for a turbulence velocity scale.
4. *Reynolds Stress Models.* These models employ several (usually seven) PDEs for all of the components of the turbulence stress tensor $(-\rho\overline{u_i'u_j'})$.
5. *Direct Numerical Simulation.* The time-dependent (3D) structure is resolved through a numerical solution of time-dependent Navier–Stokes equations.

Brief descriptions of algebraic and two-equation eddy viscosity models and of Reynolds stress models are given here. The first two models invoke the concept proposed by Boussinesq (1877). In this concept, turbulent stresses in the mean momentum equation $(-\rho\overline{u_i'u_j'})$ are assumed to be proportional to the mean rate of strain:

$$\tau_{ij} = -\rho\left(\overline{u_i'u_j'}\right) = \mu_t\left(\bar{u}_{i,j} + \bar{u}_{j,i}\right) - \tfrac{2}{3}\rho\delta_{ij}k \qquad (1.75)$$

where the constant μ_t, called "eddy" viscosity, is a spatial function of the mean velocity field $\bar{u}_i$. The physical idea behind this concept is explained in Tennekes and Lumley (1972). The stresses in laminar flows arise as a result of random molecular motion, which is conceptually similar to turbulent fluctuations. The resemblance between these two motions is somewhat superficial. Nevertheless, it is assumed that the transfer of momentum and heat by molecular motion is similar to that induced by turbulent fluctuations. The concept of "eddy viscosity" is phenomenological and has no mathematical basis. It should be emphasized that molecular viscosity is a property of *fluid* and that turbulence is a property of *flow*. Therefore, the eddy viscosity is likely to be a function of the flow properties (e.g., mean velocity) and may also be a tensorial quantity in a three-dimensional flow.

1.2.7.1 *Algebraic Eddy Viscosity Model*

Algebraic eddy viscosity models are based on the mixing length concept [Schlichting (1979), for example]. The "mixing length model" can be written as

$$\mu_t = 2\rho l^2\sqrt{S_{ij}S_{ij}} \qquad (1.76)$$

where l is termed the mixing length and is representative of the local large-scale motion in a turbulent flow.

In algebraic eddy viscosity models, an algebraic expression for the mixing length, l, is provided for closure. Several groups have developed algebraic eddy viscosity models (Cebeci and Smith, 1974; Crawford and Kays, 1975; Mellor and Herrig, 1973; Patankar and Spalding, 1970; Baldwin and Lomax, 1978).

One of the most widely used models for the engineering calculation of boundary layers is that due to Cebeci and Smith (1974), later modified by Baldwin and Lomax (1978). The model proposed by Baldwin and Lomax has wider applications and avoids the necessity of finding the edge of the boundary layer. Both of these models incorporate empirical relations. The Baldwin and Lomax model is given by

$$\mu_t = \begin{cases} 0.16\rho y^2 [1 - \exp(-y^+/A^+)]^2 |\omega|, & 0 \le y \le y' \\ 0.02688\,\rho F_w \left[1 + 5.5\left(\dfrac{0.3y}{y_{max}}\right)^6\right]^{-1}, & y \ge y' \end{cases} \qquad (1.77)$$

where y is the distance normal to the wall, 0 to y' is the wall region, and y' represents the smallest value of y at which μ_t in the inner and outer layer are equal.

$$y^+ = \frac{\sqrt{\rho_w \tau_w}\, y}{\mu_w}, \; A^+ = 26$$

τ_w, μ_w = shear stress and molecular viscosity at the wall

$$|\omega| = \sqrt{\left(\frac{\partial u}{\partial y} - \frac{\partial v}{\partial x}\right)^2 + \left(\frac{\partial v}{\partial z} - \frac{\partial w}{\partial y}\right)^2 + \left(\frac{\partial w}{\partial x} - \frac{\partial u}{\partial z}\right)^2} \qquad (1.78)$$

$$F_w = \left\{ \frac{0.25 y_{max} \left[\left(\sqrt{u^2 + v^2 + w^2}\right)_{max} - \left(\sqrt{(u^2 + v^2 + w^2)}\right)_{min}\right]^2}{F_{max}} \right\}$$

or $F_w = y_{max} F_{max}$

Smaller of the two values for F_w is taken

Here, F_{max} is the maximum value of the relation

$$F(y) = y|\omega|[1 - \exp(-y^+/A^+)] \qquad (1.79)$$

and y_{max} is y at that point. The value of $(u^2 + v^2 + w^2)_{min} = 0$ for the boundary layer.

The outer formulation in Eq. 1.77 could be used for the wake as well as for the boundary layer. For the wake, only the outer formulation is used. For the boundary layer, both formulations in Eq. 1.77 are employed.

1.2.7.2 Two-Equation Models In the algebraic eddy viscosity model, local equilibrium of the turbulence is assumed, and the eddy viscosity is assumed to be scalar (the same in all directions), mainly dependent on the mean velocity field. The characteristics of the turbulence field itself never enter into this formulation. The two-equation model is intended to rectify this situation by relating the eddy viscosity to some gross local properties of turbulence. These are intended for application in situations where turbulence is not in equilibrium. The most widely used model in this category is the Jones–Launder k–ϵ model (1972, 1973). The Wilcox–Rubesin model (1980), which has been specifically developed for boundary layer flows with large pressure gradients, is also used by many investigators. Because there is considerable similarity (in physical concept) between these two models, only the equations due to Jones and Launder are presented here.

The two-equation model employs equations governing dynamics or gross properties of turbulence, and it relates them to eddy viscosity through the equation

$$\mu_t = \mu C_\mu f_\mu R_T \tag{1.80}$$

where

$$C_\mu = \text{constant}$$

$$f_\mu = \exp\left[-2.5/(1 + R_T/50)\right] \tag{1.81}$$

$$R_T = \text{turbulence Reynolds number} = \frac{\rho k^2}{\mu \epsilon}$$

$$k = \text{kinetic energy of turbulence}$$

$$= \overline{u_i' u_i'}/2 = \left[\overline{(u')^2} + \overline{(v')^2} + \overline{(w')^2}\right]/2 \tag{1.82}$$

$$\epsilon = \text{turbulent energy dissipation rate } (2\nu S_{ij}' S_{ij}')$$

This concept introduces additional physics (local values of k and ϵ) in the formulation and has been very successful in computing two-dimensional flows.

An equation for the kinetic energy of turbulence (k) can be derived from the Navier–Stokes equation for velocity ($u_i(t)$), by multiplying the momentum equation by $u_i(t)$, taking the time average of all terms (similar to Eq. 1.68), and subtracting the energy equation for the mean flow. The resulting turbulent kinetic energy equation is given by

$$\frac{\partial k}{\partial t} + \bar{u}_j \frac{\partial}{\partial x_j}(k) = -\frac{\partial}{\partial x_j}\left(\frac{1}{\rho}\overline{u_j' p'} + \frac{1}{2}\overline{u_i' u_i' u_j'} - 2\nu\overline{u_i' S_{ij}'}\right)$$
$$-\overline{u_j' u_i'} S_{ij} - 2\nu\overline{S_{ij}' S_{ij}'} \tag{1.83}$$

The term on the left-hand side is the convection of k; the first term (in brackets) on the right-hand side represents diffusion of turbulence due to pressure fluctuations, velocity fluctuations, and viscous stresses, respectively. The second term on the right side of Eq. 1.83 is the production of k by mean flow velocity gradients. This normally involves transfer of energy from mean flow to turbulence. The last term represents the rate at which viscous stresses dissipate the turbulent kinetic energy. This is called *viscous dissipation* and is denoted by $\epsilon = 2\nu \overline{S'_{ij} S'_{ij}}$. The diffusion terms must be modeled to provide a proper closure equation. This has been carried out by many investigators, including Jones and Launder, who provided the following equation for kinetic energy (for incompressible flow):

$$\frac{\partial}{\partial t}(k) + \frac{\partial}{\partial x_j}(k\bar{u}_j) = \frac{\partial}{\partial x_j}\left(\left(\nu + \frac{\nu_t}{C_k}\right)\frac{\partial k}{\partial x_j}\right) - \overline{u'_i u'_j}\,\bar{u}_{i,j} - \epsilon \quad (1.84)$$

$$(1) \qquad\qquad\qquad (2) \qquad\qquad\qquad (3) \quad\quad (4)$$

where, once again, the terms on the right-hand side are, respectively, diffusion, production, and dissipation. The term on the left-hand side includes the local and convective change in k. C_k is an effective Prandtl–Schmidt number (normally assumed to be unity). Equation 1.75 is employed for $\overline{u'_i u'_j}$.

An additional transport equation for dissipation (ϵ) is needed to close these equations. This is derived from the Navier–Stokes equation through manipulation. The dissipation equation is derived in a manner similar to that employed in deriving Eq. 1.83. Tennekes and Lumley (1972, p. 87), provide derivation and interpretation of this equation. The expression so derived can be identified as follows

$$\text{Convection} = \text{generation by mean flow}$$

$$+ \text{ Generation by stretching action of turbulence}$$

$$+ \text{ Diffusive transport} + \text{Viscous destruction} \qquad (1.85)$$

Several investigators have attempted modeling the terms in the exact dissipation equation. A review and critique of the various models proposed is given by Lumley (1980). Some of the major modeling efforts are due to Launder and his group (Jones and Launder, 1972, 1973; Hanjalic and Launder, 1972; Launder et al., 1975). The modeled equation for dissipation (for incompressible flow) can be written as

$$\frac{\partial}{\partial t}(\epsilon) + \frac{\partial(\epsilon \bar{u}_j)}{\partial x_j} = \frac{\partial}{\partial x_j}\left(\left(\nu + \frac{\nu_t}{C_\epsilon}\right)\frac{\partial \epsilon}{\partial x_j}\right) - C_{\epsilon_2}\frac{\epsilon^2}{k} - C_{\epsilon_1}\frac{\epsilon}{k}\overline{u'_i u'_j}\,\bar{u}_{i,j} \quad (1.86)$$

The three terms on the right-hand side of Eq. 1.86 are, respectively, diffusive transport, viscous destruction, and production. The constants in

Eqs. 1.84 and 1.86, used by Launder's group, are $C_k = 1.0$, $C_{\epsilon_1} = 1.45$, $C_{\epsilon_2} = 1.90$, and $C_\epsilon = 1.3$, $C_\mu = 0.09$.

Equations 1.80, 1.84, and 1.86 are solved simultaneously along with the mean momentum equation to derive the values of mean velocities, k, ϵ, and stresses. Equations 1.84 and 1.86 can also be recast in conservative form (ρk, $\rho \epsilon$) and expressed in a form similar to Eq. 1.18. In such a case, the turbulence transport equations for k and ϵ can be incorporated into vectors $q[\rho, \rho u, \rho v, \rho w, \rho e_0, \rho k, \rho \epsilon]$, E, F, G, H, and T. This provides a major simplification in the numerical analysis, because the numerical scheme used for the mean momentum equations could also be used for the solution of the turbulent flow field.

Various modifications have been made to the basic two-equation (k–ϵ) model. These include compressibility effect, wall vicinity effect, and rotation and curvature effects [see Lakshminarayana (1986), for a review]. One widely used modification is the wall vicinity effect. The high-Reynolds-number model (Eqs. 1.84 and 1.86) fails near the wall where the Reynolds numbers are low. One of the successful modifications due to this effect is the Chien model (1982). In this model, the following additional terms are added to the k and ϵ equations (right-hand side as source term) to satisfy the physical constraints near the wall:

$$k \text{ Equation (Eq. 1.84):} \quad -2\nu k/(y_w)^2$$

$$\epsilon \text{ Equation (Eq. 1.86):} \quad -2\nu\epsilon[\exp(-0.5y^+)]/y_w^2$$

where y_w is the distance from the wall and y^+ is the wall variable (Reynolds number) defined by

$$y^+ = \frac{\rho y_w}{\mu}\sqrt{\frac{\tau_w}{\rho}}$$

In addition, the constants in Eq. 1.86, C_{ϵ_2} and C_{ϵ_1}, are modified to include the wall vicinity effect.

There is a controversy as to whether the constants in the k–ϵ model are universally valid. It has been suggested by Kline et al. (1982) that one may have to resort to different modeling for different zones. For example, either the constants or the modeling may vary between the boundary layer and the wake. It should be remarked here that most turbulence models presently in use are for stationary cases. The rotation is likely to alter the turbulence structure, as indicated by data and analysis of Lakshminarayana and Reynolds (1980).

1.2.7.3 *Reynolds Stress Models* The eddy viscosity models discussed so far have some major limitations, one of which is the assumption that the eddy viscosity is isotropic. Rodi (1982), in a review article, listed the major limitations of these models. In many practical situations, eddy viscosity approaches are not valid. Highly three-dimensional flows, rotating flows, and flows with large curvatures are examples where these models perform poorly. Reynolds stress modeling abandons the eddy viscosity concept. Reynolds stress models, called *second-order closure models*, employ transport equations governing the individual stress components. Models based on the transport of individual Reynolds stresses are under extensive development.

Solution of the complete set of Reynolds stress equations is a formidable task. For example, in three-dimensional, compressible, turbulent flow, the use of Reynolds stress transport models result in at least 7 additional nonlinear partial differential equations.

The Reynolds stress transport equations can be derived by multiplying the fluctuating momentum equation (for u'_i) by u'_j, adding this to the product of u'_i and the u'_j momentum equation and time averaging this sum. The resulting exact Reynolds stress equations for three-dimensional flow in a coordinate system rotating with constant angular velocity is given by (Launder et al., 1975; Galmes and Lakshminarayana, 1984).

$$\underset{(1)}{\left(\rho \overline{u'_i u'_j} \bar{u}_k \right)_{,k}} = \underset{(2a)}{\left(-\overline{p' u'_i} \delta_{kj}} - \underset{(2b)}{\overline{p' u'_j} \delta_{ik}} - \underset{}{\rho \overline{u'_i u'_j u'_k}} + \underset{(2c)}{\mu (\overline{u'_i u'_j})_{,k}} \right)_{,k}}$$

$$+ \underset{(3)}{\left(\overline{p' u'_{i,j}} + \overline{p' u'_{j,i}} \right)} - \underset{(4)}{\left(\overline{\rho u'_i u'_k} \bar{u}_{j,k} + \overline{\rho u'_j u'_k} \bar{u}_{i,k} \right)}$$

$$\underset{(5)}{-2 \mu \overline{u'_{i,k} u'_{j,k}}} - \underset{(6)}{2 \rho \Omega^P (\epsilon_{ipk} \overline{u'_j u'_k} + \epsilon_{jpk} \overline{u'_i u'_k})} \qquad (1.87)$$

where

$$\Omega^P = \text{angular velocity of reference frame}$$
$$\epsilon_{ipj} = \text{the permutation tensor}$$

The physical meaning of each of the terms in Eq. 1.87 is as follows:

Term 1	Convection of Reynolds stresses and turbulence intensities ($i = j$) by the mean flow
Terms 2a, 2b, 2c	Diffusion of $\overline{u'_i u'_j}$ by pressure gradient, turbulent fluctuations, and viscous effects, respectively
Term 3	Pressure−strain interaction
Term 4	Production of $\overline{u'_i u'_j}$ by mean strain gradients
Term 5	Viscous dissipation of $\overline{u'_i u'_j}$
Term 6	Redistribution by Coriolis forces

The main effect of rotation is in the redistribution of the turbulence intensities and shear stresses. It is relatively easy to prove, by collapsing the indices in Eq. 1.87, that rotation terms do not appear in the exact transport equation for the kinetic energy of turbulence ($k = \frac{1}{2}\overline{u_i'u_i'}$).

Terms 1, 4, and 6 in Eq. 1.87 need not be modeled because they contain only the Reynolds stresses, gradients of the mean velocity, and the rotation rate (all of which are "knowns" at the second-order closure level). Terms 2, 3, and 5 must be modeled. It is beyond the scope of this book to go through the details of the modeling procedures. However, in general, most models for the diffusion, pressure–strain, and dissipation terms are of similar form. The most widely used Reynolds stress model for incompressible flow is due to Launder et al. (1975) and is given by

$$\left(\overline{u_i'u_j'}\,\overline{u}_k\right)_{,k} = \underset{(1)}{C_S\left\{\frac{k}{\epsilon}\left[\overline{u_k'u_l'}\left(\overline{u_i'u_j'}\right)_{,l}\right]\right\}_{,k}} \underset{(2)}{} - \underset{(3a)}{C_1\frac{\epsilon}{k}(\overline{u_i'u_j'} - \tfrac{2}{3}\delta_{ij}k)} - \underset{(3b)}{C_2\left(P_{ij} - \tfrac{2}{3}\delta_{ij}P\right)}$$

$$\underset{(4)}{-\left[\overline{u_i'u_k'}\,\overline{u}_{j,k} + \overline{u_j'u_k'}\,\overline{u}_{i,k}\right]} \underset{(5)}{- \tfrac{2}{3}\epsilon\delta_{ij}} \underset{(6)}{- 2\Omega^P\left(\epsilon_{ipk}\,\overline{u_j'u_k'} + \epsilon_{jpk}\,\overline{u_i'u_k'}\right)} \tag{1.88}$$

The significance of various terms in Eq. 1.88 is as follows. Term 2 represents net diffusive transport, based on modeling of the terms 2a, 2b, and 2c in Eq. 1.87. The diffusion is expressed by a simple gradient transport model. The pressure–strain term consists of two parts. The first part (3a) represents the interaction of fluctuating quantities (the so-called "return to isotropy" term), and the second term (3b) represents the interaction of mean strain and fluctuating quantities (the so-called "rapid term"). The fifth term represents dissipation and is modeled based on the assumption of local isotropy of the dissipation rate tensor. The introduction of this isotropic dissipation variable requires the solution of an additional transport equation for ϵ. Details of the derivation and modeling of the dissipation rate equation are available in Hanjalic and Launder (1972). The constants suggested for this model are $C_S = 0.25, C_1 = 1.5, C_2 = 0.6$.

If a full Reynolds stress model is employed, a total of 12 transport equations must be solved. These include continuity, three momentum, mean energy, six Reynolds stress, and dissipation rate equations. Models employing transport equations for the individual turbulent stresses constitute a large number of nonlinear differential equations. Several authors (Launder, 1971; Rodi, 1976; Galmes and Lakshminarayana, 1984) have made various simplifying assumptions to reduce the PDEs governing the Reynolds stress transport equations to algebraic equations. This provides a substantial simplification of the turbulence equations. One successful method is the technique due to Rodi (1976). In this technique, the net transport of $\overline{u_i'u_j'}$ is assumed to be locally proportional to the net transport (convection and diffusion) of k, turbulent kinetic energy. This model was later modified by Galmes and Lakshminarayana (1984) to include the effect of rotation. Their model is given by

$$\frac{\overline{u_i' u_j'}}{k} = \frac{2}{3}\delta_{ij} + \frac{R_{ij}\left(1 - \frac{C_2}{2}\right) + \left(P_{ij} - \frac{2}{3}\delta_{ij}P\right)(1 - C_2)}{P + \rho\epsilon(C_1 - 1)} \tag{1.89}$$

where

$$P_{ij} = -\rho\left(\overline{u_i'u_k'}\,\overline{u}_{j,k} + \overline{u_j'u_k'}\,\overline{u}_{i,k}\right)$$

$$R_{ij} = -2\rho\Omega^P\left(\epsilon_{ipk}\,\overline{u_j'u_k'} + \epsilon_{jpk}\,\overline{u_i'u_k'}\right)$$

$$P = -\rho\overline{u_i'u_j'}\,\overline{u}_{i,j}\;(\text{production of }k) = P_{kk}$$

The governing equations now consist of (a) the continuity, momentum, and energy equations for the mean flow and (b) the turbulence closure equations, Eqs. 1.84, 1.86, and 1.89. Equation 1.89 is a coupled set of six nonlinear algebraic equations, and Eqs. 1.84 and 1.86 are transport equations for k and ϵ.

In general, the complexity of the mean flow dictates the types of turbulence models to be used. Participants in several workshops (Hines, 1982; Kline et al., 1982) have made attempts to address the question of validity, universality, and accuracy of various models. There has been no consensus on this matter. Lakshminarayana (1986) has provided a critical review of various turbulence models and their application to complex flows. In the author's view, computation of turbulent flow in compressor and turbine rotor blades, as well as annulus wall boundary layer growth in both single and multistage compressors, requires Reynolds stress models to provide a physically realistic and accurate prediction of the flow field. As computer memory and speed continue to increase, utilization of the second moment closure models will become more common.

2

FUNDAMENTAL PRINCIPLES, ANALYSIS, AND PERFORMANCE OF TURBOMACHINERY

This chapter deals with one-dimensional analysis and overall performance, efficiency, nondimensional representation, bladings, and so on, used in various types of turbomachinery. The first four sections deal with a unified approach to one-dimensional analysis, performance, and nondimensional groups, followed by one section each on axial flow compressors and fans, propulsors and prop fans, centrifugal compressors and fans, axial and centrifugal pumps, axial flow turbines, and radial and mixed flow turbines. Hydraulic turbines are dealt with in the section on radial turbines. Because pumps have some unique features, they are covered in a separate section. There are many specialized books in each of these areas, and they are listed at the end of the book.

2.1 PHYSICAL PROCESSES INSIDE A TURBOMACHINERY PASSAGE

In addition to the classifications shown in Fig. 1.1, turbomachinery can be further classified as multistage compressors (or pumps) and multistage turbines. Because the pressure (or enthalpy) rise or drop per stage is limited, serial operation of several stages provides a convenient means of attaining high-pressure ratios or high-energy output.

It is important to recognize the physical processes occurring within the blade passage before we proceed to carry out an analysis. To illustrate this process, two representative turbomachines are chosen: a centrifugal pump or compressor and an axial flow turbine.

Let us consider a centrifugal compressor as shown schematically in Fig. 2.1 and Figs. 1.4 and 1.12. The purpose of the inducer is to turn the relative flow as it enters the impeller and smoothly guide it into the radial direction. Some

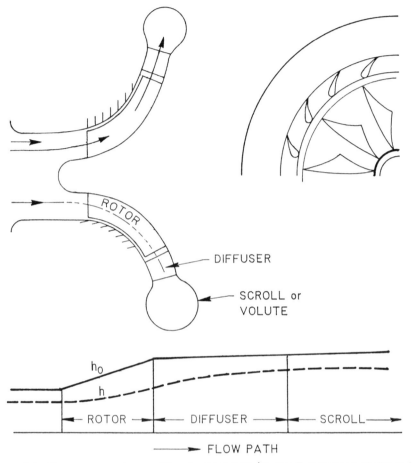

Figure 2.1   Flow processes in a centrifugal compressor. (P_o and T_o variation is similar to h_o; p and T variation is similar to h.)

pressure rise does occur in the inducer. The inducer extends axially from entry to the point where the flow has begun to move radially outward. Impellers without inducers are noisy due to flow separation and violent mixing near the leading edge. Examples of centrifugal compressors (and pumps) can be seen in Figs. 1.4–1.6, 1.9, and 1.12. They usually have a much higher pressure ratio per stage than an axial flow compressor stage due to reasons mentioned later. A typical multistage centrifugal compressor and pump can be seen in Figs. 1.4 and 1.6, respectively. In all these devices, the mechanical work from the shaft is converted to stagnation pressure rise (or head rise), angular momentum increase, and enthalpy rise. It is important to understand how efficiently this energy conversion takes place.

The flow processes occurring in a centrifugal compressor are shown in Fig. 2.1. The stagnation and static enthalpy (temperature) and pressure all

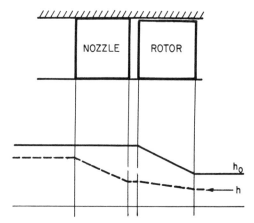

Figure 2.2   Processes occurring in an axial flow turbine. (P_o and T_o variation is similar to h_o; p and T variation is similar to h.)

increase across the rotor. The diffuser (vaned or vaneless) diffuses the high-velocity air and increases the static pressure and temperature further, while the stagnation properties (except for heat loss and viscous losses) remain nearly the same. The scroll collects the fluid from the diffuser and exhausts into a pipe. The major challenge facing a designer is to design the rotor blading for maximum pressure rise and efficiency.

In an axial turbine (Figs. 1.2 and 1.4–1.6), high-pressure and high-temperature fluid (as in a jet engine) impinges on a set of blades. Fluid energy is thus converted into mechanical energy through the rotation of the blades. The stagnation pressure (or head) and the stagnation temperature as well as all the properties change as the fluid goes through a turbine. The processes occurring in a turbine are shown in Fig. 2.2. The high-temperature/pressure/velocity fluid is expanded through stationary vanes called *nozzles*. Static pressure and enthalpy (or head) drop occurs through these nozzles, which guide the flow smoothly into a turbine rotor. All the drop in the stagnation enthalpy and pressure (or head) occurs through the rotor as shown. In order to convert the fluid energy efficiently, it is essential to guide the flow smoothly using streamlined blades.

Considerable research has been directed toward efficient energy conversion through aerodynamic optimization of flow path and blading. Early designs were crude, resulting in very large flow losses. Considerable advances in the design of streamlined bodies (e.g., aircraft wing and fuselage) have enabled turbomachinery designers to advance the state of the art substantially to provide very efficient energy conversion processes in compressors, pumps, fans, and turbines.

This chapter is concerned with the overall performance and elementary (one-dimensional) analysis of all types of turbomachinery. The purpose of one-dimensional theory is to relate the performance, such as stagnation temperature and pressure rise and the shaft power and efficiency, to the average fluid properties at the inlet and exit and the geometry of the passages.

The assumptions made in developing these theories are as follows:

1. A control volume approach (Eqs. 1.6–1.8) is used to relate the performance to flow properties far upstream and far downstream of the blade row.
2. The flow is inviscid and steady (relative to the blades), and the upstream and downstream flows are uniform.
3. The flow is axisymmetric; that is, the fluid properties are circumferentially uniform across the entire passage at the exit and inlet.

2.2   ENERGY AND ANGULAR MOMENTUM TRANSFER IN TURBOMACHINERY

As indicated earlier, turbomachines are devices in which energy is transferred either to or from a continuously flowing fluid by dynamic action of one or more moving blade rows. The energy is transferred from/to a rotor or to/from a fluid. This also changes the energy and angular momentum of the fluid, and the mass flow is conserved. Some of the most basic relationships relating the energy input/output to fluid properties will be derived in this section.

Consider the turbomachine (say a centrifugal compressor) shown in Fig. 2.3. To apply the integral equations valid for a control volume (Eqs. 1.2–1.5), a control volume enclosing part of the blade span, as shown in Fig. 2.3, is chosen. The inner and outer surfaces of the axisymmetric control volume are stream surfaces; therefore, there is no mass, momentum, or energy transfer through the inner (marked I) and outer (marked O) surfaces of the control volume. The inlet and exit surfaces, where the flow enters and exits, have mean radii of r_1 and r_2, respectively, measured from the axis. The inlet velocity has an axial component V_{z_1}, a tangential component V_{θ_1}, and the velocity vector $\mathbf{V}_1$, as shown in the plan view. The inlet and exit surfaces are taken far away from the blade, so that the properties can be assumed to be axisymmetric and uniform along the height of the stream tube. Let the height of the stream tube be dr_1 and dr_2 at the inlet and exit, respectively. Let other properties at the inlet be $\mathbf{V}_1, h_1, T_1, \rho_1, h_{o1}, T_{o1}$, all uniform in the tangential direction. Likewise, the properties of the flow at the exit are $V_{z_2}, V_{r_2}, V_{\theta_2}, \mathbf{V}_2, h_2, T_2, \rho_2, h_{o2}, T_{o2}$. The flow is assumed to be steady.

The mechanism by which energy transfer takes place is through shaft power input (from an electric motor as in an industrial compressor or from a turbine as in a jet engine), which is in the form of torque ($F_\theta \times r$), where F_θ is the tangential blade force. Input shaft power is transmitted to the control volume as torque on the blade element enclosed within the control volume.

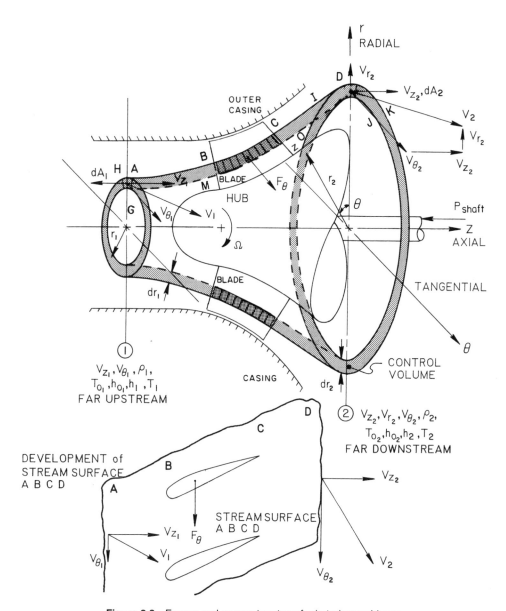

Figure 2.3   Energy and momentum transfer in turbomachinery.

The moving blade exerts a tangential force, F_θ, on the fluid surrounding it (as shown in the plan view, F_θ is the resultant force from the entire blade). This tangential force is then responsible for increasing (or decreasing in the case of a turbine) the angular momentum (rV_θ) of the fluid. Hence, $|\mathbf{V}_2| > |\mathbf{V}_1|$, $V_{\theta_2} > V_{\theta_1}$ for a compressor. The passage can be designed to diffuse the flow, thus providing not only an increase in kinetic energy $(V^2/2)$ but also an increase in the static pressure rise. Thus, all the static and stagnation flow properties change across a turbomachine rotor blade row.

A quantitative analysis of energy and momentum transfer can be made using the control volume equations (Eqs. 1.2–1.5). It should be emphasized that there is no energy, mass, and momentum transfer through control surfaces I and O (axisymmetric stream surfaces), with the exception of input shaft power transmitted to the blade in the control volume as torque. Therefore, the integration has to be performed only at the inlet (station 1, face GH) and exit (station 2, face JK) surfaces. The continuity equation is given by

$$\iint_{\text{ring GH}} \rho_1 \mathbf{V}_1 \cdot d\mathbf{A}_1 = \iint_{\text{ring JK}} \rho_2 \mathbf{V}_2 \cdot d\mathbf{A}_2 \tag{2.1}$$

The angular momentum is given by Eq. 1.4, applied in the θ-direction. The surface forces in the tangential direction on the control surfaces are assumed to be negligible. Also, we have body force, $B = F_\theta$, and the second integration on the left-hand side of Eq. 1.4 includes the moment of all the body forces $(r \times F_\theta)$ due to all blades enclosed within the control volume. Thus Eq. 1.4, applied to the control volume in Fig. 2.3, is given by (tangential component):

$$\iiint_{\text{c.v.}} rF_\theta \rho \, dv = \iint_{\text{ring JK}} (r_2 V_{\theta_2})(\rho_2 \mathbf{V}_2 \cdot d\mathbf{A}_2) - \iint_{\text{ring GH}} (r_1 V_{\theta_1})(\rho_1 \mathbf{V}_1 \cdot d\mathbf{A}_1)$$

$$\tag{2.2}$$

If the thickness of the stream tube is assumed to be small, assumption of uniform entry and exit velocity to the control volume reduces Eq. 2.2 to

$$\iiint rF_\theta \rho \, dv = d\dot{m}(r_2 V_{\theta_2} - r_1 V_{\theta_1}) \tag{2.3}$$

On the left-hand side of the equation, F_θ is the body force (or blade force) per unit mass. Thus the integration represents the sum of the moments of all body forces exerted inside the control volume. This must be equal to the torque transmitted by the shaft through the axisymmetric control surface, past the location where the blade is cut (e.g., MZ and BC in Fig. 2.3).

Denoting this torque by dT, we obtain

$$dT = d\dot{m}\left(r_2 V_{\theta_2} - r_1 V_{\theta_1}\right) \tag{2.4}$$

where dT is the torque applied to the axisymmetric stream tube by the shaft, and $d\dot{m}$ is the mass flow through the control volume. The term in parentheses represents the change in *angular momentum* of the absolute flow. The shaft power input to the control volume ($dp_{\text{shaft}} = \Omega\, dT$) is given by (negative into the control volume)

$$-dP_{\text{shaft}} = \Omega(d\dot{m})\left(r_2 V_{\theta_2} - r_1 V_{\theta_1}\right) \tag{2.5}$$

Hence the general Euler equation for turbomachinery relating the input shaft power to the change in angular momentum can be written as

$$-P_{\text{shaft}} = \dot{m}\Omega\left(r_2 V_{\theta_2} - r_1 V_{\theta_1}\right) \tag{2.6}$$

It should be emphasized here that this equation is valid only for a two-dimensional strip (small dr_1) shown in Fig. 2.3. In most turbomachinery, V_{θ_2} and V_{θ_1} are functions of radius, and Eq. 2.6 can be used if V_{θ_2} and V_{θ_1} are mass-averaged values derived from Eq. 2.5.

For an axial flow turbomachine (pump/compressor, fan) $r_2 = r_1$; therefore,

$$-P_{\text{shaft}} = \dot{m}U\left(V_{\theta_2} - V_{\theta_1}\right)$$

For a compressor, pump, or fan, exit tangential velocities (absolute component) are higher than that at the inlet and P_{shaft} is (negative) into the control volume. The situation is reversed ($V_{\theta_2} < V_{\theta_1}$) for a turbine, resulting in power output. For a centrifugal turbomachine, we have

$$-P_{\text{shaft}} = \dot{m}\left(U_2 V_{\theta_2} - U_1 V_{\theta_1}\right) \tag{2.7}$$

where U_1 and U_2 are blade speeds at the exit and inlet, respectively. Because $U_2 > U_1$ in a centrifugal turbomachine, for the same change in tangential velocity, a centrifugal turbomachine has a higher input/output of shaft power, resulting in higher pressure rise or drop. This will be discussed in detail later. It is clear from these equations that change in angular momentum or tangential velocity is directly proportional to the shaft power. Higher blade speeds result in higher input/output of shaft power.

The energy equation (Eq. 1.8) applied to the control volume in Fig. 2.3 is given by

$$\dot{q} = P_{\text{shaft}} + P_{\text{shear}} + \dot{m}(h_{o2} - h_{o1})$$

where $\mathbf{B} \cdot \mathbf{V} = 0$ because body forces are exerted by the blade and $\mathbf{B} \cdot \mathbf{V} = 0$ for both inviscid and viscous flow. Once again the flow is assumed to be steady and one-dimensional. Neglecting the shear forces exerted on the control surface and assuming adiabatic flow, we obtain

$$\frac{P_{shaft}}{\dot{m}} = h_{o1} - h_{o2} \tag{2.8}$$

It should be recognized that this equation is valid for viscous, compressible flows and includes all losses associated with viscous flows over blades. Only the shear stresses on the control surface (due to shear layers in the r direction) are neglected. If the upper and lower control surfaces coincide with the outer and inner walls of the turbomachine (where $\mathbf{V} = 0$), $P_{shear} = \boldsymbol{\tau} \cdot \mathbf{V}$ is identically zero ($\boldsymbol{\tau} \neq 0$, but $\mathbf{V} = 0$) and Eq. 2.8 is exact.

It is thus clear that shaft power input to a compressor, pump, or fan or shaft power output from a turbine is directly related to the change in total enthalpy of the fluid. For a compressor (or fan or pump), $h_{o2} > h_{o1}$; therefore P_{shaft} is negative into the control volume and vice versa for the turbine.

Equation 2.5, derived from the angular momentum equation, and Eq. 2.8, derived from the energy equation, relate P_{shaft} to the flow and thermodynamic properties. Combining these, we obtain

$$-\frac{P_{shaft}}{\dot{m}} = (h_{o2} - h_{o1}) = (U_2 V_{\theta_2} - U_1 V_{\theta_1}) \tag{2.9}$$

For liquid handling turbomachine, we have

$$P_{shaft}/\dot{m} = \frac{p_{o2} - p_{o1}}{\rho} = g(H_2 - H_1) \tag{2.10}$$

Hence,

$$-\frac{P_{shaft}}{\dot{m}} = g(H_2 - H_1) = (U_2 V_{\theta_2} - U_1 V_{\theta_1}) \tag{2.11}$$

where $(H_2 - H_1)$ = ideal head rise in units of length. Similarly the equation for a turbine is given by

$$\frac{P_{shaft}}{\dot{m}} = (h_{o1} - h_{o2}) = (U_1 V_{\theta_1} - U_2 V_{\theta_2}) \tag{2.12}$$

It is evident that the stagnation pressure and enthalpy rise (or drop) and the head rise (or drop) in a turbomachine are directly proportional to the change in tangential velocity and blade speed. This is one of the most basic equations and principles of turbomachinery.

It should be noted here that Eqs. 2.4–2.12 are valid for a stream tube or in a situation where the flow is uniform along the radii, or when the hub/tip ratio is very large. These equations can also be used at various radii to determine the local blade element performance.

2.3 TURBOMACHINERY PROCESS REPRESENTATION ON AN h–s DIAGRAM AND EFFICIENCIES

The flow process in a thermal turbomachine can be represented on an h–s diagram and compared with an idealized process to define an efficiency. The flow through a compressor or fan is represented by a compression process, and the flow through a turbine is represented by an expansion process.

2.3.1 Compressor / Pump Process and Efficiency

The compression process occurring in a single-stage compressor (rotor-stator) is shown in Fig. 2.4. The rotor compresses the fluid from p_1 (or P_{o1}) to p_2 (or P_{o2}). Most configurations (both axial and centrifugal) consist of a rotor followed by a stator (diffuser) as in Figs. 1.2 and 1.4. In the case of a centrifugal pump/compressor, the diffuser or stator could be radial vanes or

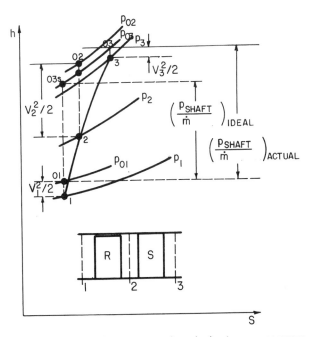

Figure 2.4 Compression processes in a single-stage compressor.

a vaneless diffuser (Fig. 1.4). In all these cases, the stator process can be represented by the compression process shown in Fig. 2.4. The static pressure would increase from p_2 to p_3 along the line 2–3, and the stagnation pressure would decrease from p_{02} to p_{03} due to viscous losses. The purpose of the diffuser is to convert (decrease) kinetic energy from $V_2^2/2$ to $V_3^2/2$ and to further increase the static pressure. A large increase in velocity at the exit of the stage is thus avoided. The diffuser also serves the purpose of guiding the flow smoothly into the next rotor or ducting.

Efficiency is based on a comparison of the actual compressor to an ideal one that is operating with the same mass flow and pressure rise as shown in Fig. 2.4. Performance is represented in terms of work input to an ideal and an actual compressor operating with the same mass flow and pressure rise:

$$\eta_c = \frac{\text{Work input to an isentropic compressor}}{\text{Work input to an actual compressor}} = \frac{h_{o3s} - h_{o1}}{h_{o3} - h_{o1}} \quad (2.13)$$

If the temperature rise is not large, c_p is nearly the same at the inlet and exit; thus

$$\eta_c = T_{o1}\left[(p_{o3}/p_{o1})^{(\gamma-1)/\gamma} - 1\right]/(T_{o3} - T_{o1}) \quad (2.14)$$

If the flow is adiabatic, $T_{o3} = T_{o2}$. Therefore, the efficiency of a compressor can be determined by measuring the stagnation pressure and the temperature rise across a stage. This is usually done by taking measurements in a reservoir downstream.

Efficiency can also be expressed in terms of shaft power/unit mass. Substitution of Eq. 2.8 in Eq. 2.13 results in the following expression (considering that P_{shaft} into the compressor is negative and $T_{o2} = T_{o3}$):

$$-\frac{P_{\text{shaft}}}{\dot{m}} = \frac{c_p T_{o1}}{\eta_c}\left[\left(\frac{P_{o3}}{P_{o1}}\right)^{(\gamma-1)/\gamma} - 1\right] \quad (2.15)$$

The shaft power depends on pressure rise and inlet stagnation temperature and is inversely proportional to the efficiency. Efficiency can also be determined by the direct measurement of P_{shaft} through a torque meter, pressure rise, and entry temperature.

When dealing with pumps, the definition of efficiency is based on actual power delivered to the fluid to the power delivered to the pump. The overall efficiency of a pump is defined as (for the single-stage pump similar to the insert shown in Fig. 2.4)

$$\eta_p = \frac{(H_3 - H_1)g}{\text{work input/unit mass}} = \frac{\rho Q g(H_3 - H_1)}{\text{BHP}} \quad (2.16)$$

where BHP is the brake horse power required to drive the pump, ρQ is the mass flow, and $g(H_3 - H_1)$ is the actual head rise. The hydraulic or Euler efficiency can be defined as

$$\eta_H = \frac{\text{actual head rise}}{\text{ideal head rise}} = \frac{g(H_3 - H_1)}{(U_2 V_{\theta 2} - U_1 V_{\theta 1})}$$

It should be remarked here that the definition of efficiency of a compressor differs from that of a pump. The compressor efficiency compares the work input to an isentropic and an actual compressor for the same mass flow and pressure rise, while the pump efficiency is based on the actual head (or pressure) rise to the work input. The overall and hydraulic efficiency would be the same if there are no transmission losses (mechanical efficiency) and leakage losses in the system.

In many instances, where large pressure rise is required, the compressors and pumps are arranged in series. Such units are called multistage compressors and pumps, respectively. Schematics of multistage axial compressors are shown in Fig. 1.2. A two-stage centrifugal compressor is shown in Fig. 1.4, and a three-stage centrifugal pump used in the space shuttle is shown in Fig. 1.6.

In aerospace propulsion applications, where thrust-to-weight ratio is an important parameter, it is desirable to achieve high pressure rise across each stage to reduce the number of stages needed for the total pressure rise across the entire compressor. This is one of the major research thrusts and is achieved through high blade speed, high inlet Mach number, and improved blading design.

The compression process occurring in a multistage compressor is shown in Fig. 2.5. For simplicity, the process is shown between the stagnation pressures at the inlet and exit of each stage. For example, p_{o3} is the pressure at the exit of the first stage. The detailed processes between $o1$ and $o3$ are represented in Fig. 2.4. The actual process in a single-stage or a multistage compressor is much more complicated than that shown in Figs. 2.4 and 2.5. There is a stagnation and static pressure loss and entropy rise from rotor exit to stator inlet (duct losses) and from one stage to the other; therefore, the process cannot be represented by a continuous line as shown. But for simplicity and overall performance evaluation, this representation is adequate.

In a multistage compression, the isentropic work supplied from $o1$ to $o7_s$ (ABCD) is less than the work input in a series of isentropic compressions (AB, EF, GH) due to diverging pressure lines on an h–s diagram. Therefore, the isentropic work input to the second stage ($o3$–$o5$) is EF and not BC. The overall compressor efficiency ($o1$–$o7$) is defined as

$$\eta_c = \frac{h_{o7_{ss}} - h_{o1}}{h_{o7} - h_{o1}} \tag{2.17}$$

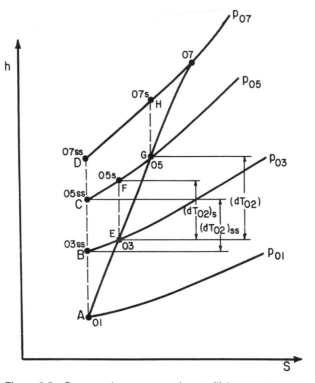

Figure 2.5   Compression processes in a multistage compressor.

and this has to be related to individual stage efficiency (or polytropic efficiency) which is defined as (e.g., second stage)

$$\eta_p = \frac{h_{o5_s} - h_{o3}}{h_{o5} - h_{o3}} \tag{2.18}$$

The constant pressure lines are diverging in an enthalpy–entropy diagram; hence, EF > BC, GH > CD. The isentropic work input to the entire stage (AB + BC + CD) is less than the sum of the isentropic work input to individual stages (AB + EF + GH). Hence, the individual stage efficiencies would be higher than the efficiency of the entire multistage compressor.

Relation Between Polytropic Efficiency and Overall Efficiency.   The infinitesimal compression process can be represented in an *h–s* diagram as shown in Fig. 2.6. For an isentropic process we have

$$dh_{os} = \frac{dP_o}{\rho_o} = \eta_p \, dh_o$$

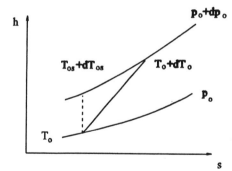

Figure 2.6 Infinitesimal compression process.

where η_p is the efficiency of the infinitesimal compression process. Therefore,

$$\frac{dP_o}{\rho_o c_p T_o} = \eta_p \frac{dh_o}{h_o} = \frac{\eta_p \, dT_o}{T_o}$$

Using the equation of state for a perfect gas and the relationship between specific heats for a perfect gas ($p/\rho = RT, c_p - c_v = R$), the above equation can be written as

$$\frac{dT_o}{T_o} = \frac{dP_o}{\rho_o c_p T_o \eta_p} = \frac{\gamma - 1}{\gamma \eta_p} \frac{dP_o}{P_o} \tag{2.19}$$

Integrating between 1 and 2 (initial and final state), we obtain

$$\log \frac{T_{o2}}{T_{o1}} = \frac{\gamma - 1}{\gamma \eta_p} \log \frac{P_{o2}}{P_{o1}}$$

$$\frac{T_{o2}}{T_{o1}} = \left(\frac{P_{o2}}{P_{o1}} \right)^{(\gamma - 1)/\gamma \eta_p}$$

Thus we get

$$\frac{T_{o2}}{T_{o1}} - 1 = r^{(\gamma-1)/\gamma \eta_p} - 1 \qquad \text{where } r = \frac{P_{o2}}{P_{o1}}$$

and

$$\eta_c = \frac{\dfrac{T_{o2s}}{T_{o1}} - 1}{\dfrac{T_{o2}}{T_{o1}} - 1} = \frac{r^{(\gamma-1)/\gamma} - 1}{r^{(\gamma-1)/\gamma \eta_p} - 1} \tag{2.20}$$

The proof as to why η_c is $< \eta_p$ is given below. Let us consider an n-stage compression process, then

$$\eta_c = \frac{(dT_{o1})_{ss} + (dT_{o2})_{ss} + \ldots (dT_{on})_{ss}}{dT_{o1} + dT_{o2} + \ldots dT_{on}}$$

where, for example, $(dT_{o2})_{ss}$ is the isentropic temperature rise across the second stage (see Fig. 2.5). Let us assume $dT_{o1} = dT_{o2} = \cdots = dT_o$. Then

$$\eta_c = \frac{\sum\limits_{n=1}^{n} (dT_{on})_{ss}}{n \, dT_o}, \qquad \eta_{pn} = \frac{(dT_{on})_s}{dT_{on}} = \frac{(dT_{on})_s}{dT_o}$$

$$dT_o = \frac{(dT_{on})_s}{\eta_{pn}}, \qquad \eta_c = \frac{\sum\limits_{n=1}^{n} (dT_{on})_{ss}}{\dfrac{n}{\eta_{pn}}(dT_{on})_s} = \eta_{pn} \frac{\sum\limits_{n=1}^{n} (dT_{on})_{ss}}{n(dT_{on})_s}$$

$$(dT_{on})_{ss} < (dT_{on})_s$$

or

$$\eta_c < \eta_{pn}$$

2.3.2 Turbine Expansion Process and Efficiency

The expansion process in a turbine is represented in Fig. 2.7. A nozzle is used at the inlet to provide partial expansion of the gas as well as to guide the flow smoothly into a rotor. The process through nozzle vanes is shown in Fig. 2.7. Point $o1$ represents inlet stagnation condition. Nozzle expansion occurs along curve 1–2. Change in stagnation pressure $(P_{o2} - P_{o1})$ is due to viscous effects only, because there is no work extraction from the fluid inside the nozzle vanes. The process along curve 2–3 represents the expansion through the rotor, the exit stagnation pressure being P_{o3}. If the flow is isentropic in the nozzle, condition $2s$ is achieved. If the flow is isentropic only in the rotor, condition $o3_s$ or 3_s is achieved. Points $o3_{ss}$ and 3_{ss} represent stagnation and static conditions of flow, when isentropic conditions exist in the entire stage, including the gap between the nozzle and the rotor.

Two types of definitions are used for turbine efficiency, depending on the application. In many situations, energy at the exit $(V_3^2/2)$ is considered a loss (e.g., power plants) and the design philosophy is to achieve as low a velocity at the exit as possible. In such situations a "total-static" efficiency is defined

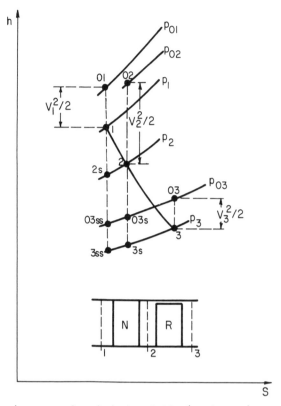

Figure 2.7 Expansion process in a single-stage turbine (nozzle, rotor).

as follows (assuming constant c_p across the stage):

$$\eta_{ts} = \frac{\text{Shaft power output of an actual turbine}}{(\text{Shaft power output} + \text{Power wasted in exhaust kinetic energy}) \text{ of an ideal turbine}}$$

$$= \frac{c_p(T_{o1} - T_{o3})}{c_p\left(T_{o1} - T_{o3_{ss}}\right) + c_p\left(T_{o3_{ss}} - T_{3_{ss}}\right)}$$

$$= \frac{T_{o1} - T_{o3}}{T_{o1} - T_{3_{ss}}}$$

which can be expressed in terms of pressures as follows (because $o1$–3_{ss} is an isentropic process):

$$\eta_{ts} = \frac{1 - (T_{o3}/T_{o1})}{1 - (p_3/P_{o1})^{(\gamma-1)/\gamma}} \tag{2.21}$$

In most aeronautical applications, the definition of efficiency is based on stagnation conditions, because the jet velocity is used to derive useful power (e.g., thrust power in turbojets, turbo fans, etc.). The so-called total-to-total efficiency, which is defined as isentropic efficiency in this book, is given by

$$\eta_t = \frac{\text{Actual work output}}{\text{Ideal work output}} = \frac{T_{o1} - T_{o3}}{T_{o1} - T_{o3_{ss}}}$$

$$= \frac{1 - T_{o3}/T_{o1}}{1 - (P_{o3}/P_{o1})^{(\gamma-1)/\gamma}} \tag{2.22}$$

The actual work output is measured either by a torque meter to derive the brake horse power, or by measuring P_{o3} and T_{o3} in a reservoir downstream of the turbine. The latter definition should be used in a multistage turbine, where the kinetic energy at the exit of a stage (except the last one) is not lost.

For a multistage turbine, as in a multistage compressor, polytropic efficiency for each stage can be related to overall efficiency of the turbine through Eq. 2.19. An expression similar to Eq. 2.20 for a turbine can be derived as follows:

$$\eta_t = \frac{1 - (r)^{(\gamma-1)\eta_p/\gamma}}{1 - r^{(\gamma-1/\gamma)}} \tag{2.23}$$

where $r = P_{o3}/P_{o1}$ is the overall pressure ratio.

For liquid handling turbines (e.g., hydraulic turbine shown in Fig. 1.10), the efficiency is defined as

$$\eta_t = \frac{\text{actual work output}}{\text{fluid power input}} = \frac{\text{BHP}}{\rho Q g (H_1 - H_3)}$$

where $H_1 - H_3$ is the actual head drop across the turbine.

It should be remarked here that in many instances, the temperature drop in a turbine or the temperature rise in a compressor is large. For accurate assessment of efficiency, it is necessary to include the variation c_p and γ in many of the expressions developed in this section. For example in deriving Eqs. 2.21 and 2.22, it is assumed that $c_{p3} = c_{p1}$, which is not valid for modern high-speed turbines. In many situations, allowance must be made for humidity in the air. For accurate calculations, it is essential to use the actual values of thermodynamic properties rather than the idealized (e.g., perfect gas) values. The polyotospic efficiency can be derived accurately by integrating Eq. 2.19 by allowing for continuous variation in thermodynamic properties. Cumpsty (1989) provides a review of the procedure used for accurate estimation of efficiency for compressors allowing for real gas effects. This becomes critical in such areas as refrigeration compressors, very high-pressure ratio compressors, and very high-pressure ratio and temperature ratio turbines.

Table 2.1   Summary of performance equations for turbomachinery[a]

	Turbine	Compressor/Pump
$P_{\text{shaft}}/\dot{m}$	$U_1 V_{\theta 1} - U_2 V_{\theta 2}$ $h_{o1} - h_{o2}$	$U_2 V_{\theta 2} - U_1 V_{\theta 1}$ $h_{o2} - h_{o1}$
(Power input for compressor, output for turbine)	$c_p T_{o1} \eta_t \left[1 - \left(\dfrac{P_{o3}}{P_{o1}} \right)^{(\gamma-1)/\gamma} \right]$	$\dfrac{c_p T_{o1}}{\eta_c} \left[\left(\dfrac{P_{o3}}{P_{o1}} \right)^{(\gamma-1)/\gamma} - 1 \right]$
Efficiency (η)	$\dfrac{h_{o1} - h_{o3}}{h_{o1} - h_{o3s}}$	$\dfrac{h_{o3s} - h_{o1}}{h_{o3} - h_{o1}}$
Efficiency(Liquid)	$\dfrac{\text{BHP}}{\rho Q(H_1 - H_3)g}$	$\dfrac{\rho Q(H_3 - H_1)g}{\text{BHP}}; H = \dfrac{P_o}{\rho g}$
η_{pn} (stage efficiency)	$\dfrac{h_{o(n-1)} - h_{on}}{h_{o(n-1)} - (h_{on})_s}$	$\dfrac{(h_{on})_s - h_{o(n-1)}}{h_{on} - h_{o(n-1)}}$
η (efficiency for multistage)	$\dfrac{1 - r^{\eta_p(\gamma-1)/\gamma}}{1 - r^{(\gamma-1)/\gamma}}$	$\dfrac{r^{(\gamma-1)/\gamma} - 1}{r^{(\gamma-1)/\gamma\eta_p} - 1}$
	[η_p is assumed constant for every stage]	
Pressure rise (or drop) in terms of temperature rise or drop	$\dfrac{P_{o3}}{P_{o1}} = \left(1 - \dfrac{\Delta T_o}{\eta_t T_{o1}} \right)^{\gamma/(\gamma-1)}$	$\dfrac{P_{o3}}{P_{o1}} = \left[1 + \dfrac{\eta_c \, \Delta T_o}{T_{o1}} \right]^{\gamma/(\gamma-1)}$
	ΔT_o = temperature rise or drop	

[a]Notations are based on Figs. 2.4–2.7 and Sections 2.2 and 2.3.

The aerospace industry has evolved very accurate and consistent methods of assessing efficiency through sophisticated measurement techniques. For example, temperatures are measured to an accuracy of a fraction of a degree. This is essential in view of the fact that the aero engine customer (air frame manufacturer) requires a very accurate estimate of the specific fuel consumption, which requires very accurate specifications of the component efficiencies, including those of compressors and turbines. The range of the aircraft and fuel carried for an aircraft mission is estimated based on these efficiencies.

A summary of various relationships for energy transfer and efficiencies derived in this chapter are given in Table 2.1.

2.4 NONDIMENSIONAL REPRESENTATION

By means of dimensional analysis, a group of variables representing some physical situation is reduced into a smaller number of dimensionless groups. This enables a unique representation of certain classes of machines based on pressure rise (or drop) and mass flow. Most importantly, it enables reduction of laboratory testing effort by reducing the number of variables. Specifically, the following can be accomplished:

1. Prediction of a prototype performance from tests conducted on a scaled model (similitude).
2. Unique representation of the performance (e.g., altitude, Mach number, Reynolds number effect).
3. Determination of a best machine on the basis of efficiencyfor specific head, speed, and flow rate.

Let us consider a machine with the following geometric variables: diameter (D), height of the blade (l_1) and chord length (l_2). Let the operating condition be inlet stagnation pressure (P_{o1}), inlet stagnation temperature (T_{o1}), mass flow rate $(\dot{m})$, and rotational speed (N). Fluid properties are the universal gas constant (R), kinematic viscosity (ν), and the ratio of specific heats (γ). Results of many tests and analyses (see Table 2.1) have indicated that, for the normal operating range, the performance of compressors, pumps and turbines may be represented by

$$P_{o2} = f(\dot{m}, P_{o1}, R, T_{o1}, \gamma, N, \nu, D, l_1, l_2)$$
$$\eta = f(\dot{m}, P_{o1}, R, T_{o1}, \gamma, N, \nu, D, l_1, l_2)$$

Because R and T_{o1} are the only variables which include temperature, they could be combined as RT_{o1}. For liquid handling machinery, this variable is not relevant, but an additional variable, vapor pressure, will be important. This will be explained later. Hence,

$$P_{o2} = f(\dot{m}, P_{o1}, RT_{o1}, \gamma, N, \nu, D, l_1, l_2)$$

with units Mass = (M), Length = (L), Time (t)

$$P_{o1} = (ML^{-1}t^{-2}), \qquad R = p/\rho T = (L^2 t^{-2}\theta^{-1}), \qquad \dot{m} = (Mt^{-1})$$
$$T_{o1} = (\theta), \qquad N = (t^{-1}), \qquad \nu = (L^2 t^{-1}), \qquad D, l_1 \text{ and } l_2 = (L)$$

Function f may be written as

$$0 = f(\dot{m}, P_{o1}, P_{o2}, RT_{o1}, \gamma, N, \nu, D, l_1, l_2)$$

According to the Buckingham π theorem, if there are n variables and m fundamental units (such as length L, mass M, and time t), the equations relating to the variables can be collected and expressed as $(n - m)$ dimensionless numbers. In this procedure, select m variables which include a group of all fundamental units from the list of n variables, say P_{o1}, RT_{o1}, and D.

Then set up a dimensionless equation combining these variables with each of the other remaining variables $(m, \gamma, N, \nu, l_1, l_2)$. Example: We select P_{o1}, RT_{o1}, and D and combine them, say, with $\dot{m}$. According to the Buckingham π theorem, we have

$$(P_{o1})^x (RT_{o1})^y (D)^z \dot{m} = \pi_1$$

$$\left(\frac{M}{Lt^2}\right)^x \left(\frac{L^2}{t^2}\right)^y (L)^z \left(\frac{M}{t}\right) = \pi_1$$

Therefore,

$$x = -1, \qquad y = 1/2, \qquad z = -2$$

and we obtain one of the groups:

$$\pi_1 = \frac{\dot{m}\sqrt{RT_{o1}}}{P_{o1}D^2}$$

Similarly, we obtain

$$\pi_2 = P_{o2}/P_{o1}, \qquad \pi_3 = \frac{ND}{\sqrt{\gamma RT_{o1}}}, \qquad \pi_4 = \gamma$$

$$\pi_5 = l_1/D, \qquad \pi_6 = l_2/D, \qquad \pi_7 = (ND^2)/\nu$$

By dimensional analysis, f can therefore be written as

$$0 = f\left(\frac{P_{o2}}{P_{o1}}, \frac{\dot{m}\sqrt{RT_{o1}}}{P_{o1}D^2}, \frac{ND}{\sqrt{\gamma RT_{o1}}}, \gamma, \frac{ND^2}{\nu}, \frac{l_1}{D}, \frac{l_2}{D}\right)$$

Thus, 10 variables are reduced to seven groups. For geometrically similar machines, $l_1/D, l_2/D$ would be nearly constant. Hence,

$$\frac{P_{o2}}{P_{o1}} = f\left(\frac{\dot{m}\sqrt{RT_{o1}}}{P_{o1}D^2}, \frac{ND}{\sqrt{\gamma RT_{o1}}}, \gamma, \frac{ND^2}{\nu}\right) \qquad (2.24)$$

Similarly,

$$\eta = f\left(\frac{\dot{m}\sqrt{RT_{o1}}}{P_{o1}D^2}, \frac{ND}{\sqrt{\gamma RT_{o1}}}, \gamma, \frac{ND^2}{\nu}\right) \qquad (2.25)$$

It is interesting to note that some of these nondimensional groups could have been written directly from basic fluid/gas dynamic principles. For example, it is known from one-dimensional gas dynamics that $\dot{m}\sqrt{RT_o}\,/$ $P_oD^2 = f(M, \gamma)$; and compressibility effect should depend on characteristic velocity (ND) and speed of sound, which leads to the second group; the last group is the ratio of inertial force to viscous force, based on the characteristic velocity ND.

The pressure (or head) rise or drop in turbomachines, then, is a function of a mass flow parameter, nondimensional speed, γ, and the Reynolds number. Equation 2.24 clearly reflects the consequence of the energy and Euler equations in turbomachines. This relationship was, of course, specified in Eq. 2.9, which indicates that the pressure rise (or drop) or temperature rise (or drop) is proportional to the mass flow rate and the blade speed. Equation 2.8 incorporates the effect of viscosity. Therefore, the pressure or temperature rise is a function of efficiency, which is controlled mainly by viscous effects including boundary layer growth on the blade and wall surfaces and by shock–boundary-layer interaction. The compressibility effect is controlled by the local speed of sound, and this appears in both the mass flow parameter and the blade speed. The Reynolds number effect depends on the flow regime and has appreciable influence in machinery which operates in a wide range of conditions (e.g., altitude, speed, gas, etc.). The Reynolds number has to change by nearly an order of magnitude, before its influence is felt. Thus the most important nondimensional variables are mass flow parameter and rotational speed (or Mach number based on rotational speed). Changes in pressure or temperature caused by Reynolds number changes are not as significant as those caused by mass flow and speed parameters, even though the Reynolds number effect is significant in efficiency prediction.

Equation 2.24 is valid for all compressible flow turbomachinery (centrifugal and axial compressors and turbines). The corresponding relationship for incompressible flow will be derived later in this chapter. It is often convenient to present the product

$$\frac{\dot{m}\sqrt{RT_{o1}}}{P_{o1}D^2} \times \frac{\Delta T_{o1}}{T_{o1}} = \frac{\dot{m}\Delta T_{o1}\sqrt{R}}{P_{o1}D^2\sqrt{T_{o1}}} \sim \frac{P_{shaft}(\gamma - 1)}{P_{o1}\sqrt{RT_{o1}}\,D^2\gamma}$$

where ΔT_o is the stagnation temperature change. Then, for a given gas (γ = constant)

$$\frac{P_{o2}}{P_{o1}} = f\left\{ \frac{P_{shaft}}{P_{o1}D^2\sqrt{RT_{o1}}}, \frac{ND}{\sqrt{\gamma RT_{o1}}} \right\}$$

Thus, the pressure rise versus nondimensional shaft power curve represents a unique relationship for a fixed value of $ND/\sqrt{\gamma RT_{o1}}$.

2.4.1 Blade Loading Coefficientand Flow CoefficientRelationship

Some turbomachinery designers prefer to use $(\Delta T_{o1}/T_{o1})$ instead of P_{o2}/P_{o1} to represent the performance. They are related by

$$\frac{P_{o2}}{P_{o1}} = \left(1 + \eta_c \frac{\Delta T_o}{T_{o1}}\right)^{\gamma/(\gamma-1)} \qquad \text{Compressor}$$

$$\frac{P_{o2}}{P_{o1}} = \left(1 - \frac{1}{\eta_t}\frac{\Delta T_o}{T_{o1}}\right)^{\gamma/(\gamma-1)} \qquad \text{Turbine}$$

From the previous analysis, it is clear that

$$\frac{\Delta T_{o1}}{T_{o1}} = f\left(\frac{\dot{m}\sqrt{RT_{o1}}}{D^2 P_{o1}}, \frac{ND}{\sqrt{\gamma RT_{o1}}}, \gamma, \frac{ND^2}{\nu}\right)$$

The efficiencies in Eq. 2.25 are included in the last group (Reynolds number). Dividing $\Delta T_{o1}/T_{o1}$ by $(ND/\sqrt{\gamma RT_{o1}})^2$ and $(\dot{m}\sqrt{RT_{o1}}/D^2 P_{o1})$ by $ND/\sqrt{\gamma RT_{o1}}$, we obtain

$$\frac{\gamma R \Delta T_{o1}}{N^2 D^2} = f\left(\frac{\dot{m} RT_{o1}\sqrt{\gamma}}{P_{o1} ND^3}, \frac{ND^2}{\nu}, \frac{ND}{\sqrt{\gamma RT_{o1}}}, \gamma\right)$$

Because γ itself is a group, the temperature rise and mass flow parameters can be divided by $\gamma - 1$ and $\sqrt{\gamma}$, respectively, to prove

$$\frac{P_{\text{shaft}}/\dot{m}}{N^2 D^2} = c_p \frac{\Delta T_{o1}}{N^2 D^2} = f\left(\frac{\dot{m}}{\rho_{o1} ND^3}, \frac{ND}{\sqrt{\gamma RT_{o1}}}, \gamma, \frac{ND^2}{\nu}\right)$$

This equation relates the mass flow parameter to the blade speed. Hence, $ND/\sqrt{\gamma RT_{o1}}$ is now a weaker parameter and mainly controls the compressibility effect, and viscous effects are controlled by the parameter ND^2/ν. If these effects are not large (subsonic flow without shock waves), one can write

$$\psi = \frac{P_{\text{shaft}}/\dot{m}}{U^2} = \frac{c_p \Delta T_{o1}}{U^2} = f\left(\frac{\rho_{o1} V_z D^2}{\rho_{o1} ND^3}\right) = f\left(\frac{V_z}{ND\pi}\right) = f(\phi) \quad (2.26)$$

Similarly

$$\eta = f(\phi), \qquad \text{where } \phi = V_z/U$$

The coefficient ψ is called the *loading coefficient* and reflects the pressure/temperature rise across a compressor or drop across a turbine. Th coefficient ϕ, called the *flow coefficient*, is now the parameter reflecting the effect of the mass flow as well as the blade speed.

2.4.2 Nondimensional Relationship for a Fixed Turbomachinery Geometry and Gas

Equations 2.24–2.25 can be considerably simplified for a given turbomachine, whose performance is evaluated at various flow conditions (e.g., P_{o1}, T_{o1}, and N). In this case the groups are nondimensionalized by some reference pressures and temperatures, usually with sea level values. The mass flow parameter is then given by

$$\frac{\dot{m}\sqrt{RT_{o1}}}{P_{o1}D^2} \sim \frac{\dot{m}\sqrt{\theta}}{\delta}$$

where $\theta = T_{o1}/T_a$ and $\delta = P_{o1}/P_a$. The speed parameter is given by

$$\frac{ND}{\sqrt{\gamma RT_{o1}}} = \frac{N}{\sqrt{\theta}}$$

Hence,

$$\frac{P_{o2}}{P_{o1}} \quad \text{or} \quad \frac{T_{o2}}{T_{o1}} \quad \text{or} \quad \eta = f\left(\frac{\dot{m}\sqrt{\theta}}{\delta}, \frac{N}{\sqrt{\theta}}, \text{Re}\right) \qquad (2.27)$$

The new groupings are called *corrected mass flow* and *corrected speed*, respectively.

2.4.3 Relationships for Incompressible Machines

For incompressible machines, the relationships derived earlier are further simplified for liquid-handling turbomachinery and in situations where the flow is considered incompressible.

1. Density is constant. The rate of mass flow ($\dot{m}$) can be replaced by the rate of volume flow $\dot{Q}$.
2. Because the compressibility effect is small, the speed of sound and the parameter $ND/\sqrt{\gamma RT_{o1}}$ are not relevant. The effect of blade speed is included in the flow coefficientparameter.
3. The pressure ratio P_{o2}/P_{o1} can be replaced by the pressure difference across the machine—that is, by $\Delta P_o/P_{o1}$, where $\Delta P_o = \rho g \Delta H$.

Pressure difference, efficiency, and the stagnation pressure can be represented by

$$\Delta H = f(Q, N, D, \mu, \rho, l_1, l_2)$$

$$\eta = f(Q, N, D, \mu, \rho, l_1, l_2)$$

Let us consider Eq. 2.26:

$$\psi = \frac{P_{shaft}/\dot{m}}{U^2} \cong \frac{g(H_2 - H_1)}{U^2} = f\left(\frac{V_z}{ND}\right) = f(\phi) \qquad (2.28)$$

The stage loading coefficient in this case is called the *head rise coefficient*. As mentioned earlier, the Reynolds number dependency is weak, so for a given turbomachine the "loading or head rise coefficient" is uniquely related to the flow coefficient. Other ways of representing the performance are based on the definition of a volume flow coefficient and elimination of blade speed, N, and diameter, D, as follows:

$$\phi = \frac{V_z}{U} \sim \frac{V_z D^2}{D^2 ND} \sim \frac{\dot{Q}}{ND^3}$$

$$\psi = \frac{g\Delta H}{U^2} \sim \frac{g\Delta H}{N^2 D^2}$$

Therefore, the following relationships can also be used instead of Eq. 2.28:

$$C_H = \frac{g\Delta H}{N^2 D^2} = f\left(\frac{Q}{ND^3}\right) = f(C_Q) \qquad (2.29)$$

$$C_p = \frac{P_{shaft}/\dot{m}}{N^2 D^2} \sim \frac{P_{shaft}}{\rho N^3 D^5} = f\left(\frac{Q}{ND^3}\right) = f(C_Q) \qquad (2.30)$$

where C_H is called the *head coefficient* and C_p is called the *power coefficient*; C_Q is called the *flow rate (or volume rate) coefficient*. Similarly, $\eta = f(C_Q)$.

The performance of an incompressible, liquid handling turbomachine can therefore be represented by Eq. 2.28, 2.29, or 2.30. Detailed interpretation of the performance of turbomachinery based on these nondimensional groups will be discussed later in this chapter.

One of the major by-products of nondimensional analysis is the definition of a group that can be used to classify various types of turbomachinery. This is accomplished through the definition of "specific speed." Let us consider Eq. 2.29. The two groups in this equation could be used to eliminate the dimension D to define a "specific speed" given by

$$\overline{N} = \frac{\left(\dfrac{Q}{ND^3}\right)^{1/2}}{\left(\dfrac{g\Delta H}{N^2 D^2}\right)^{3/4}} = \frac{N\sqrt{Q}}{(g\Delta H)^{3/4}} \tag{2.31}$$

The specific speed is a combination of operating conditions. Holding this constant ensures (within the limits of the assumptions made) similar flow conditions in geometrically similar machines. Equation 2.31 is a nondimensional combination, only when consistent units are used (e.g., N is radian/sec, Q is cubic feet per second or cubic meter per sec, H is feet or meter). But in practice, g is assumed to be constant, providing

$$S = \frac{N\sqrt{Q}}{(\Delta H)^{3/4}} \tag{2.32}$$

This is a dimensional number, and the most commonly used units are (rpm, gallons per minute, feet) or (rpm, m^3/sec, m).

Turbine designers usually prefer to work with a parameter called "power specific speed." This is derived by eliminating the size of the turbine through the following combination:

$$\overline{N}_p = \frac{C_p^{1/2}}{C_H^{5/4}} = \frac{N\sqrt{\text{BHP}}}{\sqrt{\rho}\,(g\Delta H)^{5/4}}$$

This is a dimensional number.

The specific speed $(\overline{N})$ is a significant parameter in the initial selection of turbomachinery. There is a wide variation in specific speed for turbomachinery with appreciable compressibility effect. The range is somewhat better defined for hydraulic turbines and pumps. The concept of "specific speed" is widely used by liquid handling turbomachinery designers but is much less useful for modern gas compressors and turbines. The concept of "power specific speed" and "specific speed" is extremely useful in model testing. For example, a prototype with a prescribed N, Q, ΔH, and BHP can be modeled by keeping these nondimensional parameters constant to achieve dynamically similar flow.

The suitability ranges, based on specific speed are given in Fig. 2.8. The initial design of turbomachinery depends on the parameters $\overline{N}$ and $\overline{N}_p$. Each achieves maximum efficiency in a certain range of values of $\overline{N}$ as shown in Fig. 2.9. For example, if $\overline{N}$ is between 2 and 4, an axial compressor is superior to a centrifugal compressor.

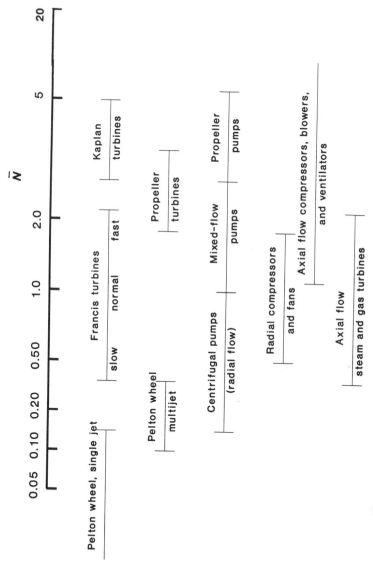

Figure 2.8 Correlation of rotor design and specific speed suitability ranges of various designs (Csanady, "Theory of Turbomachinery" 1964; reprinted with permission from McGraw-Hill.)

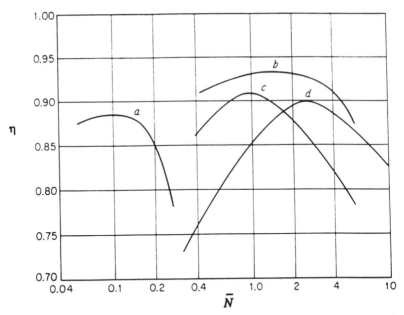

Figure 2.9 Operating range and typical efficiencies of various turbomachinery based on specific speed parameter. (a) Pelton wheel. (b) Francis propeller turbine. (c) Large single-stage centrifugal pump. (d) Axial compressors. (stage efficiency) (From Csanady, "Theory of Turbomachines," 1964; reprinted with permission from McGraw-Hill).

The usefulness of the concept of specific speed lies in deciding whether to use an axial or a centrifugal type of turbomachine for a compressor/pump or turbine. A centrifugal compressor/pump would be more suitable for low-volume flow rate with high head rise. For high volume and low head rise, an axial compressor/pump is more suitable. The peak efficiencies of each type of machine occur in different ranges of specific speed values $\overline{N}$ (see Fig. 2.9; Csanady, 1964). When we examine the cases illustrated in Fig. 2.9, we note that centrifugal compressor efficiency will peak around $\overline{N} = 1$, while the peak efficiencies of an axial flow compressor will occur at much higher values of $\overline{N}$.

2.4.4 Relationship to Cavitation in Pumps

In liquid handling machinery (hydraulic pumps, turbines, rocket pumps, etc.), a widely encountered phenomena is "cavitation." Cavitation is the formation of vapor bubbles when the static pressure is reduced locally below the fluid vapor pressure. Cavitation in turbomachinery may occur on the suction surface and near the blade tip region due to the presence of a vortex and also at other locations where the fluid pressure falls below the vapor pressure. The occurrence of cavitation is followed by a decrease in the pressure change and a drop in efficiency. One of the most important effects is the subsequent collapse of the bubble in the region of pressure rise. This is accompanied by erosion and material damage. Further treatment of this subject will be covered later. The cavitation performance of a turbomachine can be represented by a cavitation parameter, commonly known as the *Thoma parameter*, given by

$$\tau = \frac{2(P_{o1} - p_v)}{\rho V_1^2}$$

where p_v is the vapor pressure. The other form of the Thoma parameter is the net positive suction head (NPSH)

$$\tau = \frac{H_{sv}}{H} \tag{2.33}$$

where H_{sv} is the *net positive suction head* (*NPSH*) ($= H_s - h_v$), H_s is the total head at the inlet, which includes both the kinetic and potential energy, and h_v is the head corresponding to vapor pressure (p_v/ρ), H is the total head rise.

The suction head is one of the major parameters used in evaluating the cavitation behavior of pumps. The cavitation behavior at a given speed and volume flow is a function of H_s. By normalizing H_{sv} by the total head of the machine H (which depends on Q and N), the dependency of the speed and the volume rate can be eliminated. Therefore, τ is a similarity parameter for cavitation, and it is possible to determine experimentally the cavitation characteristics at one speed and one volume rate and scale it up to determine the cavitation characteristics at a different value of Q and N.

One other method of evaluating the cavitation performance is through the definition of a "*suction specific speed.*" This was first introduced by Bergeron (1940). Wislicenus (1965) provided a more comprehensive treatment on this subject. The suction specific speed is defined as

$$SS = \frac{N\sqrt{Q}}{(H_{sv})^{3/4}} \tag{2.34}$$

This is again a dimensional number and consistent units have to be employed. Suction specific speed is clearly a grouping of operating conditions that permit similar flow conditions to exist in similar liquid handling turbomachines. Constant values of both τ and SS, or of SS alone, will result in similar inlet flow and cavitation conditions. The suction specific speed of centrifugal pumps (Wislicenus, 1965) range from 7000 to 10,000 (rpm, gpm, ft), while the low-pressure axial turbopump inducer (similar to the rocket fuel pump inducer shown in Fig. 1.7) used in the space shuttle has a specific speed of 70,000 and operates at $\phi = 0.07$. The pumps usually cavitate at SS values above 7000–8000. Hence the SS is a design variable and a proper choice of variables N, Q, and H_{sv} will ensure cavitation free operation. Rocket pump inducers are used in front of the main pump to break the bubbles and ensure cavitation free operation in the main pump. Therefore, rocket pump inducers operate at high speeds to minimize the weight and size of the system. The long, narrow passages (Fig. 1.7) provide the time and space for the collapse of the cavitation bubbles. These inducers operate at high suction specific speeds.

2.5   AXIAL FLOW COMPRESSORS AND FANS

Compressors and fans increase the pressure of the fluid. If pressure rise is small and mass flow is large (as in ventilating fans, fans used in a turbofan engine, etc.) the device is called a *fan*, whereas if the pressure rise is high, the device is called a *compressor*. Sometimes a middle-range pressure rise device is termed a *blower*. The pressure rise normally encountered may be as low as a fraction of an inch of water gauge (say a hundredth of a psi) or as high as 300–400 psi (as in a multistage compressor of a modern jet engine). This chapter deals with the one-dimensional analysis, performance, and type of blading used in these devices. Several specialized books [e.g., Horlock (1973), Cumpsty (1989), Bolces and Suter (1986), Wallis (1983), Eck (1973), and Eckert and Schnell (1980)] provide very detailed treatment of compressors and fans.

As indicated earlier, several types of compressors are used for propulsion, energy, and industrial applications. Both the centrifugal and the axial types are widely used. A comparison between the two follows:

1. The two machines are comparable in weight.
2. Aerodynamically, the axial machine has better flow characteristics. Real fluid effects in centrifugal compressors are much larger, and therefore these devices have a lower efficiency.
3. For aircraft applications, the frontal area is very important. For the same mass flow, the centrifugal machine has a higher frontal area and therefore larger drag.
4. The advantage of the centrifugal compressor lies in the simplicity of manufacture, especially when the turbomachine is miniaturized (e.g., space power plants).
5. For moderate stagnation pressure ratios and low mass flow, centrifugal compressors are superior to the axial type. The multistaging of centrifugals involves larger aerodynamic or hydrodynamic losses; thus for high pressure ratios and large mass flow, an axial machine is preferred. The selection criteria is based on the specific speed range described earlier.

6. Recent interest in centrifugals for use in (a) small gas turbine units in cars, (b) space power plants, (c) rocket engines, and (d) artificial heart pumps has revived the aerodynamic investigation of this machine.

7. For industrial applications, such as natural gas pumping and rocket engines, safety and reliability are more important and thus centrifugal compressors are sometimes preferred.

8. Jumbo jets, where large mass flow and large pressure rise are encountered, invariably utilize axial flow compressors (Fig. 1.2), while centrifugals are common in turbo shaft and gas turbines used in aircraft and helicopters with small engines or thrust power (Fig. 1.4).

The list above is by no means comprehensive; other differences will be dealt with in later chapters.

Fans which have low pressure rise are usually single stage, while compressors can be single stage or multistage depending on the application. A stage consists of a guide vane and rotor, or a stator/rotor combination. A schematic of the blading and the flow processes occurring within an axial flow compressor blade row is shown in Fig. 2.10. A multistage compressor employs more than one stage. For example, some of the high thrust turbofans/jets have as many as 15–20 stages (Fig. 1.2).

The purpose of the inlet guide vane (IGV) is to guide the air smoothly into a rotor, which is very sensitive to any small change in incidence in flow angle or nonuniformity in velocity. Guide vanes also serve the purpose of preventing the injection of foreign objects into the engine (e.g., birds) and of breaking up large turbulent eddies in the atmosphere. Air is turned through

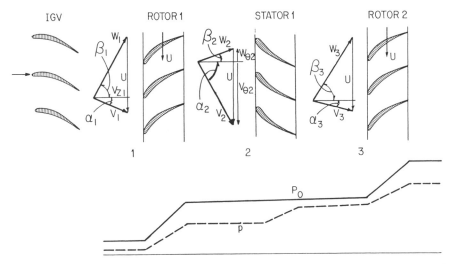

Figure 2.10 Axial flow compressor velocity triangles. (h_o and T_o distribution are similar to P_o distribution; h and T distribution are similar to p distribution.)

the proper angle by the IGV before it impinges on the rotor blade of the first stage. The velocity triangle shown for the IGV (Fig. 2.10) indicates that the flow is accelerated through the IGV, resulting in a decrease in static pressure. Energy is added to the fluid by the rotor blades, thereby increasing its stagnation pressure, temperature, and kinetic energy. The fluid is discharged at a proper angle of attack to stator blades where the static pressure is further increased by flow diffusion. The stagnation pressure remains nearly the same through the stator (except for losses), but the static pressure and temperature increase while the kinetic energy decreases. The air is directed to the second-stage rotor, and the process repeats itself. The last stage usually has a guide vane or stator that turns the fluid in the axial direction. In a multistage compressor, it is not desirable to operate all stages at the same speed. Because of changes in velocity, density, pressure, and hub/tip ratio, operating all of them at the same speed will produce poor performance. Therefore, these stages are usually divided into two segments, the low-pressure compressor (usually the first half of the stages) and the high-pressure compressor (the last half of the stages), and are operated at differing speeds.

The purpose of this section is to introduce the reader to some basic fundamentals and to include a state-of-the-art review of axial compressor technology. In a turbomachinery rotor, the flow may be viewed from two frames of reference. One is the absolute or stationary frame fixed to the ground, and the second one is the rotating or relative frame fixed to the rotor. The difference between the rotative velocity vector (**W**) and the absolute velocity vector (**V**) is the blade velocity vector $\mathbf{\Omega} \times \mathbf{r}$:

$$\mathbf{V} = \mathbf{U} + \mathbf{W}$$

Velocity triangles can be constructed as shown in Figs. 2.10 and 2.11. The inlet absolute velocity to Rotor 1 is V_1. The relative velocity is $\mathbf{W}_1$ and should align closely with the rotor blade angle at the inlet. The exit relative velocity ($\mathbf{W}_2$) is nearly parallel to the blade at the exit. Subtraction of **U** will result in an absolute velocity $\mathbf{V}_2$, which should line up with the stator blade setting. For a compressor, $V_2 > V_1$ and $W_2 < W_1$; in the absolute frame, kinetic energy is added by the shaft, and in the relative frame the compressor acts like a diffuser. For incompressible inviscid flows, $p_1 + \frac{1}{2}\rho W_1^2 = p_2 + \frac{1}{2}\rho W_2^2 = P_{oR}$. The rotor blade turns the relative velocity $\mathbf{W}_1$ to $\mathbf{W}_2$, thereby imparting angular momentum to the air and thus increasing the absolute tangential velocity. If the flow is incompressible, the axial velocity remains the same and the velocity triangles can be superimposed as in Fig. 2.11. The definition of angles used in compressor practice is shown in Fig. 2.11 and defined under nomenclature.

In an axial machine (pump/compressor/turbine), if the streamlines are parallel to the rotation axis, the stagnation enthalpy of the relative flow is

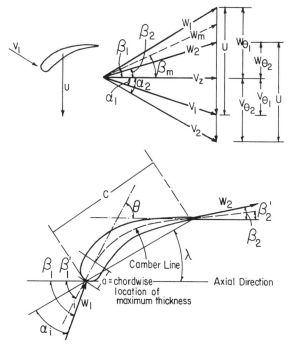

Figure 2.11 Blade velocity triangles and nomenclature. (Velocity triangle shown is for incompressible flow.)

constant. This follows from the manipulation of Eq. 2.9.

$$h_2 - h_1 = U(W_{\theta 1} - W_{\theta 2}) - \frac{(U - W_{\theta 2})^2 + V_{z2}^2}{2} + \frac{(U - W_{\theta 1})^2 + V_{z1}^2}{2}$$

$$h_2 + \frac{W_2^2}{2} = h_1 + \frac{W_1^2}{2} = (h_o)_R = \text{constant}$$

For incompressible, inviscid flow, we have

$$h_{oR} = \frac{P_{oR}}{\rho} = \frac{p_1}{\rho} + \frac{W_1^2}{2} = \frac{p_2}{\rho} + \frac{W_2^2}{2} = C_1(r) \tag{2.35}$$

Thus, the Bernoulli equation is valid for relative flow along a streamline in an axial machine for an inviscid, incompressible flow. In a more generalized case, where the streamlines are not parallel to the axis (mixed and centrifugal turbomachinery, as well as axial turbomachinery with flared annuli and hub),

Eq. 2.9, valid along a streamline, can be manipulated to give

$$h_2 - h_1 = (U_2 V_{\theta 2} - U_1 V_{\theta 1})$$
$$+ \frac{(U_1 - W_{\theta 1})^2 + V_{z1}^2 + V_{r1}^2}{2} - \frac{(U_2 - W_{\theta 2})^2 + V_{z2}^2 + V_{r2}^2}{2}$$

or

$$(h_{o2})_R - U_2^2/2 = (h_{o1})_R - U_1^2/2 \qquad (2.36)$$

or

$$I = h_{oR} - \frac{U^2}{2} = h + \frac{W^2}{2} - \frac{U^2}{2} = \text{constant along a streamline}$$

where I is called the *rothalpy*.

There are certain restrictions on the validity of this equation. It is valid for steady, adiabatic flows (including the viscous regions, except the rotor casing). Equation 2.36, which is based on Eq. 2.9 is a simplified version of Eq. 1.5 or 1.8. The assumptions made in deriving Eq. 2.9 from 1.5 are that (1) the flow is steady in relative frame, (2) work done by body forces are zero, and (3) there is no heat transfer into or from the fluid. In addition, the work done by shear forces in Eq. 1.5 or P_{shear} in Eq. 1.8 should be zero. This is not true in the outer casing region (where the wall is stationary and the blade is rotating). In this region ($\mathbf{W} = -\mathbf{U}$) and $P_{\text{shear}} = \int \boldsymbol{\tau} \cdot \mathbf{W} \, ds$ (see Fig. 1.18 for notations) is non-zero. Hence the concept of constant rothalpy is not valid in the casing region of a rotor, where the work done by shear forces may not be negligible. Thus the concept of constant rothalpy is valid everywhere for steady adiabatic, irreversible flow except the casing region.

For incompressible, inviscid flow ($\nabla h_o = \nabla P/\rho$) flow, $[p + \rho(W^2/2) - \rho(U^2/2)]$ is constant along a streamline. This equation is strictly valid for inviscid flows.

2.5.1 Aerodynamic Shapes and Type of Blading Used

Aerodynamic or streamline shaped airfoils are used to achieve pressure rise efficiently. The side of the blade leading the rotation is the pressure side. The flow velocity is decelerated (or pressure is raised) on the pressure side as shown in Fig. 2.12. On the suction side, the flow is accelerated rapidly (say, up to B in Fig. 2.12) and then decelerated to the same level as the pressure side at the trailing edge. Detailed physics of this phenomena will be explained in Chapter 3. In Fig. 2.12, increased Mach number represents decreased static pressure and vice versa.

The nature and type of blading employed in compressors depends on the application and the Mach number range. Subsonic bladings usually consist of circular arcs, parabolic arcs, or combinations thereof. The blade surfaces are designed to provide smooth entry and exit flow, with minimum loss and

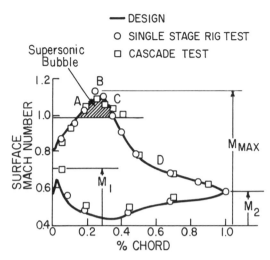

Figure 2.12 Cascade and single-stage rig airfoil surface static pressure measurements verify the CDA blade element design system (Hobbs and Weingold, 1984; Reprinted with permission from ASME; Behlke, 1986). (A) Continuous acceleration to boundary layer transition. (B) Supersonic bubble. (C) Shock-free deceleration. (D) Diffusion controlled to prevent separation.

maximum pressure rise. The nomenclature used in practice is given in Fig. 2.11. Many countries have developed their own profiles (e.g., United States, Britain, Japan, West Germany, etc.) for subsonic flows. Early British designs are due to Howell (1942), and the American designs are due to Herrig et al. (1957). Typical NACA 65 series blades are shown in Fig. 2.13. The NACA 65 series designation is as follows: NACA 65 $(x)y$, where x is 10 times the design lift coefficient of an isolated airfoil and y is the maximum thickness in percent of chord. The profiles shown in Fig. 2.13a are in ascending order of loading.

There has been considerable research in recent years to design "shock free" and "controlled diffusion" airfoils for high-speed as well as multistage compressor applications. Controlled diffusion airfoils (Fig. 2.13b) are designed and optimized (Hobbs and Weingold, 1984; Behlke, 1986; Bauer et al. 1977; Sanz, 1983) specifically for subsonic and transonic applications, by minimizing boundary layer separation and by diffusing the flow from supersonic to subsonic velocities without a shock wave (see Section 3.4.3). The profile of the control diffusion airfoil is considerably more robust at the leading and trailing edges than that of the double circular airfoil. At the trailing edge, it is more robust than the NACA 65 series blading. The profiles in Fig. 2.13b are widely used for high subsonic, transonic flows (Hobbs and Weingold, 1984), whereas the NACA 65 series or similar British profiles are used in low subsonic flows (Johnsen and Bullock, 1965). The present trend is toward the use of custom-tailored airfoils rather than the standardized series

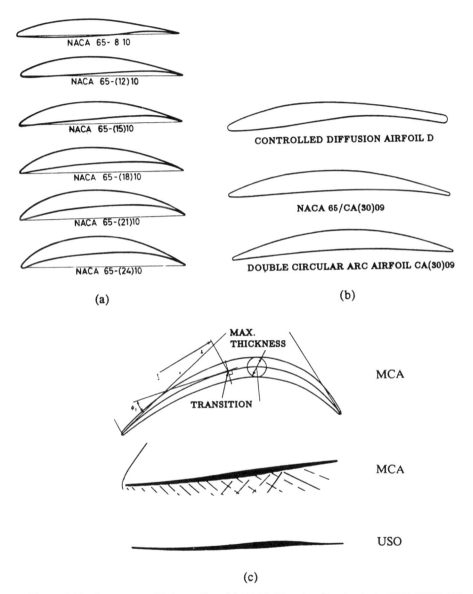

Figure 2.13 Compressor blade profiles. (a) NACA 65 series (Herrig et al., 1957; NACA TN 3916). (b) Recent trends. (Copyright © Hobbs and Weingold, 1984; Reprinted with permission from ASME.) (c) Supersonic profiles.

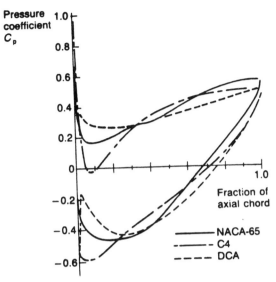

Figure 2.14 The predicted pressure distribution about three common blade profiles in cascade. Stagger 33.6°, camber 27.3°, solidity 1.25, thickness / chord ratio 0.10, $M_1 = 0.6$, incidence 0° (Cumpsty, "Compressor Aerodynamics," 1989, p. 142 reprinted with permission from Longman Group UK).

of bladings. The surface pressure distribution (or Mach number) is specified for a particular application as shown in Fig. 2.12, and the airfoil shape that provides this distribution is derived (e.g., Fig. 2.13b). Many of the aircraft engine companies have succeeded in deriving such blade shapes, resulting in improved efficiency.

Cumpsty (1989, p. 142) has compared the blade pressure distribution of NACA 65, C4, and DCA (double circular arc) blades, all of which have differing thickness distribution. The predicted blade pressure distribution for the cascades with identical parameters is shown in Fig. 2.14. Several important conclusions are drawn from these plots. The conclusions drawn are very specific to this cascade and may not be true for all operating conditions. Nevertheless, it provides a guidance on the selection criteria.

1. NACA 65 series gives the highest pressure rise.
2. Both the NACA 65 series and C4 profiles, which were widely used in early stages of compressor development, suffer from low pressures on the pressure surface leading edge. Sudden acceleration and deceleration are undesirable from the point of view of boundary layer growth.
3. DCA has much better pressure distribution, but the suction peak is aft of the leading edge and the adverse pressure gradient on the suction side toward the trailing edge is much larger, an undesirable feature from the point of view of boundary layer separation.

Hence in selecting the profile, close attention should be paid to the following:

1. Loading distribution and lift coefficient, which directly affect the pressure rise.
2. The location and magnitude of suction peak, which control the boundary layer growth on the suction side.
3. One should select a blade shape to avoid flow separation and reduce the boundary layer growth.
4. One should provide a smooth pressure distribution on the pressure side, thereby avoiding sudden acceleration or deceleration near the leading edge.

For blades operating in supersonic flows, blade leading and trailing edges are very thin and blade thickness is very small (Fig. 2.13c). For supersonic inlet flow and smooth entry, proper control of supersonic and subsonic turning is essential. With such control, the loss can be minimized. The multiple circular arc (MCA) airfoil was used in early designs (Fig. 2.13c) (Morris et al., 1972). The MCA profile is suited for started mode, while the unstarted mode requires a different profile (Fig. 2.13c) designated the unstarted strong oblique shock profile (USO). The entrance regions in MCA and USO airfoils are straight.

2.5.2 Pressure and Temperature Rise in Terms of Velocity and Flow Angles

The velocity triangles for an axial flow compressor/fan/blower are shown in Figs. 2.10 and 2.11. Let us relate the performance characteristics to these velocity triangles. Equation 2.9 and the velocity triangle, shown in Fig. 2.10, can be used to express temperature/pressure rise in terms of velocities and flow angles (assuming c_p change across the blade row is small).

$$T_{o2} - T_{o1} = \frac{U}{c_p}(V_{z2} \tan \alpha_2 - V_{z1} \tan \alpha_1) = \frac{U}{c_p}(V_{z1} \tan \beta_1 - V_{z2} \tan \beta_2)$$

$$= \frac{U}{c_p}[U - V_{z2} \tan \beta_2 - V_{z1} \tan \alpha_1] \tag{2.37}$$

$$\psi = \frac{c_p(T_{o2} - T_{o1})}{U^2} = \frac{c_p \Delta T_o}{U^2} = [1 - \phi(\tan \alpha_1 + (\text{AVR}) \tan \beta_2)] \tag{2.38}$$

where axial velocity ratio AVR $= V_{z2}/V_{z1}$. In most cases, this is usually close to unity.

In terms of Mach numbers, Eq. 2.37 can be written as (for an adiabatic compressor)

$$
\frac{T_{o3}}{T_{o1}} = \frac{T_{o2}}{T_{o1}} = 1 + \frac{(\gamma - 1) M_b^2}{1 + \dfrac{\gamma - 1}{2} M_1^2} \left[1 - \frac{M_{z1}}{M_b} (\tan \alpha_1 + \text{AVR} \tan \beta_2) \right]
$$

$$(2.39)$$

where $(M_b = U/\sqrt{\gamma R T_1}, \; M_{z1} = V_{z1}/\sqrt{\gamma R T_1}, \; M_1 = W_1/\sqrt{\gamma R T_1})$. Equation 2.39 can be substituted in Eq. 2.14 to derive the following expression for the pressure rise:

$$
\frac{P_{o3}}{P_{o1}} = \left[1 + \eta_c \frac{M_b^2 (\gamma - 1)}{1 + \dfrac{\gamma - 1}{2} M_1^2} \left[1 - \frac{M_{z1}}{M_b} (\tan \alpha_1 + \text{AVR} \tan \beta_2) \right] \right]^{\gamma/(\gamma-1)}
$$

$$(2.40)$$

It is evident that both stagnation temperature rise and pressure rise are strong functions of blade speed (U, M_b), axial velocity or axial Mach number (or mass flow), inlet and outlet flow angles, and, more specifically, the absolute or relative turning angles. For example, according to Eq. 2.38, for a given blade (β_2) and inlet angle (α_1), the pressure and temperature rise depend strongly on the flow coefficient. Furthermore, the pressure rise depends on the efficiency as well as the flow coefficient.

It is clear that for a compressor with α_1, β_2, and U held constant, the temperature and pressure rise decrease with an increase in the mass flow rate, or ϕ. On the other hand, if α_1, β_2, and V_z, are held constant, the pressure and temperature rise increase in direct proportion to U. This equation is consistent with the relationship derived from nondimensional analysis (Eqs. 2.24 and 2.26). Therefore, high blade speed and low mass flow contribute to high pressure or temperature rise for a given blade row and fixed power input. Furthermore, higher flow turning ($\beta_1 - \beta_2$) or ($\alpha_1 - \alpha_2$) contributes to higher pressure and temperature rise. There is a limiting value beyond which higher turning leads to boundary flow separation. The flow turning angle is directly related to the pressure gradient in the streamwise direction, and almost invariably adverse (increasing) pressure gradients exist near the suction surface of the blade near the trailing edge (Fig. 2.12). Depending on the location of the maximum velocity on the suction surface and its relation to the trailing edge, the flow may separate, leading to decreased performance. Lower efficiency will decrease the pressure rise, as indicated by Eq. 2.40.

If all the properties at the inlet $(T_{o1}, T_1, \alpha_1, \beta_1, V_z, U, P_{o1}, h_{o1}, \rho_1)$ and the exit angle β_2 are known, the gas dynamic equations can then be used to predict all properties at the exit. If β_2 is not known, the analysis outlined in the next chapter can be used to predict β_2 accurately for inviscid flows. There is always a slight variation between the blade outlet angle (β_2') and flow angle (β_2), called the *deviation angle*. It is also assumed that η_c is known. This can be estimated from the loss correlations presented in Chapter 6, or a reasonable value can be assumed based on experience with a similar machine.

If the flow is incompressible and inviscid, the above relationships can be simplified considerably as follows (the stagnation pressure of the relative flow is therefore constant):

$$p_1 + \tfrac{1}{2}\rho W_1^2 = p_2 + \tfrac{1}{2}\rho W_2^2 \quad \text{and} \quad V_{z2} = V_{z1}$$

Static pressure rise across the rotor is given by

$$\frac{p_2 - p_1}{\rho} = \frac{W_1^2 - W_2^2}{2} = \frac{V_z^2}{2}(\tan \beta_1 - \tan \beta_2)(\tan \beta_1 + \tan \beta_2)$$

The static pressure rise across the stator is given by (Fig. 2.10)

$$\frac{p_3 - p_2}{\rho} = \frac{1}{2}(V_2^2 - V_3^2) = \frac{V_z^2}{2}(\tan \alpha_2 - \tan \alpha_3)(\tan \alpha_2 + \tan \alpha_3)$$

From velocity triangles we know that

$$V_z(\tan \beta_1 - \tan \beta_2) = V_z(\tan \alpha_2 - \tan \alpha_1)$$

Therefore, if the stages are identical $\alpha_1 = \alpha_3$ the static pressure rise (Δp_{13}) is given by

$$\Delta p_{12} + \Delta p_{23} = \Delta p_{13} = \rho V_z U(\tan \beta_1 - \tan \beta_2)$$

Because $V_1 = V_3$, $P_{o3} - P_{o1} = p_3 - p_1$.

2.5.3 Pressure and Temperature Rise in Terms of Lift and Drag

It is much more convenient to express pressure and temperature rise in terms of the lift and drag coefficients for the blade. These aerodynamic parameters are readily available for many of the standard profiles (e.g., Herrig et al., 1957).

The aerodynamic forces exerted by the fluid on the blade and by the blade on the fluid are shown in Fig. 2.15. These are equivalent to "action" and "reaction" forces. In the case of a propeller/fan/compressor/pump, the

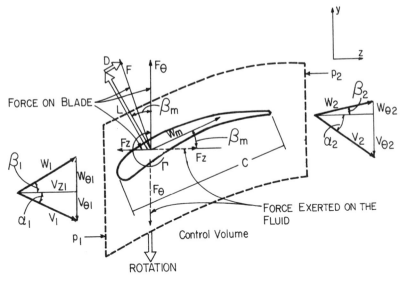

Figure 2.15   Stage loading in terms of lift and drag.

rotation of the blade (caused by mechanical drive) introduces blade forces F_θ and F_z (resultant lift force $= L$ and drag force $= D$). In this case, the pressure surface leads the blade motion. The tangential force accelerates the fluid by increasing the tangential momentum. In the case of a turbine/windmill, the fluid forces F_θ and F_z exerted on the blade impart angular momentum to the blade (transfer of energy from fluid to a rotor). The torque thus imparted to the blade is converted into mechanical/electrical power. A circulation is set up around the blade, resulting in aerodynamic forces D and L in mean cascade flow direction and normal direction, respectively.

Let us define the aerodynamic forces acting on a blade. The lift and drag forces for a compressor blade are defined with reference to a mean velocity $(\mathbf{W}_1 + \mathbf{W}_2)/2$, and a mean angle β_m shown in Fig. 2.15.

By defining the aerodynamic parameters

$$C_L = \frac{L}{\frac{1}{2}\rho W_m^2 C} \quad \text{and} \quad C_D = \frac{D}{\frac{1}{2}\rho W_m^2 C} \tag{2.41}$$

it can be proved that (Fig. 2.15)

$$F_\theta = L \cos \beta_m + D \sin \beta_m = \frac{\rho W_m^2 C}{2} C_L \cos \beta_m \left(1 + \frac{C_D}{C_L} \tan \beta_m \right) \tag{2.42}$$

Because $F_\theta U = C_p \, \Delta T_o \, \rho S V_z$

$$\psi = \frac{V_z}{2U} \sec \beta_m C_L \frac{C}{S} \left[1 + \frac{C_D}{C_L} \tan \beta_m \right] \tag{2.43}$$

It is evident from Eq. 2.43 that the loading coefficient is directly proportional to the lift coefficient, which should be obvious from physical considerations. Expressions for lift and drag are derived in the next chapter. Change in axial velocity (or mass flow) for a given blade geometry affects C_L through changes in incidence, V_z and sec β_m. As indicated earlier (Eq. 2.38 or 2.40), a decrease in the mass flow would increase the pressure or the temperature rise. It is also evident that the loading coefficient increases with an increase in the drag coefficient. The viscous effects (or drag) increase the temperature rise due to viscous dissipation. But the pressure rise will be lower due to decrease in efficiency as per Eq. 2.15.

2.5.4   Reaction

Because the designer has a choice in deciding what fraction of the stage static enthalpy/pressure rise occurs through the rotor or stator, one of the design variables used in practice is the reaction, defined by (Fig. 2.10):

$$R = \frac{h_2 - h_1}{h_3 - h_1} = \frac{\text{Static enthalpy rise across rotor}}{\text{Static enthalpy rise across stage}} \tag{2.44}$$

For isentropic flow (assuming streamlines are parallel to the axis and $V_1 = V_3$) we have

$$R = \frac{W_1^2 - W_2^2}{\left(W_1^2 - W_2^2\right) + \left(V_2^2 - V_3^2\right)} = \frac{W_1^2 - W_2^2}{2U(V_{\theta 2} - V_{\theta 1})} \tag{2.45}$$

For incompressible flow and inviscid flow, we have

$$R = \frac{p_2 - p_1}{p_3 - p_1} = \frac{W_1^2 - W_2^2}{2U(V_{\theta 2} - V_{\theta 1})} \tag{2.46}$$

For a constant axial velocity, Eq. 2.46 can be simplified to provide

$$R = \frac{1}{2} - \frac{V_z}{U} \left(\frac{\tan \alpha_1 - \tan \beta_2}{2} \right)$$

Because flow decelerates through the compressor, adverse static pressure gradients exist in the streamwise direction inside the passage. Performance may be limited by boundary layer separation due to this adverse pressure gradient. From this point of view, it is desirable to split the static pressure (or enthalpy) rise equally between the two rows. This is the preferred procedure by most designers, but not necessarily an optimum one. Radial variations give rise to higher reaction near the tip and a lower (sometimes negative) reaction near the hub. Therefore, 50% reaction can be maintained at only one radial location, usually the midradius.

Depending on the choice of the reaction, one can determine β_2 (for a given α_1 and V_z/U), thus deciding on turning angle at the mid-radius. It should be noted that for $\alpha_1 = \beta_2$, the velocity triangles are symmetrical and $R = 0.5$. If $\beta_2 > \alpha_1$, the rotor pressure rise will be greater, and vice versa, for $\beta_2 < \alpha_1$.

2.5.5 Performance

The validity of similarity relationships, such as Eq. 2.24, 2.26, 2.27, 2.29, or 2.30, has been established, so no attempt will be made here to present these results. It is clear from these equations, as well as from Eq. 2.38, that the loading coefficient (or pressure rise) is a linear function of mass flow parameter, or V_z/U, if α_1 and β_2 remain unchanged (this is true for a fixed geometry compressor) as shown in Fig. 2.16. Small changes in α_1 and β_1 do not affect the value of β_2. The flow is smoothly guided by the blade and exits at an angle close to the blade outlet angle. The difference between the blade outlet angle and the flow angle is called *deviation angle*. The deviation angle is dependent on the cascade geometry and pressure gradients. For a rotor operating in normal design range the change in β_2 is not large. If we assume that α_1 and β_2 are fixed, then from Eq. 2.38, $\psi = 1 - K\phi$, where K is nearly constant in the nominal operating range. The actual performance drops rapidly at flow coefficients above and below the design operating condition. Design operation is usually at the peak efficiency condition. In an off-design mode, the viscous effects result in larger boundary layer growth and, in some instances, flow separation resulting in the observed trend.

The measured performance of a low-speed single-stage compressor operating at Penn State is shown in Fig. 2.17. Details of the compressor are given in Lakshminarayana (1980). The compressor has a hub/tip ratio of 0.5, with an annulus diameter of 0.932 m. The inlet guide vane consists of 43 blades, followed by a rotor with 21 blades and stator with 25 blades. The inlet axial velocity is 29 m/s. The design flow coefficient (based on tip speed) is 0.56, and the corresponding pressure rise coefficient (ψ_p) is 0.486. The blade element data at the tip are NASA 65 series: chord 15.41 cm, spacing 14.2 cm, maximum thickness 5.10% of chord, stagger angle 45°. The design turning angles vary from 16° near the tip to 30° at the hub. $\bar{\psi}_p$, $\bar{\phi}$, and $\bar{\eta}$, in Fig. 2.17, are mass-averaged values for the entire passage. As mass flow is decreased, the pressure or temperature rise will increase along AB (Fig. 2.17), as per Eq. 2.38 (α_1 and β_2 held nearly constant). The ideal pressure rise will be

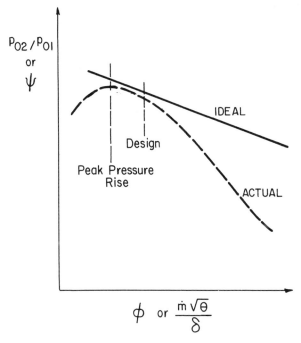

Figure 2.16   Actual and ideal performance of a compressor stage.

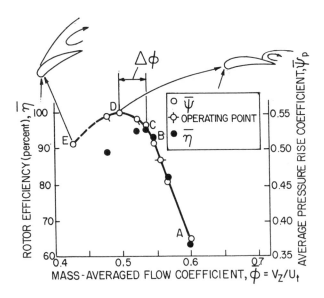

Figure 2.17   Overall performance of a compressor rotor.

higher. The measured characteristic is nearly linear along AB and is consistent with Eq. 2.38. Beyond point B, if the mass flow is decreased further, the rate of increase in the pressure rise will not be as high and the characteristic is nonlinear. Equation 2.38, based on inviscid considerations (no losses), is no longer valid. Therefore, the rate of increase in pressure rise is decreased, but the pressure rise still increases with a decrease in the mass flow until it reaches a point such as D. The incidence (and pressure gradient) effects on the viscous flow on the blade and wall surfaces influence the pressure rise substantially beyond point C, leading to the onset of stall at point D. Any further decrease in mass flow will result in a substantial decrease in the pressure rise due to massive flow separation on the blades and walls as shown in the insert of Fig. 2.17. The inception of stall may lead to "rotating stall," which results from separation on one or more compressor blades. Rotating stall is typified by alternate stalling and unstalling of the blades, resulting in a separated zone (shed out of the blade trailing edge) moving in the circumferential direction at a fraction of the blade speed. If the rotor is part of a multistage system, there could be a flow stoppage or reversal in the entire compressor system. Alternate loading and unloading of several stages will result in a pressure wave propagating through the compressor system, often resulting in mechanical failure. Under such circumstances, the compressor is said to "surge." The phenomena of rotating stall and surge are covered in Sections 2.5.7 and 2.5.8, respectively.

As indicated in Fig. 2.17, efficiency increases up to a point where the viscous effects tend to become appreciable (e.g., point C in Fig. 2.17). Beyond this point, the flow is dominated by viscous effects resulting in a decrease in the efficiency. Compressor designers allow sufficient margin between the stall point and the operating point (e.g., points D and C).

The performance of a high-speed multistage compressor used in the energy-efficient engine (E^3) is shown in Fig. 2.18. This compressor has 10 stages, a 23:1 pressure ratio, and a mass flow rate of 54.4 kg/s. Detailed performance characteristics of this stage are given in Hosny et al. (1985). The performance of the first six stages (pressure ratio of 9.5), is shown in Fig. 2.18. The plot is based on the similarity relationship represented by Eq. 2.27. Because the front stages of these compressors operate supersonically near the tip, the performance curves are steep. The supersonic sections have to operate at a unique incidence beyond which the losses will be higher, resulting in a steep curve as shown in Fig. 2.18. The pressure rise characteristic is different for each corrected speed ($N\sqrt{\theta}/\delta$), plotted as a percentage of the design speed. As explained earlier, the operating line is below the stall line to prevent surge. It should be noted that the compressor is operated at the peak efficiency point and the efficiencies are maximum at the design speed.

At higher Mach numbers, the performance (mean flow versus pressure rise) is very steep due to the following reasons. The compressibility effect has

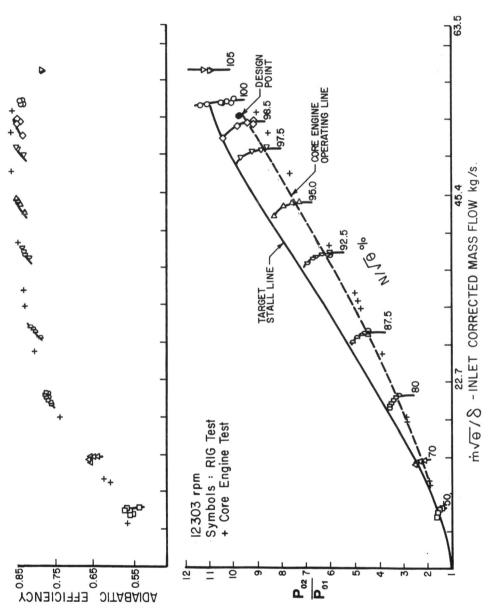

Figure 2.18 Performance of energy efficientengine (E[3]) compressor (Hosny et al., 1985; NASA CR 174955).

a dual effect as the mass flow is increased. Let us examine Eq. 2.40. At fixed M_b, as the mass flow is increased, M_{Z_1}/M_b increases and the term $(1 + [(\gamma - 1)/2]M_1^2)$ increases and the efficiency decreases rapidly due to off-design shock structure, thus giving rise to rapid decrease of P_{o2}/P_{o1} at higher mass flows. Thus the pressure rise decreases rapidly as the mass flow is increased at higher Mach numbers. This is clear from a comparison of the performance of a low-speed compressor shown in Fig. 2.17 and the characteristics of a high-speed compressor shown in Fig. 2.18.

2.5.6 Pressure Rise Limitations; Incidence and Deviation Angles

Because the compressor acts as a diffuser, an adverse pressure gradient exists on the hub and annulus wall surfaces and on the blade. The boundary layer growth on the surfaces of the blade and the walls of the compressor limit the pressure rise. The value of the pressure coefficient at which stall occurs depends on the condition of the boundary layer (whether separated or unseparated on the wall and blade surfaces), the presence of shock waves, and the Reynolds number, as well as on compressor parameters including aspect ratio, stagger angle, chord length, blade spacing, and so on. Because the boundary layers encountered in compressors are much more complex than idealized geometries for which analysis is available, the prediction of inception of stall and other phenomena is usually empirical in nature. The blade boundary layers can be predicted reasonably well for two-dimensional flows and isolated blade rows.

If the compressor is viewed as a diffuser, relative flow diffusion occurs in the case of rotor and absolute flow diffusion occurs in the case of a stator. The diffusion ratios which control the boundary layer behavior are W_2/W_1 and V_2/V_1 for the rotor and stator, respectively. The value of these ratios, as well as the inviscid velocity variation along the chord length, are important.

The boundary layer growth in the region of adverse pressure gradient controls the static pressure rise. It can be shown that the boundary layer thickness at the trailing edge is controlled by the parameter

$$D_{\text{loc}} = \frac{W_{\text{max}} - W_2}{W_{\text{max}}}$$

where W_{max} is the peak suction velocity as shown in Fig. 2.12. This is related to the limiting pressure rise of an equivalent diffuser with inlet velocity of W_{max} and exit velocity of W_2, given by $(C_p) = 1 - (W_2/W_{\text{max}})^2$. Because the magnitude of W_{max} involves a knowledge of the flow field, it is better to use a parameter that does not require a knowledge of W_{max}. Hence, Lieblein

(1965) suggested a diffusion factor defined by

$$D \sim \frac{W_{max} - W_2}{W_1} \sim \frac{W_1 + f\left(\dfrac{W_{\theta 1} - W_{\theta 2}}{\sigma}\right) - W_2}{W_1} = 1 - \frac{W_2}{W_1} + \frac{W_{\theta 1} - W_{\theta 2}}{2\sigma W_1}$$

$$(2.47)$$

The equation is valid for both rotor flows (compressible and two-dimensional) and stator flows. For stator or cascade flows, relative velocities are replaced by corresponding absolute velocities.

The author's reinterpretation of Eq. 2.47 is that maximum velocity depends on the local circulation. Consider a "lifting line analysis" where the blade is replaced by a vortex line of strength, Γ, at the location of minimum C_p. The maximum velocity induced by this vortex line is $+u$ on the suction side and $-u$ on the pressure side, and in the limiting process (Γ/C is the vortex strength per unit length) we have

$$u = \frac{\Gamma}{2C} = \frac{S(W_{\theta 1} - W_{\theta 2})}{2C}$$

This expression includes not only the overall diffusion ratio (W_2/W_1), but also localized effects due to maximum velocity.

A plot of Eq. 2.47 against losses measured in various cascade tests shows a very close correlation between losses and the diffusion factor (Fig. 2.19). The loss coefficient $\bar{\zeta}$ is the passage averaged losses. This represents loss in

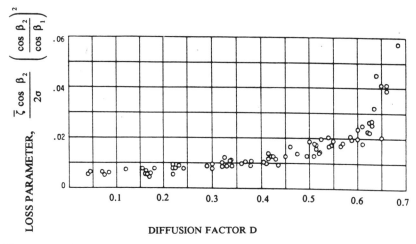

Figure 2.19   Loss parameter versus diffusion factor for low-speed NACA 65-(A_{10}) 10 cascade blades (Lieblein, 1965).

stagnation pressure due to boundary layer and wakes. The losses increase considerably beyond a diffusion factor of 0.6.

A word of caution is in order. The diffusion factor, shown in Fig. 2.19, is strictly valid for two-dimensional incompressible flows. The effect of compressibility and three-dimensionality on the diffusion factor needs further exploration. Koch and Smith (1976) have provided a modified empirical expression for the diffusion factor valid for high subsonic flow and is more general than Lieblein's expression. Furthermore, the peak pressure rise attainable in a compressor is controlled by the diffusion of the blade flow as well as by the behavior of the wall boundary layers. Stall may occur initially due to wall boundary layer separation followed by blade boundary layer separation. These mechanisms will be discussed in a later chapter.

The selection of incidence (Fig. 2.11) depends on type of airfoil, Mach number and Reynolds number ranges, leading edge radius, solidity, and the camber. The main criterion used for the selection of incidence is the same as that used by aircraft wing aerodynamicists. The incidence is chosen to maximize the lift-drag or pressure rise ratio and efficiency. A detailed discussion of various methods (often conflicting) of choosing the incidence to the blade is given by Johnsen and Bullock (1965). There is a range of incidence angles (or flow coefficient) in which the loss is minimum. Beyond this incidence angle range, the losses increase. Hence the optimum incidence angle is determined from this loss versus incidence (or flow coefficient) and pressure rise versus incidence plots. For example, the ideal flow coefficient (or incidence) for the Penn State compressor lies near C (where the efficiency is maximum) in Fig. 2.17.

The selection of incidence is based on cascade data correlation or testing. The blade profiles are tested in a cascade wind tunnel to generate typical plots as shown in Fig. 2.20. This plot, reproduced from Lieblein (1965), shows performance of British C_4 circular arc, $P4$ parabolic arc profiles with maximum thickness of 10% chord. The turning angle is the same for all of them. This figure clearly shows the effect of blade shape, incidence, and Mach number on loss coefficient ζ. At any given Mach number, there is an optimum incidence or range of incidence where the losses are minimum. The design incidence is usually chosen for minimum loss. Such a plot is usually referred to by practitioners as *loss bucket*. The losses increase with an increase in Mach number, increasing dramatically at off-design incidence conditions. This is caused by adverse pressure gradient and the associated increase in the boundary layer growth on the suction side. Some profiles (e.g., DCA) are less sensitive to Mach number changes than others. A typical deflection versus incidence curve for a blade row is also shown in Fig. 2.20(d). As the incidence increases, the turning angle also increases, reaching the stall point beyond which the loading and flow turning fall off rapidly. This is the region of flow separation. The nominal incidence at which the compressor designer chooses to operate the compressor is based on plots such as those in Fig. 2.20(d).

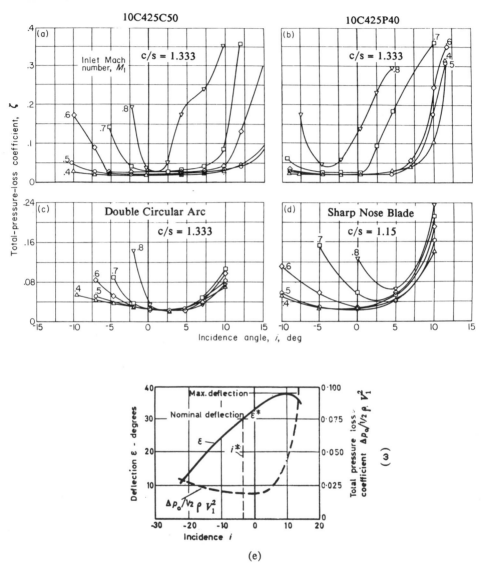

Figure 2.20 Cascade performance, effect of incidence, Mach number, and blade profile. (a–d) Typical loss distribution for various blade profiles ($\alpha_1 = 55°$, $\alpha_2 = 30°$) C4 circular arc, P4 parabolic arc and double circular arc have camber angle of 25°, sharp nose blade on a camber angle of 27.5°. (Johnsen and Bullock, 1965; NASA SP 36). (e) Cascade characteristics (11C1 / 45 / C30; $s/c = 0.9$, $\alpha_1 = 44.5°$, $\alpha_2 = 22°$, $R_e = 2 \times 10^5$). (From Horlock, 1973; AXIAL FLOW COMPRESSORS; reprinted with permission from Kreiger Publications).

The flow at the trailing edge does not leave at the same angle as the angle of the camber line. The difference between the blade or the camber angle and the average flow angle is called the *deviation angle* (Fig. 2.11). In an inviscid flow, the flow near the trailing edge on the blade surfaces follows the surface contours exactly, but the average flow angle deviates from the blade angle due to the fact that the flow has to adjust itself to a sudden removal of transverse pressure gradient near the trailing edge. Hence the average flow angle downstream of the suction peak can be expected to deviate from the camber angle. Deviation is caused by both inviscid and viscous effects. Boundary layer growth influences the deviation angle. Most earlier designs were based on empirical correlation, and this is no longer justified due to the availability of accurate and fast computational techniques. A survey of various correlations for the deviation angle can be found in Cumpsty (1989, p. 168) and Johnsen and Bullock (1965). One of the more successful deviation rules is due to Lieblein (1965), given by

$$\delta = \delta_{uo} + m\theta$$

where δ_{uo} is the deviation angle for a similar cascade of an uncambered airfoil with the same inlet flow angle as the cambered airfoil cascade, θ is the camber angle (Fig. 2.11), and $m = m_{\sigma=1}/\sigma^b$; b varies from 0.97 for $\beta_1 = 0$ to 0.55 for $\beta_1 = 70°$. The range of values of $m_{\sigma=1}$, based on a large number of cascade testings, is given by Lieblein (1965).

2.5.7 Rotating Stall

As the blade loading or the incidence is increased, the adverse pressure gradient on the suction surfaces increases the boundary layer growth, and its eventual separation, leading to blade stall as indicated in Figs. 2.17 and 2.18. This results in decreased pressure rise and efficiency. Furthermore, closing of the exit throttle or increase in the back pressure may not result in stalling of all the blades. Likewise, not all the blade passages would stall at the same time. This would result in a phenomenon called *rotating stall*, which refers to progression of the stall pattern from blade to blade in which one or more passages are stalled; this region propagates (in time) to subsequent passages. This phenomenon is illustrated in Fig. 2.21. Let us assume that passage 3–4 is stalled due to instantaneous increase in incidence at the inlet. Because of blockage caused by this separation, the incidence in passage 4–5 decreases, thus unloading and unstalling the blade in this passage. At the same time, incidence to passage 2–3 increases, initiating stall in this passage. Thus the stall cell moves right to left, opposite to the direction of rotation. The speed of propagation of the stall cell is found to be 50–70% of the blade speed. There may be several stall cells in a blade row rotating at the same time. The development of stall cells in a typical aeroengine is shown in Fig. 2.22 (Day,

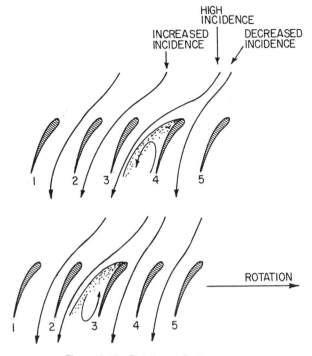

Figure 2.21 Rotating stall phenomenon.

1992). Part-span stall is usually encountered in the front stages of the compressor with large aspect ratio blades and low speed. Part span stall is usually encountered during the take-off or start-up when the mass flow is low and incidence is high. Full-span stall occurs in the medium range, and its effect is much more damaging. The cells extend all the way, leading to vibratory stresses (resonance if it coincides with the natural frequency of blades), noise, and deterioration in performance. Once again, this occurs only at peak loading and compressor never operates (under normal circumstances) in this flow regime. This may lead to low pressure and mass flow in the combustor leading to temporary shut down of the engine. The engine restart can usually be accomplished through quick response, but under no normal circumstances is such a situation allowed to occur.

The flow field of a compressor during stall is extremely complex. The computation of the inception of stall, based on linearized theories, is unsatisfactory. The next few years will see a major effort to predict stall and rotating stall numerically using complete Euler and Navier–Stokes equations. A comprehensive survey of phenomena, causes, various types of stall, correlations, and analyses available is given in the book by Pampreen (1993). The prediction of the stall performance is based on empiricism at the present

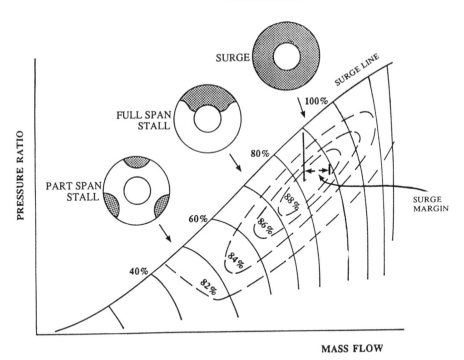

Figure 2.22 Compressor performance map. Solid lines are characteristics at various speeds, dotted lines are corresponding efficiencies.(Copyright © Day, 1992; reprinted with permission of VKI.)

time. One such approach is the diffusion factor described earlier. Computational procedures are available to predict the onset of blade separation and the endwall flow separation. The latter is the most likely mechanism for the onset of rotating stall. The present-day approach is to develop correlations and establish limitation for the peak pressure rise and thus establish stable boundaries for the operation of a compressor. One successful correlation is due to Koch (1981), described below.

Compressor blade geometry is characterized by camber angle, spacing, stagger angle, and aspect ratio (mean blade height/mean chord). All these parameters influence pressure rise and efficiency. The compressor rotor or stator could be viewed as a diffuser in the relative or absolute frame, respectively. Increasing the camber angle will increase the turning angle ($\beta_1 - \beta_2$), thereby increasing the diffusion (W_2/W_1) ratio. Increase in stagger angle has a similar effect.

Koch (1981) has examined data from over 50 compressor builds to systematically evaluate the effect of Reynolds number, tip clearance, blade geometry, and velocity triangles on the peak pressure rise. His major conclusions

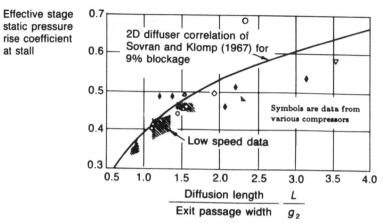

Figure 2.23 Correlation of stalling effective static pressure rise coefficients for low-speed stages. (From Cumpsty, 1989; reprinted with permission of Longman Scientific and Technical.)

are that the peak pressure rise increases with an increase in the Reynolds number, an increase in the camber, a decrease in the axial spacing of blade rows, a decrease in the tip clearance height, and an increase in the stagger angle. He has combined all these geometrical variations to define an "equivalent diffuser" with the parameter

$$\frac{L}{g_2} = \frac{\text{Diffusion Length}}{\text{Exit Passage Width}} = \frac{C\dfrac{2\pi}{360}\dfrac{\theta}{2} \bigg/ \sin\dfrac{\theta}{2}}{\left(\dfrac{A_2}{A_1}\right)\dfrac{Sh_1\cos\beta_1}{h_2}}$$

Here, h_1 and h_2 are the heights of the blade at inlet and exit, respectively, and (A_2/A_1) is the flow area (normal to flow direction) at inlet and exit, respectively. θ is the camber angle in degrees. It is clear that the ratio (L/g_2) is a function of the blade spacing, chord length, camber angle, and aspect ratio. Koch found that this combination provides an equivalent parameter representative of the diffusion process in a compressor. His correlation of the data (reproduced from Cumpsty, 1989), shown in Fig. 2.23, indicates that the combination of parameters given by the equation above provide a good representation of the performance and peak pressure rise of a compressor. The data in Fig. 2.23 include various combinations of aspect ratio, blade spacing, camber, and blade stagger angles. The data are correlated for variations in the tip clearance, Reynolds number, and blade row spacing (axial). Such plots provide guidance to the designers in avoiding stalled flow in compressors.

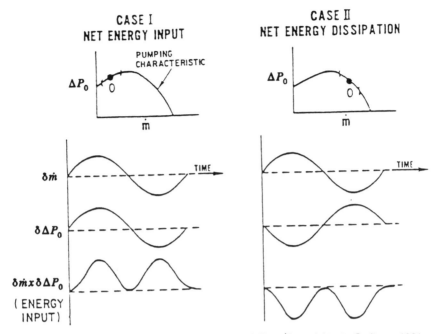

Figure 2.24 Physical mechanism for dynamic instability. (Copyright © Greitzer, 1981; reprinted with permission from ASME.)

2.5.8 Surge in Compressors

The compressor surge occurs when there is complete breakdown of the flow field in the entire system (not just the blading), and rapid changes in the mass flow lead to alternating stall and unstalled behavior resulting in violent oscillations in pressure, propagation of pressure waves, and the failure of the entire compression system. Physical mechanism responsible for the occurrence of dynamic instability in pumping systems is given by Greitzer (1981, 1985) and is shown in Fig. 2.24. A comprehensive survey of various mechanisms, causes and effects of surge as well as theories available is given in Pampreen (1993). Consider the operation of the compressor at two different operating points on either side of the characteristics (Figs. 2.18 and 2.24), one at high mass flow and one at mass flow below the peak pressure rise point (stall region). The dynamic instability is characterized by fluctuation in mass flow and pressure rise. In case II, shown in Fig. 2.24, the pressure rise decreases with an increase in the mass flow, and thus the perturbations of $\dot{m}$ and ΔP_o are out of phase. The only source of energy to sustain this wave motion is the pumping energy, which is proportional to $\delta\dot{m} \times \delta(\Delta P_o)$, whose integration over one cycle is negative. Thus the mechanical input to the pump is lower over one cycle than in the steady-state operation. The perturbations will thus decay and the stable operating condition "0" is restored. The reverse is true for the operating point of case I, which has a positive slope. The product of $\delta \dot{m} \times \delta(\Delta P_o)$ is positive over one cycle, and the mechanical energy input to the pump is higher than the steady state. This corresponds to the unstable region. This leads to instability and surge. Thus surge will occur in the positive slope region shown in Figs. 2.18, and 2.24. The phenomenon of surge has been investigated by many. A review of this can be found in Day (1992), Greitzer (1981, 1985), and Cumpsty (1989). Surge can lead to violent oscillations, noise, and mechanical failure of the compressor/pump system or the entire jet engine. One of the goals of the designer is to increase the surge margin, the difference in the mass flow between the best peak efficiency condition to surge point, as illustrated in Fig. 2.22. This is carried out through improved blading design, casing and blading treatment, blading modification near the tip, and ensuring that the flow into the compressor rotor is uniform.

The behavior of a compressor during rotating stall and surge is shown in Fig. 2.25. These data were acquired by Reiss and Blocker (1987) in a three-stage axial compressor with a tip diameter of 340 mm, running at 17,000 rpm, and a pressure rise ratio of 2.0. The compressor with high-volume downstream (6 m^3) shows intermittent rotating stall (indicated by rapid changes) followed by surge. This case has both the rotating stall and the surge, while the compressor with low-volume downstream has only rotating stall. The rotating stall is at a much higher frequency than the surge.

Inlet distortions, such as those caused by the foreign objects and blade-row wakes, can cause the rotating stall and surge to occur earlier, thus decreasing

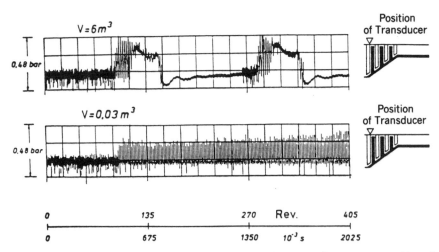

Figure 2.25 Pattern of unsteady pressure fluctuations for surge (large-pressure side volume) and full-span stall (small-pressure side volume) (Reiss and Blocker, 1987). (The original version of this material was first published by the AGARD / NATO in CP421, "Advanced Technology for Aero Gas Turbine Components," September 1987.)

the safety margin and the "surge margin" for the operation of a compressor. The stall characteristics can also be improved by movable guide vanes and stator blades. Some compressors operate with a control system built to change the incidence/inlet angle to the rotor as it approaches the stall region.

2.5.9   Effect of Mach Number

The trend in aircraft engine compressor design is to increase the tip speed and the relative Mach number. Even though the inlet absolute flow to these compressors is subsonic, future developments may include supersonic absolute flow. The pressure rise coefficient for a compressor blading is given by

$$C_p = \frac{2(p_2 - p_1)}{\rho W_1^2} = \frac{2(p_2 - p_1)}{M_{1R}^2 \gamma p_1} = \frac{2[(p_2/p_1) - 1]}{\gamma M_{1R}^2} \qquad (2.48)$$

Thus the controlling factor for the pressure rise is either the value of C_p or the diffusion coefficient introduced earlier; and the inlet relative Mach number. The value of C_p, is limited by the boundary layer phenomena. For a fixed value of C_p or D, the pressure rise can be increased by increasing the relative Mach number. This effect can be clearly shown for a rotor alone configuration using the analysis developed earlier. Consider a configuration where the entry Mach number (absolute) is axial with a value of 0.5. Let us assume that diffusion factor $(D) = 0.5$, $\sigma = 1.0$, and the exit axial Mach number is 0.5. In this case, $M_{1R} = \sqrt{0.25 + M_b^2}$, and $M_{2R} = \sqrt{0.25 + (1 - \psi)^2 M_b^2}$, $W_{\theta 1} = U$, $W_{\theta 2} = (1 - \psi)U$. Furthermore, the rotor is assumed to be of high hub/tip ratio. Therefore Eq. 2.47 can be solved for maximum loading coefficient (ψ_{max}) that can be achieved at a given blade speed (U, M_b). Knowing ψ_{max}, the pressure rise can be calculated using Eqs. 2.37 to 2.40. The calculated results for the pressure rise at various tip speeds ($M_b = M_t$) for two assumed efficiencies are shown in Fig. 2.26. The pressure rise increases rapidly with an increase in the tip speed. The gradient, $\partial(p_{o3}/p_{o1})/\partial M_t$, also increases with an increase in the Mach number. The potential advantage of higher tip speed is clearly evident from this figure. Equation 2.39 indicates that increasing blade tip speed would increase the temperature rise if the velocity triangles are kept similar. But the limitation on the diffusion factor (or flow turning) reduces the ψ_{max} that can be achieved. Therefore, the rise in pressure is less rapid, especially at lower Mach numbers. The maximum values of the loading coefficient that can be achieved for this hypothetical rotor is also shown in Fig. 2.26. ψ_{max} decreases with an increase in blade speed. Efficiencies decrease with blade tip Mach number (M_b) due to higher losses caused by shocks and shock–boundary-layer interaction. This decrease in efficiency is more dominant at higher Mach numbers.

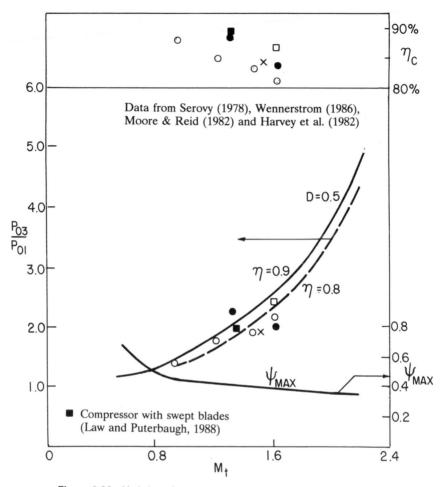

Figure 2.26   Variation of pressure rise, efficiency, and ψ_{max} with tip speed.

Some of the data from the rotor alone or stage alone (from Serovy, 1978; Wennerstrom, 1986; Moore and Reid, 1982; Harvey et al., 1982; Law and Puterbaugh, 1988) are shown in Fig. 2.26. Wherever available, the performance of the rotor alone is shown. The pressure rise follows the predicted trend, provided that proper efficiency is taken into consideration. For example, at $M_t \sim 1.6$, the data points are closer to the $\eta = 0.8$ curve. Efficiency as high as 89% has been achieved at $M_t = 1.3$ by the U.S. Air Force group (Wennerstrom, 1986; Law and Wennerstrom, 1986). Continued trends should increase M_t beyond 1.6 and achieve reasonably good efficiencies. It should be emphasized that estimated values of P_{o3}/P_{o1} are for the hypothetical rotor, and whose geometrical and flow parameters are mentioned earlier. The data correspond to differing geometries. This comparison is shown only to indicate trends.

Performance characteristics of two compressors, for which the data are shown in Fig. 2.26, are as follows:

Performance Parameter	Harvey et al. (1982)	Law and Puterbaugh (1988)
Corrected rpm	12,464	20,450
Rotor tip Mach number (M_t)	1.61	1.35
Corrected mass flow (kg/s)	78.8	28.29
Rotor total pressure ratio	2.34	2.06
Stage total pressure ratio	2.28	1.99
Rotor adiabatic efficiency	0.868	0.8967
Stage adiabatic efficiency	0.868	0.946
Tip diameter (meter)	0.8409	0.4318

Very high pressure, as well as reasonably good efficiencies, have been achieved by increasing the tip speed and the inlet Mach number.

Mass flow also increases with an increase in the blade speed. High flow rate per unit area is an important design criteria for aircraft engines, because of the obvious relationship between the frontal area, component weight, and viscous drag. Thus, an increased Mach number has the dual advantage of decreased frontal area and an increased pressure ratio. Both of these contribute to a higher thrust/weight and more compact power plant. Improved metallurgical and manufacturing technology has permitted higher tip speeds to be achieved.

2.5.10 Transonic and Supersonic Compressor

It is clear from the preceding section that higher relative velocity at the inlet results in higher temperature and pressure rise in a compressor. Major advances have been made during the last 10–20 years, resulting in compact, efficient, and lightweight compressors. When the relative Mach number becomes greater than unity, the qualitative and quantitative behavior of the flow changes substantially. Shock waves are formed at the leading edge or inside the passage, resulting in higher static and temperature rise as well as higher stagnation pressure and viscous losses. In present-day compressors, the axial and total Mach number of absolute flow at inlet is always subsonic. Because blade speed is Ωr, the blade speed is higher at higher radii. With the present trend toward higher blade speeds, the relative Mach numbers may reach supersonic ($\mathbf{W} = \mathbf{V} - \mathbf{U}$) near the outer sections of the rotor, while the relative flow near the inner radii remains subsonic. A *transonic compressor* is one where the relative flow remains subsonic ($M_R < 1$) at inner radii and supersonic ($M_R > 1$) at outer radii, with a transonic region in the middle. This is especially true for high aspect ratio or low hub/tip ratio compressors. In some low-aspect-ratio and compact compressors, there may be cases where the entire flow, hub to tip, has supersonic relative flow. Such a compressor is known as a *supersonic compressor*.

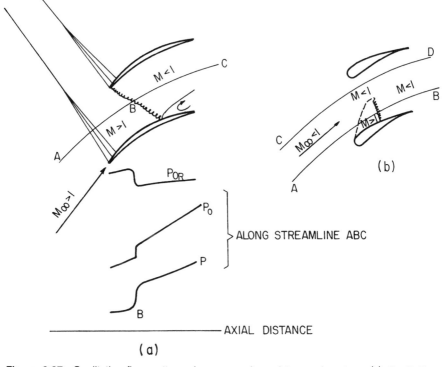

Figure 2.27 Qualitative flow patterns in supersonic and transonic rotors. (a) Qualitative variation of flow properties across a rotor with supersonic in flow and subsonic exit flow. (b) Subsonic inlet flow with supersonic flow regions inside the passage. (h_o and T_o follow the same trend as P_o, T and h follow the same trend as P, h_{oR} and T_{oR} are constant for an adiabatic compressor; jump in P_o and h_o depends on the velocity triangle at B).

There are major differences between a subsonic, transonic, and supersonic compressor. Consider the qualitative flow behavior in the passage shown in Fig. 2.27. Here are the major differences between a purely subsonic flow through a blade row, supersonic flow at the entry, and a subsonic flow at the entry with supersonic regions inside. There are a large number of variations in addition to these three items, which will be dealt with in Chapter 3.

1. The leading edge and the blade shape differ considerably when the entry flow is supersonic. To avoid detached shocks, sharp-nosed leading edges are used with supersonic entry. The flow is expanded smoothly over the suction surface (Fig. 2.27) with expansion waves; and a shock wave occurs within the passage, originating on the pressure side.

2. There is a jump in static pressure and static temperature across a shock wave, and thus the relative flow diffusion takes place suddenly across a shock wave. This is beneficial in increasing the pressure rise as compared to purely subsonic flow where the diffusion is carried out across a

finite length cascade. The disadvantage of such a diffusion is that, associated with localized large pressure rise, there are boundary layer shock interaction effects which result in most cases with flow separation. This feature, combined with the entropy rise across a shock wave, results in higher pressure losses and lower efficiency.

3. Mixed flow from hub to tip results in large radial pressure gradients (sometimes discontinuous) in the radial directions. This results in three-dimensional flow in a passage. The validity of two-dimensional analyses for turbomachinery flows with three-dimensional shocks is somewhat questionable.

4. The shocks respond to changes in downstream conditions. This brings about a drastic change in the flow behavior with variation in downstream conditions. Both the location and the strength of the shock are altered.

5. The upstream should be shock-free or should consist of alternate weak shocks and expansion waves. Otherwise, there would be large turning upstream of the flow, a condition which is unacceptable. There is only one incidence (at a given operating condition) when the flow at the entry is smoothly guided through the blade suction surface, and there would be no upstream shocks or expansion waves. Such a condition is called a *unique incidence condition* and will be dealt with in the next chapter. Operating the compressor below or above this condition results in large losses. Hence, the characteristic ($\psi - \phi$) is very steep for a supersonic flow as indicated in Fig. 2.18 (design speed and 105% speed). This is one of the most critical limitations of a supersonic compressor.

6. Associated with the shock wave is blade choking, which controls the mass flow that can be passed through a compressor.

7. The designer has to incorporate the shock losses (both direct and indirect effects) in the process.

8. Higher static pressure ratio as well as pressure jumps cause rapid area changes and blade curvature. This has to be incorporated in the design procedure.

Early transonic and supersonic compressor designs were failures. The efficiencies were poor and the reliability was not good. It was initially believed that the low efficiencies obtained were due to the shock pattern alone. It was later recognized that the losses were more a result of flow disturbances caused by shocks. Major improvements have been made in blading design, shock optimization, and hub-to-tip design. Some of the early blade designs are given in Hawthorne (1964). The present designs are shown in Figs. 2.13b and 2.13c. The most successful designs are the ones custom designed, based on the controlled diffusion airfoil and shock-free airfoil design concepts. This has enabled high efficiency to be achieved. The efficiency of a supersonic compressor (single stage) which was as low as 40%

during 1940 has been increased to nearly 90% (e.g., Fig. 2.26). The highest relative Mach number has increased to 2.4 in experimental compressor stages. This has been achieved through a careful design of blading, choice of aspect ratio, radial variation of blading shape, and other blade and flow parameters. The Mach numbers in stators have not increased as rapidly as rotor Mach numbers. A historical account of transonic and supersonic compressor development can be found in Serovy (1978) and Wennerstrom (1986). The present trend is toward low aspect ratio and higher Mach number blading, so as to achieve supersonic flow at most radial locations. The number of stages could be reduced for a given pressure ratio, thus enabling substantial savings in weight and size.

Supersonic compressor stages can be broadly classified as follows:

1. Shock-in-rotor/subsonic stator configuration
2. Shock-in-rotor/shock-in-stator configuration
3. Rotor with subsonic turning/supersonic shock-in-stator
4. Rotor with all supersonic turning/shock-in-stator
5. Rotor with all supersonic turning/subsonic stator

These classifications are self-explanatory and are shown in Fig. 2.28a. Classification 1 implies that the shock is inside the rotor and that the relative velocity at inlet to rotor is supersonic. The rotational speed of the rotor is very high. Shock-in-rotor or shock-in-stator implies that within a blade channel, subsonic flow follows a stabilized shock. The stages with fully supersonic flow and shock-in-rotor produce the highest pressure rise.

One of the major considerations of supersonic and transonic compressors is the mass flow rate. The unique incidence and the sonic choking conditions inside the passage control the mass flow through the compressor. The most efficient supersonic compressor operates with an axial Mach number of unity. From gas dynamic considerations, it can be shown that maximum mass flow per unit area is achieved at an axial Mach number of unity and the mass flow per unit area decreases as the Mach number is decreased or increased from unity. Therefore, achieving very high subsonic velocities at the inlet to a compressor, which results in very high supersonic velocities in the compressor, is the goal of designers of modern compressors for aircraft engines. Today, entry axial Mach numbers in the range of 0.4–0.6 are typical.

The change in the shock pattern in a cascade as the Mach number is increased from low supersonic to high supersonic inlet flow is shown in Fig. 2.28b. The shock pattern as well as separation streamlines are shown. Shock reflections from separated streamlines, as well as a leading edge shock at $M_{1R} = 2.52$, can be seen in the figure. The presence of stronger shocks and larger separated zones, along with the flow disturbances caused by these shocks in the radial direction, is responsible for the poor efficiency of these compressors. A method for analyzing the one-dimensional flow through such cascades is given in the next chapter. Very low efficiency, resulting from

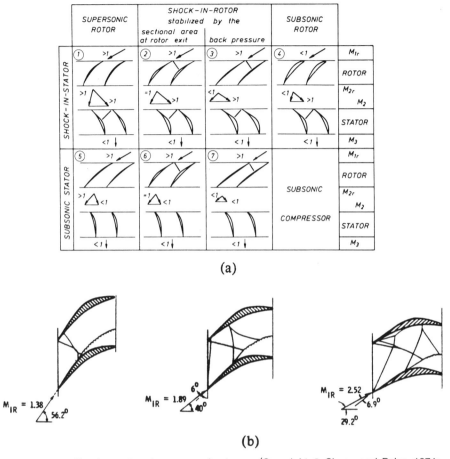

(a)

(b)

Figure 2.28 Shock structure in supersonic stages. (Copyright © Simon and Bahn, 1974; reprinted with permission from ASME.) (a) Supersonic compressor stages (subscript *r* stands for relative Mach number). (b) Change in shockwave pattern as inlet Mach number is increased.

these shock waves, may limit operating range of these supersonic compressors at very high Mach numbers.

2.5.11 Current Trends in Compressor and Fan Development

Wisler (1989) has provided a summary of current trends in aircraft engine compressor design and performance. These can be summarized as follows:

1. *Higher Speeds.* The advantages of higher speeds have already been explained. The future trend is toward increasing the tip Mach number as well as increasing the inlet Mach number (absolute flow). But the

successful design will almost invariably involve a tradeoff between efficiencyand the pressure rise.

2. *Higher Pressure Rise per Stage / Spool.* The advantage of higher pressure rise is the reduced weight and drag of the engine, resulting in improved specific fuel consumption for the power plant. The data shown in Fig. 2.26 indicate that the future trend is toward increasing the pressure ratio beyond 2 and at the same time attaining higher efficiencies.

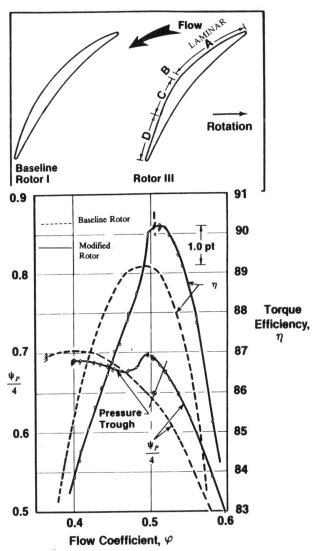

Figure 2.29 Improvement in performance with custom-tailored airfoil (region A—laminar flow; region B—transition; region C—rapid turbulent diffusion; region D—deceleration). (Copyright © Wisler, 1989; reprinted with permission of General Electric Aircraft Engines.)

Table 2.2 **Effect of diffusion factor and tip speed on ψ_{max}**
and P_{o2}/P_{o1} ($\sigma = 1$, $M_z = 0.5$, $\eta = 0.9$)

	ψ_{max}		P_{o2}/P_{o1}	
M_t	$D = 0.5$	$D = 0.6$	$D = 0.5$	$D = 0.6$
0.5	0.70	0.88	1.23	1.29
0.8	0.47	0.57	1.41	1.51
1.2	0.40	0.47	1.86	2.08
1.8	0.36	0.43	3.22	3.95
2.1	0.36	0.42	4.00	5.56
2.4	0.35	0.42	6.23	8.25

3. *Higher Aerodynamic Loading.* Achievement of higher aerodynamic loading, without separating the boundary layer, has always been the aim of the compressor designer. By controlling transition, laminar length, and boundary layer separation (through proper choice of blade profile and pressure distribution), the diffusion factor (and hence the loading coefficient C_p) for a compressor can be improved.

One such example carried out by the General Electric Company is illustrated in Fig. 2.29, where the extent of laminar flow on the airfoil was controlled. The peak suction velocity was moved from 25% chord to 38% at midspan and to 55% at the tip. The transition region (laminar to turbulent) has nearly constant axial velocity (or no pressure gradient). The growth of the turbulent boundary layer was controlled using an appropriate velocity distribution (or pressure gradient) through thickness and camber changes. The new blade section and its improved performance are shown in Fig. 2.29. In effect, such transition control increases the diffusion factor.

A calculation of maximum pressure rise (ψ_{max}), for the rotor with $\sigma = 1$, $M_z = 0.5$, $M_{\theta1} = 0$, $\eta = 0.9$ at two values of diffusion factor, shown in Table 2.2 (similar to those described earlier and shown in Fig. 2.26), reveals the advantage of increasing the diffusion factor through boundary layer control.

4. *Lower Aspect Ratio and Higher Solidities.* The trend today is toward lower aspect ratios (blade height/chord ratio) in axial compressors. The aspect ratio of compressor blades has decreased by a factor of 3 (4 to 1.4) during the last 40 years (Wisler, 1989). The advantages of low-aspect-ratio compressor blades are described in Wennerstrom (1986). There has been a gradual realization that low aspect ratios (blade height/chord) are beneficial in achieving higher loading, higher efficiency, and good stall/surge range in aircraft engine compressors. In earlier days, moderate aspect ratios were preferred for mechanical design reasons. There are indications that stall performance can be improved with lower aspect ratios. Also, the compressor is more rugged

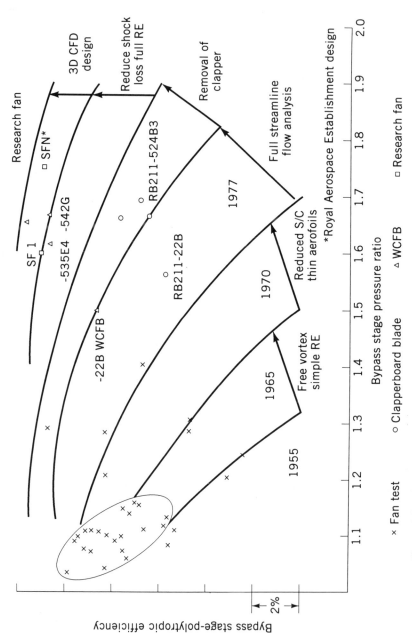

Figure 2.30 Increased fan efficiency through improved aerodynamics (RE stands for radial equilibrium equation [Chap. 4]). (Courtesy of Rolls Royce PLC.)

mechanically with lower blade aspect ratios. With the trend toward controlled diffusion airfoils, and higher pressure rise, the average solidity used in aircraft compressors has increased more than 40% (1 to 1.4) during the last 40 years (Wisler, 1989).

5. *Improved Blade Profiles.* As indicated earlier, current design trends include the use of custom designed airfoils as opposed to the use of standardized airfoils. This provides effective control of the chordwise pressure gradient and boundary layer behavior, thereby increasing the loading coefficientand the pressure rise substantially.

6. *Improved Loss Prediction / Control.* Compressor developers have been actively involved in identifying the sources of loss. Various means are used to (a) control and reduce the losses due to the boundary layer on endwalls, and (b) improve viscous behavior of the flow through a compressor and fans. Various attempts to achieve these goals have resulted in a continuous improvement in the efficienciesof compressors and fans, as shown in Figure 2.30. This has been achieved through improved analytical techniques, blading design and loss reduction techniques.

2.6 PROPULSORS / PROPELLERS / PROP FANS / UNDUCTED FANS

Propellers are used in aircraft and marine (both surface and underwater) vehicles for developing thrust. They are considered to be "extended turbomachinery." A schematic of conventional and modern propellers/prop fans/propulsors used in aircraft and marine applications is shown in Fig. 2.31. Details on the prop fans used in aircraft are given in Hawkins (1984) and Arndt (1984), those on marine propellers and propulsors are presented in Gearhart (1966), and those on aircraft propellers are given in McCormick (1979). A photograph of a ship propeller is shown in Fig. 1.8, and an aircraft prop fan is shown in Fig. 1.3.

Many of the theories available for ducted turbomachinery can be utilized to analyze the flow through propellers. Propellers essentially increase the pressure of the fluid, and this increase is much smaller than that encountered in axial flow compressors. The pressure rise is utilized to accelerate the fluid, thus providing the thrust necessary to propel the vehicle.

The recent introduction of the prop fan (Hawkins, 1984; Arndt, 1984) for aircraft propulsion has generated renewed interest in these types of turbomachines. In the prop fan (tractor configuration), the propeller is located in front of the engine and is driven by a turbine through gearing. The main difference between a conventional propeller and a prop fan is in the number of blades and the speed. A conventional propeller has very few blades (2–4), large diameter, and low pressure rise. The operating speeds in these conventional propellers are kept low to avoid shock waves in the tip region. In view

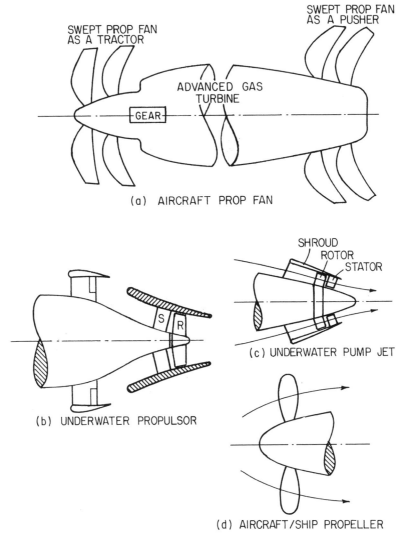

Figure 2.31 Propeller and propulsor configurations used in aerospace and marine applications.

of this, the early use of propellers was mainly confined to low speeds on short-range aircraft, ships, and marine propulsors.

The prop fan (tractor) and unducted fan (aft version or pusher prop fan) are under intensive development. These are expected to achieve high performance and efficiencies. They are likely to replace some of the turbo fans which are in wide use at the present time. Present-day turbo fans have bypass ratios as high as 8–9, with a fan pressure rise of 1.3–1.9. The propulsive

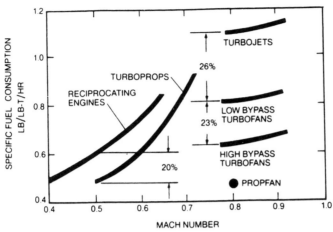

Figure 2.32 Specific fuel consumption (mass flow rate of fuel per hour / unit thrust) for various power plants. (Copyright © Hawkins, 1984; reprinted with permission of AIAA.)

efficiency increases with an increase in the bypass ratio (through decreased heat rejection in the jet) and a decrease in pressure ratio. Thus, at low speeds and short range, the turbo prop is the most efficient power plant, but its efficiency (inversely proportional to specific fuel consumption) decreases with an increase in the speed (Fig. 2.32). Moreover, the propeller power plant is bulky due to the need for a gear box. This has led to the introduction of the prop fan for aircraft. The present prop fan effort is directed toward improving the efficiencies beyond that which can be achieved with a turbo fan.

With modern advances in turbomachinery blade design, aircraft engine manufacturers have developed a propeller (called the *prop fan*) which can be operated efficiently at $M = 0.8$ that produces high thrust. This has been achieved through the use of eight to ten highly loaded blades (similar to compressor) with thin profile and sweep. The sweep reduces the onset of compressibility effects and the losses at the blade tips, thus improving efficiency over the conventional straight bladed propeller. This has enabled higher efficiencies to be achieved. The prop fan utilizes all the technology available for high-speed compressor/fan design, thereby providing a compromise between a high-bypass-ratio fan with high blade loading and a large-diameter unducted propeller, hence the name *prop fan*. It includes the best features of both turbomachines. Low specific fuel consumption (defined as the ratio of fuel consumption rate per hour per unit thrust) has been achieved for the prop fan as shown in Fig. 2.32. A major problem in the use of the prop fan is the associated noise and vibration. Various attempts are being made to reduce these.

The prop fan (Fig. 2.31) combines the best features of a variable-pitch fan, counterrotation propellers, and modern turbomachinery aerodynamic

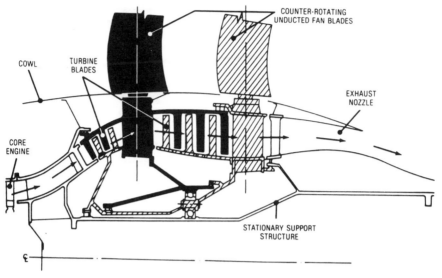

Figure 2.33 Unducted fan configuration. (Copyright © Hawkins, 1984; reprinted with permission of AIAA).

developments to achieve a fuel saving of as much as 20–30% over the conventional turbofans and turbojets. A schematic of the "pusher" type of prop fan (unducted fan, as opposed to the "tractor" type) is shown in Fig. 2.33. Unlike a "tractor" configuration, it eliminates the use of gearing because the turbine can drive the prop fan (UDF) directly. This is a considerable improvement over existing propeller-driven power plants and the turbofan. The disc loading of the General Electric design (Hawkins, 1984) is 65 hp/ft^2, and the bypass ratio is in excess of 35. UDF features include (a) an inner turbine rotor which supports and powers a rear row of variable-pitch fan blades through a rotating turbine rear frame and (b) an outer turbine rotor which supports and drives another row of variable-pitch fan blades. Variable pitch allows optimization of fan incidence angle at various flight speeds and also allows for thrust reversal. The turbine stages as well as the fan stages are counterrotating. Such a configuration has better noise and vibration characteristics. Counterrotation also eliminates swirl losses, improves efficiency, and allows higher pressure ratios. Fuel savings of up to 30% have been achieved.

2.6.1 Analysis and Performance

The major difference between the conventional propeller and the prop fans and propulsors is the pressure rise and thrust level. Basic principles are very similar to those described in the previous section and are outlined below.

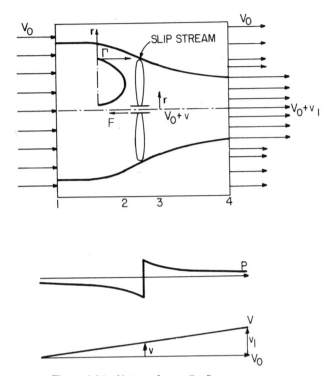

Figure 2.34   Nature of propeller flow processes.

The principles on which propellers operate can be explained from the actuator disk theory, illustrated in Fig. 2.34. Let us assume that the flow is incompressible, the pressure and velocity are uniform far upstream and downstream and the slip stream is well-defined. The static pressure decreases as it approaches the disk, increases across the blade, and decreases again to ambient conditions downstream. The pressure difference Δp provides the flow acceleration for the generation of thrust. Using the Bernoulli equation across stations 1 and 2 and across 3 and 4, along with Eqs. 1.2 and 1.3 for the control volume, it can be shown that the thrust developed is given by (see McCormick, 1979)

$$F = 2\rho A(V_o + v)v \quad \text{and} \quad v_1 = 2v$$

where A = disk area.

It should be noted that $\rho A(V_o + v)$ is the mass flow through the propeller plane and $2v$ is the change in velocity between far downstream and far upstream. The actuator disk analysis/concept does not reveal the mechanism generating the pressure rise. This mechanism is best illustrated by using a blade element analysis, similar to the one described for the axial flow compressor/fan in Section 2.5.

Consider the blade forces shown in Fig. 2.15. The difference in angle $(\beta_1 - \beta_2)$ is small for a propeller, because the camber is small. Let us write $\beta_m = \phi$. The angle $(90° - \phi)$ is called the *effective pitch angle* of the propeller, and $(90° - \lambda)$ is called the *geometrical pitch angle* of the propeller, since the propeller aerodynamicists measure all angles with reference to the tangential direction. Consider a unit spanwise height of the propeller at radius r, where the chord length is C. From Eq. 2.42, the infinitesimal torque $dT = rF_\theta$ and infinitesimal thrust $dF = F_z$ due to unit blade height are given by

$$dT = \tfrac{1}{2}\rho W_m^2 Cr[C_L \cos\phi + C_D \sin\phi]$$
$$dF = \tfrac{1}{2}\rho W_m^2 C[C_L \sin\phi - C_D \cos\phi]$$

(2.49)

For a propeller with no camber, we have

$$W_m \approx W_1 \approx W_2 = \frac{W_z}{\cos\phi}$$

Therefore the total thrust and torque are given by

$$F = \frac{1}{2}\rho W_z^2 B \int_{hub}^{tip} \frac{C(r)}{\cos^2\phi}[C_L \sin\phi - C_D \cos\phi]\, dr$$

$$T = \frac{1}{2}\rho W_z^2 B \int_{hub}^{tip} \frac{C(r)r}{\cos^2\phi}[C_L \cos\phi + C_D \sin\phi]\, dr$$

(2.50)

where B is the number of blades, and $W_z = V_o$, the flight speed (Figure 2.34). The propeller efficiency is given by,

$$\eta_p = \frac{\text{Thrust Power Output}}{\text{Power Input}} = \frac{FV_o}{2\pi NT}$$

(2.51)

Knowing the section C_L and C_D, blade geometry, and flight speed (W_z), the torque and efficiency can be calculated from Eqs. 2.50 and 2.51. Methods for calculating section C_L and C_D from a knowledge of airfoil/blade profile thickness and incidence angle are described in Dommasch et al. (1967) or McCormick (1979) and are usually based on thin airfoil theory.

A dimensional analysis, similar to that described in Section 2.4, can be carried out to prove that the nondimensional groups are given by

Torque coefficient: $C_T = T/\rho N^2 D^5$

Thrust coefficient: $C_F = \dfrac{F}{\rho N^2 D^4}$

Power coefficient: $C_p = \dfrac{P_{shaft}}{\rho N^3 D^5} = 2\pi C_T$

Hence

$$\eta_p = \frac{C_F}{C_p}\left(\frac{V_o}{ND}\right) \tag{2.52}$$

The parameter $J = V_o/ND$ is similar to the flow coefficient for a compressor ·, .d is a measure of the blade stagger angle. It is called the *advance ratio*. I ˖rformance is usually represented as C_p versus V_o/ND and C_T versus V_o/ND. Detailed performance maps and design charts can be found in McCormick (1979) and Dommasch et al. (1967) for aircraft propellers and in Gearhart (1966) and Dickmann and Weissinger (1955) for marine propellers and propulsors.

Prop fan and unducted fan analysis and design differ from those outlined above. The blade sections near the root (mid-radius to root) have large camber and resemble a compressor or a fan blade, while the outer (mid-radius to tip) sections resemble a propeller with small camber and thickness. Methods of analysis and design are still evolving. One of the methods (Metzger and Rohrbach, 1979) is to analyze the mid-radius to root section using blade element analysis, as described in Section 2.5 (Eqs. 2.37–2.43). It should be noted here that W_1 and W_2, as well as β_1 and β_2, differ considerably. The sectional pressure rise, thrust, and torque are all influenced by the camber angle, incidence, stagger angle, and blade thickness. The static pressure rise can be derived from the analysis of Metzger and Rohrbach (1979), and the thrust developed can be derived from the known pressure rise. From midradius to tip, the equations given in this section can be utilized to derive thrust and torque (Eq. 2.50). If the C_L and C_D of the sections are known, the thrust developed by the blade section can be determined. For high speed flows, it is necessary to include density variation $\rho(r)$ in Equations 2.50 and 2.51. Metzger and Rohrbach utilized this approach and analyzed the root sections using various methods available for turbomachinery (cascade theory; see Chapter 3). The outer sections were designed using Goldstein's classical vortex theory (see McCormick, 1979), which is a more advanced theory than the blade element theory described in this section. Their results for the eight-bladed $M = 0.8$ rotor are shown in Fig. 2.35. The agreement between the predicted and measured values is good.

The measured performance of a scale model of a counter rotating unducted fan (Fig. 2.33) designed for $M = 0.72$, altitude = 10,668 m, $C_p' = 4.17$ ($= 550\ \text{SHP}/A\rho N^3 D^3$, where A is the annulus area), and tip speed = 237 m/s, radius ratio = 0.42, is shown in Fig. 2.36 (Sullivan, 1987). The values of C_p and η are much higher than those encountered in conventional propellers (see McCormick, 1979). Considering that V_o/ND is analogous to flow coefficient, this performance characteristic is very similar to those of axial flow compressors. The power loading coefficient decreases with an

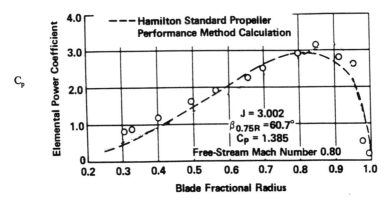

Figure 2.35   Comparison of predicted and measured power loading distribution. (Copyright © Metzger and Rohrbach, 1979; reprinted with permission of AIAA.)

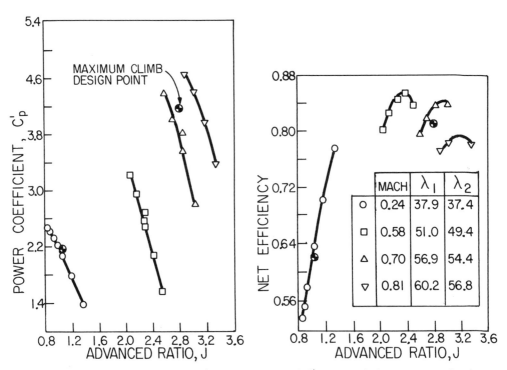

Figure 2.36   Performance of UDF (λ_1 is the stagger of 1st rotor, λ_2 is the stagger angle of the 2nd rotor at 3/4 radius). (Copyright © Sullivan, 1987; reprinted with permission of General Electric Co.).

increase in advance ratio (increased flow coefficient) and increases with Mach number, which is similar to the ψ–ϕ relationship for a compressor. The peak value of C_p occurs close to stall. Decreasing the advance ratio further from this peak results in blade stall. The UDF exhibits a high pressure coefficient and high efficiency. The efficiency peaks at 85% at $M = .65$, and drops to 77% at $M = .81$. As improved design methods become available, efficiencies are likely to reach higher values than those shown in Fig. 2.36. Sullivan (1987) also found that efficiency and C_p are sensitive to rotor-rotor spacing. An increase in spacing decreases η and C_p. This is caused by mixing of the flow in the space between two rotors.

Three-dimensional and compressibility effects are appreciable in these prop fans. Therefore, simple analyses presented in this section should only be used to provide qualitative trends. The cascade theories presented in the next chapter should provide an accurate prediction of the flow from hub to midspan of the prop fan, and the outer half of the blade is best handled by numerical methods or by two-dimensional strip theories corrected or three-dimensional effects caused by vortex shedding (McCormick, 1979). Circulation, lift, and pressure rise all go to zero gradually toward the tip, and the analysis of the tip section region becomes very complicated. Smith (1987) provides a method of incorporating three-dimensional effects into the design of the prop fan/UDF.

2.7 CENTRIFUGAL COMPRESSOR AND FANS

Centrifugal compressors are used in situations where mass flow requirements are low to moderate and the pressure rise is high. For the same frontal area, the centrifugal turbomachine has lower mass flow rate than an axial machine, because the frontal flow area is a fraction of the total frontal area (Figs. 1.4, 1.5, 1.9 and 1.12). Early turbojet engines had centrifugals, but because of their low cycle pressure ratio, they have been mostly replaced by axials (except in small engines). Most low-to-moderate thrust turboshaft, turbo prop, and turbo jet engines have centrifugal compressors. They have also found widespread use in industrial, automotive, commuter aircraft, rocket engine (pumps), and other applications. Pressure ratio as high as 12–14 and speeds in excess of 100,000 rpm have been achieved. The centrifugals are basically simpler, more compact, and less costly but less efficient than the axials.

Let us consider Eq. 2.36 and draw some conclusions regarding the mechanism causing pressure/enthalpy/temperature rise in axial and centrifugal compressors. In a purely axial machine, $U_1 = U_2$, and thus the static enthalpy/pressure rise can be achieved only by decelerating the relative flow. Likewise, the stagnation enthalpy rise of the absolute flow can be increased only by the flow turning (Eq. 2.9). The rotor passages act like diffuser in relative frame. The centrifugal compressor achieves part of its pressure rise

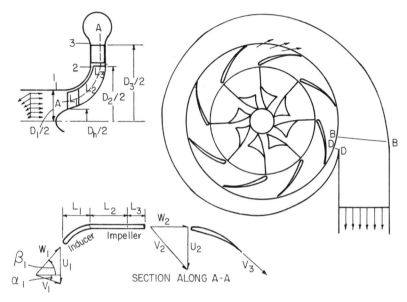

Figure 2.37 Centrifugal compressor.

from the centrifugal and Coriolis forces due to rotation and the change of radii. This is in addition to the pressure rise achieved through flow turning as in an axial compressor, where the energy interchange is through change of tangential momentum. In centrifugals, the change in radii provides additional pressure and enthalpy rise as indicated by Eq. 2.9. In view of the large pressure rise and the presence of centrifugal and Coriolis forces, the flow through centrifugal compressor passages is affected adversely by large boundary layer growth, flow separation, and secondary flow. In view of these features, centrifugal compressors have lower efficiency than axials.

A schematic of the centrifugal impeller is shown in Fig. 2.37. A photograph is shown in Fig. 1.12. The flow, entering axially, is turned towards the radial direction by an "inducer." Centrifugal fans and compressors without inducers are noisy and inefficient due to large incidence and separation that occurs at the impeller leading edge. The temperature and pressure rise across an inducer can be calculated using Eqs. 2.37–2.40. A review of the three-dimensional viscous effects and nature of flow field in inducer passages used in rocket pumps (Fig. 1.7) is given in Lakshiminarayana (1982).

The flow from an inducer goes through either a "mixed-flow path" or a radial path depending on the design. The former is called a *mixed-flow compressor* and the latter is called a *radial compressor*. The pressure rise in this section is due to both centrifugal and Coriolis forces and to some extent is also due to blade camber. A high-velocity jet comes out radially at the exit of the rotor, and it is then diffused in a stator or a vaneless diffuser. Because of compressibility effects, the height of the blade (normal distance between the shrouds) decreases continuously from the leading edge to the trailing edge. The flow is then collected in a scroll as shown in Fig. 2.37 or connected to a second stage through a crossover duct as shown in Figs. 1.4 and 1.6. Multistaging is much more difficult for centrifugals and involves considerable mixing losses through the duct connecting the two stages. For this reason, multistaging of centrifugal compressors is not very common.

In many instances, the diffuser vanes are eliminated, and the flow is "dumped" into a volute. This reduces the tendency of the compressor to stall and surge, but the result is achieved at considerable loss in stagnation pressure. Stall and surge usually occur at the exit of the impeller and at low flow rates, as in an axial flow compressor. At very low flow rates, unsteady separated flows are observed at the intake of a centrifugal compressor. The vaned diffuser is more sensitive to stall and surge, so many have adopted a "vaneless diffuser" (Fig. 1.4), though this does not completely eliminate the tendency to stall and surge.

There has been a trend recently to combine the best features of centrifugals and axials to design a hybrid compressor. This may include a centrifugal stage in the back or front end, followed or preceded by an axial multistage compressor. Schweitzer and Fairbanks (1981) designed and built a combined centrifugal back end with a low aspect ratio/highly loaded axial compressor stage to achieve a design pressure ratio of 18:1 in only seven stages, with a polytropic efficiency of 91.5%. A hybrid compressor called the *Axi Fuge* has been developed by Solar Turbines Inc. (Wiggins, 1986) and has achieved high efficiency and performance. The Axi Fuge (which can also be classified as a mixed-flow compressor) has a typical centrifugal compressor annulus, but the individual stages of rotor and stator are similar to those of an axial compressor. The Axi Fuge greatly improved on the efficiency of a centrifugal compressor, while preserving the compactness of these machines. A comparison of the three types of compressors operating with the same pressure rise is given in Fig. 2.38. Also included in this figure is a schematic of the Axi Fuge

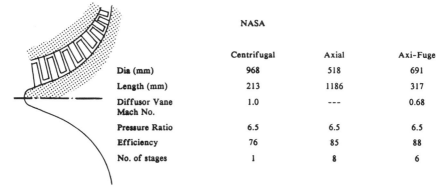

	NASA		
	Centrifugal	Axial	Axi-Fuge
Dia (mm)	968	518	691
Length (mm)	213	1186	317
Diffusor Vane Mach No.	1.0	---	0.68
Pressure Ratio	6.5	6.5	6.5
Efficiency	76	85	88
No. of stages	1	8	6

Figure 2.38 Comparison between Axi Fuge, axial, and centrifugal compressors. (Copyright © Wiggins, 1986; reprinted with permission of ASME.)

compressor, built and tested at Solar (Wiggins, 1986). It is clear that the Axi Fuge achieves the best features of both types of compressors.

2.7.1 Analysis and Performance

The analysis of centrifugal compressor flow is complicated. Because the flow passages are narrow, viscous and rotational effects play a major role. Nevertheless, one-dimensional inviscid analysis provides some qualitative trends.

There are three types of blading used. They are illustrated in Fig. 2.39 along with the corresponding velocity triangles. Let us examine the nature of the flow field as it enters the impeller and exits the scroll. As shown in Fig. 2.37, the flow enters the inducer with or without swirl at an inlet relative angle of β_1. The inducer provides some pressure rise. The streamline path L_2 is partly axial and partly radial, terminating in a radial path L_3. The

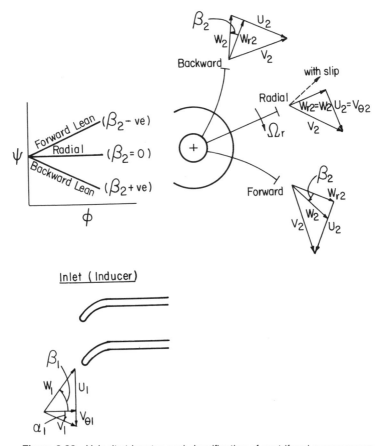

Figure 2.39 Velocity triangles and classification of centrifugal compressors.

relative velocity at the exit W_2 differs depending on the type of blading used (Figs. 2.37 and 2.39). The absolute velocity V_2 is then diffused to V_3 through a radial vane diffuser or a vaneless diffuser. In the vaned diffusers, the diffusion is achieved partly by flow turning and partly by area changes (higher radius has higher annulus area). In the vaneless diffuser, the diffusion is purely a result of area change. The exit velocity V_3 has a large swirl component. The exit flow is collected in a volute (notice the area change in the circumferential direction) and exits in a pipe. Clearly, the flow path is much more complex than in an axial flow compressor.

In a centrifugal compressor or pump, the pressure rise or head rise occurs not only due to flow turning, but also due to change in radii (due to centrifugal and coriolis forces). Consider Equation 2.36,

$$(h_2 - h_1) = \frac{W_1^2 - W_2^2}{2} - \frac{U_1^2 - U_2^2}{2}$$

Thus part of the static enthalpy or pressure or temperature or head rise occurs due to diffusion of relative velocity and part of it is due to the centrifugal effect. The pressure rise in an axial flow compressor occurs only due to relative flow diffusion. In a centrifugal machine, the contribution from these sources can be nearly 50% each. This is precisely the reason, the centrifugal compressor (for the same mass flow and relative velocity diffusion) can produce a much higher pressure rise than a corresponding axial flow compressor. The relative velocity diffusion in a centrifugal compressor is caused by the flow turning and area changes. In addition, the shear and turbulent stresses cause additional decrease in relative velocity, resulting in further static pressure/temperature rise. The viscous or shear pumping effect is very inefficient, and associated with this is a decrease in efficiency and increase in losses. A detailed treatment of shear pumping effect is covered in Lakshminarayana (1978).

If the flow is assumed to be inviscid and axisymmetric upstream of the rotor, downstream of the rotor, and downstream of the diffuser, Eqs. 2.7 and 2.8 can be used to determine the pressure rise (considering the velocity triangles shown in Fig. 2.39).

The temperature rise is given by

$$C_p(T_{o2} - T_{o1}) = U_2^2 - (W_{r_2} \tan \beta_2)U_2 - U_1(U_1 - W_{z_1} \tan \beta_1)$$

$$\psi = \frac{C_p(T_{o2} - T_{o1})}{U_2^2} = \left(1 - \frac{W_{r_2}}{U_2} \tan \beta_2\right)$$

$$-\left(\frac{r_1}{r_2}\right)^2 \left(1 - \frac{W_{z_1}}{U_1} \tan \beta_1\right)$$

$$(2.53)$$

where β_1 and β_2 are positive in the direction opposite to rotation shown in

Fig. 2.39. β_1 is positive, and β_2 is positive for backward leaning blades and negative for forward-leaning blades, and zero for radial blades.

Equation 2.53 can also be expressed as

$$
\frac{T_{o2}}{T_{o1}} = 1 + \frac{(\gamma - 1)M_{b2}^2}{1 + \dfrac{\gamma - 1}{2}M_1^2} \left\{ \left(1 - \frac{M_{r2}}{M_{b2}} \tan \beta_2 \right) - \left(\frac{r_1}{r_2} \right)^2 \left[1 - \frac{M_{z1}}{M_{b1}} \tan \beta_1 \right] \right\}
$$

$$(2.54)$$

where

$$
M_{b2} = \frac{U_2}{\sqrt{\gamma R T_1}}, \quad M_{z1} = \frac{W_{z1}}{\sqrt{\gamma R T_1}}, \quad M_{r2} = \frac{W_{r2}}{\sqrt{\gamma R T_1}}, \quad M_1 = \frac{V_1}{\sqrt{\gamma R T_1}}
$$

If $r_1 = r_2$, Eq. 2.54 ($M_{b2} = M_{b1}$, $M_{r2} = M_{z2}$) reduces to Eq. 2.39. Pressure rise can also be expressed in terms of Mach numbers and angles, similar to Eq. 2.40.

Thus very high pressure (or temperature) ratios can be achieved at very high blade speed, and large r_2/r_1. When the entry flow is axial (as is usually the case), we obtain

$$
\tan \beta_1 = \frac{M_{b1}}{M_{z1}}
$$

Hence

$$
\frac{T_{o2}}{T_{o1}} = 1 + \frac{(\gamma - 1)M_{b2}^2}{1 + \dfrac{\gamma - 1}{2}M_1^2} \left[1 - \frac{M_{r_2}}{M_{b_2}} \tan \beta_2 \right]
$$

In Eq. 2.53, W_{r_2}/U_2 or W_{z_1}/U_1 correspond to flow coefficients.

For a fixed β_1, β_2, Eq. 2.53 satisfies the similarity relationship given by Eqs. 2.24 and 2.26. The ideal performance characteristics of three types of rotors with zero inlet swirl velocity ($\psi = 1 - \phi \tan \beta_2$) are shown schematically in Fig. 2.39. Here, $\phi = W_{r2}/U_2$. The radial-bladed impeller has constant stagnation temperature and pressure rise at all mass flows. ψ increases with increased mass flow for the forward-leaning blade. The backward-leaning blade exhibits characteristics similar to those of an axial flow compressor (i.e., decrease in ψ with increase in ϕ).

Forward-leaning blades produce the highest pressure rise for a given ϕ and speed, but they have the disadvantage of producing very high velocities at the exit. The rising pressure characteristics may lead to unstable operation. Another disadvantage to the forward-leaning configuration is higher blade stress, because the centrifugal force on curved blades induces bending. Hence they are rarely used in practical application.

For the same mass flow and blade tip speed, the absolute velocity at the exit of a radial impeller is higher than that of backward-leaning blades ($V_2 = \sqrt{(U_2 - W_2 \sin \beta_2)^2 + W_{r2}^2}$, where β_2 is positive). Hence at very high pressure ratios, the exit absolute flow for a radial impeller becomes supersonic earlier than a backward-leaning impeller. Under these circumstances, backward-leaning blades become attractive. It is not uncommon to encounter lean as high as $-30°$ in a modern centrifugal compressor for aerospace and automotive applications.

Knowing the temperature (or absolute stagnation enthalpy) rise, the static enthalpy can be calculated from Eq. 2.36. Equation 2.14 can be used to determine the stagnation pressure rise. The temperature and pressure rise can also be expressed in terms of Mach numbers in much the same way as Eq. 2.40 derived for an axial flow compressor. These equations can be simplified considerably for a radial compressor. The expression for reaction is given by Eq. 2.44. This quantity can be expressed in terms of velocity triangles for the impeller and the diffuser.

For incompressible flow, we have

$$\psi = \frac{g\Delta H}{U_2^2}$$

and Eq. 2.53 can then be used to determine the head rise (ΔH). If the entry

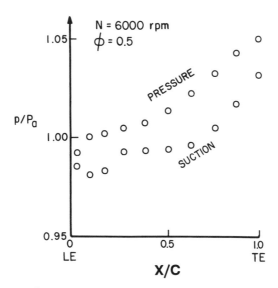

Figure 2.40 Pressure distribution on a rotor blade along the chord at midspan. (Copyright © Mizuki et al., 1980; reprinted with permission of ASME).

flow is axial, this equation reduces to

$$\frac{g\Delta H}{U_2^2} = 1 - \frac{Q}{U_2}\frac{\tan \beta_2}{2\pi r_2 b_2} = 1 - \phi \tan \beta_2 \qquad (2.55)$$

where b_2 is the height of the blade at the exit. For a radial blade, the head rise is constant. For forward-leaning blades (β_2 is negative) the head rise increases with the volume flow, whereas for backward-leaning blades there will be a decrease in head rise with an increase in the volume flow. The performance characteristics are as shown in Fig. 2.39. Equation 2.55 is a confirmation of nondimensional analysis carried out in Section 2.4.3 and represented by Eq. 2.28.

The pressure distribution on a rotor blade (inducer + impeller) measured by Mizuki et al. (1980), shown in Fig. 2.40, clearly reveals the pressure rise process occurring within the rotor. Most of the pressure rise occurs within the curved part of the duct, clearly revealing the effect of centrifugal forces in achieving the pressure rise in centrifugal pumps/compressors.

Eckardt (1980) carried out systematic measurements of the performance and flow field in high-speed centrifugal compressors. The performance of radial blades (rotor O) and backward-leaning blades (rotor A, $\beta_2 = 30°$) are shown in Fig. 2.41. Both rotors had identical tip diameter (400 mm), tip width (26 mm), shroud contour axial length (130 mm), and number of blades (20). Rotor O shows nearly constant ψ over a reasonable mass flow range, except in the region of surge and very high mass flow, with a peak efficiency near 90%, at 10,000 rpm. As expected, the efficiency decreases at higher speed due to Mach number effects. The backward-leaning vane rotor shows falling ψ characteristics, similar to those of an axial compressor, but the slopes are not as great as those of an axial flow machine. This clearly shows that the centrifugal compressor is less sensitive to mass flow changes than an axial flow compressor. An axial flow compressor achieves *all* of its pressure rise through flow turning (influenced by camber, incidence, stagger angle), while the centrifugal achieves its pressure rise through a combination of the effects mentioned above, with the pressure rise due to incidence and camber being a fraction of the total pressure rise. Surge and stall in a centrifugal compressor occurs, as in an axial flow stage, when the mass flow is decreased below a critical value. High incidence to the inducer tends to stall the blades as massive flow separation occurs inside the blade passages.

If the relative Mach number is fixed, there is an optimum diameter ratio at which maximum mass flow can be passed through the inlet. An expression for the mass flow as function of the inlet diameter can be derived by using the continuity equation:

$$\dot{m} = A\rho_1 M_{1\,rel} a \cos \beta$$

where $M_{1\,rel} = W_1/\sqrt{\gamma R T_1}$ is the relative inlet Mach number, and β_1 is the relative inlet angle at the tip, A is the inlet area. The analysis below assumes

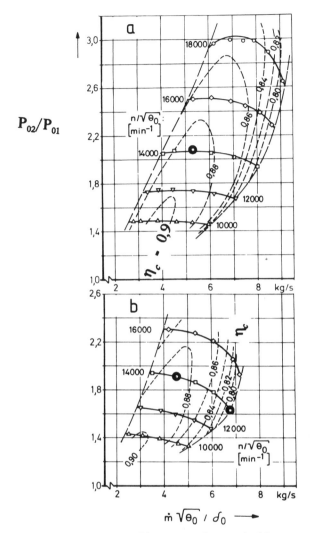

Figure 2.41 Compressor performance: (a) Rotor O ($\beta_2 = 0°$); (b) Rotor A ($\beta_2 = 30°$). (Dotted lines are iso-efficiency contours, $\theta_0 = T_{o1} / T_a$, $\delta_0 = P_{o1} / p_a$). Design conditions are indicated by bold hollow circles. (Copyright © Eckhardt, 1980; reprinted with permission of ASME.)

that $M_{1\,rel} a \cos \beta_1$, which is the absolute inlet velocity (V_1) for axial entry, is constant from hub to tip. Using the equation of state and the notations of Fig. 2.37, mass flow can be written as

$$\dot{m} = \frac{\pi}{4} D_1^2 \left[1 - \frac{D_h^2}{D_1^2} \right] \frac{P_{o1} \sqrt{\gamma} M_{1\,rel} \cos \beta}{\left[1 + \frac{\gamma - 1}{2} M_1^2 \right]^{\gamma(\gamma-1)} \sqrt{RT_1}}$$

Hence,

$$\frac{\dot{m}\sqrt{RT_{o1}}}{P_{o1}D_2^2} = \sqrt{\gamma}\,\frac{\pi}{4}\,\frac{D_1^2}{D_2^2}\left[1 - \frac{D_h^2}{D_1^2}\right]\frac{M_{1\,\text{rel}}\cos\beta}{\left(1 + \dfrac{\gamma - 1}{2}M_1^2\right)^{(\gamma+1)/2(\gamma-1)}} \quad (2.56)$$

Thus, the mass flow parameter (introduced earlier) depends on M_1, $M_{1\,\text{rel}}$, β, D_1, D_h, and D_2. Let us consider a case where there is no inlet swirl ($M_{1\text{rel}}\cos\beta = M_1$).

$$M_1^2 = M_{1\,\text{rel}}^2 - M_{b2}^2\frac{D_1^2}{D_2^2} \quad (2.57)$$

where M_{b2} is the impeller tip Mach number, D_1 is the impeller "eye" outer diameter, and D_2 is the impeller tip diameter (Figure 2.37).

With values of M_{b2}, $M_{1\,\text{rel}}$ fixed, a plot of mass flow parameter versus D_1/D_2 can be derived as shown in Fig. 2.42. This plot clearly indicates that for a given $M_{1\,\text{rel}}$ and M_{b2}, there is a radius ratio which results in maximum mass flow. Details of analyses for zero swirl flow and inlet swirling flow can be found in Hawthorne (1964, p. 565) and Shepherd (1956). The two extremes, where the mass flow is zero, correspond to zero area and zero absolute Mach number, respectively. It is also clear from Fig. 2.42 that for a given mass flow, there is a radius ratio where relative Mach number is minimum. In many high-speed applications, the relative inlet Mach number is a critical parameter, high values of which would result in a decrease in efficiency and an increase in losses. It is also clear from Fig. 2.42b that a positive swirl in absolute flow would increase the mass flow for the same relative velocity or decrease relative velocity for the same mass flow.

2.7.2 Slip Factor

As in an axial flow compressor, the flow does not follow the blade exit angle and there will be some deviation between the blade exit angle and the flow angle as illustrated in Fig. 2.39. The difference in exit flow and blade angles is the deviation angle and is caused by both inviscid and viscous effects.

A widely used correlation parameter, the slip factor, is defined by the ratio of actual swirl velocity to ideal swirl velocity given by

$$K = (V_{\theta_2})/(V_{\theta_2})_b$$

This is similar to the deviation rule used in axial flow compressor practice, and it is a measure of the deviation between the blade exit angle and flow exit angle (Fig. 2.11). The deviation of this value K from unity correlates with the decrease in pressure rise. A review of various correlations available for

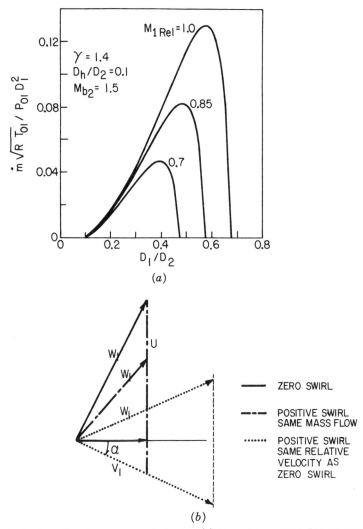

Figure 2.42 Centrifugal compressor behavior. (a) Mass flow versus Diameter ratio (axial entry). (b) Change of velocity triangles with change in inlet swirl.

slip factor is given by Wiesner (1967). His relationship for slip factor is given by

$$K = \frac{(V_{\theta_2})}{(V_{\theta_2})_b} = 1 - \frac{U_2}{(V_{\theta_2})_b} \frac{\sqrt{\cos \beta_2'}}{B^{0.7}} = 1 - \left[\frac{1}{1 - \frac{W_{r2}}{U_2} \tan \beta_2'} \right] \frac{\sqrt{\cos \beta_2'}}{B^{0.7}}$$

$$(2.58)$$

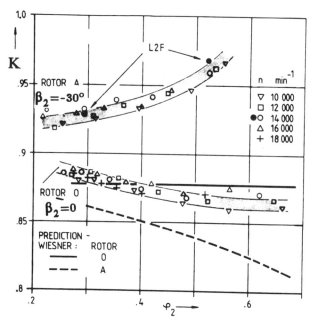

Figure 2.43 Variation of slip factor K with flow coefficient ϕ (Rotor 0, $\beta_2 = 0°$; Rotor A, $\beta_2 = 30°$). (Copyright © Eckardt, 1980; reprinted with permission of ASME.)

where B is the number of blades, $W_{r2}/U_2 = \phi_2$. This correlation can be incorporated in Eq. 2.53 (note that $W_{r_2} \tan \beta_2 = U_2 - V_{\theta_2}$) to derive the actual pressure and the temperature rise in a centrifugal compressor. A comparison of slip factors predicted and measured by Eckardt (1980) for rotors O ($\beta_2 = 0°$) and A ($\beta_2 = 30°$) is shown in Fig. 2.43. The performance of these compressors are shown in Figure 2.41. It is clear from the figure that predicted slip factors are reasonably close to measured values for the radial impeller, but for the backward curved vane, the departure is considerable. The empirical slip factor (Eq. 2.58) is constant for a radial impeller, whereas for backward-swept vanes the empirical slip factor decreases with an increase in the flow coefficient (W_{r2}/U_2). The measured slip factor shows the opposite trend for a backward swept vane. Eckardt attributes this to the curvature of the blade. The centrifugal force due to curvature of the vane increases with the flow coefficient (W^2/R), and this has a tendency to align the flow in the direction of the channel shape. The slip factor approaches unity for choking conditions.

A word of caution is in order in the use of slip factor. The empirical correlations are derived for a certain class of centrifugal compressors (e.g., incompressible flow), and caution should be exercised in using these slip factors for high-speed flow. Nevertheless, the slip factor provides a method of estimating the actual pressure rise.

As in axial flow compressors, the pressure rise in centrifugal compressors increase with an increase in the number of blades (increase in solidity), but the losses increase much more rapidly beyond a certain number of blades. Hence, empirical correlations have been derived for the number of blades. For example, Eckert and Schnell (1961) have provided the following equation for the choice of number of blades:

$$B = \frac{2\pi \cos \beta_m}{C_1 \ln D_2/D_1}$$

where $\beta_m = (\beta_1 + \beta_2)/2$ and $C_1 = 0.35$–0.45. A more detailed treatment on the choice of number blades can be found in Turton (1984).

2.7.3 Diffusers

As shown in Fig. 2.37, absolute velocity and kinetic energy increase substantially from inlet to exit across an impeller. Diffusers have been used in many flow devices to convert the high kinetic energy at the exit of the rotor into static pressure rise. For example, the stator vanes in an axial flow compressor act as a diffuser. In centrifugal machines, several types of diffusers are employed. These are shown schematically in Fig. 2.44. The diffuser serves the purpose of (a) converting the kinetic energy into pressure rise and (b) collecting the fluid and guiding it smoothly into either the next stage or into a piping system. The volute chamber is common to all these systems. The cross-sectional area increases from a minimum along section DD to a maximum along section BB, as shown in Fig. 2.37. The section DD is usually designed to follow the streamline at that location. The flow in these devices is

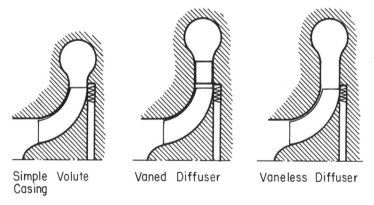

Simple Volute   Vaned Diffuser   Vaneless Diffuser
Casing

Figure 2.44 Diffusers used in centrifugal compressors (see Fig. 2.37 for details of vaned diffusers).

complex, and viscous effects dominate the flow. Nevertheless, simple analyses can be used to obtain overall performance.

For inviscid, compressible flow through a diffuser, the one-dimensional continuity and momentum equations can be used to derive the static pressure rise coefficient.

The basic equations (one-dimensional inviscid) governing the flow in a vaneless diffuser is given by (Fig. 2.44)

$$rV_\theta = \text{constant}$$

$$V_r 2\pi rh = \text{constant}$$

where h is the width of the duct. This enables the flow angle to be calculated.

The dramatic effect of adding a diffuser section is shown in Fig. 2.45. Abdelhamid and Bertrand (1979) performed experiments in a simple blower with and without a vaneless diffuser. The data without the diffuser were taken with the flow discharging into a room. The authors found the flow to be steady without the diffuser and the pressure rise improved substantially with the addition of a diffuser, but the presence of unsteadiness (rotating stall, described earlier) was observed. The critical flow coefficient at which the diffuser stalls depends on diameter ratio as well as diffuser width as shown in Fig. 2.45.

The flow field and the pressure rise are very sensitive to diffuser area changes as in two-dimensional and conical diffusers. The results, shown in Fig. 2.46, clearly indicate this effect. The data were acquired with $U_b = 250$ m/s, $R_e = 4 \times 10^6$, $b_3/b_1 = 0.83$, $R_2/R_1 = 1.65$, b_2/b_1 varies from 1.17 for configuration 1 to 0.5 for configuration 9. The diffuser wall angle has considerable influence on area change and the resulting pressure gradient. It is clear that even a small change in wall angle has a major influence on

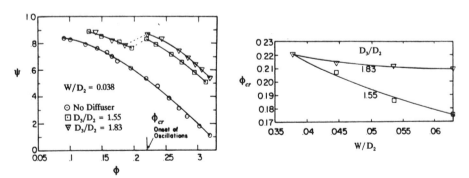

Figure 2.45 Head coefficient and flow coefficient and the onset of instability (at ϕ_{cr}) as a function of geometry (W/D_2 = diffuser width to inlet diameter; D_3/D_2 = diffuser outer to inner diameter). See Fig. 2.37 for notations. (Copyright © Abdelhamid and Bertrand, 1979; reprinted with permission of ASME.)

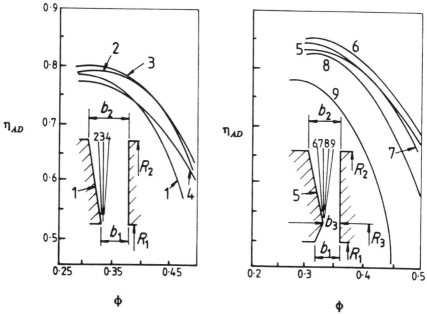

Figure 2.46 Effect of geometry on vaneless diffuser performance. [(Turton (1984); reprinted with permission of Chapman & Hall Ltd.)]

efficiency, η_{Ad}, ratio of actual to ideal pressure rise. Geometries 3 and 6 have the best performance. As indicated earlier, the major effect in a diffuser is the adverse pressure gradient that controls the boundary layer growth and separation. A one-dimensional inviscid analysis followed by a simple boundary layer calculation may provide important information on wall contouring.

Vaned diffusers with streamline-shaped (airfoil or cascade) blades should improve the pressure rise characteristics of centrifugal compressors through increased area ratio as well as flow turning. Japikse (1984) provides a review of the state of the art. The first two configurations, shown in Fig. 2.47a, are known as cascade diffusers as they are a cascade of blades located in an annulus. Type b is called a channel diffuser. This type is most suitable for high speed applications. Experimental measurement of the performance of cascade (circular arc) diffusers (Fig. 2.47a) was carried out by Sakurai (1975) and is shown in Fig. 2.48. He tested a large number of configurations, varying area ratio A_r, length ratio C/S_1, (where C is the camber-length, S_1 is the channel width at the inlet, Fig. 2.47), and equivalent diffuser angle ($2\theta = $ arctan $(S_2 - S_1)/2C$), and he plotted the pressure recovery for the diffuser (C_{pR}). Pressure recovery is based on the static pressure rise across the diffuser, normalized by the average inlet dynamic head. The optimum values of variables at which C_{pR} and η_{AD} are maximum are indicated. The essential features and pressure recovery for a channel type of diffuser are given by Japikse (1984).

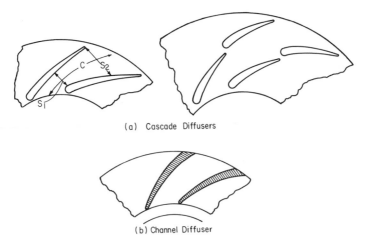

(a) Cascade Diffusers

(b) Channel Diffuser

Figure 2.47   Types of vaned diffusers.

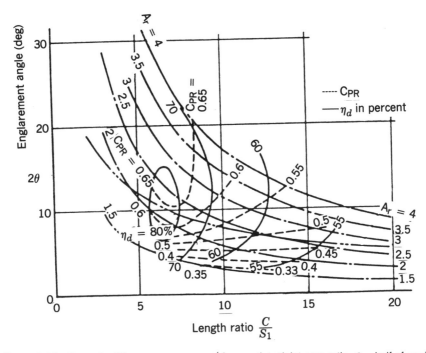

Figure 2.48   Cascade diffuser recovery map (A_r = outlet / inlet area ratio, θ = half of equivalent conical diffuser angle, η_d = actual / ideal pressure recovery in percent. (Copyright © Sakurai, 1975; reprinted with permission of ASME.)

One recent approach has been to vary the inlet swirl (through the inlet guide vane angle) and adjust the vaned diffuser angle to maximize the pressure recovery. Simon et al. (1987) carried out a systematic experimental study and showed, by matched adjustment of inlet guide vanes and diffuser vanes, that the operating range of a backward curved impeller can be extended. The optimum combination is a function of the flow coefficient and pressure ratio. Improved efficiencies were also observed.

In view of the presence of substantial viscous effects, adverse pressure gradients, and (in some cases) compressibility effect(s), the flow field in these diffusers deviate considerably from those derived from one-dimensional inviscid consideration. The performance can be estimated using empirical correlations (e.g., Japikse, 1984) or through computation of the flow field using techniques described in Chapter 5.

2.8   AXIAL AND CENTRIFUGAL PUMPS

Liquid handling turbomachinery where energy is imported to the fluid are called *pumps*. They can be classified as axial, centrifugal, and mixed-flow pumps. The analyses of these pumps are similar to those described in Sections 2.5 and 2.7 (axial and centrifugal compressors). Because the density is constant in these devices, the analysis becomes much simpler. This section deals with additional analysis/performance related to pumps.

One of the major limitations in the performance of liquid handling turbomachinery is the phenomena known as *cavitation*, described in Section 2.4.4. Some of the major effects of the cavitation bubble on liquid handling machinery are (a) decreased performance and efficiency and (b) damage to the structure resulting from the collapse of the bubbles and from the pressure waves resulting from this collapse. The latter phenomenon is responsible for noise, vibration, and possible failure. The mixing of the vapor with the liquid phase causes a loss in pressure rise for pumps, decreased pressure drop in turbines, and increases in flow losses, resulting in decreased efficiency. The cavitation acts as flow blockage, accelerating the flow in other regions of the pump and thus decreasing the static pressure rise. Significant regions of cavitation may lead to sudden decrease in the pressure or head rise in the pump and a large decrease in the efficiency.

The treatment of pumps will be covered under noncavitating and cavitating performance. It is beyond the scope of this book to cover physics of cavitation. The reader is referred to the book by Knapp et al. (1970) and Brennen (1995) for basic treatment on this subject.

2.8.1   Noncavitating Performance

The noncavitating performance of pumps is similar to the performance of low-speed compressors (incompressible), and the relevant equations are given in Sections 2.3–2.5 and 2.7. A typical axial flow pump looks similar to an axial flow compressor, even though the criteria used for blading design may be different. The pump designer tends to move the suction peak toward the trailing edge to delay cavitation. The centrifugal pump looks similar to a centrifugal compressor. An example of a centrifugal pump used in the space shuttle main engine is shown in Fig. 1.6.

The Euler equation for a pump is given by Eqs. 2.7 and 2.11, and the efficiency is given by Eq. 2.16. The performance representation is given by Eq. 2.28. Additional performance parameters for pumps are given by specific speed (Eq. 2.31 or 2.32), cavitation parameter (Eq. 2.33), and suction specific speed (Eq. 2.34).

The axial pump blading nomenclature and velocity triangles are shown in Fig. 2.10. The loading coefficient in terms of flow angles is given by Eq. 2.38:

$$\psi = 1 - \phi(\tan \alpha_1 + \tan \beta_2), \text{ where } \psi = g(H_2 - H_1)/U^2$$

$$\frac{(P_{o2} - P_{o1})}{\rho U^2} = \frac{g\Delta H}{U^2} = \eta_H \psi$$

where ΔH is the actual head rise and ϕ is the flow coefficient. The hydraulic efficiency, η_H, in the ratio of (actual/ideal) head rise. The loading coefficient in terms of lift and drag is given by Eq. 2.43. The performance of an axial flow pump is similar to that shown in Figs. 2.16 and 2.17 for a compressor and is explained in Section 2.5.

Multistage pumps have been successfully built and used in cryogenic pumping systems employed in liquid rocket engines (e.g., J2, Apollo Saturn II, second and third stages). One such application (tested in water) (Crouse and Sandercock, 1967) is shown in Fig. 2.49. The two-stage pump consisted of an inducer, followed by an axial flow rotor–stator, terminated by a second-stage rotor. The inducer usually consists of three or four helical blades of large chord length at large stagger angle (they wrap around the circumference almost 300–360° as shown in Fig. 1.7). They accept low pressure from a tank and increase the head sufficiently to keep the downstream (main pump) from cavitating. The inducers almost invariably operate with cavitation. The long and narrow passages are designed to gradually and gently collapse these bubbles before they enter the main pump. It is also designed to provide smooth entry velocity to the rotor. The inducer is followed by a rotor–stator–rotor combination in this particular configuration. The blade-turning angle varied from a low value of 2.3° (diffusion factor of 0.41) at the tip of the transition rotor to a high value of 66° (diffusion factor of 0.7853) at the root of the second stage rotor. The inlet diameter was 15.6 cm. The cumulative head rise curve for noncavitating performance is shown in Fig. 2.49. The pump was designed for a suction specific speed of 30,000 (Eq. 2.34, RPM, GPM, ft), a hub/tip ratio of 0.4, a flow coefficient of 0.108, $H_{sv} = 75$ m, and $U_t = 38$ ms. The inducer (ψ_{1-2}, η_{1-2}) provides a small head rise ($\psi_{1-2} = 0.14$) at design condition and at a low efficiency. Most of the

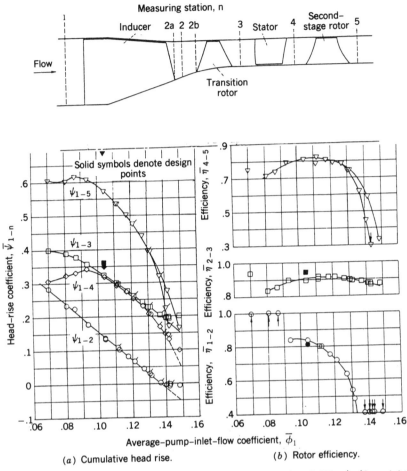

Figure 2.49 Noncavitating performance (Net positive suction head, 75 m). (Copyright © Crouse and Sandercock, 1967; NASA TND 3962.)

head rise occurs through transition rotor and second-stage rotor, achieving an overall efficiency of about 81.1% at the design condition.

Wislicenus (1986) provides detailed design philosophy as well as criteria used in the blading design. The choice of the type of turbomachinery is based on a chart such as that shown in Fig. 2.9. The blading design is usually based on the loading coefficient and the turning angle and is optimized to avoid or delay cavitation. Some designers have employed NACA 65 Series blades and double circular blades shown in Fig. 2.13. Most aircraft axial flow compressor bladings are designed to have a high suction peak (or large acceleration) near the leading edge. This is based on the criteria to avoid flow separation (decreased adverse pressure gradient near the trailing edge of the suction

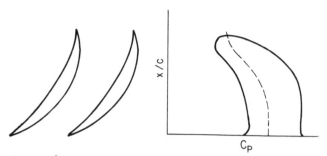

Figure 2.50 Typical axial flow pump blading designed from assumed pressure distribution (Wislicenus, 1986; NASA RP 1170).

surface). These suction peaks (or $C_{p\,min}$) are too high for pump application. A more appropriate choice would be a pressure distribution such as that shown in Fig. 2.50, where the suction side near the leading edge (up to midchord) has a nearly constant pressure and velocity and most of the turning or loading takes place beyond the midchord. This improves the cavitation characteristics, thus increasing the head rise, efficiency, and life of the pump.

When comparing the pressure distribution sought for an axial flow pump blade (Fig. 2.50) with that sought for an axial flow compressor blade (Figs. 2.12, 2.14), it is clear that the criterion used in a compressor is to achieve a moderate adverse pressure gradient to prevent separation (by loading the blade near the leading edge) and that the criterion used in a pump is to reduce the minimum pressure and load the blade toward the trailing edge, thus reducing both the flow separation and the cavitation.

The diffusion factor, shown in Fig. 2.19 (and Eq. 2.47), is employed to minimize flow separation, and cavitation criteria (explained later) is used to select the $C_{p\,min}$. Some of the recent trends in compressor development (Section 2.5.11) can be used to improve the pump design to achieve higher efficiency, pressure rise, and mechanical life.

The analysis and performance of noncavitating mixed flow and centrifugal pumps are similar to those described in Section 2.7. The geometry is shown in Figs. 2.37 and 2.39, head rise ($\psi = (g\Delta H/U_2^2)$) is given by Eq. 2.53, and slip factor is given in Section 2.7.2. The diffuser analysis is identical to that described in Section 2.7.2. Detailed blading design and geometrical design of passages, volute, and the return passages for a multistage pump can be found in Wislicenus (1986).

Centrifugal pumps are widely used in spacecraft as well as in industrial application. One typical example is shown in Fig. 1.6. This is a high-pressure fuel pump used in the space shuttle's main engine, and it has the following performance characteristics: $\dot{m} = 67.6$ kg/s, $P_{o1} = 1.547$ MPa, $P_{o2} = 42.1$ MPa, and $\eta_p = 0.763$. It is driven by a turbine with a flow rate of 71.9 kg/s, inlet temperature of 996 K, pressure ratio of 1.411, $\eta_t = 0.839$, $N = 35,000$ rpm, and an output of 45,785 kW. This is a three-stage pump, and each stage consists of an inducer and impeller combination.

The performance of a small centrifugal pump, tested by the Japanese Space Agency (Kamijo and Hirata, 1985), is shown in Fig. 2.51. This is a single-stage pump (inducer and impeller configuration similar to first-stage pump shown in Fig. 1.6) with an inducer tip diameter of 6.5 cm, an impeller

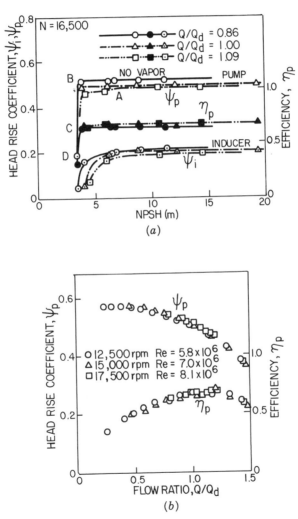

Figure 2.51 Performance of a rocket pump. (a) Cavitating performance. (b) Noncavitating performance. (Q_d is the design volume flow, ψ_i is the head-rise coefficient for inducer, ψ_p is the head-rise coefficient for pump). (Copyright © Kamijo and Hirata, 1985; reprinted with permission of ASME.)

diameter of 13.4 cm, an inducer inlet tip blade angle of 89.7°, an impeller inlet blade angle of 76.2°, an impeller outlet blade angle of 65°, and six blades. Because the number of blades is small and the outlet angle is large (small swirl with a backward lean), the head rise decreases with an increase in the flow rate, but not as rapidly as an axial flow pump (compare Figs. 2.51 and 2.49). The noncavitating characteristics are very similar to those of the compressor shown in Fig. 2.41. Furthermore, the performance trend follows Eq. 2.53 as well as the similarity relationship (Eq. 2.28). The peak efficiency is only 68%, which is reasonable for a small pump with an inducer. As indicated earlier, the inducer efficiency is low.

2.8.2   Cavitating Performance

Cavitation is a dynamic phenomenon in liquid handling machinery that severely limits their performance. With current trend toward high-speed machinery and propulsion, the cavitation phenomenon plays a major role in further advances in liquid-handling machinery. As indicated earlier, when the local pressure is reduced below the vapor pressure, cavitation occurs. Some of the locations in a pump susceptible to cavitation are shown in Fig. 2.52. Cavitation occurs on blade surfaces where static pressure is below vapor pressure ($C_{p \, min}$ location), and it also occurs near the blunt trailing edge, at corners (such as the intersection of the blade and the wall), near the axial and the radial gaps, in vortex regions such as the tip vortex, in the secondary vortex, in the scraping vortex, and in the trailing edge vortex. The static pressure inside a vortex reaches low values, resulting in cavitation. Under extreme breakdown conditions, the entire passage may be dominated by cavitation bubbles (e.g., supercavitating bladerows and cascades). The inception and development of cavitation is controlled by the properties of the fluid (vapor pressure, surface tension, air content, purity, etc.) and the geometry of the body (curvature, bluntness, suction peak, sharp corners, roughness, etc.). A reduction in ambient or inlet pressure or an increase in the temperature-dependent vapor pressure leads to cavity formation. Another source of cavitation is vortex cavitation, caused by the low pressure at the core of the vortex. Hence the cavitation in turbomachinery can be broadly classified as surface cavitation (vapor bubbles at or near the blade surface) and vortex cavitation (tip vortex, trailing vortex, secondary vortex, etc.). The major source of vortex cavitation is due to tip vortex near the tip clearance region of a rotor.

The major consequence of cavitation is loss in performance and efficiency, material damage, noise, and severe vibrations. For example, the inducers used in early rockets (J2 rocket engines) had a life of 2–3 minutes. In many instances, cavitation cannot be avoided, and thus material selection and proper operating conditions become extremely crucial.

For machines with short or intermittent service, such as rocket pumps and storm water pumps, the eroding effects of cavitation are not important; but in

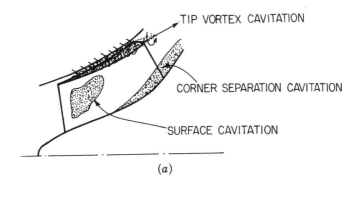

TIP VORTEX CAVITATION

CORNER SEPARATION CAVITATION

SURFACE CAVITATION

(*a*)

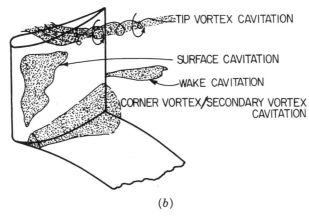

TIP VORTEX CAVITATION

SURFACE CAVITATION

WAKE CAVITATION

CORNER VORTEX/SECONDARY VORTEX CAVITATION

(*b*)

Figure 2.52 Location of cavitation in (a) centrifugal pump and (b) axial flow pump. Partial cavitation. Under extreme breakdown condition the entire passage may be dominated by cavitation bubbles.

continuous operation or reusable rocket pumps, stable and cavitation-free operation is important. Most pumps operate with some cavitation, but appreciable cavitation bubbles have pronounced damaging effects on the performance, efficiency, vibration, and noise. As explained earlier, the vapor pockets consist of a large volume of vapor; this affects the flow path through large blockage, resulting in flow acceleration and reduced pressure rise (or decreased absolute pressures). The resulting pressure reduction causes decreased head or pressure rise. When the pump operates with substantial cavitation, the pressure reduction will be dramatic, resulting in a breakdown of performance. This is shown in Fig. 2.53 for the pump configuration described earlier (Fig. 2.49). The net positive suction head is defined by Eq. 2.33 as $H_{sv} = H_s - h_v$. Let us consider the case of $H_{sv} = 4.57$ m in Fig. 2.53. At low flow coefficients (0.08–0.105), the inlet velocity is low and the static

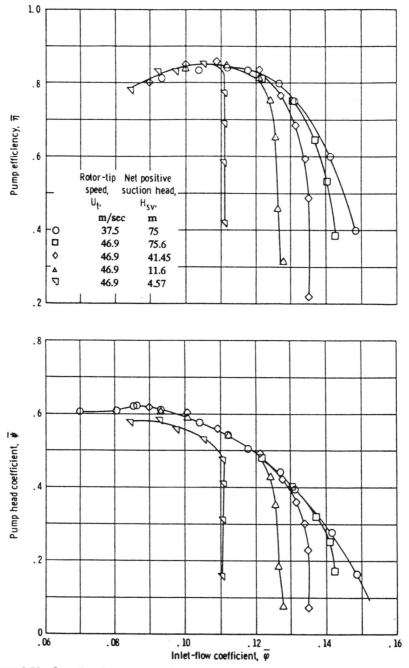

Figure 2.53 Overall performance of axial flow pump under cavitating conditions (Crouse and Sandercock, 1967; NASA TND 3962).

pressure rise is high. Consider the equation

$$H_{sv} = h_s + \frac{V^2}{2} - h_v = \frac{p_s}{\rho} + \frac{V^2}{2} - \frac{p_v}{\rho}$$

Because p_s/ρ is high, the cavitation is kept to a minimum. As the flow coefficient is reduced to say 0.11 at $H_{sv} = 4.57$ m, h_s is very low, giving rise to extensive cavitation. This results in complete breakdown of the performance as measured by head rise and efficiency.

A plot of H_{sv} (NPSH) versus head rise and efficiency at the same volume (or flow coefficient) shown in Fig. 2.51 for the rocket pump described earlier shows the dramatic effect of decreasing the NPSH. Appreciable cavitation occurs below NPSH = 5 m, causing breakdown in the performance of the inducer–impeller combination. In this plot, the total suction head (NPSH) is gradually reduced, keeping all other operating conditions the same. The head rise and efficiency remain nearly the same as NPSH is reduced. At some point of reduced NPSH (in this case at NPSH < 6.25 m for inducer and NPSH < 4 m for pump), a change in head is evident. This is the condition (shown as B) under which complete flow breakdown occurs. It should be noted that the inducer breakdown occurs first, followed by the breakdown of the flow in the impeller. Both tip vortex and surface cavitation may have occurred, even though some cavitation may be present even above this point. This is the region which has very low pressures on the suction side of the blade, resulting in vapor pockets in substantial portions of the blade passage from the leading to the trailing edge.

The sequence of events leading to the flow breakdown is shown in Fig. 2.51a. Inception of cavitation occurs near point A. This is followed by point B, where a substantial part of the passage is cavitating. Region C is the supercavitating region. This is followed by a dramatic decrease in head at D, the breakdown point. Dynamic instabilities leading to violent vibration and failure can occur in regions C and D (Fig. 2.51). Even though stall and surge is not involved in this phenomenon as in axial flow compressors (Sections 2.5.7 and 2.5.8), the instability is connected with the unsteady cavitation. This is known to be the cause of many pump failures.

Cavitation performance curves of both an axial flow pump and a centrifugal pump are similar. For example, a ψ–ϕ curve shown for an axial flow pump in Fig. 2.53 is similar for a centrifugal pump, and the NPSH versus ψ curve shown in Fig. 2.51 for an inducer–centrifugal-pump combination is similar to that of an axial flow pump.

There have been many attempts made to reduce cavitation damage in the main pump, by incorporating an inducer upstream. These are high-solidity, large-stagger-angle (80–85°), axial flow pumps consisting of fewer blades (typically three or four) and operating with cavitation. There is a small pressure rise in these inducers as the fluid goes through these long and narrow passages, which enable the cavitation bubble to break down, thus enabling the main pump to operate nearly cavitation-free. A review of the complex flow field in these devices is given by Lakshminarayana (1982).

Wislicenus (1986) provides a systematic approach to the similarity representation of the performance of cavitating pumps. The cavitation parameter is represented either by a Thoma parameter (Eq. 2.33) or, more generally, in terms of suction specific speed (Eq. 2.34). The cavitation phenomena depends on H_{sv}, N, and Q (Figs. 2.53 and 2.51). These can be grouped together to form a parameter which provides a unique representation of the cavitating pump under all operating conditions (for the same fluid and pump). A unique representation of specific speed and the Thoma cavitation parameter, developed by Wislicenus (1986), is shown in Fig. 2.54. The data shown are for commercial pumps (excluding rocket pumps). This figure shows that suction speed of commercial pumps is limited to 12,000. Rocket pumps operate up to a suction specific speed of about 50,000. For example, data at $H_{sv} = 4.57$ m, shown in Fig. 2.53, represents SS (RPM, GPM, ft) $\approx 26,500$. The design suction specific speed of a low-pressure oxidizer pump inducer used in the space shuttle main engine is 70,000 and $\phi = 0.07$. Not all points shown in Fig. 2.54 represent cavitation-free operation. Most data points for suction specific speed (SS) above 7000 represent operation with some local cavitation (near the suction peak and the tip vortex cavitation).

A simple analytical technique for determining the minimum achievable NPSH is given in Cooper (1989; see chapter by Furst and Desclaux). Cooper (1989, 1993) and Swift et al. (1983) contain recent advances.

The design process for pumps usually starts with the selection of suction speed (e.g., Fig. 2.54), followed by the selection of the type of pump, specific speed, and other nondimensional groups. Details of design procedure can be found in Wislicenus (1986) and Balje (1981).

The relationships governing similarity in flow are given by (Eq. 2.29)

$$\phi = \frac{Q}{ND^3} = \text{const.}, \qquad \psi = \frac{g\Delta H}{N^2 D^2} = \text{const.}$$

In addition to the specific speed ($\overline{N}$), given by Eq. 2.31, one other parameter that is useful in the choice of the best machine for a particular application (for a given ΔH and Q) is given by the specific diameter defined as

$$d_s = \frac{(\psi)^{1/4}}{\sqrt{\phi}} = \frac{D(g\Delta H)^{1/4}}{\sqrt{Q}}$$

It should be noted here that for a given ΔH and Q, the above equation is a function of D only and Eq. 2.31 is a function of N only. Thus, we have two additional equations for the choice of N and D.

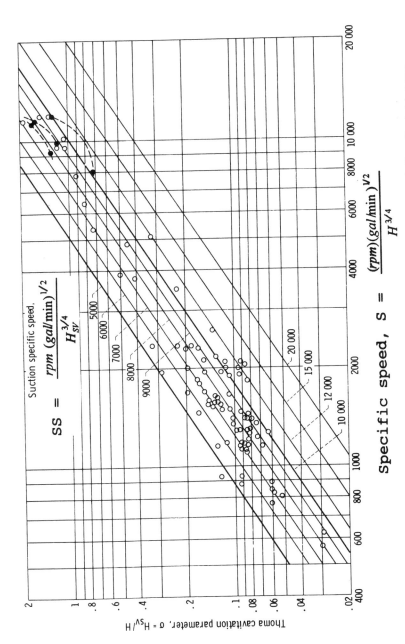

Figure 2.54 Points of acceptable pump performance with respect to cavitation. For total pump head H (ft) for multistage pumps, only first stage is considered. Total inlet head H_{sv} above vapor pressure of fluid pumped is referred to centerline of impeller. For double suction pumps, one-half of total capacity of pump is used for calculating SS and S (Wislicenus, 1986; NASA RP 1170).

141

The application of basic specific speed is limited to the values of N, Q, and H at which a particular machine has its best efficiency point (e.g., Figs. 2.51 and 2.53). It is expected that this condition always exists under similar flow conditions—that is, at only one value of Q/ND^3. Thus there is only one basic specific speed at the best efficiency point associated with one class of geometrically similar machine, and it is called the *specific speed* of that class of machine.

The specific speed and specific diameter can be used to choose the best pump. This procedure, suggested by Balje (1981), is shown in Fig. 2.55. This plot contains large amounts of data from various centrifugal pumps. For given values of H and Q, either a speed or a diameter can be chosen to determine the optimum value of d_s or $\bar{N}$ for best efficiency. For example, best efficiency at $\bar{N} = 0.3$ occurs at $d_s = 8$. Best efficiency is achieved at higher specific speed. The best operating range for a centrifugal pump is from $\bar{N} = 0.1$–0.8, which is consistent with Figs. 2.8 and 2.55. The best range of $\bar{N}$ for axial flow pumps is 1–20. Balje (1981) has given a plot similar to Fig. 2.55 for axial flow pumps, and this plot indicates that the best efficiency is achieved in the range $\bar{N} = 2$–6 and $d_s = 1.5$–3.

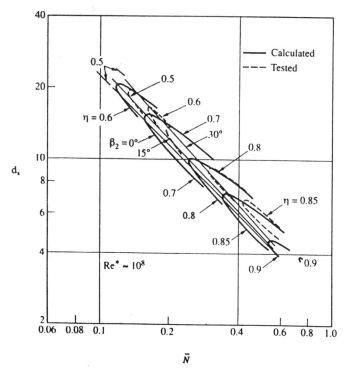

Figure 2.55 Comparison of calculated efficiency contours with test data on centrifugal pumps. (From Balje, 1981; reprinted with permission of John Wiley & Sons, Inc., *Turbomachines; a Guide to Design, Selection and Theory*.)

2.9 AXIAL FLOW TURBINES

The function of a turbine is to extract energy from the fluid and convert it to mechanical/electrical energy. A classification of turbines is shown in Fig. 1.1. Enclosed turbines can be classified as radial (Fig. 1.9), axial (Fig. 1.2), and mixed (Fig. 1.10), as in compressors. This section is intended to provide a brief overview of one-dimensional analysis and performance. More detailed treatment can be found in the book by Horlock (1973a), which is exclusively devoted to axial flow turbines.

An example of extended turbomachinery is a windmill. A windmill has very few blades (two to four) with large spacing; therefore, the interference due to adjoining blades is very small. The analysis of the windmill is based on airfoil theory, in much the same way as the propeller theory described earlier (Eqs. 2.50 and 2.51). The reader is referred to aerodynamics textbooks such as Bertin and Smith (1989) or McCormick (1979) for a method of calculating (a) C_L and C_D from a known profile (thickness and camber) and (b) incidence of the airfoil. The forces exerted by the fluid on the blade (windmill) are similar to that of the compressor or propellor blade shown in Fig. 2.15. The difference between a windmill (turbine) and a gas/steam turbine is that the flow turning in a windmill is very small and the flow upstream and downstream of the windmill rotor is exposed to the atmosphere, while the flow turning in gas and steam turbines can be substantial, varying from $50°$ to $180°$ (Pelton wheel). Radial gas turbines are similar to centrifugal compressors. The flow path in this case will be from outer to inner radius. These are used only in small power/propulsion plants. The axial type is widely used and is covered in detail in this section. Description and analysis of radial turbines and hydraulic turbines are covered in the next section.

The history of turbines dates back several hundred years, starting with windmills, steam turbines, and water turbines and eventually leading to gas turbines. The gas turbine is a relative newcomer. The flow in turbines, unlike that in compressors, is much better behaved, in view of the favorable pressure gradients that exist. Therefore, they are easier to design. Large flow turning and pressure drop can be accomplished because of the desirable characteristics brought about the pressure drop along the flow path. Large pressure drop can be achieved per stage without the danger of separating the flow. Therefore, fewer stages are required in operating turbines as compared to compressors to achieve the same change in pressure. Therefore, much higher efficiencies can be achieved in a turbine stage. The turbine has other problems that are not present in compressors. For instance, the presence of high-temperature gases introduces higher stresses and decreases life. Also, the presence of particles in combustion products results in erosion of blading.

In the early days of jet engine development, poor efficiencies in compressors and high efficiencies in turbines (due largely to experience gained from steam turbine practice) led to a large number of compressor stages (as many as 12–17) being driven by one or two single turbine stages. Considerable effort was extended from 1945 to 1980 in improving compressor efficiency, which has led to a decrease in the ratio of (compressor/turbine) stages. With the advent of the turbo fan and prop fan, improvement in turbine efficiency and performance has become much more critical. Figure 2.33 shows the complexities introduced by the prop fan. The turbine design/operation is becoming much more critical in these power plants than in earlier versions of turbo jets. Even in the energy-field (steam and hydraulic turbines), the efficiencies are becoming critical due to increased demand for energy and decreased resources.

A schematic of a multistage turbine is shown in Fig. 2.56 (also see Fig. 1.2). The front stages where the pressures are high are called the *high-pressure* (*HP*) *turbine*, while the aft stages where the pressures are low are called the *low-pressure* (*LP*) *turbine*. The HP and LP stages rotate at different speeds. The design philosophies of these stages also differ. The purpose of a nozzle is to accelerate the flow and guide it smoothly into the

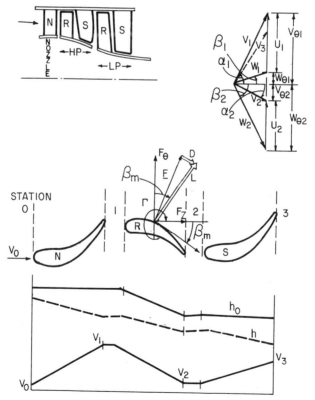

Figure 2.56 Turbine configurations, velocity triangles, and flow processes. (P_o and T_o variation is similar to h_o; p and T variation is similar to h; $1\frac{1}{2}$ stages shown.)

first-stage rotor. A certain amount of static pressure drop is achieved through the nozzle. The flow then passes through the rotor where all the stagnation and static properties change. If there is no static enthalpy and static pressure drop in a rotor, the turbine is called an *impulse turbine*. In such configurations, all of the static pressure/enthalpy drop occurs only in the stator. The distribution of flow properties along the flow path is also shown in Fig. 2.56. These trends are typical, but may vary from one type of design to the other.

The velocity triangles at the exit of the nozzle (flow is usually axial at inlet to nozzle), exit of the rotor, and exit of the stator are also shown in Fig. 2.56. The absolute velocity increases from V_0 to V_1 across the nozzle, decreases (extraction of kinetic energy) from V_1 to V_2 across the rotor, and increases again from V_2 to V_3 across the stator. Because the density is decreasing, the annulus area is increased to avoid excessively high velocities.

The following section covers fundamentals of turbine theory, performance, physical process, and so on. for a very detailed treatment, the reader is referred to Horlock (1973a) and Glassman (1972).

2.9.1 Expressions for Pressure and Temperature Drop

The analysis of a turbine stage is very similar to that of a compressor stage. Many of the concepts, equations, and physics developed are common to both, with the exception of major differences in the property changes listed below.

1. In a turbine the parameters P_o, h_o, and T_o decrease through the rotor, while they increase across a compressor.
2. The compressor usually operates at lower temperatures (500–1000°R) than a turbine (1000–3500°R).
3. The flow turning is much higher (50–180°) in a turbine than in a compressor (20–35°).
4. The blade thickness and profiles differ considerably because of the reasons mentioned above.
5. The annulus area decreases in a multistage compressor and increases in a multistage turbine.
6. Material considerations (high temperature and stress) are much more critical in a turbine.

The concept of velocity triangles and some of the relationships (e.g., Eqs. 2.35 and 2.36) are the same for both systems. The compressor terminology (Fig. 2.11) 1–2 for rotor and 2–3 for stator is very similar to that used for a turbine (Fig. 2.56). Therefore, the equations are almost identical. Because the flow turning is large in turbines, care should be exercised in taking proper signs for V_{θ_1}, V_{θ_2}, W_{θ_1}, and W_{θ_2}. The velocity triangles for a turbine are shown in Fig. 2.56. For clarity, the axial velocity is shown constant in this figure, but the analysis allows for its variation along the flow path. In this figure, V_{θ_1}, W_{θ_1}, α_1 and β_1 are positive; V_{θ_2}, W_{θ_2}, α_2, and β_2 are negative, and U is in the positive direction. In the case of turbine (this and the next section), the convention chosen is that the absolute and relative angles and velocities are positive when measured in the direction of rotation. The temperature drop is given by the following equation (similar to Eqs. 2.37 and 2.38 for a compressor), assuming $U_1 = U_2$:

$$\psi = c_p \frac{(T_{o1} - T_{o2})}{U^2} = \frac{c_p \Delta T_o}{U^2} = \left[\frac{V_{z_1}}{U} \tan \alpha_1 - \frac{V_{z_2}}{U} \tan \alpha_2 \right] \quad (2.59)$$

But $(V_{z_2}/U) \tan \alpha_2 = (V_{z_2}/U) \tan \beta_2 + 1$. It should be noted here that β_2 is negative and U is positive in Fig. 2.56, so $V_{z_2} \tan \alpha_2$ will be negative, because V_{θ_2} is negative. The temperature drop will be proportional to $(|V_{\theta_1}| + |V_{\theta_2}|)$, where $|V_{\theta_1}|$ and $|V_{\theta_2}|$ are absolute values of tangential velocities in the absolute frame. Thus,

$$\psi = \left[\phi(\tan \alpha_1 - AVR \tan \beta_2) - 1 \right]$$

Similarly, the temperature ratio drop corresponding to Eq. 2.39 is given by

$$\frac{T_{o2}}{T_{o1}} = 1 - \frac{(\gamma - 1) M_b^2}{1 + \frac{\gamma - 1}{2} M_1^2} \left[\frac{M_{z1}}{M_b} (\tan \alpha_1 - AVR \tan \beta_2) - 1 \right] \quad (2.60)$$

where M_b, M_1, M_{z1} are based on $a = \sqrt{\gamma RT_1}$, and the pressure ratio is given by

$$\frac{P_{o2}}{P_{o1}} = \left[1 - \frac{(\gamma - 1) M_b^2}{\eta_t \left(1 + \frac{\gamma - 1}{2} M_1^2 \right)} \left\{ \frac{M_{z1}}{M_b} (\tan \alpha_1 - AVR \tan \beta_2) - 1 \right\} \right]^{\gamma/\gamma - 1}$$

$$(2.61)$$

In most turbine cases, β_2 is negative, so $\tan \alpha_1$ and $\tan \beta_2$ terms are additive.

Like compressors, the pressure drop and temperature drop in a turbine are strongly dependent on the blade speed (M_b, U), axial velocity or mass flow, inlet and outlet flow angles, and, more specifically, the absolute ($\alpha_1 - \alpha_2$) or relative flow turning angles ($\beta_1 - \beta_2$). Higher turning angles produce larger temperature and pressure drops, and thus a higher work output. Unlike compressors, large flow turning can be accomplished without separating the flow. The effect of mass flow (or flow coefficient) is opposite to that of a compressor. With α_1 and β_2 fixed and blade speed held constant, higher mass flow produces higher temperature and pressure drops.

If α_1, β_2 and $\dot{m}$ are held constant, higher speeds contribute to higher pressure or temperature drops and higher work output per stage. Higher speeds thus result in more compact power plants. This is why aerospace turbomachines operate at the highest possible speed allowed by stress considerations. It is clear from Eq. 2.61 that an increase in mass flow or flow coefficient would decrease P_{o2}/P_{o1} and increase the pressure drop. Higher speeds involve higher shaft power, so higher pressure and temperature drops have to occur.

Turbine performance can also be expressed in terms of lift and drag as in a compressor. The forces exerted by the fluid on a blade are shown in Fig. 2.56. The tangential force, F_θ, and the axial force, F_Z, can be resolved into a lift and drag as in compressors. Defining the lift and drag coefficient as per Eq. 2.41 (Fig. 2.15), the expression for the tangential force is given by Eq. 2.42. It should be noted here that β_m is negative for this case illustrated in Fig. 2.56; therefore, F_θ, the tangential force, is reduced by the drag. The equation for the turbine loading coefficient is the same as Eq. 2.43.

An explanation is in order. For both turbines and compressors, the loading coefficient is directly proportional to the lift coefficient. For a compressor, the loading coefficient increases with an increase in the drag coefficient. The viscous drag increases the internal energy of the fluid through a temperature rise, resulting in an increase in ψ. For a turbine, an increase in the drag coefficient results in a decrease in the loading coefficient. Because the viscous drag tends to dissipate energy and increase temperature, T_{o2} is higher in a real turbine than in an ideal turbine, with higher values of C_D or viscous losses. Therefore, ψ or $(T_{o1} - T_{o2})$ is lowered by drag forces.

2.9.2 Degree of Reaction

The degree of reaction in a turbine is given by

$$R = \frac{h_1 - h_2}{h_1 - h_3} \qquad (2.62)$$

This is similar to the expression given for a compressor (Eq. 2.44).

Equations 2.45 and 2.46, derived for a compressor, are also valid for a turbine. Detailed velocity triangles should be known before the reaction can be determined. Simple expressions for reaction can be derived for turbines with repeating stages, where the flow at the exit of each stage is identical and where the axial velocity is constant. Equation 2.62 can be simplified to express the reaction in terms of the rotor flow inlet and outlet angles only. The relative stagnation enthalpy is unchanged across an axial flow turbine rotor in an adiabatic flow; therefore,

$$h_1 - h_2 = \frac{W_2^2 - W_1^2}{2}$$

$$h_1 - h_3 = h_{o1} - h_{o3} = h_{o1} - h_{o2} = U(V_{\theta_1} - V_{\theta_2})$$

Hence,

$$R = \frac{V_z}{2U}(\tan \beta_2 + \tan \beta_1) = (tan \, \beta_m)\frac{V_z}{U} \qquad (2.63)$$

A turbine stage designed for zero reaction, where all the static pressure drop occurs in a nozzle or stator, is called an *impulse stage*. Using Eqs. 2.59 and 2.63 and noting that $V_z \tan \alpha_1 = V_z \tan \beta_1 + U$, the following expression can be derived for repeating stages with constant axial velocity (AVR = 1):

$$\psi = 2(\phi \tan \beta_1 - R) \qquad (2.64)$$

The temperature and pressure ratios, likewise, can be expressed in terms of reaction by replacing the term $[\phi(\tan \alpha_1 - \text{AVR} \tan \beta_2) - 1]$ by $2(\phi \tan \beta_1 - R)$ in Eqs. 2.60 and 2.61. It is clear that low- and zero-reaction turbines have a higher temperature drop (work output) and lower temperature ratio (T_{o2}/T_{o1}) and pressure ratio (P_{o2}/P_{o1}) than does a corresponding high-reaction turbine for the same mass flow, blade speed, and relative inlet angle. When these repeating stages are designed to have axial exit flow with no swirl ($\phi \tan \beta_3 = 1$, i.e., "repeating" stages with constant axial velocity and axial exhaust), the loading coefficient is given by

$$\psi = 2(1 - R) \qquad (2.65)$$

Therefore, a zero-reaction turbine (impulse turbine) produces twice as much work as a 50%-reaction turbine, both operating under the conditions mentioned above.

Because the temperature drop is higher for low/zero-reaction turbines, the stators of subsequent stages have lower temperatures and, therefore, lower cooling requirements. It should be noted that the rotor surface senses the *relative* stagnation temperature, and the stator surface senses the *absolute* stagnation temperature. Both these quantities are lower for an impulse rotor and stator. The nozzle is usually the same for both high- and low-reaction turbines; only the temperature drop across rotors and subsequent stators differs. Thus, an impulse stage (rotor–stator) is likely to operate at lower skin temperatures and thus require less cooling. A disadvantage of an impulse stage is that all the acceleration occurs within a stator passage, and this is likely to increase losses. Thus, an impulse stage usually has lower efficiency than a corresponding reaction turbine.

2.9.3 Characteristics of Impulse and Reaction Turbines

In practice, the reaction varies from hub to tip. Some stages are designed to operate at 50% reaction at midspan, with outer and inner radii operating at higher and lower reactions, respectively. If the reaction is below zero (which

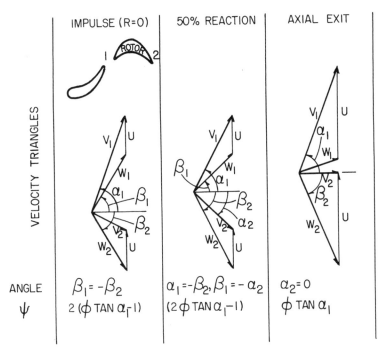

Figure 2.57 Properties of various axial flow turbine stages.

may occur near the hub), the relative flow will decelerate, which is an undesirable situation. It is not possible to cover all possibilities in this section. Several cases, which provide insight into major differences in the design, are dealt with here. These are 0%, 50%, and 100% reaction, as well as axial exit velocity (Fig. 2.57). A designer may vary the reaction appreciably along the span (say 20–70%) and modify the design and provide acceptable blade shapes from material, stress, and manufacturing considerations.

Let us consider an impulse stage. The velocity triangles for this case are shown in Fig. 2.57. Because the reaction is zero, $W_1 = W_2$, $h_1 = h_2$, $(h_{o1})_R = (h_{o2})_R$, and $\beta_1 = -\beta_2$ and all *static* pressure drop occurs in the nozzle or the stator and all stagnation pressure drop occurs in a rotor. Equation 2.59 in this case (for AVR = 1) can be simplified to

$$\psi = 2(\phi \tan \alpha_1 - 1) \tag{2.66}$$

For a 50% reaction, it can be proven that the velocity triangle is symmetrical, as shown in Fig. 2.57. Therefore,

$$\psi = (2\phi \tan \alpha_1 - 1) \tag{2.67}$$

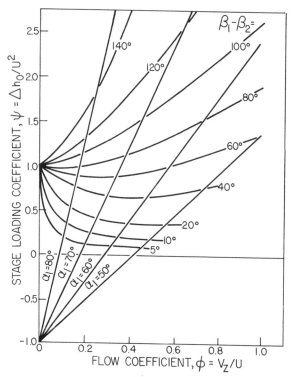

Figure 2.58 Fifty-percent reaction stage. (Copyright © Horlock, 1973a; Kreiger Publishing Co.).

Horlock (1973a) has given a detailed description of the design and off-design performances of impulse, 50% reaction, and axial exit flow turbines. A design plot which is based on Eq. 2.67 for 50% reaction is shown in Fig. 2.58. It is clear from this plot that, at a given α_1, ψ increases linearly with ϕ. When the flow coefficient is fixed, an increase in α_1 results in increased ψ. For example, at $\phi_1 = 0.5$, increasing α_1 from 50° to 70° increases ψ from 0.2 to 1.75, an increase of nearly ninefold. The present trend in the design of the nozzle is to use as high an α_1 as possible. But it should be realized that increasing α_1 increases W_1 for a given blade speed, and thus the flow is likely to reach supersonic speeds and limit the mass flow. Therefore, the designer has to vary α_1, U, R, and V_z, (or ϕ) to arrive at an optimum design for a given turbine inlet temperature (T_{o1}).

It should be emphasized here that the curves shown in Fig. 2.58 are for ideal conditions; no viscous losses or shock losses or three-dimensional effects are included. These effects become important at operating conditions other than the design condition (off-design conditions). It is possible to estimate the performance at off-design conditions (varying α_1 and ϕ) by utilizing Eqs. 2.59–2.61 with an assumed correlation for the efficiency. Estimation of efficiency is critical, especially for the calculation of the pressure drop from Eq. 2.61. The efficiencies vary considerably at off-design conditions. They are sensitive to changes in mass flow, Mach number, blade speed, and Reynolds number. Decrease in efficiency is mainly caused by viscous effects, including shock–boundary-layer interaction. Methods of calculating these effects are covered in Chapter 6.

2.9.4 Total-to-Static Efficiency

In many turbines (especially power turbines), the kinetic energy in the exhaust should be as small as possible, because this represents aerodynamic loss. The exhaust energy is utilized for thrust generation in a turbo prop, turbo fan, and turbo jet. But in automotive gas turbines, hydraulic turbines, and all other open-cycle systems, exit kinetic energy is considered a loss. Therefore, a more appropriate definition to represent the performance of these turbines is the total-to-static efficiency, given by Eq. 2.21. The total-to-static efficiency can be related through a velocity ratio as follows (refer to Fig. 2.56; note the differences in designation of stations between Fig. 2.56 and Fig. 2.7):

$$\eta_{ts} = \frac{h_{o1} - h_{o3}}{h_{o1} - h_{3ss}} = \frac{2\psi U^2}{V_{is}^2} \qquad (2.68)$$

where $h_{o1} - h_{3ss} = V_{is}^2/2$ is the isentropic velocity that could be achieved through isentropic expansion of the flow from h_{o1} to h_{3ss}. The velocity ratio U/V_{is} is one of the design parameters in such a case. Maximum η_{ts} is achieved when

$$\psi = V_{is}^2/2U^2$$

Another useful performance measure of a rotor is the efficiency based on the kinetic energy at inlet to the rotor (Fig. 2.56) given by

$$\eta_{KE} = \frac{h_{o1} - h_{o2}}{V_1^2/2} \qquad (2.69)$$

It should be noted here that $V_1^2 \neq V_{is}^2$ (where $V_{is} = h_{o1} - h_{2ss}$), except in the case of a reversible impulse turbine. The efficiency based on inlet kinetic energy represents how well the rotor converts potential and kinetic energy into mechanical power. For practical purposes, the kinetic energy at the exit cannot be zero. Hence $\eta_{KE} = 1$ cannot be achieved in practice. An optimum value for η_{KE} can be derived for an isentropic turbine as follows:

$$\eta_{KE} = \frac{\psi U^2}{V_1^2/2} = \frac{4(\phi \tan\alpha_1 - 1)}{\sigma^2} \qquad \text{(Impulse)}$$

$$= 2\frac{(2\phi \tan\alpha_1 - 1)}{\sigma^2} \qquad \text{(50\% reaction stage)}$$

where $\sigma = V_1/U$ is the velocity ratio, which has to be optimized.

For a reversible impulse stage (from velocity triangles in Fig. 2.57) we have

$$\eta_{KE} = 4\frac{(\sigma \sin \alpha_1 - 1)}{\sigma^2}$$

For a given α_1, we have $\partial \eta_{KE}/\partial \sigma = 0$ when $\sigma = V_1/U = 2/\sin \alpha_1$. Similarly, for a 50% reversible reaction stage we have

$$\eta_{KE} = \frac{2(2\sigma \sin \alpha_1 - 1)}{\sigma^2}$$

Therefore, for this stage, η_{KE} is maximum when $\sigma = 1/\sin \alpha_1$. Because α_1 is usually large, the best η_{KE} is achieved when V_1/U is close to unity for a 50% reaction stage and close to 2 for an impulse turbine. For the same enthalpy drop in a stage $(h_{o1} - h_{o2})$, an optimum reaction turbine requires a higher blade speed than does an optimum impulse turbine. If the blade speed is fixed, the present analysis as well as the earlier analysis (Eq. 2.65) indicates that a reaction turbine requires more stages (for the same total output power) than does an impulse turbine.

It is also interesting to note that $V_1 = V_{is}$ for an isentropic impulse turbine and $V_1 = 0.707V_{is}$ for a 50%-reaction stage. In the latter case, the kinetic energy drop $V_1^2/2$ is split evenly between the rotor and the stator $(V_1^2/4$ each).

2.9.5 Performance

The performance of turbines can be expressed in terms of nondimensional parameters (Eq. 2.24). It is clear that Eqs. 2.59 to 2.61 follow these similarity relationships. The performance of a typical two-stage turbine is shown in Figs. 2.59 and 2.60. The data are from a two-stage turbine (Whitney et al., 1972). The turbine had a mean diameter of 66 cm, with a mean blade height of 10 cm, and operated at an overall pressure ratio of 1.4–4.0 and at 40–108% speed at an inlet temperature of 378° K. The total pressure ratio (P_{o1}/P_{o2}) versus mass flow parameter $\left(\dot{m}\sqrt{\theta_{cr}}/\delta\right)$ indicates, as explained earlier, that the pressure ratio increases linearly with mass flow up to a certain point (denoted by A), beyond which the increase in pressure ratio is much more rapid. Compressibility effects beyond point A limit the mass flow through the turbine passage. The mass flow per unit area for compressible flow is given by

$$\frac{\dot{m}}{A} = \frac{P_o\sqrt{\gamma}}{\sqrt{RT_o}}M\left(\frac{1}{1 + \frac{\gamma - 1}{2}M^2}\right)^{(\gamma+1)/2(\gamma-1)} \tag{2.70}$$

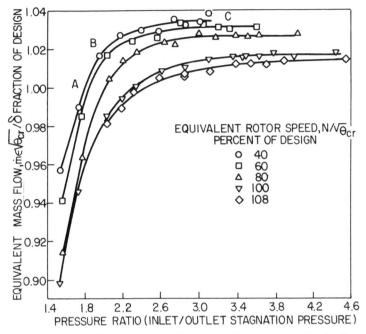

Figure 2.59 Performance of a two-stage turbine; θ_{cr} is the squared ratio of critical velocity (at $M = 1$) at turbine inlet temperature to critical velocity at standard sea-level. (Whitney et al., 1972; NASA TND 6960).

Because the term in parentheses is close to unity at low Mach numbers, the mass flow increases linearly with an increase in inlet velocity (say up to point A). At points such as B, choking conditions are reached and the mass flow through the passage is fixed.

The pressure drop characteristics can be explained on the basis of Eq. 2.61. At a low Mach number and at fixed $N/\sqrt{\theta_{cr}}$, the term $(1 + (\gamma - 1/2)M_1^2)$ is close to unity, while M_{z1} increases linearly with mass flow parameter. This accounts for the behavior of the performance up to A in Fig. 2.59. At higher mass flow, the rapid increase in pressure drop is caused by compressibility effects. The density and temperature are decreasing through the blade row, and hence any increase in mass flow will bring about a rapid increase in the axial velocity (because $\dot{m} = \rho V_z A$; V_z increases faster compared to incompressible flow as ρ is also decreasing due to expansion) and a very rapid increase in the axial Mach number, because the speed of sound is also decreasing as a result of decrease in temperatures. Even though M_1 is also increasing, the direct effect of an increase in M_z is substantial in Eqs. 2.60 and 2.61. Hence, at higher mass flow and at high subsonic Mach numbers, the pressure drop is much more rapid (from A to B in Fig. 2.59). Near sonic and choking conditions, M_z increases very rapidly and the pressure drop and mass flow characteristics are very steep as indicated from

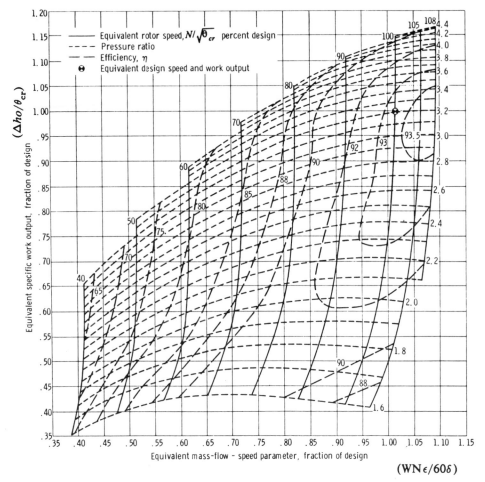

Figure 2.60 Performance map. Design value of specific work output, 76,818 joules per kilogram; design value of mass flow − speed parameter, 79,764 kilogram rpm per second (Whitney et al., 1972; NASA TND 6960).

B to C in Fig. 2.59. The curve B to C in Fig. 2.59 probably represents the presence of supersonic flow, where the pressure drop is much more rapid and the nozzle and/or the blade row is choked.

A composite performance map of the two-stage turbine (Whitney et al., 1972) is shown in Fig. 2.60. Performance variations at various operating conditions are shown in this plot. The parameters used in this plot can be derived from Eq. 2.27 (for a fixed machine and gas). The first and the second group are multiplied to derive a new group, $\dot{m}N/\delta$. Hence, the functional relationship (Eq. 2.27) can be written as

$$\frac{\Delta h_o}{\theta_{cr}} = f\left(\frac{\dot{m}N}{\delta}, \frac{N}{\sqrt{\theta}}, \text{Re}\right)$$

One has to achieve an order of magnitude variation in the Reynolds number in the turbulent regime before it has any major effect on the pressure rise and the efficiency. Therefore, the first two variables in the above equation are more important. In Fig. 2.60, $\Delta h_o / \theta_{cr}$ is plotted against $\dot{m} N / \delta$ (weight flow $W N \epsilon / 60 \delta$ is used instead of this parameter) for various values of $N \sqrt{\theta}$. One can also plot $\Delta h_o / \theta_{cr}$ versus $\dot{m} \sqrt{\theta} / \delta$ at various values of $N / \sqrt{\theta}$. The efficiency plots in Fig. 2.60 reveal that peak efficiencies are achieved near design operating conditions. The efficiencies are low at off-design conditions and for supersonic flows, for which passage will be choked.

Several conclusions can be drawn from a plot such as the one shown in Figs. 2.59 and 2.60:

1. For a fixed pressure ratio, there exists an optimum equivalent weight-speed parameter at which the efficiency is highest. Efficiency decreases below or above this value.

2. The maximum attainable value of $\Delta h_o / \theta_{cr}$ is not necessarily the maximum efficiency point. (This is true of a compressor also; see Fig. 2.17).

3. By selecting an optimum value of $N / \sqrt{\theta}$ and $\dot{m} N / \delta$, an efficient turbine can be designed.

Kacker and Okapuu (1982) correlated more than 100 sets of data from 33 turbines and provided a comprehensive assessment of turbine enthalpy drop ($\Delta h_o / U^2$) as a function of the flow coefficient (V_z / U). Their correlation is shown in Fig. 2.61. The efficiency of a turbine depends strongly on the loading coefficient and the flow coefficient. The loading coefficient influences the pressure gradient in the passage, and this increases the losses. The flow coefficient is a direct measure of the mass flow, for a given speed and machine. Higher flow coefficient, and hence higher mass flow, results in a higher pressure drop (Fig. 2.59), and the corresponding losses also increase. Thus it is not surprising that highest efficiencies occur at low loading and low flow coefficient. For aircraft application, where the major criterion is the weight, higher loading results in fewer stages, thus reducing the weight of the power plant. An optimization study has to be carried out to derive an optimum pressure ratio and weight of an aircraft turbine. For a land-based turbine, it is preferrable to operate at lower loading and low flow coefficient to achieve higher efficiency. Moustapha et al. (1987) built a turbine to operate at a very high pressure ratio of 3.76, $\Delta h_o / U^2 = 2.47$, $V_z / U = 0.64$, and a stage efficiency of 82%.

A plot of efficiency versus pressure ratio, at two different mass-averaged reaction levels, obtained for the turbine in the energy efficient engine (Thulin et al., 1982 and Leach, 1983) is shown in Fig. 2.62. The high-pressure turbine of rotor diameter 82 cm is designed for a pressure ratio of 4, a loading coefficient (mean radius) $\Delta h_o / U^2$ of 1.62 at a flow coefficient (V_z / U) of 0.351, and a mean reaction of 43% at speed of 13,232 rpm. The higher

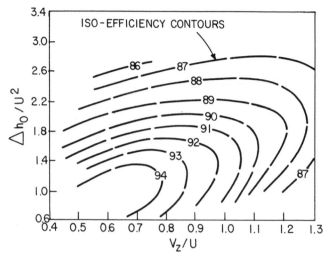

Figure 2.61 Correlation of stage efficiency in contour form as a function of stage loading and flow coefficient. (Copyright © Kacker and Okapuu, 1982; Reprinted with permission of ASME.)

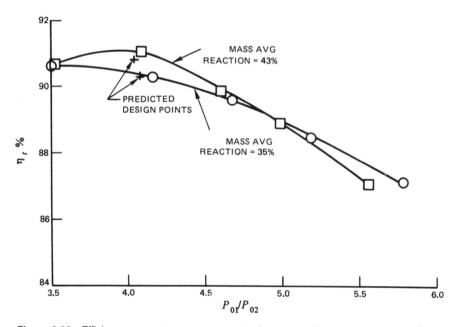

Figure 2.62 Efficiency versus stage pressure ratio for energy-efficient engine turbine (Thulin et al., 1982; NASA CR 165608).

pressure ratio gives rise to large acceleration of relative flow and the presence of supersonic exit flow. Therefore, both the shock losses and viscous losses increase with an increase in the pressure ratio. Efficiencies at two levels of reaction indicate that higher-reaction turbines produce higher efficiency. The reasons for this can be explained by comparing an impulse with a reaction turbine. The results obtained from the test program demonstrate that appreciable efficiency gains could be achieved by designing a single-stage turbine to operate at a low flow coefficient with an attendant high blade stress and reaction level.

It should be remarked here that high efficiencies can be achieved in turbines as a result of favorable pressure gradients which control the boundary layer growth. It is clear from the plots and equations presented earlier that the limitation in turbine performance is brought about by the following factors:

1. Compressibility effects limit the mass flow that can pass through the turbine. Therefore, the temperature and pressure drops are limited by choking conditions on the mass flow (Figs. 2.59 and 2.60).
2. There is a limit on the maximum inlet temperature because of material and stress considerations.
3. There is a limit on maximum blade speed U_b, because of stress considerations.

As the inlet temperature and speed increases, the combined stresses increase, so there must be a compromise between the maximum inlet temperature and the maximum blade speed. Recent research activities have been focused in the area of new materials that can withstand higher temperatures and higher stresses at the same time. This would enable improvement in cycle efficiencies, decrease the number of turbine stages, and thereby improve weight/power output of the turbine. This will result in increased thrust/weight ratio for an aircraft, spacecraft, or power turbines. Considerable effort has also been spent in developing efficient cooling techniques and surface coatings, so that gas temperatures can be increased.

2.9.6 Typical Blading

Blade shapes used for turbines differ considerably from those used for compressors. Large turning and stresses encountered in practice result in the blade shapes shown in Fig. 2.63. Typical blade shapes used for subsonic inlet flow (subsonic or supersonic exit) and supersonic inlet and exit flow are shown in Fig. 2.63. Profiles shown in Fig. 2.63a are earlier profiles derived by NACA. Profiles shown in Fig. 2.63b are typical blades used for inlet subsonic flow and exit supersonic flow. The thickness distributions, suction surface curvature, and trailing edge shape are varied for particular applications. The

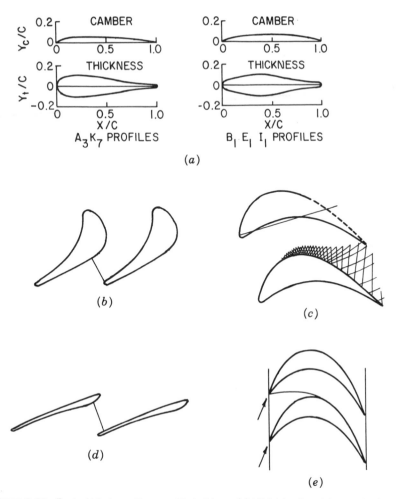

Figure 2.63 Typical blade profiles used in turbines. (a) NACA basic turbine profile (Dunavant and Erwin, 1956, NASA TN 3802). (b) Profiles for subsonic inlet and supersonic exit. (Copyright © Sieverding, 1976; reprinted with permission of VKI.) (c) Shock-free airfoil design ($M_1 = 0.39$, $M_2 = 1.15$, and $\theta = 79°$). (Copyright © Sobieczky and Dulikravich, 1982; reprinted with permission from ASME.) (d) Steam turbine tip section. (Copyright © Sieverding, 1976; reprinted with permission of VKI.) (e) Supersonic inlet and supersonic exit flow.

design of the blade section depends on the passage Mach number distribution, stress levels, and various other factors. Profiles can be generally classified as follows:

1. Profiles derived by various agencies, such as NACA and NGTE (e.g., A_3K_7 profiles).
2. Profiles with circular arc and parabolic arc camber specified much the same way as compressor blading.

3. Profiles derived graphically or empirically from a specified pressure or Mach number distribution for optimum performance.
4. Each industry (steam, gas, small engine, large engine) has developed their own profiles over the years, modified them continuously (by trial and error), and arrived at optimum blade shapes. These are modified to develop new sections for new applications.
5. The recent trend is toward custom-designed or custom-tailored airfoils. This concept was introduced in the section on compressors (Figs. 2.12, 2.13 and 2.29).

The aircraft industry has developed a series of computer programs to optimize the profile in terms of shock and viscous losses, location of shock and expansion waves, location of transition and smooth acceleration. Profiles shown in Figs. 2.63c and 2.63e are some examples of theoretically derived airfoil shapes. Profile c was derived from a numerical code which uses a finite volume method and the method of characteristics. Profile e was derived from a potential analysis, modified to satisfy the unique incidence condition. Sieverding (1976) has provided a comparison of pressure distribution and shock location for profiles shown in Fig. 2.63b as well as for some close derivatives of these blade shapes. He has provided a detailed interpretation, including advantages and disadvantages, of these blade shapes.

The profiles shown in Fig. 2.64 are some of the latest designs used in the E^3 (energy-efficient engine) engine. High pressure ratio and efficiencies are achieved through the incorporation of various aerodynamic improvements. A good example of the custom design airfoil is the one developed for a spacecraft turbine (two stage) by Huber et al. (1992), shown in Fig. 2.65. A very high specific work requirement per stage (442 BTU/lb$_m$), $\psi = 4.15$,

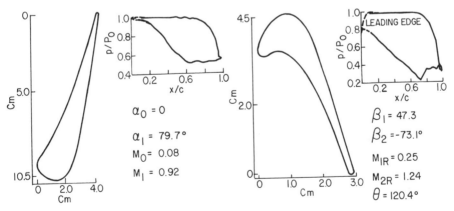

Figure 2.64 Vane and blade mean radius profile characteristics of energy-efficient engine turbine blading (Thulin, 1982; NASA CR 165608).

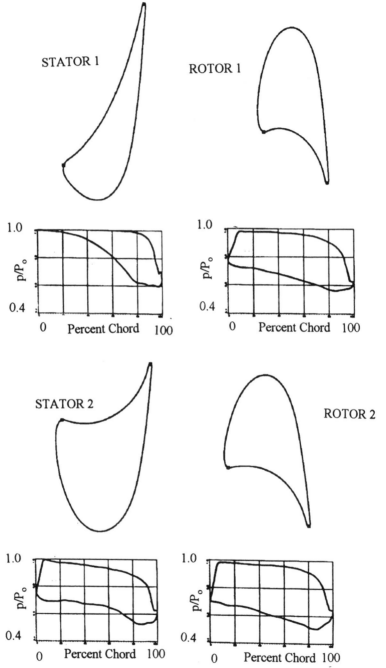

Figure 2.65 Blade geometry used in advanced turbopump drive (ATD) turbine for space shuttle (midspan airfoil sections for the two stage turbine (nozzle at the inlet), $N = 24{,}000$ rpm). (Copyright © Huber et al., 1992; reprinted with permission of AIAA.)

resulted in unconventional blades with large flow turning ($\sim 160°$). The profile for the two-stage turbine and the corresponding pressure distribution derived from Euler/Navier–Stokes methods show thick blades with large flow turning. The total pressure ratio for the turbine is 3.93 and operates with hydrogen/oxygen gas mixture and runs at 24,000 rpm.

All the shapes, except the one shown as Fig. 2.63e, are suitable for subsonic entry flow. For supersonic inlet, as in a compressor, the leading edge dictates the unique incidence at which the blading would operate.

The spacing of blades is one of the critical parameters in turbomachinery. Close spacing involves a larger number of blades, larger weight, and increased losses due to increased frictional surfaces. The force per blade can be kept small if a large number of blades are employed. On the other hand, with larger spacing, the forces/blade should be large because there are fewer blades; friction surface, weight, and losses are all minimized. Hence an optimum value of blade spacing exists that provides reasonably good efficiency as well as tangential momentum extraction (output). Early turbine designers widely employed an empirical criterion (based on a large number of experiments) due to Zwifel (1945), based on the nondimensional force in a cascade.

$$Z = \frac{2F_\theta}{\rho V_2^2 C_z}$$

where C_z is the axial chord, V_2 is the exit velocity, and F_θ is the blade force. Equating the work done to the enthalpy rise, we obtain

$$F_\theta U = c_p(T_{o1} - T_{o2})\rho S V_z$$

Substituting Eq. 2.59 in the above expression and assuming that the flow is incompressible ($V_{z_1} = V_{z_2}$), it can be proven that

$$Z = \frac{2F_\theta}{\rho V_2^2 C_z} = 2\cos^2 \alpha_2 \frac{S}{C_z}(\tan \alpha_1 - \tan \alpha_2)$$

It should be remarked here that α_2 is usually negative for a turbine, and thus the terms in the parentheses are additive. Zwifel suggested a value of 0.8 for the coefficient Z. The maximum turning allowed is dictated by the viscous effects and the Mach number consideration. It is clear from the above equation that good performance can be obtained with either (a) very high turning and fewer number of blades or (b) small turning with a larger number of blades. This criterion can be used to determine S/C_z (or number of blades) for a known value of α_1 and an assumed value of α_2. This criterion is used with some success, even when the flow is compressible.

2.9.7   Blade Pressure Distribution

The nature of the pressure distributions on a typical nozzle and rotor blades are shown in Figs. 2.64 and 2.65. These are very heavily loaded blades. The static pressure decreases in both the nozzle and the rotor blade. This is brought about by acceleration of the absolute flow in the nozzle and relative flow in the rotor. Because of this acceleration, the flow which is subsonic at the inlet may reach supersonic velocities near the trailing edge. The pressure distribution for the rotor in Fig. 2.64 indicates that sonic conditions are reached near 70% chord, resulting in a shock wave in this region. The static pressure increases behind the shock wave.

On the pressure side of rotor blade (Figs. 2.64 and 2.65), large deceleration occurs near the leading edge, followed by nearly flat pressure distribution; the relative flow is accelerated during the aft 20–50% of chord. On the suction side, however, the flow is accelerated rapidly near the leading edge, followed by moderate acceleration during the remainder of the chord. It is only in the aft 10–15% of chord that the pressure rises again to reach the same level as the pressure side.

Comparing this with the pressure distribution for a typical compressor (Fig. 2.12), several differences become clear:

1. The pressure drop in a turbine is much higher than the corresponding pressure rise in a compressor. While the turning angle in a compressor varies from 20° to 35°, the corresponding turning angle in a turbine can be as high as 160°.

2. The design of the pressure surface is similar in both cases; it is usually designed for nearly constant Mach number or static pressure.

3. In the case of a compressor suction surface, the relative flow is accelerated first and then decelerated. The separation criterion prevents rapid deceleration (Fig. 2.12B–D). The turbine flow, on the other hand, is accelerated rapidly all the way from the leading edge to the trailing edge (Figs. 2.64 and 2.65).

4. The turbine designer (in high subsonic cases) usually delays the formation of shock losses. For this reason, inlet subsonic and exit supersonic flows are not uncommon in aircraft turbines. The compressor designer, on the other hand, is dealing with the deceleration of high Mach number flow and has to allow for shock waves within the passage. This results in higher losses and lower efficiency.A transonic compressor is less efficientthan a transonic turbine.

5. In the case of a turbine, the flow acceleration near the pressure surface is much more pronounced toward the trailing edge.

Design flow turning, flow inlet and exit angles, and inlet and exit Mach numbers for both the nozzle and the rotor blades of the E^3 turbine and the

ATD turbine are also shown in Figs. 2.64 and 2.65. Industrial gas turbines and power steam turbines are not designed for such high blade loading.

2.9.8 Effect of Mach Number

As indicated earlier, higher inlet Mach numbers produce greater pressure and temperature drops and higher work output, resulting in compact turbines suitable for aerospace applications. Therefore, the trend in turbine development for aircraft and spacecraft (e.g., space shuttle) has been to increase the speed and the relative Mach number at the tip. High Mach numbers have been in use in both steam turbines and aircraft gas turbines for a long time. The application of turbines in rocket engines and hypersonic vehicles has resulted in a renewed interest in high-Mach-number operation.

In turbines, the work-output-per-unit mass increases as the nozzle exit tangential velocity increases (Eq. 2.6). The mass-flow-per-unit cross-sectional area increases as the nozzle exit axial Mach number nears unity. Thus high exit nozzle swirl velocity and high mass flow produce higher work output. Oates (1984) has analyzed the limitation of the compressibility effect on the maximum attainable swirl in a turbine nozzle. Consider uniform entry flow, with zero swirl and isentropic flow (Fig. 2.56). Let the Mach numbers at inlet and exit be M_0 and M_1, respectively. Using the one-dimensional continuity equation ($\rho_0 V_0 A_0 = \rho_1 V_1 \cos \alpha_1 A_1$) and isentropic relationship between density and temperature and noting $T_{o0} = T_{o1}$, the following relationship between the inlet and exit Mach number can be derived:

$$\frac{M_1}{M_0} = \left[\frac{1 + \dfrac{\gamma - 1}{2} M_1^2}{1 + \dfrac{\gamma - 1}{2} M_0^2}\right]^{(\gamma+1)/2(\gamma-1)} \frac{A_0}{A_1} \frac{1}{\cos \alpha_1}$$

The swirl velocity is given by

$$V_1^2 \sin^2 \alpha_1 = V_1^2\left(1 - \cos^2 \alpha_1\right) = M_1^2 a_1^2 \left(1 - \cos^2 \alpha_1\right)$$

$$\frac{V_1^2}{a_0^2} \sin^2 \alpha_1 = \frac{T_1}{T_0} M_1^2 \left(1 - \cos^2 \alpha_1\right)$$

Hence

$$\frac{V_1^2 \sin^2 \alpha_1}{a_0^2} = \frac{1 + \dfrac{\gamma - 1}{2} M_0^2}{1 + \dfrac{\gamma - 1}{2} M_1^2} M_1^2 - \left[\frac{1 + \dfrac{\gamma - 1}{2} M_1^2}{1 + \dfrac{\gamma - 1}{2} M_0^2}\right]^{2/(\gamma-1)} M_0^2 \left(\frac{A_0}{A_1}\right)^2$$

$$(2.71)$$

The derivative of this equation with respect to M_1^2 may now be taken and equated to zero.

$$M_0^2 \left(\frac{A_0}{A_1}\right)^2 \left[\frac{1 + \dfrac{\gamma - 1}{2} M_1^2}{1 + \dfrac{\gamma - 1}{2} M_0^2}\right]^{(\gamma+1)/(\gamma-1)} = 1 \qquad (2.72)$$

Hence Eqs. 2.71 and 2.72 give the maximum possible swirl when $M_1 \cos \alpha_1 = 1$. This occurs when the axial Mach number is unity or when the nozzle is choked.

In the case of a compressor, the limiting factor is C_p, which influences boundary layer growth. For a given C_p, a higher Mach number produces higher pressure rise as indicated by Eq. 2.48. In a turbine, the limiting factor on C_p is the choking condition. Here again, if the pressure coefficient is held fixed, the higher inlet Mach number produces a larger pressure drop (as per Eq. 2.48). This is also clear from Eqs. 2.59–2.61 and comments made earlier.

The inlet Mach number at which the throat will choke is called the *choking Mach number*. Based on Eq. 2.70, the ratio of the inlet area ($S \cos \beta_1$; see Fig. 2.66) and throat area (A^*) for one-dimensional flow is given by

$$\frac{A}{A^*} = \frac{S \cos \beta_1}{A^*} = \frac{1}{M_0} \left[\frac{2}{\gamma + 1}\left(1 + \frac{\gamma - 1}{2} M_0^2\right)\right]^{(\gamma+1)/2(\gamma-1)} \qquad (2.73)$$

For a given β_1 and throat area, this relationship provides a means of determining the inlet choking Mach number (M_0) that results in sonic conditions at the throat. As this ratio is increased, the inlet Mach number at which the choking condition occurs also decreases. Dramatic changes occur as the inlet Mach number is increased gradually. At $M_0 = M_{cr}$ a shock wave forms on the suction side. Kynast (1960) and Sieverding (1976) have measured pressure distributions at various inlet Mach numbers, with all other variables held constant and for a fixed geometry. Dring and Heiser (1978) have provided a clear explanation of the phenomenon that occurs as the flow progresses, from the subsonic to the supersonic regimes. Fig. 2.66 illustrates this effect. At exit $M = 0.7$, the flow is subsonic everywhere, the pressure distribution is smooth on both surfaces, and the pressure drop is moderate. At $M = 0.9$, a small supersonic region appears on the suction side of the blade. As explained earlier, higher Mach numbers result in large pressure drops. This is clear from Fig. 2.66. This operating condition, which has a large pressure drop, a very small supersonic region, and a weak shock wave, probably has the best flow features and peak efficiency. Therefore, this would be a typical operating condition for most aircraft turbines. As the inlet Mach is further increased, the supersonic region would extend all the way across the passage, resulting in a shock shown in Fig. 2.66. As the Mach number is further increased (or the back pressure is decreased), the shock wave moves further downstream (as in a convergent divergent nozzle) until

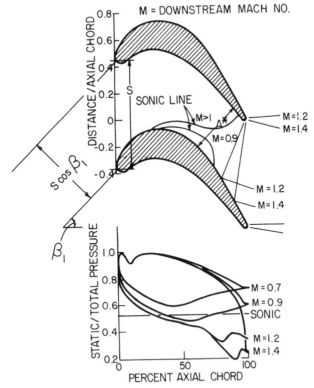

Figure 2.66   Transonic cascade flow (Dring and Heiser, 1978; AFAPL TR 78-52).

the shock wave is formed at the trailing edge. This condition is the "limit loading" condition, because it represents the maximum loading possible. It is clear from Fig. 2.66 that a large change in the pressure distribution for $M > 1$ occurs near the suction of the trailing edge. The measurements carried out by Kynast (1960) and Sieverding (1976) at various inlet Mach numbers (0.22–0.85) clearly reveal this effect.

2.9.9   Transonic and Supersonic Turbine

The classification of high-speed turbines is very similar to that of compressors described earlier. A transonic turbine is one where the relative flow remains subsonic at inner radii, and supersonic at outer radii, with a transonic region in the middle. When the entire flow from hub to tip is supersonic, the turbine is classified as a supersonic turbine. The qualitative flow features for these cases are shown in Fig. 2.66. The static pressure increases across the shock wave, resulting in a dip as shown for $M = 1.2$ and 1.4 in Fig. 2.66. In this region, the flow would decelerate and then accelerate again. The stagnation enthalpy in a nozzle and stator remains unchanged across the shock wave in adiabatic flow, but static enthalpy and static temperatures increase across the shock wave. Some of the other features described under "transonic and supersonic compressors" are valid in cases of high-speed turbines as well.

There are essentially three types of turbines, characterized by the inlet and exit flow:

1. Subsonic inlet and exit, with supersonic flow regions inside the passage
2. Subsonic at inlet, supersonic at exit
3. Supersonic at inlet and at exit

The flow phenomena in high-speed turbine rotors and stators are very similar to nozzle flows. Vanes and rotors are usually designed for shock-free operation. Because of nonuniformity and changes in design conditions, blade rows normally operate at back pressures other than for which they were designed. The phenomena in the blade rows at various back-pressure conditions can be best illustrated using a classical nozzle flow as an example. Following Liepmann and Roshko (1957), the nature of nozzle flow can be illustrated as shown in Fig. 2.67. Point J represents an ideal operating

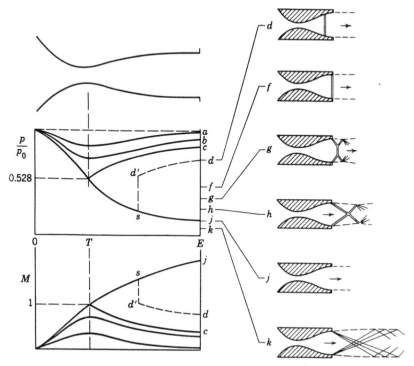

Figure 2.67 Effect of pressure ratio on flow in a Laval nozzle (From Liepmann and Roshko, 1957; reprinted with permission of John Wiley & Sons, Inc.)

condition at the design back pressure. The nozzle exit pressure is equal to the downstream pressure. When the back pressure is increased (or when the expansion is lower than it is designed for), the transition from the nozzle exit pressure, which is lower than the downstream pressure, has to occur through a shock wave as shown in Fig. 2.67d–h. The location of the shock wave depends on the ratio of actual to design pressure. The shock, which is originally located inside the passage, moves downstream as the back pressure is decreased. If the back pressure is less than the nozzle exit pressure (under expansion), further expansion will occur through expansion waves as shown in Fig. 2.67k.

A schematic of design, overexpanded, and underexpanded flow in a nozzle blade row is illustrated in Fig. 2.68. Horlock (1973a) provides a clear explanation of the conditions under which a supersonic nozzle would operate with overexpanded and underexpanded conditions shown in Fig. 2.68. If the rotor blade speed is increased (from the design value), since the relative flow angle (β_1) should remain the same, the deflection of the flow through the

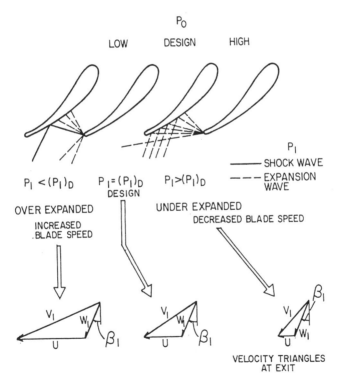

Figure 2.68 Nozzle flow field in the presence of rotor downstream at low, design, and high pressure ratios.

nozzle must increase; resulting in a shock pattern, as shown. The flow passes through the trailing edge shock, through its reflection, and through the expansion wave emanating from the other blade. If the rotor blade speed is decreased, the nozzle deflection decreases the expansion wave pattern will be as shown in Fig. 2.68.

The shock wave patterns that exists in some typical bladings are shown in Figs. 2.64 and 2.66. The E^3 engine turbine blade shock wave is located near the trailing edge, aft of the chordwise location $X/C = 0.70$. The performance of transonic turbines (subsonic inlet and supersonic exit) is shown in Figs. 2.59 and 2.60. The condition of operation from B to C in Fig. 2.59 corresponds to transonic turbines. A supersonic turbine (Fig. 2.63, case e) can only operate at one condition (unique incidence). Efficiency and performance will be very poor away from this unique incidence condition.

The method of estimating the performance of these turbines is very similar to the analysis presented earlier. Equations 2.59–2.61 are valid with the exception of the fact that η_t for a supersonic or transonic turbine is lower than that for a subsonic turbine. The isentropic relationships for the determination of static properties are valid only if the shocks are weak and nearly isentropic. A state-of-the-art review on transonic turbomachines is given by Bolcs and Suter (1986).

2.9.10 Current Trends in Turbine Development

Several authors (Rohlik, 1983; Rosen and Facey, 1987; Swihart, 1987; Hauser et al., 1979; Tanaka, 1983; Clifford, 1985) have provided historical perspective and predicted trends in turbine development. Major improvements anticipated in the area of aircraft gas turbines will be to increase the turbine inlet temperature, develop new materials to withstand higher temperatures, increase the turbine pressure ratio, achieve higher blade loading and efficiency, improve cooling techniques, and improve design and computational analysis (3D viscous, high-temperature, heat-transfer, turbulent flow field prediction). The last topic is covered in a Chapter 5, and turbine cooling techniques are covered in Chapter 7.

Turbine Inlet Temperature and Materials   It is known from the gas turbine cycle analysis that higher cycle temperature produces a larger amount of work per unit mass flow and improves the (weight/power) ratio of the gas turbine. A dramatic improvement that can be achieved in the horsepower of a gas turbine engine/unit mass flow with an increase in inlet temperature is shown in Fig. 2.69. The method of calculating thrust power from a cycle analysis can be found in most aerospace propulsion books (see Hill and Peterson, 1992, for example). The specific horsepower (hp/lb/s) can be doubled from the current levels of 200 at 1950 K to 450 at 2500 K. This capability compares with an ideal specific horsepower of 540 that does not

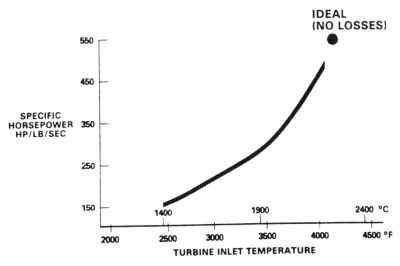

Figure 2.69 Advances in core specific power possible with high-temperature lightweight materials. (Copyright © Rosen and Facey, 1987; reprinted with permission of AIAA.)

include any losses. This improvement is one of the major goals of aerospace engine designers.

A major problem associated in achieving this increased performance is the availability of material that can withstand such high temperatures and combined stresses (due to temperature, rotation, and aerodynamic loading). Temperature capabilities of various materials drawn from Rohlik (1983) and Swihart (1987) are shown in Fig. 2.70. At the present time, gas temperatures as high as 1950 K have been achieved through material improvement and turbine cooling. Several U.S. agencies have undertaken the development of integrated high-performance turbine engine technology (IHPTET). To achieve the goals of IHPTET, the material should be capable of operating at temperatures in excess of 600 K above the best present-day turbines. The trend to achieve this goal is illustrated in Fig. 2.70. The rate of progress during the next decade must be much higher than that achieved during the 1970s and 1980s to attain this goal. The types of materials that are presently under consideration are also shown. Ceramics, which show promise of operating to 1600 K without cooling, have had problems with brittleness in rotating components. The composites that show promise for high temperature include metal matrix, intermetallics, and carbon. The high strength/weight ratios of these materials will allow for major improvements in aerodynamics and heat transfer.

The progress made in increasing the turbine inlet temperature is shown in Fig. 2.71. Engines still in the development stage (it usually takes about 10 years to develop an engine from conception to the operational stage) are expected to achieve turbine inlet temperatures of 2200 K by the year 2000,

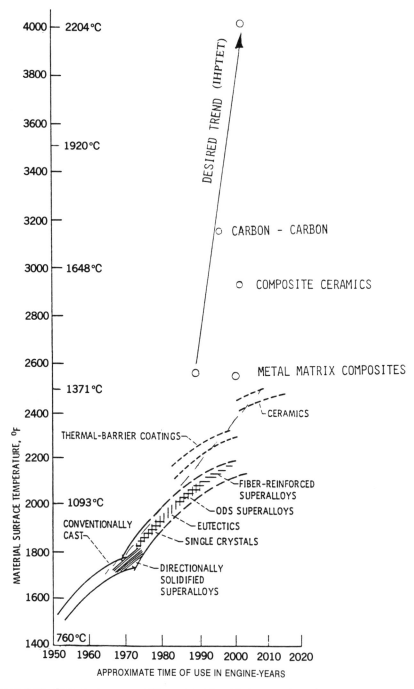

Figure 2.70 Temperature capabilities of turbine blade materials. (Modified from Rohlik, 1983.

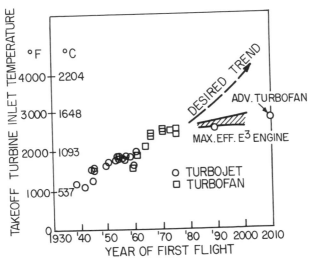

Figure 2.71 Progress in turbine inlet temperature (Rohlik, 1983; MODIFIED).

but the desired trend, shown in Fig. 2.70 for IHPTET, is 2500 K by the year 2000. One has to clearly distinguish between achievable goals (based on present knowledge and material) and desired goals (based on possible new material and development). During the first 30 years of development (1940–1970), the temperature increased by nearly 800 K. If the same rate of progress continues, one can hope to achieve 2200 K by the year 2000. As mentioned earlier, this will be achieved through a combination of developments, including improvements in combustion, new material, new methods for cooling the blades, and better flow and heat-transfer predictive capabilities.

Turbine Cooling   Turbine cooling is employed in most modern gas turbines. Chapter 7 is devoted to this advanced and fast developing topic. Turbine cooling enables high turbine inlet temperatures to be achieved for a given turbine material. Film and convection cooling allow the blade temperature to be maintained at values much lower than those encountered without cooling. Advances in this area have given rise to considerable improvement in turbine inlet temperature, and thereby the specific power (Fig. 2.69), as shown in Fig. 2.72. This figure is based on Clifford (1985). The author has included developments since 1950 and projections beyond 1987. Considerable basic research is presently underway in the area of materials and cooling technology, and should yield major improvements in turbine inlet temperature in the 1990s and beyond.

Pressure Ratio and Loading Coefficient   The need for advances in increasing the pressure ratio across a turbine is associated with a desire to increase the stagnation pressure ratio (maximum pressure ratio of the cycle,

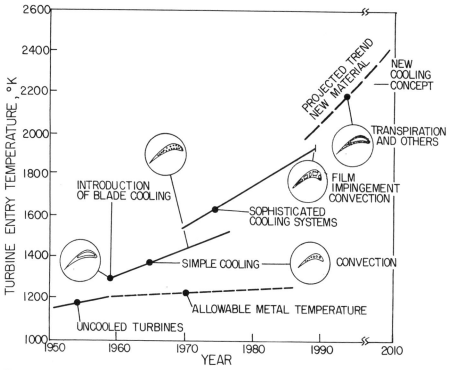

Figure 2.72 Variation of engine TET over recent years. (Copyright © Rolls Royce, plc, from Clifford, 1985; AGARD CP 390.)

ratio of pressure at compressor exit to engine inlet) of the core engine cycle. The cycle pressure ratio will continue to increase to a level of 60 by the year 2010, from the present level of 35 (Fig. 2.73). This will result in an improved thrust/mass ratio, thrust/weight ratio, and thermal efficiency of the cycle. This can only be achieved through improved turbine design, including higher pressure ratio, inlet temperature, and cooling. Hence, the aerodynamic and heat-transfer problems become much more severe. The pressure ratio per stage has increased continuously through better aerodynamic design tools.

A summary of the present state-of-the-art pressure ratios and loading coefficients is shown in Fig. 2.74. The trends differ between a large commercial engine such as the E^3 and prop fan engines, and small engines for use in helicopters and turbo props. The former usually consists of several stages as described earlier, while the small engines incorporate single stage turbines for reasons of compactness and simplicity. These small turbines have to achieve high pressure and temperature drop, resulting in high blade speed. Two examples of very heavily loaded turbines, shown in Fig. 2.74, are due to Moustapha (1987) and Bryce et al. (1985). All the data shown in Fig. 2.74 are for small engine turbines, with the exception of the E^3 engine (Leach, 1983;

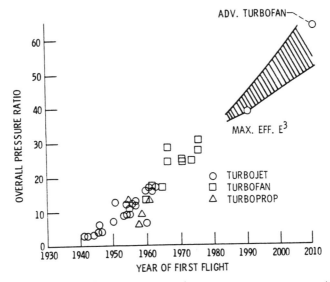

Figure 2.73   Progress in pressure ratio (Rohlik, 1983; NASA TM 83414).

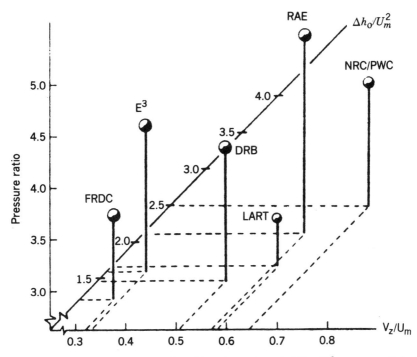

Figure 2.74   High-pressure research turbine stage characteristics [E^3—energy efficient engine (Leach, 1983) performance in Figs. 2.62 and 2.64; FRDC—Ewen et al. (1973); DRB—Okapuu (1974); LART—Liu et al. (1979); RAE—Bryce et al. (1985); NRC / PWC—Moustapha et al. (1987)]. (Copyright © Moustapha et al., 1987; reprinted with permission of ASME.)

Thulin et al., 1982) and the RAE turbine (Bryce et al., 1985) engine. In the small turbine field, the maximum loading coefficient ($\Delta h_o/U_m^2$) achieved is 2.5, with a pressure ratio of 3.76 and an efficiency of 82%. In the large turbine area, the maximum pressure ratio (per stage) achieved is 4.5, with an aerodynamic loading coefficient of 2.1 and an efficiency of 87%. High loading and high pressure ratio result in supersonic exit flow, so the efficiencies are generally lower. Attempts are being made to achieve high loading and high pressure ratio at reduced blade speeds. The turbine for the E^3 engine achieves a pressure ratio of 4.0, $\psi = 1.62$ at an efficiency of 90% (Fig. 2.62).

It is conceivable that turbines for large aircraft engines beyond the year 2000 will be operating at 2200 K, inlet pressures of 700–800 psia, while the small turbines may reach a temperature of 2000 K, pressures of 250–500 psia, and efficiencies of 89–91% (Rohlik, 1983). A great deal of research is underway in achieving these goals. The increased emphasis will be on high temperature materials, high turbine inlet temperature, high blade loading, high pressure ratio, and increased efficiency. These will be achieved through complex cooling systems, many innovative aerodynamic concepts such as contoured endwalls, low-aspect-ratio blading, reduced blade-row spacing, various endwall trenching and grooving, active tip clearance control systems, and variable nozzle and stator geometry. In addition, computational techniques for design and analysis will enable major design improvements.

The trend in turbine design for space turbines is shown in Fig. 2.75, including earlier rocket engine turbines (RL10, J2, F1, SSME). The recent design for advanced space shuttle main engine fuel pump (G^3T) and oxidizer pump (GGOT), due to Huber et al. (1992), achieves the highest loading and turning. The blading used in the GGOT turbine is similar to that shown in Fig. 2.65. The oxidizer pump turbine (GGOT) is single stage with shaft speed

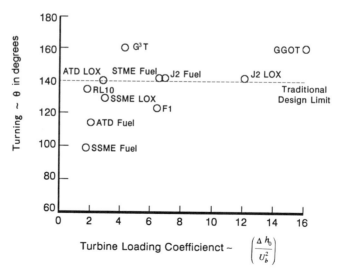

Figure 2.75 Current status of design technology for space turbines. (Copyright © Huber et al., 1992; reprinted with permission of AIAA.)

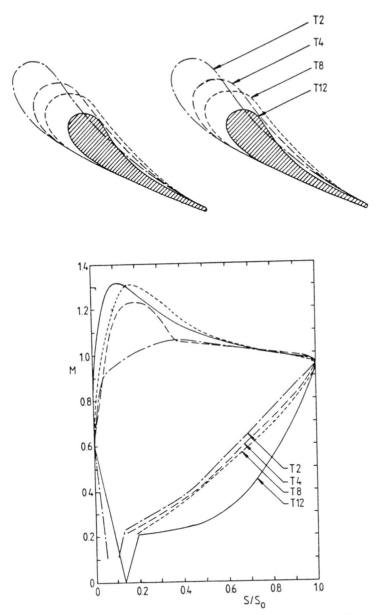

Figure 2.76 Controlled acceleration Airfoil (CAA). (Copyright © Baines et al., 1985; reprinted with permission of ASME.)

of 7880 rpm, a pressure ratio of 1.6, and a loading coefficient of 16.1. It is designed to deliver 14,168 kW at a flow rate of 27.39 kg/s. The nozzle accelerates the flow from $M_0 = 0.35$ to $M_1 = 1.05$ through a combination of turning and endwall convergence.

Blading Design The controlled acceleration airfoil (as in controlled diffusion airfoil for a compressor) and computer-aided design will eventually supplement or replace empirical methods. A recent example of this is an effort by Baines et al. (1986). A systematic development of airfoil, from 1969 technology (T2) to present technology based on a time-marching numerical technique and boundary layer optimization, is shown in Fig. 2.76. The objective was to increase the blade loading and reduce the blade solidity without incurring additional losses. Both the suction surface and the pressure surface contours were extensively modified to achieve this goal. The blade T12, with shorter chord, thin leading edge achieved the best (theoretical) performance. These techniques and others, including shock-free airfoil design, optimization of pressure distribution in the radial, chordwise, blade-to-blade direction, optimization for the reduction of losses (e.g. endwall contours), and adoption of variable geometry nozzles to account for off-design operation (e.g., Takagi, 1986) will become common practice in industry.

2.10 RADIAL AND MIXED-FLOW TURBINES

Because most of the hydraulic turbines used in practice are radial or mixed flow, radial gas turbines and hydraulic turbines are covered in one section in a unified manner.

Hydraulic turbines have been in existence for more than a century, whereas the radial steam turbine was introduced around the turn of the century. Hence, the axial turbine is much older than its radial counterpart and has been, and still is, the target of considerable development work. The material to be covered in this chapter is fairly classical, because research and development work in this area has been slow compared to the effort spent on axial flow compressors and turbines. Dixon (1984) and Von Der Nuell (1964) provide a good introduction to overall performance and analysis. The design aspects are covered in Whitfield and Baines (1990). The latest developments on cooling, computation, and design, covered in later chapters, are also applicable to radial turbines.

Because of manufacturing ease, robustness, and erosion resistance, radial gas turbines are used in many applications. Inward radial gas turbines have been used in automotive turbochargers and gas turbines (Fig. 1.9), aircraft auxiliary power units, aircraft engine starters, cogeneration gas turbines, enhanced oil recovery systems, and many mechanical drive systems (e.g., automotive torque converter). There are large radial turbines operated by blast furnaces. There has been renewed interest in improving radial turbines for use in automotive engines for passenger cars and trucks (Boyd et al.,

1983; Mulloy and Weber, 1982; Lane, 1981; Rohlik, 1983; Ewing et al., 1980). The advanced gas turbine (AGT 101), built by Allied Signal and Ford Co. (Fig. 1.9), has achieved inlet temperatures in excess of 1650° K through the use of ceramics. Efficiencies in excess of 84% have also been achieved. There are several types of radial gas turbines, but two of the most commonly used radial gas turbines are shown in Fig. 2.77. The radial inflow turbine is very similar to the centrifugal compressor (Fig. 2.37) except that the flow path is reversed. As in centrifugal compressors, there are three types of blading: radial, backward-curved and forward-curved vanes (Fig. 2.39). In the cantilevered turbine, the blade chord length is small. Cantilevered turbines are usually of the impulse type ($W_2 = W_3$) as shown in Fig. 2.77b. The reaction type ($W_3 > W_2$) suffers from the fact that the exit area must increase to keep velocity within a reasonable limit. Decrease in area due to decrease in radius along the flow path is one of the major limitations on the increase in the relative velocity.

The analysis of this type of turbomachinery is very similar to that of an axial turbine. The entire rotor and nozzle flow path is in the radial direction with a change in radius. Therefore, the analysis is identical to that of the axial turbine covered earlier, with the exception of the fact that $U_1 \neq U_2$.

Hydraulic turbines, as mentioned earlier, have been in existence for a long time. The Francis type of radial turbine (shown in Figs. 2.77c and 1.10) and the mixed type (the entry flow is partly radial and partly axial) are widely used in hydraulic powerplants. For example, the Grand Coolee installation has 12 units, each running at 72 rpm with a capacity of 600 MW, a 10-m diameter, and $Q = 850$ m^3/s. Francis turbines are usually designed for medium head installations (25–500 m) and specific speed $\overline{N}$ of 0.3–2.0 (Eq. 2.31 and Figs. 2.8 and 2.9). The Francis inward radial turbine has flow entering through the wicket gates and then on to the rotor as shown in Fig. 2.77c. The radial velocity is gradually turned to axial at the exit.

There are two types of propeller hydraulic turbines, one with a fixed blade setting and the other with a variable blade setting. The latter, which can be operated at differing head and mass flow, is called the *Kaplan turbine*. In a propeller turbine, the fluid passing through the wicket gates usually achieves a free vortex distribution, changing its path from predominantly radial to predominantly axial. The absolute tangential velocity decreases as shown in Fig. 2.77d. The blades are fewer in number (three to eight) and are usually of airfoil shape, with little or no camber. They are particularly suited for low head (30 m) applications, with maximum efficiencies of around 94%. They are usually designed at a specific speed, $\overline{N}$, of 2–5 (Fig. 2.8 and 2.9).

2.10.1 Analysis and Performance of Radial Gas Turbines

The one-dimensional analysis of radial turbines is very similar to that of an axial turbine and a centrifugal compressor, provided that the nature of the velocity triangles is taken into account.

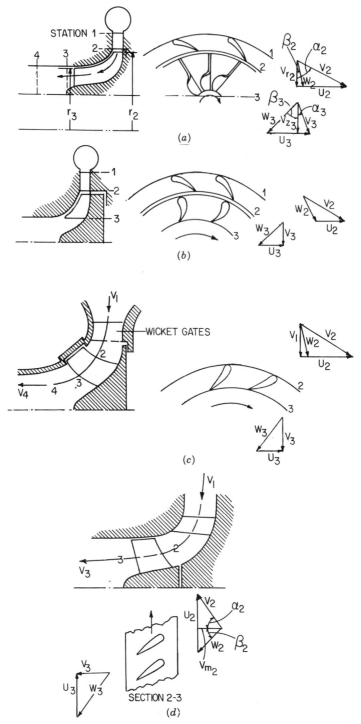

Figure 2.77 Radial gas and hydraulic turbines and velocity triangles (a) Radial inflow, axial outflow. (b) Cantilevered. (c) Francis turbine. (d) Propeller turbine.

The expansion through a radial turbine can be represented on an h–s diagram as shown in Fig. 2.7. The work output is given by Eq. 2.12, and the efficiency is given by Eq. 2.21 or 2.22. The analysis of a cantilevered turbine is very similar to that of an axial flow turbine (Eqs. 2.59–2.67), except that the inlet blade speed is different from the exit blade speed and the radial velocity (corresponding to axial velocity) changes continuously due to radius change and the compressibility effect. The former effect is the dominant one.

The more widely used turbine configuration is the inward (radial to axial) configuration shown in Fig. 2.77a. One can write the one-dimensional continuity and momentum equation across each of the components, inlet nozzle, and rotor. The energy equations (for adiabatic flow) given below provide expressions for the loading coefficient, temperature, and pressure ratios:

Nozzle: $\quad\quad\quad\quad h_{o1} = h_{o2}, \quad h_1 - h_2 = \dfrac{V_2^2}{2} - \dfrac{V_1^2}{2}$

Rotor $\quad\quad\quad\quad (P_{\text{shaft}}/\dot{m}) = h_{o2} - h_{o3} = U_2 V_{\theta_2} - U_3 V_{\theta_3}$ $\quad\quad$ (2.74)

Diffuser $\quad\quad\quad h_{o3} = h_{o4}, \quad h_4 - h_3 = \frac{1}{2}(V_3^2 - V_4^2)$

The loading coefficient is derived in a fashion similar to Eq. 2.59, for an axial turbine, and is given by

$$
\begin{aligned}
\psi &= \frac{C_p(T_{o2} - T_{o3})}{U_2^2} = \frac{U_2 V_{\theta_2} - U_3 V_{\theta_3}}{U_2^2} \\
&= \frac{(U_2 V_{r_2} \tan \alpha_2 - U_3 V_{z_3} \tan \alpha_3)}{U_2^2} \\
&= \frac{V_{r_2}}{U_2} \tan \alpha_2 - \frac{U_3}{U_2}\left(\frac{V_{z_3}}{U_2} \tan \beta_3 + \frac{U_3}{U_2}\right)
\end{aligned}
\quad (2.75)
$$

The appropriate velocity triangles are shown in Fig. 2.77a. Note that β_3 (which is negative) refers to the configuration shown in Fig. 2.77a. Denoting $V_{r_2}/U_2 = \phi$, the flow coefficient, this equation reduces to

$$
\psi = \phi\left[\tan \alpha_2 - \frac{r_3}{r_2}\left(\frac{V_{z_3}}{V_{r_2}} \tan \beta_3\right)\right] - \frac{r_3^2}{r_2^2}
\quad (2.76)
$$

The temperature ratio can be expressed in terms of Mach numbers following the procedure used in deriving Eq. 2.60.

$$
\frac{T_{o3}}{T_{o1}} = \frac{T_{o3}}{T_{o2}} = 1 - \frac{(\gamma - 1)M_{b_2^2}}{1 + \dfrac{\gamma - 1}{2}M_2^2}[\psi]
\quad (2.77)
$$

And an expression for P_{o3}/P_{o2} is given by the equation in Table 2.1:

$$\frac{P_{o3}}{P_{o2}} = \left[1 - \frac{1}{\eta_t}\frac{(\gamma-1)M_{b_2}^2}{1+\dfrac{\gamma-1}{2}M_2^2}\psi\right]^{\gamma/(\gamma-1)} \tag{2.78}$$

$$M_2 = \frac{V_2}{\sqrt{\gamma R T_2}}, \qquad M_{b_2} = U_2/\sqrt{\gamma R T_2}$$

For the configuration shown in Fig. 2.77a β_2 and α_2 are both in the same direction and are positive. For $r_3 = r_2$, $V_{z_3}/V_{r_2} = $ AVR, Eqs. 2.75–2.78 reduce to Eqs. 2.59–2.61 for an axial turbine. The reaction is defined by Eq. 2.62 as

$$R = \frac{h_2 - h_3}{h_1 - h_3} = \frac{\frac{1}{2}\left([U_2^2 - U_3^2] - [W_2^2 - W_3^2]\right)}{(U_2 V_{\theta_2} - U_3 V_{\theta_3}) + \left(\dfrac{V_3^2 - V_1^2}{2}\right)} \tag{2.79}$$

Unlike an axial turbine, the reaction is now a function of inner and outer radii and velocities at stations 1, 2, and 3. For a given blade geometry, the velocities can be evaluated as described earlier to determine the reaction.

For a specific case, when $\alpha_3 = 0$, $\beta_2 = 0$, $V_{\theta_1} = 0$, $V_{\theta_2} = U_2$, and $V_{\theta_3} = 0$ (radial turbine with axial exit) the reaction can be written as

$$R = \frac{U_2^2 - \dfrac{V_2^2}{2} + \dfrac{V_3^2}{2}}{U_2^2 + \dfrac{V_3^2 - V_1^2}{2}} = \frac{U_2^2 - W_2^2 + V_3^2}{2(U_2^2) + V_3^2 - V_1^2} \tag{2.80}$$

If $V_1 = W_2 = V_3$, which is a very special case, the reaction is 50%.

Equation 2.76 is the general expression for the work done. If the geometry of the runner, rpm, and blade setting are known and β_2, β_3, r_3, r_2, and V_{r_2} are all known, then P_{shaft} can be calculated. It should be mentioned here that the 2D analysis, considered in the next chapter on cascade analysis, will provide predictions for α_2 and β_2. Likewise, the prediction of the rotor flow, which has to be essentially based on 3D analysis, will provide an estimate for β_3. In many approximate calculations, β_3 is assumed to be equal to the blade outlet angle, or a correlation is employed to derive its value. Knowing (or assuming) η and T_{o1}, the pressure drop can be calculated from the equation in Table 2.1.

As in a centrifugal compressor, the advantages of radii change are clear from Eq. 2.76. The enthalpy and pressure drop occur through a decrease in kinetic energy as well as through centrifugal and Coriolis forces. The explanation given for a centrifugal compressor is valid for the radial turbine case also, and the reader is referred to that discussion. The advantages of radial turbines lie in extracting larger work in a single stage. The radial path and centrifugal forces play an important role in achieving this goal. Larger pressure and temperature drops can be achieved in a single stage compared to an axial turbine. It is this feature that is attractive in its application to small power turbines such as automotive gas turbines and turbochargers.

Many of the conclusions and observations made for an axial turbine, based on performance equations, are valid for a radial turbine also and will not be repeated.

The performance of an efficient high-temperature turbine designed and tested by Allison (Ewing et al., 1980) (1258 HP, $N = 55,000$ rpm, turbine inlet temperature $= 1530$ K, mass flow $= 5.2$ lb/s) is shown in Fig. 2.78. The performance testing was carried out at low temperatures. Comparing this with the performance of an axial turbine (Fig. 2.60), similarities are clearly evident. The interpretations are similar to those for an axial flow turbine. The enthalpy drop increases with an increase in mass flow (at constant N), and the efficiencies achieved are lower than those for an axial turbine stage. The following conclusions are generally valid:

1. The radial turbine is suitable for low mass flow, high pressure drop, and low power application.
2. The radial turbine is more robust and is more resistant to corrosion and erosion. It is more easily maintained than an axial turbine.
3. The cooling of a radial turbine passage is more difficultthan the cooling of an axial stage (Large and Meyer, 1982).

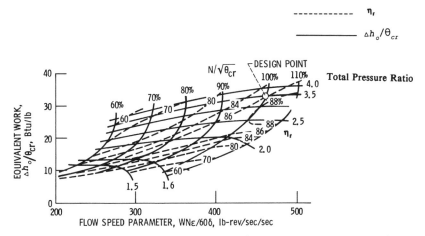

Figure 2.78 Radial gas turbine performance (Rohlik, 1983; NASA TM 83414).

Recent investigations by Okapuu (AGARD, 1987) indicate that the best feature of a radial and an axial turbine could be incorporated into a "hybrid"- or "mixed"-flow turbine, with a substantially higher efficiency than a corresponding axial stage and a smaller frontal area than a corresponding radial turbine. In aircraft (e.g., turbo prop) applications, both efficiency and frontal areas are important and selection is a compromise between the highest efficiency that can be achieved (or low specific fuel consumption) and lowest frontal area (or drag).

One of the most commonly employed configurations is the radial turbine with radial entry flow ($\beta_2 = 0$) and axial exit flow ($\alpha_3 = 0$). The relative velocity comes in radially, and the blade is therefore radial at the inlet. Equation 2.75 reduces to (because $\phi = \cot \alpha_2$)

$$\psi = 1 \tag{2.81}$$

The temperature and pressure ratios are given by Eqs. 2.77 and 2.78, respectively, with $\psi = 1$.

In a radial turbine, where the exhaust kinetic energy is not utilized for thrust power generation (e.g., turboshaft), the recovery of exit kinetic energy V_3^2 and the static properties at the exit (p_3, h_3) are important in evaluating the turbine performance. The values of h_3, p_3, and so on, for a turbine with radial entry ($\beta_2 = 0$) and axial exit ($\alpha_3 = 0$) can be derived as follows. Along a pitch line, we have

$$h_2 - h_3 = (h_{o2} - h_{o3}) - \frac{V_2^2}{2} + \frac{V_3^2}{2}$$
$$= U_2^2 - \frac{V_2^2}{2} + \frac{V_3^2}{2} \tag{2.82}$$

Hence

$$\frac{T_3}{T_2} = 1 - \frac{1}{c_p T_2}\left[U_2^2 - \frac{V_2^2}{2} + \frac{V_3^2}{2}\right]$$
$$= 1 - \frac{M_{b_2}^2(\gamma - 1)}{2}\left[2 - \left(\frac{V_2}{U_2}\right)^2 + \left(\frac{V_3}{U_2}\right)^2\right]$$
$$= 1 - \frac{M_{b_2}^2(\gamma - 1)}{2}\left[1 - \cot^2 \alpha_2 + \left(\frac{r_3}{r_2}\right)^2 \cot^2 \beta_3\right] \tag{2.83}$$

It is interesting to examine an ideal situation and derive a condition for optimum operation. The ideal situation is one where the exhaust kinetic energy is nearly zero (no heat or energy loss) and the turbine process is isentropic. In such a case, by examining Fig. 2.7 (see Section 2.9.4), it is clear that

$$h_{o1} - h_{o3ss} = h_{o1} - h_{3ss} = \frac{V_{is}^2}{2} \tag{2.84}$$

where $V_{is}^2/2$ is the ideal kinetic energy available for a turbine operating in the pressure range of P_{o1} and P_{o3}. Therefore, the performance of an actual and an ideal turbine would be optimum when ($V_{\theta 3} = 0$, $V_{\theta_2} = U_2$)

$$h_{o1} - h_{o3} = U_2^2 = \frac{V_{is}^2}{2}$$

or

$$U_2 = 0.707 V_{is} \tag{2.85}$$

Therefore, efficiency will be maximum at or near this operating condition.

Recent advances and projected developments in radial gas turbine technology are similar to those described in Section 2.9. Rohlik (1983) has provided a summary of the current development work as well as projected improvements. The major emphasis will be on new materials, higher turbine inlet temperature, higher pressure drop, better designs, and higher efficiency. The following are some major developments:

1. *Ceramic Turbine.* The small gas turbine is an ideal candidate for ceramic blades. The potential benefits are low cost, low weight, corrosion and erosion resistance and increased turbine temperatures. The projected and present temperature ranges are similar to those for axial turbines (Figs. 2.70 and 2.71).

2. *Incorporation of Cooling Devices to Increase the Turbine Inlet Temperature.* Some progress has already been made (Lane, 1981; Large and Meyer, 1982), and major advances in this area will continue in much the same way as an axial turbine.

3. *Improvement in Off-Design Performance.* This has been achieved through varying the inlet nozzle angle ($\pm 10°$ variation in α_2; see Large and Meyer, 1982) and translating the nozzle sidewall (Boyd et al., 1983).

4. *Improved Design Techniques to Achieve High Pressure Drop and Efficiency.* Based on the current aerodynamic advances and material availability, a radial turbine has been designed for an inlet temperature of 1260 K, a cycle pressure ratio of 14:1 and an η_t of 86% and incorporates complex cooling arrangements (Snyder and Roelke, 1990). Whitfield (1990) has provided a design method to minimize inlet and exit Mach numbers resulting in reduced losses.

The analysis of the flow field in a volute for radial turbines is similar to those outlined in Section 2.7.3. Bhinder (1970) and Rohlik (1983) have provided details of one-dimensional analysis. Considerable work has been carried out by the Cincinnati group (e.g., Eroglu and Tabakoff, 1989) to understand the flow field in these radial turbine components through measurement and numerical analysis.

2.10.2   Analysis and Performance of Hydraulic Turbines

The analysis of a Francis type of hydraulic turbine (Fig. 2.77c) is very similar to that for a radial gas turbine (Fig. 2.77a), but the analysis is simplified considerably due to constant density. The configuration and velocity triangles are shown in Fig. 2.77c. The loading coefficient, expressed in terms of head drop $[\psi = g(H_2 - H_3)/U_2^2$, Eq. 2.28] is given by Eq. 2.76.

If the blade setting, blade speed, and the exit flow angle from the wicket gate are known, then α_2, β_2, and β_3 are known. Therefore, the head-rise coefficient, ψ, can be calculated from Eq. 2.76. Knowing the radii at 2 and 3 and the corresponding through flow velocities, V_{m_2} and V_{m_3} are derived from continuity considerations.

If the relative flow at entry is radial ($\beta_2 = 0 = W_{\theta_2}$), and the exit direction is axial without any absolute swirl ($V_{\theta_3} = 0$), Eq. 2.75 simplifies to

$$\psi = \phi \tan \alpha_2 = 1$$

This is the ideal head rise, and the loading coefficient is constant for all mass flows for such a configuration ($\beta_2 = \alpha_3 = 0$).

The hydraulic turbine characteristics are represented through the similarity relationship given by Eqs. 2.29 and 2.30. Efficiency (see Section 2.3.2) is given by

$$\eta = \frac{\text{Actual power output in bhp}}{\rho g Q(H_2 - H_3)} \tag{2.86}$$

From Eqs. 2.28–2.30,

$$\eta = f(C_Q) = f(C_p) \tag{2.87}$$

The performance of a small Francis turbine is shown in Fig. 2.79 (power-specific speed, Section 2.4.3, of 29). The constant head characteristic is clearly evident, and the head coefficient C_H is related to the loading coefficient ψ through the equation

$$C_H = \frac{g(H_2 - H_3)}{N^2 D^2} = \pi^2 \frac{g(H_2 - H_3)}{U_2^2}$$

$$= 9.85\psi \quad \text{for } \beta_2 = \alpha_3 = 0 \tag{2.88}$$

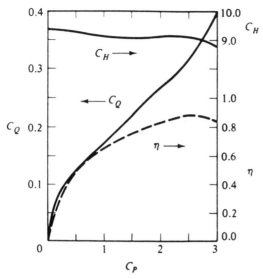

Figure 2.79 Performance curves for a Francis turbine (N = 600 rpm, D = 0.68 m, $\bar{N}_p$ = 29 = $\dfrac{N\sqrt{BHP}}{H^{5/4}}$ (RPM, BHP, Ft)). (From White, 1979; reprinted with permission of McGraw-Hill.)

The measured C_H is about 9.03 and is nearly constant for all values of C_p and C_Q. Efficiency and power increase with an increase in C_Q. This property is similar to that of radial and axial turbines presented earlier. For this particular turbine (Fig. 2.79), the maximum efficiency is 89% at C_p = 2.70, C_Q = 0.34, and C_H = 9.03.

The static head, h_2 and h_3, and velocities can be determined by using the inviscid equation (along a streamline)

$$h_2 + \frac{W_2^2}{2g} - \frac{U_2^2}{2g} = h_3 + \frac{W_3^2}{2g} - \frac{U_3^2}{2g} \tag{2.89}$$

where h is the static head in meters. Across the wicket gate we have

$$h_1 + \frac{V_1^2}{2} = h_2 + \frac{V_2^2}{2} \tag{2.90}$$

Quantities h_1 and V_1 are known from the entry conditions, and h_2 and V_2 can be determined from the wicket gate exit angle and the continuity equation. The blade setting enables determination of β_2, β_3, W_2, and W_3. The meridional velocities V_{m_2} and V_{m_3} are calculated from the continuity equation. Therefore, the static head or pressure at the exit can be calculated using Eqs. 2.89 and 2.90.

The analysis of a propeller turbine (Fig. 2.77d) is similar to that of axial and radial turbines. The property change across the wicket gate 1–2 can be determined using the Bernoulli equation and the continuity equation to provide an estimate for α_2, V_m and V_{z_3}. The blade setting will enable the determination of velocity triangles. The loading coefficient can then be determined from Eq. 2.76 (usually $r_2 = r_3$ for a propeller turbine) with $V_{r_2} = V_{m_2}$. The value of V_{m_2} is determined from continuity considerations. The propeller turbines are usually designed for zero exit swirl.

A comparison of Pelton wheel (with 180° turning), Francis, and Kaplan turbine performance is shown in Fig. 2.80. The efficiency characteristics of the Kaplan (adjustable blade) and Francis turbines are very similar, even though the range of operation is different as indicated in Figs. 2.8 and 2.9. The fixed-blade propeller turbine is more sensitive to change in mass flow and power. This is clear from Fig. 2.80. The efficiency of the fixed-blade propeller turbine at off-design conditions is poor, because the change in mass flow would affect the incidence and the angle of the relative flow into the rotor. This introduces considerable profile losses due to larger blade boundary layer and separation. The Francis turbine with $\beta_2 = 0$ and $\alpha_3 = 0$ is less

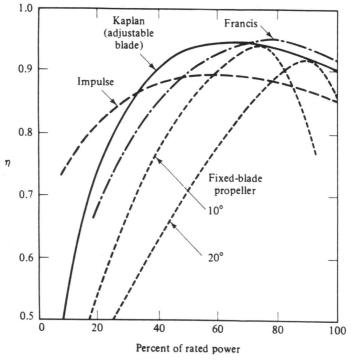

Figure 2.80   Efficiency versus power level for various turbine designs at constant speed and head. (From White, 1979; reprinted with permission of McGraw-Hill.)

sensitive to changes in operating conditions and is therefore more attractive. As described in Section 2.8.2, the cavitation seriously affects performance of all hydraulic machines, both pumps and turbines. Both surface and vortex cavitation are present in hydraulic turbines. As the NPSH is decreased, the inception of cavitation occurs followed by a complete breakdown.

2.10.3 Operating Range of Axial and Radial Turbines

It will be useful to look at the performance range of all axial and radial turbomachinery (both gas and liquid handling machines) to generalize the range of operation. A very useful plot of loading coefficient versus flow coefficient due to Horlock (1973a) is shown in Fig. 2.81. It is clear from this plot as well as Fig. 2.74 that most highly efficient axial flow turbines have loading coefficients in the range 1–4, whereas the radial turbines have optimum loading coefficients of around 1 (as indicated earlier, for $\beta_2 = \alpha_3 = 0$ and $\psi = 1$). The Kaplan turbine and the windmill, both of which have very little camber and hence very low turning, have the lowest loading. The Pelton

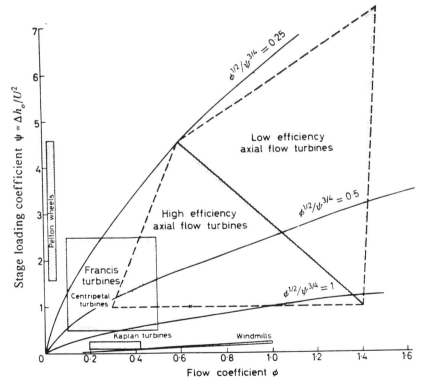

Figure 2.81 Loading coefficients, flow coefficients, and shape parameters for various types of turbines. (Copyright © Horlock, 1973a, Kreiger Publishing Co.).

wheel, however, has the highest loading coefficient of all hydraulic turbines, but it suffers from low efficiencies and is therefore very rarely used. The Pelton wheel was a major power plant in early days of hydraulic turbine development. Also shown in this figure is the isoline of constant $\phi^{1/2}/\psi^{3/4}$ which is specific speed similar to Eq. 2.31. As indicated in Fig. 2.8, the Kaplan turbine has the highest specific speed and the Pelton wheel has the lowest. Most gas turbines fall in the moderate range. In the selection of a suitable turbine, a complete system study should be carried out for the specific application before a final decision is made.

3

CASCADE INVISCID FLOWS

3.1 INTRODUCTION

The successful development of axial flow turbomachinery can be attributed to the use of of two-dimensional cascade data and analyses. If radial velocities are negligible and stream surfaces are cylindrical, the development of a cylindrical section, such as the one shown in Fig. 3.1a, would result in a rectilinear cascade of infinite number of blades. In situations where there are appreciable radial flows due to radius variations of hub or casing walls, the blading enclosed in a stream tube such as A'B'C'ABC could be developed into a two-dimensional cascade plane, as shown in Fig. 3.1b. The equations of motion as well as the geometry become simple. This transformation assumes that the streamlines lie on an axisymmetric surface. A meridional coordinate (direction of resultant axial and radial velocities, m) and a tangential coordinate (θ, y) are defined and the flow equations are developed in (m, θ) or (m, y) coordinate systems. The governing equation for the flow in the stream tube A'B'C'ABC can be shown to be a two-dimensional equation. Details of the assumption made and the governing equations are given in Chapter 4. There exists a large amount of analytical as well as experimental information on this configuration. Cascade data and analysis have been widely used for the design and analysis of (a) a wide range of turbomachinery for subsonic, transonic, and supersonic flows; and liquid handling machinery. The configuration is a simple one, making it possible to investigate such properties as lift, drag, pressure distribution, transition and separation, boundary layer growth, shock–viscid interaction, regions of vortex and cavity flows, and boundary layer control. The cascade model is an extremely useful tool in carrying out a parametric study of the variables, either through experimentation or through

189

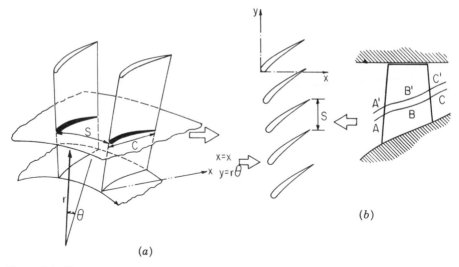

Figure 3.1 Development of cascade of airfoils. (a) Cylindrical stream surface. (b) Noncylindrical stream surface.

analysis, in arriving at an optimum blade geometry. The emergence of highly efficient axial flow compressors, pumps, and turbines used in marine, aircraft, and land vehicles is largely due to the use of cascade data and analysis.

The physical nature of the flow processes that occur in both compressor and turbine cascades is shown in Figs. 3.2 and 3.3, respectively. In a compressor or pump cascade, the flow (corresponding to the relative flow in a compressor rotor) decelerates and $p_2 > p_1$. Typical blade pressure distribu-

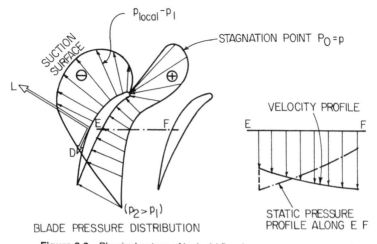

Figure 3.2 Physical nature of inviscid flow in a compressor cascade.

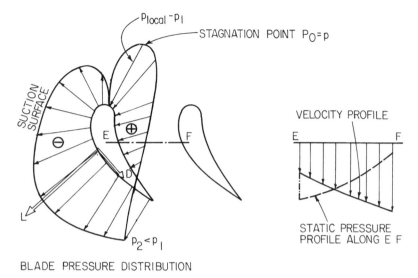

BLADE PRESSURE DISTRIBUTION

Figure 3.3 Physical nature of inviscid flow in a turbine cascade.

tion (positive towards surface) is shown here as well as in Figs. 2.14 and 2.12. In a turbine cascade, the flow (corresponding to the relative flow in a turbine rotor) accelerates and the passage mean static pressure decreases. Typical blade pressure distributions are shown in Figs 3.3 and 2.64–2.66. The lift force (L), shown in Figs. 3.2 and 3.3, is the resultant of all the aerodynamic forces acting on the blade normal to the mean velocity vector $\mathbf{V}_m (= \mathbf{V}_1 + \mathbf{V}_2)$, and the drag force is the resultant aerodynamic force parallel to the mean velocity vector $\mathbf{V}_m$. Static pressure varies continuously across the passage, along the streamline, and along the blade surface, as shown in Figs. 3.2 and 3.3, which also shows the velocity variations, across the passage, in inviscid flow. The velocities are uniform far upstream and far downstream of the cascade.

Incompressible, compressible, transonic, and supersonic cascade flows will be dealt with in this chapter. The reader is referred to the books by Scholz (1965) and Gostelow (1984) for additional reading.

3.2 AERODYNAMIC FORCES AND GOVERNING EQUATIONS

3.2.1 Velocity Triangles; Lift and Drag Forces

All the equations derived in this section are valid for far upstream and downstream (usually one chord from the nearest edge), where the flow is uniform. Typical velocity triangles for both compressible and incompressible

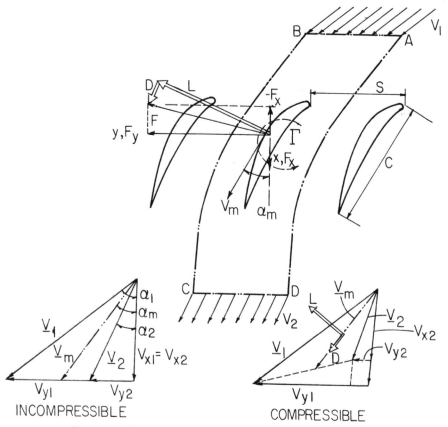

Figure 3.4 Velocity triangles and forces for compressor cascade.

flow and the forces are shown in Figs. 2.15 and 3.4 for a compressor cascade
and Figs. 2.56 and 3.5 for a turbine cascade. A Cartesian coordinate system is
used in this chapter, with x representing the axial direction and y the
tangential direction, V_x and V_y are the corresponding velocity components.

In incompressible flow, the axial velocity remains unchanged between inlet
and outlet. In compressible flow, the axial velocity decreases (for constant
inlet and exit areas) for a compressor cascade (because $\rho_2 > \rho_1$) and increases for a turbine cascade (because $\rho_2 < \rho_1$). The change in axial velocity
(V_x) is governed by the equation (for constant inlet and exit areas)

$$\frac{\Delta V_x}{V_{x_1}} = \left(\frac{V_{x_2}}{V_{x_1}} - 1\right) = \left(\frac{V_2 \cos \alpha_2}{V_1 \cos \alpha_1} - 1\right) = \left(\frac{\rho_1}{\rho_2} - 1\right) \qquad (3.1)$$

where α_1 and α_2 are inlet and outlet flow angles, respectively (Figs. 3.4 and

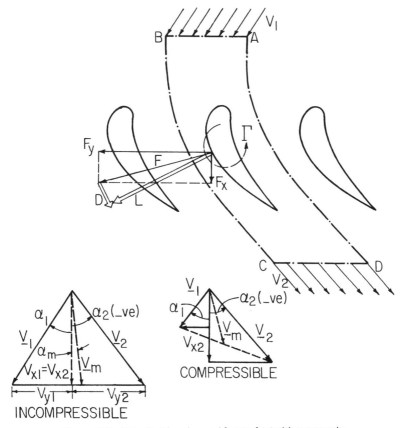

Figure 3.5 Velocity triangles and forces for turbine cascade.

3.5). The above equation utilizes the continuity equation

$$\rho_1 V_1 \cos \alpha_1 = \rho_2 V_2 \cos \alpha_2 \tag{3.2}$$

It is normal practice to design the rotor and stator blade row such that the axial velocity is constant along the flow path, and this is accomplished by varying the annulus area.

General expressions for the forces acting on a cascade of blades in compressible flow will be derived before proceeding to the theoretical analysis. A control surface ABCD is chosen (Figs. 3.4 and 3.5) midway between the passages such that the flow properties along AD and BC are identical. The control surfaces AB and CD are located far upstream and downstream, respectively, such that the flow properties are uniform along these two surfaces. The sign convention adopted in this chapter is that all the angles are measured with respect to the positive x axis, the clockwise direction being positive. The axes (x, y) are shown in Figs. 3.4 and 3.5. For example, α_2 is positive for the compressor cascade (Fig 3.4) and negative for the turbine cascade (Fig. 3.5).

Applying the integral form of the momentum equation, we get the forces exerted by the fluid on the cascade blade in the x and y directions. These forces can then be resolved into lift and drag components, respectively. The following vectorial mean quantities are utilized in the analysis:

$$\frac{\mathbf{V}_1 + \mathbf{V}_2}{2} = \mathbf{V}_m, \qquad \frac{V_{x_1} + V_{x_2}}{2} = V_{x_m}, \qquad \frac{V_{y1} + V_{y2}}{2} = V_{y_m} \qquad (3.3)$$

and

$$\tan \alpha_m = \frac{V_{y_1} + V_{y_2}}{V_{x_1} + V_{x_2}} = \frac{\tan \alpha_1 + \dfrac{\rho_1}{\rho_2} \tan \alpha_2}{1 + \dfrac{\rho_1}{\rho_2}} \qquad (3.4)$$

If ρ_m is the density satisfying the continuing equation for V_m, we obtain

$$\frac{1}{\rho_m} = \frac{1}{2}\left(\frac{1}{\rho_1} + \frac{1}{\rho_2} \right) \quad \text{or} \quad \rho_m = \frac{2\rho_1 \rho_2}{\rho_1 + \rho_2} \qquad (3.5)$$

Referring to Figs. 3.4 and 3.5 the force exerted by the fluid on the body (this is equal and opposite to the forces exerted by the blade on the fluid inside the control volume) in the x direction can be derived by using the momentum equation (Eq. 1.3) applied to the control volume ABCD. It can be proven that

$$-F_x = S(p_2 - p_1) + \rho_1 V_{x_1}^2 S\left(\frac{\rho_1}{\rho_2} - 1 \right) \qquad (3.6)$$

Similarly, the force in the y direction is given by

$$F_y = \rho_1 V_{x1} S(V_{y_1} - V_{y_2}) = \rho_1 V_{x_1}^2 S\left(\tan \alpha_1 - \frac{\rho_1}{\rho_2} \tan \alpha_2 \right) \qquad (3.7)$$

where α_1 and α_2 are inlet and outlet air angles, respectively.

It should be remarked here that in a compressor cascade, where $p_2 > p_1$, the force exerted by the fluid on the blade in the axial direction (F_x) is negative, as shown in Fig. 3.4. for a turbine cascade, where $p_2 < p_1$, it is positive. The direction of the force in the y direction depends on the change in momentum in the y direction. If $(V_{y_1} - V_{y_2})$ is positive, it is in the positive y direction, as shown in Figs. 3.4 and 3.5.

The lift and drag forces are forces acting parallel and perpendicular to the mean velocity (V_m), respectively. Expressions for these forces can be derived by resolving F_x and F_y in the direction of V_m as well as perpendicular to it:

$$
L = -F_x \sin \alpha_m + F_y \cos \alpha_m = -\left[S(p_1 - p_2) + \rho_1 V_{x_1}^2 S \left(1 - \frac{\rho_1}{\rho_2} \right) \right] \sin \alpha_m
$$
$$
+ \left[\rho_1 V_{x_1}^2 S \left(\tan \alpha_1 - \frac{\rho_1}{\rho_2} \tan \alpha_2 \right) \right] \cos \alpha_m
$$

$$(3.8)$$

For the compressor cascade shown in Fig. 3.4, the force F_x is in the negative x direction and α_m is positive. For the turbine cascade shown in Fig 3.5, the force F_x is positive in the x direction and α_m is negative. The force F_y is in the positive y direction for both cases.

Similarly, the drag force is given by

$$
D = F_y \sin \alpha_m + F_x \cos \alpha_m = \rho_1 V_{x_1}^2 S \left(\tan \alpha_1 - \frac{\rho_1}{\rho_2} \tan \alpha_2 \right) \sin \alpha_m
$$
$$
+ \left[S(p_1 - p_2) + \rho_1 V_{x_1}^2 S \left(1 - \frac{\rho_1}{\rho_2} \right) \right] \cos \alpha_m
$$

$$(3.9)$$

For supersonic isentropic flow, static pressures and densities can be related to the inlet and exit mach numbers. Under these conditions, the generalized equation (Eq. 1.3) should be employed to evaluate the lift and drag from the known inlet and outlet flow properties. For incompressible and inviscid flow, $\rho_1 = \rho_2$ and p_1 and p_2 are expressed in terms of inlet and exit velocities, using the Bernoulli equation. Hence, the lift is uniquely determined once the values of S, V_{x_1}, α_1 and α_2 are known.

For subsonic cascades, it is convenient to express the lift and drag in terms of inlet properties and flow angles. To carry this out, the static pressure difference ($p_1 - p_2$) can be expressed (Liepmann and Roshko, 1957) in terms of M_1 and M_2, and neglecting the fourth-order terms (e.g., M_1^4 and M_2^4), the static pressure difference can be expressed as (Borisenko, 1962, pp. 697–798)

$$
p_1 - p_2 = \rho_m \frac{(V_2^2 - V_1^2)}{2}
$$

$$(3.10)$$

The viscous effects can be allowed for (approximately) by writing

$$
p_1 - p_2 = \rho_m \left(\frac{V_2^2 - V_1^2}{2} \right) + \Delta P_o
$$

$$(3.11)$$

where ΔP_o is the stagnation pressure loss. This could include all the losses encountered in a blade row: profile loss, secondary flow loss, wake mixing loss, and shock boundary layer interaction loss. A description of these losses is given in Chapter 6.

An expression for the lift coefficient can be derived by substituting Eq. 3.11 into Eq. 3.8. The following terms appear in the equation, which can be simplified using cascade relationships Eqs. 3.2– 3.8:

$$\frac{V_2^2 - V_1^2}{2V_m^2}\tan \alpha_m = \frac{V_{x_2}^2 \sec^2\alpha_2 - V_{x_1}^2 \sec^2\alpha_1}{2V_m^2} \tan \alpha_m$$

$$= \frac{V_{x_1}^2}{V_m^2}\frac{\rho_1}{\rho_m}\left\{-\tan^2\alpha_m\left(\tan \alpha_1 - \frac{\rho_1}{\rho_2}\tan \alpha_2\right)\right.$$

$$\left. + \left(\frac{\rho_1}{\rho_2} - 1\right)\tan \alpha_m\right\} \quad (3.12)$$

and

$$\frac{\rho_1 V_{x_1}^2}{\rho_m V_m^2} = \frac{\rho_m}{\rho_1}\cos^2\alpha_m = \left(\frac{2}{1 + (\rho_1/\rho_2)}\right)\cos^2\alpha_m \quad (3.13)$$

The final expression for the lift coefficient is given by

$$C_L = \frac{L}{\frac{1}{2}\rho V_m^2 C} = 2\frac{S}{C}\cos \alpha_m \left[\frac{2}{1 + (\rho_1/\rho_2)}\tan \alpha_1 - \frac{2}{1 + (\rho_2/\rho_1)}\tan \alpha_2\right]$$

$$- \frac{S}{C}\frac{\sin \alpha_m \cos^2 \alpha_m}{\cos^2 \alpha_1}\left(\frac{\Delta P_o}{\frac{1}{2}\rho V_1^2}\right)\left(\frac{2}{(\rho_1/\rho_2) + 1}\right) \quad (3.14)$$

Similarly, an expression for the drag coefficient can be written as (using Eq. 3.9)

$$C_D = \frac{D}{\frac{1}{2}\rho_m V_m^2 C} = 2\frac{S}{C}\cos^3 \alpha_m \frac{\Delta P_o}{\rho_1 V_1^2 \cos^2 \alpha_1}\left(\frac{2}{1 + (\rho_1/\rho_2)}\right) \quad (3.15)$$

Substituting this equation into Eq. 3.14, we get

$$C_L = 2\frac{S}{C}\cos \alpha_m \left[\frac{2\tan \alpha_1}{1 + (\rho_1/\rho_2)} - \frac{2\tan \alpha_2}{1 + (\rho_2/\rho_1)}\right] - C_D \tan \alpha_m \quad (3.16)$$

The expression for C_L and C_D involve the density ratio (ρ_2/ρ_1). The equation of state, isentropic relationships, and the inviscid energy equation can be used to express ρ_2/ρ_1, in terms of P_{o1}, P_{o2}, V_1, V_2, and M_1 as follows (Scholz, 1965):

$$\frac{\rho_2}{\rho_1} = \left[1 + \left(\frac{M_1^2}{\frac{\gamma + 1}{\gamma - 1} - M_1^2}\right)\left(1 - \frac{V_2^2}{V_1^2}\right)\right]^{1/\gamma - 1} \left(\frac{P_{o2}}{P_{o1}}\right) \quad (3.17)$$

The lift coefficient for subsonic flow can be calculated if S/C, α_1, α_2, C_D, M_1, P_{o1}, and P_{o2} are known (because V_2/V_1 in Eq. 3.1 can be expressed in terms of ρ_2/ρ_1, α_1, and α_2). If the loss coefficient $[\zeta_p = (\Delta P_o/\frac{1}{2}\rho_1 V_1^2)]$ is known, the drag coefficient can be calculated from Eq. 3.15.

When the flow is incompressible, $V_{x_1} = V_{x_2}$ and $\rho_1 = \rho_2$. The expressions for the lift and drag coefficients are simplified considerably. Equations 3.15 and 3.16, respectively, reduce to

$$C_L = 2\frac{S}{C}\cos\alpha_m[\tan\alpha_1 - \tan\alpha_2] - C_D \tan\alpha_m \quad (3.18)$$

$$C_D = 2\frac{S}{C}\cos^3\alpha_m\left(\frac{\Delta P_o}{\rho_1 V_1^2 \cos^2\alpha_1}\right) = \frac{S}{C}\zeta_p\frac{\cos^3\alpha_m}{\cos^2\alpha_1} \quad (3.19)$$

For a turbine blade, α_2 and α_m are usually negative and hence the lift coefficient is much higher than a compressor blade. (Also see Section 2.9.1.)

3.2.2 Kutta–Joukowski Theorem

Assuming $\Delta P_o = 0$ in Eq. 3.14, we have

$$L = C_L \frac{1}{2}\rho_m V_m^2 C = \rho_m V_m\left[SV_{x_1}\frac{\rho_1}{\rho_m}\left(\frac{2\rho_2}{\rho_1 + \rho_2}\tan\alpha_1 - \frac{2\rho_1}{\rho_1 + \rho_2}\tan\alpha_2\right)\right] \quad (3.20)$$

Because

$$\Gamma = S(V_{y_1} - V_{y_2}) = SV_{x_1}\left(\tan\alpha_1 - \frac{\rho_1}{\rho_2}\tan\alpha_2\right)$$

Eq. 3.20 reduces to

$$L = \rho_m V_m \Gamma \quad (3.21)$$

Eq. 3.21 establishes the validity of the Kutta–Joukowski theorem, which can be stated as follows: The force acting per unit height of a cascade blade is equal to $\rho_m \mathbf{V}_m \times \boldsymbol{\Gamma}$, where $\mathbf{V}_m$ is the vectorial mean velocity and ρ_m is the average density as defined by Eq. 3.5.

3.2.3 Governing Equations

The coordinate system most commonly employed in cascade practice is shown in Fig. 3.6, where x and y are aligned in the cascade axial and tangential directions, respectively, and X and Y are aligned in the direction of chord length and in normal direction, respectively. The corresponding velocity components are shown in Fig. 3.6.

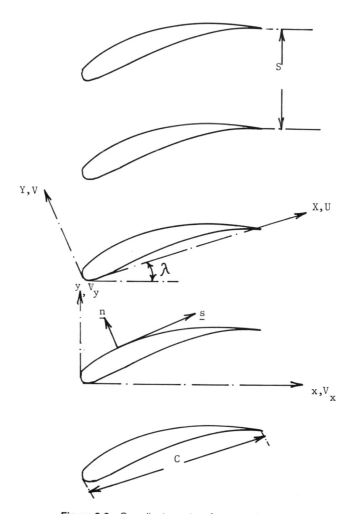

Figure 3.6 Coordinate system for cascade analysis.

The most general equations of motion for inviscid flow are given by Eq. 1.18 (with $G = S = w = O$, $Re \rightarrow \infty$ and $u = V_x$, $v = V_y$). The boundary conditions to be satisfied by these equations are as follows: (1) The flow is uniform far upstream and downstream (i.e., $V_x = V_{x_1}$, $V_y = V_{y_1}$ at far upstream), (2) $\mathbf{V} \cdot \mathbf{n} = 0$ on the solid surface, and (3) the Kutta–Joukowski condition at the trailing edge. The Kutta–Joukowski condition ensures that the flow leaves the sharp trailing edge smoothly and that the velocity there is finite. In an inviscid flow, this is possible only when there is a stagnation point at the trailing edge and the pressure difference across the trailing edge is zero.

These are the most general equations valid for unsteady compressible inviscid flow. The equations are nonlinear and the boundary conditions are complicated. In view of these complications, no analytical solution exists for these generalized equations. The time marching solution for transonic flow through a cascade is based on these equations.

The two-dimensional flow equation in terms of stream function are given by Eqs. 1.58–1.60 (with $u = V_x$, $v = V_y$). The boundary conditions to be satisfied in this equation are as follows:

1. $(\partial \Psi / \partial s)|_{surface} = \nabla \psi \cdot \mathbf{s} = \mathbf{V} \cdot \mathbf{n} = 0$, where s is the unit vector parallel to the surface, as shown in Fig. 3.6.

2. ψ_x and ψ_y are specified far upstream and the Kutta–Joukowski condition is enforced at the trailing edge.

For incompressible flow, the governing equation is given by Eq. 1.63 with the boundary conditions same as above.

If the flow is irrotational, the governing equations are given by Eqs. 1.52 and 1.57, with $\phi_z = 0$, $\phi_x = V_x$, and $\phi_y = V_y$.

The boundary conditions to be satisfied by these equations are as follows:

1. $\partial \phi / \partial n = \nabla \phi \cdot \mathbf{n} = 0$ (3.22)

    On the blade surface, $\mathbf{n}$ is the unit normal vector shown in Fig. 3.6.

2. $\phi_x = V_{x_1}$ and $\phi_y = V_{y_1}$ at far upstream, and the Kutta–Joukowski condition is specified at the trailing edge.

The right-hand side of the equation for the boundary condition (Eq. 3.22) is nonzero for cascade blades with normal velocity, such as those caused by coolant injection or boundary layer blowing.

For incompressible flow, $\phi_x \ll a$ and the governing equation is Eq. 1.55, with $\phi_z = 0$.

Methods of solving these equations for incompressible, compressible, subsonic, transonic, and supersonic flows are described in the subsequent sections.

Some remarks on the trailing edge condition for cascades follow. In a subsonic inviscid flow, the discharge flow angle, and hence the lift, can be determined for a sharp-pointed (trailing edge) airfoil in a cascade. The sharp point, or discontinuity, would lead to an infinite velocity and therefore must be the location of the singularity or stagnation point. For cascade airfoils having no discontinuity in surface profiles (the most common case is the rounded trailing edge) there is no preferred location for the downstream stagnation point in an inviscid flow. There are, therefore, an infinite number of possible solutions which exist. The only possible approach in inviscid theory for blading having a rounded trailing edge is to specify the outlet angle, or the lift, as well as the inlet angle.

As is discussed in Gostelow (1984), the flow around rounded trailing edges cannot be uniquely solved until the role of viscosity is introduced. Thwaites (1960) and Spence and Beasley (1960) provide techniques of incorporating the role of viscosity in the specification of trailing edge condition. In practice, the inviscid pressure distribution curves a lot as the trailing edge is approached, but the viscous solution does not. For a conventional cascade wake with no appreciable curvature, the static pressures at the trailing edge on suction and pressure surfaces should be equal. Therefore, the one stagnation point for the potential flow (the uniquely defined discharge angle) must be chosen so as to give no loading across the wake at the trailing edge. A curved wake can sustain a transverse gradient, and thus static pressures are not equal on either side of the trailing edge. This can be calculated using the approach suggested by Spence and Beasley (1960).

3.3 INCOMPRESSIBLE, INVISCID IRROTATIONAL CASCADE FLOW THEORIES

When the flow is incompressible and irrotational, the cascade flow is governed by the following equations:

$$\psi_{xx} + \psi_{yy} = 0 \qquad (3.23)$$

$$\phi_{xx} + \phi_{yy} = 0 \qquad (3.24)$$

The boundary conditions to be satisfied by these equations are that the flow velocity and pressure far upstream are equal to the values in the undisturbed free stream (hence, ψ_x, ψ_y or ϕ_x, ϕ_y are specified upstream) and that the blade surfaces are streamlines.

The flow represented by these equations (Eqs. 3.23 and 3.24) and their solutions belong to a class termed "potential flow," which has historically been the most widely explored and most developed field in the subject area of fluid mechanics. This has led to such useful and practical solutions as flow around an aircraft wing, cascades of blades, and flow around other streamlined bodies. Robertson (1965), Milne-Thompson (1960), and Karamachetti (1966) provide a very good coverage of this field as it relates to a single body.

The methods available for solving the potential flow through a cascade or designing a cascade for a prescribed pressure distribution can be briefly classified as follows:

1. *Conformal Transformation Method.* In this method, the flow around a cascade of blades is transformed into the flow around a cylinder. Because the flow in the latter plane is known exactly, the reverse transformation provides the flow in the cascade plane. The procedure is reversed for a design problem.

2. *Method of Singularities.* This is an approximate method (exact in the limit where the blade is replaced by large number of surface singularities) where the blade is replaced by a set of singularities such as sources, sinks, and vortices. In the method of surface singularities or panel method, the blade surface is replaced by panels of source or vortex sheets of infinitesimal length.

3. *Numerical Method.* In this method, equations are solved numerically using a relaxation solution, finite difference technique, or finite element method.

4. *Graphical Method.* In the method developed by Wislicenus (1965), the deviation between the camberline and the mean streamline is derived empirically using cascade experimental data. The mean streamline and thickness effects are determined from an assumed pressure distribution.

A good description of the first two methods as applicable to a flow around a single body (airfoil, cylinder, etc.) is given by Robertson (1965).

The methods described above could be utilized either for the design of a blade profile with a given pressure distribution or for the analysis of the flow field for a given blade profile and entry conditions. The method of singularities (including the panel method) is one of the widely used techniques because of its accuracy as well as the ease with which it can be adapted to the present-day computer. No attempt will be made in this section to describe all the techniques. Only those techniques that are either very basic or widely used will be described here. Numerical methods are described in Chapter 5. The graphical method, which is mostly based on experimental data for the conventional profiles, is mainly due to Wislicenus (1965). This method will not be described here. For other techniques, the reader is referred to Scholz (1965), Gostelow (1984), Borisenko (1962), Robertson (1965), and Johnsen and Bullock (1965).

3.3.1 Conformal Transformation Methods

3.3.1.1 Complex Potential Because the stream function (ψ) in Eq. 1.39 and the potential function (ϕ) in Eq. 1.50 satisfy the Cauchy–Riemann condition, $\psi_x = -\phi_y$ and $\psi_y = \phi_x$, they can be grouped together to define a complex potential (which is analytic) as follows:

$$F = \phi + i\psi \quad \text{and} \quad W = \frac{dF}{dz} = V_x - iV_y \tag{3.25}$$

where W is the "complex velocity" and F is the "complex potential."

This enables representation of irrotational flow by complex potential and complex velocity. Some common examples of the plane flow are:

1. Uniform flow parallel ($V = V_x$) to the x axis:

$$\phi = Vx, \quad \psi = Vy, \quad F = Vz \tag{3.26}$$

2. Uniform flow with components V_x in the x direction and V_y in the y direction:

$$\phi = V_x x + V_y y, \quad \psi = V_x y - V_y x, \quad F = (V_x - iV_y)z \tag{3.27}$$

3. Source (Q) and sink ($-Q$) located at $z = z_o = x_o + iy_o$:

$$\phi = \pm \frac{Q}{2\pi} \ln r, \quad \psi = \pm \frac{Q}{2\pi} \theta, \quad F = \pm \frac{Q}{2\pi} \ln(z - z_o) \tag{3.28}$$

where

$$r = \sqrt{(x - x_o)^2 + (y - y_o)^2} \quad \text{and} \quad \theta = \tan^{-1}\left(\frac{y - y_o}{x - x_o}\right)$$

4. Vortex (Γ) located at $z = z_o$ (positive Γ for anticlockwise circulation):

$$\phi = \frac{\Gamma}{2\pi} \theta, \quad \psi = -\frac{\Gamma}{2\pi} \ln r, \quad F = -\frac{i\Gamma}{2\pi} \ln(z - z_o) \tag{3.29}$$

5. Complex singularity (combination of flows in items 3 and 4 above):

$$A = Q - i\Gamma, \quad F = \frac{Q - i\Gamma}{2\pi} \ln(z - z_o) \tag{3.30}$$

6. Doublet (source and sink pair) of strength μ located $z = z_o$ and oriented at an angle α to the x axis:

$$\phi = -\frac{\mu \cos(\alpha - \theta)}{2\pi r}, \qquad \psi = -\frac{\mu \sin(\alpha - \theta)}{2\pi r}, \qquad F = -\frac{\mu}{2\pi}\frac{e^{i\alpha}}{z - z_o}$$

$$(3.31)$$

7. Flow around a cylinder of radius a, with clockwise circulation Γ, in a uniform flow of velocity V:

$$F(z) = \underset{\substack{\text{Uniform}\\\text{flow}}}{Vz} + \underset{\substack{\text{Doublet}\\\text{simulating}\\\text{thickness}}}{V\frac{a^2}{z}} + \underset{\substack{\text{Circulatory}\\\text{flow}}}{\frac{i\Gamma}{2\pi}\ln\frac{z}{a}} \qquad (3.32)$$

and the complex velocity is given by

$$W(z) = V\left(1 - \frac{a^2}{z^2}\right) + i\frac{\Gamma}{2\pi z} \qquad (3.33)$$

Because Laplace's equation (Eq. 3.23 or 3.24) governing the inviscid, incompressible, irrotational flow is linear, the solution of complex flows can be obtained as a superposition of simple flows. It can be proven (Karamachetti, 1966) that the solution of Laplace's equation $\nabla^2 A = 0$, can be expressed as

$$A = f_1(\bar{z}) + f_2(z)$$

where f_1 and f_2 are arbitrary complex functions and this is the general solution. The real and imaginary parts of $f_1(z)$ and $f_1(\bar{z})$ must separately satisfy the equation. Hence, the theory of the complex variable is an extremely useful and powerful mathematical tool in the solution of potential flows.

3.3.1.2 Conformal Transformation

The physical interpretation of the function $\zeta = \zeta(z)$ is that for each value of z in the z complex plane, there is a corresponding value of ζ in the complex plane of the variable ζ. Hence, $\zeta(z)$ is considered as a "transformation" or "mapping" function. A transformation is said to be "conformal" when the angle between any two curves passing through any point z(at which $d\zeta/dz \neq 0$) is preserved and its sense also remains unchanged. It can be proven (Robertson, 1965; Milne-Thompson, 1960) that the mapping through "analytic functions" (defined as those functions that are differentiable) is always conformal. The points in the

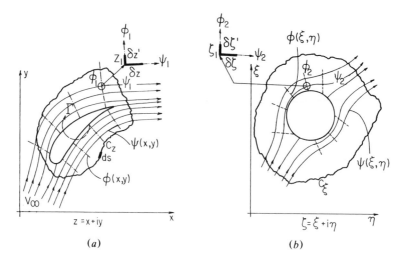

Figure 3.7 Illustration of conformal mapping technique.

complex domain where the function $\zeta(z)$ is analytic are called "regular" points of the function; and where the function is not analytic, the function is said to be "singular."

The concept of conformal transformation is extremely useful in the study of irrotational and incompressible flow around bodies such as airfoils, cascades, cylinders, and cavity flows. Let us take, for example, the flow around a single airfoil. The physical plane (z) shown in Fig. 3.7 has the airfoil as well as the corresponding network of potential lines and streamlines. Consider the curves ψ_1 and ϕ_1 intersecting at right angles. Let the corresponding curves in the ζ plane be ψ_2 and ϕ_2, respectively, as shown in Fig. 3.7. If δz and $\delta z'$ are infinitesimal lengths along curves ψ_1 and ϕ_1, respectively, at z_1 in the z plane, then the corresponding lengths at ζ_1 in the ζ plane are $\delta \zeta$ and $\delta \zeta'$ along curves ψ_2 and ϕ_2, respectively, as shown in Fig. 3.7. The elements $\delta \zeta$ and δz are related by

$$\delta \zeta = \frac{d\zeta(z)}{dz} \delta z, \qquad \arg \delta \zeta = \arg\left(\frac{d\zeta}{dz}\right)_{z_1} + \arg \delta z \qquad (3.34)$$

Here $(d\zeta/dz)_{z_1}$ is the local scaling factor; and because the function $\zeta(z)$ is analytic, it is independent of the direction of δz. Therefore, all infinitesimal elements passing through a point are scaled by the same factor. Similarly, it can be proven that the angle between the two lines ϕ_1 and ψ_1 is retained in the ζ plane (Milne-Thompson, 1960). Hence, the transformation is said to be "conformal."

Let us now consider the transformation of an airfoil profile, shown in Fig. 3.7, into a standard shape about which the flow is known. Let the mapping function $\zeta(z)$, yet unknown, transform the airfoil into a circle. Consider the arbitrary flow in the physical plane given by

$$F(z) = \phi(x, y) + i\psi(x, y) \tag{3.35}$$

The streamlines and potential lines constitute an orthogonal network of curves. Let this flow be transformed into ζ plane, where the complex potential is given by

$$F[z(\zeta)] = \phi[x(\xi, \eta), y(\xi, \eta)] + i\psi[x(\xi, \eta), y(\xi, \eta)] \tag{3.36}$$

Hence, the streamlines and the equipotential lines in the ζ plane will also be orthogonal and the new function (Eq. 3.36) will represent the complex potential of the flow which is obtained by the transformation of the flow in the physical plane.

The mapping function $\zeta(z)$ is said to perform conformal mapping of the flow in the z plane to a corresponding flow in the ζ plane. Denoting the complex velocities in the z plane and ζ plane as W_z and W_ζ, respectively, we have

$$V_x - iV_y = W_z = \frac{dF}{dz}, \qquad V_x' - iV_y' = W_\zeta = \frac{dF}{d\zeta},$$

$$\frac{W_\zeta}{W_z} = \frac{dz}{d\zeta} = 1/(d\zeta/dz) \tag{3.37}$$

The last expression scales the velocities. It can be proven (Karamachetti, 1966) that the vorticity and circulation remain invariant during the transformation.

The remainder of this section will be devoted to the analysis of the cascade flow by the conformal transformation method.

3.3.1.3 Conformal Transformation Method for Cascade Flows The procedure used for the analysis of the cascade flow by the conformal transformation method usually consists of the following steps:

1. The infinite number of profiles of a cascade in the physical plane are transformed into a single profile in the imaginary plane. This step is usually simple, but often results (depending on the profile and solidity) in an awkward shape.
2. A single profile is transformed into a simple (usually circular) profile. This is achieved through several steps and is the most complicated part.
3. Singularities (vortices, sources, and sinks) are incorporated in the final mapped plane so as to simulate the flow in the physical plane. The

equations for the complex potential, and hence the complex velocity, can now be written.

4. Knowing the flow in the final transformed plane and the overall transformation function, the flow in the physical plane can now be determined. The velocity, blade pressures, deviation angles, and other aerodynamic parameters can now be calculated.

The physical plane of a cascade is mathematically a multiply connected region. A multiply connected region is a "connected region in which there are irreconcilable paths, or, equivalently, irreducible circuits" (Karamachetti, 1966, p. 256). The transformation of such a profile can be carried out using a multivalued function. Any analytic function $\zeta = \zeta(z)$ is multivalued, if for each value of z in the domain of the definition, there corresponds more than one value of $\zeta = \zeta(z)$. Common examples of such functions are $\zeta = \sqrt{z}$, $\ln z$, e^z, and so on.

The concept of Riemann surface is useful in dealing with the mutlivalued function. Consider the transformation $z = \ln \zeta$ or $\zeta = e^z$:

$$\zeta = re^{i\theta} = r(\cos \theta + i \sin \theta) \quad \text{and} \quad z = \ln r + i\theta \tag{3.38}$$

Thus, the circle in the ζ plane (Fig. 3.8) maps into one of the strips:

$$2n\pi \le y \le 2(n + 1)\pi, \qquad n = 0, \pm 1, \pm 2 \ldots$$

in the z plane. Each point in the ζ plane (e.g., ζ_1 in Fig. 3.8) corresponds to an infinite number of points in the z plane (shown as $\ldots z_1^{-2}, z_1^{-1}, z_1^0, z_1^1, z_1^2$, etc.). Inversely, the infinite number of strips in the z plane can be transformed into a single profile in the ζ plane using this transformation. The contour (circle of radius r) in the ζ plane can be imagined to be made of an infinite number of sheets cut along the positive ξ axis, as shown in Fig. 3.8. For example, contour S_0 (ABCDE) consists of the strip starting from the positive ξ axis and makes one revolution ending at the cut DE. In Eq. 3.38, θ of this strip ranges 0 to 2π and is transformed into A'B'C'D'E' (hatched part in the z plane). The second sheet S_1 ranges from $\theta = 2\pi$ to 4π. These sheets are said to form a Riemann surface. Thus, an infinite cascade of flat plates of spacing 2π is transformed into infinite Riemann surfaces (single sheets or profiles, in reality) in the ζ plane.

The conformal transformation methods are exact for potential flows, but the transformation of an arbitrary geometry (e.g., large camber, thick blades) into a circle is very cumbersome. In view of this and the availability of efficient panel methods, the conformal transformation methods are rarely used at the present time. The flat plate cascade analysis is widely used to validate approximate analysis and numerical techniques, because it is the only exact solution available. The generalized conformal transformation methods and their limitations are listed in Table 3.1. The method's usefulness

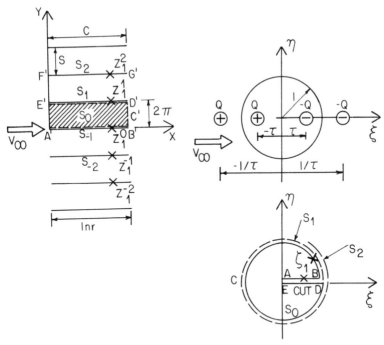

Figure 3.8 Conformal mapping of a flat-plate cascade.

lies in validating the computational techniques and codes for accuracy. Gostelow (1964) developed a new family of airfoils using these techniques. His scheme, valid for large camber (as high as 110°), is useful in validating panel and computational fluid dynamic codes for such geometry. The method of analyzing the flow through flat and cambered plate cascades are discussed below.

3.3.1.4 Flat Plate Cascade   Weinig (1935, 1964) carried out an analysis of the flow through a cascade of flat plates at an incidence. The basic principle on which all cascade transformations and subsequent analysis are based is the same. Hence, only Weinig's analysis will be described in detail here. More-generalized procedures are summarized in Scholz (1965), Johnsen and Bullock (1965), and Horlock (1973, 1973a).

Let us consider the basic principles on which the flat plate cascade transformation (or mapping function) is based before presenting Weinig's analysis. Consider the flat-plate cascade (z plane) and the cylinder (ζ plane), shown in Fig. 3.8. The uniform flow past a cylinder of radius unity in the ζ plane can be represented by placing a source (Q) and the sink ($-Q$) upstream and downstream, respectively, at a distance $1/\tau$ on either side of the origin. In order to make the circumference of the cylinder a streamline,

Table 3.1 Limitation of the conformal transformation methods

Method	Limit on C/S	Limit on f/C	Limit on t_{max}/C	Remarks
Weinig (1935, 1964)	None	0	0	Flat plate cascade
Weinig (1957)	None	< 0.08	0	Circular arc cascade
Howell (1948)	< 1.5	< 0.15	< 0.15	Compressor cascade
Garrick (1944)	< 1.5	< 0.2	< 0.2	
Traupel (1945, 1948)	> 0.5	None	> 0.15	Turbine cascade
Shirakura (1972)	Not available	None	None	
Kraft (1958)	Not available	None	None	Turbine cascade
Cantrell and Fowler (1959)	Not available	None	None	Design (impulse)

the source (Q) and sink ($-Q$) are placed at inverse points inside the circle at distances ($-\tau$) and (τ), as shown in Fig. 3.8. τ is a transformation parameter to be discussed later. On the basis of Eq. 3.28, the complex potential for the singularity combination shown in the ζ plane is given (by linear superposition of complex potential) by

$$F(\zeta) = \frac{V_\infty S}{2\pi}\left[\ln\frac{\zeta + 1/\tau}{\zeta - 1/\tau} + \ln\frac{\zeta + \tau}{\zeta - \tau}\right] \qquad (3.39)$$

where $V_\infty S$ represents the source (Q) strength and S is a length, yet unknown. For uniform flow in the z plane, the complex potential is $V_\infty z$. Hence, the mapping function which transforms the flat plate cascade at zero incidence into the exterior of a circle in the ζ plane is $[F(\zeta)/V_\infty]$ and is given by Eq. 3.39. The points $\zeta = \pm 1/\tau$ are located at $\pm\infty$ in the z plane. Because the transformation is conformal, the numerical values of the potential and stream functions are conserved and so are the characteristics of the singularities. On passing around any external singularity (e.g., source at $\zeta = -1/\tau$), we obtain the intensity $Q = V_\infty S$. This corresponds to the flow through a section of height S (blade spacing) on the x axis, which thus represents the volume flow in the passage bounded by planes $A'B'$ and $D'E'$. Passing around the second time raises the intensity by Q, resulting in the second passage bounded by planes $D'E'$ and $F'G'$. It is convenient to imagine the process with an infinite sheeted surface, shown in Fig. 3.8. Every time ζ crosses the cut DE, it passes from one sheet to the next succeeding sheet,

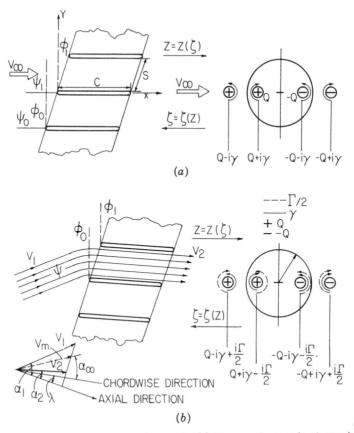

Figure 3.9 Flow through a staggered cascade. (a) Staggered cascade at zero incidence. (b) Staggered cascade at incidence.

depending on whether ζ is increasing or decreasing. These sheets range from 0 to 2π, 2π to 4π, and so on. This step is basic to all cascade transformations and involves the transformation of an infinite number of profiles (equally spaced passages with identical flows) into a Riemann surface in the ζ plane. This concept applies not only to the sources but also to the vortices located exterior to the circle.

If the cascade were staggered, there would be a jump not only in ψ but also in ϕ values from the leading edge of one blade to the leading edge of the subsequent blade. Denoting the reference blade by zero and the adjoining (positive y direction) blade by 1, the differences in stream function and potential function between the two blades are given by (Fig. 3.9a)

$$\psi_1 - \psi_0 = SV_\infty \cos \lambda = Q$$
$$\phi_1 - \phi_0 = SV_\infty \sin \lambda = \gamma \tag{3.40}$$

This flow can be generated by locating a vortex of strength γ in addition to a source or sink of strength Q in the exterior and interior of the circle (Eq. 3.40), shown in Fig. 3.9a. Hence, the resulting flow distribution in the ζ plane (circle plane) will have a complex singularity of $Q - i\gamma (= SV_\infty e^{-i\lambda})$, $Q + i\gamma$, $-(Q + i\gamma)$, $-(Q - i\gamma)$ located at $-1/\tau$, $-\tau$, τ, $1/\tau$, respectively, from the center of the circle, as shown in Fig. 3.9a. This will make the circle a streamline. The transformation of an infinite cascade of flat plates at a stagger angle of λ and at zero incidence to a cylinder, using the concepts outlined before for the zero stagger case, is therefore given by

$$z = \frac{S}{2\pi}\left[e^{-i\lambda}\ln\frac{\zeta + 1/\tau}{\zeta - 1/\tau} + e^{i\lambda}\ln\frac{\zeta + \tau}{\zeta - \tau}\right] \qquad (3.41)$$

The transformation parameter τ can be determined by equating the known cascade chord length in the z plane to the corresponding value in the ζ plane. The leading and trailing edges of a flat plate cascade are singular and $dz/d\zeta = 0$ at these locations. With $\zeta = e^{i\theta}$, and evaluating $dz/d\zeta$ from Eq. 3.41 and equating it to zero, the following expressions are obtained for the coordinates of the leading (l) and trailing (t) edges:

$$\tan\theta_l = \frac{1 - \tau^2}{1 + \tau^2}\tan\lambda, \qquad \theta_l = \theta_t + \pi, \qquad z_t - z_l = x_t - x_l = C \quad (3.42)$$

Then Eq. 3.42 provides the following relationship between solidity (C/S), stagger angle, and the transformation parameter τ:

$$\frac{C}{S} = \frac{2}{\pi}\left\{\cos\lambda\ln\frac{\sqrt{1 + 2\tau^2\cos 2\lambda + \tau^4} + 2\tau\cos\lambda}{1 - \tau^2}\right.$$

$$\left. + \sin\lambda\tan^{-1}\frac{2\tau\sin\lambda}{\sqrt{1 + 2\tau^2\cos 2\lambda + \tau^4}}\right\} \qquad (3.43)$$

Hither to, we have dealt with the case of a cascade of flat plates at zero incidence. The effect of incidence is to generate a net circulation of Γ around the cascade blade, its value being determined by the Kutta–Joukowski condition described earlier. The additional effects due to incidence (or generation of circulation around a cascade blade) can be simulated by locating a complex source (vortices) of strength $i\Gamma/2$ at $x = 1/\tau$ and $-1/\tau$ and $-i\Gamma/2$ at $x = \tau$ and $-\tau$. The first pair of vortices $(i\Gamma/2)$ generate a circulation Γ in the exterior of the circle, and the second pair $(-i\Gamma/2)$ makes the outer contour (circle) a streamline. This singularity combination satisfies the governing equation as well as the boundary condition described in Section 3.2.3. The complex potential due to the combined source, sink, and vortex singularities in the ζ plane, shown in Fig. 3.9b, is given by

$$F(\zeta) = \frac{V_m S}{2\pi} \left\{ e^{-i(\lambda + \alpha_\infty)} \ln\left[\frac{\zeta + 1/\tau}{\zeta - 1/\tau}\right] + e^{i(\lambda + \alpha_\infty)} \ln\left[\frac{\zeta + \tau}{\zeta - \tau}\right] \right\}$$

$$+ \frac{i\Gamma}{4\pi} \ln \frac{\zeta^2 - 1/\tau^2}{\zeta^2 - \tau^2} \tag{3.44}$$

The modifications carried out above to include the incidence effect do not change the equation for the cascade transformation given by Eq. 3.41. In Eq. 3.44, α_∞ is the mean incidence and is unknown for a cascade. The problem is thus reduced to the solution of the flow around a cylinder, as shown in fig. 3.9b. The solution is expressed as a complex potential (Eq. 3.44). Let us now evaluate properties of the cascade, including circulation, lift, lift coefficient, local velocity, blade pressure distribution, and flow outlet angle.

At the blade trailing edge,

$$\left(\frac{dF}{dz}\right)_{z=z_t} = \left(\frac{dF}{d\zeta}\right)_{\zeta=\zeta_t} \left(\frac{d\zeta}{dz}\right)_{\zeta=\zeta_t} \tag{3.45}$$

In the physical plane, the Kutta–Joukowski condition is satisfied if: $(dF/dz)_{z=z_t} = 0$. However, $(d\zeta/dz)_{\zeta=\zeta_t} = \infty$; therefore, $(dF/d\zeta)_{\zeta=\zeta_t} = 0$. Hence, taking the derivative of Eq. 3.44 and evaluating its value at $\zeta = \zeta_t$ (Eq. 3.42), the value of circulation Γ can be proven to be

$$\Gamma = \frac{4V_m S\tau \sin \alpha_\infty}{\sqrt{1 + 2\tau^2 \cos 2\lambda + \tau^4}} \tag{3.46}$$

From Eq. 3.21, the lift is therefore given by

$$L = \frac{4\rho V_m^2 S\tau \sin \alpha_\infty}{\sqrt{1 + 2\tau^2 \cos 2\lambda + \tau^4}} \tag{3.47}$$

The lift coefficient is therefore

$$C_L = \frac{L}{\frac{1}{2}\rho V_m^2 C} = 8\frac{S}{C}\frac{\tau \sin \alpha_\infty}{\sqrt{1 + 2\tau^2 \cos 2\lambda + \tau^4}} \tag{3.48}$$

If we know C_L, the outlet angle (α_2) can be calculated using Eq. 3.18. One of the most useful parameters is the ratio of the lift in a cascade blade and the corresponding isolated airfoil both at the same incidence (α_∞). The ratio, referred to here as *Weinig's coefficient* (K), is given by

$$C_{L\text{ isolated}} = 2\pi \sin \alpha_\infty$$

$$K = \frac{C_{L_{\text{cascade}}}}{C_{L_{\text{isolated}}}} = \frac{4}{\pi} \frac{S}{C} \frac{\tau}{\sqrt{1 + 2\tau^2 \cos 2\lambda + \tau^4}} \tag{3.49}$$

For low space/chord ratio, Eq. 3.49 can be simplified to

$$K = \frac{2}{\pi} \frac{S}{C} \frac{1}{\cos \lambda} \tag{3.50}$$

This approximation is good up to $S/C = 0.7$.

The above expression can also be derived from Eq. 3.18 by assuming $C_D = 0$. For a closely packed blade row (small S/C), $\alpha_2 = \lambda$ and $\alpha_m = \lambda + \alpha_\infty$. Hence, from Eq. 3.18 we obtain (for small flow turning),

$$C_L = 2\frac{S}{C}[\tan \alpha_1 - \tan \alpha_2] \cos \alpha_m$$

$$= 2\frac{S}{C}[2(\tan \alpha_m - \tan \alpha_2)] \cos \alpha_m = 4\frac{S}{C}\frac{\sin \alpha_\infty}{\cos \lambda}$$

This provides the same expression as Eq. 3.50 for K.

Wislienus (1965) has provided a comprehensive plot of Weinig's coefficient (K) derived from Eqs. 3.49 and 3.43, which is reproduced in Fig. 3.10. If we know the incidence, S/C, and λ, the values of K and C_L for the cascade blade $(C_L = 2K\pi \sin \alpha_\infty)$ can be determined. The slope of the lift–incidence

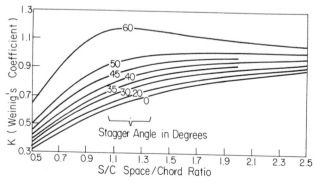

Figure 3.10 Weinig's coefficient (K) for flat-plate cascade. (From Wislicenus 1965; reprinted with permission of Dover Publications, Inc.)

curve is given by

$$\frac{dC_L}{d\alpha_\infty} = 2\pi K \cos \alpha_\infty \tag{3.51}$$

The local velocity can be determined from the following equation:

$$\frac{dF}{dz} = V_x - iV_y = \frac{dF}{d\zeta}\frac{d\zeta}{dz} \tag{3.52}$$

The value of $dF/d\zeta$ can be derived from Eq. 3.44, and the derivative of the mapping function $(d\zeta/dz)$ can be obtained from Eq. 3.41. If the interest lies only in the velocity on the blade surface, then only the real part of Eq. 3.52 need be considered.

The local pressure coefficient can be determined using the Bernoulli equation,

$$C_p = \frac{p_l - p_1}{\frac{1}{2}\rho V_1^2} = 1 - \left(\frac{V_l}{V_1}\right)^2 \tag{3.53}$$

where V_l is the local velocity.

Some of the conclusions that can be drawn from Weinig's analysis for a flat plate cascade are as follows:

1. The lift coefficient of a cascade blade is less than that of an isolated blade for the same chord length and incidence at moderate stagger angles. This is caused by the interference of the adjoining blades on the flow around a blade.

2. For a flat plate cascade at moderate values of S/C, the lift coefficient is linearly proportional to (S/C) and inversely proportional to the cosine of the stagger angle.

Fan designers would find Weinig's analysis very useful from the point of view of preliminary design. If loading or pressure rise is prescribed, Eqs. 3.18 and 2.43 and Fig. 3.10 can be used to select the preliminary geometry of blading.

It should be emphasized here that there are many practical limitations in the use of Weinig's analysis. The analysis is valid for a very thin airfoil without camber. If the camber and thickness effects are large, as in a compressor or turbine, Weinig's analysis provides only a qualitative (and preliminary) estimate of the cascade properties. Nevertheless, Weinig's analysis is the only closed analytical solution available for the cascade flow and is extremely valuable in understanding, theoretically, the effects of various cascade parameters such as S/C, λ, and α_m on the overall performance.

3.3.1.5 *Slightly Cambered Circular Arc Profiles* In the case of an isolated cambered airfoil (very thin), the effect of camber is to displace the center of the cylinder in the transformed plane from the center of the coordinate system. There are major difficulties, mathematical as well as physical, in extending such a technique to a cascade of cambered plates because of the stagger angle as well as the adjoining blades. Instead of this technique, Weinig (1957, 1964) suggests a simpler and approximate technique for the solution of flow through a cascade of cambered and thin airfoils. The approximation introduced by Weinig is equivalent to the simplified singularity method, where the singularities are distributed along the chord length; the tangency condition valid for a cambered blade (or streamline of camber shape) is also satisfied on the chord length. The technique suggested by Weinig uses the same transformation as a flat plate cascade (Eq. 3.41), but it modifies the complex potential of the flow in the ζ plane by introducing additional complex sources to yield the streamline of camber shape at the blade location. A good interpretation of this theory can be found in Scholz (1965).

The camber effect is generated by placing additional complex sources within the image circle (Fig. 3.9) at $-\tau$ and $+\tau$ of strength equal to $\pm i S V_m e^{i\lambda}/R$, respectively. Here, R is the radius of curvature, shown in Fig. 3.11. The additional term in the complex potential (Eq. 3.44) for the flow in the complex plane due to this added singularity is given by

$$F(\zeta) = \frac{i V_m S}{2\pi} \frac{e^{i\lambda}}{R} \ln\left(\frac{\zeta + \tau}{\zeta - \tau}\right)$$

$$= \phi' + i\psi' \tag{3.54}$$

It can be verified (Scholz, 1965) that the real part of this complex potential is equal to $\ln(V/V_m)$, and the imaginary part is due to camber. Because the total complex potential (Eqs. 3.44 and 3.54) is known, the velocity can be determined in the cascade plane. Weinig (1957, 1964) has provided expressions for local velocity as well as the expression for the ratio of blade to flow turning angle, both valid for a cascade of thin airfoils with small circular arc cambers. The analysis is valid for smooth shockless entry flow. This is defined as the condition at which the streamline joins the camber at the entrance with a continuous slope. The ratio of flow turning angle (ϵ) to blade turning angle ($\theta = \alpha_1' - \alpha_2'$) is given by

$$\mu = \frac{\epsilon}{\theta} = \frac{2S}{\pi C} \cos \lambda \ln \frac{\tau^2 + 1}{\tau^2 - 1} \tag{3.55}$$

The value of τ is derived from Eq. 3.43. In this analysis, the deviation of the

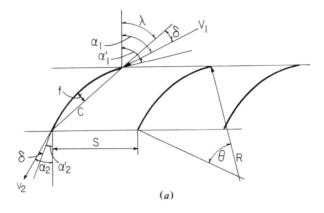

(a)

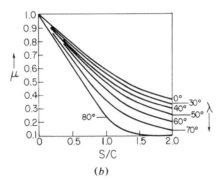

(b)

Figure 3.11 Aerodynamic characteristics of a cascade of circular arc cambered plates. (a) Circular arc cascade. (b) Variation of μ with S / C and λ. (From Eck, 1973; reprinted with permission of Pergamon Press.)

flow from the profile at inlet and outlet are equal; hence,

$$\delta = \alpha_2 - \alpha_2' = \alpha_1' - \alpha_1, \qquad \delta = \frac{\theta - \epsilon}{2} = \frac{\theta}{2}(1 - \mu) \qquad (3.56)$$

Weinig's theory can be utilized in many ways. If the blade row has shock-free entry and the camber is small, the theory could be utilized either for the design or analysis problem. If the entry is not shock-free, the theory can be utilized to derive the value of the deviation angle (as a first approximation) and hence the lift coefficient. Both of these cases will be illustrated through examples later. The values of μ for various space chord ratios derived from Eqs. 3.55 and 3.43, as calculated by Eck (1973), are shown in Fig. 3.11. This plot is extremely useful in the preliminary design or analysis of a lightly loaded (or small camber) blade. Some relationships valid for a circular arc blade cascade (with moderate camber) are listed below. The notations are defined in Fig. 3.11.

$$\lambda = (\alpha'_1 + \alpha'_2)/2, \qquad C = R(\alpha'_1 - \alpha'_2) = R\theta,$$

$$\frac{f}{C} = \frac{R}{C} - \sqrt{\left(\frac{R}{C}\right)^2 - \frac{1}{4}} \qquad (3.57)$$

Because the theory applies only to a small turning angle, the flow angles for shock-free entry are

$$\alpha_1 = \lambda + \delta, \qquad \alpha_2 = \lambda - \delta \qquad (3.58)$$

and the following expression for the lift coefficient can be derived by using Eq. 3.18 (with $C_D = 0$):

$$C_L = 2\frac{S}{C}\cos\lambda\frac{\sin 2\delta}{\cos^2\lambda - \sin^2\delta} \qquad (3.59)$$

The corresponding lift for an isolated airfoil is (with shock-free entry)

$$C_L = 4\pi(f/C) \qquad (3.60)$$

where f/C is the maximum camber/chord ratio. Hence, the following equation for the ratio of lift coefficient of a cascade to that of an isolated airfoil, with shock-free entry, can be derived (assuming $\sin^2\delta \ll \cos^2\lambda$, $\sin 2\delta = 2\delta$) in Eq. 3.59 and by using Eqs. 3.55–3.60:

$$K_c = \frac{C_L}{C_{L_{\text{isolated}}}} = \frac{8}{\pi^2}\left(\frac{S}{C}\right)^2\ln\left(\frac{1+\tau^2}{1-\tau^2}\right) \qquad (3.61)$$

Values of K_c derived from Eqs. 3.61 and 3.43 are shown plotted in Fig. 3.12.

Example 3.1. *Analysis:* Consider a cascade of blades with a circular arc camber and $\lambda = 36°$, $\alpha_1 = 51°$, $\theta = 30°$, $\alpha'_2 = 21°$, and $S/C = 1.0$. Find the lift coefficient.

From Fig. 3.11, $\mu = 0.55$, and from Eq. 3.56, $\delta = 6.75°$. Thus, $\alpha_1 = 51°$ and $\alpha_2 = 27.75°$. The lift coefficient based on these angles and Eq. 3.18 (with $C_D = 0$) is 1.064. The lift coefficient based on the exact panel method (inviscid) is 1.035. Thus, the approximate theory provides the value of the lift coefficient reasonably accurately for circular arc blades of small camber. This example illustrates the usefulness of Weinig's analysis to determine the deviation and hence the outlet angle.

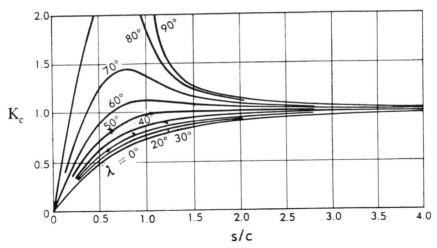

Figure 3.12   Weinig's coefficient (K_c) for a circular arc profile with smooth inflow conditions (Scholz, 1965). (The original version of this material was first published by AGARD / NATO in AGARDograph AG 220.)

Example 3.2. *Design Problem:* The approximate method due to Weinig provides a quick means of arriving at a preliminary design, especially for a lightly loaded axial fan, pump, and compressor rotor. An illustration of this is given below in the design of an axial flow fan to provide the following characteristics: $Q = 18.83$ m^3/s, $r_h = 0.229$ m, $r_t = 0.458$ m, $N = 1750$ rpm, $\Delta P_o = 7.62$ cm of water. The hub, midradius, and tip sections will be designed using Weinig's cascade theory (neglecting all the radial velocities). The entry flow is assumed to be axial, and the blades have smooth entry flow. Annulus area $= 0.49 m^2$, $V_{x_1} = V_{x_2} = 38.1$ m/s, $\tan \beta_1 = \Omega r/V_x = 1.464$ r (values at three radii are tabulated in Table 3.2), $U_t = 84$ m/s (0.43 m of H$_2$O at standard atmosphere). Assuming 85% aerodynamic efficiency at all the radial locations, we obtain

$$\psi = \frac{\Delta P_o}{\frac{1}{2}\rho U_t^2} = \frac{7.62}{43} = 0.1775 = \frac{UV_{\theta_2}}{U_t^2} \times 0.85$$

Hence, $V_{\theta_2} = 17.4/(r/r_t)$, for constant ψ at all locations.

Knowledge of V_{θ_2} enables the outlet velocity triangle to be established at each radius. The outlet angles calculated are tabulated in Table 3.2. At mid-radius, choose $S/C = 1$; hence,

$$C_L = 2(S/C)\cos\beta_m(\tan\beta_1 - \tan\beta_2) = 0.7$$

Choosing the same chord length at all radii will result in $S/C = 1.33$ at the

Table 3.2 Design characteristics of a fan rotor[a]

Location	r	Ωr	V_{X_1}	V_{θ_2}	W_{θ_2}	β_1	β_2	ϵ	β_m	C_L	C	S	μ	δ	β'_1	β'_2	θ	λ	f/C
Hub	22.9	42	38.1	34.7	7.05	48	10	38	29	1.11	15.24	10.36	0.7	7	55	3	52	29	0.113
Mid-radius	34.35	62.4	38.1	23.16	39.24	58	46	12	52	0.7	15.24	15.24	0.49	6.4	64.4	39.6	25	52	0.055
Tip	45.8	84	38.1	17.37	66.13	65	60	5	63	0.5	15.24	20.25	0.3	5.8	70.8	54.2	16.6	62.5	0.036

[a] All lengths are in centimeters, velocity in meters per second, angles in degrees.

tip and 0.68 at the root. The corresponding values of C_L are 0.50 and 1.11, respectively.

The vane angles can now be determined using Fig. 3.11. At this stage, the designer can choose either a flat plate cascade or a cascade with cambered blades. For the sake of convenience, circular arc blades are selected and the thickness effect is neglected. It should be emphasized that the thickness effect is neglected here to avoid the complexity in the design, because the purpose of this example is to illustrate the basic principles. [Eck (1973) gives methods of incorporating the thickness effect empirically.]

Choose an aspect ratio of 2; hence, $C = 15.2$ cm. The corresponding spacing at tip, mid-radius, and hub are 20.25 cm, 15.2 cm, and 10.36 cm, respectively. The number of blades based on blade spacing at the tip is 14.17; hence, choose 14 blades with blade spacing and chord length as shown in Table 3.2. If λ is assumed to be equal to β_m, Fig. 3.11 can be used to derive the value of μ at various radii. This provides the value of δ from the following equation (because ϵ is known):

$$\delta = \frac{\epsilon \left(\dfrac{1}{\mu} - 1 \right)}{2}$$

Knowing δ, the inlet and outlet blade angles as well as the camber angle can be determined from Eq. 3.56. The stagger angle and the camber can then be determined using Eq. 3.57. This completes the preliminary design. Design characteristics of the rotor are tabulated in Table 3.2. Note that λ and β_m are nearly the same; hence, the approximation made earlier is valid. In Table 3.2, β is used to denote the blade and flow angles in the relative frame. W_{θ_2} is the relative tangential velocity.

3.3.1.6 *Application of Conformal Transformation Methods: Comparison and Limitations*   The range of applicability of the theories, such as the limitations on C/S, f/C, t_{max}/C, and so on, are listed in Table 3.1.

As mentioned earlier, Weinig's analysis is useful for thin blades with small or no camber. The main application of this method lies in low-pressure-rise rotors such as fan blades. It is also useful as a test case for many of the unsteady as well as steady flow theories. This is the only analysis available that has its solution in analytical form; hence, it is very useful as a test case in the development of both steady, two-dimensional flow theories for cascades. Howell and Garrick's transformations are generally suitable for compressor cascades of solidity of less than 1.5, with no more than 15–20% (based on chord) camber and thickness. These are found to agree well with the data from a compressor cascade (see Pollard and Wordsworth, 1962; Hall and Thwaites, 1964). Shirakura's method seems to work well for all the turbine (accelerated flow) test cases he has compared. He shows good agreement

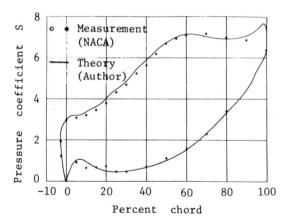

Figure 3.13 Comparison of measured and predicted pressure distribution on a turbine cascade with $\theta = 110°$, $t_{max}/C = 0.15$, $\sigma = 1.8$, $\lambda = 35°30'$, and $\alpha_1 = 30°$, $S = (1 - C_p)$ (Eq. 3.53). (Copyright © Shirakura, 1972; reprinted with permission of JSME.)

with the experimental data due to Dunavant and Erwin (1956) for solidities varying from 1.5 to 1.8, t_{max}/C from 0.1 to 0.15, and the camber θ from 65° to 110°. One extreme case is shown in Fig. 3.13. The agreement is excellent and the predicted deviation is $-3°.05'$, while the measured value is $-2°.58'$.

3.3.2 Cascade Analysis by Method of Surface Singularity: The Panel Method

It can be proven that the solution of the Laplace equation (3.23) or (3.24) is unique if the boundary conditions at infinity and on the body are satisfied, and, in addition, if the magnitude of the circulation around the body is specified. Hence, the solution of the Laplace equation is unique if the boundary conditions given by Eq. 3.22 and others are satisfied. Thus, the kinematical problem of finding the flow pattern of an incompressible irrotational flow through a cascade is reduced to a mathematical problem of deriving a particular solution of the Laplace equation.

One of the most widely used techniques of finding the solution to a Laplace equation, which is linear, utilizes the principle of "superposition" in deriving the combined solution. If ϕ_1, ϕ_2, and ϕ_3 are solutions of Eq. 3.24, the resulting flow represented by

$$\phi = \phi_1 + \phi_2 + \phi_3 \tag{3.62}$$

is also a solution of the Laplace equation. This can easily be shown by substituting Eq. 3.62 in Eq. 3.24. The boundary condition satisfied by the combined flow should be the same as that of the flow represented by individual solutions. Hence, the combined flow also represents the irrotational and incompressible flow. If, for example, ϕ_1, ϕ_2, and ϕ_3 represent the source, sink, and vortex flows, respectively, then the combined flow represented by $\phi_1 + \phi_2 + \phi_3$ is also irrotational and satisfies the same boundary conditions at infinity as the individual singularities. This concept led to the development of an "equivalent" method of solving the flow around the body by replacing it with sources, sinks, and vortices.

3.3.2.1 *Generalized Method*

The surface singularity method is the most accurate method for the inviscid incompressible flow, both in two-dimensional and three-dimensional flows, and approaches the exact solution in the limit. A clear distinction must be made between approximate methods (i.e., thin airfoil theory, Karamachetti, 1966) and exact methods (e.g., conformal transformation and surface singularity method). Approximate solutions introduce analytical approximations (such as small camber, small thickness, etc.) into the formulation, and hence place a limit on the accuracy regardless of the numerical procedure used. In the exact formulation, the numerical approximations are introduced for the purpose of calculation, which can be made as accurate as possible by using a refined numerical technique.

The method of surface singularities is based on the transformation of the differential equation governing the flow (Laplace's equation) into an integral equation and is valid for irrotational inviscid flow (Korn, 1899; Lamb, 1945). The method due to Prager (1928) and others (e.g., Kellog, 1953; Martensen, 1959) utilizes vorticity distribution around the surface, whereas the method due to Hess and Smith (1966) and others (Giesing, 1964; Johnston and Rubbert, 1975; Bristow, 1976) utilizes source and doublet distribution around the surface. The unknown distribution of source or vortices replacing the surface of the body is calculated employing an integral equation derived from the boundary condition on the surface, superposing the potential due to these sources with the undisturbed flow field. The method was applied to a cascade of blades by Martensen (1959), Giesing (1964), Minassian (1975), McFarland (1982, 1984), and others. Generalized features of these techniques are described below.

As shown by Korn (1899) and Karamachetti (1966), the solution for the irrotational flow can be represented as the influence of source and doublet (or vortex sheets) on the boundary surface. The resulting expression, based on Green's theorem, is given by

$$\Phi = \Phi_\infty + \phi = \Phi_\infty + \iint_S q(Q)\phi_s(Q, P)\, dS + \iint_S \mu(Q)\phi_d(Q, P)\, dS$$

(3.63)

where Φ is total potential at P, Φ_∞ is the potential due to undisturbed flowfield, and ϕ is the perturbation caused by the body. ϕ_s and ϕ_d are velocity potential due to a unit source and a unit doublet, respectively,

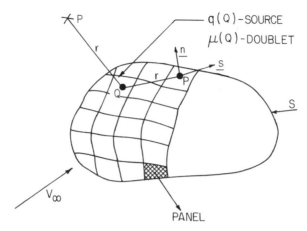

Figure 3.14   Generalized panel method.

located at Q on the surface, which is at a distance r from the point P (Fig. 3.14). The values of ϕ_s and ϕ_d for various types of panels are tabulated in Table 3.3. The integration is taken over all the panels replacing the body. The first integration represents the contribution from all the sources $q(Q)$ located on the body. The second integration is the contribution from all the doublets of strength $\mu(Q)$. It can be proven (Karamachetti, 1966, p. 344) that the doublet potential ϕ_d can be generated by differentiating the source potential (ϕ_s) with respect to a certain chosen direction $(\phi_d = -(\partial/\partial n_Q)\phi_s)$, where n_Q is the axis of the doublet at Q. The potential $(d\phi_s)$ due to a source $(q\,dS)$ (q is the source strength per unit area) at Q is given by

$$d\phi_s = -\frac{q(Q)\,dS}{4\pi r}$$

where dS is the element of the surface. Hence, the total potential due to all sources $q(Q)$ is given by

$$\phi_s = -\iint_S \left(\frac{q(Q)\,dS}{4\pi r}\right)$$

A similar equation can be derived for the doublet or vortex distribution. A summary of the properties of source, doublet, and vortex panels is given in Table 3.3. Expressions for the velocity potential are also given in the table. For detailed derivation of these equations, the reader is referred to Karamachetti (1966) and Katz and Plotkin (1991).

There are two types of flow problems, depending on the specification of the boundary condition. In the Dirichlet or first boundary value problem, the

Table 3.3 Properties of source, doublet and vortex panels

	Schematic	Properties	$\phi(P)$ - Potential function
Point source (3D)		Radial streamlines	$\phi(P) = -\dfrac{q(Q)}{4\pi r}$ r is the distance between P and Q
Source panel (3D)		Normal velocity jump across panel and the tangential velocity is continuous	$\phi(P) = \displaystyle\iint_{panel} q(Q)\left(-\dfrac{1}{4\pi r}\right)ds$ $q(Q)$ = Strength of source per unit area
Source panel (2D)		Normal velocity jump across panel and the tangential velocity is continuous	$\phi(P) = \dfrac{1}{2\pi}\displaystyle\int_{l} q(Q)\,(lnr)dl$ $q(Q)$ = Strength of source per unit length
Point doublet (3D)		Near the axis, streamlines are in the direction of doublet axis	$\phi(P) = \mu(Q)\,\dfrac{\partial}{\partial n_Q}\left(\dfrac{1}{4\pi r}\right) = -\dfrac{\mu(Q)}{4\pi}\,\dfrac{cos\theta_\mu}{r^2}$ n_Q is along doublet axis at Q
Doublet panel (3D)		Tangential velocity jump across panel, normal velocity continuous	$\phi(P) = \displaystyle\iint_{panel}\mu(Q)\,\dfrac{\partial}{\partial n_Q}\left(\dfrac{1}{4\pi r}\right)ds = -\iint_{panel}\dfrac{\mu(Q)}{4\pi}\,\dfrac{cos\theta_\mu}{r^2}\,ds$ $\mu(Q)$ = double strength per unit area
Doublet panel (2D)		Tangential velocity jump across panel, normal velocity continuous	$\phi(P) = -\displaystyle\int_{panel}\mu(Q)\,\dfrac{\partial}{\partial n_Q}\,(lnr)ds = -\int_{panel}\dfrac{\mu(Q)}{2\pi}\,\dfrac{cos\theta_\mu}{r}\,ds$ $\mu(Q)$ = double strength per unit length
Vortex panel (3D)		Tangential velocity jump across panel	$\phi(P) = \dfrac{1}{4\pi}\displaystyle\iint_{panel}\gamma(Q)\,\dfrac{cos\theta_v}{r^2}\,ds$ $\gamma(Q)$ = Strength of vortex per unit area

223

value of the potential (ϕ) (or stream function, ψ, because they are interchangeable) is prescribed. In the Neumann or second boundary value problem, the derivative of the potential, $\partial\phi/\partial n$, is prescribed. The technique due to Hess and Smith (1966) and others belongs to this class of problem.

The methods of Hess and Smith (1966), Giesing (1964), Bristow (1976), and others utilize source and doublet distribution, whereas those due to Prager (1928) and Martensen (1959) utilize the vortex distribution to represent the flow around an arbitrary body. It is clear from Table 3.3 that the vortex sheet is equivalent to a doublet sheet. Hence, the distribution of a doublet over a closed surface, the axis directed along normals, may be replaced by a system of vortex filaments lying on the surface (see Lamb, 1945, p. 212; Milne-Thompson, 1960, p. 549). The boundary condition (see Fig. 3.14) is,

$$\mathbf{V} \cdot \mathbf{n}|_s = 0 \text{ on the surface, where } \mathbf{V} = \mathbf{V}_\infty + \mathbf{v} \tag{3.64}$$

where v is the perturbation in velocity due to the body.

Hence, from Eq. 3.63, we obtain

$$-\mathbf{V}_\infty \cdot \mathbf{n}(P)\big|_p = \mathbf{v} \cdot \mathbf{n}(P)\big|_p = \frac{\partial\phi}{\partial n}\bigg|_p = \iint_S q(Q)\frac{\partial}{\partial n}[\phi_s(Q,P)]\,dS$$

$$+ \iint_S \mu(Q)\frac{\partial}{\partial n}[\phi_d(Q,P)]\,dS \tag{3.65}$$

This equation is valid on the surface (both P & Q are located on the surface of the body). Similar equations can be written for stream function. The stream function and potential functions are related by the Cauchy–Riemann relationship $\partial\phi/\partial s = \partial\psi/\partial n$; $\partial\phi/\partial n = -\partial\psi/\partial s$. In Eq. 3.65, $\partial/\partial n$ is the differentiation (in outward normal) at the surface at point P, as shown in Fig. 3.14. Subscript p refers to the surface point. Equation 3.65 is the principal equation in the surface singularity method or the panel method. This boundary condition is suitable for analysis purposes, where the body shape is specified. For a design problem, the following equation based on the tangential derivative is needed.

$$(\mathbf{V}_\infty + \mathbf{v}) \cdot \mathbf{s}|_p = \iint_S q(Q)\frac{\partial}{\partial s}[\phi_s(Q,P)]\,dS$$

$$+ \iint_S \mu(Q)\frac{\partial}{\partial s}[\phi_d(Q,P)]\,dS + \frac{\partial(\Phi_\infty)}{\partial s} \tag{3.66}$$

Equations 3.65 and 3.66 can be used to express local velocity (tangential to the surface) and the local blade or body shape in terms of source and doublet distribution. This equation can then be solved to derive the body shape for a given (optimum) distribution of sources and doublets or vortices.

A continuous distribution of sources (or sinks) and doublet panels (vortex lattices) is applied to the surface of the body, as shown in Fig. 3.14, and the strength is adjusted such that the boundary condition (Eq. 3.65) is satisfied. While Hess and Smith (1966) and Giesing (1964) prescribe the boundary condition on the body surface (i.e., Eq. 3.65), Prager (1928) and Martensen (1959) apply the condition such that the total tangential velocity (perturbation and the undisturbed flow) is zero on the inside profile of the body. The integral equation derived by Prager and others are similar due to the reasons mentioned earlier. Either the doublet or the vortex distribution can be used to simulate the circulation.

Numerical solution of Eq. 3.65 or 3.66 is usually accomplished by panel-type influence coefficient methods (Johnston and Rubbert, 1975). The surface is divided into a finite number of panels, each having a control point (such as Q in Fig. 3.14) at which Eq. 3.65 (the boundary condition) is satisfied. The integral equation can then be expressed as a set of algebraic equations. The strength of the source/doublet/vortex panel could be assumed to be constant or allowed to vary across each panel as follows:

$$q_J \text{ (or } \mu_J) = q(Q) \text{ [or } \mu(Q)] = \sum_{k=1}^{N} a_{kJ}\tau_k \qquad (3.67)$$

where τ_k is a parameter representing the distribution of singularity strength of a particular panel, N is the number of panel segments in panel J. If the strength of the panel is constant, then $N = 1$, $a_{kJ}\tau_k$ is the strength of a particular panel (k). If we approximate the boundary condition by M discrete points, Eq. 3.65 can be written for each control point (i.e., $i = 1, \ldots, M$) as

$$-(\mathbf{V}_\infty \cdot \mathbf{n})_i = \sum_{k=1}^{M} \tau_k A_{ik} \qquad (3.68)$$

where

$$A_{ik} = \sum_{(k)} \iint_{\text{Panel } J} a_{kJ}(K_{iJ})\, dS \qquad (3.69)$$
$$\text{All panel segments having non zero } a_{kJ}$$

where K_{iJ} is the kernel function (induced velocity at i due to a unit singularity located on panel J).

If each panel is of constant source or doublet or vortex strength τ_k (strength is uniform across each panel but varies from panel to panel), then the summation is over k only (Eq. 3.68). A_{ik} is the induced velocity at i due to a source of unit strength of the kth panel.

The unknown strength, τ_k in Eq. 3.68, can be obtained by solving this set of linear algebraic equations using the methods described in Appendix B. Knowing the values of τ_k, the desired flow properties can be obtained by the derivatives of the potential function (Eq. 3.63). This equation can also be expressed as a set of algebraic equations:

$$\Phi(P) = \sum_{k=1}^{M} C_{pk}\tau_k \tag{3.70}$$

where C_{pk} is similar to A_{ik} in Eq. 3.68.

If we know $\Phi(P)$, the velocities and hence the pressure on the surface can be calculated using the equation

$$C_p = \frac{P_{\text{local}} - P_1}{\frac{1}{2}\rho V_1^2} = 1 - \left(\frac{V_{\text{local}}}{V_1}\right)^2 \tag{3.71}$$

Detailed discussion of the panel method and its implementation to isolated airfoils and wings can be found in Katz and Plotkin (1991).

A brief review of various panel methods applicable to potential flow through cascades, and the limitations of these methods, is given by Minassian (1975). The earlier surface singularity method due to Martensen (1959) used a discrete vortex distribution, while the technique used by Minassian utilizes a piecewise continuous distribution. Martensen placed a large number of discrete vortices around the airfoil surface. The method is very sensitive to coordinate distribution and has difficulty near the trailing edge. In the "piecewise" continuous type of singularities, these difficulties are overcome by using singularity elements of small length. The strength distribution over each element may either be constant or vary in a prescribed manner. Giesing's (1964) method utilizes source and sink distribution with piecewise continuous distribution of singularities for the solution of cascade flow. This approach has since been extended by McFarland (1982, 1984) to include source and doublet distribution. A classification of the surface singularity method for a cascade is given in Table 3.4.

Minassian's method is an extension of the techniques developed by Martensen (1959) and Mavriplis (1971). Minassian's method utilizes vortex elements of constant strength and superimposes solutions for two linear onset flows. The circulation is generated by introducing a uniform vorticity distribution on the internal surface. Minassian claims that the vortex method has some advantage over the corresponding source treatment. Because the strength of the elements is identically equal to the required surface velocity

Table 3.4   Classification of panel (surface singularity) methods for cascades

Source	Boundary Condition	Nature of Singularities	Functional Approximation for q, μ, and γ	Application
Riegels (1961)	Neumann	Vortex lattice	$\gamma(Q)$ constant over panel	Incompressible cascade 2D flow
Martensen (1959)	Neumann	Vortex lattice	$\gamma(Q)$ = constant over the panel	Cascade of blades (2D)—incompressible flow
Giesing (1964)	Neumann	Source lattice and a vortex within the profile to generate circulation	$q(Q)$ = constant over the panel	Cascade of blades (2D)—incompressible flow
Minassian (1975)	Neumann	Vortex lattice	Piecewise constant distribution of vorticity over the cascade airfoil	Multielement (tandem) cascades, including compressibility effect
McFarland (1982, 1984)	Neumann	Sources doublets and vortices	$\mu(Q)$ and $q(Q)$ varying linearly across the panel	Cascade of airfoils; quasi-3D blade row including compressibility effect

locally, the summing process for individual vortex elements is eliminated, resulting in less computational time. This argument may not be valid in view of the insignificant computational time required for the panel method. Hence, the two methods should prove to be equally accurate and attractive.

3.3.2.2 Vortex Panel Methods for a Cascade

This method is originally due to Martensen (1959) and was extended by Minassian (1975) to include piecewise continuous distribution and this method is described in this section. The Neumann boundary condition applied externally on the body contour results in a Fredholm's equation of the first kind, which is very difficult to solve. To overcome this difficulty, the condition of zero tangential velocity on the internal body contour of the cascade blade is imposed to derive an integral equation governing the flow. Martensen (1959) has proven that the solutions derived by both methods (the tangential condition on the internal surface and the Neumann condition on the external surface) are identical. The potential function at any point on a cascade is given by the

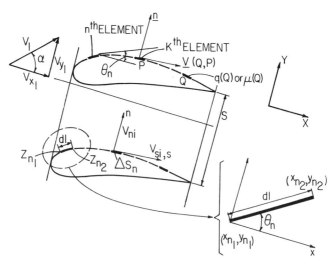

Figure 3.15 Panel method for cascade.

Fredholm's integral of the second kind (based on Eq. 3.63):

$$\Phi(x, y) = \iint_S \gamma(Q)\,\phi_\nu(Q, P)\,dS + V_m(x \cos \alpha_m + y \sin \alpha_m) \quad (3.72)$$

where ϕ_ν is the potential function at P due to a vortex panel located at Q on the body (Fig. 3.15). $V_m \cos \alpha_m$ and $V_m \sin \alpha_m$ are onset flows (mean) in x and y directions, respectively. $V_m \cos \alpha_m$ (axial velocity) is constant for incompressible flow through the cascade (Figs. 3.4 and 3.5), and $V_m \sin \alpha_m = (V_{y_1} + V_{y_2})/2$. Likewise, α_m is the flow angle defined by Eq. 3.4, with $\rho_1 = \rho_2$. The integral term is the perturbation due to the cascade. The stream function and potential function satisfy the Cauchy–Riemann condition and hence can be used interchangeably. The streamwise derivative of Eq. 3.72 and the internal boundary condition results in,

$$-\frac{\gamma(P)}{2} + \iint \gamma(Q)B(Q, P)\,dS = -V_m\left[\left(\frac{dx}{ds}\right)_p \cos \alpha_m + \left(\frac{dy}{ds}\right)_p \sin \alpha_m\right]$$

$$(3.73)$$

where $-\gamma(P)/2$ (γ is positive anticlockwise) is the internal tangential velocity due to vorticity in the neighborhood of P (Panel enclosing point P) on the profile, the integral represents the contribution due to all other vortex panels replacing blades in a cascade, and $B(Q, P)$ is the induced velocity $[(\partial\phi/\partial s$ or $\partial\psi/\partial n)]$ at P due to a unit vortex placed at Q. The blade surface is approximated by a large number of surface elements (ΔS_n). For the purpose of solution, the integral equation (Eq. 3.73) is expressed as a system of linear equations:

$$-\frac{\gamma_k}{2} + \sum_{n=1}^{M} A_{kn}\gamma_n = R_k \qquad (3.74)$$

where $A_{kn} = B_{kn}(\Delta S)_n$, ΔS_n is the length of surface element, and B_{kn} is the kernel (induced velocity) at the kth element due to a unit vortex located at n. R_k is the right-hand side of Eq. 3.73 and consists of the tangential (to the blade surface) components of three distinct onset flows: (1) $V_m \cos \alpha_m$ is the mean axial velocity, (2) $V_m \sin \alpha_m$ is the mean tangential velocity, and (3) V_Γ is the component of velocity induced by the circulation generating a uniform vorticity distribution (γ_c) placed in the vicinity of the body along the inner contour. Hence, Eq. 3.74 can be expressed as a set of three linear equations:

$$-\frac{\gamma_k^i}{2} + \sum_{n=1}^{M} A_{kn}\gamma_n^i = R_k^i \qquad (i = 1, 2, 3)$$

where $R_k^1 = -V_m[(dx/ds)_p \cos \alpha_m]$ and $R_k^2 = -V_m[(dy/ds)_p \sin \alpha_m]$. The last condition ($R_k^3$) prescribes the circulation around the airfoil. Minassian (1975), following Mavriplis (1971), proves that this step results (for a constant vortex strength) in an additional term in the expression for R_k given by

$$-\frac{1}{2} - \sum_{n=1}^{M} A_{kn}$$

Hence, for a given surface location, Eq. 3.74 consists of three distinct additive parts. The Kutta condition is satisfied by finding a circulation that will force the rear stagnation point to occur at the trailing edge.

The kernel function A_{kn} is the tangential (to the surface element k) component of induced velocity at k due to an infinite row of unit vortex elements n equally spaced (S), replacing the blades (Fig. 3.15). This can be calculated using the potential solution of a row of infinite vortices (of strength $\gamma\, dx$) spaced equally (S apart) at a stagger angle of λ [see Milne-Thompson (1960, p. 375) for proof], given by

$$du_\gamma - idv_\gamma = \frac{i\gamma dx}{2S} e^{i\lambda} \coth\left[\frac{\pi(z - z_0)}{S} e^{i\lambda}\right] \qquad (3.75)$$

where z_0 ($x_0 + iy_0$) is the location of the vortex elements, and du_γ and dv_γ are the velocities induced by the nth vortex in x and y directions, respectively.

Equation 3.75 can be used to find the kernel function A_{kn}. Let us consider the vortex panel n of length dl shown in Fig. 3.15. Integrating Eq. 3.75

between the limits z_{n_1} (leading edge of the panel) and z_{n_2} (trailing edge of the panel), the kernel function due to nth element at point p of the kth element can be proved to be (Minassian, 1975)

$$du_{kn} - idv_{kn} = \frac{\exp(-i\theta_n)}{2\pi i} \ln \frac{\sinh \pi \left(\bar{z}_k - \bar{z}_{n_1}\right)/S}{\sinh \pi \left(\bar{z}_k - \bar{z}_{n_2}\right)/S} \qquad (3.76)$$

where $\bar{z}_k$ is the conjugate complex coordinate of the control point of element k, θ_n is the angle the element makes with the x axis, and du_{kn} and dv_{kn} are the velocity components parallel and perpendicular to the x axis, respectively, induced by the element n at the control point of k. $\bar{z}_{n_1}$ and $\bar{z}_{n_s}$ are, respectively, the first and last points of elements n (Fig. 3.15). The value of A_{kn} is derived by resolving du_{kn} and dv_{kn} in a direction tangential to the element k (v_{s_i}).

Hence, Eqs. 3.74 and 3.76 constitute a complete set of equations for the solution of γ_k. For a given singularity element, the term R_k on the right-hand side of Eq. 3.74 consists of three distinctive additive parts. The equation is solved for each of these flows and superposed as described earlier and in Martensen (1959) and Minassian (1975). If only inlet conditions are prescribed, the onset flow (V_m) has to be determined from Eqs. 3.74–3.76 using Kutta condition iteratively. The algebraic equations are solved to derive values of γ_k. Local velocities as well as pressure can then be calculated using Eqs. 3.70 and 3.71, respectively. In most applications, 50 panels ($M = 50$) are adequate. For the analysis, it is necessary to specify the inlet flow angle, solidity, and blade coordinates. Martensen excluded the presence of a sharp trailing edge in his calculation by stipulating that the surface should be integrable. Pal (1965) later extended Martensen's analysis to include the trailing edge effect as well as the effect of the downstream and the upstream rows, as in a tandem cascade. Excellent agreement between the theory and the experiment, except in the trailing region where the viscous effect is dominant, is shown. This method has also been extended to the tandem cascade by Minassian (1975). Comparison of the predictions from the surface singularity analysis and the exact conformal transformation method for a 70° camber airfoil in a cascade, shown in Fig. 3.16, indicate excellent agreement. Martensen (1959) shows excellent agreement between the predicted and the measured pressure distribution for several turbine cascades with large flow-turning angles.

There is no limitation on the applicability of the vortex panel method (or the source panel method). It is valid for all solidities, blade thicknesses, and cambers and can be extended to include compressibility effects. The method could be reversed to derive the blade profile from an assumed velocity distribution. Lewis (1982) has attempted such an inverse technique using the vortex singularity method. The vortex panel method could be inaccurate for

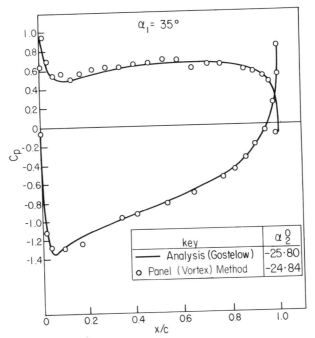

Figure 3.16 Pressure distribution for 70° camber airfoil in cascade 10c4 / 70c50, $S / C =$ 0.9. (From Gostelow, 1964, 1965; reprinted with permission of the University of Liverpool.)

very small thicknesses as the distance between singularities on the pressure and suction surfaces become small.

3.3.2.3 Source–Vortex (Doublet) Method for a Cascade The source–vortex (doublet) technique is similar to the vortex lattice technique. The elements of the panel in this case consist of sources and doublets. The earlier solution due to Giesing (1964) employed only sources and sinks (constant across the panel), with a unit vortex placed within the profile to generate circulation around the profile. The latest method due to McFarland (1982, 1984) employs both source (sink) and doublet distribution around the profile. The earlier solution due to Giesing employed constant source strengths, while the method due to McFarland employs a higher-order method developed by Bristow (1976). The technique is very similar to the one described in Section 3.3.2.1.

It can be proven (Hess and Smith, 1966) that the induced velocity at point P (Fig 3.15) due to a two-dimensional source panel at P is $q(P)/2$ and that the two-dimensional form of Eq. 3.65 with source distribution valid for an infinite cascade of blades, shown in Fig. 3.15, is given by

$$\frac{q(P)}{2} + \frac{1}{2\pi} \iint q(Q) A(Q, P) \, dS = -\mathbf{V}_m \cdot \mathbf{n}(P) \qquad (3.77)$$

where $A(Q, P) = \mathbf{n} \cdot \mathbf{V}(Q, P)$ and $\mathbf{V}(Q, P)$ is the velocity induced at P due to an infinite row of source panels $q(Q)$ (Fig. 3.15) that replace the blades that are spaced at S apart. The first term is the contribution to the outward normal velocity at point P on the boundary due to the source density in the neighborhood of P. The integral represents the contribution of the source density on the remainder of the boundary surface (including all the blades) to the outward normal velocity.

The integral in Eq. 3.77 is solved for three "basic" flows: flow at a zero angle of incidence, flow at a 90° angle of incidence, and a circulatory flow around the cascade blade. The solution procedure is very similar to that outlined for vorticity distribution earlier in this chapter. The integral in Eq. 3.77 is expressed as a system of linear algebraic equations (similar to Eq. 3.74). The kernel function $A(Q, P)$ is calculated from the induced velocity due to an infinite row of source panels (derived by multiplying the right-hand side of Eqs. 3.75 and 3.76 by i). The three "basic flow" solutions are superposed such that correct angle of attack is obtained and the Kutta–Joukowski condition is satisfied. Giesing (1964) has shown excellent agreement between his predictions, the exact method due to Garrick (1944), and the experimental data for the NACA 65010 cascade tested by Herrig et al. (1957).

Bristow (1976) introduced modifications to the surface singularity technique. The main feature of this modification was the introduction of a mean-square singularity density minimization. Both vortex and source distributions are employed, and each element (Fig. 3.15) is modeled by connected line segments on which piecewise source density of uniform strength and piecewise vortex density with linearly varying strength are positioned. The flow tangency condition on the surface and the Kutta–Joukowski condition at the trailing edge (for a sharp trailing edge blade) provide only one equation for each of the two unknowns q and μ (or γ) in Eq. 3.65. This is overcome by introducing a closure condition. Bristow (1976) also introduced the minimization of the mean-square density strength. The system of equations and technique employed by McFarland (1982, 1984) are similar to those of Bristow (1977). These equations are generated by using normal velocity surface boundary conditions, far stream boundary conditions and a tangential velocity error minimization equation. Bristow (1977) also developed a technique for bodies with curved panels, such as the leading edge of an airfoil.

McFarland (1982) utilized this technique to derive solutions for incompressible and compressible flow through cascades, and he found excellent agreement between the exact hodograph results of Gostelow (1964) and the finite difference techniques. The panel code has been used in many turbomachinery applications. One such application is shown in Fig. 3.17. The measured blade pressure distribution is compared with the predictions from McFarland's (1984) panel code for an automotive torque converter stator at midspan (By and Lakshminarayana, 1991). The flow-turning angles were approximately 123°, 54°, and 7°, respectively, for speed ratio, SR, of 0, 0.6,

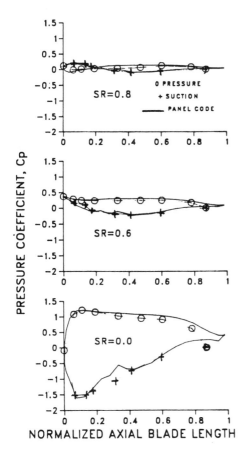

Figure 3.17 Comparison of C_p distribution predicted by the panel code with the measured values at the blade midspan section of a torque converter stator (By Lakshminarayana, 1991.) Reproduced by permission of SAE.

and 0.8, respectively. The speed ratio is the ratio of the speed of the turbine to the speed of the pump. The agreement is quite good considering that the flow is very complex. An extension of the source panel method to quasi-three-dimensional flow (e.g., contracting side walls in a cascade) was attempted by Renken (1976). He shows good agreement between his solutions and those based on the finite difference solutions.

3.4 SUBSONIC INVISCID CASCADE FLOWS

In common with many other types of turbomachinery, modern jet engines operate at very high ranges of speed, where the compressibility effect is substantial. The compressibility effect results in a change of density throughout the flow medium. The advantage of high-Mach-number operation is that the mass flow per unit area as well as the pressure ratio across the stage are high, resulting in a compact power or propulsion plant. From the definition of the pressure rise (or drop) coefficient across a blade row, it can be proven that (Eq. 2.48)

$$\frac{p_2}{p_1} = 1 + C_p \frac{\gamma M_{1R}^2}{2} \tag{3.78}$$

Thus, for a given C_p, the pressure rise increases with the Mach number. However, excessive Mach number results in shock waves, thus decreasing the efficiency of the turbomachinery owing to entropy production. Increasing the Mach number results in an augmentation of the pressure difference acting on a solid body, and it causes the effect of the body to be felt at greater distances. Hence, the interference effects are much greater at higher Mach numbers.

It is well known that the velocity change in a stream tube is given by

$$\frac{1}{V} \frac{dV}{ds} = \frac{1}{(M^2 - 1) A} \frac{dA}{ds}$$

where V is the velocity in the streamwise (s) direction, and A is the area change. It is evident from this equation that the Mach number effect on velocity (for a fixed dA/ds) is relatively greater at high Mach numbers. The flow acceleration on the suction side of a blade (where the Mach numbers are higher) is much larger than those on the pressure side. Thus, increase in inlet Mach number (subsonic) changes suction surface pressures substantially. This is evident from the turbine cascade data shown in Fig. 2.66.

In high subsonic flows, two inflow Mach numbers are of practical interest. The *critical mach number* ($M_{1\,cr}$) is defined as the Mach number for which sonic conditions are first reached locally in the flow field. In such case, there will be no shocks. When the Mach number is increased further, the mass flow per unit area is maximum, and sonic conditions are achieved at the throat cross section of the cascade. In this case, downstream influences are not felt upstream, and the back pressure has no influence on the upstream flow field. The cascade is said to be "choked," and the upstream Mach number is called the *choking Mach number* (M^*). Supercritical flow conditions for compressor and turbine cascades are shown schematically in Fig. 3.18. Using gas dynamic equations, expressions for the "critical" ($M_{1\,cr}$) and "choking" (M^*) Mach numbers can be derived and are given by (Scholz, 1965), respectively, (see Fig. 3.18 for notations)

$$M_{1\,cr} = \sqrt{\frac{\gamma + 1}{\gamma - 1}\left\{1 - \left[\frac{\left(\frac{2}{\gamma + 1}\right)^{\gamma/\gamma - 1} - C_{p\,min}}{1 - C_{p\,min}}\right]^{(\gamma-1)/\gamma}\right\}} \tag{3.79}$$

$$\frac{A^*}{S \cos \alpha_1} = M^* \left\{\frac{\gamma + 1}{2}\left(1 - \frac{\gamma - 1}{\gamma + 1} M^{*2}\right)\right\}^{1/(\gamma-1)}$$

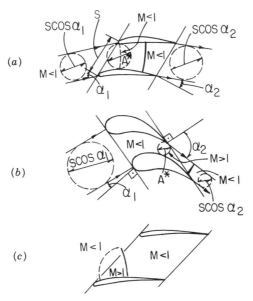

Figure 3.18 Supercritical choked flow through a cascade (Scholz, 1965). (The original version of this material was first published by AGARD / NATO in AGARDograph AG 220.) (a) Decelerating or a compressor cascade. (b) Accelerating or turbine cascade. (c) Supercritical compressor cascade at design conditions.

For example, if $C_{p\,\min} = -3$, $A^*/S \cos \alpha_1 = 0.8$, and $\gamma = 1.4$, then $M_{1\,\mathrm{cr}} = 0.45$ and $M^* = 0.6$.

Hence, depending on the inlet Mach number, the subsonic flow through a cascade can be classified as a

1. Subcritical cascade $(M_1 < M_{\mathrm{cr}})$
2. Supercritical cascade $(M_{\mathrm{cr}} < M_1 < M^*)$ or $(M_{\mathrm{cr}} < M_1 < 1.0)$
3. Choked cascade $(M_1 = M^*)$

For shock-free operation and minimum profile loss, the cascade must be operated below M_{cr}.

The Mach number range to achieve lower profile losses can be extended beyond M_{cr} by carefully designing the blade shape (e.g. supercritical blades or controlled diffusion blades), as shown in Figs. 3.19, 2.12, and 2.76. Whether shock-free operation beyond M_{cr} is possible or not is still open for discussion. The shock-free designs showed weak oscillating shock waves which had, however, little or no influence on losses and boundary layer behavior.

Several methods are available for the solution of compressible flow through a cascade. Gostelow (1984) has provided a review of these techniques. Some of the earlier techniques are described in Scholz (1965). The following

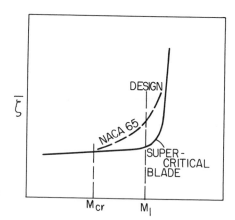

Figure 3.19 Supercritical or controlled diffusion blades.

methods are probably among the most useful:

1. Solution via incompressible flow field
2. Analytical solution via transformation
3. Numerical solution: Finite difference and finite element techniques
4. Numerical solution: Streamline curvature method
5. Numerical solution: Design problem

Techniques 1 and 2 are valid only for potential flows, while the numerical techniques, 3 and 5, have general validity.

Because the flow perturbations in a cascade are not small, linearization (or small perturbation approximation) is not applicable to practical situations. Hence, treatment of these analyses is not included here, except for brief remarks. The reader is referred to books on gas dynamics (Borisenko, 1962; Shapiro, 1953; Schreier, 1982; Sears, 1960; Von Mises, 1958) for linearized solutions for wings and other configurations where the perturbations are small.

Linear theories are based on the assumption that the perturbations in the streamwise and the normal directions are small (X, Y in Fig. 3.6). Hence, the equation governing this flow is given by

$$(1 - M_\infty^2)\phi_{XX} + \phi_{YY} = 0 \qquad (3.80)$$

where ϕ is the perturbation potential and M_∞ is the incident Mach number. This equation can be reduced to a Laplace equation by the transformation of the coordinate $X_i = X$, $Y_i = Y\sqrt{1 - M_\infty^2}$. The resulting equation is the same as the equation governing the incompressible flow through a cascade of blades. It can be proven that the coordinates X_i and Y_i correspond to an equivalent cascade in an incompressible flow. Likewise, the stagger angle as well as the spacing of the blades are changed. The pressure coefficient in the two cases are related by

$$C_p = \frac{1}{\sqrt{1 - M_\infty^2}} C_{pi} \tag{3.81}$$

where C_{pi} is the pressure coefficient of an equivalent incompressible cascade, the solution of which is obtained by any of the methods described in Section 3.3. This transformation, called the *Prandtl–Glauert transformation*, has been widely used for flow around single airfoils (Scholz, 1965; Shapiro, 1953; Schreier, 1982; Sears, 1960; Von Mises, 1958). It has been extended to a cascade of blades by several investigators. These are described in great detail in Scholz (1965). The application of this rule may be justified only for widely spaced blades. The blades behave as isolated airfoils only when the interference effects are small.

Because linearization in the physical plane may introduce inaccuracies which limit its practical application, several investigators have tried to overcome this problem by linearizing the equations in the hodograph plane. Brief reviews of various techniques of doing this are given in Scholz (1965) and Gostelow (1984). One of the most widely used transformations is that due to Von Karman (1941) and Tsien (1939). Their transformation results in the following relationship between the compressible and incompressible case (Karman-Tsien pressure correction formula):

$$C_p = \frac{C_{pi}}{\sqrt{1 - M_\infty^2} + \frac{C_{pi}}{2}\left(\frac{M_\infty^2}{1 + \sqrt{1 - M_\infty^2}}\right)} \tag{3.82}$$

This technique, which has been extended to cascades by Adams (1956), is not very accurate, for the reasons mentioned earlier. Both of these methods are still restricted to slender profiles with large spacing.

3.4.1 Solution via an Incompressible Flow Field

Several techniques have been developed that provide a semitheoretical or theoretical transformation of compressible flow through a cascade to a corresponding incompressible flow. These techniques do not utilize the small perturbation assumption, which is not valid for a cascade.

Among the simplest of these is the technique proposed by Lieblein and Stockman (1972). The correction was obtained from the exact numerical solutions derived from the computer program developed by Katsanis (1969). The following correction factor was derived from these exact solutions:

$$V = V_i \left(\frac{\rho_i}{\bar{\rho}}\right)^{V_i/\bar{V}_i} \tag{3.83}$$

where V is the corrected compressible velocity, V_i is the local velocity for the cascade in incompressible flow, ρ_i is the incompressible density (taken to be ρ_o), $\bar{\rho}$ is the average compressible density, and $\bar{V}_i$ is the average incompressible velocity across the passage. The average density $\bar{\rho}$ is obtained from the following equation, based on a one-dimensional continuity equation and isentropic relations:

$$\frac{\bar{\rho}}{\rho_o}\left\{\frac{\gamma+1}{\gamma-1}\left[1-\left(\frac{\bar{\rho}}{\rho_o}\right)^{\gamma-1}\right]\right\}^{1/2} = \frac{\bar{V}_i}{V^*} \tag{3.84}$$

where V^* is the critical velocity.

The incompressible velocity V can be derived from any one of the techniques described in Section 3.3. McFarland (1982) utilized Eqs. 3.83 and 3.84 in combination with the panel method described in Section 3.3. The pressure distribution measured in a high-turning turbine cascade by Schwab (1982) is shown compared with the predicted values in Fig. 3.20 for a nozzle with 75° turning and solidity based on axial chord of 0.716. The agreement is

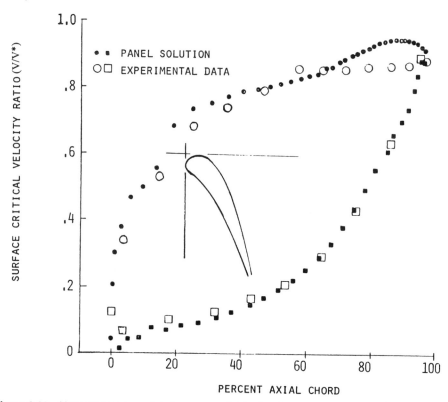

Figure 3.20 Measured and predicted pressure distribution for a turbine cascade (Schwab, NASA TM 82894,1982).

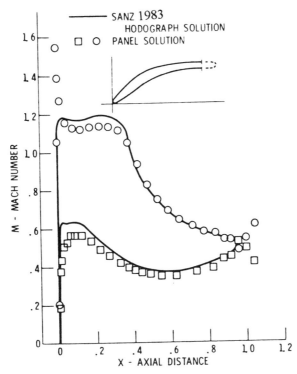

Figure 3.21 Measured and predicted Mach distribution for a compressor stator. (Copyright © McFarland, 1982; reprinted with permission of ASME.)

excellent except near the suction peak, where the Mach numbers are very high subsonic. The discrepancy near the suction peak may be due to the viscous effect, which is not included in this analysis. The prediction for a compressor cascade blade is shown in Fig. 3.21. The shock-free cascade blade was designed by Sanz (1983) using the method of complex characteristics due to Garabedian and Korn (1976) described in Section 3.4.3. The agreement between the panel method and the hodograph solution is very good, although the panel method predicts a velocity "spike" on the suction surface at the leading edge. It is encouraging that a simple relationship such as Eqs. 3.83 and 3.84 provides such excellent agreement. It is necessary to validate this method fully for a wide range of cascades geometries.

Oswatitsch and Rhyming (1957) employed the governing equations written in terms of potential and streamlines. In many situations, the equations cast in $\phi-\psi$ (Potential and Stream functions) coordinates are extremely useful. This is especially true for the indirect problem, where the physical boundaries $(x, y,$ or $s, n)$ are not known. The use of these coordinates for the indirect problem (or design problem) will be dealt with later. Using the definition of the stream function, we can write the continuity equation and the irrotationality condition as

$$\frac{1}{\rho V}\frac{\partial}{\partial \phi}(\rho V) + \frac{\partial \theta}{\partial \psi} = 0 \qquad (3.85)$$

$$\frac{1}{V}\frac{\partial V}{\partial \psi} - \frac{\partial \theta}{\partial \phi} = 0 \qquad (3.86)$$

where θ is the direction of the velocity vector and V is the velocity. The corresponding equation for the incompressible flow is given by

$$\frac{1}{V_i}\frac{\partial V_i}{\partial \phi_i} + \frac{\partial \theta}{\partial \psi_i} = 0 \qquad (3.87)$$

$$\frac{1}{V_i}\frac{\partial V_i}{\partial \psi_i} - \frac{\partial \theta}{\partial \phi_i} = 0 \qquad (3.88)$$

where subscript i represents the values for incompressible flow.

Integration of Eqs. 3.85 and 3.87 along a streamline (ψ) results in the following equation,

$$\frac{\rho V}{\rho_1 V_1} = \frac{V_i}{V_1} \qquad (3.89)$$

because $\rho = \rho_1$ for the incompressible flow case.

Hence, the constant of integration is chosen so that the densities and velocities of the compressible and incompressible cases are identical.

Similarly, by integrating Eqs. 3.86 and 3.88 along the constant ϕ lines, it can be proven that

$$\frac{V}{V_b} = \frac{V_i}{V_{ib}} \qquad (3.90)$$

where the subscript b refers to values along the contour of the blade.

Equations 3.89 and 3.90 constitute a set of equations relating the incompressible flow to the compressible flow. Equation 3.89 is valid exactly (within the approximations made) in the center of the channel, according to stream filament theory. Hence, this equation is used to determine the compressible flow velocity along the mean streamline (midpassage) by Oswatitsch and Rhyming (1957). If we know the values of V_{ib} from the incompressible flow theory, Eq. 3.90 can be used along a potential line to determine the compressible flow velocity along the blade contour. Oswatitsch and Rhyming (1957) computed the flow field for a turbine cascade at $M = 0.42$, and they showed good agreement between their results and the numerical solutions of Wu and Brown (1952). Kynast (1960) later utilized this method to compute the flow field in a turbine cascade at axial Mach numbers of 0.22, 0.40, 0.60,

0.75, and 0.85. The agreement is quite good up to $M = 0.6$, but small deviations are observed near the leading edge of the suction surface for $M = 0.75$ and 0.85. Except for this, the agreement is very good, even for the peak suction surface pressures. Errors near thick leading edges (as in a turbine cascade) exist in most of the solutions available.

3.4.2   Analytical Solution via Transformation

In this method, the equations of motion relating the potential and stream functions, which are nonlinear, are transferred into a linear equation either in the physical plane or in the hodograph plane. In the latter case, the dependent variables are changed into independent variables, which result in a linear equation. This method is attributed to Chapyglin (1944).

If an analytical solution is sought, the most convenient choice of coordinates are the intrinsic coordinates, with s being the streamwise and n the principal normal directions. The compressible equations valid in this system are derived in Liepmann and Roshko (1957) and in Von Mises (1958). In an intrinsic coordinate system, the equations of motion reduce to (because $M_n = 0$)

$$(1 - M^2)\phi_{ss} + \phi_{nn} = 0 \tag{3.91}$$

This equation is nonlinear because M is the local Mach number, $M(s, n)$.

One of the methods of solving Eq. 3.91 in the physical plane is due to Lakomy (1971, 1974). This method is accurate and seems to provide very good agreement with the experimental data. Lakomy transforms the nonlinear equation (Eq. 3.91) into a linear equation through the transformation of dependent variables. The transformed equation is analogous to the incompressible flow equations. By comparing the flow fields of incompressible and compressible flows in the hodograph plane, he was able to derive the compressible flow solution from a known incompressible flow solution. It should be emphasized here that this method is exact, unlike the approximate methods described earlier, which also employed a similar technique (i.e., to derive compressible flow solutions from a known or equivalent incompressible flow solution).

Lakomy (1971, 1974) transforms the nonlinear Eq. 3.91 into a linear equation using the transformation

$$ds^i = (V/V^i)\, ds, \qquad dn^i = (V/V^i)\sqrt{1 - M^2}\, dn \tag{3.92}$$

where the superscript i denotes the quantities in the transformed plane. In these and all subsequent equations, the velocities are nondimensionalized by inlet velocity, and the density is nondimensionalized by its inlet value. The resulting equation is

$$\phi^i_{s^i s^i} + \phi^i_{n^i n^i} = 0 \tag{3.93}$$

This equation represents the incompressible flow through an equivalent cascade. The transformation is not yet complete, because the resulting "equivalent" cascade may not satisfy the boundary conditions (closure of the profile). To accomplish this, Lakomy introduces another transformation into a hypothetical plane (denoted by a tilde) such that $d\tilde{\phi}^i = d\bar{\phi}$ and $d\tilde{\psi}^i = d\bar{\psi}$; and on the airfoil surface, $\tilde{V}(s) = \bar{V}^i(s)$ and $\tilde{\theta}^i(s) = \bar{\theta}(s)$. This completes the transformation of the equations as well as the boundary condition to an equivalent incompressible flow cascade. Equations 3.92–3.93 are employed to derive solutions for a cascade at high subsonic Mach numbers (shock-free). The reader is referred to Lakomy (1971, 1974) for details.

The predictions from this theory agree very well with experimental data for compressor and turbine cascades.

3.4.3 Design of Supercritical Airfoils via Hodograph Solution

The choice of the hodograph coordinate system, in which the velocity components $(U, V), (Q, \theta)$, are used as independent variables, are attractive in designing subcritical and supercritical airfoils (shock-free). The velocities U, V are in X and Y directions (e.g. Fig. 3.6), and Q and θ are the total velocity and the streamline angle, respectively. Hodograph methods for the analysis and the design of cascades in incompressible flow have been studied by many investigators and these are reviewed by Scholz (1965). For example, Cantrell and Fowler (1959) developed a very useful technique for designing turbine cascades from a prescribed velocity distribution using the hodograph method. Hobson (1979) developed a method of designing shock-free transonic flow in a turbomachinery cascade. Many numerical analysts use this exact method and the profile to validate their computer codes.

The hodograph technique developed by Bauer et al. (1977) for the design of supercritical shock-free airfoils has been extended by McIntyre (1976), Stephens (1978), and Sanz (1987) to design shock-free cascades. The conventional operation in transonic regime has supersonic patches on the suction surface (e.g., Fig. 3.18c) resulting in strong shocks and losses associated with these shocks. This results in reduced performance, decreased pressure rise for a compressor, and a decreased pressure drop for a turbine. The efficiency is also decreased. Hence, design of shock-free airfoils is very attractive in the transonic flow range.

In the inverse design method, either the velocity or the pressure is specified on the blade and the blade shape is derived by solving the potential equation (Eq. 1.52) through the use of complex characteristics, after mapping the flow field into a velocity $(U, V$ or $Q, \theta)$ plane. The viscous corrections can be included in the design using the methods outlined in Chapter 5.

A brief outline of the shock-free cascade design method due to Bauer et al. (1977), McIntyre (1976), and Sanz (1987) is given below. The system of equations used are the potential equations (Eqs. 1.52 and 1.57) which are valid for steady, inviscid and irrotational flows. Equation 1.52 and the irrotational condition in two dimensions are transformed into (U, V) coordinates with X and Y as dependent variables, through the Legendre transformation, resulting in the following equations (where $X_U = \partial X / \partial U$ etc.):

$$(a^2 - U^2)Y_V + UV(X_V + Y_U) + (a^2 - V^2)X_U = 0 \quad \text{and} \quad X_V - Y_U = 0$$
$$(3.94)$$

Thus the nonlinear equation (Eq. 1.52) is transformed into a linear system of first-order partial differential equations by considering $X(U, V)$ and $Y(U, V)$. These equations can be solved by method of characteristics. If ξ and η are characteristic coordinates, these equations can be reduced to a set of four characteristic equations given by

$$V_\xi = \tau_+ U_\xi, \qquad V_\eta = \tau_- U_\eta, \qquad X_\xi + \tau_- Y_\xi = 0, \qquad X_\eta + \tau_+ Y_\eta = 0$$

where $\tau_\pm = (UV \pm a\sqrt{Q^2 - a^2})/(a^2 - V^2)$. These characteristic equations are solved simultaneously from the initial point (ξ_0, η_0) in the four-dimensional space. Garabedian and Korn (1976) developed a technique whereby the flow is mapped into a unit circle (similar to incompressible flow shown in Figs. 3.8 and 3.9 for a flat-plate cascade) by conformal transformation.

Sanz (1987) has developed a computer code for the design of shock-free cascade airfoils. One example of a supercritical compressor cascade is shown in Fig. 3.21. The design (prescribed) and the predicted Mach number distribution on the final derived shape is in good agreement. A second example, the 160°-turning turbine rotor blade geometry for a space craft turbine, is shown in Fig. 2.65. The preliminary shape was derived using the Sanz (1987) code. It was subsequently modified to include three-dimensional and viscous effects.

The hodograph technique is extremely useful in deriving baseline or preliminary shapes and can substantially reduce the development cycle for a turbomachinery system. Stephens (1978) designed and successfully tested a supercritical compressor cascade. His results clearly demonstrate improved performance, reduced losses, and, most importantly, successful implementation of this technique to turbomachinery application.

3.4.4 Numerical Solution

Numerical methods for solving the equations governing the two-dimensional inviscid compressible flow through a cascade can be broadly classified as

1. Finite difference and finite element methods
2. Relaxation technique
3. Streamline curvature method

The numerical methods developed for viscous, shocked flows are equally applicable to subsonic, shock-free flows. These methods are covered in detail in Chapter 5.

3.4.5 Concluding Remarks

It can be concluded from the analyses presented in this section that the solution of inviscid, subsonic, and incompressible flow through a cascade has essentially been completed, with the exception of further refinements needed for the accurate solution of the flow near the leading and trailing edges. For the general case of a blade with a rounded trailing edge, inviscid solutions are indeterminate and viscosity must be invoked, if only in its simplest form of making the suction surface and pressure surface velocities equal at the trailing edge. The method of singularities, solution via analytical transformation, and many of the numerical methods described later accurately predict the inviscid flow (subsonic and incompressible flows) through cascades for the direct problem. Iterative procedures converge slowly as the sonic condition is approached. For high subsonic regions, the most appropriate method would involve a numerical solution of compressible potential or stream function equations described in Chapter 5.

3.5 NATURE OF TRANSONIC AND SUPERSONIC FLOWS AND SOME ANALYTICAL METHODS

3.5.1 Transonic and Supersonic Flows

The definition of supersonic and transonic compressors was introduced in Section 2.5.10. The terminology "supersonic" and "transonic" cascade is less obvious. In general, the flow is classified as follows:

Transonic compressor cascade: $M_{cr} < M_1 < 1.0$
Transonic turbine cascade: $M_{cr} < M_2 < 1.0$
Supersonic compressor cascade: $M_1 \geq 1.0$
Supersonic turbine cascade: $M_2 \geq 1.0$

In many cases, the terminology is based on inlet flow (compressor) or exit flow (turbine). For example, supersonic inlet flow has $M_1 > 1$. The development of the flow as the entry Mach number is increased from subsonic to supersonic flow is illustrated in Fig. 3.22 for a compressor cascade and in Fig. 2.66 for a turbine cascade. Referring to Fig. 3.22, a supersonic bubble may exist near the leading edge or inside the passage or both, depending on the inlet Mach number and the rate of acceleration of the flow near the suction surface. As M_1 approaches M_{cr}, change in suction peak pressures occur in a transonic cascade. In Fig. 3.22a, the sonic region is initiated very

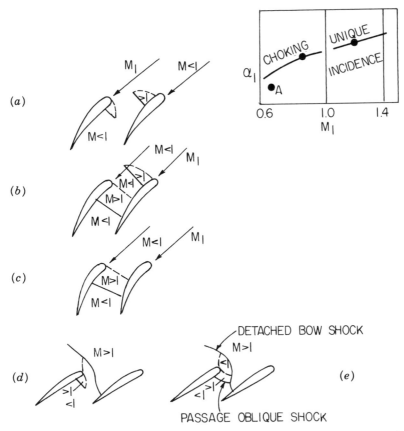

Figure 3.22 Transition from subsonic to supersonic flow in a compressor cascade (modified and adapted from Chauvin et al., 1969).

near the leading edge, and the shock wave is located further downstream. As the Mach number is increased further (without changing the flow inlet angle), transition takes place from (a) to (d) and then to (e). The condition at (b) represents a choked condition, and further increase in M_1 from (a) to (b) or (c) is possible only with change in inlet angle (α_1). The limiting conditions for the operation of a cascade in subsonic and supersonic flow are shown as insert in Fig. 3.22. At subsonic conditions, the limiting condition is achieved when the cascade is choked. At supersonic Mach numbers, the limiting condition is the unique incidence, described later. Cascade tests and analysis are extremely useful in deriving information on the shock structure, shock losses, and blade pressure distribution. Schreiber and Starken (1984) have carried out detailed tests in a transonic rotor and a corresponding cascade to prove the validity of cascade tests in a rotor environment. Across a shock wave, the static pressure and temperatures increase and the Mach number decreases. The static pressure coefficient increases and the Mach number becomes subsonic downstream of the shock wave. When the free-stream Mach number is increased continuously, the transonic range begins when the highest local Mach number reaches unity and ends when the inlet (compressor) or exit (turbine) Mach number reaches unity. In the latter case, the cascade is "supersonic".

In a "supersonic cascade flow," the approaching flow (compressor) or the exit flow (turbine) is supersonic. As the inlet Mach number to the cascade is increased to supersonic values, shock wave formation occurs ahead of the cascade, as shown in Fig. 3.22e. Various possible flow configurations in a turbomachine cascade (both turbine and compressor) are shown in Fig. 3.23, reproduced from Starken (1993). The nature of the flow field depends on inlet and exit Mach numbers and on inlet and exit flow angles. Their interdependency is illustrated in Fig. 3.23 for a compressor cascade, but they apply also to turbine cascades (Fig. 2.66). There exists a "unique incidence" or inlet flow and "unique deviation" or exit flow for certain operating ranges. For the sake of completeness, these cases are presented here and described in detail later. The cascade flow can be classified (Starken, 1993) as follows:

Subsonic Flow. At subsonic inlet and exit flow condition, the inlet velocity and inlet flow angle can be specified independently and the exit flow angle depends on the Kutta–Joukowski condition or viscous considerations.

Subsonic Choked Inlet. The subsonic choked inlet flow condition is the limiting operating range of a compressor cascade. It is also the operating point of transonic and supersonic turbine cascades. At this condition, there exists only one inlet flow angle for a fixed inlet Mach number. The variation in the exit flow conditions ranging from subsonic to axial supersonic velocity are achieved by varying the downstream back pressure. The axial supersonic exit conditions are achieved at unique deviation angle (described later) for a given blade geometry and inlet flow.

Transonic Inlet. At transonic inlet flow conditions, the inlet flow angle is specified by the back pressure at subsonic exit flow conditions and by the unique incidence at transonic and supersonic exit flow conditions.

Transonic/Supersonic Choked Inlet. If the flow experiences a minimum cross section within the blade passage, either due to the blade geometry or due to boundary layer separation at supersonic inlet velocity, it may also choke. This inlet condition is therefore called *transonic/supersonic choked*. It is identical to the so-called "unstarting" of transonic or supersonic cascades.

Supersonic (Axial Subsonic) Inlet. Starting is generally achieved by increasing the Mach number and the inlet flow angle. They are characterized by the unique incidence condition for the inlet flow which means blade

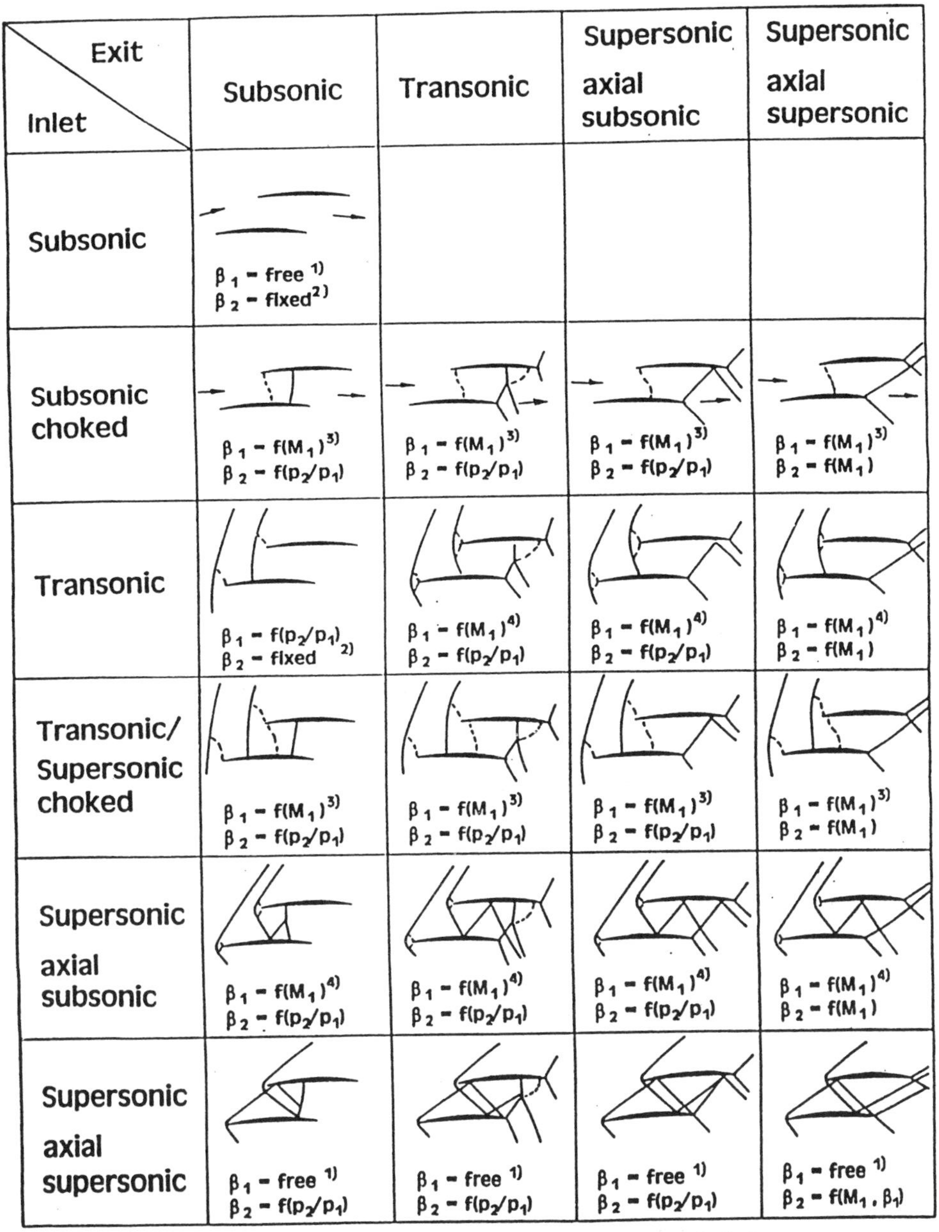

Inlet \ Exit	Subsonic	Transonic	Supersonic axial subsonic	Supersonic axial supersonic
Subsonic	β_1 = free [1] β_2 = fixed [2]			
Subsonic choked	β_1 = f(M_1) [3] β_2 = f(p_2/p_1)	β_1 = f(M_1) [3] β_2 = f(p_2/p_1)	β_1 = f(M_1) [3] β_2 = f(p_2/p_1)	β_1 = f(M_1) [3] β_2 = f(M_1)
Transonic	β_1 = f(p_2/p_1) [2] β_2 = fixed [2]	β_1 = f(M_1) [4] β_2 = f(p_2/p_1)	β_1 = f(M_1) [4] β_2 = f(p_2/p_1)	β_1 = f(M_1) [4] β_2 = f(M_1)
Transonic/ Supersonic choked	β_1 = f(M_1) [3] β_2 = f(p_2/p_1)	β_1 = f(M_1) [3] β_2 = f(p_2/p_1)	β_1 = f(M_1) [3] β_2 = f(p_2/p_1)	β_1 = f(M_1) [3] β_2 = f(M_1)
Supersonic axial subsonic	β_1 = f(M_1) [4] β_2 = f(p_2/p_1)	β_1 = f(M_1) [4] β_2 = f(p_2/p_1)	β_1 = f(M_1) [4] β_2 = f(p_2/p_1)	β_1 = f(M_1) [4] β_2 = f(M_1)
Supersonic axial supersonic	β_1 = free [1] β_2 = f(p_2/p_1)	β_1 = free [1] β_2 = f(p_2/p_1)	β_1 = free [1] β_2 = f(p_2/p_1)	β_1 = free [1] β_2 = f(M_1, β_1)

1) free in the meaning of an independent variable parameter between certain limits
2) fixed due to viscous effects
3) determined by passage throat
4) determined by unique incidence

Figure 3.23 Flow pattern and related boundary conditions of possible cascade flow configurations, where β_1 is the inlet flow angle and β_2 is the exit flow angle (Starken, 1993). (The original version of this material was first published by AGARD AG 328.)

geometry and inlet Mach number determine the "unique incidence" or the inlet flow angle. The exit flow again depends on the back pressure at axial subsonic exit conditions and on the blade geometry and inlet flow at axial supersonic exit conditions.

Axial Supersonic Inlet. If the axial component of the inlet flow is increased to supersonic values, the inlet flow angle, between certain limits, is again independent of the inlet Mach number and a free parameter. The exit flow behavior is identical to the axial subsonic case described before.

Physical laws governing subsonic and supersonic inviscid flows are different. The former is governed by an elliptic equation (Eq. 1.52 or 1.60, $M < 1$), whereas the equation for the supersonic case is hyperbolic. The mathematical difficulty in solving the "mixed" flows is so great that there are very few analytical solutions. If shock waves are present inside the passage, an additional difficulty arises due to jump in the flow properties across the shock. Such discontinuities, and the resulting shock boundary layer interaction, are beyond the scope of most analyses. There are other difficulties involved in the solution of these flows. One of them is due to the dual value of velocities for the same mass flow parameter. There are two velocities which give the same value of ρV, one in the subsonic region and the other in the supersonic region. If a numerical analysis (e.g., finite difference) is employed (where the calculation of ρ lags by one iteration), the equation governing the flow (e.g., Eq. 1.60) does not provide any indication as to whether the flow is supersonic or subsonic at any grid point.

In external aerodynamics, several theories are available for the prediction of the flow field around a wing. These employ a "small perturbation" assumption, which reduces Eq. 1.52 or 1.60 into a linear equation for perturbation potential function or stream function, respectively. Various methods (parabolic method, parametric differentiation, method of local linearization, or integral methods) have been proposed to solve such flows, especially over a thin slender body. The two-dimensional solutions available from such flows are fairly accurate for a single airfoil. A review of these techniques can be found in Guderly (1962). Most analytical techniques are not valid for cascade flows, because the governing equations (Eqs. 1.60 or 1.52) are nonlinear and the perturbations are not small. Hence, the linearized and small perturbation theories are not valid for such flows, except for small-turning blade rows. The methods that are applicable to cascade flows are mostly numerical and are described in Chapter 5.

The qualitative nature of flow inside the passage of a supersonic compressor cascade is shown in Fig. 3.24. This particular case has a detached bow shock at the entrance, a passage oblique shock and a series of reflected shocks. The shock–boundary-layer interaction results in rapid boundary layer growth and thick wakes. The nature of measured shock structure and the blade pressure distribution in a supersonic compressor cascade ($M_1 = 1.53$,

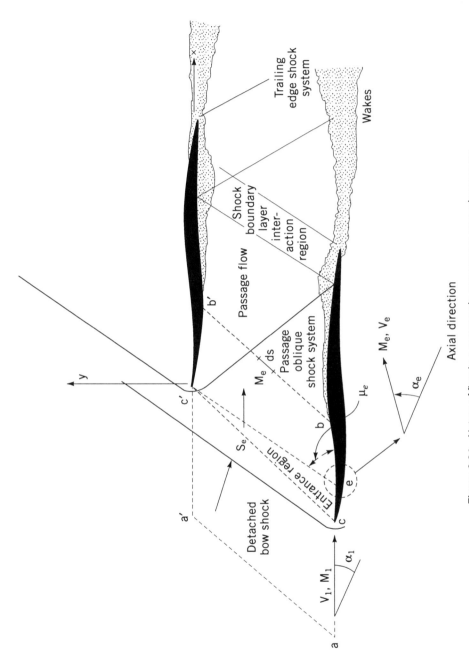

Figure 3.24 Nature of flow in a supersonic compressor cascade passage.

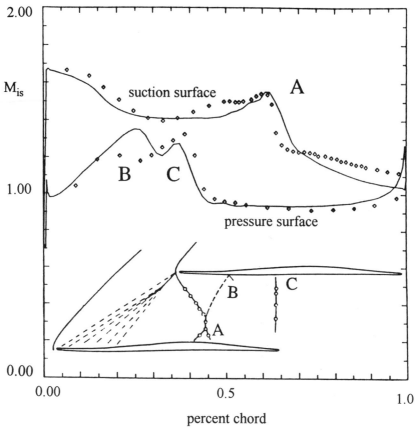

Figure 3.25 Isentropic blade surface Mach numbers and shock structure identification for PAV-1.5 cascade computation. Calculated (*solid line*) and experimental (*symbols*) values. The bottom half of the plot shows the shock-wave pattern deduced from flow visualization and L2F measurements. (Adapted from Schreiber, 1988 AGARD CP 401.)

$p_2/p_1 = 2.13$, Re $= 2.7 \times 10^6$), shown in Fig. 3.25, clearly reveals the complex flow features that exist in supersonic compressors. The predictions are achieved through a full Navier–Stokes solution (Kunz and Lakshminarayana, 1992c), described in Chapter 5. The features labeled A, B, and C in Fig. 3.25 correspond, respectively, to the local compression region where the bow shock impinges on the suction surface, the Mach reflection impinges on the pressure surface, and the passage shock impinges on the pressure surfaces. These regions have substantial streamwise pressure gradients.

Several excellent reviews on the supersonic cascade are available (e.g., Chauvin et al., 1969, 1970; Lichtfuss and Starken, 1974; Bolcs and Suter, 1986).

3.5.1.1 Supersonic Inlet Flow   The most common example of this case occurs when the supersonic relative inlet flow (subsonic axial velocity) is decelerated through a cascade to subsonic exit flow. Many modern aircraft engine fans (outer radii) operate in this mode. The physical nature of the flow is shown (cases d and e in Fig. 3.22) in Fig. 3.24. The deceleration is accomplished through a normal shock. Detailed discussion of this case can be found in Scholz (1965), Lichtfuss and Starken (1974), and Chauvin et al. (1969). Depending on the Mach number and the leading-edge shape, the shock waves could be either attached to or detached from the leading edge. For a thick leading edge, the shock waves are detached, as shown in Figs. 3.22 and 3.24. The upper part of the shock wave continues as an oblique shock, while the lower part remains as an oblique shock or normal shock depending on the back pressure.

The changes in the shock structure that occur as the Mach number is continuously increased are shown in Fig. 3.22 (cases d and e). At slightly supersonic Mach number, the shock wave is detached and no choking occurs (case d). As the Mach number is increased further the supersonic region of the pressure side is connected with that of the suction side and a passage shock is generated (case e, started condition). The exit flow may be subsonic.

In started supersonic cascade flow, the approach flow Mach number and the inlet flow angle are not independent of each other. There is one particular incidence, called the *unique incidence*, at which cascade operation is possible. The flow upstream of the blade row must be periodic and hence a unique relationship exists between the inlet Mach number and the inlet flow angle. The unique incidence condition establishes a particular mass flow. This phenomenon is sometimes referred to as the *entrance problem*. Lichtfuss and Starken (1974) provide a very detailed explanation of this phenomenon, a summary of which is given below. The explanation is for cambered blades with attached shocks. The axial Mach number is a controlling factor in deciding the nature of the inlet flow. Only the infinite cascade, representative of the rotor flow, is discussed. For a discussion of the finite cascade, as in a cascade in a wind tunnel, refer to Starken (1978) and to Lichtfuss and Starken (1974).

The concept of "unique incidence" and operating boundaries of a supersonic cascade are illustrated in Fig. 3.26, based on the example of a circular arc cascade treated by Lichtfuss and Starken (1974). The cascade blades are at a stagger angle of $60°$, $S/C = 1$, and the chamber line has a radius of 5 chord lengths. To the left of the branch point s, the axial Mach number is subsonic. The "unique incidence" line is a function of the inlet Mach number and the cascade geometry. Consider the high angle of attack, shown at point B. As the inset indicates, the blade has a series of expansion fans emanating from each leading edge. These propagate upstream of the flow. The upstream flow has to pass through a large number of expansion waves from various blades before reaching the leading edge of the blade. The axial component of velocity increases until the inlet flow direction is parallel to the

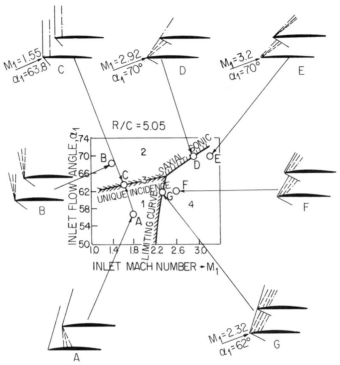

Figure 3.26 Supersonic inlet flow boundaries for a circular arc cascade ($\lambda = 60°$, $S/C = 1$, $R/C = 5$). (Modified and adapted from Lichtfuss and Starken, 1974.)

suction side, and the operating point moves to the "unique incidence" condition as shown (point C). If the angle of attack is negative as shown (point A and inset A in Fig. 3.26), the upstream flow has to pass through a series of shock waves before reaching the leading edge. This decreases the axial Mach number, resulting in an increase in incidence, and the operating point moves toward the unique incidence point (C) shown in Fig. 3.26. The "unique incidence" is realized when the expansion lines upstream of the cascade cancel the effects of the oblique shock upstream. Such a condition is realizable along the unique incidence line, and this operation condition is shown as inset C in Fig. 3.26. Other situations where the upstream flow is not affected are discussed in Lichtfuss and Starken (1974). The "unique incidence line" shown in Fig. 3.26 depends on the stagger angle, space chord ratio, and suction surface profile. A method for using the inlet flow field solution to derive this line will be described later.

The unique incidence curve ends at s, which corresponds to sonic axial velocity. If the axial Mach number is supersonic, operation is possible in the region bounded by the axial sonic line and the limiting curve, as shown in Fig. 3.26. The shock configuration at E (and inset E) in Fig. 3.26 indicates that the expansion and shock waves do not influence the upstream flow, and hence remain undisturbed. If the Mach number is reduced, while keeping the axial component supersonic, a situation such as that shown in inset D, point D, in Fig. 3.26 occurs. In this case, the first expansion wave just goes through the leading edge of the following blade. If the Mach number is reduced further, the disturbances originating from the blades would run in front of the cascade, resulting in subsonic axial Mach numbers. Thus, no continuous transition from axial supersonic to axial subsonic velocities is possible at this inlet flow angle. A similar situation occurs at incidences below s and along the limiting curve (cases F and G). Operation at supersonic axial velocities is therefore possible only along and to the right of the lines bounding the "axial sonic" and "limiting curve" marked in Fig. 3.26 (Region 4). At axial subsonic velocities, operation is confined to the "unique incidence" line (marked 3) and is not possible in the shaded region (regions marked 1 and 2). Thus, at high subsonic axial Mach numbers, a unique relationship between the incident Mach number and the air-flow angle exists and is a function of the cascade geometry and the inlet Mach number.

3.5.1.2 Analyses for the Supersonic Inlet Flow

Several analyses are available for predicting the inlet flow in a supersonic cascade. Unlike a subsonic cascade, the inlet conditions in this case are uniquely determined by the blade geometry and the shock-wave pattern. There are three methods available for the calculation of the entrance flow: (1) The irrotational or simple wave model yields an approximate solution to a class of entrance flows with sharp leading edge blades and attached shocks, as shown in Fig. 3.24. This approach is described in Levine (1957), Yamaguchi (1964), and Lawaczeck (1972). (2) Semiempirical models are used when detached shock waves are present in the flow. The shock waves are generally detached when the leading edge has a finite radius. The analysis on this is due to Kantrowitz (1950), Starken (1971), and Novak (1967). (3) An inviscid solution, valid for rotational flows, was pursued by York and Woodard (1976). In this technique, the inviscid equations of motion and energy equations are solved numerically by the method of characteristics.

The simple wave analysis is due to Levine (1957), Yamaguchi (1964), Starken (1971), and Lichtfuss and Starken (1974). A description of Levine's (1957) method is given below. The objective is to determine the incidence (i) or the inlet angle (α_1), at a prescribed inlet Mach number M_1. For simplification, the flow condition at one arbitrary point, such as e in Fig. 3.24, is assumed arbitrarily. The flow direction is parallel to the blade surface at this location as shown in Fig. 3.24. The Mach number is determined such that the left-running Mach line through this point meets the leading edge of the adjoining blade. The flow is supersonic and uniform upstream of the expansion wave ec' (before it enters the blade passage). The flow direction and the Mach number are constant along the characteristic ec'. The flow angle α_e, and the wave angle μ_e are known at this location. The control volume $aa'c'e$

is chosen to satisfy the continuity and the Prandtl-Meyer relationship. Because the flow is of the simple-wave type, the following Prandtl–Meyer relationship holds (see Fig. 3.24 for notations):

$$\alpha + \nu = \text{const}$$

$$\nu(M) = \frac{\sqrt{\gamma + 1}}{\sqrt{\gamma - 1}} \tan^{-1} \sqrt{\frac{\gamma - 1}{\gamma + 1}(M^2 - 1)} - \tan^{-1} \sqrt{M^2 - 1}$$

Hence,

$$\alpha_1 + \nu(M_1) = \alpha_e + \nu(M_e) \tag{3.95}$$

where the subscript 1 represents the upstream condition, and e represents the conditions at the arbitrarily chosen point e (Fig. 3.24). This can be combined with the continuity equation to provide

$$\rho_1 V_1 S \cos \alpha_1 = \rho_e V_e S_e \sin \mu_e \tag{3.96}$$

where S_e is the distance between point e and the leading edge of the next blade, as shown in Fig. 3.24. The value of S_e can be determined from the cascade geometry to be

$$\frac{S_e}{S} = \sqrt{\left(\cos \lambda - \frac{y_e}{S}\right)^2 + \left(-\sin \lambda - \frac{x_e}{S}\right)^2} \tag{3.97}$$

where x_e and y_e are coordinates of point e.

The inlet Mach number is given by [assuming isoenergetic ($T_o = \text{const}$) and homentropic ($P_{o1} = P_{oe}$) flow]

$$\frac{M_1 \cos \alpha_1}{\left[1 + \dfrac{\gamma - 1}{2}M_1^2\right]^{(\gamma+1)/2(\gamma-1)}} = \left\{\frac{S_e}{S}\right\} \frac{M_e \sin \mu_e}{\left[1 + \dfrac{\gamma - 1}{2}M_e^2\right]^{(\gamma+1)/2(\gamma-1)}} \tag{3.98}$$

Eqs. 3.95 and 3.98 are to be satisfied by the flow in the entrance region ($aa'c'e$). The values of μ_e, α_e, S_e, S and γ are known. Equations 3.95 and 3.98 thus represent the "unique incidence condition." For example, if the value of M_e is fixed and M_1 is assumed known, there is only one value of α_1 (unique incidence) that satisfies these equations. Thus, for a chosen value of M_e, M_1 and α_1 are uniquely related (unlike a subsonic cascade). By moving point e along the contour from the blade leading edge, the unique incidence curve, such as that shown in Fig. 3.26, can be obtained. The axial supersonic limiting curve in Fig. 3.26 represents a second solution of Eqs. 3.98 and 3.95.

More thorough discussion of the physical nature of the flow, the effects of leading edge thickness and camber, and the maximum thickness can be found in Bolcs and Suter (1986) and in the chapter by Erwin and Ferri (Hawthorne, 1964).

In several instances, the supersonic regions are either small or a detached shock wave exists. The latter occurs with finite-radius leading edges. Many investigators (Moeckel, 1949; York and Woodard, 1976; Starken et al., 1985; Freeman and Cumpsty, 1989) have provided approximate solutions for this problem. Most of these methods utilize the continuity, momentum, and energy equations, integrated across the blade passage and a control volume approach. The governing equations are written for the control volume between the inlet (e.g., aa' in Fig. 3.24) and an arbitrary line bb' in the entrance region. The lateral sides of the control volume are taken as consecutive stagnation streamline (e.g., ac, a'c' in Fig. 3.24). The continuity, momentum, and energy equations for the control volume abb'a' are written and solved iteratively.

A knowledge of the unique incidence angle is important for an under-standing of the flow through a supersonic cascade flow. The real fluid effects and nonlinearity of the governing equations make the results derived from the simple analysis described above to deviate from the actual achievable inlet angles. A more accurate range can be determined by using the numeri-cal methods that solve the inviscid Euler or viscous Navier–Stokes equa-tions (with grid clustering and minimum artificial dissipation) described in Chapter 5.

In actual rotors, the real fluid effects have an appreciable influence on the achievable range of operation. The compressors generally operate at condi-tions other than the unique incidence as shown in Fig. 3.27, due to Schreiber and Starken (1984). Tests were carried out both in a rotor and in a corresponding cascade. The transonic compressor rotor was operated at different speeds (70–100%). The design pressure ratio was 1.51 at a relative inlet (tip) Mach number of 1.38. The choking and unique incidence curves for the cascade were derived from the analysis and the testing. The limiting conditions for the rotor illustrates that the range of operation is higher than those predicted by cascade analysis and that it reaches zero operating range (unique incidence) condition somewhere between $M_1 = 1.2$ and $M_1 = 1.5$.

3.5.1.3 *Supersonic Exit Flow*

Supersonic exit velocities may occur in the case of transonic or supersonic turbines. They may also occur in some supersonic compressors, which have high supersonic inlet velocities. Even though the concept of supersonic inlet flow is relatively new, steam turbines with supersonic exit velocities were in existence long ago (Stodola, 1924).

In what follows, the physical phenomenon with supersonic exit flow is described using a simplified inviscid approach. Lichtfuss and Starken (1974) provide a comprehensive treatment of the flow through a supersonic cascade, including exit flow. The description given below is mainly drawn from their

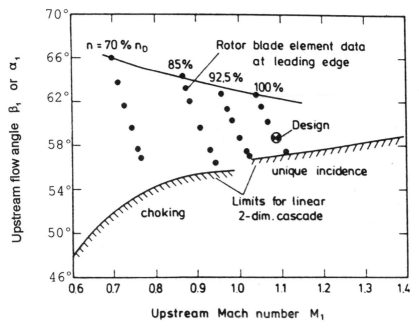

Figure 3.27 Range of inlet flow conditions for axial flow compressor rotor (β_1) and cascade (α_1) blades (at midspan) n_D is the design speed. (Copyright © Schreiber and Starken, 1984; reprinted with permission of ASME.)

article. The phenomenon is described (Lichtfuss and Starken, 1974) on the basis of the inviscid flow, even though the boundary layer growth and shock–boundary-layer interaction have major influence on this flow. The real fluid effects can only be captured from the numerical solution of the exact Navier–Stokes equation. The phenomena and analyses described should be considered qualitative in nature.

Lichtfuss and Starken (1974) deal with the cascades of flat plates and cambered profiles with sharp trailing edges. Because the latter configuration is more practical, treatment of this case is given in detail. The flow at the inlet is assumed to be supersonic. Detailed treatment of this case is given in Lichtfuss and Starken (1974), only a brief description follows here.

It should be emphasized that the flow field and the operating boundaries described below vary from cascade to cascade, and hence are very dependent on the cascade geometries and profiles and the inlet flow. With the super-sonic exit flow, the back pressure is an additional parameter that can be chosen independently of the inlet flow. At supersonic exit axial velocities, however, the back pressure cannot be considered as a free parameter, but is determined solely by the inlet flow and the cascade geometry.

The limiting boundaries of the cascade exit operation are shown in Fig. 3.28. The cascade geometry is the same as that used for describing the

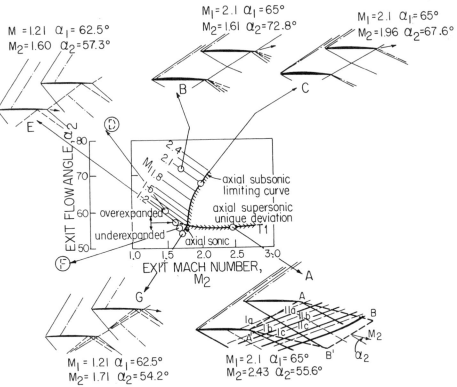

Figure 3.28 Supersonic exit flow boundaries and throttle curve of a circular arc blade cascade. Cases A, B, and C: High subsonic axial inlet; cases E and G: Low subsonic axial inlet. (Modified from Lichtfuss and Starken, 1974.)

inlet flow (Fig. 3.26). The branch marked "axial supersonic" corresponds to the "unique deviation" line and is the only possible operating condition (for a given inlet flow and the cascade geometry) at supersonic axial exit velocity. Consider the operation at (A) and the inset corresponding to this point. The axial component of the inlet flow is subsonic, and the unique incidence relationship, described earlier, is valid. The flow in regions Ia and IIa (inset A in Fig. 3.28) can be computed by the method of characteristics, thereby fixing the flow conditions at the trailing edge. Right-hand- and left-hand-running tail shocks are generated. The shock-wave angle is determined from the flow conditions on the pressure and suction surfaces and from the Kutta condition, because the static pressure and the outlet angle are identical behind both shocks at the trailing-edge region. This will facilitate the computation of the flow in regions Ib and all other regions downstream. The shock waves emanating from the trailing edge are attenuated by their interferences with the Mach lines. Far downstream, only one left-hand-running Mach line and only one right-hand-running Mach line of each passage remain, and these do not interfere with the shocks. The whole flow region downstream of the cascade is covered by the left-hand-running Mach lines originating from the suction surface and the right-hand-running Mach lines originating from the pressure surface. The flow is uniform far downstream. Hence, the flow condition is determined by the inlet flow and the geometry of the cascade; the back pressure is not an independent parameter, and the downstream condition does not influence the cascade flow. As long as the inlet flow is fixed and the outlet axial flow is supersonic, only one deviation condition is possible. Hence, this is referred to as the "unique deviation" concept. Thus, for a given inlet flow and geometry, operation is possible only along the branch T1 (Fig. 3.28) for the supersonic exit axial flow case.

This is not the only case possible for this inlet flow. If the exit axial Mach number is subsonic, but with the same inlet flow as before, several possibilities exist for the exit flow. Such a condition is shown as point B and inset B in Fig. 3.28. This condition can be achieved by an increase in the back pressure from the case shown earlier. The inlet condition and the passage flow (up to the left-hand-running trailing-edge shock wave) at this operating point are identical to the case shown in A, but the important difference is the reflection of the shock wave by the pressure surface, which cancels all the left-hand-running characteristics originating from the suction surface. Because all left-hand-running Mach lines behind the shock must originate in a region infinitely far downstream, the expansion is a simple-wave type. The right-hand-running shock waves of each blade are attenuated by the expansion waves originating from the trailing edge. A uniform exit flow is achieved, and this flow field is thus dependent on (or can be calculated from) the exit Mach number, flow angle, or pressure, which has to be specified. The computation for the flow field is simplified substantially if the outlet angle at the trailing edge is specified. If the back pressure is increased further, an upstream movement of the shock results in a subsonic flow field inside the passage, and the method of characteristics fails, but this is not a limiting condition for the operation of the cascade.

On the other hand, if the back pressure is decreased, the trailing edge shock wave moves downstream; and at a certain value of the exit pressure, Mach number, or outlet angle, the shock wave just goes through the trailing edge, as shown in inset C in Fig. 3.28. The operating line is along BC. A simple wave flow exists downstream, and all left-hand-running Mach lines downstream of the trailing-edge shock must come from the exit region.

As the back pressure is decreased further, the left-hand-running shock wave originating from the trailing edge leaves the passage and the exit flow is no more controlled by the back pressure. This operation is possible only along the "unique deviation" line marked in Fig. 3.28. The transition between points such as C and A is not continuous. Thus, at constant inlet flow, the supersonic outlet flow can exist only at the unique deviation condition, but subsonic exit flow operation is possible along operating lines such as BC. The exit flow in such a case is controlled by the exit pressure.

In the example considered so far, the left-hand-running characteristics originate near the suction surface and leave the blade passage at low back pressure. Let us now consider the operation at a different inlet Mach number (e.g., $M_1 = 1.21$). The operation at this condition is at points D, E, F, and G in Fig. 3.28. At E, the suction surface characteristics are reflected by the pressure surface; hence, the flow downstream should be of the simple-wave type with an axial subsonic component. The flow angle at the exit is chosen to be parallel to the suction surface. This is an adapted flow condition (i.e., neither expansion nor compression near the trailing edge). If the back pressure is increased, an overexpansion occurs, point D in Fig. 3.28. The shock pattern is similar to case B ($M_1 = 2.1$). If the back pressure is decreased, shown as case F in Fig. 3.28, the right-hand-running shock waves are attenuated by the expansion waves. The maximum exit Mach number is achieved (case G) when the last Mach line from the trailing edge just meets the trailing edge of the adjacent blade. This condition corresponds to an axial sonic velocity component.

Hence, at low inlet Mach numbers and low back pressures, the throttle curves end at the axial sonic limiting curve. At higher inlet Mach numbers, a limiting condition is achieved at the axial subsonic exit flow velocity (case C). The operating limits of the exit flow shown plotted as M_2 versus α_2 (Fig. 3.28) provide the boundaries within which the cascade operation is possible. Lichtfuss and Starken (1972) have demonstrated the existence of the upper boundary curve and the "unique deviation" condition.

3.5.1.4 *Analysis of the Supersonic Exit Flow*

Several analyses are available for predicting the exit flow in a supersonic cascade. These are reviewed in Lichtfuss and Starken (1974). One of the simplest methods of calculating the exit flow is by the use of a relationship such as that shown in Eq. 3.95. For axial subsonic outlet Mach numbers, neglecting shock losses, the exit region has a simple wave flow as described earlier, and hence the following equation is valid:

$$\alpha - \nu = \alpha_2 - \nu_2 = \alpha_{TE} - \nu_{TE} \qquad (3.99)$$

where the subscript 2 refers to the far downstream condition, and TE refers to the trailing edge condition. Either the back pressure or the trailing edge flow angle α_{TE} can be chosen as the variable. If expansion occurs at the suction side of the trailing edge, the following equation can be used for the right-hand-running Mach lines:

$$\alpha_s + \nu_s = \alpha_{TE} + \nu_{TE} \qquad (3.100)$$

Here, the subscript s denotes the values at the suction side. The Mach number M_s and ν_s are independent of the outlet exit flow condition. Hence,

(from Eqs. 3.99 and 3.100)

$$\alpha_2 - \nu_2 = 2\alpha_{TE} - (\alpha_s + \nu_s) \qquad (3.101)$$

$(\alpha_s + \nu_s)$ is constant for a given inlet flow and cascade geometry, and this can be determined from the procedure outlined earlier. α_{TE} depends on back pressure imposed. Hence, there are two unknowns $(\alpha_2$ and $\nu_2)$ for the determination of the far downstream flow. Equation 3.101 and the one-dimensional form of the continuity equation (similar to Eq. 3.96) can be used for the determination of the angles α_2 and ν_2. These equations have to be solved iteratively, and the procedure is similar to that described for the inlet supersonic flow. Here again, the real fluid effects (e.g., flow separation near the trailing edge) make the analysis somewhat qualitative. Accurate solution can only be achieved from Navier–Stokes solution, with grids clustered near the trailing edge.

The reader is referred to the papers quoted in this chapter and the book by Bolcs and Suter (1986) for a more comprehensive treatment of transonic and supersonic cascades and turbomachinery, including the performance and the experimental data.

4

THREE-DIMENSIONAL INVISCID AND QUASI-VISCOUS FLOW FIELD

In present day turbomachinery, the three-dimensional effects are hardly negligible and their incorporation into design or analysis is essential for accurate prediction of the performance or improved design of the machinery.

This chapter deals with three-dimensional inviscid and quasi-viscous effects. The three-dimensionality is caused by both the viscous and inviscid effects. The quasi-viscous methods, which incorporate the real fluids effects in an approximate or global manner, are included in this chapter. For example, secondary flow arises due to viscous boundary layer, but the analysis of its effect is treated inviscidly. The viscous effects and their methods of analyses are covered in Chapters 5 and 6. Some of the three-dimensional inviscid effects are due to:

1. Compressibility, and radial density and pressure gradients
2. Radial variation in blade thickness and geometry
3. Presence of finite hub and annulus walls, annulus area changes, flaring, curvature and rotation
4. Radially varying work input or output
5. Presence of two phase flow, and coolant injection
6. Radial component of blade force and the effects of blade skew, sweep, lean and twist
7. Leakage flow due to tip clearance and axial gaps
8. Nonuniform inlet flow and presence of upstream and downstream blade rows
9. Mixed-flow (subsonic, supersonic, and transonic flow) regions along the blade height with shock–boundary-layer interaction
10. Secondary flow caused by inlet velocity/stagnation pressure gradient and flow turning

261

Most of these are caused by inviscid effects, which can be treated by the use of inviscid equations of motion. The secondary flow is caused primarily by the presence of viscous layers on the walls. The dominant influence here is the velocity gradient normal to the wall or the presence of normal (to the streamline) vorticity upstream of the blade row. Three-dimensional inviscid (or secondary flow) theories have been developed to predict the nature of three-dimensional flow away from the walls. These theories account for effects of the velocity gradients upstream of the blade row. In many cases, the inviscid and viscous effects augment each other. For example, even though the leakage flow arises due to blade unloading, its subsequent roll-up and diffusion is controlled by viscous effects. Likewise, two- and three-dimensional shocks are often associated with separation and hence the viscous interaction effects cannot be ignored.

The three-dimensional effects are illustrated in Figs. 1.15, 1.16, and 4.1. Most of the inviscid effects are schematically shown in Fig. 4.1. The radius change of the annulus and the hub walls, as well as change in stream tube height due to area changes, results in spanwise flows. The radial variation of blockage (and area changes in the radial direction) gives rise to radial or spanwise flows. Because the rotor blade is usually thicker at the root than at the tip (from structural considerations), radial shifts in the streamlines occur, thus resulting in radial flows. Likewise, radially varying work input or output

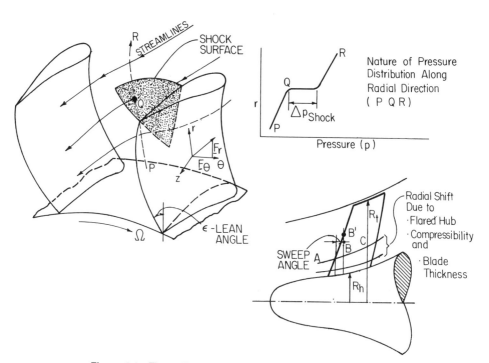

Figure 4.1 Three-dimensional inviscid effects in turbomachinery.

(Δh_o) provides an additional mechanism for spanwise flow generation. This will be explained later. The presence of two-phase flow (usually cavitation bubbles in a pump or hydraulic turbine) results in density discontinuity and deviation from the radial equilibrium equation. The compressibility effect, likewise, induces radial flows inside the blade passage, even in situations where the flow is in simple radial equilibrium upstream and downstream of the blade row, governed by the equation

$$\frac{\partial p}{\partial r} = \rho \frac{V_\theta^2}{r} \tag{4.1}$$

This equation indicates that the centrifugal force is directly proportional to density, which changes along the streamline inside the passage, resulting in an imbalance between the radial pressure gradient and the centrifugal force. This introduces additional acceleration terms (extended form of Eq. 4.1). Thus additional radial flows arise due to the compressibility effect. A similar argument can be put forth to explain the presence of spanwise flows in a passage with a three-dimensional (skewed) shock structure as shown in Fig. 4.1. In this case, the pressure jump ($\partial p/\partial r$) and streamline divergence across a shock (say Q in Fig. 4.1) disturbs the simple radial equilibrium, giving rise to imbalance between the centrifugal forces and the radial pressure gradient. To achieve a new equilibrium condition (say from Q to R), radial acceleration ($\rho DV_r/Dt$) and radial flows have to develop. Additional three-dimensional effects arise due to sudden change in flow turning and the associated streamline divergence as well as jump in density across the shock. This disturbs the radial equilibrium and introduces three-dimensionality including radial flows. Measurements taken in transonic and supersonic turbomachinery reveal the presence of large radial flows due to three-dimensional shock structure. Even though these flows can be treated by inviscid theories, away from blade and walls, the viscous effects and flow separation due to three-dimensional shock–boundary-layer interaction and the resulting radial flows in these separated regions make it necessary to include the three dimensional viscous effects. Likewise, the blade sweep and radial blade force introduce three-dimensionality. For example, the blade sweep shown in Fig. 4.1 introduces large pressure gradients along BB'. Because of the absence of blade beyond B', the radial line spans the blade passage up to B' and the free stream beyond B'. This introduces appreciable radial pressure gradient and associated radial flows. Likewise, the radial blade force, F_r, introduces radial acceleration.

The tip leakage flow results from the unloading of the blade, causing a very complex three-dimensional and vortex flow field at the tip. This will be dealt with in a separate section. The upstream nonuniform flow (radial or circumferential) results in radially varying output/input or circumferentially nonperiodic flow, both of which provide an additional mechanism for the generation of spanwise flows.

The equations of motion governing the turbulent flow through turbomachinery are highly nonlinear, and hence most of the analytical solutions available are for simple flows. These solutions involve several assumptions, depending on the type of machinery and the geometry of blade-row and flow parameters. Attempts to solve the equations (numerically) governing the flow field started in the late 1960s (Cooper and Bosch, 1966; Marsh, 1968). The classical three-dimensional design or analysis is based on an iterative solution of axisymmetric equations and blade-to-blade formulations (cascade solution) discussed earlier. The following techniques can be classified as axisymmetric solutions:

1. Simplified radial equilibrium analysis (SRE)
2. Actuator disk theories (ADT)
3. Passage averaged equations and their solutions (PAE)

The nonaxisymmetric solutions can generally be classified as follows:

1. Lifting line and lifting surface approach
2. Quasi-three-dimensional methods
3. Numerical solutions of exact equations (potential, Euler, and Navier Stokes)

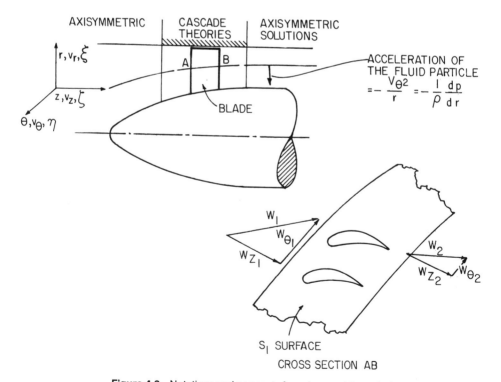

Figure 4.2   Notations and concepts for axisymmetric analysis.

The axisymmetric solution is used to predict radial or spanwise variation of properties of flow far upstream and downstream of the blade rows. Once the local values of flow parameters are known, the cascade theories, described in Chapter 3, can be used to predict the blade-to-blade variation of flow properties. This procedure is illustrated in Fig. 4.2. This technique of combining the axisymmetric theory with the cascade theory is limited to axial turbomachinery. The lifting line and lifting surface theories are mainly used for the analysis of axial turbomachinery, especially propellers. In the quasi-three-dimensional methods, the blade-to-blade flow field and hub-to-tip flow are suitably coupled to derive a composite solution. The numerical solution of exact governing equations will be covered in Chapter 5.

4.1 AXISYMMETRIC SOLUTIONS

In axisymmetric theories, the flow upstream and downstream of the blade row is assumed to be axisymmetric and the governing equations are solved in these regions. Once the radial variation of properties (V_r, V_θ, V_z) is established, the cascade solutions described in Chapter 3 can be used to solve the blade-to-blade flow (S₁ surface in Fig. 4.2).

The equations governing the axisymmetric, inviscid and steady flow upstream and downstream of the blade row in the cylindrical coordinate system $(\partial/\partial\theta = \mu = \lambda = 0$ in Eqs. 1.9–1.10) are given by (Fig. 4.2)

Continuity:
$$\frac{\partial}{\partial r}(\rho r V_r) + \frac{\partial}{\partial z}(\rho r V_z) = 0 \qquad (4.2)$$

Radial momentum:
$$\frac{\partial V_r}{\partial t} + V_r\frac{\partial V_r}{\partial r} + V_z\frac{\partial V_r}{\partial z} - \frac{V_\theta^2}{r}$$
$$= -\frac{1}{\rho}\frac{\partial p}{\partial r} + F_r \qquad (4.3)$$

Tangential momentum:
$$\frac{\partial V_\theta}{\partial t} + V_r\frac{\partial V_\theta}{\partial r} + V_z\frac{\partial V_\theta}{\partial z} + \frac{V_\theta V_r}{r} = F_\theta \qquad (4.4)$$

Axial momentum:
$$\frac{\partial V_z}{\partial t} + V_r\frac{\partial V_z}{\partial r} + V_z\frac{\partial V_z}{\partial z} = -\frac{1}{\rho}\frac{\partial p}{\partial z} + F_z \qquad (4.5)$$

The expressions for vorticity are given by
$$\xi = -\frac{\partial V_\theta}{\partial z}, \qquad \eta = \frac{\partial V_r}{\partial z} - \frac{\partial V_z}{\partial r}, \qquad \zeta = \frac{1}{r}\frac{\partial}{\partial r}(rV_\theta) \qquad (4.6)$$

The forces F_r, F_θ, and F_z are body forces or viscous forces in the quasi-viscous analysis.

4.1.1 Simplified Radial Equilibrium Equation: Analytical Solutions

For steady flow with cylindrical stream surfaces (radial velocity is zero), Eqs. 4.2–4.6 are simplified considerably, and the resulting radial momentum equation is called the *simplified radial equilibrium equation* (SRE). These equations are strictly valid away from the blade row. Only the radial equilibrium equation is relevant because the flow does not vary in either the tangential or the axial direction, since in a steady inviscid flow in a cylindrical annulus without any body force, V_θ, V_z, h, and $h_o = f(r)$ only and $\xi = \partial V_\theta / \partial z = 0$. Therefore, the radial component of the momentum equation is given by

$$\frac{V_\theta^2}{r} = \frac{1}{\rho}\frac{dp}{dr} = \frac{dh}{dr} - T\frac{ds}{dr} \qquad (4.7)$$

In the equation above, the following thermodynamic relationship for entropy is used:

$$T\,ds = dh - \frac{dp}{\rho}$$

Because the curvature effect has a major influence on the flow field in turbomachinery, it is essential to provide a physical explanation of the forces exerted in simple flows and to establish a sign convention. In this particular flow field, the curvature due to swirling flow (due to machine radius r) introduces a centripetal acceleration $(-V_\theta^2/r)$ in the negative r direction, which keeps the flow moving in a swirling motion along a cylindrical section in axial flow turbomachinery. This acceleration is balanced by the radial pressure gradient, $-(1/\rho)(\partial p/\partial r)$, which is acting downward or in the negative r direction (see Fig. 4.2). Thus, $\partial p/\partial r$ is always positive. Hence, the static pressure in a swirling flow in an annulus always increases with an increase in the radius.

The static enthalpy in Eq. 4.7 can be replaced by stagnation enthalpy using the equations

$$h_o = h + \frac{V^2}{2} = h + \frac{V_\theta^2}{2} + \frac{V_z^2}{2}$$

$$\frac{dh_o}{dr} - T\frac{ds}{dr} = \frac{V_\theta^2}{r} + V_\theta\frac{dV_\theta}{dr} + V_z\frac{dV_z}{dr} = \frac{V_\theta}{r}\frac{d}{dr}(rV_\theta) + V_z\frac{dV_z}{dr} \qquad (4.8)$$

Equation 4.8 is an ordinary differential equation and can be easily solved when all the variables, with exception of one, are known. For example, for isentropic flow, if V_θ and h_o are prescribed or known, then Eq. 4.8 can be solved for $V_z(r)$. This is usually the design problem. If V_z and h_o are prescribed, then $V_\theta(r)$ can be determined.

Design Example. *Case A:* Let us consider a case with $h_o =$ constant and $s =$ constant, with different distributions for V_θ. If $rV_\theta =$ constant, it is clear from Eq. 4.8 that V_z = constant. Thus from Eq. 4.6, $\eta = \zeta = 0$. This design case is called a "free-vortex" design. The advantage of this design is that the work input/output and circulation are constant at all radii, and there is no shed vortex from the blade. Most of the early designs were based on this free-vortex design concept. The tangential velocity is prescribed to vary inversely with the radii, and the blading is designed accordingly (see example 3.2 in Section 3.3.1.5). This usually results in very large tangential velocity and flow turning near the hub, and the blading is highly twisted because both the inlet and exit angles as well as stagger angles vary considerably along the radius. Even though the losses due to shed vortex (due to $d\Gamma/dr$) are eliminated, the hub wall losses are likely to be higher.

If the tangential velocity is prescribed according to $V_\theta = kr$, with k, h_o and s constant, then the solution of Eq. 4.8 can be proven to be $V_z = \sqrt{C_1 - 2k^2r^2}$. This results in large tangential velocity in the tip region. Thus, the flow turning is very large in this region, and the axial component of vorticity ζ equals $2k$. Because this represents a solid body rotation, it is called a "forced-vortex" design. The nature of the axial and tangential velocity distribution for both the free- and forced-vortex cases is shown in Fig. 4.3. Very rarely machines are designed for this type of distribution, because the tip losses would be large as well as the losses due to the trailing vortex system. Because circulation around the airfoil is increasing in proportion to the radius, this will result in a strong trailing vortex system at the exit within the blade wake.

The designer usually prescribes velocities somewhere between a forced-vortex and a free-vortex design according to the equation

$$V_\theta = ar \pm \frac{b}{r}$$

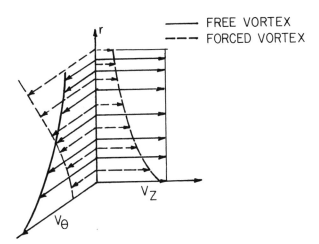

Figure 4.3   Free- and forced-vortex velocity distribution.

where a and b can be varied to suit a particular design. This is called *exponential design.*

If h_o = constant, s = constant, and the outlet angle (α_2) is specified, then Eq. 4.8 reduces to

$$\frac{dV}{dr} + \frac{V \sin^2 \alpha}{r} = 0$$

and the solution is given by

$$\frac{V}{V_m} = \exp\left(-\int_{r_m}^{r} \frac{\sin^2 \alpha}{r} dr\right) \qquad (4.9)$$

where the boundary condition $V = V_m$ at $r = r_m$ (mid-radius) is specified.

Design Example. *Case B:* A more general case is when $h_o = h_o(r)$, s = constant, and $\alpha = \alpha(r)$ is specified. The designer has a much wider choice in this case. A solution for such a case can be derived as follows.

The radial equilibrium equation is given by

$$\frac{V_\theta}{r} \frac{d}{dr}(rV_\theta) + V_z \frac{dV_z}{dr} = \frac{dh_o}{dr}$$

Substituting $V_z = V \cos \alpha$ and $V_\theta = V \sin \alpha$, we get

$$V \frac{\partial V}{\partial r} + \frac{V^2 \sin \alpha}{r} = \frac{dh_o}{dr} \qquad (4.10)$$

The solution to Eq. (4.10) is given by

$$V^2 = \frac{2 \int_{o}^{r} [e^M]\left(\frac{dh_o}{dr}\right) dr + C_1}{e^M} \qquad (4.11)$$

where

$$M = 2 \int^{r} \frac{\sin^2 \alpha}{r} dr$$

If $V = V_m$ at $r = r_m$ (normally the midradius), the constant in the above equation can be evaluated knowing that

$$\int_o^r (\ )\, dr - \int_o^{r_m} (\ )\, dr = \int_{r_m}^r (\ )\, dr \tag{4.12}$$

The final expression for V is given by

$$\frac{V^2}{V_m^2} = \frac{2\int_{r_m}^r \left[\exp\left(2\int \frac{\sin^2 \alpha}{r}\, dr \right) \right] \left(\dfrac{dh_o}{dr} \right) dr}{\exp\left(2\int_{r_m}^r \frac{\sin^2 \alpha}{r}\, dr \right)} + \exp\left(-2\int_{r_m}^r \frac{\sin^2 \alpha}{r}\, dr \right) \tag{4.13}$$

Note that for $dh_o/dr = 0$, Eq. 4.13 reduces to Eq. 4.9. If $\alpha =$ constant and $dh_o/dr = 0$, Eq. 4.13 simplifies to

$$\frac{V}{V_m} = \left(\frac{r}{r_m} \right)^{-\sin^2 \alpha}$$

In many cases, the flow outlet angle is designed to be constant (as in nozzle flows or the last stage stator of a jet engine). Hence such simple analyses are extremely useful.

The analyses presented in this section are widely used in the preliminary design (first cut) of blading. In recent years, many of the computer codes utilize these analyses to prescribe the inlet or exit boundary conditions. For example, in time iterative techniques, the static pressure on the hub wall is prescribed and the simplified radial equilibrium equation is used to prescribe $p(r)$ after each iterative step, thereby updating the boundary conditions as the computation progresses toward a steady state. It is also used by experimentalists to determine the static pressure away from the walls from the measured wall and hub static pressures. Static pressure is one of the most difficult quantities to measure in a three-dimensional flow, but recent development of optical measurements provides very accurate velocity data. These can be used to derive the static pressure distribution from hub-to-tip from known (and accurate) wall static pressures downstream. This will also enable determination of stagnation pressure distribution from gas dynamic equations.

Horlock (1973) provides a summary of design distributions employed in early designs based on this simplified radial equilibrium concept. Two examples of low hub-tip ratio designs are illustrated in Fig. 4.4. The design with 60% reaction (Fig. 4.4a) at the mean section and a free-vortex distribution could result in a very low reaction at the root, a heavily loaded stator root section, and a very high stagger angle at the rotor and stator tip. This also

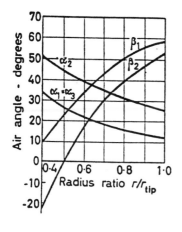
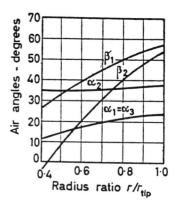

(a)   (b)

Figure 4.4 Various types of vortex design for axial flow compressor (copyright © Horlock (1973), Kreiger Publishing Co.). (See Fig. 2.10 for notations.) (a) Free-vortex design (60% reaction mean). (b) Exponential design (60% reaction mean).

depends on the hub/tip ratio. The case in Fig. 4.4b is a compromise, where the following distribution (called exponential design) is used:

$$V_\theta = a \pm \frac{b}{r}$$

The loading as well as the blade twists are moderate, and the reaction is nearly uniform along the span.

The choice of distribution of $V_\theta(r)$ depends on the type of turbomachinery, the entry Mach number, the hub/tip ratio, and the desired exit velocity distribution. Because modern design techniques are based on a computer-aided design system and computational fluid dynamics software, optimum blading is sought. The blading may have twist, lean, skew, and sweep, and thus the choice becomes much more complicated. A recent trend is toward low-aspect-ratio blading. The myth that the free-vortex blading has the lowest losses has been replaced by a more systematic optimization, which includes (a) reduction of blade boundary layer and annulus wall boundary layer growth and (b) reduction of shock losses. Hence, the choice of tangential velocity distribution at the exit for practical design may be more complicated than those outlined in this section.

4.1.2 Actuator Disk Theories

This is a mathematical model where the blade row is replaced by an appropriate plane (or disk) of discontinuity. The pressure rise or drop is assumed to be concentrated at the plane of the discontinuity. All the flow turning is assumed to take place at the actuator plane. The available theories can be broadly classified as:

1. *Linearized Theories.* In this theory the perturbation due to blade row is assumed to be small. Hawthorne and Horlock (1962) developed a theory for the incompressible flow case with cylindrical walls (constant hub/tip ratio). Lewis and Hill (1971) investigated the effect of sweep and dihedral. A comprehensive review of actuator disk theories is given by Marble (1964) and Horlock (1977).
2. *Nonlinear Theories.* The theoretical analysis of the flow through a turbomachine, when large perturbations are present, has been investigated by Oates (1972) and Marble (1964).

The actuator disk theories are applicable in situations where loading and perturbations due to blading are small, such as fans and ventilators. Nevertheless, they provide a qualitative estimate of the three-dimensional effects due to blading, including a prediction of radial variation in flow properties. No attempts will be made here to provide comprehensive developments. The reader is referred to the book by Horlock (1977) on this subject. Only the linearized analysis for incompressible flow will be presented to expose the reader to the principles.

The tangential velocity as well as p and P_o have a jump condition at the disk, and the radial and axial velocities are continuous across the disk.

The following assumptions are made in the development of linear theories:

1. Flow is steady and axisymmetric.
2. The entropy and stagnation enthalpy are constant along the streamlines, upstream and downstream of the blade row.
3. The radial velocity and axial vorticity are small everywhere.
4. Free-vortex conditions exist far upstream and downstream of the actuator disk.

The analysis presented below is due to Hawthorne and Horlock (1962). We may express the velocity components as

$$V_r(r, z) = v_r(r, z)$$
$$V_\theta(r, z) = V_{\theta\infty}(r) + v_\theta(r, z)$$
$$V_z(r, z) = V_{z\infty}(r) + v_z(r, z) \qquad (4.14)$$

The subscript ∞ refers to the basic free-vortex flow (far upstream and downstream), and v_r, v_θ, and v_z are the perturbations due to the disk. Using equations of motion and the Helmholtz vorticity equation (Eqs. 1.40 and 4.6), it can easily be proven, on the basis of the assumptions made, that the perturbed flow is also irrotational. It should be noted here that the equations are solved upstream and downstream of the disk and not across the disk.

By defining a potential function such that $\partial\phi/\partial r = v_r$ and $\partial\phi/\partial z = v_z$ and substituting these in the linearized continuity equation, it can be shown that the governing equation for the perturbed flow is given by ($v = \text{grad } \phi$)

$$\frac{\partial^2\phi}{\partial r^2} + \frac{1}{r}\frac{\partial\phi}{\partial r} + \frac{\partial^2\phi}{\partial z^2} = 0 \tag{4.15}$$

Equation 4.15 can be solved by separation of variables to prove that

$$(v_z)_\pm = \pm\left\{\frac{(V_{z\infty})_2 - (V_{z\infty})_1}{2}e^{\pm(k_1/h)z}\right\} \tag{4.16}$$

where the plus sign refers to values upstream of the disk and the minus sign refers to the values downstream, $h = r_t - r_h$, and k_1 is the first Bessel function solution and is approximately equal to π. Thus,

$$V_{z_{\text{upstream}}} = (V_{z\infty})_1 + \frac{(V_{z\infty})_2 - (V_{z\infty})_1}{2}e^{(\pi/h)z} \tag{4.17}$$

$$V_{z_{\text{downstream}}} = (V_{z\infty})_2 + \frac{(V_{z\infty})_1 - (V_{z\infty})_2}{2}e^{-(\pi/h)z} \tag{4.18}$$

where the subscripts 1 and 2 refer to values far upstream and far downstream, derived from the simplified radial equilibrium equation (Eq. 4.8) described earlier. Hawthorne and Horlock obtained good agreement between this theory and their experiment in an isolated row of guide vanes.

4.2   QUASI-THREE-DIMENSIONAL AND THREE-DIMENSIONAL THEORIES

4.2.1   Governing Equations in Rotating Coordinate System and Interpretation

The most convenient coordinate system for use in turbomachinery rotors is the relative or rotating coordinate system. For stationary three-dimensional flow (nozzles, stators, guide vanes), the equations can be simplified by substituting $\Omega = 0$. The advantages of a relative system are as follows: The relative flow is steady in most cases; boundary conditions are easier to apply; and the velocity profiles and boundary layers are similar to those observed in a stationary system. The advantages of a relative system are clear from

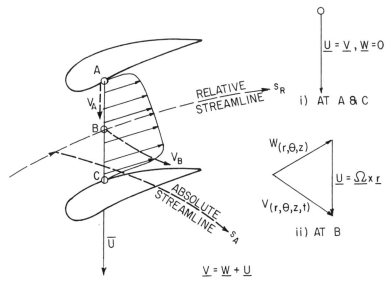

Figure 4.5 Absolute and relative velocity profiles in a rotor blade row (viscous boundary conditions on the surface).

Fig. 4.5. The absolute velocity profile is highly skewed, and the absolute velocities on the surface of the blade is not zero. Furthermore, it can be proven (Dean, 1959) that the absolute velocity in a rotor has to be unsteady for the pressure rise or drop to occur across a rotor. Consider the absolute streamline s_A (total velocity, V) in Fig. 4.5. Along the streamline s_A, for an adiabatic inviscid flow we have

$$\left(\frac{DV}{Dt} \right) = \frac{\partial V}{\partial t} + V \frac{\partial V}{\partial s_A} = -\frac{1}{\rho} \frac{\partial p}{\partial s_A} = -\frac{\partial h}{\partial s_A} \qquad (4.19)$$

Hence

$$\left(\frac{\partial V}{\partial t} \right) = -\frac{\partial h_o}{\partial s_A}$$

Thus the absolute velocity has to be unsteady to produce a stagnation enthalpy change. Similarly, it can be proven that the static pressure in a stationary (fixed point) frame has to be unsteady. Let us consider the streamline s_A (fixed frame), for an inviscid flow:

$$\frac{Dh_o}{Dt} = \frac{\partial h_o}{\partial t} + V \frac{\partial h_o}{\partial s_A} = \frac{\partial h_o}{\partial t} - V \frac{\partial V}{\partial t} = \frac{\partial h}{\partial t} = \frac{1}{\rho} \frac{\partial p}{\partial t}$$

Hence the stagnation enthalpy change across a blade can be achieved only when the pressure at a fixed point is unsteady. Thus for a rotor, it is essential to include time-dependent terms in the absolute reference frame. It is necessary to include the time-dependent terms in the relative frame of reference only when the inflow (absolute or relative) velocity is circumferentially nonuniform or time-dependent. If the absolute flow is uniform at inlet to a rotor, the relative flow is usually steady.

The governing equation in a relative (noninertial) frame of reference can be derived as follows. As indicated in Chapter 2 (Section 2.5 and Figs. 2.10 and 2.11), the velocity (**V**) in an absolute frame can be related to the velocity (**W**) in a relative frame through the equation

$$\mathbf{V} = \mathbf{W} + \mathbf{\Omega} \times \mathbf{r} \tag{4.20}$$

where **r** is the radius vector. In turbomachinery, the rotation $\mathbf{\Omega}$ is about z axis. Hence the absolute velocities are related to the relative velocities as follows:

$$\mathbf{i}_\theta \cdot \mathbf{V} = \mathbf{i}_\theta \cdot \mathbf{W} + \mathbf{i}_\theta \cdot (\mathbf{\Omega} \times \mathbf{r}), \qquad V_z = W_z, \quad V_r = W_r \tag{4.21}$$

where $\mathbf{i}_\theta$ is the unit vector in the tangential direction.

In addition, the absolute acceleration (in a fixed frame) **a** of a particle can be related to the displacement **r** and the velocity **W** relative to the moving system through the equation (see Greenwood, 1988, or Vavra, 1960, for detailed derivation).

$$\mathbf{a} = \left(\frac{D\mathbf{V}}{Dt} \right)_A = \left(\frac{D\mathbf{W}}{Dt} \right)_R + \mathbf{\Omega} \times \mathbf{\Omega} \times \mathbf{r} + 2\mathbf{\Omega} \times \mathbf{W} \tag{4.22}$$

It is assumed that $\mathbf{\Omega}$ is steady. The subscripts A and R stand for absolute (fixed) and relative frame of reference, respectively. Substituting these expressions in Eqs. 1.9 and 1.10 results in the following continuity and momentum equations in the relative system:

$$\frac{\partial \rho}{\partial t} + \nabla \cdot (\rho \mathbf{W}) = 0 \tag{4.23}$$

$$\underset{(1)}{\frac{D\mathbf{W}}{Dt}} + \underset{(2)}{2\mathbf{\Omega} \times \mathbf{W}} + \underset{(3)}{\mathbf{\Omega} \times \mathbf{\Omega} \times \mathbf{r}}$$

$$= -\underset{(4)}{\frac{\nabla p}{\rho}} + \underset{(5)}{\frac{\mathbf{F}}{\rho}}$$

$$+ \frac{1}{\rho} \frac{\partial}{\partial x_j} \left[\mu \underset{(6)}{\left(\frac{\partial W_i}{\partial x_j} + \frac{\partial W_j}{\partial x_i} \right)} + \underset{(7)}{\delta_{ij} \lambda \, \mathrm{div}\, \mathbf{W}} \right] \tag{4.24}$$

where **F** is the body force.

To avoid confusion, the subscript R will be dropped from now on. In equations in the rotating reference frame, all derivations (spatial and time) refer to the rotating coordinate system.

It is evident that the equation of motion of fluid in a rotating system is identical (except that $\mathbf{V}$ is replaced by $\mathbf{W}$) in form to the equation in the absolute frame of reference, provided a fictitious body force (per unit mass) equal to $-(2\boldsymbol{\Omega} \times \mathbf{W} + \boldsymbol{\Omega} \times \boldsymbol{\Omega} \times \mathbf{r})$ acts on the fluid in addition to the body and surface forces. The term $2\boldsymbol{\Omega} \times \mathbf{W}$ is the Coriolis force, and $\boldsymbol{\Omega} \times \boldsymbol{\Omega} \times \mathbf{r}$ is the centrifugal forces due to system rotation. Batchelor (1967, p. 555) provides detailed interpretation of the effect of Coriolis force on a fluid element. Considering the inviscid, incompressible flow, Eq. 4.24 can be written as

$$\frac{D\mathbf{W}}{Dt} + 2\boldsymbol{\Omega} \times \mathbf{W} = -\nabla\left[p/\rho - (\boldsymbol{\Omega} \times \mathbf{r})^2/2\right]$$

The term in brackets on the right side of the preceding equation is called reduced pressure (p^*). Thus, the centrifugal force is equivalent in its effect to the contribution to the pressure. The Coriolis force ($-2\boldsymbol{\Omega} \times \mathbf{W}$) acts in a plane normal to $\boldsymbol{\Omega}$ and $\mathbf{W}$; thus, the Coriolis force tends to change the direction of the velocity component $\mathbf{W}$ in the plane normal to $\boldsymbol{\Omega}$. It is thus clear that the Coriolis force has no effect on the inviscid flow field in an axial flow turbomachinery (Fig. 4.6) but has a major effect on the flow field in centrifugal and radial turbomachinery (Fig. 4.7).

The physical significance of various terms in Eq. 4.24 is illustrated in Figs. 4.6 and 4.7, respectively, for an axial and a centrifugal machine. Vavra (1960) has provided a very detailed interpretation and derivation of Eqs. 4.22–4.24. The first term in Eq. 4.24 consists of both the local and the convective acceleration of the relative flow. The fourth term is the static pressure gradient. Both of these ($D\mathbf{W}/Dt, \nabla p/\rho$) vary in the streamwise, radial, and tangential directions and are influenced by the flow turning, compressibility, rotation, curvature, thickness distribution, blade geometry, and flow geometry including flow angles, incidence, Mach number, and Reynolds number.

The relative frame or rotating coordinate system introduces two additional terms: term 2 in Eq. 4.24 (Coriolis acceleration) and term 3 (centripetal acceleration). The Coriolis acceleration is produced by two equal and cumulative effects as indicated by Eq. 4.22; ($\boldsymbol{\Omega} \times \mathbf{W}$) arises due to (a) substantial derivative of ($\boldsymbol{\Omega} \times \mathbf{r}$) in the relative system and (b) a component of equal magnitude $\boldsymbol{\Omega} \times \mathbf{W}$ due to the system rotation. The direction of Coriolis acceleration is approximately in the radial direction for axial turbomachinery (Fig. 4.6) and approximately in the tangential direction for centrifugal turbomachinery (Fig. 4.7). The Coriolis acceleration has a major influence in centrifugal turbomachinery; it influences the pressure rise/drop as well as losses. For example, in an inviscid flow, even in the absence of $D\mathbf{W}/Dt$, and

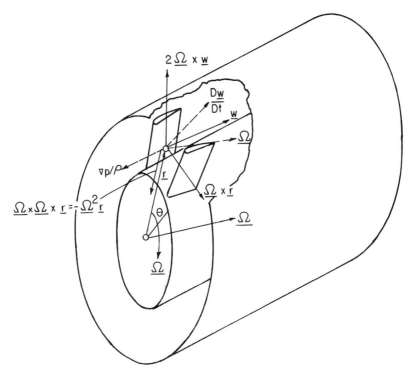

Figure 4.6 Forces and acceleration in an axial turbomachine rotor. (Directions for DW/Dt and $\nabla p/\rho$ are arbitrary.)

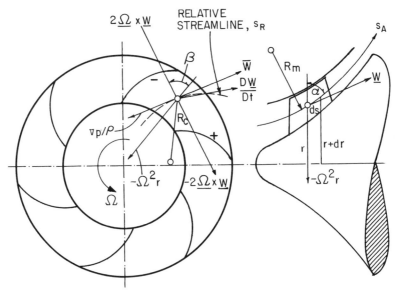

Figure 4.7 Forces and acceleration in a centrifugal machine. (Directions for DW/Dt and $\nabla p/\rho$ are arbitrary.)

F, a pressure gradient is set up by the Coriolis force in the tangential direction, given by (for radial-bladed turbomachinery)

$$-\frac{1}{\rho r}\frac{\partial p}{\partial \theta} = 2\Omega W \tag{4.25}$$

It can be proven, using Eq. 4.25, that the Coriolis force sets up a velocity gradient in the blade-to-blade direction in a centrifugal impeller. The static pressure gradient in Eq. 4.25 can be written as

$$P_{OR} = p + \rho W^2/2$$

where P_{OR} is the relative stagnation pressure. If the fluid is inviscid (no losses) $(dP_{OR}/rd\theta) = 0$; hence, Eq. 4.25 can be written as

$$dW/d\theta = 2\Omega r$$

For a linear variation in velocity, the difference between the velocity on the suction side (W_s) and the pressure side (W_p) is given by

$$W_s - W_p = 2\Omega r(\theta_s - \theta_p)$$

where $\theta_s - \theta_p$ is the angular extent of the passage. This provides a quick estimate of the blade loading on a centrifugal impeller blade.

It is thus clear that the Coriolis force introduces pressure gradient as well as the velocity gradient in the tangential direction in centrifugal turbomachinery, even in the absence of flow turning. In addition, the Coriolis acceleration tends to have a major influence on the blade boundary layer and losses; this will be dealt with in Chapter 5. The Coriolis force in axial turbomachinery is mainly in the radial direction. The Coriolis acceleration introduces radial outward flows inside the blade boundary layers, resulting in skewed boundary layers. This is explained in Chapter 5.

The centripetal acceleration term in Eq. 4.24 (due to rotating frame of reference), $\Omega \times \Omega \times \mathbf{r}$, is in the radial direction for both axial and radial turbomachinery. It should be noted that this acceleration is, in addition to the centrifugal acceleration, arising from the streamline curvature (blade camber, R_c; meridional flow curvature, R_m; as shown in Fig. 4.7) and the swirling flow (W_θ^2/r). The centrifugal acceleration due to rotation influences (a) the pressure rise or drop and the viscous effects in a centrifugal machine and (b) the viscous effects in an axial flow turbomachine.

The additional terms in Eq. 4.24 are blade and gravitational forces (term 5 in Eq. 4.24) and viscous terms (terms 6 and 7). Several variations of Eq. 4.24 are employed in practice. The viscous terms are usually the same in all these formulations. Hence, only the inviscid form of Eq. 4.24 will be presented below.

The acceleration $D\mathbf{W}/Dt$ can be expressed in two different forms:

$$\frac{D\mathbf{W}}{Dt} = \frac{\partial \mathbf{W}}{\partial t} + (\mathbf{W} \cdot \nabla)\mathbf{W} = \frac{\partial}{\partial t}\mathbf{W} + \nabla\left(\frac{W^2}{2}\right) - (\mathbf{W} \times \nabla \times \mathbf{W}) \quad (4.26)$$

Hence Eq. 4.24 can be expressed as

$$\frac{\partial \mathbf{W}}{\partial t} + (\mathbf{W} \cdot \nabla)\mathbf{W} + 2\mathbf{\Omega} \times \mathbf{W} + \mathbf{\Omega} \times \mathbf{\Omega} \times \mathbf{r} = -\frac{\nabla p}{\rho} + \frac{\mathbf{F}}{\rho} \quad (4.27)$$

or

$$\frac{\partial \mathbf{W}}{\partial t} - (\mathbf{W} \times \nabla \times \mathbf{W}) + 2\mathbf{\Omega} \times \mathbf{W} + \mathbf{\Omega} \times \mathbf{\Omega} \times \mathbf{r} = -\frac{\nabla p}{\rho} - \nabla\left(\frac{W^2}{2}\right) + \frac{\mathbf{F}}{\rho}$$
$$(4.28)$$

The static pressure in Eq. 4.28 can be eliminated by the use of the following relationships:

$$-\frac{\nabla p}{\rho} = T \nabla s - \nabla h \quad (4.29)$$

and

$$-\frac{\nabla p}{\rho} - \frac{\nabla W^2}{2} + \Omega^2 r = -\nabla I + T \nabla s \quad (4.30)$$

where

$$I = h + W^2/2 - \frac{\Omega^2 r^2}{2} \quad (4.31)$$

The gravitational force, which may be important in some liquid-handling machinery, can be absorbed in the expression for I,

$$I = h + \frac{W^2}{2} - \frac{\Omega^2 r^2}{2} + gz_h$$

distance z_h must be measured vertically upward from a horizontal plane of reference.

The quantity I was first introduced by Wu (1952) and is termed *rothalpy*. $I = h_o - V_\theta U$ and is nearly constant along a streamline in an adiabatic and steady flow.

The equations of motion (Eq. 4.28) for inviscid flow can now be expressed (using Eq. 4.30) as

$$\frac{\partial}{\partial t}(\mathbf{W}) - (\mathbf{W} \times \nabla \times \mathbf{W}) + 2\boldsymbol{\Omega} \times \mathbf{W} = -\nabla I + T \nabla s + \frac{\mathbf{F}}{\rho} \quad (4.32)$$

Using the first law of thermodynamics, the energy equation for inviscid flow can be expressed as

$$\frac{DI}{Dt} = \dot{Q} + \frac{1}{\rho}\frac{\partial p}{\partial t} \quad (4.33)$$

Lyman (1993) provides detailed derivation and interpretation of the transport equation for rothalpy. If the flow is viscous and body forces exist (note $\mathbf{F} \cdot \mathbf{W} = 0$; hence, work done by blade forces do not appear in Eq. 4.33), then an additional term $(\nabla \cdot \boldsymbol{\tau} + \rho \mathbf{B}) \cdot \mathbf{W}$ (where $\boldsymbol{\tau}$ is the viscous and turbulent stress vector) is added to the right side of Eq. 4.33. Hence, the change in rothalpy in inviscid flow is caused by (1) pressure fluctuation in rotating frame of reference, (2) work done by body forces on the relative flow, and (3) heat transfer, which is important only in cooled and high-temperature turbines. The viscous effects, as explained in Section 2.5 (below Eq. 2.36), arises due to work done by viscous forces in the casing region (where $\mathbf{W} = -\mathbf{U}$) and $(\nabla \cdot \boldsymbol{\tau}) \cdot \mathbf{W} \neq 0$.

The entropy equation is

$$T\frac{Ds}{Dt} = \dot{Q} \quad (4.34)$$

where $\dot{Q}$ is the rate of heat transfer.

The equation of state relating the pressure, density, and temperature is given by

$$p = \rho R T \quad (4.35)$$

It can be proven that I is nearly constant along a relative streamline (s_R in Fig. 4.7). Consider Eq. 4.32, and take the dot product of this equation along the streamwise directions s_R. For inviscid (isentropic) flow with $\mathbf{F} = 0$, this streamwise momentum equation reduces to

$$\frac{\partial W}{\partial t} = -\frac{\partial I}{\partial s_R} \quad (4.36)$$

Hence, for steady relative flow, I is constant along a relative streamline. Eq. 4.36 is a more generalized version of Eq. 2.35 and 2.36 derived from one-dimensional analysis. The fact that I is constant along a relative streamline provides a major simplification in the analysis as well as computation. For example, in the numerical analysis, the grids could be placed approximately along the streamwise (body-fitted coordinate system) direction, in which case the change in the variable I, along the streamwise coordinate, is small. This provides improved accuracy in the numerical solution.

The equations of continuity, momentum, energy, entropy, and state in scalar form for steady, inviscid flow in a rotating cylindrical coordinate system are given below (based on Eq. 4.32). θ is measured from pressure surfaces in the direction of rotation. In this notation, W_θ is negative for a compressor.

Continuity:
$$\frac{1}{r}\frac{\partial}{\partial r}(\rho r W_r) + \frac{1}{r}\frac{\partial}{\partial \theta}(\rho W_\theta) + \frac{\partial}{\partial z}(\rho W_z) = 0 \qquad (4.37)$$

r Momentum:
$$-\frac{W_\theta^2}{r} - W_\theta\frac{\partial W_\theta}{\partial r} + \frac{W_\theta}{r}\frac{\partial W_r}{\partial \theta} + W_z\left(\frac{\partial W_r}{\partial z} - \frac{\partial W_z}{\partial r}\right) - 2\Omega W_\theta$$

$$= -\frac{\partial I}{\partial r} + T\frac{\partial s}{\partial r} + \frac{F_r}{\rho} \qquad (4.38)$$

θ Momentum:
$$\frac{W_r W_\theta}{r} + W_r\frac{\partial W_\theta}{\partial r} - \frac{W_r}{r}\frac{\partial W_r}{\partial \theta} - W_z\left(\frac{1}{r}\frac{\partial W_z}{\partial \theta} - \frac{\partial W_\theta}{\partial z}\right) + 2\Omega W_r$$

$$= -\frac{1}{r}\frac{\partial I}{\partial \theta} + \frac{T}{r}\frac{\partial s}{\partial \theta} + \frac{F_\theta}{\rho} \qquad (4.39)$$

z Momentum:
$$-W_r\left(\frac{\partial W_r}{\partial z} - \frac{\partial W_z}{\partial r}\right) + W_\theta\left(\frac{1}{r}\frac{\partial W_z}{\partial \theta} - \frac{\partial W_\theta}{\partial z}\right)$$

$$= -\frac{\partial I}{\partial z} + T\frac{\partial s}{\partial z} + \frac{F_z}{\rho} \qquad (4.40)$$

Energy:
$$W_r\frac{\partial I}{\partial r} + W_\theta\frac{\partial I}{r\,\partial \theta} + W_z\frac{\partial I}{\partial z} = \dot{Q} \qquad (4.41)$$

Entropy:
$$W_r\frac{\partial s}{\partial r} + \frac{W_\theta}{r}\frac{\partial s}{\partial \theta} + W_z\frac{\partial s}{\partial z} = \frac{\dot{Q}}{T} \qquad (4.42)$$

State:
$$p = \rho R T \qquad (4.43)$$

The conservative form (Eq. 1.18) of continuity and momentum equations for inviscid unsteady flow are given, respectively, by (based on Eq. 4.27)

$$\frac{\partial \rho}{\partial t} + \frac{1}{r}\frac{\partial}{\partial r}(\rho r W_r) + \frac{1}{r}\frac{\partial}{\partial \theta}(\rho W_\theta) + \frac{\partial}{\partial z}(\rho W_z) = 0 \qquad (4.44)$$

$$\frac{\partial}{\partial t}(\rho W_r) + \frac{1}{r}\frac{\partial}{\partial r}(\rho r W_r W_r) + \frac{1}{r}\frac{\partial}{\partial \theta}(\rho W_r W_\theta) + \frac{\partial}{\partial z}(\rho W_r W_z)$$

$$-\frac{\rho W_\theta^2}{r} - 2\rho \Omega W_\theta - \rho \Omega^2 r = -\frac{\partial p}{\partial r} + F_r \qquad (4.45)$$

$$\frac{\partial}{\partial t}(\rho W_\theta) + \frac{1}{r}\frac{\partial}{\partial r}(\rho r W_r W_\theta) + \frac{1}{r}\frac{\partial}{\partial \theta}(\rho W_\theta W_\theta) + \frac{\partial}{\partial z}(\rho W_z W_\theta)$$

$$+2\rho \Omega W_r = -\frac{\partial p}{r\,\partial \theta} + F_\theta \qquad (4.46)$$

$$\frac{\partial}{\partial t}(\rho W_z) + \frac{1}{r}\frac{\partial}{\partial r}(\rho r W_r W_z) + \frac{1}{r}\frac{\partial}{\partial \theta}(\rho W_\theta W_z) + \frac{\partial}{\partial z}(\rho W_z W_z) = -\frac{\partial p}{\partial z} + F_z \qquad (4.47)$$

Many investigators have used Eqs. 4.37–4.43 or 4.41–4.47 for quasi-viscous analysis, by including viscous forces in F_r, F_θ, and F_z. In such a case, the energy equation (Eq. 4.41) should include work done by the viscous forces $\mathbf{F} \cdot \mathbf{W}$. The entropy production (Eqs. 4.34 and 4.42) should include viscous dissipation (ϕ/ρ, Eq. 1.12), as well as entropy change due to heat transfer as in cooled turbines ($\dot{Q}$).

Equations 4.37–4.43 or 4.41–4.47 represent a complete set of equations which are to be solved simultaneously to provide the variation of flow properties W_r, W_θ, W_z, p, I, s, and T inside a turbomachinery passage. In view of the highly nonlinear nature of these equations, very few attempts have been made in deriving an analytical solution for these equations. Various methods of solving these equations numerically will be described in Chapter 5.

4.2.1.1 *Irrotational Flow* When the flow is isentropic and steady and the *absolute flow* is irrotational, a simpler equation (similar to Eq. 1.52) can be derived by introducing the potential function:

$$\phi_r = V_r$$

$$\phi_\theta = rV_\theta$$

$$\phi_z = V_z \qquad (4.48)$$

The continuity equation and irrotationality condition can be combined to give

$$\left(1 - M_r^2\right)\phi_{rr} + \left(1 - M_\theta^2\right)\frac{\phi_{\theta\theta}}{r^2} + \left(1 - M_z^2\right)\phi_{zz} - 2M_zM_r\phi_{rz}$$
$$-2\frac{M_rM_\theta}{r}\phi_{r\theta} - 2\frac{M_\theta M_z\phi_{\theta z}}{r} + \frac{1}{r}\left(1 + M_\theta^2\right)\phi_r = 0 \qquad (4.49)$$

Analytical solutions are available for only such flows.

For an incompressible flow case, Eq. 4.49 reduces to

$$\phi_{rr} + \frac{\phi_r}{r} + \frac{1}{r^2}\phi_{\theta\theta} + \phi_{zz} = 0 \qquad (4.50)$$

It should be emphasized here that these equations are restricted to absolute irrotational flow and thus are applicable to only a certain class of turbomachinery flows.

4.2.2 Lifting Line and Surface Theories

The concept behind the lifting line and surface theory for both enclosed and extended turbomachinery is similar to that of the Prandtl lifting line theory. These theories are analytical and are valid for free-vortex or nearly free-vortex flows. It is widely used by hydrodynamicists and aerodynamicists to predict the thrust and torque developed by ship propellers and aircraft propellers, respectively. Recently it has been extended to include the prop fan described in Section 2.6. The analyses can be classified according to Mach number regime. The solutions are somewhat simpler for incompressible flow and utilize Eq. 4.50. Because the applicability of lifting line and surface theories are limited to restricted classes of turbomachines (e.g., prop fan, ship propeller, ventilating fans, propulsor, and other lightly loaded blades), only the rudiments of the approach will be presented. The reader is referred to quoted references for detailed mathematical treatment, extensions to transonic machines, inverse design technique, and so on. The principle on which the theories are based is similar to the singularity method for cascade. Singularities such as vortices and pressure dipoles are used to simulate and solve Eqs. 4.49 and 4.50.

4.2.2.1 *Incompressible Lifting Line Theory*   The governing equation for incompressible flow is given by Eq. 4.50. In this technique, the blade is replaced by singularity (vortex, source, and sinks) lines replacing the blade. These are located along the camber and span the entire annulus. Tyson (1952) derived the solution for a single vortex spanning an annulus. This was extended by Tamura and Lakshminarayana (1975, 1976) for a blade row (in free-vortex flow) to include both the source and the vortices distributed along

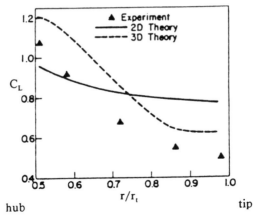

Figure 4.8 Experimental and theoretical distributions of lift coefficient (C_L) for propeller pump rotor for $\phi = 0.13$. (Copyright © Howells and Lakshminarayana, 1977; Reprinted with permission of ASME.)

the camber of the three-dimensional blade. Tamura and Lakshminarayana (1976) have provided a detailed analysis and solution for the flow (ϕ) and simulated the flow in a hypothetical rotor and studied the three-dimensional inviscid effects due to hub/tip ratio, stagger angle, and number of blades and found that the three-dimensional effects are considerable for low hub/tip ratio. Howells and Lakshminarayana (1977) designed and built a free-vortex rotor (propeller pump) and carried out a validation of the analysis. Detailed blade pressure distribution at various radii, dihedral angles, and flow coefficients were measured and compared with the above three-dimensional analysis as well as the two-dimensional panel method (Chapter 3). Their result, shown in Fig. 4.8, clearly indicates the inadequacy of two-dimensional theories. The three-dimensional trends are predicted well. Falcao (1975) has provided a solution of the Laplace equation for varying circulation $(\Gamma(r))$, including the trailing vortex system in the wake. As indicated earlier, lifting line and surface theories are widely used in analyzing three-dimensional flow fields in marine and aircraft propellers. The reader is referred to Tsakonas (1979) for details.

4.2.2.2 Compressible Flow Theories The compressible flow solutions were derived by Okurounmu and McCune (1974), Namba (1974), Lordi and Homicz (1981), and Hanson (1982, 1983). More recently, Hawthorne et al. (1984) made attempts to apply the singularity method for the design of large deflection blades. These analyses are usually based on the assumption that the flow far upstream and downstream are uniform, and the perturbations are small compared with the undisturbed flow. As mentioned earlier, linearization makes the analysis invalid for modern highly loaded compressor or turbine blades, but the theory is valid for a prop fan (Section 2.6), where the perturbations are small and the interference effects due to adjoining blades are not significant. The theory due to Okurounmu and McCune (1974) utilizes the velocity potential, similar to but more complicated than the analysis developed for incompressible flow. The theory of Namba (1974) is based on the solution of the pressure equation (for steady flow), derived first by taking the divergence of the momentum equation and then eliminating the density and velocity from the inviscid irrotational equations in a rotating cylindrical coordinate system. The resulting equation is given by

$$\left(1 - M_{z_0}^2\right)\frac{\partial^2 p}{\partial z^2} + \frac{1}{r^2}\left(1 - M_{z_0}^2\left(\Omega^*\right)^2 r^2\right)\frac{\partial^2 p}{\partial \theta^2}$$

$$- 2M_{z_0}^2\Omega^*\frac{\partial^2 p}{\partial z\,\partial\theta} + \frac{1}{r}\frac{\partial}{\partial r}\left(r\frac{\partial p}{\partial r}\right) = 0 \tag{4.51}$$

where M_{z_0} is the undisturbed axial Mach number, $\Omega^* = \Omega r_t/W_{z_0}$ and all lengths (z, r) are normalized by r_t.

Namba (1974) solved Eq. (4.51) by representing the blade by a sheet of pressure dipoles and by the method of superposition. Lordi and Homicz (1981) have included the thickness distribution by solving this equation as well as the perturbation potential equation (similar to Eq. 4.49) due to sources and sinks simulating the thickness effect. Thus, Okurounmu and McCune use a vortex representation of the lifting surface, Namba uses pressure dipoles, and Lordi and Homicz use pressure dipoles to generate lift and sources to simulate the thickness effect. None of the authors have shown comparison with data from a high-speed compressor. As mentioned earlier, modern compressors and turbines are very heavily loaded and applicability of these theories are somewhat limited to an understanding of the qualitative nature of three-dimensional effects. In spite of the limitations mentioned above, the theories are valid for enclosed turbomachinery with low loading (fans) and unshrouded turbomachinery (e.g., marine propellers, windmills, aircraft propellers, prop fans).

There has been a recent renewal of interest in these theories due to the introduction of prop fans and unducted fans (UDFs), shown in Figs. 1.3 and 2.33. Lifting line and surface have been powerful tools in analyzing flow over finite wings and aircraft propellers for more than half a century. The prop fan poses a significantly more difficult problem due to the large number of blades and higher cruise Mach numbers ($M = 0.80$), with relative Mach numbers in the tip reaching transonic speeds. Hanson (1982, 1983) developed a theory for the prediction of pressure and lift on these prop fan blades using approaches similar to those described in this section and those used in the propeller analysis. This is a linear theory and valid for compressible flows. Hanson provided a unified theory for predicting the acoustic field as well as the aerodynamic field of a prop fan by including the time-dependent terms in

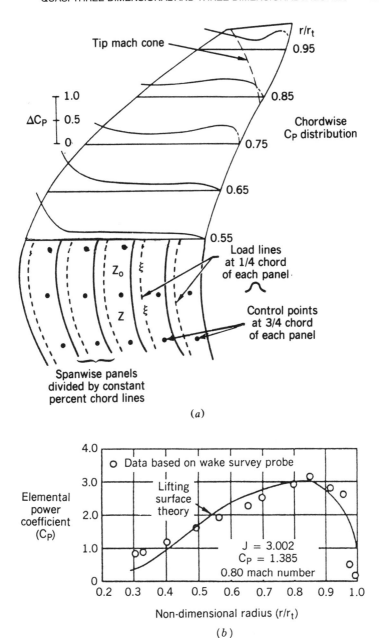

Figure 4.9 Lifting surface theory for prop fan (helical coordinate system): (a) Panel location and pressure distribution. (b) Comparison between experimental and theoretical distribution of power coefficient. (Copyright © Hanson, 1982, 1983; reprinted with permission of AIAA.)

Eq. 4.51. He employs helical lifting surfaces using the pressure potential (dipole) method. An expression for the pressure disturbance caused by helically convected thickness and loading sources was developed from the acoustic analogy using the free-space Green's function. The blades are assumed to be thin, so mass source is not utilized. The method is illustrated in Fig. 4.9. The lift distribution is centered at 1/4 chord of each panel. The blade surface is divided into a number of spanwise panels. It is not necessary to resort to the simple division as shown in Fig. 4.9. One can resort to an arbitrary, unequal spacing and division as well. The radial distribution of local power coefficient (C_P) predicted by Hanson and shown in Fig. 4.9, indicates excellent agreement with the data. It should be noted here that, even though completely numerical techniques are evolving for the prediction of pressure on a high-speed prop fan, the analysis given here would be extremely useful in resolving the flow field (as in zonal method with analytical zones) in the wake and near the tip of the blade.

4.2.3 Passage Averaged Equations

In view of the great difficulties involved in the solution of the equations of motion, energy and state, several simplifying flow models have been proposed for the solution of the "hub-to-tip flow" or meridional through flow.

Many researchers have attempted to develop passage averaged equations and solve them by analytical or simplified numerical technique. The motivation for this is that there are very accurate and efficient techniques for solving the blade-to-blade equations. To obtain a more accurate coupling between the blade-to-blade and hub-to-tip directions, it is necessary to include some gross nonaxisymmetric effects in predicting the radial distribution of properties upstream of, inside, and exiting the blade row. One of the powerful techniques is to solve for an "average" flow (average properties across the blade), thus satisfying the equations in a global fashion. This techniques is analogous to the momentum integral technique used in the boundary layer theory, where gross properties of boundary layer (momentum thickness, etc.) can be predicted accurately using reasonable assumptions for the velocity distribution. In this technique, the equations of motion are averaged from blade to blade, thus eliminating the terms $\partial/\partial\theta$. Elimination of one independent variable simplifies the equations considerably.

One of the earlier analyses is due to Smith (1968), who adopted an averaging technique to develop an approximate form of radial equilibrium equation. Extensions of this concept are due to Horlock and Marsh (1971), Hirsch and Warzee (1976, 1979) Sehra and Kerrebrock (1981), Jennions and Stow (1985, 1986), Adamczyk (1985), and Hirsch and Dring (1987). Generalized analysis will be presented in detail, and later it will be shown that the axisymmetric flow analysis is a special case of the generalized technique.

There are three types of averaging used in practice. They can be classified as follows:

1. *Algebraic averaging:* This is similar to Reynolds averaging used for Navier–Stokes equation.
2. *Density-weighted (ρ) averaging:* This is similar to Favre averaging used for Navier–Stokes equation.
3. *Mass-weighted averaging (ρV_z):* This approach introduces blockage due to blade as well as viscous layers (boundary layers and wakes) on the flow field and is used in the design process.

The first two averaging techniques introduce correlations or interaction terms due to asymmetry in the tangential direction. These terms are similar to those arising from time averaging of Navier–Stokes equation (Reynolds or Favre averaging). In the latter case, the terms arise due to time averaging, while in the former case, correlation terms arise due to blade-to-blade (spatial) averaging. The physical meaning of these terms are totally different, and this is where the analogy between passage averaging and Reynolds averaging ends. The approach followed in this section is to introduce the density weighted averaging procedure due to Smith (1968), Hirsch and Warzee (1976, 1979), and Jennions and Stow (1985).

The concept is illustrated in Fig. 4.10. The flow field can be divided into a series of blade-to-blade surfaces (S_1) and one hub-to-tip surfaces (S_2). This concept was originally proposed by Wu (1952). The solution on the S_1 surface (with known inlet properties) can be determined from the cascade theories presented in Chapter 3. The inlet properties, which vary from hub to tip, are governed by the radial and axial momentum equations. Therefore, a solution of the radial and axial momentum equations is necessary to establish the initial conditions for the cascade solution. In simplified radial equilibrium and actuator disk theories, these are determined upstream and downstream and are valid only for specialized cases (inviscid flows). In a more general case, the streamlines are curved as shown in Fig. 4.1. The inlet conditions should be based on the solution of the entire radial momentum equation.

The quasi-three-dimensional techniques, employing passage averaged equations, involve the solution of the passage averaged radial equilibrium equation on a surface such as a mean S_2 surface and exact solution of tangential and axial momentum equations along several two-dimensional blade-to-blade surfaces such as S_1 (Fig. 4.10). The technique provides accurate values of inlet and exit boundary conditions for solution of the S_1 surface. The solution of flow properties on the S_1 surface is based on cascade theories described in Chapter 3. A description of the averaging procedure and various forms of passage averaged radial equilibrium equations used are given below.

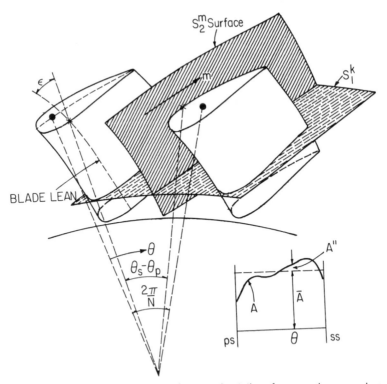

Figure 4.10 S_1 and S_2 stream surfaces and notations for averaging procedure.

The assumptions made in deriving the passage averaged equation for a mean S_2 surface are as follows:

1. The flow is steady in the relative frame of reference. All the analyses presented in this section assume inviscid flow, with slip conditions at the surface, and allow for viscous effects through entropy changes.
2. The viscous effects are allowed through the force vector **F** and the entropy changes. The heat transfer effects are allowed through entropy changes. Many authors employ loss correlation and relate it to ∇s.
3. Other assumptions made, which are specific to some cases, are stated in the proper context.

4.2.3.1 Passage Averaging Technique and Averaged Continuity Equation
Let us consider a blade row with a lean ($\epsilon(r, z)$), sweep ($\tau(r)$), and N number of blades as shown in Figs. 4.10 and 4.11. The angular extent (surface to surface) $\theta_s - \theta_p$ is a function of r and z.

A brief summary of the averaging procedure and averaged equations is presented here. Details can be found in Smith (1968), Hirsch and Warzee (1976, 1979), Horlock and Marsh (1971), Sehra and Kerrebrock (1981), and

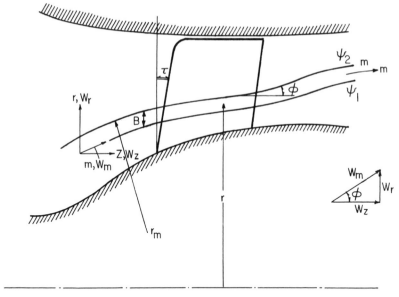

Figure 4.11   Quasi-three-dimensional analysis, geometry and notations.

Jennions and Stow (1985, 1986). The averaging technique presented here is based on Hirsch and Warzee (1979) and Jennions and Stow (1985).

We can denote any quantity $A(W_r, W_\theta, W_z, I, p, s, T, \rho)$ by

$$A = \bar{A} + A'', \qquad \text{with } \overline{A''} = 0 \tag{4.52}$$

$$A = \hat{A} + A', \qquad \text{with } \overline{A'} \neq 0, \overline{\rho A'} = 0 \tag{4.53}$$

where $\bar{A}$ and $\hat{A}$ are algebraic and density averaged values (similar to Reynolds and Favre's density averaging, respectively) given by (illustration in Fig. 4.10, algebraic average used for scalar quantities only)

$$\bar{A} = \frac{\int_{\theta_p}^{\theta_s} A \, d\theta}{\theta_s - \theta_p}, \qquad \hat{A} = \frac{\int_{\theta_p}^{\theta_s} \rho A \, d\theta}{\int_{\theta_p}^{\theta_s} \rho \, d\theta} = \frac{\overline{\rho A}}{\bar{\rho}} \tag{4.54}$$

where

$$\theta_s - \theta_p = \frac{2\pi}{N}\left(1 - \frac{t}{S}\right) = \frac{2\pi b}{N} \tag{4.55}$$

The blockage t/S is due to blade thickness only. But it could be generalized to include the displacement thickness due to blade boundary layers as well. Furthermore, from the above relationship we can prove the following:

$$(\hat{A} - \bar{A}) = \overline{\rho' A''}/\bar{\rho} = -\bar{A}' \tag{4.56}$$

The lean angle (ϵ) is defined by

$$\tan \epsilon = -r\frac{\partial \theta}{\partial r} \tag{4.57}$$

and the blade angle is

$$\tan \beta' = r\frac{\partial \theta}{\partial z} \tag{4.58}$$

Along a mean line m, between the pressure and suction side (Fig. 4.10), we have

$$\theta_p = \theta_m + (1 - b)\frac{\pi}{N} \tag{4.59}$$

$$\theta_s = \theta_m - (1 - b)\frac{\pi}{N} + \frac{2\pi}{N} \tag{4.60}$$

Hence,

$$\tan \beta' = \tan \beta + (\tan \phi)\tan \epsilon \tag{4.61}$$

where

$$\tan \phi = W_r/W_z \quad \text{and} \quad \tan \beta = W_\theta/W_z \tag{4.62}$$

It is implicitly assumed here that the flow and blade angles are identical on the blade surface within the passage. This is true in most turbomachinery, with the exception being separated flows and blades with film cooling.

The integration of any arbitrary quantity A can be written as

$$\frac{\overline{\partial \rho A}}{\partial r} = \frac{N}{2\pi b}\int_{\theta_p}^{\theta_s}\frac{\partial \rho A}{\partial r}d\theta = \frac{N}{2\pi b}\frac{\partial}{\partial r}\int_{\theta_p}^{\theta_s}\rho A\,d\theta - \frac{N}{2\pi b}\left[(\rho A)_s\frac{\partial \theta_s}{\partial r} - (\rho A)_p\frac{\partial \theta_p}{\partial r}\right]$$

$$= \frac{1}{b}\frac{\partial}{\partial r}(b\bar{\rho}\hat{A}) - \frac{N}{2\pi b}\left[(\rho A)_s\frac{\partial \theta_s}{\partial r} - (\rho A)_p\frac{\partial \theta_p}{\partial r}\right] \tag{4.63}$$

Similarly,

$$\frac{\overline{\partial(\rho A)}}{r\partial\theta} = \frac{N}{2\pi br}\left[(\rho A)_s - (\rho A)_p\right] \tag{4.64}$$

and

$$\frac{\overline{\partial(\rho A)}}{\partial z} = \frac{1}{b}\frac{\partial}{\partial z}(b\bar{\rho}\hat{A}) - \frac{N}{2\pi b}\left[(\rho A)_s\frac{\partial\theta_s}{\partial z} - (\rho A)_p\frac{\partial\theta_p}{\partial z}\right] \tag{4.65}$$

In the above equations subscript p and s refer to values on the pressure and suction surfaces, respectively. The average of the continuity equation (Eq. 4.37 or 4.44), derived from substituting these equations (Eq. 4.56–4.65), is given by

$$\frac{1}{br}\frac{\partial}{\partial r}(b\bar{\rho}r\hat{W}_r) + \frac{1}{b}\frac{\partial}{\partial z}(b\bar{\rho}\hat{W}_z) - \frac{N}{2\pi br}\left[(\rho\hat{W}_z)_s(-\tan\phi_s\tan\epsilon_s + \tan\beta_s')\right.$$
$$\left. + (\rho W_z)_p(\tan\phi_p\tan\epsilon_p - \tan\beta_p') - (\rho W_\theta)_s + (\rho W_\theta)_p\right] = 0 \quad (4.66)$$

The last term in square brackets is zero (from relationships 4.57–4.62). The term is zero in both the inviscid flow (no slip condition) and viscous flow ($\mathbf{W} = 0$). Therefore, the averaged continuity equation is given by

$$\frac{\partial}{\partial r}(rb\,\bar{\rho}\hat{W}_r) + \frac{\partial}{\partial z}(rb\bar{\rho}\hat{W}_z) = 0 \tag{4.67}$$

This is an exact equation, and no assumptions are made in deriving this equation. A physical explanation of Eq. 4.67 is as follows. Consider a one-dimensional axisymmetric stream tube bounded by streamlines ψ_1 and ψ_2 (Fig. 4.11). The mass flux through this stream tube is $(2\pi rB\rho W_z)$, and this should be constant as shown by Eq. 4.67. Such simplifications are made in many quasi-three-dimensional analyses. For example, in analyzing the flow in a contracting two-dimensional rectilinear cascade, the following equation derived from Eq. 4.67 is used:

$$\frac{\partial}{\partial x}(\rho hW_x) + \frac{\partial}{\partial y}(\rho hW_y) = 0 \tag{4.68}$$

where $h = h(x)$ is the height of the channel in the z direction, the spanwise flow is assumed to be small. W_x and W_y are axial and tangential components, respectively, in a cascade.

4.2.3.2 Passage Averaged Radial Equilibrium Equation (S_2 Surface)
Let us consider the average of the radial momentum equation, which is the principal equation governing the three-dimensional flow from hub to tip. It is more convenient to work with the static pressure because it represents blade

loading. Therefore, Eq. 4.45 can be written as (for steady, inviscid flow)

$$\frac{1}{r}\frac{\partial}{\partial r}(\,prW_rW_r\,) + \frac{1}{r}\frac{\partial}{\partial \theta}(\,\rho W_rW_\theta\,) + \frac{\partial}{\partial z}(\,\rho W_rW_z\,)$$

$$-\frac{\rho W_\theta^2}{r} - \rho\,\Omega^2 r - 2\rho\,\Omega W_\theta = -\frac{\partial p}{\partial r} + F_r \qquad (4.69)$$

Averaging the convection terms in Eq. 4.69 is very similar to that carried out for the continuity equation. The average radial pressure gradient is given by

$$\overline{\frac{\partial p}{\partial r}} = \frac{\partial \bar{p}}{\partial r} + \frac{1}{b}\frac{\partial b}{\partial r}\left(\bar{p} - \frac{p_s + p_p}{2}\right) - \frac{p_s - p_p}{bS}\tan\epsilon_m \qquad (4.70)$$

where S is the blade spacing and ϵ_m is the mean blade lean angle. The term $F_b = [(p_s - p_p)/bS]$ is proportional to blade forces and, hence, the last term in Eq. 4.70 represents its component in the radial direction. For radial blades, this force is purely tangential and is zero in the radial direction.

Typical convective terms in Eq. 4.69 are (with $A = \rho W_r W_z$)

$$\overline{\frac{\partial}{\partial z}\rho W_r W_z} = \frac{1}{b}\frac{\partial}{\partial z}\left(b\overline{\rho W_r W_z}\right) - \frac{N}{2\pi b}\left[(\,\rho W_r W_z\,)_s\frac{\partial \theta_s}{\partial z} - (\,\rho W_r W_z\,)_p\frac{\partial \theta_p}{\partial z}\right] \qquad (4.71)$$

and writing

$$\overline{\rho W_r W_z} = \overline{\rho\left(\hat{W}_r + W_r'\right)\left(\hat{W}_z + W_z'\right)} = \bar{\rho}\hat{W}_r\hat{W}_z + \bar{\rho}\overline{W_r' W_z'} \qquad (4.72)$$

Here, W_z' and W_r' represent the spatial variation of the mean flow (Fig. 4.10) and not the fluctuating (time-dependent) quantities. Substitution of Eq. 4.72 in Eq. 4.71 results in the following expression:

$$\overline{\frac{\partial}{\partial z}\rho W_r W_z} = \frac{1}{b}\frac{\partial}{\partial z}\left(b\bar{\rho}\hat{W}_z\hat{W}_r\right) + \frac{1}{b}\frac{\partial}{\partial z}\left(b\bar{\rho}\overline{W_z' W_r'}\right)$$

$$-\frac{N}{2\pi b}\left[(\,\rho W_r W_z\,)_s\frac{\partial \theta_s}{\partial z} - (\,\rho W_r W_z\,)_p\frac{\partial \theta_p}{\partial z}\right] \qquad (4.73)$$

When all the terms in Eq. 4.69 are averaged, the terms that involve $\partial\theta_s/\partial z$, $\partial\theta_s/\partial r$, and so on, drop out as in continuity Eq. 4.66.

Such averaging can be carried out for Eq. 4.38, 4.45, or 4.69. Therefore, several forms of average radial equilibrium equations are available. By carrying out the averaging procedure described above for Eq. 4.38, we can derive the following passage averaged radial equilibrium equation (Hirsch and Warzee, 1979):

$$\hat{W}_z\left(\frac{\partial \hat{W}_r}{\partial z} - \frac{\partial \hat{W}_z}{\partial r}\right) - \frac{\hat{W}_\theta}{r}\frac{\partial}{\partial r}\left(r\hat{V}_\theta\right) = \hat{T}\frac{\partial \hat{s}}{\partial r} - \frac{\partial \hat{I}}{\partial r} + \frac{\overline{F}_r}{\overline{\rho}} + \frac{F_{b,r}}{\overline{\rho}} - \sum_{i=1}^{5} G_i/\overline{\rho}$$

(4.74)

where

$$F_{b,r} = \left[\left(\hat{T}s'_s - h'_s\right)\rho_s - \left(\hat{T}s'_p - h'_p\right)\rho_p\right]\frac{N}{2\pi b}\tan\epsilon_m = \frac{p_s - p_p}{bS}\tan\epsilon_m$$

$F_{b,r}$ is the component of the blade force in the radial direction, given by the last term in Eq. 4.70. Subscript p and s refer to values on pressure and suction surfaces, respectively. G_i are the fluctuating terms due to blade-to-blade variation of flow properties given by

$$G_1 = \frac{1}{br}\frac{\partial}{\partial r}\left(br\overline{\rho W'_r W'_r}\right)$$

$$G_2 = \frac{1}{br}\frac{\partial}{\partial z}\left(br\overline{\rho W'_r W'_z}\right)$$

$$G_3 = -\frac{1}{r}\overline{\rho W'_\theta W'_\theta}$$

$$G_4 = -\frac{1}{2b}\frac{\partial b}{\partial r}\left[\left(\hat{T}s'_s - h'_s\right)\rho_s + \left(\hat{T}s'_p - h'_p\right)\rho_p\right]$$

$$G_5 = -\overline{\left(h' - \hat{T}s'\right)\frac{\partial \rho}{\partial r}} - \overline{\rho T'\frac{\partial s'}{\partial r}}$$

Equation 4.69 can also be expressed in nonconservative form by expanding the terms and subtracting the continuity equation. The averaged form of the equation is in nonconservative form and is given by

$$\overline{\rho}\hat{W}_r\frac{\partial}{\partial r}\hat{W}_r + \overline{\rho}\hat{W}_z\frac{\partial}{\partial z}\hat{W}_r - \overline{\rho}\frac{\hat{W}_\theta\hat{W}_\theta}{r} - \overline{\rho}r\Omega^2 - 2\Omega\overline{\rho}\hat{W}_\theta$$

$$= -\frac{\partial \overline{p}}{\partial r} + F_{b,r} + \overline{F}_r - \sum_{i=1}^{4} G_i$$

(4.75)

where G_1, G_2, and G_3 are the same as before, and G_4 is the second term on the right-hand side of Eq. 4.70.

Let us interpret the passage average radial equilibrium equations (Eqs. 4.74 and 4.75), which are the principal equations for the meridional solution of the flow. $\hat{W}_z$, $\hat{W}_r$, $\hat{V}_\theta$, $\hat{T}$, $\hat{s}$, and $\hat{I}$ are passaged averaged values as per Eq. 4.54, $\overline{\rho}\hat{W}_\theta\hat{W}_\theta/r$ is the average centrifugal force term due to swirl, $\overline{\rho}r\Omega^2$ is the average centrifugal force due to noninertial frame of reference, $\partial \bar{p} / \partial r$ is the average radial pressure gradient, and $F_{b,r}/\rho$ is the blade force in the radial direction. This is a direct consequence of averaging. The blade-to-blade variation of properties, which arises due to blade forces, is now partly replaced by the blade force. The remaining part is in G_i functions. This quantity has to be evaluated from a blade-to-blade solution (Chapter 3) of the θ and z momentum equations. $\bar{F}_r$ is the dissipative and body forces. This accounts for entropy increases in an adiabatic flow and is an approximate method of including the viscous effects. It should be remarked that for purely axisymmetric flow, $\bar{F}_r$, $F_{b,r}$ and G_i are all zero. Therefore, these terms in effect represent the deviation from the axisymmetric solution and are source terms in the partial differential equations.

The static enthalpy change from the mean (h'_s, h'_p) and entropy changes (s'_s, s'_p) should also be derived from the blade-to-blade or cascade solution. For example, $h'_s = h_s - \hat{h}$, where h_s is derived from the blade-to-blade solution and $\hat{h}$ is derived from the passage averaged equation.

For meridional through-flow calculation, a mean stream function ψ_m can be defined by

$$\hat{W}_z = \frac{1}{\overline{\rho} r b} \frac{\partial \psi_m}{\partial r} \tag{4.76}$$

$$\hat{W}_r = -\frac{1}{\overline{\rho} r b} \frac{\partial \psi_m}{\partial z} \tag{4.77}$$

These satisfy the continuity equation (Eq. 4.67) exactly. Equation 4.74, along with Eqs. 4.76 and 4.77, reduces to

$$\frac{\partial}{\partial r}\left(\frac{1}{\overline{\rho} r b}\frac{\partial \psi_m}{\partial r}\right) + \frac{\partial}{\partial z}\left(\frac{1}{\overline{\rho} r b}\frac{\partial \psi_m}{\partial z}\right)$$

$$= \frac{1}{\hat{W}_z}\left(\frac{\partial \hat{I}}{\partial r} - \hat{T}\frac{\partial \hat{s}}{\partial r} - \frac{\hat{W}_\theta}{r}\frac{\partial}{\partial r}\left(r\hat{V}_\theta\right) - \frac{F_{b,r}}{\overline{\rho}} - \frac{\overline{\overline{F}}_r}{\rho} + \sum_{i=1}^{5}\frac{G_i}{\overline{\rho}}\right) \tag{4.78}$$

Therefore, the passage-averaged governing equations for the meridional flow are given by Eqs. 4.67 and 4.74 or 4.75 in terms of velocity as a dependent variable, or by Eqs. 4.76–4.78 in terms of stream function. Because the solution technique is based on two-dimensional flow equations, it is more convenient to solve the meridional flow in terms of stream function ψ_m. A method of solving Eq. 4.78 is described later.

Let us now consider the corresponding equations governing the blade-to-blade flow on an axisymmetric surface.

4.2.3.3 Equation for Blade-to-Blade Surface (S₁ Surface)   The equation for the S_1 surface (Fig. 4.10), governed by the tangential and axial momentum equations, is best solved using a coordinate system based on the

meridional flow direction m (resultant of axial and radial velocity) and θ as shown in Figs. 4.10 and 4.11. The solution on the axisymmetric surface S_1 is carried out using quasi-two-dimensional equations. This replaces the x and y directions used in cascade theories.

The meridional velocity can be defined as (Fig 4.11)

$$W_m^2 = W_z^2 + W_r^2 \tag{4.79}$$

$$W_r = W_m \sin \phi$$

$$W_z = W_m \cos \phi \tag{4.80}$$

and the total velocity W is given by

$$W^2 = W_m^2 + W_\theta^2 \tag{4.81}$$

Thus, the velocity components are W_m and W_θ in m and θ directions, respectively. The transformation of gradients is given by

$$\frac{\partial}{\partial m} = \cos \phi \frac{\partial}{\partial z} + \sin \phi \frac{\partial}{\partial r} \tag{4.82}$$

or

$$W_m \frac{\partial}{\partial m} = W_z \frac{\partial}{\partial z} + W_r \frac{\partial}{\partial r} \tag{4.83}$$

The principal equation in the $m-\theta$ plane is the tangential momentum equation. By using Eqs. 4.79–4.82, we can express Eq. 4.39 as

$$\underset{(1)}{\frac{W_\theta W_m \sin \phi}{r}} \quad - \quad \underset{(2)}{\frac{W_m \sin \phi}{r} \frac{\partial}{\partial \theta}(W_m \sin \phi)}$$

$$+ \underset{(3)}{W_m \frac{\partial W_\theta}{\partial m}} \quad - \quad \underset{(4)}{\frac{W_m \cos \phi}{r} \frac{\partial}{\partial \theta}(W_m \cos \phi)} + \underset{(5)}{2\Omega W_m \sin \phi}$$

$$= -\underset{(6)}{\frac{1}{r} \frac{\partial I}{\partial \theta}} + \underset{(7)}{\frac{T}{r} \frac{\partial s}{\partial \theta}} + \frac{F_\theta}{\rho} \tag{4.84}$$

Because the surfaces are axisymmetric, $\partial \phi / \partial \theta = 0$, therefore terms 2 and 4 in the above equation can be combined to provide $(W_m/r)(\partial/\partial\theta)W_m$. Therefore, Eq. 4.84 can be simplified to

$$
W_m \frac{\partial W_\theta}{\partial m} - W_m \frac{\partial W_m}{r\partial\theta} + \frac{W_\theta W_m \sin\phi}{r} + 2\Omega W_m \sin\phi
$$
$$
\text{(1)} \qquad\qquad \text{(2)} \qquad\qquad \text{(3)} \qquad\qquad \text{(4)}
$$
$$
= -\frac{1}{r}\frac{\partial I}{\partial\theta} + \frac{T}{r}\frac{\partial s}{\partial\theta} + \frac{F_\theta}{\rho} \tag{4.85}
$$

The above equation contains only two independent variables (m, θ) and two dependent variables for velocity W_m, W_θ and is much more convenient to solve. There is no blade force, and thus F_θ represents only viscous forces.

The transformation of the continuity equation, which is needed to satisfy mass conservation, is much more difficult to obtain in a compact form because ϕ is a function of r and z, and derivatives such as $\partial\phi/\partial r$ and $\partial\phi/\partial z$ appear in the transformed equation. The continuity equation can, therefore, be averaged over the stream tube, bounded by say ψ_1, ψ_2 (Figs. 4.10 and 4.11) with a radial stream tube thickness of $B(r, z)$ or $B(m)$. The averaging process is similar to that employed in deriving Eqs. 4.67 and 4.74 and is defined by

$$
\overline{\rho A} = \frac{1}{B}\int_r^{r+B} \rho A\, dr \tag{4.86}
$$

The average of partial derivatives is given by

$$
\overline{\frac{\partial(\rho A)}{\partial r}} = \frac{1}{B}\left[(\rho A)_{\psi_1} - (\rho A)_{\psi_2}\right] \tag{4.87}
$$

$$
\overline{\frac{\partial(\rho A)}{\partial m}} = \frac{1}{B}\frac{\partial}{\partial m}(B\overline{\rho A}) - \frac{1}{B}\left[(\rho A)_{\psi_1}\left(\frac{\partial r}{\partial m}\right)_{\psi_1} - (\rho A)_{\psi_2}\left(\frac{\partial r}{\partial m}\right)_{\psi_2}\right]
$$
$$
\tag{4.88}
$$

By using Eqs. 4.79–4.83 and 4.86–4.88, we can integrate the continuity equation (Eq. 4.37) across the stream tube bounded by streamlines ψ_1 and ψ_2 (Fig. 4.11) with thickness $B(m)$ to provide the following continuity equation:

$$
\frac{1}{r}\frac{\partial}{\partial m}(r\overline{\rho W_m}B) + \frac{\partial}{r\partial\theta}(\overline{\rho W_\theta}B) = 0 \tag{4.89}
$$

where $\overline{\rho W_m}$ is average value across the stream tube. In the above equation, the terms containing $(\rho A)_{\psi_1}$ and $(\rho A)_{\psi_2}$ (values along ψ_1 and ψ_2, respectively) drop out because the flow domain is bounded by streamlines. This is very similar to terms appearing in square brackets in Eq. 4.66. The procedure and steps are similar to those used in deriving Eq. 4.67 and will not be repeated here. Similarity between Eq. 4.67 and Eq. 4.89 should be noted. The superscript in Eq. 4.89 will be dropped, and future reference to W_θ, W_m and ρ refers to mean values in the stream tube. Defining a stream function (which exactly satisfies the continuity equation, Eq. 4.89), given by

$$\rho W_\theta = -\frac{1}{B}\frac{\partial \psi_b}{\partial m}, \qquad \rho W_m = \frac{1}{rB}\frac{\partial \psi_b}{\partial \theta} \qquad (4.90)$$

and substituting this equation in the tangential momentum equation (Eq. 4.85) yields the following expression in terms of stream function ψ_b (Hirsch and Warzee, 1979)

$$W_m \frac{\partial}{\partial m}\left(\frac{1}{\rho B}\frac{\partial \psi_b}{\partial m}\right) + \frac{W_m}{r}\frac{\partial}{\partial \theta}\left(\frac{1}{\rho B r}\frac{\partial \psi_b}{\partial \theta}\right) - W_m \sin\phi\left(\frac{W_\theta}{r} + 2\Omega\right)$$

$$= \left(\frac{1}{r}\frac{\partial I}{\partial \theta} - \frac{T \partial s}{r \partial \theta}\right) - \frac{F_\theta}{\rho} \qquad (4.91)$$

This is the governing equation for the blade-to-blade surface S_1. For inviscid flow through a rectilinear cascade, I, s = const, $\Omega = \phi = 0$, $r \to \infty$, $m = x$, $r\theta = y$, and B = constant, and this equation reduces to Eq. 1.60 (with $W_m = u$, $W_\theta = v$). In addition, ρ = constant for incompressible flow and we recover the Laplace equation (Eq. 1.63). Thus, the three-dimensionality is contained in $B(m)$ and ϕ.

The governing equations (refer to Figs. 4.10 and 4.11) are Eq. 4.74 or 4.78 for the meridional through-flow (S_2 surface) and Eq. 4.91 for blade-to-blade or S_1 surface. The continuity equation is satisfied exactly in both cases. The variables that couple the two surfaces are b, B, ϕ, $F_{b,r}$, ΣG_i, and so on. The terms containing these unknowns are considered source terms. For example, $F_{b,r}$ and ΣG_i in Eq. 4.74 or 4.78 have to be derived from a solution of Eq. 4.91. Likewise, B and ϕ in Eq. 4.91 have to come from a solution of Eq. 4.74 or 4.78. It should be noted here that these are two two-dimensional partial differential equations (PDEs) compared to four three-dimensional momentum and continuity equations (Eqs. 4.37–4.40) needed for an exact three-dimensional solution. This provides considerable simplification of the governing equations and appreciable reduction in computation time, but the coupling of S_1 and S_2 equations through the source terms represents difficulty in the interaction process. The solution technique could be either the finite difference or finite element method described in Chapter 5. Equations can also be solved by the streamline curvature technique described later in this chapter.

4.2.3.4 Quasi-Three-Dimensional Solution Technique

The procedure for solving the quasi-three-dimensional flow using an iterative procedure is as follows (Hirsch and Warzee, 1979):

1. Solve the meridional flow equations (Eqs. 4.74, 4.75, or 4.78), assuming $F_{b,r} = \Sigma G_i = 0$.

2. Define several S_1 surfaces, from hub to tip and estimate $B(m)$ and $\phi(m)$ for each of the S_1 surfaces. Because radial derivatives are present in Eqs. 4.74, 4.75, 4.78 and (ΣG_i), it is necessary to have an adequate number of S_1 surfaces depending on the radial variation of properties (velocity, pressure, density).

3. Solve Eq. 4.91 on each of these S_1 surfaces (m, θ). This solution should provide an estimate of correlations such as $\overline{W_r' W_r'}$, and so on, and enable a first approximation for ΣG_i and $F_{b,r}$. This should be calculated at every grid point if it is a finite difference or finite element technique or at selected streamlines if it is a streamline curvature technique. Curve fitting to smooth these functions (for continuous variation of G_i and $F_{b,r}$) along m and θ surfaces may be necessary to avoid numerical stability problems.

4. Solve Eqs. 4.74, 4.75, or 4.78 a second time to derive a second approximation to B and ϕ.

5. Repeat procedures 1–4 until there is convergence and until values of B, ϕ, ΣG_i, and $F_{b,r}$ do not vary from $(n-1)$th to nth iteration. Hirsch and Warzee (1979) report that two or three iterations are sufficient to obtain reasonable convergence for an axial compressor (where B and ϕ are weak functions of m) and that more iterations are needed for a centrifugal compressor where variations in ϕ (due to axial–radial path) and B (converging passages) are substantial.

The results for the S_1 surface for a DFVLR transonic compressor rotor (28 rotor blades, 60 stator blades, $\eta = 85\%$, tip speed 14,000 rpm, mass flow 11.7 kg/s) at 18% span from the hub, stream tube 2, is shown compared with predictions from Hirsch and Warzee (1979) in Fig. 4.12. The agreement is quite good.

It is now recognized that the coupling of S_1 and S_2 surfaces is a highly unstable process and that the S_1–S_2 composite technique is replaced by complete Euler or Navier–Stokes solver. Nevertheless, the blade-to-blade (S_1) and hub-to-tip (S_2) solutions (uncoupled) are widely used in industry for analysis and design and during the development cycle.

4.2.3.5 Equations in Streamline Line Coordinates (r_m, ϕ)

Many authors (Smith, 1968; Jennions and Stow, 1985, 1986) prefer to work with the streamline curvature method for both the S_1 and S_2 surfaces. For example,

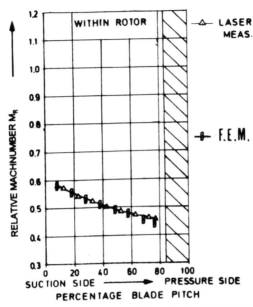

Figure 4.12 Comparison between data and predictions for the DFVLR Axial Flow compressor at 39 percent chord, 18 percent span from the hub based on Finite Element Method (FEM). (Copyright © Hirsch and Warzee, 1979; reprinted with permission of ASME.)

Wilkinson (1972) has solved the blade-to-blade equation using the streamline curvature method, and Smith (1968) and Jennions and Stow (1985, 1986) have solved the radial equilibrium equation using the streamline curvature method.

Jennions and Stow (1985) used the averaged axial momentum equation (Eq. 4.40 averaged over the passage) and Eq. 4.75 and manipulated the equations (using transformation such as Eqs. 4.79–4.83) to derive the following radial equilibrium equation in terms of $\bar{\rho}$, $\hat{W}_z$, ϕ, β, G_i, F_r, and F_z.

$$\underbrace{\frac{\sec^2 \phi}{\bar{\rho}} \frac{\partial \bar{p}}{\partial r}}_{(1)} = -\hat{W}_z^2 \underbrace{\left(\frac{\partial}{\partial z} + \tan \phi \frac{\partial}{\partial r} \right) \tan \phi}_{(1)} + \underbrace{\frac{\left(\hat{W}_z \tan \beta + \Omega r \right)^2}{r}}_{(2)}$$

$$+ \underbrace{\frac{\tan \phi}{\bar{\rho}} \left(\frac{\partial}{\partial z} + \tan \phi \frac{\partial}{\partial r} \right) \bar{p}}_{(3)} + \underbrace{\frac{\left(\bar{F}_{b,r} - \bar{F}_{b,z} \tan \phi \right)}{\bar{\rho}}}_{(4)}$$

$$+ \underbrace{\frac{\left(\bar{F}_r - \bar{F}_z \tan \phi \right)}{\bar{\rho}}}_{(5)} + \underbrace{\left[\sum_{i=1}^{4} \frac{G_i}{\bar{\rho}} - \sum_{i=1}^{5} \frac{G_{iz} \tan \phi}{\bar{\rho}} \right]}_{(6)} \quad (4.92)$$

This equation is derived from multiplying the average axial momentum equation by $\tan \phi$ and subtracting it from the radial momentum equation (Eq. 4.75). This equation provides an important coupling between the axial and radial momentum equations. This, in combination with the tangential momentum equation and continuity equation satisfies all the equations. The equation in this format also provides a convenient means of accessing the importance of various terms.

The first term on the right-hand side of Eq. 4.92 is the centripetal acceleration due to meridional curvature ϕ, r_m shown in Fig. 4.11. The second term, (V_θ^2/r), is the centripetal acceleration due to swirling flow. The third term is the contribution due to axial and radial pressure gradients. The fourth term consists of the blade force in the radial direction (Eq. 4.70), plus an additional component due to the axial blade force given by

$$F_{b,z} = -F_b r \frac{\partial \theta_c}{\partial z} + \frac{1}{\bar{\rho}} \frac{1}{b} \frac{\partial b}{\partial z} \left(\bar{p} - \frac{p_s + p_p}{2} \right)$$

where θ_c is the angle of the mean camber surface. The fifth term, $F_z \tan \phi$, arises from the axial force in the axial momentum equation. The last term is the effect on nonaxisymmetry in the flow field. The expressions for G_{iz} are similar to G_1, G_2, G_3 except the correlations such as $\overline{W_r' W_z'}$, $\overline{W_\theta' W_z'}$ and $\overline{W_z' W_z'}$ appear in these equations. In most axial turbomachinery, the variation of W_r in the blade-to-blade direction can be neglected (except when three-dimensional shocks are present) and the perturbation or correlation terms G_3, G_4, and G_5 are retained. Thus, only the axisymmetry in W_θ and W_z are retained in this formulation. The modeling of loss terms or friction terms $F_r - F_z \tan \phi$, carried out by various authors, is somewhat arbitrary and is included as loss coefficient. This will be covered in a later chapter. Many authors neglect G_i terms and it should be cautioned here that $G_i = 0$ only when the flow is axisymmetric.

Equations 4.91 and 4.92 represent a complete set of equations for the quasi-three-dimensional flow solution. Both of these equations can be solved by the streamline curvature method, described later in the chapter. The coupling procedure is very similar to the procedure described earlier. Jennions and Stow (1985) computed the flow through a highly loaded nozzle vane and showed excellent agreement between the measured and the predicted blade pressure distribution as well as the exit Mach number distribution. In the latter case, the core flow (hub-to-tip flow, with the exception of annulus and hub wall boundary layer) is predicted very well and the accuracy is as good as that of the fully three-dimensional (numerical) inviscid solution. The streamline curvature program should include grids within the blade row (not just in between) to obtain an accurate solution.

4.2.3.6 Axisymmetric Radial Equilibrium Equation

If the flow upstream and downstream of the blade is axisymmetric and inviscid, Eq. 4.92 is simplified considerably. In this case, terms 4, 5, and 6 in Eq. 4.92 are zero. Equations 4.80–4.83 can be used to simplify Eq. 4.92. The local quantities are the same as averaged quantities, and the absolute system is used ($W = V$). Furthermore,

$$dm = -(d\phi)r_m, \qquad V_r = V_m \sin \phi = W_r, \qquad W_z = V_z \qquad (4.93)$$

$$\partial z/\partial m = \cos \phi, \qquad \Omega = 0, \qquad W_z \tan \beta = V_\theta \qquad (4.94)$$

Hence,

$$-\frac{\tan \phi}{\rho} \frac{\partial}{\partial z} p = \tan \phi \left(V_z \frac{\partial V_z}{\partial z} + V_r \frac{\partial V_z}{\partial r} \right) = V_m \tan \phi \frac{\partial}{\partial m}(V_m \cos \phi)$$

$$= V_m \sin \phi \frac{\partial V_m}{\partial m} + \frac{V_m^2 \sin^2 \phi}{r_m \cos \phi} \qquad (4.95)$$

$$V_z^2 \left[\frac{\partial}{\partial z} + \tan \phi \frac{\partial}{\partial r} \right] \tan \phi = -\frac{V_m^2}{r_m \cos \phi} \qquad (4.96)$$

Therefore, Eq. 4.92 for axisymmetric, inviscid, compressible flow reduces to

$$\underset{(1)}{\frac{1}{\rho} \frac{\partial p}{\partial r}} = \underset{(2)}{\frac{V_m^2 \cos \phi}{r_m}} - \underset{(3)}{V_m \sin \phi \frac{\partial V_m}{\partial m}} + \underset{(4)}{\frac{V_\theta^2}{r}} \qquad (4.97)$$

Equation 4.97 can also be derived directly by substituting Eqs. 4.83, 4.93, and 4.94 into Eq. 4.3. It should be remarked here that the sign convention used for r_m in Eq. 4.93 is positive (since $\partial\phi/\partial m$ is negative) when the streamline curvature projection is concave downward, as shown in Fig. 4.11 and 4.13 (streamline DE).

Physical interpretation of this equation is shown in Fig. 4.13. The first term is the radial pressure gradient, and the last term is the centripetal acceleration $(-V_\theta^2/r)$, and these exist even when $r_m = \infty$ and $\phi = 0$. The second term is the radial component of the centripetal acceleration $(-V_m^2/r_m)$ due to streamline curvature and the third term is the radial component of convective acceleration of the meridional flow. Equation 4.97 is a modified version of simplified radial equilibrium equation (Eq. 4.8) to include the meridional streamline curvature effect. Equation 4.97 can be solved iteratively for the streamline curvature, provided that either p or V_θ are specified. Many investigators employ a form which does not have the pressure gradient.

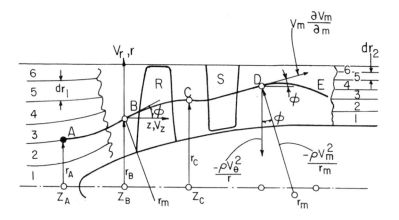

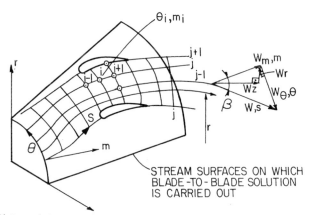

Figure 4.13 Nature of stream surfaces and notations used (sparse grid shown for clarity).

This can be derived by writing

$$\frac{1}{\rho}\frac{\partial p}{\partial r} = \frac{\partial h}{\partial r} - T\frac{\partial s}{\partial r} = \frac{\partial h_o}{\partial r} - T\frac{\partial s}{\partial r} - \frac{\partial}{\partial r}\left(\frac{V_m^2}{2}\right) - V_\theta\frac{\partial V_\theta}{\partial r} \quad (4.98)$$

By substituting this equation in Eq. 4.97, we obtain the following equation:

$$\frac{\partial h_o}{\partial r} - T\frac{\partial s}{\partial r} = \frac{V_m^2 \cos\phi}{r_m} - V_m \sin\phi\frac{\partial V_m}{\partial m} + V_\theta\frac{\partial V_\theta}{\partial r} + \frac{\partial}{\partial r}\left(\frac{V_m^2}{2}\right) + \frac{V_\theta^2}{r}$$

$$(4.99)$$

Equation 4.98 is a more generalized form than the simplified radial equilibrium equation (Eq. 4.8) and is widely employed for the solution of axisymmetric turbomachinery flows. Equation 4.99 can be rearranged to give

$$\frac{\partial}{\partial r}(V_m^2) = -2V_m^2\left[\frac{\cos\phi}{r_m} - \frac{1}{V_m}\sin\phi\frac{\partial V_m}{\partial m}\right] + 2\left[\frac{dh_o}{dr} - T\frac{\partial s}{\partial r} - \frac{V_\theta}{r}\frac{\partial(rV_\theta)}{\partial r}\right]$$

$$= 2V_m^2 F(r) + 2T(r) \tag{4.100}$$

where $T(r)$ and $F(r)$ are functions of radius only and are usually derived by an iterative process in the streamline curvature method. Novak (1976) derived and made extensive use of Eqs. 4.99 and 4.100 to predict a large class of axisymmetric flows by the streamline curvature method.

An extension of the meridional flow equation to two-phase flow to compute the effect of wetness in a stream turbine has been carried out by Yeoh and Young (1982). The thermodynamic relationship for entropy and enthalpy for two-phase flow can be written as

$$h_o = (1 - y)h_g + yh_l + \frac{V^2}{2} \tag{4.101}$$

$$s = (1 - y)s_g + ys_l \tag{4.102}$$

This assumes that the droplets (l) and vapor (g) have the same velocity, where g and l refer to gas and liquid, respectively; y is the wetness fraction. Furthermore,

$$\frac{1}{\rho_g}\nabla p = \nabla h_g - T_g\nabla s_g \tag{4.103}$$

$$0 \approx \nabla h_l - T_l\,\nabla s_l \tag{4.104}$$

If we substitute the above equations and assume that the flow is in equilibrium, the additional term arising due to two-phase flow can be proven to be

$$T(r) = 2\left[\frac{dh_o}{dr} - T_g\frac{ds_g}{dr} - \frac{V_\theta}{r}\frac{\partial}{\partial r}(V_\theta r) + \frac{L}{T_s}(T_s - T_g)\frac{dy}{dr}\right] \tag{4.105}$$

where L is the enthalpy of evaporation at saturation condition, and T_s is the saturation temperature. Yeoh and Young (1982) have solved these equations for a 14-stage steam turbine using the streamline curvature technique described later.

4.2.4 Streamline Curvature Method

As mentioned earlier, the equations governing a blade-to-blade surface and passage averaged S_2 equations can be solved either by the finite difference, finite element, or streamline curvature method. One of the efficient methods of solving these equations (either S_1 or S_2) is the streamline curvature technique. It has been in use for a long time in fluid mechanics, but its application to turbomachinery started in the mid-1960s (Novak, 1976). This is (computationally) a very efficient technique based on iteration of the streamline defined by ϕ and r_m. The technique will be illustrated for simple cases. An excellent review of the streamline curvature technique can be found in Novak (1976) and Hearsey (1986). Wilkinson (1972) provides a systematic and mathematical approach to this technique. Hearsey's (1986) technique will be described in this section.

The solution of the passage averaged and axisymmetric equations will be covered first, followed by the solution for the blade-to-blade flow.

4.2.4.1 *Meridional Through-Flow Solution*   The equation governing the average meridional through-flow in nonaxisymmetric flow is given by Eq. 4.92, and the axisymmetric through-flow is given by Eq. 4.99 or 4.100. The former requires input from blade-to-blade solution and the latter is a stand-alone equation, except for the specification of V_θ distribution downstream of the blade row. To illustrate the principle, only the axisymmetric equation is considered (Eq. 4.100). The procedure involves establishing the streamline by an iterative procedure by satisfying the continuity (in a global fashion) and the radial momentum equations. The following are the steps involved.

1. Divide the annulus into a number of stream tubes as shown in Fig. 4.13 using the continuity equation for each stream tube. It could be done with either equal mass flow ($\dot{m}/6$ in Fig. 4.13) or equal spacing. The radii of each stream tube is established from the equation $\rho_1 V_{z_1} \pi r_1 \, dr_1 = \rho_2 V_{z_2} \pi r_2 \, dr_2$. The axial stations are usually chosen to be between the blade rows (e.g., A, B, C, D, and E in Fig. 4.13). In this equation, subscripts 1 and 2 represent the inlet and exit stations. Initial approximation for V_{z_2} can be obtained from the simplified radial equilibrium or the actuator disk theory described earlier in this chapter. The value of dr_2 can be established from this equation and the global mass conservation. This is a one-dimensional approximation.

2. Once the initial streamlines are in place, the streamline curvature can be determined from the equation

$$\frac{1}{r_m} = \frac{d^2r/dz^2}{\left[1 + \left(\dfrac{dr}{dz}\right)^2\right]^{3/2}} \tag{4.106}$$

where (e.g. derivatives at point B in Fig 4.13)

$$\frac{dr}{dz} = \frac{1}{2}\left[\frac{r_c - r_B}{z_c - z_B} + \frac{r_B - r_A}{z_B - z_A}\right]$$

and

$$\frac{d^2r}{dz^2} = \frac{2}{(z_c - z_A)}\left[\frac{r_c - r_B}{z_c - z_B} - \frac{r_B - r_A}{z_B - z_A}\right]$$

Likewise, the streamline angle and r_m can also be computed.

3. For the analysis problem, either the relative velocity angle or the rV_θ distribution is specified. Hence, $V_\theta = V_m \tan \beta$. The values of V_θ or β can be derived from a solution of the blade-to-blade solution also. This also provides an estimate for h_o for the rotor. The entropy distribution can be specified from a loss correlation (radial distribution to include profile, annulus wall and hub wall boundary layers, and tip leakage losses). These are passage-averaged values. This provides a first estimate for $T(r)$ in Eq. 4.100.

4. An initial estimate is made for V_m distribution. Because V_z is known from the simplified radial equilibrium equation or from the actuator disk theory, V_r is estimated from ϕ and V_m is known. This provides an estimate for $F(r)$.

5. Equation 4.100 is then solved, starting from midradius on toward the inner and outer radii. The velocities (V_m) computed enable an estimate of densities at each calculation point.

6. The mass flow is computed across the annulus. This mass flow, in general, will not equal inlet mass. The values of V_m at midradius is adjusted in proportion to the error in mass flow computed. This process is repeated at each axial station until continuity is established. The mass flow in each stream tube is computed. In general, each stream tube will not contain proper percentage of the flow, because the streamlines are based on first approximation. This error is stored in each computing location.

7. The new values of V_m will not enable modification of the streamline. The new radii of each streamline (for example, A, B, C, D, and E), based on the new mass flow, is computed and stored at each computing station for each of the streamlines.

8. Almost invariably, the new streamlines will not be continuous and any further iteration will lead more often to a divergent solution. A smoothing or relaxation factor is needed to control the instability. In early computations a spline fit was employed to obtain a continuous streamline. Wilkinson (1972) provides a more systematic approach given

below:

$$r_{new}^{n-1} = r^{n-1} + R(r^n - r^{n-1}) \qquad (4.107)$$

where r_{new}^{n-1} is the streamline radius to be used with new iteration, r^{n-1} is the streamline radius in the previous $(n - 1)$th iteration, and r^n is the value derived from the nth iteration (after satisfying mass flow conservation). Wilkinson made a rigorous mathematical attempt to derive an expression for the relaxation factor. His expression, after modification by Hearsey (1986), is given by

$$R = \frac{1}{1 + \dfrac{k}{8}\left(\dfrac{\Delta r}{\Delta z}\right)^2} \qquad (4.108)$$

where (subscript mR stands for values at midradius)

$$k = \frac{1}{(\rho)_{m\text{R}}}\left[\frac{\partial(\rho V_m)}{\partial V_m}\right]_{m\text{R}} \approx (1 - M_m^2)_{m\text{R}}$$

9. After establishing the new "damped" radius, the new values of ϕ, r_m, $T(r)$, and $F(r)$ can be derived and procedures 5–8 repeated until convergence.

Novak (1976) and Wilkinson (1972) provide a very detailed step-by-step procedure as well as many practical hints in implementing the procedure.

The solution of the passage averaged equation (Eq. 4.92) is very similar to the one just described. In this case, terms 4–6 have to be calculated from a blade-to-blade solution (described later), and pressure and friction terms can be replaced by h_o and s using the thermodynamic relationship. Jennions and Stow (1985, 1986) solve this equation.

One of the best examples where the streamline curvature method is more suitable and efficient is the axisymmetric flow through an annulus with struts or stator blade rows. Davis and Miller (1975) have computed nonswirling flow through a duct (shown in Fig. 4.14), solving Eq. 4.100 by streamline curvature and Eq. 4.99 by the finite difference method. An allowance for strut can be made in Eq. 4.100 by including a blockage factor due to strut thickness in step 6 of the streamline curvature method. Even though this is a simple case, it is not a trivial one, and no analytical solution is available. Some analytical solutions for the swirling flow through an axisymmetric duct (no struts or blades) can be found in Batchelor (1967). Davis and Miller (1975) chose 16 axial computing stations and 10 radial locations for the streamline curvature (using spline-fit technique) and 38 axial locations and 9 radial locations for the finite difference technique. The results show good agreement. Further-

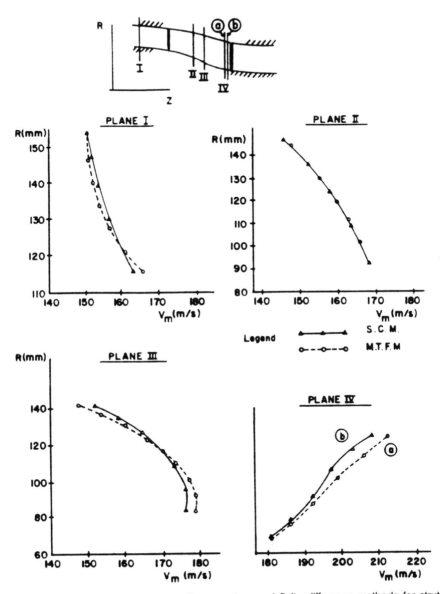

Figure 4.14 Predictions from streamline curvature and finite difference methods for strut spanning an annulus. (S.C.M. streamline curvature method, M.T.F.M. finite difference technique). (Copyright © Davis and Miller, 1975; reprinted with permission of ASME.)

more, they employed the spline-fit technique to determine streamline curvature, which is not as accurate as the one described earlier due to Wilkinson (1972). The streamline curvature method converged in about half the computer time as compared to the finite difference method.

The streamline curvature method is widely used in industry for baseline design, which can then be validated against more advanced computational codes. It is relatively easy to incorporate empirical loss correlations. For example, the $\partial s / \partial r$ distribution in Eq. 4.100 or 4.92—arising from blade profile losses, losses due to annulus and hub wall boundary layers, and secondary flow and tip leakage losses—can be specified. In such a case, this technique becomes a quasi-viscous method. The formulation becomes inconsistent when the losses (empirical) are introduced through the term $\partial s / \partial r$, without the sources (viscous terms) which generate these losses. Horlock (1971) has provided a detailed discussion on this inconsistency and has shown that the effect of viscous terms (introduced through body force term F_r) is small near the design point operation of a turbomachine and is significant at off-design conditions.

The streamline curvature method is usually restricted to shock or cavitation free flows. For flows with shocks and cavitation, numerical techniques described in Chapter 5 are more attractive. The streamline curvature technique and the code can be used with the Euler/Navier–Stokes solver to provide boundary conditions, blade forces, and source terms in the Euler equations, as well as initial conditions for the entire domain. This will provide a substantial reduction in computation time (Hale et al., 1994). The streamline curvature is still a powerful technique for simple cases (blade-to-blade or axisymmetric or passage averaged equations) because it can be easily programmed and is computationally more efficient. This technique has been widely used in the design process (Jennions and Stow, 1985, 1986; Novak, 1976), where the program is utilized for a large number of designs to arrive at an optimum blade geometry. With the availability of faster computers with larger storage capacities, the finite difference technique of solving these equations, in an inverse mode, may become a common practice.

4.2.4.2 *Blade-to-Blade Solution*

For the blade-to-blade solution, Eq. 4.85 (or Eq. 4.91) has to be solved. This equation can be expressed in streamline curvature form. There are two forms available and used.

In the first form, used by Katsanis (1966, 1968), the equation is expressed in terms of local blade angle (β), its derivative in the streamwise direction ($\partial \beta / \partial s$), streamline angle ($\phi$), and the total velocity (W). See Fig. 4.13 for notations and physical ideas behind the concept. In the second form, used by Wilkinson (1972), the equations are expressed in terms of streamline angle ϕ, β (or $d\theta / dm$), $d\beta / dm$ (or $d^2\theta / dm^2$), and W_m. Both of these equations are identical and are valid for inviscid, compressible, rotational flows. The choice of form of variables chosen depends on the method of solution.

The form used by Wilkinson (1972) and Novak and Hearsey (1977) can be derived from Eq. 4.85. This equation can be expressed in terms of streamline curvature. For example (Fig. 4.13),

$$W_\theta = W_m \tan \beta = W_m \frac{r d\theta}{dm} \tag{4.109}$$

$$W_m = W \cos \beta = W \frac{dm}{ds} \tag{4.110}$$

The dependent variables (W_m, W_θ, ϕ) in Eq. 4.85 can be expressed in terms of $(W, d^2\theta/dm^2, d\theta/dm)$, where $W = \sqrt{W_m^2 + W_\theta^2}$. The notations used are shown in Fig. 4.13. The equation, in terms of streamline curvature, is given by

$$\frac{\partial W^2}{\partial \theta} = A \frac{d^2\theta}{dm^2} + B \frac{d\theta}{dm} + C + D \tag{4.111}$$

where $d\theta/dm$, and $d^2\theta/dm^2$ are slopes and curvature of streamlines.

$$A = 2(rW_m)^2$$

$$B = 2W_m \frac{d}{dm}(r^2 W_m)$$

$$C = 4\Omega r W_m \sin \phi$$

$$D = 2\left(\frac{\partial I}{\partial \theta} - T \frac{\partial s}{\partial \theta} - \frac{F_\theta r}{\rho} \right)$$

Most authors neglect the viscous terms, F_θ, and assume that the entry flow is uniform ($\partial I/\partial \theta = \partial s/\partial \theta = 0$, hence $D = 0$). Wilkinson and Katsanis both solved the equations by the streamline curvature method. The method due to Novak and Hearsey (1977), who solved the equations for the streamline by using the finite difference technique, is given below and is more accurate than the original Wilkinson's method, which determined the shapes of stagnation streamlines iteratively. The gradients $d\theta/dm, d^2\theta/dm^2$ are the slopes and curvature of the streamline, respectively, on the blade-to-blade surface. (These are analogous to the slopes of the streamline in meridional streamline $\partial r/\partial z, d^2 r/dz^2$.) The procedure used in the solution of Eq. 4.111 by the streamline curvature method is very similar to the technique described earlier for the solution of meridional flow. Let us denote the coordinates of one of the streamlines at station i, along jth streamline by m_i, θ_i, and so on, as shown in Fig. 4.13. The slopes between $(n + 1)$th and nth iteration can be written as

$$k_i = \frac{d\theta^{n+1}}{dm} = \frac{d\theta^n}{dm} + \frac{1}{2}\left[\frac{\theta_{i+1} - \theta_i}{m_{i+1} - m_i} + \frac{\theta_i - \theta_{i-1}}{m_i - m_{i-1}}\right]$$

$$N_i = \frac{d^2\theta^{n+1}}{dm^2} = \frac{d^2\theta^n}{dm^2} + 2\left[\frac{\dfrac{\theta_{i+1} - \theta_i}{m_{i+1} - m_i} - \dfrac{\theta_i - \theta_{i-1}}{m_i - m_{i-1}}}{m_{i+1} - m_{i-1}}\right]$$

(4.112)

The velocity gradient can be written as

$$M_i = \frac{d(W^2)}{d\theta} = \left[\frac{W_{j+1}^2 - W_{j-1}^2}{\theta_{j+1} - \theta_{j-1}}\right]_i$$

(4.113)

where j refers to the streamline and i refers to the station. Equations 4.112 and 4.113 can be substituted into Eq. 4.111 to derive the following matrix equation

$$[X]\{\theta\} = \{M\}$$

(4.114)

where M and θ are row matrices containing values at $i = 1$ to $i = N$, the number of stations along the streamline, and X contains coefficients of difference equations.

The step-by-step procedure is similar to the procedure outlined for the meridional solution, with the exception that Eq. 4.100 is now replaced by difference equations for $d^2\theta/dm^2$ and $d\theta/dm$. The initial streamlines can be established by a one-dimensional continuity equation. The values of W_m, B, C, and A can then be estimated to start the solution. Equation 4.111 is solved for W^2, updating A, B, and C continuously from the previous pass. The mass flow can be estimated from

$$\dot{m} = \int_{\theta_s}^{\theta_p} \rho W \cos \beta (r\, d\theta)$$

and W_m is adjusted to satisfy the global continuity equation at each calculation station J.

The method due to Abdallah and Henderson (1987) is similar to that of Novak and Hearsey (1977), except that the velocities are calculated from the continuity equation (Eq. 4.89) with $B = $ constant.

Wilkinson (1972), Novak and Hearsey (1977), and Abdallah and Henderson (1987) have computed flows through various turbine and compressor cascades and have shown excellent agreement with the measured values as well as the exact solutions.

4.2.4.3 Composite Solution The composite solution of quasi-three-dimensional flow involves iterative processes involving solutions of Eq. 4.92 or Eqs. 4.100 and 4.111. Novak and Hearsey (1977) illustrate this procedure for a turbine nozzle and a centrifugal compressor. The procedure to be followed involves the information available for the blade row, and whether a design or an analysis problem is sought. For the analysis problem, the blade geometry and the inlet flow are specified and the outlet angles or V_θ distribution is estimated from a one-dimensional consideration or cascade correlation. This provides an estimate for $T(r)$ in Eq. 4.100, and the solution of the passage averaged equation can be carried out to predict V_m, ϕ from Eq. 4.100. It is economical to iterate on this solution, because the initial approximation for V_θ and h_o is not accurate. At this stage, the blade-to-blade equation (Eq. 4.111) is solved, with initial values of A, B, C, and D being derived from a solution of Eq. 4.100. This provides a better estimate for V_θ, and h_o, and the solution of Eq. 4.100 will now provide a better estimate for W_m. This procedure is repeated until convergence. The iterative solution, based on the streamline curvature method, is somewhat cumbersome and may not converge easily. The composite solution is best carried out by the finite difference or element technique, described later, using the entire set of equations.

Nevertheless, the streamline curvature technique is very economical and most suited for the blade-to-blade or meridional streamline solutions. As mentioned earlier, it is a powerful tool in the design of turbomachinery. References (Jennions and Stow, 1985, 1986; AGARD, 1989; Hearsey, 1986) provide excellent examples of such procedures.

4.2.5 Three-Dimensional Flow Analysis

A general theory for three-dimensional flow in subsonic and supersonic turbomachinery (axial, mixed, and radial types) was developed by Wu as early as 1952 (Wu, 1952). This inviscid theory is valid for all design and off-design conditions. The flow is assumed to be steady, and the equations of motion, energy, and state are satisfied on two sets of intersecting stream surfaces (Fig. 4.10), known as S_1 and S_2 surfaces. The family of S_1^k surfaces are from blade-to-blade and S_2^m surfaces are from hub-to-tip. A complete solution for the three-dimensional flow should be obtained by an iterative process by solving the equations for S_2 and S_1 surfaces simultaneously. Wu introduces two stream functions for the three-dimensional flow in a turbomachinery passage, one stream function for the S_1 surface and a second one for the S_2 surface. These are designated as ψ_b and ψ_m in the passage averaged analysis. The stream functions for the S_1 and S_2 surfaces are different and distinct. It should be emphasized here that the solutions S_1 and S_2 are interconnected, and for a complete three-dimensional solution these two solutions should be derived simultaneously.

Wu's technique consists of defining two sets of stream function ψ_b and ψ_m for S_1 and S_2 surfaces, respectively. The governing equation for the S_1 surfaces is similar to Eq. 4.91 and the governing equation for the S_2 surfaces is similar to Eq. 4.78, with the exception that the properties ψ_m, I, s, T, W_θ and $F_{b,r}$ now refer to the values on the local S_2 surface, $G_i = 0$, and b is the distance between the two S_2 stream surfaces. These two equations are solved iteratively until convergence is achieved. The governing equations are derived for the stream surfaces (S_1 and S_2), and hence the shape of the surface appears explicitly in these equations. The coupling of the equation is through the source terms on the right-hand side of Eqs. 4.91 and 4.78, respectively. For example, ψ_m in Eq. 4.78 is solved assuming $b =$ constant and by approximating the source term (terms on the right-hand side). For example, $r\hat{V}_\theta$ and $\hat{I}$ are specified from the one-dimensional analysis. A solution is obtained for ψ_m (and hence W_z, W_r). This is then used in Eq. 4.91 to solve equations for the S_1 surface, again using assumed or approximate values for the source terms due to the S_2 surface. This provides a first estimate for the source term in Eq. 4.78, which is again solved, and these steps are repeated until convergence is achieved. In view of the efficient solution techniques available for the Euler solution, this technique of solving the three-dimensional flow through turbomachinery is no longer attractive. Wu's formulation may be computationally more efficient, but it is more difficult to code than the Euler equations. The formulation is most useful in the blade-to-blade (S_1) and average hub-to-tip (S_2) surfaces (quasi-three-dimensional solutions). It is also useful in the design mode, because losses, pressure, and velocity field can be specified to derive the blade camber, lean, sweep, and so on, through the establishment of S_1 and S_2 surfaces. These analyses (uncoupled S_1 and S_2 solutions) can provide solutions which are nearly as accurate and as useful as the fully three-dimensional solution, and this can be carried out in much less computational time. The analysis is not valid when the stream surfaces are highly warped (as in shocked flows). Hence, the application of this technique is limited to subsonic turbomachinery, where this approach is as accurate and numerically more efficient than full Euler solutions.

There has been certain confusion in literature regarding the definition of S_2 surface. A recent paper by Zhu and Wang (1987) addresses some of these issues. The terminology used and the governing equations employed are classified below.

1. *Pitch-Averaged S_2 Surface.* The equations for the pitch-averaged S_2 surface are given by Eq. 4.74, 4.75 or 4.78. This pitch-averaged surface is a mathematical concept and cannot be identified geometrically. Furthermore, this surface is not a geometrical mean (parallel to camber surface at the midpassage). It represents a surface on which the properties are a true average between the two blade surfaces at a specified r, z location. This surface may be highly warped.

2. *Mean S_2 Surface.* This is a surface located midway between the pressure and the suction surface and parallel to the camber surface. This is the mean S_2 surface on which Novak and Hearsey (1977) solved the centrifugal compressor flow field. Most use this approach in quasi-three-dimensional solutions as described later. Equation 4.78 or 4.92, neglecting fluctuating terms, is employed in this formulation.

3. *Local S_2 Surface.* These are S_2^m surfaces (a large number of them) from pressure to suction side.

In addition to the various definitions used for the S_2 surface, the techniques of solution of S_1 and S_2 surface equations also differ widely depending on the assumptions made, accuracy desired, and whether a design or analysis problem is solved. Here is a classification of solution techniques.

4.2.5.1 Solution Techniques and Predictions Very few attempts have been made for the simultaneous solution of equations governing S_1 and S_2 surfaces. Most of the solutions are based on one S_2 surface solution (midpassage) and several S_1 surfaces. Even though a complete and composite solution of S_1 and S_2 surfaces has been carried out (Krimmerman and Adler, 1978; Wang and Yu, 1988), in the opinion of the author it is not an attractive alternative to a numerical solution of the entire Euler equation. Wu's formulation is most attractive for one mean or passage averaged S_2 surface and several S_1 surface solutions. This is computationally efficient and reasonably accurate for most design/analysis purposes.

Most methods of solving the S_1 and S_2 equations are based on the finite difference method. There are essentially three techniques for solving these equations. The streamline curvature method was explained earlier. Hirsch and Warzee (1976, 1979) solved one passage-averaged S_2 and several S_1 surfaces using the finite element method. Krimmerman and Adler (1978) also employed the finite element method for a complete solution. Most numerical solutions are based on the finite difference technique and solve either the S_1 surface equation or the S_2 surface equation, with some iteration between the two. The details of the finite difference techniques will be described in Chapter 5, and only a brief summary of the methodology and typical results will be presented here.

Solution of S_1 Surface Equations. The principal equation used for the quasi-three-dimensional solution in the blade-to-blade direction is given by Eq. 4.85 or 4.91 and is expressed in difference form. Many investigators (Smith and Frost, 1970; Katsanis and McNally, 1977; Davis and Miller, 1972; Raukhman, 1971; Chen and Zhang, 1987; Wang et al., 1986; Wu et al., 1985; and Wu and Wang, 1984) have solved the equation for governing blade-to-blade direction numerically using the finite difference technique.

Most authors specify outlet angle, and this is not satisfactory because the outlet angle should be predicted from equations to satisfy the Kutta–

Joukowski condition at the trailing edge for inviscid flows (see Section 3.2.3). Wu and Wang (1984) used the criterion developed by Joukowsky (1954). This criterion equates the velocities on the pressure and suction sides at the blade cutoff points. Wu and Wang used the body-fitted coordinate system and solved the finite difference equations using the LU decomposition scheme (Appendix B.3). Relaxation factors are utilized to iterate ψ and density. Additional stability was achieved by increasing the inlet Mach number gradually to the required value. The correct outlet angle, as well as ratio of mass flow in a tandem cascade, was predicted.

Satisfactory solution of S_1 surface equations has been achieved. The methods have been used in a wide variety of turbomachinery, both in the design and analysis mode. It is a standard design/analysis program in most aircraft engine industries. If complete viscous flow is sought, it is better to solve the full Navier–Stokes equations or resort to viscous/inviscid coupling programs.

Solution of S_2 Surface Equations. The governing equations for a passage-averaged S_2 are Eqs. 4.74, 4.75, and 4.78. The method of solution is very similar to those described for the S_1 surface. Some of the earlier solutions are due to Marsh (1968) and Davis and Miller (1972). Marsh has also developed ways of allowing for irreversibility effects by introducing the local polytropic efficiency in the computer program.

One of the most widely used computer codes is that due to Katsanis and McNally (1977), who modified S_2 surface equations by expressing entropy in terms of stagnation temperature and pressure of the relative flow which are nearly constant along a streamline. This is achieved by writing

$$ds = c_p \frac{dT_{oR}}{T_{oR}} - R \frac{dP_{oR}}{P_{oR}} \tag{4.115}$$

$$I = c_p T_{o1} - \Omega(rV_{\theta_1}) \tag{4.116}$$

The resulting equation for an S_2 surface is given by (similar to Eqs. 4.76–4.78)

$$\frac{\partial^2 \psi_m}{\partial r^2} + \frac{\partial^2 \psi_m}{\partial z^2} = \frac{\partial \psi_m}{\partial r} \frac{\partial (\ln \rho b r)}{\partial r} + \frac{\partial \psi_m}{\partial z} \frac{\partial (\ln \rho b r)}{\partial z}$$
$$- \frac{rb\rho}{W_z} \left[\frac{W_\theta}{r} \frac{\partial}{\partial r}(rV_\theta) + \xi W^2 + \zeta + F_{b,r}/\rho \right] \tag{4.117}$$

where

$$\xi = \frac{1}{2} \left[c_p \frac{R}{P_{oR}} \frac{\partial P_{oR}}{\partial r} - \frac{1}{T_{oR}} \frac{\partial T_{oR}}{\partial r} \right]$$

$$\zeta = \Omega^2 r - \frac{RT_{oR}}{P_{oR}} \frac{\partial P_{oR}}{\partial r}$$

Katsanis and McNally (1977) utilized Eq. 4.117 and solved the flow on the mean S_2 surface (midchannel) using quasi-orthogonals. Equation 4.117 was expressed in streamline (m) and quasi-orthogonal coordinates (n), and the resulting equation is expressed as

$$\nabla^2 \psi_m = G(m, n)$$

The equations are expressed in finite difference form and solved by the successive overrelaxation technique. When the flow is supersonic, the method fails. Katsanis and McNally resort to the velocity gradient equation and solve the equation by streamline curvature method. The velocity gradient equation for the S_2 surface is similar to Eq. 4.85 for the S_1 surface. Katsanis and McNally (1977) use the quasi-orthogonal or velocity gradient method employing s and t coordinate directions shown in Fig. 4.15. (s, t), (m, n), and (z, r) are orthogonal coordinate systems. If the angle between s and the axial direction (z), is α, and the angle between m and the axial direction is ϕ, then the streamlines make an angle of $(\phi - \alpha)$ with the new axis, s. If $\alpha = \phi$, then s and m coincide. The principal equation along the quasi-normal direction, t, is given by

$$
\underbrace{W\frac{dW}{dt}}_{} = \underbrace{\frac{W^2 \cos^2 \beta \cos(\phi - \alpha)}{r_m}}_{(1)} + \underbrace{W \cos \beta \frac{dW_m}{dm} \sin(\phi - \alpha)}_{(2)}
$$

$$
- \underbrace{\left(\frac{W^2 \sin^2 \beta \cos \alpha}{r} \right)}_{(3)} - \underbrace{2\Omega W \sin \beta \cos \alpha}_{(4)} + \underbrace{Wc\frac{d\theta}{dt}}_{(5)}
$$

$$
+ \left(c_p \frac{dT_o}{dt} - \Omega \frac{d(rV_{\theta_1})}{dt} - T\frac{\partial s}{\partial t} \right) \tag{4.118}
$$

where $c = W \sin \phi \sin \beta \cos \beta + r \cos \beta (\partial W_\theta / \partial m + 2\Omega \sin \phi)$, W is the total velocity, and β is the flow angle.

The significance of each of the terms in Eq. 4.118 can be explained with reference to Fig. 4.15 and Eq. 4.45. The term on the left-hand side is the convective acceleration of particle in t direction, and the last term on the right-hand side contains the static pressure gradient in the t direction as well as the centripetal acceleration due to rotation $(-\Omega^2 r)$. The other terms on the right-hand side are, respectively:

1. The component of centripetal acceleration due to meridional streamline curvature $(-W_m^2 / r_m)$ in the t direction.
2. The t component of meridional flow acceleration in the t direction given by

$$
\left(W_m \frac{dW_m}{dm} \right) \sin(\phi - \alpha) = W \cos \beta \frac{dW_m}{dm} \sin(\phi - \alpha)
$$

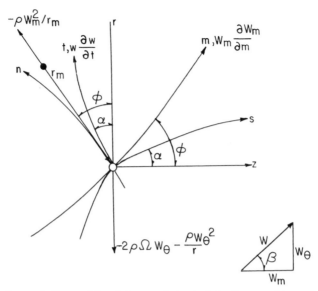

Figure 4.15   Forces in the meridional plane and physical significance of terms in Eq. 4.118.

3. The component of centripetal acceleration due to swirling flow $(-\rho W_\theta^2/r)$ in the t direction.
4. The component of Coriolis force $(-\rho 2\Omega W_\theta)$ in the t direction.
5. The t component of the acceleration in the θ direction. It should be recognized here that r, θ, z is an orthogonal system, and thus θ and t are not orthogonal. Hence, the acceleration in the θ direction is c and its component in t direction is $c\, d\theta/dt$.

The notations used in S_1 and S_2 analysis have been kept consistent throughout this book. The α, ϕ angles defined by Katsanis and McNally (1977) are ϕ and α, respectively, in this book.

In the supersonic region, Eq. 4.118 is solved as an initial value problem. Velocity W is specified at the hub $(t = 0)$ and this value is varied during the iteration until the mass conservation equation given below is satisfied:

$$\int_{hub}^{tip} \rho Wrb \cos(\phi - \alpha)\cos \beta\, dt = \dot{m}$$

Katsanis and McNally's program has been used to solve a large class of turbomachinery flow fields.

This technique, once again, is useful in subsonic and/or mildly transonic flows. The limited success they had in supersonic flows was due to the introduction of numerical dissipation in the technique. Convergence is not assured in the mixed flow region due to the dual value for density for a given mass flow parameter as explained in Section 3.5.1.

Recently, Chen and Dai (1987), and Zhao (1986) have made attempts to incorporate recent developments in computational fluid dynamics and solve the transonic flow (with shock wave) using only one set of equations. Zhao solves these equations by the artificial density technique, which consists of modifying the density so as to introduce numerical dissipation in the supersonic region. The equations are discretized and solved by the successive line overrelaxation method. Upwind differencing is used in the supersonic region. Because the equations are in nonconservative form, the authors had considerable difficulty with the calculation of density because it is not a unique function of mass flux (Section 3.5.1). They have employed the velocity gradient equation (similar to Eq. 4.118) in the stream function coordinate system and obtained velocities directly. The stream functions are derived from the principal equation, including the shock region. Knowing ψ and $\mathbf{W}$, the densities can be calculated accurately. They computed the flow in a DFVLR compressor and compared the predictions with data from Dunker et al. (1977). The results, shown in Fig. 4.16, indicate good agreement with data at the leading edge and fair agreement at the trailing edge. The shock–boundary-layer interactions are large, and the flow downstream of the shock wave can only be captured using a full Navier–Stokes equation. The three-dimensional shock structure is also predicted from this analysis, as shown in Fig. 4.16b.

Quasi-Three-Dimensional Solution. As explained earlier, the quasi-three-dimensional solution is a technique where three-dimensional effects are approximated. A coupled solution of one S_2 (mean- or passage-averaged) and several S_1 surfaces would provide a reasonable solution for the blade-to-blade flow as well as for the meridional through-flow. One such solution by Hirsch and Warzee (1976, 1979) was described earlier (Section 4.2.3.4). They utilized the passage averaged S_2 equation and local S_1 surface equations and solved them by the finite element method.

Quasi-three-dimensional solutions have been obtained by Katsanis and McNally (1977), Zhu and Wang (1987), and Wang, Zhu, and Wu (1985), but none of the authors provide a comparison with the experimental data. The solution by Hirsch and Warzee, shown in Fig. 4.12, is a good example of the quasi-three-dimensional solution. Perhaps the most widely used quasi-three-dimensional solution is due to Katsanis (1968) and Katsanis and McNally (1977). Pouagare et al. (1983) computed the flow field in a low-speed axial compressor using the code developed by Katsanis (1968) and Katsanis and McNally (1977) and compared them with the data acquired from a miniature rotating five-hole probe. This is a low-speed compressor at Penn State and has a hub/tip ratio of 0.5 and a tip diameter of 0.932 and is operated at 1088 rpm, with $\psi_p = 0.486$, and $\phi = 0.56$. (See Section 2.5.5 and Fig. 2.17 for further details on this compressor.) The results shown in Fig. 4.17 indicate

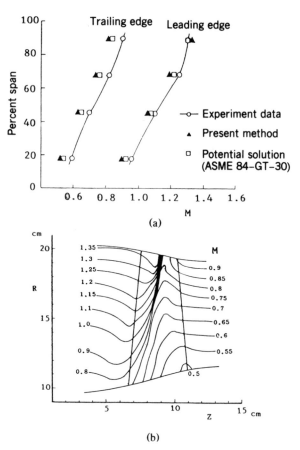

Figure 4.16 Predicted Mach number distribution for DFVLR transonic axial flow compressor $\dot{m}$ = 17.3 kg / s, N = 20,260 rpm, T_o = 288.2 K (Dunker et al., 1977). (a) Comparison of radial Mach number distribution on mean S_2 surface. (b) Mach number contours on S_2 surface. (Copyright © Zhao, 1986; reprinted with permission of ASME.)

that the agreement is good at most locations, and the departure near the blade tip is attributed to the annulus wall boundary layer and the tip clearance effects. The wake at Z = 1.07 is not predicted because the code used is an inviscid code.

Three-Dimensional Solution. The full three-dimensional solution includes utilization of several S_2 and S_1 surfaces, and iterating the solution between these two surfaces to obtain a converged solution. If the flow is inviscid and the relative flow is steady, simultaneous solution of S_1 and S_2 surfaces should provide a nearly accurate solution for moderately loaded blades. The method is not accurate when viscous flow effects and shock–boundary-layer interaction are dominant.

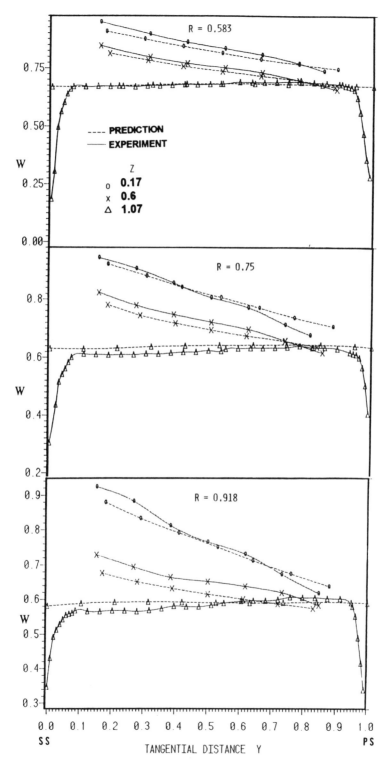

Figure 4.17 Measured and predicted blade-to-blade distribution of velocity, *W*, normalized by blade tip speed for Penn State Axial Flow Compressor (Pouagare et al., 1983; reprinted with permission of AIAA.)

Some of the investigators have made attempts for a complete solution of the equations governing S_1 and S_2 surfaces (Krimmerman and Adler, 1978; Wang and Yu, 1988; Liu and Ji, 1988). Krimmerman and Adler's (1978) solution is based on the finite element method, while those in Wang, Zhu and Wu 1985; and Wang and Yu (1988) are based on the finite difference technique.

Wang, Zhu and Wu (1985) used eleven S_1 surfaces, nine S_2 surfaces (including annulus and hub walls), and 80 grids in the streamwise or axial directions for computing the entire flow in an axial compressor. Wang and Yu (1988) also computed the transonic flow in a centrifugal compressor in which Eckardt (1976) obtained detailed flow data using a laser two-focus system. Good agreement up to about midchord is obtained, with considerable deviations near the trailing edge. The flow field beyond the midchord is highly three-dimensional, turbulent, and subjected to rotation and curvature effects. It is not surprising that an inviscid code was not able to predict the jet–wake feature and separated flows that exist in these regions.

There have been several attempts to incorporate viscous effects through loss correlation (using the equation $T \nabla s = \nabla h - \partial p / \rho$). Quasi-viscous methods have found wide application in the design process, but they are not accurate in predicting the local flow properties. For example, the viscous loss distribution from hub to tip is included in the mean S_2 surface equation to define gross effects of viscosity on the hub-to-tip flow field, which is then used in the analysis of the S_1 surface. Such an approach fails for fully 3D flows, or for S_2 and S_1 surfaces which are highly warped; coupling the two solutions becomes difficult, if not impossible. A detailed assessment of the Katsanis (1968) and Katsanis and McNally's (1977) code, modified to include boundary layer growth, has been carried out by Povinelli (1984). The prediction of outlet angle and loss were good for rectilinear and annular cascades. The loss and outlet angle predictions for endwall flow and rotor were poor.

In summary, the S_1 and S_2 surface techniques are useful for moderate three-dimensional flows and for subsonic flows and are extremely useful in the design process. The quasi-three-dimensional viscous technique with one mean hub–tip surface and several S_1 surfaces with loss correlations included in the S_2 surface equations is widely used in the design and analysis procedure in industry. It will remain as a design tool for the foreseeable future because these are well-developed codes and incorporate a wide array of loss correlations. They can also easily be modified to include new loss correlations. The quasi-three-dimensional (both viscous and inviscid) solutions are not very accurate for shocked transonic or supersonic rotor flows. In these cases, the stream surfaces are highly warped, and a composite solution by Navier–Stokes code provides a more accurate solution. The potential solution methods are restricted to cascade flows.

4.2.5.2 *Multistage Turbomachinery Analysis*
In a multistage turbomachinery, nonuniformities and unsteadiness due to rotor-stator, rotor-rotor, stator-stator interaction introduce major complexity in the analysis of the

flow field. The current practice is to use the passage averaged equations developed in Section 4.2.3 and assume the flow field is axisymmetric at the inlet of each blade row, and solve the through flow equations by streamline curvature or other technique. Accurate numerical solution of the entire multistage turbomachinery flow field (as a coupled system) using the unsteady Navier–Stokes equations requires several million grid points (at least 0.5 to 1 million grid point for each blade row for steady flow calculation) and involves intensive computing (Rai, 1987, 1989). This is prohibitively expensive and is beyond the capability of the present-day computers.

One of the successful methods of analyzing this flow is due to the work of Adamczyk (1985) and Adamczyk et al. (1990). An average-passage equation (the equation governing an average passage in the blade row as opposed to passage-average equation, Section 4.2.3, which is the axisymmetric flow field) is derived by filtering the Navier–Stokes equation in both the space and the time to remove all information except those associated with the time-averaged flow field within a typical passage of a multistage configuration. For example, the flow field due a blade row immediately preceding the blade row under consideration can be represented by Eqs. 4.74, 4.75, or 4.78, and Eqs. 4.84 or 4.91. These equations include the nonuniformity (or asymmetry) in the upstream blade row through apparent stress and blade force terms ΣG_i and $F_{b,r}$. Implicit in this equation is the assumption that flow is periodic and that the flow field is identical in every passage at every instant of time. This is not true in a multistage turbomachinery. This problem can be overcome by decomposing the flow variable $A(W_r, W_\theta, W_z, I, p, s, T, \rho)$ into fluctuations arising from unsteadiness, nonperiodicity, and nonuniformity due to all upstream blade rows.

For example, Adamczyk (1985) decomposed velocity into deterministic and nondeterministic fluctuations. The deterministic fluctuation consists of all nonuniformities and unsteadiness clocked with the shaft speed and the blade-passing frequency. This includes all fluctuations that repeat every revolution and the blade-to-blade variation; the latter component is included in ΣG_i and $F_{b,r}$ functions in Eq. 4.74. The nondeterministic fluctuation includes random turbulence (Section 1.2.6) and variation in flow field from revolution to revolution.

Based on this concept, Eq. 4.53 (density averaging) can be written as (Adamczyk, 1985; Suryavamshi et al., 1994)

$$A = \hat{A} + \tilde{A}_{BP} + \tilde{A}_{BA} + \tilde{A}_R + a$$

where A is the instantaneous value at any given location and $\hat{A}$ is the time-averaged or axisymmetric value. This is the same for each blade row and is the passage-averaged value $\hat{A}$ in Eq. 4.53. $\hat{A}_{BP}$ is the blade-to-blade periodic fluctuating component, derived from averaging all of the passages in a blade row, and this is equivalent to A' in Eq. 4.53 for an isolated blade row. $\hat{A}_{BA}$ is the blade aperiodic fluctuating component. This represents the variation in passage-averaged properties from one blade passage to the other.

This arises from differing stator-rotor blade counts, differing flow in each blade row (due to nonperiodic incidence, inviscid flow field, tip clearance, secondary flows, and wakes). A_R represents all other unsteadiness (deterministic) and is clocked with the shaft frequency. The quantity 'a' is the nondeterministic fluctuation, associated with turbulence, and revolution-to-revolution variation is flow field. Experimentally, these quantities can be derived by ensemble-averaging (time-average) the data at constant (r, θ, z) at once per revolution. For example, if 400 revolutions of data are acquired, the ensemble-averaged value includes 400 samples. The difference between the ensemble-averaged value $A_E(r, \theta, z)$ and the instantaneous value $A(r, \theta, z, t)$ represents the nondeterministic fluctuation (a). The ensemble-averaged values are then averaged over all of the passages to derive a blade-to-blade periodic value $(\hat{A} + \tilde{A}_{BP})$. A phase-locked averaging will then decouple the blade periodic $(\hat{A} + A_{BP})$ component from the aperiodic (deterministic) fluctuating component $(A_{BA} + A_R)$. A procedure similar to that carried out in Section 4.2.3 results in additional "apparent or mixing stresses," given by

$$R_{ij} = G_{ij} + \overline{\left(\tilde{A}_{BA}\right)_i \left(\tilde{A}_{BA}\right)_j} + \overline{\left(\tilde{A}_R\right)_i \left(\tilde{A}_R\right)_j} + \overline{a_i a_j}$$

where subscript i and j correspond to components. For example, G_i is the stress arising from nonaxisymmetry in an isolated blade row and is given by Eq. 4.74; $i = r$, $j = r$, is the G_1 value. Similarly, stress correlations due to blade aperiodic and deterministic component results in additional "apparent stresses." The stress due to nondeterministic fluctuations $\overline{(a_i a_j)}$ includes the Reynolds stress due to turbulent fluctuations (Eqs. 1.68 and 1.69) and other fluctuations not clocked with the shaft frequency.

The value of G_{ij} can be derived from blade-to-blade flow analysis. The stress $\overline{a_i a_j}$, if dominated by turbulence, can be modeled using the turbulence models described in Section 1.2.7. Other components in the preceding equation has to be modeled based on experiment or analysis. Attempts are underway to model these terms.

4.2.6 Secondary Flow and Vorticity: Inviscid Effects and Theories

In many instances, the flow and the thermal field entering a blade row are nonuniform. Some typical distributions are shown in Fig. 4.18. The radial gradient in velocity, stagnation pressure, and stagnation temperature may be caused by casing/hub wall boundary layers (Figs. 4.18a and 4.18e), or by the presence of an upstream blade row (as in Figs 4.18b and 4.18f). As indicated earlier, the blade row causes radially nonuniform velocity, stagnation pressure, and stagnation enthalpy profiles, especially in a non-free-vortex blade row. Furthermore, the combustion chamber preceding the turbine blade row produces a nonuniform temperature profile, both in the radial direction (as in Fig. 4.8d) and in the circumferential direction. In addition, the multistage

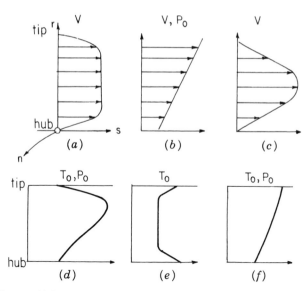

Figure 4.18 Nature of inlet stagnation temperature, stagnation pressure and velocity profiles.

turbomachinery may encounter fully developed flow in the embedded stages (as in Fig. 4.18c).

Even though some of this nonuniformity arises from viscous effects, many of these arise even in inviscid flow (e.g., Figs. 4.18b and 4.18f). Even though a complete analysis of this is beyond the scope of analytical techniques, the discipline known as "secondary flow theories" has emerged as a powerful tool in calculating the effects of nonuniformity in the flow field on the blade row performance.

Let us consider a two-dimensional cascade blade row and inlet velocity profile shown in Fig. 4.19. AAA is the streamline in a cascade in the uniform flow region, and BBB is the streamline in the shear layer. If we neglect viscous effects and the velocity variation in the n direction and assume the flow to be incompressible (with constant density) and steady, the pressure gradient in the normal direction (n) is balanced by the centripetal acceleration at point A along the streamline AAA shown in Fig. 4.19. Hence

$$\left(\frac{\partial p}{\partial n}\right)_A = \frac{\rho u_A^2}{R_A} \tag{4.119}$$

where R_A is the radius of the streamline AAA at point A, where the total streamwise velocity is u_A.

If the boundary layer approximation is invoked, the pressure gradient normal to the side wall for the streamlines AAA and BBB would be the

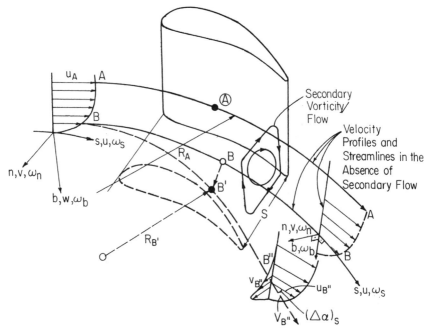

Figure 4.19 Secondary flow phenomenon and notations used (only one blade shown for clarity).

same, and thus we obtain

$$\left(\frac{\partial p}{\partial n}\right)_A = \left(\frac{\partial p}{\partial n}\right)_B = \frac{\rho u_A^2}{R_A} > \frac{\rho u_B^2}{R_B}$$

Because $u_B < u_A$, $R_B = R_A$, there is an imbalance between the normal pressure gradient and the centripetal acceleration and thus streamline BBB in the shear layer will deflect more, thereby developing a cross-flow toward the suction side. The fluid particle originating at B would follow the path BB′B″ as shown in Fig. 4.19 with $R_{B'} < R_A$. The cross-flow (v), which is a deviation from the design or main or primary flow, is called the *secondary flow*. From continuity consideration, there should be spanwise velocities w. It can be proven that for a turning duct along a mean line (neglecting streamwise pressure gradient), $\partial u/\partial s = 0$; hence $\partial w/\partial b = -\partial v/\partial n$. This simple explanation provides a clear physical reasoning for the occurrence of the secondary flow in a curved bend or a blade passage. The secondary flow gives rise to secondary vortices. These deviations in primary flow ($v_{B'}, w_{B''}$) result in vorticity: $\omega_s = -\partial v/\partial b + \partial w/\partial n$. In the illustration shown, the normal component of vorticity ω_n is in the negative n direction at inlet, and the streamwise vorticity component ω_s is in the negative s direction at exit. The nature of secondary flow in a rotor blade row is shown in Figs. 1.15 and 1.16.

Secondary flow is thus generated whenever a shear layer (or normal vorticity) is turned through a duct or cascade. Secondary flow can also be viewed as the development of a three-dimensional or skewed boundary layer, if it is caused by viscous effect in the wall region. Some other examples where secondary flow and vorticity are developed are as follows:

1. A sheared flow passing around an obstacle (e.g., the flow past a bridge pier or wall and blade leading edge intersection region)
2. Turning of stratified flow in a bend or blade row
3. Vortex motions induced by atmospheric and oceans currents by the earth's rotation

A simple quantitative estimate of the secondary vorticity can be made using a purely kinematical relationship. Consider a constant cross-sectional area curved duct as shown in Fig. 4.20. The scale in this figure is exaggerated to illustrate the point. Let the inlet normal vorticity due to shear layer (e.g., Fig. 4.19) be $\omega_n \ (= \partial u / \partial b)$. Because of the curvature of the duct, there will be a velocity differential between the two streamlines, with velocity $u - \Delta u$ and u, along PP'P″ and QQ', respectively. If the velocity is constant ($\Delta u = 0$) and the turning is small, convection and rotation of the vorticity vector by the flow gives rise to a component of vorticity, ω_{s_1}, in the streamwise direction. The streamwise distance PP″ is equal to the streamwise distance QQ'. The vorticity vector is now along Q'P″, giving rise to a streamwise component vorticity (for small flow turning) $\omega_{s_1} \approx -\omega_n \, d\theta$, assuming that ω_n is unchanged. Let us now consider additional effects due to velocity differential Δu. Let Q'P' be the new location of the vortex filament (QP). The differential distance ds_2(P'P″) due to velocity differential can be estimated from simple flow considerations. For example, assuming vorticity in the spanwise

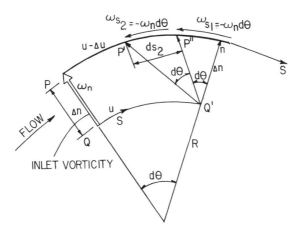

Figure 4.20   Estimate of secondary vorticity.

direction (normal to paper) is zero,

$$\frac{du}{dn} = -\frac{u}{R}, \qquad ds_2 = (\Delta u)(\Delta t) = -\left(\frac{u}{R}\,\Delta n\right)\left(\frac{R\,d\theta}{u}\right)$$

Thus the vorticity vector QP (assuming no distortion) will be convected downstream and is inclined at an angle (vector Q'P') as shown in Fig. 4.20. This develops an additional streamwise vorticity $\omega_{s_2} = -\omega_n\,d\theta$.

Hence, the total streamwise vorticity developed is given by

$$\omega_s = \omega_{s_1} + \omega_{s_2} = -2\omega_n\,d\theta \qquad (4.120)$$

Thus, half the secondary vorticity is generated due to velocity differential between the streamlines, with the remainder due to the translation of the vorticity vector. This simple secondary vorticity expression (Eq. 4.120) is due to Squire and Winter (1951), who derived this equation through a systematic approximation to the partial differential equations governing the flow field. The derivation given here is much simpler.

In the sign convention adopted in this book, s is the streamwise direction, n is the principal normal (positive being directed toward the center of curvature), and b is the binormal direction. The s, n, b system follows the right-hand rule. For example, in Fig. 4.19, b is directed downward as shown. ω_n at the inlet is in the negative direction and ω_s at the exit is in the negative direction. According to this sign conversion, ω_n in Fig. 4.20 and Eq. 4.120 is negative and hence Eq. 4.120 should read $\omega_s = +2\omega_n\,d\theta$. Some researchers use the n direction opposite to those in this book. Hence ω_n in their notation is positive in Fig. 4.19, and thus Eq. 4.120 has the negative sign on the right-hand side.

Extensive literature is available on secondary flow. It is beyond the scope of this book to cover all aspects of secondary flow. The reader is referred to periodic reviews written on the subject (Lakshminarayana and Horlock, 1963; Hawthorne, 1966; Horlock and Lakshminarayana, 1973; Ucer et al., 1985, articles by Serovy, Sieverding, and Braembussche in Vol. 2). Only the analytical aspect of the secondary flow is covered in this section; the experimental aspects, including its effect on losses and efficiency, are covered in Chapter 6.

Secondary flows can have significant effects on the performance of turbomachinery. The effect of secondary flow is summarized below:

1. It introduces cross-flow (v and w) velocity components, which result in three-dimensionality in the flow field.

2. These secondary flows tend to form a vortex, which will eventually initiate a separation region near the suction surface of the wall. Their effect on the overall performance is thus substantial.

3. The secondary flows have appreciable effect on the flow turning (usually overturning in the wall regions and underturning outside the wall region), thus affecting the pressure rise or drop in turbomachinery.

4. The resulting loss decreases efficiency. The endwall flow losses, including secondary flow, accounts for approximately 2–4% drop in efficiency of turbomachines.

5. The secondary flow introduces off-design conditions in the downstream row (incidence changes). The rotor–stator interaction, due to secondary flow, results in unsteady flow/pressure field in subsequent blade rows. This interaction may result in vibration, flutter, and noise.

6. Secondary flow affects the temperature field as well as the cooling requirement in a turbine.

7. It is also responsible for cavitation (vortex cavitation) and resulting deterioration in performance and damage in liquid handling machinery.

4.2.6.1 Generalized Expressions for Secondary Vorticity A more generalized secondary vorticity expression for nonrotating curved flows is derived by Hawthorne (1951, 1955a) and for the rotor flow by Smith (1957). Lakshminarayana and Horlock (1973) provided generalized equations for the development of vorticity, valid for compressible, stratified, rotating and nonrotating, and viscous flows. Consider the coordinate system s, n, b (orthogonal locally) shown in Fig. 4.19. The unit vectors s, n, b are along the streamwise, principal normal, and bi-normal directions, respectively (Fig. 4.19). Taking the scalar product of Eq. 1.36, the following governing equations can be derived for the streamwise vorticity (ω_s), normal vorticity (ω_n), and vorticity in the bi-normal direction (ω_b). The expressions for ω_s, ω_n, and ω_b are given by, respectively (Lakshminarayana and Horlock, 1973),

$$\frac{\partial}{\partial s}\left(\frac{\omega_s}{\rho u}\right) = \frac{2\omega_n}{\rho u R} - \frac{1}{u^2\rho^3}\left[\frac{\partial p}{\partial n}\frac{\partial \rho}{\partial b} - \frac{\partial \rho}{\partial n}\frac{\partial p}{\partial b}\right] + \frac{\mu}{\rho^2 u^2}\mathbf{s}\cdot\nabla^2\boldsymbol{\omega}$$

$$+ \frac{\mathbf{s}}{\rho u^2}\cdot\nabla\times\left(\frac{\mathbf{B}}{\rho}\right) \tag{4.121}$$

$$\frac{\partial}{\partial s}(\omega_n u) = \frac{u\omega_b}{\tau} + \frac{\omega_n u}{a_n}\frac{\partial a_n}{\partial s} - \left(\omega_b u\frac{\partial \alpha_b}{\partial s}\right) + \frac{1}{\rho^2}\left[\frac{\partial p}{\partial s}\frac{\partial \rho}{\partial b} - \frac{\partial p}{\partial b}\frac{\partial \rho}{\partial s}\right]$$

$$+ \mathbf{n}\cdot\nabla\times\left(\frac{\mathbf{B}}{\rho}\right) + \mathbf{n}\cdot\frac{\mu}{\rho}\nabla^2\boldsymbol{\omega} \tag{4.122}$$

$$\frac{\partial}{\partial s}\left(\frac{\omega_b}{\rho}\right) = \frac{\omega_n}{\rho R} + \frac{\omega_n}{\rho}\frac{\partial \alpha_n}{\partial s} + \frac{\omega_b}{\rho a_b}\frac{\partial a_b}{\partial s} + \frac{1}{\rho^3 u}\left(\frac{\partial \rho}{\partial s}\frac{\partial p}{\partial n} - \frac{\partial \rho}{\partial n}\frac{\partial p}{\partial s}\right)$$

$$+ \frac{\mathbf{b}\cdot\nabla\times\left(\dfrac{\mathbf{B}}{\rho}\right)}{\rho u} + \frac{\mu}{\rho^2 u}\mathbf{b}\cdot\nabla^2\boldsymbol{\omega} \tag{4.123}$$

These expressions are exact, except for the viscous terms. The effect of compressibility on viscous terms is neglected. This assumption is not very critical, because the majority of the flows in turbomachinery are turbulent and this term would be approximated by an empirical eddy viscosity (turbulent). The expression for vorticity is given by

$$\nabla \times \mathbf{V} = \mathbf{s}\omega_s + \mathbf{n}\omega_n + \mathbf{b}\omega_b$$

where $\mathbf{V} = \mathbf{s}u$, $\omega_n = \partial u/\partial n$. R is the principal radius of curvature (Fig. 4.19), and τ is the radius of torsion defined by Frenet-Serret's formulae:

$$\frac{\partial \mathbf{s}}{\partial s} = \frac{\mathbf{n}}{R}, \quad \frac{\partial \mathbf{b}}{\partial s} = -\frac{\mathbf{n}}{\tau}, \quad \frac{\partial \mathbf{n}}{\partial s} = \frac{\mathbf{b}}{\tau} - \frac{\mathbf{s}}{R}$$

a_n and a_b are the distances between adjacent streamlines in n and b directions, respectively. $\mathbf{B}$ is the body force. In the case of a turbomachinery blade row, it represents the blade force, $\partial\alpha_n/\partial s$ is the rate of angular rotation of $\mathbf{n}$ vector along the streamwise direction, and $\partial\alpha_b/\partial s$ is the rate of angular rotation of $\mathbf{b}$ vector along the streamwise direction. In most situations, ω_b is normally small, and only Eqs. 4.121 and 4.122 are relevant. Furthermore, ω_b and $\partial\alpha_b/\partial s$ terms in Eq. 4.122 can also be neglected in many axial turbomachines, see James (1987) for additional discussion.

The appearance of factor 2 in the second term in Eq. 4.121 arises from equal contributions from the substantial derivative $D\omega/Dt$ and the term $(\omega \cdot \nabla)\mathbf{V}$, respectively. Examination of the inviscid terms in Eq. 4.121 shows how the secondary vorticity is developed (a) when there is a normal component of vorticity (ω_n) in a flow of radius of curvature R (Fig. 4.19) and (b) when density and pressure gradients exist in the b and n surfaces in mutually perpendicular directions [because $\nabla(\rho^{-1})$ and ∇p are normal, respectively, to surfaces of constant density and constant pressure, the vector $\nabla(\rho^{-1}) \times \nabla p$ is tangential to the curve of the intersection of these surfaces]. Furthermore, even in the absence of these effects, an existing secondary vorticity will change (due to compressibility and velocity changes) through the term $\partial(\omega_s/\rho u)/\partial s$ caused by the vortex stretching, and by diffusion and dissipation through viscosity.

There has been no analytical solution of Eqs. 4.121–4.123 for the generalized case, but there have been some attempts made to solve the streamwise vorticity equation numerically. One such attempt is due to Briley and McDonald (1984). They decomposed the velocity into primary and secondary velocity and employed a set of momentum equations and the transport equation for secondary vorticity (Eq. 4.121) and obtained good predictions for both the primary and secondary flows, including viscous effects for the flow through a curved duct.

In many practical applications, a simplified from of Eqs. 4.121–4.123 is employed. The value of τ is usually large and other terms in Eq. 4.123 are small, and thus ω_b can be assumed to be constant. Likewise, the change of ω_n is brought about mainly by compressibility effect, viscous diffusion, and wall boundary layer growth through the blade row. For inviscid and compressible flow, Eq. 4.121 can be written as (for $\partial p/\partial b = 0$; $\partial p/\partial n = -\rho u^2/R$, in the s, n, b notation used here, n is toward the center of curvature)

$$\frac{\partial}{\partial s}\left(\frac{\omega_s}{\rho u}\right) = \frac{2\omega_n}{\rho u R} + \frac{1}{\rho^2 R}\frac{\partial \rho}{\partial b} = \frac{2\omega_n}{\rho u R} - \frac{1}{\rho R T}\frac{\partial T}{\partial b} \qquad (4.124)$$

which can also be expressed as

$$\frac{\partial}{\partial s}\left(\frac{\omega_s}{\rho u}\right) = \frac{2}{\rho \rho_o u^2 R}\frac{\partial P_o}{\partial b} \qquad (4.125)$$

This is one of the basic equations for secondary vorticity developed by Hawthorne (1955a).

In turbomachinery applications, the static temperature changes through the blade row, while the stagnation temperature for a stator and relative stagnation temperature for a rotor remain constant along a streamline in an adiabatic flow. Therefore, it is useful to express secondary vorticity in terms of T_o. The static temperature in Eq. 4.124 can be eliminated to prove (Lakshminarayana, 1975):

$$\frac{\partial}{\partial s}\left(\frac{\omega_s}{\rho u}\right) = \frac{2\omega_n}{\rho u R}\left(1 + \frac{\gamma-1}{2}M^2\right) - \frac{|\nabla T_o|}{\rho T R}\cos\beta \qquad (4.126)$$

where β is the angle between ∇T_o and the coordinate b. The angle β represents the rotation of isothermal surfaces as the flow progresses through the blade row.

It is clear from simplified secondary vorticity expressions, Eqs. 4.124–4.126, that the secondary flow can develop under the following situations:

1. When there are normal components of vorticity such as those caused by wall boundary layers or shear gradient in the free stream due to upstream flow (Figs. 4.18 and 4.19).
2. When there is a radial temperature (static or stagnation) gradient at the inlet such as those caused by combustion chambers preceding a nozzle blade row (Fig. 4.18). In either case, it is essential that a stagnation pressure gradient be present as per Eq. 4.125.

Equations 4.124 and 4.125 suggest a mechanism for reducing the secondary flow. The shear velocity gradient (ω_n) and temperature gradient can be properly chosen or manipulated so as to minimize the stagnation pressure gradient, thus reducing the secondary flow development.

Lakshminarayana (1975) provided a solution to Eq. 4.126, when only the stagnation temperature gradient is present ($\omega_n = 0$). His expression for the rotation of isothermal surfaces is given by

$$\beta = \frac{R|\nabla T_o|}{2T_o a}\left\{\frac{\ln(1 + a\epsilon)}{a} - \epsilon\right\} \tag{4.127}$$

where R is the radius curvature of the passage, $a = (R/u)(du/ds)$, ϵ is the flow-turning angle. The thermal transport and the presence of hot and cold spots on a turbine blade can be predicted quantitatively using this analysis [see Lakshminarayana (1975) for details].

In aircraft engines, it is not uncommon to encounter high temperatures at inlet near the midspan (due to the preceding combustor) as shown in Fig. 4.18d. Such hot spots can be transported by secondary flows toward the blade tip, causing considerable blade damage. The qualitative trend (warping of the isothermal surface) can be predicted using Eq. 4.127.

4.2.6.2 *Calculation of Secondary Velocities and Change in Outlet Angle in a Cascade*   Let us consider the cascade blade row in Fig. 4.19, with inlet vorticity ω_n and exit secondary vorticity ω_s. The secondary flow in this instance is defined as the deviation of primary flow from the primary flow direction (s). Let us denote the velocity vector by

$$\mathbf{V} = s\mathbf{u} + \mathbf{n}v + \mathbf{b}w \tag{4.128}$$

where u is the primary velocity, and v and w are secondary velocities.

The secondary vorticity ω_s is given by

$$\frac{\partial w}{\partial n} - \frac{\partial v}{\partial b} = \omega_s \tag{4.129}$$

By defining a secondary stream function ψ_s (streamlines in the bn plane) given by

$$w = -\frac{\partial \psi_s}{\partial n}, \qquad v = \frac{\partial \psi_s}{\partial b} \tag{4.130}$$

we obtain the following Poisson equation:

$$\nabla^2\psi_s = -\omega_s \tag{4.131}$$

This equation can be solved numerically or analytically using the appropriate expression for ω_s. The only analytical solution available is for the incompressible flow, with Squire and Winter's (1951) equation for secondary vorticity ($\omega_s = 2\epsilon\omega_n$). Hawthorne (1955b) has provided an analytical solution of Eq. 4.131 for incompressible flow. This solution can be used to calculate secondary velocities v and w. These can then be averaged over the blade to determine $\bar{v}$ and $\bar{w}$, which will then provide an estimate for the deviation in exit angle ($\Delta\bar{\alpha}_2$) due to secondary flow and the deviation in flow angle from its primary value. Typical predictions from this analysis are shown later in this section.

4.2.6.3 *Secondary Flow in Rotors: Effect of Rotation*   An expression for secondary vorticity in a blade row rotating with angular velocity Ω can be derived by taking curl of Eq. 4.24. For simplicity, all viscous terms are neglected. The generalized expressions are given in Lakshminarayana and Horlock (1973). An intrinsic coordinate system is used to derive expressions for the secondary vorticity (see Fig. 4.21). s', n', b' are coordinate systems along the relative velocity (W) direction, principal normal direction, and bi-normal direction, respectively. The relationships between the component of absolute vorticity (ω) and the relative vorticity, ζ, along s', n', b' are given

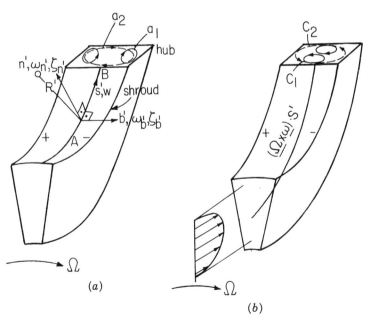

Figure 4.21 Nature of secondary flow in centrifugal compressors / pumps. (a) Effect of meridional curvature (b) Effect of rotation.

by (based on Eq. 4.20)

$$\zeta = \omega - 2\Omega$$

$$\zeta_{s'} = \omega_{s'} - 2\Omega \cdot s'$$

$$\zeta_{n'} = \omega_{n'} - 2\Omega \cdot n' = \frac{\partial W}{\partial b'}$$

$$\zeta_{b'} = \omega_{b'} - 2\Omega \cdot b' = -\frac{\partial W}{\partial n'} + \frac{W}{R'} \qquad (4.132)$$

where R' is the principal radius of curvature of the relative streamline. In turbomachinery application, the component of *absolute* vorticity (ω) along the *relative flow* ($\omega_{s'}$) direction is of interest. The expression for $\omega_{s'}$ for inviscid flow is given by (Lakshminarayana and Horlock, 1973)

$$\frac{\partial}{\partial s'}\left(\frac{\omega_{s'}}{\rho W}\right) = \frac{2\omega_{n'}}{2\rho WR'} - \frac{2\Omega \times \omega}{\rho W^2} \cdot s' - \frac{1}{\rho^3}\frac{1}{W^2}\left(\frac{\partial \rho}{\partial b'}\frac{\partial p}{\partial n'} - \frac{\partial \rho}{\partial n'}\frac{\partial p}{\partial b'}\right) \qquad (4.133)$$

Similarity between this equation and Eq. 4.121 is evident. Additional secondary vorticity arises due to rotation effects, when $\Omega \times \omega$ has a component in the relative streamwise direction. For a purely axial flow rotor, ω is usually in the (sn) plane, and the component of $\Omega \times \omega$ in the s' direction is negligibly small. Hence the rotation-induced secondary flow is small. It should be emphasized that in modern axial flow turbomachinery, with large viscous effects and annulus wall and hub wall flaring, this contribution may be appreciable. On the other hand, for radial turbomachinery, the secondary vorticity produced by the rotation term (($2\Omega \times \omega) \cdot s'$) may dominate those produced by the flow curvature R' (Lakshminarayana and Horlock, 1973).

The development of secondary vorticity and flow in turbomachinery rotors can be interpreted using a simplified form of Eq. 4.133. Let us consider incompressible, inviscid, homogeneous flow. The secondary flow arises due to $\omega_{n'}$ and $\Omega \times \omega$.

The secondary vorticity is given by

$$\left(\frac{\omega_{s'}}{\rho W}\right)_1^2 = \int_1^2 \frac{2\omega_{n'}}{\rho WR'}\,ds' - \int_1^2 \frac{2(\Omega \times \omega) \cdot s'}{\rho W^2}\,ds' \qquad (4.134)$$

where 1 and 2 represent the inlet and local (or exit) flow conditions, and $\omega_{s'}$ is the absolute vorticity in the relative streamwise direction. For axial turbomachinery, if the secondary flow induced by the annulus and hub wall boundary is of interest, $\Omega \times \omega \cdot s'$ is nearly zero, Eq. 4.134 can be simplified to (for ρ, W, and R' const; see Lakshminarayana and Horlock, 1973)

$$\omega_{s'_2} - \omega_{s'_1} = 2\,\omega_{n'}\,\epsilon'$$

where ϵ' is the turning angle of the relative flow and $\omega_{n'}$ is the component of absolute vorticity in the principal normal direction (n') of the relative flow.

Let us consider a centrifugal/radial turbomachinery. Three sources of secondary flow can be recognized from Eq. 4.134:

1. The first source of secondary flow is the meridional curvature [change of the flow path from the axial to the radial direction (Fig. 4.21a)] in the presence of $\omega_{n'}$.
2. The second source of secondary flow is the blade camber, again in the presence of $\omega_{n'}$, similar to that shown in Fig. 4.19.
3. The third source is caused by the direct effect of rotation (last term in Eq. 4.134).

Let us consider a radial compressor. In this case, s' is in the radial direction, n' is in the axial direction, and b' is in the tangential direction. For irrotational absolute flow $\zeta = -2\Omega$ and $\zeta_{n'} = -2\Omega$. Hence, the axial vorticity spans the entire passage from inlet to exit, and blade-to-blade and is in a direction opposite to rotation and is called "relative eddy."

The nature of secondary flow in predominantly axial flow path of a centrifugal turbomachinery (as in centrifugal compressor/pump inducers) may be different from that along the radial flow path of the impeller. This is clear by examining Eq. 4.134. This may result in major changes in the nature and magnitude of secondary flow as the flow travels from the inlet to the exit of a centrifugal compressor or pump or turbine.

Let us consider the secondary flow induced in the initial flow path (axial to mildly radial) of a centrifugal compressor or pump. The curvature of the meridional streamline is R' (as shown in Fig. 4.21a), and $\mathbf{\Omega} \cdot \mathbf{n}'$ and $\partial W/\partial b$ are all small (because blade boundary layer growth is small). Hence, no significant secondary flow can be expected from this source. Similarly, it can be proven that the development of secondary flow is not significant due to the blade camber (which is small, Fig. 1.12). Furthermore, at inlet the value of $(\mathbf{\Omega} \times \mathbf{\omega}) \cdot \mathbf{s}'$ is nearly zero and its contribution to the secondary flow in the nearly axial flow path of a centrifugal pump/compressor/radial turbine should be small. Hence, no significant secondary flow is expected from any of these sources until the flow starts to deviate substantially from the axial direction. This is clear from laser doppler measurements reported by Hathaway et al. (1993) at a nondimensional shroud meridional distance of 0.475. Only significant deviations from the primary flow observed by Hathaway et al. at this location are due to leakage flow.

Let us now consider the region where the flow path changes from predominantly axial to predominantly radial (e.g., flow path AB marked in

Fig. 4.21a). Let us consider the secondary flow induced by all three sources identified earlier:

1. *The Secondary Flow Induced by Meridional Curvature and* $\omega_{n'}(\partial W / \partial b'$ $+ 2\mathbf{\Omega} \cdot \mathbf{n}')$. The contribution from each of these terms depends on the type of machinery. Significant blade boundary layer growth is expected at this location, and thus $\omega_{n'}$ may be significant. The secondary flow due to $\partial W/\partial b'$ is confined to the surface boundary layers, shown as "a1" (toward the shroud) in Fig. 4.21a. The contribution due to $2\mathbf{\Omega}' \cdot \mathbf{n}'(-2\mathbf{\Omega}$ at the exit) covers the entire passage. The nature of this secondary flow is shown by a dashed line and marked as "a2" in Fig. 4.21a. The experimental data due to Hathaway et al. (1993) (Fig. 4.22) shows evidence of flow toward the shroud inside the blade boundary layer (secondary flow a1), but the contribution due to $(-2\mathbf{\Omega})$ is not observed. As shown later, all other contributions to secondary flow are in a direction opposite to the secondary flow due to this case. Hence, secondary flow due to this source may have been overshadowed by other sources.

2. *The Source Due to Blade Camber.* In this case, n' is now toward the center of curvature of the blade. Because n' is nearly in the tangential direction, $\mathbf{\Omega} \cdot \mathbf{n}'$ is very small. Hence, the dominant source for this case is the velocity gradient $(\partial W/\partial b'$, where b' is in the spanwise direction) inside the hub and shroud wall boundary layers. The secondary flow is from pressure to suction surface on both the hub and the shroud surfaces (shown as "C1" in Fig. 4.21b) and is similar to that shown for an axial flow compressor/turbine in Fig. 4.19.

3. *Direct Effect of Rotation.* The secondary flow due to the term $-2(\mathbf{\Omega} \times \boldsymbol{\omega}) \cdot \mathbf{s}'/\rho W^2$ is the direct rotation effect. This term is equal to $-2(\mathbf{\Omega} \times \boldsymbol{\zeta}) \cdot \mathbf{s}'/\rho W^2$ (Eq. 4.134) and the main contribution to this term comes from the vorticity, $\zeta_{b'}$, as shown below (Fig. 4.21b). The second term on the right-hand side of Eq. 4.134 can be written as (for predominantly radial flow path, $\mathbf{\Omega} \approx -\mathbf{n}\Omega$)

$$-2\big(\mathbf{\Omega} \times (\mathbf{b}'\zeta_{b'} + \mathbf{n}'\zeta_{n'} + \mathbf{s}'\zeta_{s'})\big) \cdot \mathbf{s}'/\rho W^2 \approx 2\Omega\zeta_{b'}/\rho W^2$$

$$\approx \frac{2\Omega}{\rho W^2}\left(\frac{-\partial W}{\partial n'} + \frac{W}{R'}\right)$$

The first term in the above expression induces secondary flow from the pressure to the suction surface on both the hub and the shroud, whereas the second term (due to curvature of the meridional streamline) induces global secondary flow (not confined to wall boundary layers) in the entire passage. These are shown in Fig. 4.21b as "C1" and "C2", respectively. Inside the wall boundary layers, the term $\Omega(\partial W/\partial n')$ is likely to dominate. Outside the wall boundary layer, the secondary flow due to meridional curvature will cover the whole passage, with flow from the suction to the pressure surface near the hub and from the pressure to the suction surface near the shroud.

To support the various features revealed by the secondary flow theories, experimental data obtained by Hathaway et al. (1993) and Ubaldi et al. (1993) will be used. Hathaway's data were obtained in a large-scale, low-speed, backswept impeller with a tip speed of 153 m/s. The impeller had 20 blades with a back sweep of 55°. The inlet diameter was 0.870 m and the exit diameter was 1.524 m. The mass flow was 30 kg/s at 1862 rpm. The data were acquired with a laser-doppler velocimeter at the peak efficiency. Ubaldi et al.'s (1993) data were also obtained in a low-speed, backswept contrifugal impeller with an outlet blade angle of 67°, seven blades with inlet shroud radius of 122 mm, and an exit diameter of 200 mm. The data were acquired with a hotwire anemometer and fast response transducer. The secondary velocity near the exit of both of these impellers is shown in Fig. 4.22.

Evidence of some of the sources and magnitude of secondary flows discussed earlier are clear from the data acquired by Ubaldi et al. (1993). All the sources mentioned and shown in Fig. 4.21, with the exception of the last source (due to $\Omega W/R'$), drive the flow from the pressure to the suction surface inside the wall boundary layers and from the hub to the shroud inside blade boundary layers. This feature is clear from Ubaldi et al.'s data. The vortex marked C2 in Fig. 4.22 (corresponding to C2 in Fig. 4.21b) is due to the combined effect of meridional curvature and rotation ($\Omega W/R'$). The measured vortex is in the direction predicted by the analysis. The data due to Hathaway et al. (1993) also confirms the presence of secondary flows. Hathaway et al. do not have data in the nearwall region of the hub and thus the nature of the secondary flow in this region is not confirmed. However, the nature of secondary flow from suction to pressure surface away from the hub is as predicted by the hypothesis. Secondary flow in the blade region measured by Hathaway et al. is consistent with the author's hypothesis. The data from Hathaway et al. and Ubaldi et al. indicate that the flow field in the shroud region, especially near the pressure surface, is dominated by the leakage flow effects, and the secondary flow effects are overshadowed by the leakage flow. This will be dealt with in the next section. For a radial impeller, only the contribution due to $\Omega \, \partial W/\partial n'$ is relevant as $R' \to \infty$ and the camber is zero.

It is interesting to assess the relative order of magnitude of the curvature effect and the rotation effect in the generation of secondary flow. The ratio of the last two terms in Eq. 4.134 can be written approximately as $W/R'\Omega$. If W is approximated as Ωr, this ratio is equal to r/R'. Hence, secondary flow due to meridional curvature dominates if the ratio (r/R') is large and the rotation dominates the secondary flow development if this ratio is small. It is possible to derive a more rigorous parameter, Richardson number, investigated by Bradshaw (1973), for the combined effects of the curvature and the

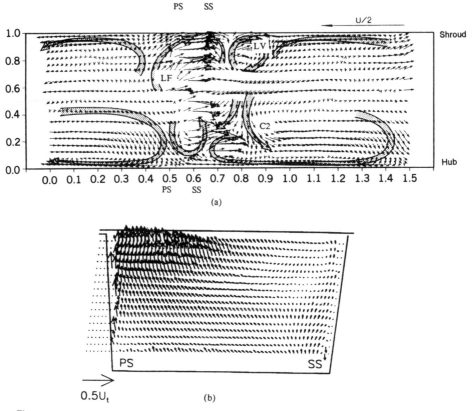

Figure 4.22 Secondary velocity vectors for backward swept centrifugal compressor. (a) At the exit. (Copyright © Ubaldi et al., 1993; Reprinted with permission of ASME.) (b) Near the trailing edge. (Copyright © Hathaway et al., 1993; reprinted with permission of ASME.)

rotation, but his definition of Richardson number is valid only when the plane of the curvature of the streamline and plane of rotation of the axes coincide.

4.2.6.4 *Effects of Mach Number*   The effect of compressibility is included in the generalized secondary vorticity expressions presented earlier. The compressibility also affects the inlet vorticity ω_n and the static temperature distribution $(\partial T/\partial b)$. In most incompressible flow applications, ω_n is assumed to be frozen at its inlet value. An approximate expression for the change in ω_n due to compressibility effect can be derived, neglecting the viscous effect and the change in $\partial \rho/\partial b$ and a_b and assuming $\tau = \infty$ (Lakshminarayana, 1975). The expression is given by

$$\omega_n u = \omega_{no} u_o + \frac{1}{2T}\frac{\partial T}{\partial b}\left(u^2 - u_o^2\right) \tag{4.135}$$

where u_o is the inlet streamwise velocity. This equation can be substituted in Eq. 4.126 to derive the following expression, which is valid for inviscid and compressible flows:

$$\frac{\partial}{\partial s}\left(\frac{\omega_s}{\rho u}\right) = \frac{2\,\omega_{no}}{\rho u R}\left[1 + \frac{\gamma - 1}{2}\frac{M^2}{C}\left(\frac{u_o}{u}\right)^2\right]\frac{u_o}{u}$$

$$-\frac{1}{\rho RT}\frac{|\nabla T_o|\cos \beta}{C}\left(\frac{u_o}{u}\right)^2 \qquad (4.136)$$

where

$$C \approx 1 + \frac{\gamma - 1}{2}M^2\left[1 - \left(\frac{M_o}{M}\right)^2\right]$$

where subscript o refers to inlet conditions, β is the angle between ∇T_o and the coordinate b (Fig. 4.19). It is clear from Eq. 4.136 that the compressibility effect, flow turning, and the loading on the blade have a substantial effect on secondary vorticity development. The local values of u, ρ, T, and M can be derived from inviscid theories described in Chapter 3 and in this chapter (blade-to-blade solution). Equation 4.136 can be integrated along a streamline to derive values of ω_s at the exit. This can then be substituted in the stream function equation to solve for secondary velocities.

4.2.6.5 Viscous and Other Effects on Secondary Flow

The secondary flow theories neglect viscosity and turbulence effects on ω_n and ω_s development. The flow perturbations are assumed to occur mainly in the cross section of the bend or cascade, and the perturbations due to secondary flow in the streamwise direction are assumed to be small.

The viscosity and turbulence affect the boundary layer growth through the passage (hence, ω_n changes as the flow progresses through the blade row). In addition, the viscous effects in Eq. 4.121 tends to diffuse and dissipate the vortices. This results in decreased streamwise normal vorticity in the viscous region. The former was allowed for in an approximate manner by Lakshminarayana and Horlock (1967a).

To assess the viscous and turbulence effects on secondary flow, Pouagare and Lakshminarayana (1982) solved Eqs. 4.121 and 4.122 numerically (for incompressible and homogeneous flows with $\tau \to \infty$ and $F = 0$ and assuming that a_n and α_b are constant) along a mean streamline to obtain values of ω_s and ω_n distribution along the blade span. Both the laminar and the turbulent stresses were included in Eqs. 4.121 and 4.122 in the calculation. The Crank–Nicolson scheme was used to march in the streamwise direction. The stream function equation (Eq. 4.131) is then solved using the successive over

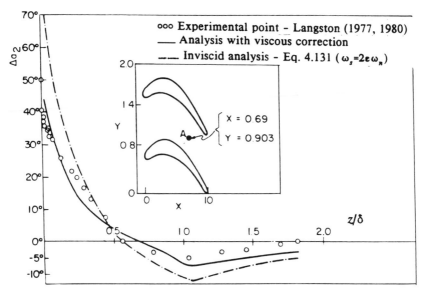

Figure 4.23 Comparison between experimental and predicted change of outlet angle for Langston's (1977, 1980) turbine cascade; $\Delta\alpha_2$ is the change in outlet angle, z is the spanwise distance (Copyright © Pouagare and Lakshminarayana, 1982; reprinted with permission of ASME.)

relaxation technique. They obtained better agreement with the data than did inviscid theories, especially near the wall region. The viscous and turbulence effects tend to reduce vorticity (both ω_n and ω_s) near the wall through dissipation; hence, vorticity and secondary velocities in the wall region are usually overpredicted from the inviscid theories. The change in outlet angle predicted for Langston's cascade (Langston 1980; Langston et al., 1977) is shown in Fig. 4.23. Details of Langston's turbine cascade are as follows; C = 281.3 mm, A = 1, S/C_x = 0.95, turning angle = 110°, α_1 = 45.3°. The agreement between the predictions and data is excellent. The effect of viscosity near the wall is appreciable. The inviscid theories tend to overpredict ω_s, v, and w in this region. Generally speaking, the inviscid theories are generally accurate for compressor cascades, IGV, and isolated compressor blade rows, but large turning in turbine cascades and blade rows requires the viscous corrections described in Pouagare and Lakshminarayana (1982).

4.2.6.6 *Application to Cascades and Turbomachinery Blade Rows*
The theories outlined above are generally valid for small flow turning. In a compressor cascade, good agreement between the inviscid theory and the experimental data has been obtained, with the exception of the nearwall region. The predictions can be improved in the wall regions using corrections for Bernoulli surface rotation (Lakshminarayana and Horlock, 1967a).

The application of the theory outlined in this section (4.2.6) is restricted to the following situations:

1. In the analysis system, where the passage averaged equations are employed with blade-to-blade solutions derived from the cascade or quasi-three-dimensional theories. In this case, the secondary flow is considered as a perturbation to the velocity field derived from these passage averaged equations. The spanwise and cross-flow velocities at the exit can be written as

$$u = u_p, \qquad v = v_p + v_s, \qquad w = w_p + w_s, \qquad \omega_{s'} = (\omega_{s'})_p + (\omega_{s'})_s$$

$$(4.137)$$

   where p refers to primary flow.
2. Blade rows with small turning and thin leading edge, with thin endwall boundary layer.
3. Viscous effects are important to endwall flows in turbine cascades with large turning, where endwall separation, horseshoe vortex, and thick boundary layers are present.
4. If Wu's solution (e.g., Kang et al., 1989) or Euler solution is employed, the secondary flow is derived as part of the solution, if the actual entry velocity, enthalpy, or temperatures are included in the formulation.
5. In liquid-handling machinery, two-phase flow may exist in the endwall region, and the theory is invalid for this case.

A detailed discussion of incorporating the secondary flow effects in a turbomachinery blade row can be found in Leboeuf and Naviere (1985), James (1987), Lakshminarayana and Horlock (1967a), Came and Marsh (1974), Smith (1955), Leboeuf et al. (1982), Lakshminarayana (1975), Dixon (1974), and Jennions and Stow (1985, 1986).

Good agreement between the predictions and data has been obtained for cascades, inlet guide vanes, and isolated rotors, but the application to the multistage compressor has not been successful. The best approach for this would be to resort to numerical solutions outlined in Chapter 5.

4.2.7 Tip Clearance and Leakage Flow Effects

4.2.7.1 Physical Nature and Experimental Evidence In almost all turbomachinery rotors, the gap between the blade and the shroud/annulus wall induces leakage flow across the gap. The leakage flow also occurs in stators with the gap near a rotating hub. The nature, magnitude, roll-up, and formation of a leakage vortex depend on the type of machinery, blade and flow parameters, and type of fluid. Figure 4.24 shows a somewhat idealized

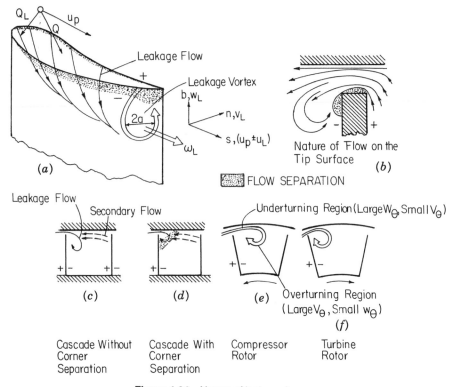

Figure 4.24 Nature of leakage flows.

picture. The phenomenon in actual machinery, where the annulus wall boundary layer, secondary flow, and scraping vortex is present (Fig. 1.15 and 1.16), is much more complex.

The leakage flow arises due to a pressure difference between the two surfaces of the blade at the tip. In a simple situation, when there is no viscous effect, the leakage velocity across the gap is given by (for incompressible flow)

$$Q_L = \sqrt{\frac{2(p_p - p_s)}{\rho}}$$

(4.138)

One of the assumptions implicit in this approximation is that the stagnation pressure across the gap (relative stagnation pressure for the rotor flow) is the same and the leakage flow arises due to static pressure difference across the gap. Thus the local leakage velocity depends on the local pressure difference, and thus the blade loading is a major parameter influencing the magnitude of the leakage flow.

The mass flow through this leakage gap, which does not participate in the energy conversion process, depends on the gap height. Furthermore, when viscous effects are present, both in the gap and in the endwall region, the height of the gap plays a very critical role in the total leakage flow. Thus if the gap is extremely small, the viscous effects within the gap as well as inviscid effects will keep this leakage flow velocity and mass flow within the gap to a minimum. As the gap is increased to practical levels, leakage flow increases, and larger mass flow goes through the gap. Hence the two most critical parameters controlling the magnitude of the leakage flow are blade clearance height and the blade loading (local pressure difference between the pressure and the suction surface).

The total velocity within the gap in the relative frame of reference can be considered to be the resultant of primary or inviscid velocity (u_p) near the blade in the absence of gap and the leakage Q_L as shown in Fig. 4.24. The leakage flow tends to roll up into a vortex. Because the leakage jet and the main flow are at differing angles, a flow discontinuity exists. This flow discontinuity eventually rolls up into the core of the vortex. The formation of a leakage vortex has been observed in cascades and in many turbomachinery, but in some cases the conditions necessary for the formation of a strong vortex may not exist because high turbulence levels, high-velocity jet, or separation zone may diffuse the leakage flow before it forms a vortex (Lakshminarayana et al., 1995). In any case, the phenomenon can be idealized to assume leakage flow and vortex to form as shown in Fig. 4.24. The viscous effects in the gap may result in flow separation on the blade tip and suction side as shown in Fig. 4.24b.

Further complications arise when there is secondary flow (which may oppose the leakage flow), scraping vortex, corner flow separation, presence of shock waves, coolant injection, and annulus wall boundary layer. For example, if the annulus wall boundary layer is much larger than the tip clearance height, viscous effects dominate, thereby decreasing the role of the leakage flow. If the coolant is ejected at the tip, this may act to diffuse the leakage flow or prevent the formation of a leakage vortex.

The secondary and leakage flows oppose each other as shown in Figs. 4.24c and 4.25. The leakage flow has a beneficial effect when corner separation is present. The leakage jet tends to "wash out" the separated region, thus improving the performance (Fig. 4.24d). The relative motion has a substantial influence on the magnitude of leakage flow, strength, and location of the leakage vortex. The rotation in the compressor tends to augment the leakage flow and move the leakage jet closer to the pressure side as shown in Fig. 4.24e. The rotation has an opposite effect in a turbine as shown in Fig. 4.24f. Furthermore, the blade loading is usually much higher for a turbine. Hence, leakage flow velocities for a turbine tend to be much higher than that encountered in a compressor.

The leakage flow and its interaction with other flow features is a very complex phenomenon. In most turbomachinery, the leakage flow has a more pronounced effect and hence is more important than the secondary flow, especially in the tip region.

The effects of leakage flow and vortex are many; the important effects are listed below.

1. The leakage flow/vortex introduces three-dimensionality to the flow field. The mixing of the leakage flow, entrainment process, vortex formation, diffusion, and convection phenomena introduces large three-dimensionality to the flow field. It is not confined to the vicinity of the tip, but spreads inward (approximately 10–30% of the span from the tip, depending on the type of turbomachinery, aspect ratio, leakage flow strength, etc.). Hence, this is an important cause of three-dimensionality in turbomachinery.

2. The dissipation and mixing of the leakage flow and vortex introduces aerodynamic losses and inefficiency. The loss in efficiency could be anywhere from 2% to 4%, or more, depending on the type of turbomachinery. This is in addition to the energy loss associated with the direct mass flow through the gap which does not contribute to the energy transfer.

3. The blade unloading at the tip, which is not confined to only the tip region, results in appreciable decrease in pressure rise/drop and less flow turning, and affects the stall and surge margin of the compressor.

4. The leakage flow and vortex in a rotor (or stator) is perceived by the succeeding stator (or rotor) blade row as inlet distortion or unsteady inflow. This causes unsteadiness in the subsequent blade row, unsteady pressure, unsteady boundary layer, unsteady transition, and noise generation.

5. The leakage flow and vortex causes appreciable change in the heat transfer in a turbine and, hence, changes the cooling flow requirements.

6. The leakage flow/vortex impinging on a subsequent blade row causes vibration, higher blade stresses, and flutter.

7. The tip vortices, with low pressure inside the core, cause cavitation in liquid-handling machinery. This result in decreased efficiency, and blade tip damage.

The purpose of this section is to describe the inviscid analyses available for predicting the effects of tip clearance on flow field (items 1 and 3 above). The other effects, including viscous effects and losses, will be dealt with in subsequent chapters. A comprehensive review of tip clearance effects in axial flow turbomachinery is given in the Von Karman Institute Lecture Series (VKI Lecture Series, 1985).

The effects of leakage flow on the flow field in a turbine cascade (Yamamoto, 1989) and in a compressor rotor (Lakshminarayana et al. (1995)

and Lakshminarayana et al. (1982)) are shown in Figs. 4.25 and 4.26, respectively. The turbine cascade turning angle was 113°, $\sigma = 1.20$, and the tip/clearance gap (τ) was 2.1% of the blade height (or 1.53% of chord). The compressor turning angle and σ were 16° and 1.09, respectively. The nature of leakage and secondary flow at the tip, as well as the nature of secondary flow at the top wall, is clearly evident from Fig. 4.25. The leakage flow interacts with the secondary flow to form a significant region where the flow is highly three-dimensional. The extent of this interaction region extends all

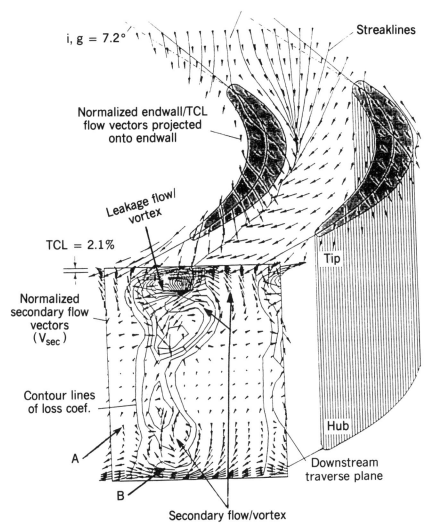

Figure 4.25 Tip clearance effects in a turbine cascade. (Copyright © Yamamoto, 1989; reprinted with permission of ASME.) (TCL stands for tip clearance, normalized secondary flow vectors are based on mass averaged flow direction.)

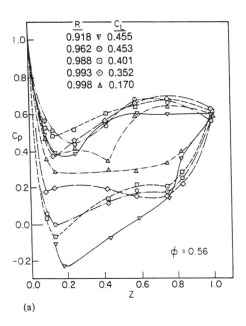

(a)

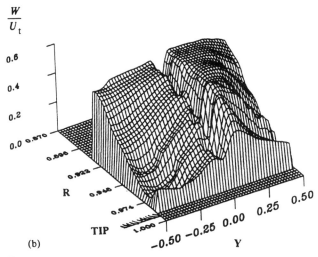

(b)

Figure 4.26 Tip clearance effects in an axial flow compressor rotor. (a) Blade pressure distribution and blade lift coefficient in the blade tip region ($\tau/h = 1.11\%$). (Copyright © Lakshminarayana et al., 1982; reprinted with permission of ASME.) (b) Relative velocity distribution at 4% chord from the trailing edge ($\tau/h = 1.52\%$). Blade at Y = 0. (Copyright © Lakshminarayana et al., 1995; reprinted with permission of ASME.)

the way to midspan in the wake region. The three-dimensionality is large, and in some locations the cross-flow and the normal components of velocity are as large as the streamwise velocity. The leakage flow field in a turbine rotor may be significantly affected by centrifuging of the blade boundary layer toward the tip and the relative motion between the wall and the blade.

The total velocity field derived from a miniature five-hole probe at the exit of the Penn State low-speed compressor (see Section 2.5.5 and Fig. 2.17 for details on this compressor) shows the major effects of leakage flow (Fig. 4.26). The leakage flow tends to move farther along the tangential direction before interacting with the main flow, as shown schematically in Fig. 4.27. This results in very low kinetic energy in the interaction zone. The interaction region corresponds to low relative velocities observed near the clearance region (closer to midpassage), as shown in Fig. 4.26a. This interaction region extends to about 15–20% of the span.

The measured chordwise distribution of the static pressure coefficient at various spanwise locations near the tip is shown in Fig. 4.26a. The distribution at the very tip of the blade ($R = 0.998$) shows maximum unloading of the blade, with substantial effect even at 8% ($R = 0.962$) of the span from the tip. The effect of the tip vortex near the blade tip is to move the suction

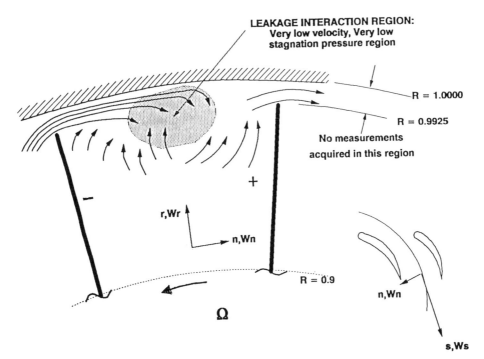

Figure 4.27 Schematic of leakage flow and its interaction with the main flow in an axial flow compressor (tip gap shown exaggerated)

peak away from the leading edge. This is especially true close to the blade tip, and this trend has been documented clearly for an isolated airfoil and a cascade (Lakshminarayana and Horlock, 1967b; Lakshminarayana and Horlock, 1962).

The leakage flow effects in a centrifugal compressor can be clearly seen in Ubaldi et al.'s (1993) data plotted in Fig. 4.22a. In this figure the flow toward the tip on the pressure side is designated as LF and the leakage vortex is designated as LV. The leakage flow has a dominant effect on the centrifugal compressor flow field as indicated by the flow field shown in Figs. 4.22a and 4.22b.

The quantitative and qualitative effects of tip clearance depend on the following variables: clearance height (τ), blade loading, blade thickness at the tip, stagger angle, incidence, annulus wall boundary layer profile and thickness, blade rotational speed, Reynolds and Mach numbers, blade boundary layer thickness at the tip, and radial pressure gradient at the tip. An excellent review of various effects, theories available, and active tip clearance control can be found in the VKI Lecture Series (1985).

4.2.7.2 Tip Leakage Flow Analyses The tip leakage flow has been analyzed by many investigators. It is beyond the scope of this book to describe all of them. These analyses can be broadly classified as follows:

1. *Leakage Flow Analyses (Rains, 1954; Vavra, 1960; Senoo and Ishida, 1986).* In this technique the leakage flow velocity is computed from the theoretical or approximated blade pressure distribution. The kinetic energy associated with the leakage flow (direct effect) is calculated to derive an expression for functional dependency of losses and efficiency. There have been several modifications made in recent years to improve the loss correlations based on this type of analysis. This type of analysis is mainly used for loss and efficiency estimate and will be described in a later chapter.

2. *Potential Tip Vortex Model (Lakshminarayana and Horlock, 1962, 1967b; Lakshminarayana, 1970).* In this model, the leakage vortex is represented by a potential vortex located at the trailing edge of the blade tip, including the tip vortices shed out from all the blades. The wall effect is included via image vortices. Hence the model consists of two infinite rows of tip vortices, spaced at 2τ apart in the spanwise gap. The induced velocity due to this system of vortices is calculated to find out induced drag. Because the vortex is a point vortex, it does not represent the real phenomenon such as that shown in Fig. 4.24a. This analysis is useful in capturing the gross effects such as efficiency and losses.

3. *Modified Potential Tip Vortex Model (Yokoyama, 1961; Lakshminarayana and Horlock, 1967b).* To allow for local chordwise effects of the leakage vortex on pressure distribution, Yokoyama (1961) and Lakshmi-

narayana and Horlock (1967b) modeled the tip vortex by assuming that the vortex leaves at an angle to the tip chord and moves in the spanwise direction. The strength of the vortex is assumed to be zero at quarter-chord point (where the leakage flow is observed to originate) and attains maximum strength at the trailing edge. The location of the center of the vortex near the suction side provides a shift in suction peak as well as a qualitative trend in the change in C_p distribution observed (similar to that shown in Fig. 4.26a). The induced flow field is calculated from the Biot–Savart's law.

4. *Combined Tip Vortex Model for Flow Prediction (Lakshminarayana, 1970)*. This model recognizes the features in a real flow, namely, the presence of a core rotating with a solid body rotation surrounded by a potential vortex. This analysis, which has the most practical features, will be covered in detail later in this section.

5. *Three-Dimensional Viscous Flow Analysis (Wu and Wu, 1954; Wadia, 1983; Hah, 1986; Bansod and Rhie, 1990; Kunz et al., 1993; Crook et al. (1993), Basson and Lakshminarayana (1993)*. There have been several attempts made to include viscosity and solve the exact and approximate form of Navier–Stokes equations in the gap region. Computational techniques, which include tip clearance as part of the formulation and flow regime, will be covered in Chapter 5.

As mentioned earlier, this section is mainly concerned with three-dimensional inviscid effects. The most appropriate models for the prediction of the flow field perturbation are models 1 and 4.

One of the earliest analyses is due to Rains (1954). He used Eq. 4.138 to calculate the energy lost due to leakage flow. He also was one of the first to analyze the roll-up of leakage flow into a vortex. Considering the deformation of the leakage flow sheet due to velocity induced by the vortex, and applying the Bernoulli equation across the gap, he proved that the radius of the vortex (a) is given by (Fig. 4.28)

$$\frac{a}{\tau} = 0.14 \left[\frac{d}{\tau} (C_L)^{1/2} \right]^{0.85} \tag{4.139}$$

where d is the distance from the leading edge and C_L is the lift coefficient. The analysis is similar (but more complicated) than the theories available for the roll-up of aircraft wing trailing vortices into two discrete vortices (Schlichting and Truckenbroadt, 1979). The analysis assumes that the blade loading and, hence, the leakage velocity are constant along the chord, and the flow is assumed to be inviscid and incompressible.

4.2.7.3 Combined Vortex Model for Flow Prediction

A potential or a point vortex, which is a mathematical simplicity, is not valid for accurate prediction of blade-to-blade flow (i.e., local outlet angles and flow field, etc.)

in the clearance region. The detailed flow pattern found in a real fluid does not resemble that of the potential vortex which explains it. The potential vortex model fails to predict the presence of a loss core near the clearance region and velocities such as those shown in Figs. 4.25–4.27.

Leakage flow originating from the tip all along the chord forms a sheet of discontinuity which roll up into a spiral to form a core of rotating fluid which lies below the suction surface and inboard of the blade tip (Fig. 4.26a). The flow visualization experiments and measurements (Gusakova, 1960; Gusakova, et al., 1960; Mehmel, 1962; Inoue et al., 1986; Rehbach, 1960; Lakshminarayana and Horlock, 1962; 1967b; Yamamoto, 1989) also confirm the validity of this model.

A theoretical model which takes into account the presence of a vortex core with solid body rotation was developed by Lakshminarayana (1970). It is assumed that the vortex core contains the shed vorticity in the field [of circulation $(1 - K)\Gamma$] so the motion outside is irrotational. K is the fraction of lift retained at the tip, a concept introduced by Lakshminarayana and Horlock (1962, 1967b). The analysis is valid for incompressible and inviscid flows, and implicit in the analysis is the vortex formation. If the flow is highly turbulent, the vortex may diffuse rapidly; the theory is not valid in such a case.

The model is illustrated in Fig. 4.28. The image vortices represent the effect of endwall, and thus the spacing between the two rows of vortices is $2(\tau + a)$. The inner region (domain R) behaves as a forced vortex and the outer region behaves as a free vortex. The boundary condition (that the normal velocities be equal to zero on the annulus or endwall adjoining the clearance region) is satisfied by using the image vortices located at equal distances from the wall. This flow model is equivalent to the solution of Poisson's equation in domain R and the Laplace equation outside the domain R (bounded by the blade surface and the annulus wall). Kinematic condition on the blade surfaces is satisfied only if there are streamwise components of vorticity lying in the wake. Because their magnitudes are likely to be very small compared to that of leakage vortices, their effect is neglected in the present analysis. A method of finding the three important parameters associated with such a vortex system (i.e., a, b, and ω in Fig. 4.28) is described first, before providing a theoretical solution for the flow field.

Radius of the Vortex Core (a). The radius of the vortex core can be determined from Rains' theory, which is based on inviscid flow and constant blade loading along the chord. The predicted values are in agreement with Rains' experiment as well as the other experiments. Hence, Rains' expression is used for computing the radius of the vortex core which is given by Eq. 4.139.

Angular Rotation of the Vortex Core (ω). The angular rotation of the vortex core can be determined once the radius of the vortex core (Eq. 4.139) and the

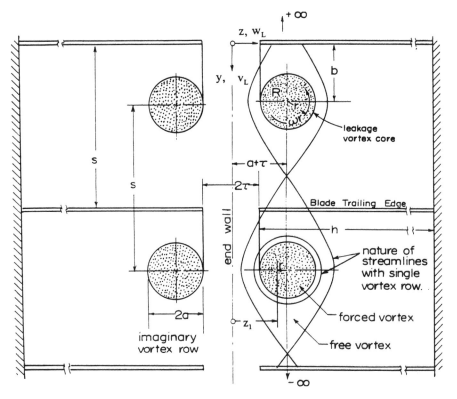

Figure 4.28 Modified clearance flow model showing the location of vortices in the tangential plane of the cascade. (Copyright © Lakshminarayana, 1970; reprinted with permission of ASME.)

strength of the shed vortex are established. ω is then given by

$$\omega = \frac{(1 - K)\Gamma}{2\pi a^2} \tag{4.140}$$

Lakshminarayana (1970) provided the following expression for the empirical coefficient (K), based on his cascade data:

$$1 - K = 0.23 + 7.45(\tau/S) \qquad \text{for } 0.01 < \tau/S < 0.1$$

This correlation would enable the calculation of both spanwise and normal velocities due to the leakage vortex. Future experiments should enable modification of this correlation for rotation and compressibility effects.

Location of the Vortex Core in the Passage (b). The induced flow due to the vortex row as shown in Fig. 4.28 tends to move the vortices away from the suction surface. If we know the induced flow field due to image vortices and assume that the vortices originated at the leading edge of the blade, we can compute the distance b at which the vortices are located at any desired axial station.

An extension of Lamb's (1945) solution for the induced flow field of an infinite row of point vortices to two infinite rows of vortices of finite radius (similar to Eq. 3.75) is carried out by Lakshminarayana (1970). The resulting expressions for induced velocities outside the domain R are v_L component (normal to V_m) and w_L component (spanwise velocity), given by:

$$v_L = \frac{(1-K)\Gamma}{2S} \left[\frac{\sinh M}{\cosh M - \cos \dfrac{2\pi(y-b)}{S}} - \frac{\sinh N}{\cosh N - \cos \dfrac{2\pi(y-b)}{S}} \right] \tag{4.141}$$

$$w_L = \frac{(1-K)\Gamma}{2S} \left[\frac{\sin \dfrac{2\pi(y-b)}{S}}{\cosh M - \cos \dfrac{2\pi(y-b)}{S}} - \frac{\sin \dfrac{2\pi(y-b)}{S}}{\cosh N - \cos \dfrac{2\pi(y-b)}{S}} \right] \tag{4.142}$$

where

$$M = \frac{2\pi(z - a - \tau)}{S}, \qquad N = \frac{2\pi(z + a + \tau)}{S} \tag{4.143}$$

Inside the vortex core, the deviation in flow-outlet angle at any spanwise location such as $z = z_1$, in Fig. 4.28, is given by

$$\tan \Delta\beta_2 = \frac{\omega r}{V_m} = \frac{(1-K)C_L C}{4\pi a} \left(\frac{r}{a} \right) \tag{4.144}$$

where $r = a + \tau - z_1$, the local radius of the vortex core. The flow is underturned when $z_1 < (a + \tau)$ and is overturned when $z_1 > (a + \tau)$.

Outside the vortex core, the change in outlet angle is given by

$$\tan \Delta\beta_2 = \frac{v_L}{V_m} = \frac{(1-K)C_L C}{4S}$$

$$\times \left[\frac{\sinh M}{\cosh M - \cos \dfrac{2\pi(y-b)}{S}} - \frac{\sinh N}{\cosh N - \cos \dfrac{2\pi(y-b)}{S}} \right] \tag{4.145}$$

because $\Gamma = C_L C V_m / 2$.

Equations 4.144 and 4.145 enable the calculation of deviations in outlet angle at all spanwise and passage locations. From the conventions adopted here (Fig. 4.28), positive $\Delta\beta_2$ indicates underturning and negative values indicate overturning. If we know the deviation angle $(\Delta\beta_2)$, we can calculate the outlet angles from the equation

$$\beta_2 = (\beta_2)_p + \Delta\beta_2$$

If $\beta_1 < (\beta_2)_p$, the flow is underturned, and if $\beta_1 > (\beta_2)_p$, the flow is overturned. β_p is the primary flow angle.

Because of the presence of a vortex core, no analytical expression can be given for the average deviation in outlet angle $(\overline{\Delta\beta_2})$ except at $z = 0$. The average outlet angle deviation at $z = 0$ can be derived by integrating Eq. 4.141 from $y = 0$ to S. The resulting expression is

$$\tan \overline{\Delta\beta_2} = \frac{(1 - K)C_L C}{2S} \tag{4.146}$$

For $\tau < z < (2a + \tau)$, it is easier to carry out the integration numerically to include the contribution due to the vortex core. For $z > (2a + \tau)$, the deviation in outlet angles due to potential vortex is small because the induced velocities due to two rows are nearly equal and opposite in nature. Depending on the strength of the vortex core, flow overturning is likely to occur for $(a + \tau) < z < (2a + \tau)$. The average overturning is small if a/S is small. The experimental results confirm these observations.

The kinetic energy associated with the induced flow field outside the vortex core gives a qualitative indication of the nature of losses, but not necessarily its magnitude.

For any spanwise position (z), the average kinetic energy associated with the leakage flow field outside the core is derived by squaring v_L and w_L from Eqs. 4.141 and 4.142, respectively, and integrating the summation over proper limits across the passage. Thus,

$$\frac{\overline{w_L^2 + v_L^2}}{V_1^2} = \left(\frac{(1 - K)C_L C V_m}{4SV_1}\right)^2 [I_1 + I_2]$$

where

$$I_1 = \int \left[\frac{\sinh M}{\cosh M - \cos \dfrac{2\pi(y - b)}{S}} - \frac{\sinh M}{\cosh N - \cos \dfrac{2\pi(y - b)}{S}} \right]^2 d\left(\frac{y}{S}\right)$$

$$I_2 = \int \left[\frac{\sin \dfrac{2\pi(y - b)}{S}}{\cosh M - \cos \dfrac{2\pi(y - b)}{S}} - \frac{\sin \dfrac{2\pi(y - b)}{S}}{\cosh N - \cos \dfrac{2\pi(y - b)}{S}} \right]^2 d\left(\frac{y}{S}\right)$$

$$\tag{4.147}$$

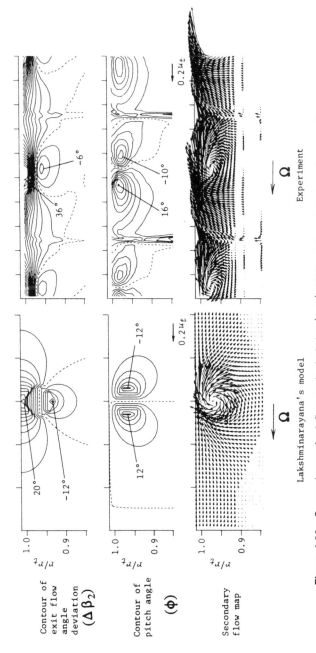

Figure 4.29 Comparisons of exit-flow angle deviation ($\Delta\beta_2$), pitch angle, and leakage flow velocity (w_L, v_L) between experiment (Inoue, 1986) and Lakshminarayana's model for a low speed axial flow compressor. (Copyright © Inoue et al., 1986; reprinted with permission of ASME.)

The limits of integration for I_1 and I_2 depends on the value of z. For example, $z = 0$, the limits of integration are 0 to 1, and hence

$$I_2 = 0, \qquad I_1 = 4\coth\frac{2\pi(a + \tau)}{S}$$

Thus, for $z = 0$ we have

$$\frac{\overline{v_L^2} + \overline{w_L^2}}{V_1^2} = \left(\frac{(1 - K)C_L V_m C}{4SV_1}\right)^2 4\coth\frac{2\pi(a + \tau)}{S} \tag{4.148}$$

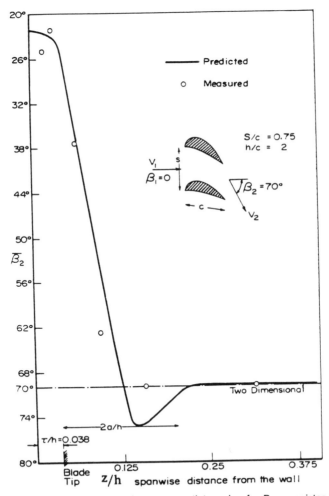

Figure 4.30 Predicted and measured average outlet angles for Bauermeister and Hubert (1962) turbine cascade ($\tau / h = 0.038$). (Copyright © Lakshminarayana, 1970; reprinted with permission of ASME.)

At other locations such as $z = a + \tau$ (Fig. 4.28), the limits of integrations are

$$0 \text{ to } \frac{b-a}{S} \quad \text{and} \quad \frac{b+a}{S} \text{ to } 1 \qquad \text{to both } I_1 \text{ and } I_2$$

This can be best handled by numerical integration.

The theory has been validated against data from a compressor rotor and a turbine cascade. Inoue et al. (1986) used hot-wire sensors to obtain detailed three-dimensional flow data in the tip region of a low-speed axial flow compressor rotor with a 22.45-cm tip radius and a tip solidity of 1.0, $C_t = 11.7$ cm. Their measurements, compared with this theory, show good agreement with the measured pitch angle (Fig. 4.29). The discrepancy in the yaw or outlet angle may be caused by the relative motion of the blade, which is not included in the analysis. The theory assumes a circular vortex, whereas the actual vortex is distorted due to the induced flow field and turbulence. The leakage flow vectors, measured and predicted, show qualitative agreement.

The predicted pitch-averaged outlet angle for a turbine cascade (Bauermeister and Hubert, 1962) with tip clearance is shown compared with data in Fig. 4.30. The average exit angles are predicted well by the theory.

The limitations of the theory are as follows:

1. The analysis has not been validated for a centrifugal compressor.
2. It is not valid for transonic rotors with shocks.
3. If viscous and turbulence effects are large, the inviscid assumption is no longer valid.
4. In many instances, the flow separates in the corner or on the blade surface near the trailing edge. The theory is not accurate for such a case.

4.2.8 Method of Incorporating the Perturbations to the Primary Flow Field by Secondary and Leakage Flows

The analyses of secondary and leakage flow described in this and earlier sections can be incorporated into three-dimensional or quasi-three-dimensional theories as follows:

1. *Passage-Averaged Radial Equilibrium Analysis.* In this case, only the passage averaged values are relevant. If the secondary and leakage flow velocities are not large, the primary flow from the passage-averaged equations can be superposed to give (at the exit plane)

$$u = u_p$$
$$v = v_p(r) + \bar{v}_s(r) + \bar{v}_L(r)$$
$$w = w_p(r) + \bar{w}_s(r) + \bar{w}_L(r) \tag{4.149}$$

where $\bar{v}_s(r)$, $\bar{w}_s(r)$ are pitch-averaged secondary velocities and $\bar{v}_L(r)$, $\bar{w}_L(r)$ are pitch-averaged velocities due to leakage flow/vortex.

2. If a quasi-three dimensional system is chosen, where u_p, v_p, and w_p are local values, then the above equations can be used with local values of v_s, v_L, w_s, and w_L.
3. If a fully three-dimensional solution is adopted, the secondary flow is derived as part of the solution. If the formulation does not have the tip clearance effect, the flow field from this analysis could be superposed on the primary viscous/inviscid flow field.

A word of caution is in order. The procedure described above assumes that there is no interaction between the primary, secondary, and leakage flow fields. Hence the method is not valid when there is strong interaction. (For example, v_s and v_L are of the same order of magnitude and are in opposite directions.)

Many successful attempts have been made to improve the flow predictions by incorporating the secondary flow, leakage flow theories, and viscous effects as perturbations to the primary flow. Such methods are widely used in industry in their design and analysis system for isolated, single, and multi-stage blade rows. One successful method is due to Adkins and Smith (1982). Recent attempts are described in Ucer and Shreeve (1992). The procedure generally involves the following:

1. The primary flow is calculated from S_2 surface equations outlined earlier; various techniques employed differ slightly in details.
2. The annulus-wall and the hub-wall boundary layers can be included to correct for the viscous effects near the walls. This is described in Chapter 5.
3. The gross viscous effects in the primary flow can be included by allowing for entropy gradient in the radial direction through loss correlations available.
4. The secondary flow and leakage velocities can be incorporated through the analyses described in this section.

Adkins and Smith (1982) utilized various approaches described in this chapter, including steps 1–4, and computed many of the complex flow fields in turbomachinery. The S_2 surface equations and secondary flow perturbations are based on the analyses by Smith (1955, 1968), and the leakage flow analysis is based on the model proposed by Lakshminarayana and Horlock (1967b). In Adkins and Smith's (1982) model, the leakage vorticity $(1 - K)\Gamma$ is distributed uniformly across the passage and varied as the first half-cycle of a sine wave in the spanwise direction. The perturbations to the flow field (v_L and w_L) are calculated to assess the effects of leakage flow on compressor flow field and performance. In addition, they allowed for endwall boundary layers and shear layer mixing through a mixing coefficient to predict the hub-to-tip distribution of velocities, pressures, temperatures, and angles at

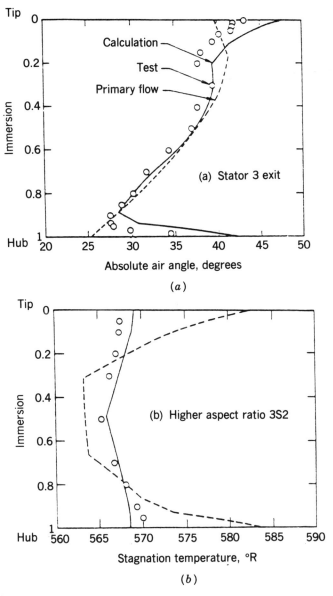

Figure 4.31 Predicted and measured averaged properties in a multistage compressor. (a) Low-speed four-stage compressor: stator 3 (exit air angle). (b) High-subsonic Three-Stage Low-Aspect-Ratio Compressor. (Copyright © Adkins and Smith, 1982; reprinted with permission of ASME.)

the exit. The predictions are shown compared with data in Fig. 4.31. The low-speed four-stage G.E. compressor (Fig. 4.31a) is comprised of four identical stages, with $A = 2.8$, $\sigma = 1.05$, $r_h/r_t = 0.7$, and $\tau/c = 0.04$. The primary flow solution consists of an S_2 inviscid solution (without any allowance for the wall boundary layer). The calculation includes the perturbations due to secondary flow, leakage flow, and annulus wall boundary layer. Considering the complexity of the flow (embedded stage), the agreement between the data and the calculation is reasonably good. They also computed flow in a high-subsonic three-stage compressor (Bursdall et al., 1979), with average aspect ratio and solidity of 1.22 and 1.10, respectively and $\tau/h = 0.016$. The stagnation temperature distribution shown compared with data in Fig. 4.31 indicates the strong influence of secondary and leakage flow in T_o distribution. The mixing in the core region is caused by the radial transport in the wake and turbulence. More detailed treatment of this subject will be covered in Chapter 6.

5

COMPUTATION OF TURBOMACHINERY FLOWS

The discipline of computational fluid dynamics (CFD) has fostered a unified approach to turbomachinery analysis and design. The practice of treating hydraulic and steam turbines, pumps, and gas-handling turbomachinery and others separately is giving way to a more integrated approach. These developments are facilitated by a common denominator: The governing equations are the same for all turbomachinery, with additional constitutive equations employed to handle special cases (e.g., two-phase flows). The boundary conditions encountered in turbomachinery are among the most complex in the CFD field.

The emergence of CFD in the 1970s provided a major impetus to solve the Euler and Navier–Stokes equations governing the flow fields in external and internal flows. Major progress has subsequently been made in the areas of technique development, grid generation, turbulence modeling, application of boundary conditions, pre- and post-data processing, and computer architecture. Most of the techniques used for the solution of the Navier–Stokes equations can be classified as finite difference, finite area/volume, finite element and spectral methods. Only the first two techniques are widely used in turbomachinery, and thus the emphasis in this book is on these two techniques. Computational techniques provide an efficient method for the analysis and design of turbomachinery. The use of CFD by turbomachinery manufacturers has increased significantly over the past decade, resulting in a shorter hardware development cycle. Combined with measurements, CFD provides a complementary tool for simulation, design, optimization, and, most importantly, analysis of complex three-dimensional flows hitherto inaccessible to the engineer. In many instances, CFD simulation provides the only detailed flow field information available, because actual testing of turbomachinery with detailed measurement in rotating passages is cumbersome, expensive, and, in many cases, impossible.

Essential ingredients for an accurate and efficient solution of the flow field are as follows:

1. Governing equations, including turbulence transport equations; validity of approximations made
2. Enforcement of proper boundary conditions
3. Adequate grid resolution and orthogonality
4. Turbulence modeling
5. Numerical technique; optimum artificial dissipation, accurate discretization, good convergence history, and proper assessment
6. Efficient code/algorithm development, including vectorization
7. Computer architecture, including parallel processing
8. Assessment of computational techniques through calibration and validation

Computational techniques widely used in turbomachinery practice can be broadly classified as follows:

1. Inviscid solvers for two-dimensional flows
2. Quasi-three-dimensional techniques
3. Boundary layer solutions, including momentum integral techniques
4. Parabolized Navier–Stokes/space marching techniques
5. Full Euler and Navier–Stokes solutions for compressible and incompressible flows

The first and second topics were covered in Chapters 3 and 4, respectively. The major thrust of this chapter is to cover topics 3–5 and to assess various techniques to numerous geometries and flow fields encountered in practice, as well as provide recommendation as to the most appropriate computational technique for the analysis and design of turbomachinery.

There have been several review articles on these topics. Adler (1980), McNally and Sockol (1981), and Lakshminarayana (1991) have provided reviews of computational methods for turbomachinery flows. Lakshminarayana (1986) has reviewed the turbulence models used in turbomachinery flow. The topics of transition in general and application to turbomachinery in particular are reviewed by Narasimha (1985) and Mayle (1991), respectively. Even though transition modeling is important (especially turbines) in turbomachinery flow prediction, the state-of-the-art knowledge is unsatisfactory for the prediction of transition in turbines, compressors, and pumps. Transition modeling is not covered in this chapter.

Major advances have been made in the numerical solution of the Navier–Stokes equations. It is neither possible nor necessary to cover all of the approaches in this book. Hence, only those techniques which are likely, in the opinion of the author, to be used widely will be covered. Most turbomachinery flows are turbulent in nature, and hence the emphasis is placed on turbulent flows.

It is assumed that the reader is familiar with the basic techniques available for the numerical solution of partial differential equations. Rudiments of techniques relevant to turbomachinery flows are introduced at the appropriate section. The reader is referred to many textbooks in the field (Smith, 1978; Anderson et al., 1984; Hoffman, 1989; Hirsch, 1990; Pantankar, 1980; Baker, 1983; Ferziger, 1981; Chapra and Canale, 1988; Shyy, 1994) for detailed coverage. A brief overview of the basic techniques useful in turbomachinery is given in Appendix B. To keep the chapter to a reasonable length, the treatment of the finite element method, details of finite difference techniques, and implementation and theoretical aspects (convergence, stability, consistency, etc.) of CFD are not covered. The fundamentals of turbulence modeling were covered in chapter 1. Application of these techniques to turbomachinery is briefly reviewed in this chapter.

In the area of design, Navier–Stokes solutions are used in the final stages of design to check problem areas (e.g., laminar and turbulent separation, adverse pressure gradient areas, shock location, tip clearance and other losses and so on), and they have begun to find widespread use in the early stages of design. It is hoped that this chapter will provide a basis for the selection of technique and code for particular applications.

5.1 GOVERNING EQUATIONS

The governing equations employed are cast either in integral form (for finite volume discretization) or in differential form (for finite difference discretization). The most appropriate coordinate system for turbomachinery application is the rotating cylindrical coordinate system. The equations in integral form can be found in most textbooks on fluid mechanics. The generalized equations in both the integral and the differential form were presented in Chapter 1 (e.g., Eqs. 1.2–1.5, Eqs. 1.9–1.12 and Eq. 1.18). In differential form, the Reynolds-averaged, compressible Navier–Stokes equations in a rotating cylindrical coordinate system are given by (the inviscid form of these equations is given by Eqs. 4.44–4.47):

$$\frac{\partial q}{\partial t} + \frac{1}{r}\frac{\partial (rE)}{\partial r} + \frac{1}{r}\frac{\partial F}{\partial \theta} + \frac{\partial G}{\partial z} = \frac{1}{r}S \qquad (5.1)$$

Conservation variable q, flux vectors E, F, and G (which now includes all the viscous and turbulence terms), and source term S are given by

$$
q = \begin{bmatrix} \rho \\ \rho W_r \\ \rho W_\theta \\ \rho W_z \\ \rho e_0 \end{bmatrix}, \qquad
E = \begin{bmatrix} \rho W_r \\ \rho W_r^2 + p - \tau_{rr} \\ \rho W_r W_\theta - \tau_{r\theta} \\ \rho W_r W_z - \tau_{rz} \\ \rho W_r h_0 - W_r \tau_{rr} - W_\theta \tau_{r\theta} - W_z \tau_{zr} + Q_r \end{bmatrix}
$$

$$
F = \begin{bmatrix} \rho W_\theta \\ \rho W_\theta W_r - \tau_{\theta r} \\ \rho W_\theta^2 + p - \tau_{\theta\theta} \\ \rho W_\theta W_z - \tau_{\theta z} \\ \rho W_\theta h_0 - W_r \tau_{\theta r} - W_\theta \tau_{\theta\theta} - W_z \tau_{\theta z} + Q_\theta \end{bmatrix},
$$

$$
G = \begin{bmatrix} \rho W_z \\ \rho W_z W_r - \tau_{zr} \\ \rho W_z W_\theta - \tau_{z\theta} \\ \rho W_z^2 + p - \tau_{zz} \\ \rho W_z h_0 - W_r \tau_{zr} - W_\theta \tau_{z\theta} - W_z \tau_{zz} + Q_z \end{bmatrix},
$$

$$
S = \begin{bmatrix} 0 \\ p + \rho W_\theta^2 + \rho \Omega^2 r^2 + 2\rho r \Omega W_\theta \\ -2\rho r \Omega W_r \\ 0 \\ 0 \end{bmatrix}
$$

If body forces are present, they can be included in the S vector as shown in Eq. 1.23. Components of shear stress tensor and heat flux are given by

$$
\tau_{\theta\theta} = \frac{2}{3}\mu\left[2\left(\frac{1}{r}\frac{\partial W_\theta}{\partial \theta} + \frac{W_\theta}{r}\right) - \frac{\partial W_z}{\partial z} - \frac{\partial W_r}{\partial r}\right] - \rho \overline{(w_\theta')^2},
$$

$$
Q_\theta = -\gamma\left(\frac{\mu}{PR}\right)\frac{1}{r}\frac{\partial e}{\partial \theta} + \overline{\rho w_\theta' e'}
$$

$$
\tau_{zz} = \frac{2}{3}\mu\left[2\frac{\partial W_z}{\partial z} - \frac{\partial W_r}{\partial r} - \frac{W_r}{r} - \frac{1}{r}\frac{\partial W_\theta}{\partial \theta}\right] - \rho \overline{(w_z')^2},
$$

$$
Q_z = -\gamma\left(\frac{\mu}{PR}\right)\frac{\partial e}{\partial z} + \overline{\rho w_z' e'}
$$

$$
\tau_{rr} = \frac{2}{3}\mu\left[2\frac{\partial W_r}{\partial r} - \frac{W_r}{r} - \frac{1}{r}\frac{\partial W_\theta}{\partial \theta} - \frac{\partial W_z}{\partial z}\right] - - \rho \overline{(w_r')^2},
$$

$$
Q_r = -\gamma\left(\frac{\mu}{PR}\right)\frac{\partial e}{\partial r} + \overline{\rho w_r' e'}
$$

$$\tau_{z\theta} = \tau_{\theta z} = \mu \left[\frac{\partial W_\theta}{\partial z} + \frac{1}{r} \frac{\partial W_z}{\partial \theta} \right] - \overline{\rho w_z' w_\theta'}$$

$$\tau_{rz} = \tau_{zr} = \mu \left[\frac{\partial W_z}{\partial r} + \frac{\partial W_r}{\partial z} \right] - \overline{\rho w_r' w_z'}$$

$$\tau_{r\theta} = \tau_{\theta r} = \mu \left[\frac{1}{r} \frac{\partial W_r}{\partial \theta} + \frac{\partial W_\theta}{\partial r} + \frac{W_\theta}{r} \right] - \overline{\rho w_r' w_\theta'}$$

where $e_o = p/(\gamma - 1)\rho + \frac{1}{2}\rho(W_r^2 + W_\theta^2 + W_z^2) =$ total energy per unit volume. Details of Reynolds-averaging procedure and approximations to these equations were introduced in Chapter 1. The dependent and Reynolds-averaged variables in Eq. 5.1 are ρW_r, ρW_θ, ρW_z, ρ, p, ρe, ρh_o, $\overline{\rho w_i' w_j'}$, $\overline{\rho w_i' e'}$. An equation of state ($p = \rho R T$), specification of molecular viscosity, and Prandtl number will close the system of equations for laminar flow. The formulation assumes Stokes hypothesis (Eq. 1.10, with $\lambda = -2\mu/3$). For turbulent flow, nine additional variables ($\overline{\rho w_i' w_j'}$, $\overline{\rho w_i' e'}$) have to be evaluated. The most general transport equations for turbulence are the Reynolds stress equations. Many of the terms in these equations have to be modeled. Modeled Reynolds stress equations are given by Launder et al. (1975) and many others.

If the turbulent heat fluxes are obtained using an eddy diffusivity model, the molecular viscosity μ is replaced by $\mu_l + \mu_t$ and μ/PR is replaced by $\mu_l/\text{PR} + \mu_t/\text{PR}_t$, where PR_t is the turbulent Prandtl number.

For application to complex general geometries, it has become standard practice to transform Eq. 5.1 into a body-fitted curvilinear coordinate system. This transformation for a Cartesian system is given in Appendix A. The transformed equations in a rotating cylindrical coordinate system are given in Govindan and Lakshminarayana (1988). The boundary conditions for time marching techniques are dealt with in Section 5.7.3.1 and these are similar for all other computational techniques.

5.2 TURBULENCE MODELING FOR TURBOMACHINERY FLOWS

Before we proceed to describe the computational techniques for viscous and turbulent flows, it is necessary to examine turbulence models applicable to turbomachinery flows, because they are an integral part of the computation. Most turbomachinery flows encountered in practice are turbulent and three-dimensional in nature. In many situations, three-dimensionality, curvature, rotation, shock–boundary-layer interaction, buoyancy, flow separation or reversal, and other effects introduce changes in the turbulence structure, thus invalidating many of the turbulence models widely used for "simple" and "mildly complex" shear layers. It becomes increasingly important to embody more physics and/or constitutive equations in providing suitable closure models for adequate prediction of these complex flows.

Bradshaw (see Kline et al., 1982) provided a basis on which shear flows can be classified as "simple" or "complex." The simple shear layer is one where the significant rate-of-strain component is $\partial u / \partial y$, the gradient normal to the wall in a two-dimensional flow. Flows subjected to significant extra rates of strain are considered "complex shear layer." It is known that even a small extra rate of strain can have a significant effect on the turbulence structure and mean velocity profiles. Turbulence structure modification has been observed experimentally when the effects of curvature, rotation, and three-dimensionality are present (e.g., Kline et al., 1982; Anand and Lakshminarayana, 1978; Johnston et al., 1972; Bradshaw, 1973). This phenomenon is discussed in Section 5.2.1. Some examples of complex shear layers in turbomachinery are shown in Fig. 5.1. These include:

1. Shear layers on rotating surfaces (Figs. 5.1d, 5.1e, and 5.1g). The boundary layers and wakes on these surfaces are almost invariably three-dimensional, thus introducing additional strains.
2. Shear layers developing on curved surfaces (such as boundary layers on turbine blades), wall jets used in cooling applications, and curved flows in turbomachinery passages (Figs. 5.1c, 5.1f, and 5.1g). The centrifugal force, arising from streamline curvature in these flows, represents an additional strain; in most cases, these shear layers are three-dimensional, introducing additional strain due to the radial or transverse components of velocity and the centrifugal acceleration.
3. Separated flows, such as those arising from a shock–boundary-layer interaction, dump diffuser processes; separated flows on stator and rotor blades; and corner or wall flow separation (Fig. 5.1b, 5.1f).
4. Swirling flows and vortices (e.g., secondary and leakage vortices, swirling flow in the annulus).
5. Interacting boundary layers (Fig. 5.1b).

5.2.1 Structural Changes of Turbulence in Three-Dimensional, Rotating, Curved, or Separated Flows

Before proceeding with a review of various turbulence models used for complex flows, it is important to understand the physical phenomena associated with these flows. Some representative cases will be covered to illustrate the effects of "extra strain" rates. This will provide a foundation for discussion of the turbulence models used in these flows.

One class of flows subjected to extra strain effects are those with streamline curvature. Bradshaw (1973) has provided a comprehensive review of this topic. It is well known that centrifugal force suppresses the turbulence on a convex surface and amplifies the turbulence on a concave surface. Often multiple curvatures are present; longitudinal or streamwise curvature as occurs over a curved body (e.g., turbomachinery blades, Fig. 4.19); and

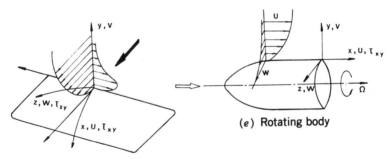

(a) Aircraft wing, helicopter blade, turbomachinery blade

(e) Rotating body

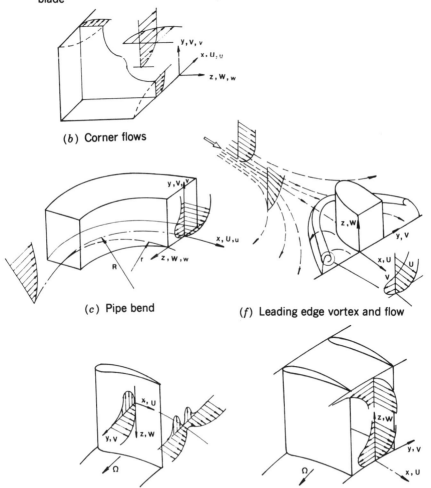

(b) Corner flows

(c) Pipe bend

(f) Leading edge vortex and flow

(d) Turbomachinery rotor blade boundary layer and wake

(g) Hub and annulus wall boundary layers in turbomachinery passages

Figure 5.1 Some examples of complex shear flows (u,v,w are velocities in x,y,z directions, respectively) (copyright © Lakshminarayana, 1986).

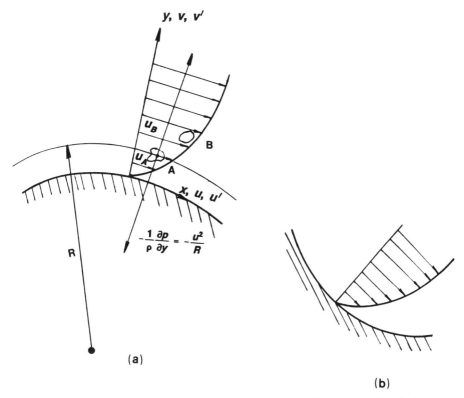

Figure 5.2 Effect of longitudinal curvature on turbulence (copyright © Lakshminarayana, 1986).

transverse or lateral curvature, as occurs in a turbulent vortex and meridional streamline curvatures (Fig. 4.11). In turbomachinery, one encounters as many as six curvatures, even though in most situations the dominant effects are caused by curvatures in the longitudinal and/or transverse direction. Examples are (a) the flow over highly cambered blades in a turbine and (b) the swirling flow through a duct, and leakage and secondary vortices.

The effects of curvature on turbulence structure are illustrated in Fig. 5.2a. The physical phenomena associated with the qualitative changes can best be explained by considering a simplified normal or radial momentum equation by ignoring the viscous effects, $\partial p/\partial y = \rho u^2/R$. The effect of centrifugal force in altering the turbulence structure can be explained using Prandtl's mixing-length concept. Let us say an eddy at location A (Fig. 5.2a) moves away from the wall where the velocity u_B is higher than u_A but retains its own momentum (u_A). At the new location (B) there is an imbalance between the local pressure gradient (higher than than at A and acting inward) and the centripetal force $(-\rho u_A^2/R)$. Thus, eddy motion away from the wall is suppressed, thereby decreasing the "mixing length." This argument, though qualitative, illustrates the "stabilizing" effect of convex curvature on turbulence, as observed by a decrease in Reynolds stress and turbulence energy levels compared to those for "simple shear layers." It has been observed, experimentally, that even mild curvature gives rise to significant turbulence amplification/suppression effects. The stabilizing effect in some of these cases becomes large enough to suppress the turbulence completely. The convex curvature affects the outward diffusion of turbulence kinetic energy from the wall region. The integral length scale as well as the relative contribution to the Reynolds stress production by large- and small-scale turbulence motions are affected significantly. These results indicate that "simple" turbulence models, which are based on "simple shear layers," fail to capture these effects. The turbulence suppression cannot be predicted by any of the approaches which utilize the zero-, one-, and two-equation turbulence models described in Chapter 1. Only a Reynolds stress transport model can predict such phenomena.

The effect of concave curvature (Fig. 5.2b) is opposite of that observed for a convex surface. In this case, the turbulence is amplified, and there is an extra production of shear stress and turbulence by the centrifugal force. The Reynolds stresses and turbulence are higher than those observed in simple flows, and the concave curvature has a destabilizing effect on the boundary layer. The development of Taylor–Görtler instabilities influence the turbulence transition and transport. Available experimental data suggest that the curvature effect on the concave and convex surfaces reduces the coupling between the wall and the outer layers, which makes turbulence transport independent of turbulence production near the wall. The longitudinal vortices observed seem to be unstable and are the dominant cause of the turbulence transport in the outer layer. Thus, the concave curvature has a strong influence on the structure of a turbulent boundary layer, and this effect cannot be captured by simple models which ignore the real physical phenomena.

The effect of rotation on the flow through a duct is shown in Fig. 5.3. The duct is rotating about a spanwise axis. This is a simplified model of flow in a centrifugal or a radial compressor, and this configuration has been investigated by Stanford group (Johnston et al., 1972; Johnston, 1974). In this flow, a normal pressure gradient is set up by the Coriolis force (Eq. 4.25), as shown in Fig. 5.3. If we apply the same argument as for a curved flow, it is evident that as the eddy tends to move away from the leading (the rotation) surface, marked +, the eddy moves into a higher pressure gradient (away from the wall inside the boundary layer) while retaining its initial momentum. Therefore, as the eddy moves from 1P to 2P on the pressure side (Fig. 5.3), the movement is amplified by the higher pressure gradient caused by the higher Coriolis force at 2P. The turbulence is amplified on this side and is destabilized. The effect is opposite on the suction, or trailing side of the channel. On this side, the turbulence is stabilized by the attenuation of turbulence. There is a negative production of turbulence and stresses. The mean velocity

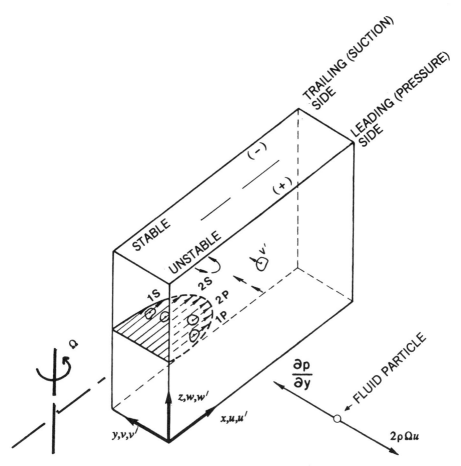

Figure 5.3   Complex shear flow in a rotating duct (copyright © Lakshminarayana, 1986).

profile, measured by Johnston (1974), clearly reveals these effects. The change in production/redistribution of turbulence due to rotation for this geometry, shown in Table 5.1, indicates that the major influence of rotation is in the redistribution of energy between the normal stresses. The $\overline{(u')}^2$ component decreases on the leading side and increases on the trailing side of the channel, and the $\overline{(v')}^2$ component increases on the leading side and decreases on the trailing side, but they have no influence on $\overline{(w')}^2$ the spanwise component. The principal stress production $\overline{(u'v')}$ increases [for $\overline{(u')}^2 > \overline{(v')}^2$] on the leading side. This is clearly evident from the wall shear stress measured by the Stanford group. The flow visualization study confirms (a) the suppression of turbulence on the trailing side and (b) the appearance of Taylor–Görtler type cells on the leading side at higher rotational speeds.

Table 5.1   Coriolis terms in the Reynolds stress equation (see Fig. 5.3 for notation) (Johnston, 1974)

Transport Equation	Production Term	
	By Mean Shear	By Rotation
$\overline{(u')^2}$	$2(-\overline{u'v'})\,\partial u/\partial y$	$-4(-\overline{u'v'})\Omega$
$\overline{(v')^2}$	0	$4(-\overline{u'v'})\Omega$
k	$(-\overline{u'v'})\,\partial u/\partial y$	0
$(-\overline{u'v'})$	$\overline{(v')^2}\,\partial u/\partial y$	$(\overline{(u')^2}-\overline{(v')^2})2\Omega$

The effect of rotation on blade boundary layers in axial flow turbomachinery is illustrated in Fig. 5.1d. The system rotates around the axis of the rotor and the dominant component of the Coriolis force $(2\,\Omega \times \mathbf{U})$ is in the radial direction; this introduces radial flow. The mean and Reynolds stress equations for such flows are given by Anand and Lakshminarayana (1978). Because the Coriolis forces are in the radial direction, turbulence intensity and stresses increase in this direction, with a corresponding decrease in the streamwise (or chordwise) direction. The effects of the Coriolis force are different for turbines and compressors, depending on the velocity triangle. The stresses as well as the turbulence intensity production are modified accordingly. The rotation effects on axial flow inducer blading have been measured and reported by Anand and Lakshminarayana (1978). Similar effects are observed in the case of free shear layers (wakes) from the blades (Lakshminarayana and Reynolds, 1980).

Separated turbulent flows represent one of the most complex flows in terms of both physical understanding and modeling. Simpson (1985) carried out a systematic study of these flows. The separated flows encountered in practice can be classified as follows: separation induced by adverse pressure gradient (e.g., suction side near trailing edge, corner separation), separation due to abrupt changes in the cross section (e.g., flow about sharp corners, blunt bodies, fence, rib), and shock-induced separation. Based on his and others' data, Simpson provided a physical model for separated flows. He argues, based on various data available, that the small-scale mean back flow does not come from downstream flow, but from the large-scale coherent eddies passing through the separated region. The inner layer in a separated flow is substantially different, phenomenologically, from that of a conventional layer. Large turbulence fluctuations occur in this region, with the energy for the flow coming mainly from the large-scale structures. Because of large-scale unsteadiness in the separated region, the mean velocity in the back flow region is the result of time-averaged fluctuation in velocity, which is large. Simpson suggests that "Reynolds stress in this region must be modeled by relating them to the turbulence structure and not to the local velocity gradients." This suggests major limitations of the conventional models based on concepts derived from the simple shear layer. The other feature that is distinctly different from the simple unseparated shear layer is that the normal stress $\left(\overline{(v')^2}, \overline{u'w'}\right)$ has a dominant effect on the mean and turbulence transport equations.

The change in turbulence in swirling flows (such as leakage or secondary vortex or flow through a turbomachinery annulus) can be attributed to curvature. Near the core of a vortex, the pressure gradient is approximately balanced by the centripetal force $(-\partial p/\partial r \approx -\rho V_\theta^2/r)$. Invoking the same arguments presented earlier (for curved flows), it is evident that the turbulence will be suppressed inside the vortex core. Data by Panton et al. (1980), confirm this trend. He measured values of effective turbulent diffusion in a vortex, which were much lower than values of molecular diffusion observed for simple shear flows. Maximum values of ν_t/ν occur near the outer edge of the vortex, away from the forced vortex core region. Initial turbulence levels in the vortex core decay, and in some cases turbulence is completely damped out in this region. These effects are brought about by centripetal force, and accordingly the curvature and rotation effects should be included in turbulence models for vortex flows. This will be clear from computations presented later in the next section.

5.2.2 Turbulence Models for Complex Flows

The equations governing the turbulent stress are given by Eq. 1.88. The production, dissipation, diffusion, and convection of turbulence stresses and energy are influenced by the Coriolis and centrifugal forces as described earlier.

Zero-equation or algebraic eddy viscosity models (e.g., Eq. 1.77) are widely used in practical engineering applications involving simple shear flows and in many of the Navier–Stokes computer codes. Two-equation models (Eqs. 1.84 and 1.86) are employed when additional details of the turbulence field, including transport effects, are important. Reynolds stress models (Eq. 1.88) have been under extensive development for decades and are often employed in complex flow situations—for example, three-dimensional flows, flow with curvature and rotation, and flows with blowing and suction. Numerical computations which utilize large-eddy simulation (LES) or the direct numerical simulation (DNS) of the time-dependent turbulent flow field are prohibitively expensive, and accordingly, their use is presently limited to very simple flows (flat-plate boundary layer, channel flows, etc.) at low Reynolds number. Lakshminarayana (1986) has reviewed various models, their limitations, and their use in turbomachinery flows.

The algebraic eddy viscosity model is valid only for two-dimensional "simple shear flows." Various data and comparisons available lead to the following conclusions:

1. The model is adequate for two-dimensional flows with mild pressure gradient, and the mean velocity field is predicted well.
2. The model is suitable for three-dimensional boundary layers with small cross flows.
3. The model is inaccurate in flows with curvature, rotation, or separation.
4. The model is not accurate for flows with pressure- or turbulence-driven secondary motions and/or when abrupt changes in mean strain are present.
5. The model cannot accurately predict shock-induced separated flow.

It should be noted that for an attached two-dimensional boundary layer with a mild pressure gradient, there is very little advantage in resorting to a higher-order model. The algebraic eddy viscosity model is adequate for the prediction of the mean velocity field.

The $k-\epsilon$ model has been successfully used to predict the two-dimensional attached flows with moderate pressure gradient. The mean and gross properties of turbulence are predicted well. Because of several shortcomings of the $k-\epsilon$ model, the model breaks down for flows with rotation, curvature, strong three-dimensionality, and separation. The constants used in these models are optimized based on well-documented turbulent flows including decaying isotropic turbulence and simple shear flows. Numerous attempts to modify and refine the $k-\epsilon$ model to include curvature, rotation, and other effects have not been successful. The effect of rotation is to redistribute the energy (term 6 in Eq. 1.88); hence, the rotation effect does not appear explicitly in transport of kinetic energy. The turbulence kinetic energy is neither destroyed nor produced by rotation effects, but the rotation has a major effect on the distribution of $\overline{w_i' w_j'}$. (Eq. 5.1)

Some of the conclusions concerning the $k-\epsilon$ models based on the results from various investigators (Lakshminarayana, 1986) are as follows:

1. The $k-\epsilon$ equations have been widely used for two-dimensional flows with moderate pressure gradient. Mean velocities are predicted fairly accurately for these cases, and gross properties of the turbulence are predicted well. The model is good for attached two-dimensional boundary layers. For these cases, the Reynolds stress model (RSM) has little advantage over the $k-\epsilon$ model.
2. The predictions from the $k-\epsilon$ model are not good for three-dimensional flows. Shortcomings are due to the modeling of the pressure–strain term, assumption of isotropy, the gradient diffusion approximation (Boussinesq Eq. 1.75), and low Reynolds number effects near walls. An isotropic eddy viscosity is adequate for three-dimen-

sional boundary layers, with very small cross flow, but fails for significant cross flows and swirl.

3. The model breaks down for flows with rotation, curvature, strong swirl, three-dimensionality, and separated flows.

When the curvature, rotation, and other extra strain effects are present, the Reynolds stress closure equations can provide a more realistic and rigorous approach to account for these complex strain fields. Some of these attempts are described below.

5.2.2.1 Algebraic Reynolds Stress Models

As indicated in Section 1.2.7.3, attempts have been made to use the simplified Reynolds stress equation to derive improved turbulence models. One such effort has resulted in the algebraic Reynolds stress equation given by Eq. 1.89.

It should be remarked here that even though the diffusion and advection of $\overline{(u_i' u_k')}$ do not appear explicitly in Eq. 1.89, they appear implicitly through the transport equations for k and ϵ. Therefore, the algebraic Reynolds stress model (ARSM) cannot be used alone, but must be employed in conjunction with the $k-\epsilon$ equations. Thus, the $k-\epsilon$ ARSM model would constitute Eqs. 1.84, 1.86, and 1.89. There have been many attempts to use the ARSM in conjunction with the $k-\epsilon$ models for the computation of turbulent flows. These attempts may be classified as follows:

1. Modified $k-\epsilon$ via C_μ from the ARSM
2. Coupled $k-\epsilon$/ARSM models

In the former models, the value of C_μ (in Eq. 1.80) is corrected locally through the ARSM. These models still employ the eddy viscosity concept, and thus they are not suitable for resolving the components of Reynolds stresses. In the latter model, $k-\epsilon$ and ARSM models are both solved in their complete form.

Warfield and Lakshminarayana (1987b) manipulated the ARSM and the individual stresses to derive an expression for C_μ in Eq. 1.80 in terms of P, ϵ, Ω, and $\partial u_i / \partial x_j$ and successfully computed and validated the data in a rotating channel (Johnston et al., 1972). The coefficient $C_{\mu i}$ $(i = 1, 2, 3)$ is nonisotropic and is a function of the local property of turbulence, mean velocity, and system rotation. The expressions for $C_{\mu i}$ are lengthy and can be found in Warfield and Lakshminarayana (1987b). The predictions compare very well with the data for a rotating channel. The asymmetry in the profile due to Coriolis force, as well as the profile near the wall, are predicted well. Thus, the major effects of rotation are captured through the ARSM.

The use of the $k-\epsilon$/ARSM model does not entail any significant additional resources because these ARSM equations are algebraic, and computational time is increased only moderately. The dramatic effect of curvature

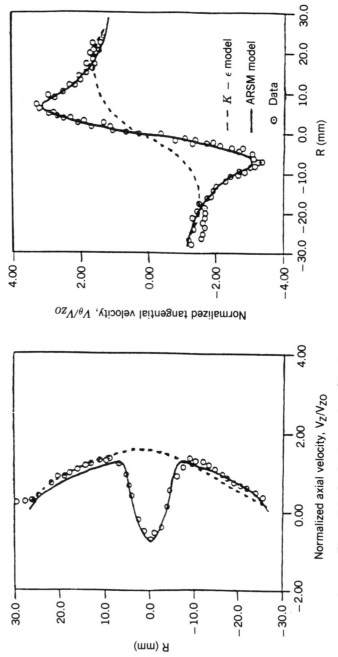

Figure 5.4 Confined vortex flow. (data from Escudier et al., 1980; computation from Hutchings and Iannuzzelli, 1987). (Copyright © Lakshminarayana, 1986; reprinted with permission.)

and the inability of the $k-\epsilon$ model in capturing these effects can be seen in predictions by Hutchings and Iannuzzelli (1987), who utilized Rodi and Scheuerer's (1983) ARSM. The predictions from the standard $k-\epsilon$ model for the swirling flow in a duct are shown compared with the experimental data in Fig. 5.4. The $k-\epsilon$ model misses the major features of the swirling flow, such as the recirculating zone along the center line, whereas $k-\epsilon/$ARSM models provide excellent agreement with the data.

Zhang and Lakshminarayana (1990) recently modified Eq. 1.89 to include the Reynolds stress model proposed by Shih and Lumley (1986) and successfully predicted the three-dimensional boundary layer profile on a compressor rotor blade (Lakshminarayana and Popovski, 1987). The comparison shown in Fig. 5.5 indicates that the prediction of mean velocity profiles, especially the radial component, is substantially improved with ARSM. The location and magnitude of maximum radial velocity is not captured well at the chordwise distance $S = 0.87$ and at the radial location $R = 0.75$. The predictions at all other radial locations are predicted well with the coupled $k-\epsilon/$ARSM model.

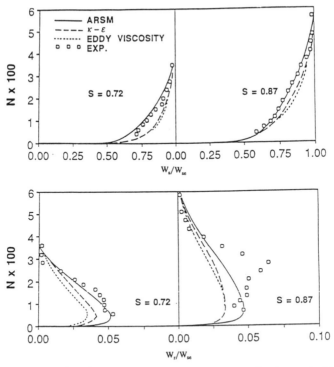

Figure 5.5 Axial Flow Compressor rotor suction surface boundary layer at midspan (Zhang and Lakshminarayana, 1990; data from Lakshminarayana and Popovski, 1987). Penn State low-speed compressor outer diameter 0.936 m, speed 1080 rpm, Reynolds number 2.8×10^5, 21 rotor blades, tip chord 15.41 cm; N is the distance normal to blade, normalized by chord length; w_s is the streamwise velocity, w_r is the radial velocity.

In summary, $k-\epsilon/\text{ARSM}$ models simulate the turbulent stresses more realistically by relating the properties to local conditions. $k-\epsilon/\text{ARSM}$ models can be implemented in an efficient manner and can predict important features of the flow. The rotation and curvature effects are captured directly instead of through modeling. Warfield and Lakshminarayana (1987b, 1989), Hah and Lakshminarayana (1980), and Luo and Lakshminarayana (1995a, b) have made extensive use of ARSM models to predict three-dimensional viscous flows inside turbomachinery including rotor wakes, turbine endwall flows, tip clearance flows, and heat transfer on turbine blades. In addition, $k-\epsilon$ and ARSM models have been extended and validated for unsteady viscous flows (Fan and Lakshminarayana, 1993; Fan et al., 1993).

5.2.2.2 *Reynolds Stress Models*   The full Reynolds stress model (RSM) (Eq. 1.88) provides a more realistic physical simulation of turbulent flow and is potentially the superior model. However, it is very complex, and it is the least tested model so far. Nevertheless, its use is likely to become more widespread during the next decade for both simple and complex flows. Details of the modeling can be found in Launder et al. (1975) and Lumley (1980).

A comparison of predictions of the flow field (Luo and Lakshminarayana, 1995b) in a 180° turnaround duct (Re = 10^6) from $k-\epsilon$, $k-\epsilon/\text{ARSM}$, and RSM models are shown compared with Monson et al.'s (1989) data in Fig. 5.6. In this figure, the convex side is at $y/H = 0$, the concave side is $y/H = 1.0$, H is the duct height, and V_m is the mean velocity. The mean velocity profile is predicted well by all the models, except very near the

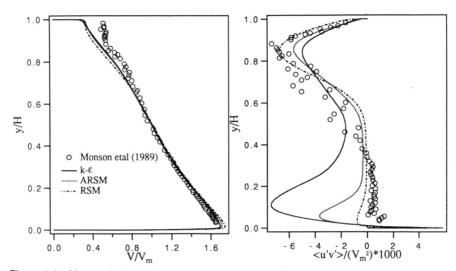

Figure 5.6 Mean velocity and turbulent shear stress at 60° location of a 180° turn around duct (Luo and Lakshminarayana, 1995b).

concave side. The $k-\epsilon$ model fails to capture the curvature effect on streamwise turbulent stress, $\overline{u'v'}$, near the convex side. The ARSM does reasonably well and better than the $k-\epsilon$ model, and the RSM provides the best agreement. It should be noted that this is a severe test case for turbulence models.

5.3 SUMMARY OF APPROXIMATIONS AND NUMERICAL TECHNIQUES

As indicated in Chapter 1, the turbomachinery flow field is complex, three-dimensional, turbulent, and, in many instances, separated. However, in many flows, simplifications to the full Navier–Stokes equations are applicable. Therefore it is not efficient to solve the full Navier–Stokes equations for all turbomachinery flow fields. For example, viscous flow through a two-dimensional cascade with attached subsonic flow can be most accurately predicted using one of the inviscid techniques described in Chapters 3 and 4, combined with a boundary layer code. It is unnecessary and inefficient to solve the full Navier–Stokes equations for this flow. Likewise, swirling flow through diffusers and ducts with large boundary layer growth can be best handled by making suitable approximations to reduce the Navier–Stokes equations to a parabolic form. A spatial marching solution should provide reasonably good results for these cases. The boundary layer procedure is approximately two orders of magnitude more efficient, and the space-marching code is about one order of magnitude more efficient (in computer time) than a full Navier–Stokes solution. Boundary layer and space-marching codes are valuable tools in design, because they enable a large number of calculations to be made in a reasonable time. For many flows, such as the ones mentioned earlier, the accuracy of the solution from any of these techniques is about the same.

Some turbomachinery flows are dominated by inviscid effects such as flow turning, blade thickness, blunt leading edge, blade lean and skew, and so on. Again it is not necessary to solve such flows using full Navier–Stokes techniques because this can be prohibitively expensive. The major physics can be captured by Euler equations, and viscous effects can be resolved from a boundary layer code for nonseparated zones. Inviscid and viscous regions are identified, the corresponding equations are solved, and the boundary between the zones is iterated until a converged solution is obtained.

Viscous dominated flow, such as the flow in a transonic compressor with strong three-dimensional shocks and associated boundary layers, and centrifugal turbomachinery flow field are best resolved by solving the Navier–Stokes equation in the entire region.

The computational techniques for turbomachinery flow analysis can be classified as shown in Table 5.2. One must clearly distinguish between

Table 5.2 Computational techniques for turbomachinery

Physical approximation ← Procedure → Solution technique

Discretization: Finite volume / finite difference / Finite element

Algebraic equations: SOR, SLOR, ADI, preconditioning, conjugate gradient, multigrid, direct method

376

physical and discretization approximations. The errors involved in solution can be classified as (1) those arising due to physical approximations, (2) those arising from discretization such as the order of accuracy of discretization, grid size, method of solution, geometry and boundary representation, artificial dissipation, factorization error, linearization, and so on, and (3) those arising from iterative solution of algebraic equations (ADI, SOR, SLOR, etc.).

The techniques, classified according to the physical approximations made, are as follows (Table 5.2):

1. *Inviscid Flow.* The viscous terms are neglected and the resulting equations are solved by any of the following techniques: panel method (incompressible, irrotational, two-dimensional flow), potential equation (irrotational, 2D, 3D), Euler equation (2D, 3D).

2. *Boundary Layer Approximations.* In this technique, streamwise diffusion terms are neglected and the pressure field is assumed to be imposed by the inviscid layer and this is prescribed from an inviscid analysis. This assumption is valid when the viscous layers are thin compared to blade spacing. It is not valid in the endwall region of the blade row or in the presence of vortices or separation.

3. *Parabolized Navier–Stokes (PNS) Equation / Space Marching Techniques.* The assumption made in this technique is similar to the boundary layer assumption, but the technique allows for transverse and normal pressure gradients. The streamwise pressure gradient is prescribed from an inviscid code, and it can be continuously updated to capture the flow physics. All of the techniques which utilize single-pass spatial marching are included in this category. A second type of inviscid approximation (with the same viscous approximation as above) does not employ any approximations for the pressure gradient term, and instead approximates the convective terms in the cross $(y, z,$ or $\eta, \zeta)$ momentum equations by decomposing the velocity into potential, rotational, and viscous components.

4. *Thin-Layer Navier–Stokes (TLNS) / Reduced Navier–Stokes (RNS) Equation.* The TLNS, where the viscous and the turbulent diffusion terms in the main stream direction are neglected, is valid for thin viscous layers. It is also a consistent simplification on computational grids that are too coarse to resolve the streamwise diffusion terms.

5. *Full Navier–Stokes and Euler Solutions.* There are two types of solutions. In the first type, the entire equation is solved in either the steady or unsteady form. In the second approximation, parabolic assumptions are made initially, and the pressure is updated by using an iterative procedure. The latter technique is not the PNS solution described earlier, but rather an initial approximation to this pressure field, which is updated to satisfy the entire Navier–Stokes equation. Therefore, the converged solution satisfies the full Navier–Stokes equation, even

though the final stage was achieved through several intermediate approximations.

6. *Zonal Technique.* The solution of the full Navier–Stokes equation in the entire domain can be prohibitively expensive. Hence we use zonal techniques, where several zones are identified and the corresponding (inviscid, boundary layer, parabolic) solutions are obtained and patched and integrated to obtain a composite solution. These are described in detail later and are very accurate, but are much less expensive techniques.

5.4   COMPUTATION OF INVISCID FLOWS

The computational techniques available for inviscid flows in turbomachinery can be classified as follows:

1. Two-dimensional and three-dimensional potential flow solutions
2. Two-dimensional, quasi-three-dimensional stream function solutions
3. Two-dimensional and three-dimensional Euler solutions

The most popular techniques used for the solution of the Euler and the Navier–Stokes equations are covered together later in this chapter because similar techniques are used to solve the two sets of equations.

The validity of potential flow solutions to turbomachinery flows is restricted to two-dimensional and cascade flows. The three-dimensional flows are almost invariably rotational, except for the cases described in Sect. 4.2.2. In view of this, the treatment of potential flows in cascades has been kept deliberately brief. The potential flow analysis is restricted to irrotational and isentropic flows. The predictions are restricted to cascades with weak shockwaves. On the other hand, the equations governing cascade flow in terms of stream function (Eqs. 1.58–1.60) and the principal equations for S_1 and S_2 surfaces (Eqs. 4.91 and 4.117) as well as the passage-averaged quasi-three-dimensional equation (Eq. 4.78) are valid for rotational flows and are more general. In addition, the approximate effects of viscosity can be included in the quasi-three dimensional formulation through empirical loss correlations or entropy gradients.

In subsonic regions, the potential flow equations are elliptic and any of the techniques described in Appendix B (SOR, SLOR, ADI, LU decomposition) can be used to solve them. If there are supersonic patches as in a transonic cascade, some special treatment is necessary to capture the shock formulation. This will be illustrated for the potential equation, and the technique is similar for the stream function equations as well. The equations can be discretized by finite difference (FD), finite element, or finite volume (FV) techniques as explained in Appendix B. Only the FD/FV technique will be described here.

The techniques available for the solution of potential and stream function equations for transonic flow (mixed subsonic, supersonic, and transonic flow) can be classified as follows: (a) strongly implicit procedure, (b) type-dependent scheme, (c) rotated difference scheme, (d) artificial time concept, (e) artificial compressibility and viscosity concept. A summary of various techniques can be found in Hirsch (1990).

Major problems arise when the cascade has mixed (subsonic, transonic, supersonic) flow in the passage. If it is purely subsonic, the relaxation method (Anderson et al., 1984) for an elliptic equation is adequate, and the panel method is the best for incompressible flow. Most of the other techniques listed above are for stabilizing the solution in the supersonic region. In the type-dependent scheme, central differencing cannot be used, because all information in this case comes from upstream. Thus, a special backward or upward differencing scheme is used in the supersonic region. In the rotated difference scheme, the coordinates are aligned in the direction of the resultant velocity, and the potential equation then reduces to Eq. 3.91.

The artificial compressibility and artificial viscosity techniques are developed to stabilize the solution in a supersonic zone through the local generation of artificial dissipation. As indicated earlier, there is no difficulty in solving these equations in the subsonic region. Instability occurs in the supersonic region and during shock formation. Some dissipative terms (nonphysical) are essential to damp these instabilities. The artificial compressibility and artificial viscosity (Jameson, 1974; Hafez et al., 1979; Dulikravich, 1988) techniques generate artificial dissipation in the locally supersonic region by modifying the density as shown below.

The equations governing the potential flow in a cascade in conservative form are given by

$$(\rho\phi_x)_x + (\rho\phi_y)_y = 0 \tag{5.2}$$

and

$$\rho = \left[1 - \frac{\gamma - 1}{2}M_1^2(\phi_x^2 + \phi_y^2 - 1)\right]^{1/(\gamma-1)}$$

where density and velocities are nondimensionalized by upstream values.

In the artificial compressibility/viscosity technique (Hafez et al., 1979), the physical density is replaced by $\tilde{\rho}$, and the governing equation is written as

$$(\tilde{\rho}\phi_x)_x + (\tilde{\rho}\phi_y)_y = 0$$

where

$$\tilde{\rho} = \rho - \mu \Delta s (\partial \rho / \partial s) = \rho - \mu \left(\frac{u}{q} \rho_x \Delta x + \frac{v}{q} \rho_y \Delta y \right) \quad (5.3)$$

and

$$q = \sqrt{u^2 + v^2}$$

u and v are velocity components in x and y directions respectively, and μ is a switching function given by (zero for subsonic flows)

$$\mu = \max \left[0, \left(1 - \frac{1}{M^2} \right) \right] \quad (5.4)$$

The additional term in Eq. 5.3 is equivalent to adding artificial viscous terms. The nature of terms added is purely to stabilize the solution in the supersonic region. $\Delta s (\partial \rho / \partial s)$ is the total differential component of the density in the flow direction. Various modifications have been suggested for Eq. 5.3, and this, as well as an analysis of artificial viscosity/compressibility effects, can be found in Anderson et al. (1984), Hafez et al. (1979), and Dulikravich (1988). In the subsonic region, $\mu = 0$, and the artificial compressibility/viscosity concept is triggered only when a supersonic flow is encountered. This technique was used by Farrel and Adamczyk (1982) and others to solve the two-dimensional and quasi-three-dimensional potential equations for a cascade, and it was used by Huo and Wu (1986) to solve the quasi-three-dimensional S_1 equation (similar to Eq. 4.91) for turbomachinery in a body-fitted coordinate system.

Huo and Wu (1986) replaced the density by Eq. 5.3 and developed a nine-point difference scheme given by

$$A_1 \psi_{j-1, k-1} + A_2 \psi_{j, k-1} + A_3 \psi_{j+1, k-1} + A_4 \psi_{j-1, k} + A_5 \psi_{j, k}$$
$$+ A_6 \psi_{j+1, k} + A_7 \psi_{j-1, k+1} + A_8 \psi_{j, k+1} + A_9 \psi_{j+1, k+1} = D_{j, k} \quad (5.5)$$

$A_1 \ldots A_9$ are coefficients which depend on the equation used. For example, for a two-dimensional potential equation they are a function of ρ only. For S_1 and S_2 stream surface equations, they depend on $\tilde{\rho}, b$, or B and other variables. $D_{j, k}$ contains all of the source functions. For a two-dimensional potential equation it is zero; for S_1 or S_2 surface equations it may include variables I, S, Ω, ϕ, F_θ, G_i, $F_{i, j}$, B, W_r, W_θ, W_z, and so on, (Chapter 4), whose values are derived from either the previous iteration of S_1 (or S_2 surface) or the present solution of S_2 (or S_1 surface).

The next step is to solve the system of algebraic equations obtained by application of Eq. 5.5 at all grid points in the solution domain using any of the techniques described in Appendix B. Huo and Wu (1986) utilized a finite difference scheme and solved these equations using a LU decomposition scheme by writing Eq. 5.5 as

$$[M][\psi] = [P] \tag{5.6}$$

or

$$[L][Q] = [P], \qquad [U][\psi] = [Q] \tag{5.7}$$

where $[\psi]$, $[P]$, and $[Q]$ are column vectors and $[M]$ is a matrix of influence coefficients in Eq. 5.5.

The pseudo-density for a body-fitted coordinate system (ξ, η) for this case is given by

$$\tilde{\rho} = \rho - \mu \left[\frac{V_\xi}{V} \rho_\xi \, \Delta \xi + \frac{V_\eta}{V} \rho_\eta \, \Delta \eta \right] \tag{5.8}$$

where V is $\sqrt{V_\xi^2 + V_\eta^2}$.

Coefficients A_i are updated and Eq. 5.7 solved iteratively until convergence is achieved.

One of the critical steps in this scheme is to determine the Mach number at every grid point and decide on the switching function (Eq. 5.4). This involves solving for the velocity by integrating the stream function along a grid line after each iteration. Therefore, the procedure for solving the flow field using S_1 equations is as follows:

1. Solve S_1 surface equation (e.g., Eq. 4.91) using the solution technique described above, or others described in Appendix B.
2. Integrate the momentum equation in η or the blade-to-blade direction (by prescribing $\psi = 0$ on one of the surface) to determine the velocity and Mach number at every grid point.
3. Prescribe the switching function.
4. Iterate equations for ψ, M, ρ until the convergence is reached.

5.5 BOUNDARY LAYER SOLUTIONS

5.5.1 Nature of Boundary Layers

In many instances, the boundary layers on turbomachinery blades are thin, and interaction effects between the viscous and inviscid regions are small enough to be negligible. Under these circumstances, an inviscid solver can be combined with a boundary layer code to predict the entire flow field and losses. This approach is valid in regions away from the endwalls, because viscous–inviscid coupling is large in the shroud wall (with or without tip clearance) and hub wall regions. Moreover, the boundary layer solutions, which assume the pressure gradient normal to the walls to be zero, are not valid in these regions. The boundary layer approximation involves neglection of streamwise diffusion terms and retention of only the dominant viscous terms (i.e., $\partial^2/\partial y^2$, where y is the direction normal to the wall). The pressure is prescribed at the edge of the boundary layer and is assumed to be constant across the viscous layer.

There are two approaches for the solution of the boundary layer equations. In the momentum integral approach, where the interest lies in global features and overall performance, the equations are integrated from the wall to the free stream, yielding differential equations for the growth of the boundary layer momentum thickness. The second approach is the differential approach, where the local flow field is resolved (including velocity profiles) through the analytical/numerical solution of partial differential equations governing the boundary layer.

The turbulent shear layers in turbomachines differ from the turbulent boundary layer on an airfoil in several respects:

1. Stronger three-dimensional effects. The flow curvature (due to camber, annulus, and hub wall radius), Coriolis and centrifugal forces, radial pressure gradient, annulus wall flaring, blade skew, and sweep induce three-dimensionality in the blade and wall boundary layers.

2. High rates of heat transfer in turbine blades.

3. Stronger acceleration and changes in flow direction.

4. High free-stream turbulence.

5. Interaction of two or more shear layers (e.g., the interaction of the pressure and suction surface boundary layers as well as the interaction of the wall and blade boundary layers).

6. Unsteadiness arising from the wakes of previous blade row and inlet distortion, blade vibration and flutter.

7. Compressibility effect and shock–boundary-layer interaction in high-speed turbomachinery.

Bradshaw (1974) estimated the strength of some of these special effects needed to change the shear stress or the distance of separation by 10%. These are tabulated in Table 5.3.

Let us consider a two-dimensional boundary layer developing on a cascade blade. The boundary layer may be laminar, transitional, turbulent, or separated. It is difficult to predict such mixed flows. Almost invariably the transition and turbulence have to be modeled. The separated flow is best solved using full Navier–Stokes equations, because elliptic effects and the

Table 5.3 Special effects needed to change C_f or distance of separation by 10% (Bradshaw, 1974)

Effect	Order of Magnitude
1. Sweep	$45°$
2. Heat transfer	$T_w/T_o = 0.7$
3. Longitudinal curvature	$\delta/R = 1/80$ or $35°$ turning angle
4. Rotation (component about blade spanwise axis, Ω)	$\Omega\delta/V_o = 1/80$
5. Free-stream turbulence (small scale)	3%
6. Free-stream unsteadiness (large scale: $\omega\delta/2\pi V_o = 0.1$)	30% (mean $\bar{c}_f$ changes by 10%)

normal (to the wall) pressure gradient effects are important and the boundary layer approximations are no longer valid.

The nature of two-dimensional boundary layer on a turbomachinery blade is shown in Fig. 5.7. In many cases, all the flow regimes (laminar, transitional, turbulent, and separated) exist and such complexity make prediction very difficult. The flow separation is usually on the suction side near the trailing edge, where the pressure gradient is adverse. The boundary layers that develop on turbomachinery blades (both rotating and stationary), and hub and annulus walls of turbomachinery are not two-dimensional. There are centrifugal and Coriolis forces, which, in addition to pressure and viscous forces, make the direction of the flow inside the boundary layer different from that outside, thus forming a three-dimensional flow configuration. Some

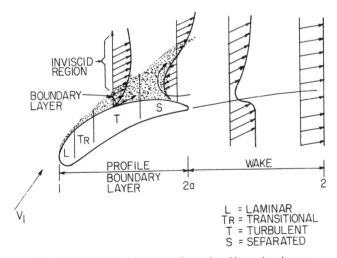

Figure 5.7 Nature of the two-dimensional boundary layer.

examples of the three-dimensional boundary layer in turbomachinery are shown in Fig. 5.1.

1. *Rotating Blade* (Fig. 5.1, Cases a and d). The nature of the three-dimensional boundary layer on an axial rotor blade and the corresponding velocity triangles are shown in Fig. 5.8. Let us assume that the flow satisfies the simplified radial equilibrium equation (Eq. 4.1) in the inviscid region. As the blade surface is approached, the absolute tangential velocity changes as indicated in the velocity triangles sketched in Fig. 5.8. Therefore, there is an imbalance between the radial pressure gradient $(-\partial p/\partial r)$ and the centripetal force $(-\rho V_\theta^2/r)$ inside the rotor blade boundary layer. This produces radial inward or outward flow depending on the nature of the velocity triangles. If the absolute and relative tangential velocities are in opposite directs (as in compressor), the centrifugal force is larger than the radial pressure gradient, and, as shown in Fig. 5.8, an outward radial velocity component will develop. If the absolute and relative tangential velocities are in the same direction (Fig. 5.8), the imbalance in forces is inward and a

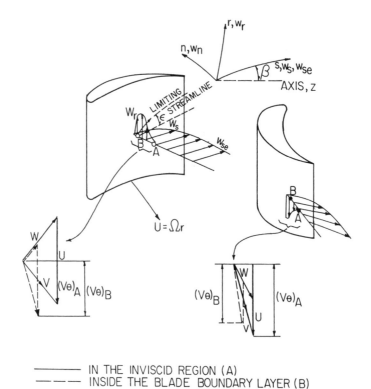

———— IN THE INVISCID REGION (A)
— — — INSIDE THE BLADE BOUNDARY LAYER (B)

Figure 5.8 Nature of radial flows inside a rotor boundary layer.

radial inward flow is produced. In a turbine, there is a wide variation in the velocity triangle from one type to another, from the leading to the trailing edge (Fig. 2.57). Therefore, both inward and outward radial flows may exist on the same blade. Similar effects exist in centrifugal turbomachinery, but are more complex due to the effects of longitudinal curvature.

2. *Stator Blade.* The centrifugal forces inside the boundary layer of a stator blade are always smaller (because V_θ decreases toward the blade surface) than the radial pressure gradient. This imbalance between the pressure forces and the centrifugal force gives rise to radial inward flow.

3. *Endwall Boundary Layer (Cases c, f, and g in Fig. 5.1).* The boundary layer that develops on the endwalls of a stationary or rotating channel is three-dimensional. Here again, the pressure gradient in the blade-to-blade direction, $\partial p/\partial y$ (in Fig. 5.1g notation), is larger than the centripetal force ($\approx -\rho u_s^2/R$, where u_s is the streamwise velocity and R is the radius of curvature, approximately equal to the radius of camber). The cross-flow that develops on the wall is toward the suction surface. This occurs in the endwalls of cascades, as well as rotating and stationary blade rows. This phenomenon is the same as secondary flow (Fig 4.19), as described in Chapter 4. The boundary layer that develops on the annulus walls of multistage turbomachinery is highly three-dimensional in nature. In addition to the effects indicated above, the following features make it very complex: (a) blade forces, (b) tip leakage flows, (c) interaction of blade and annulus wall boundary layers, and (d) sudden perturbations induced by alternate rotating and stationary blade rows.

4. *Rotating Hub-Wall Boundary Layer.* One of the important characteristics of the boundary layer on a rotating hub in a uniform stream is the skewness of the profile as shown in Fig. 5.1 (case e). The absolute velocity vector is at a 90° angle to the axial direction at the hub, and it changes its direction to the design incidence angle at the edge of the boundary layer.

The effects of three-dimensional boundary layers (described above) on the performance of turbomachinery are many:

1. The radial flows inside the profile boundary layer introduce loss gradients ($\partial s/\partial r$).

2. The wakes are three-dimensional in nature. The mixing losses are thus increased by the three-dimensionality in the blade boundary layer.

3. The rotor boundary layers (blade) are generally thin near the hub and thick at the tip. This is caused by radial outward transport of boundary layer fluid due to rotation effects as explained earlier. This has the

effect of delaying the flow separation near the hub, where there would be an increase in loading compared to the two-dimensional inviscid value. These effects will be in the opposite direction near the blade tip, where the loading decreases. The overall effect is to increase the transport of low-energy fluid toward the tip as well as to introduce large gradients in circulation and losses. This results in higher flow losses, along with larger three-dimensionality in the flow. For a stator or nozzle blade, the effects may be opposite to those of a rotor blade.

4. The turbulent mixing of the endwall boundary layer with the radial flow of the profile boundary layer produces additional losses. These are in addition to the secondary flow and the leakage flow that already exist in this region. (Chapter 4).

5. The overall cumulative effect of (1) through (4) is to make the flow highly three-dimensional in nature. The actual flow distribution in a turbomachinery blade row is significantly different from the distribution based on cascade or strip theory. Because the subsequent blade rows or stages are designed on the basis of the "design" flow distribution of the previous stage, the embedded blade row in a multistage compressor or turbine operates at flow incidences and velocities different from their design values.

6. The wall boundary layer inside the blade row (in the absence of tip clearance) is known to initiate wall stall at the intersection of the blade suction surface and the wall.

7. The importance of the annulus wall boundary layer and its effect on tip stall is generally recognized by turbomachinery designers. In addition, the boundary layer growth on the hub and annulus walls of multistage limits the stage work that can be achieved. Calculations based on two-dimensionality of the boundary layer underestimate the losses.

5.5.1.1   *Influence of Rotation and Curvature*   The rotation of the blade and hub introduces three-dimensionality and affects the stability and growth of viscous layers in turbomachinery. Let us consider a centrifugal compressor model as shown in Figs. 4.7 and 5.3. The Coriolis force tends to destabilize the boundary layer on the pressure side, thus increasing its growth on that side and stabilizing the boundary layer on the suction side. The effect of rotation appears through the Coriolis force in the momentum equations as well as implicitly through the turbulence constitutive equation, as explained earlier. Because the secondary flow and the Coriolis force effects have a major influence on the flow field in centrifugal turbomachinery, viscous effects dominate. Furthermore, long and narrow passages result in substantial boundary layer growth on the blade as well as on the endwalls (both the hub and the shroud). Thus, boundary layer approximations are not valid for such flows.

In axial flow turbomachinery, the Coriolis force is in the radial direction. This introduces radial outward flows. Thus, its effect in axial flow machinery is to introduce three-dimensionality in the viscous layers.

In a generalized curvilinear system, we can define nine curvature terms given by K_{ij} ($i = 1, 2, 3$; $j = 1, 2, 3$). In turbomachinery, all of these curvature terms are present. The dominant ones are as follows. These are given in the s, n, r system or in the m, θ, n system.

1. The radius of curvature due to blade surface/camber is given by R_c and is proportional to $\partial \beta / \partial s$. This is the principal curvature on the surface, when $n = $ constant in the direction of s.
2. The radius of curvature due to the sloping path or curvature of meridional streamline is R_m. This arises due to the annulus- and hub-wall flaring, or to radial flows. (Chapter 4).
3. The curvature due to annulus-or hub-wall radius is r_w.
4. Curvature due to swirling and rotating flows (r).

All of these curvature terms introduce centrifugal force terms in various momentum equations. The boundary layer growth is affected through additional terms that appear directly in momentum equations and have an indirect effect on turbulence as described earlier.

5.5.2 Boundary Layer Equations

In situations where the viscous layers have to be resolved accurately (including the velocity profile, shear stress and turbulence quantities), the boundary layer solutions have an advantage over Navier–Stokes solutions. Because the boundary layer equations are parabolic, a single-pass solution can be employed. Thus the computation time is minimal and a large number of grid points can be employed. Hence the boundary layer solutions, combined with a Euler/panel code, can resolve the viscous layer accurately, but coupling the two solutions to include interaction effects results in a complex coding effort. From the point of view of computational time, boundary layer solutions are very attractive for subsonic, attached, and shock-free flows.

The boundary layer equations are derived from the Navier–Stokes equation by assuming that the pressure gradient normal to the wall (or blade) is zero and the streamwise diffusion is negligible. This assumption is strictly not valid for three-dimensional boundary layers with rotation. The rotation term introduces additional forces normal to the walls, especially in centrifugal machinery. Nevertheless, for thin boundary layers, the pressure gradient normal to the wall can be assumed to be very small. The boundary layer equations for incompressible, three-dimensional flow are given by Eqs. 1.44–1.48. For turbomachinery flows (especially three-dimensional rotor/stator blade boundary layers) it is essential to include curvature and rotation terms and derive equations in a general curvilinear rotating system. Mager (1952) provided the most general equations.

The governing equation in the rotatating coordinate system for laminar flow is given by Eq. 4.27 or 4.28, where viscous terms can be represented by

$$\mathbf{F} = -\frac{\mu}{\rho}\left[\nabla \times (\nabla \times \mathbf{W}) - \frac{4}{3}\nabla(\nabla \cdot \mathbf{W})\right]$$

A complete derivation of three-dimensional boundary layer equations for incompressible, nonrotating flow is given by Nash and Patel (1972) and by Cebeci et al. (1977). Anand and Lakshminarayana (1975) extended these equations to include rotation; they also presented equations valid for turbomachinery flows, identifying explicitly the curvature and rotation terms. Lakshminarayana and Govindan (1981) extended the equations to compressible rotating flow; Thompkins and Usab (1982), Vatsa (1985), and Anderson (1987) extended these for a body-fitted coordinate system; and Karimipanath and Olsson (1993) have incorporated a transition model and studied the effect of rotation and compressibility.

Let s, n, r be the coordinate system shown in Fig. 5.8. By assuming that the normal coordinate (n) is straight (with $h_2 = 1$), we can write the arc length dS as (Nash and Patel, 1972)

$$(dS)^2 = (h_1\, ds)^2 + (dn)^2 + (h_3\, dr)^2$$

where h_1 and h_3 are locally varying functions of the position. The boundary layer equations in a orthogonal rotating coordinate system for compressible flow is given by (retaining only the dominant Reynolds stresses):

Continuity

$$\frac{\partial}{\partial s}(h_3\,\rho W_s) + \frac{\partial}{\partial n}(\rho W_n h_1 h_3) + \frac{\partial}{\partial r}(\rho h_1 W_r) = 0 \qquad (5.9)$$

s Momentum

$$\frac{\partial}{\partial s}\left(\rho h_3\left(W_s^2\right)\right) + \left(\frac{\partial}{\partial n} + \frac{1}{h_1}\frac{\partial h_1}{\partial n}\right)h_1 h_3\,\rho W_s W_n + \left(\frac{\partial}{\partial r} + \frac{1}{h_1}\frac{\partial h_1}{\partial r}\right)\rho h_1 W_s W_r$$

$$- \rho W_r^2 \frac{\partial h_3}{\partial s} + h_3\frac{\partial p}{\partial s} + 2(\rho\Omega_n W_r - \rho\Omega_r W_n) - h_3\,\rho\Omega^2 R\frac{dR}{ds}$$

$$= h_1 h_3 \frac{\partial}{\partial n}\left(\mu\frac{\partial W_s}{\partial n} - \overline{\rho w_s' w_n'}\right) \qquad (5.10)$$

r Momentum

$$
\left(\frac{\partial}{\partial s} + \frac{1}{h_3} \frac{\partial h_3}{\partial s} \right) h_3 \, \rho W_s W_r + \frac{\partial}{\partial r} \left(h_1 \, \rho (W_r^2) \right) + \left(\frac{\partial}{\partial n} + \frac{1}{h_3} \frac{\partial h_3}{\partial n} \right) h_1 h_3 \, \rho W_n W_r
$$

$$
- \rho W_s^2 \frac{\partial h_1}{\partial r} + h_1 \frac{\partial p}{\partial r} - 2\rho (\Omega_n W_s - \Omega_s W_n) - h_1 \, \rho \Omega^2 R \frac{\partial R}{\partial r}
$$

$$
= h_1 h_3 \frac{\partial}{\partial n} \left(\mu \frac{\partial W_r}{\partial n} - \overline{\rho w'_n w'_r} \right) \tag{5.11}
$$

Energy Equation

$$
\frac{\rho W_s}{h_1} \frac{\partial I}{\partial s} + \rho W_n \frac{\partial I}{\partial n} + \frac{\rho W_r}{h_3} \frac{\partial I}{\partial r} - \dot{Q}
$$

$$
= \frac{\partial}{\partial n} \left[\frac{\mu}{PR} \frac{\partial h_{oR}}{\partial n} + \mu \left(1 - \frac{1}{PR} \right) \left(W_s \frac{\partial W_s}{\partial n} + W_r \frac{\partial W_r}{\partial n} \right) - \overline{\rho h'_{oR} w'_n} \right] \tag{5.12}
$$

where $\Omega (= \mathbf{s}\Omega_s + \mathbf{n}\Omega_n + \mathbf{r}\Omega_r)$ is the angular velocity, R is the perpendicular distance from the axis of rotation, h_{oR} = stagnation enthalpy of the relative flow, and I = rothalpy $(h_{oR} - \Omega^2 R^2/2)$.

In the normal momentum equation, all terms are small as per the boundary layer assumption. The Coriolis acceleration term is $2(\Omega_r W_s - \Omega_s W_r)$ $(= \mathbf{n} \cdot 2(\Omega \times \mathbf{W}))$. This term is usually small for axial flow turbomachinery, but can be substantial for radial and centrifugal types of machinery. Therefore, as indicated earlier, the boundary layer approximations are not valid for the centrifugal type of machinery.

For a more generalized case, the values of scale factors, h_1, h_3, and their derivatives can be determined from the coordinate transformation and the Jacobian. In most turbomachinery cases, these coefficients can be expressed explicitly.

It is useful to examine these equations for several specific cases.

1. Cartesian, nonrotating, incompressible, laminar flow: $h_1 = h_2 = h_3 = 1$, $\Omega = 0$, $s = x$, $n = y$, $r = z$. These equations reduce to Eqs. 1.44–1.48, because the Bernoulli equations (Eqs. 1.47 and 1.48) can be used in the inviscid region.

2. Two-dimensional cascade flow $h_3 = 1$, $W_r = 0$, $\Omega = 0$. For a compressor cascade, the camber is usually small; therefore, $h_1 = 1$. For a turbine cascade, it is essential to retain the curvature terms due to the camber $h_1(s)$; $\partial h_1/\partial n$ is small due to the boundary layer approximation.

3. For a rotor blade with nearly radial blades, with small camber, it can be proved that

$$\frac{1}{h_1 h_3} \frac{\partial h_1}{\partial r} = \frac{\sin^2 \beta}{r}$$

where β is the blade camber angle.

The parameters that influence the growth of the boundary layer in turbomachinery are the streamwise pressure gradient, Mach number, Reynolds number, surface roughness, curvature, and rotation. It is very difficult to derive universal nondimensional parameters for rotation and curvature effects valid for all cases, because their influence on mean velocity and turbulence structure is dependent on the type of machinery and specific curvature terms. Specific cases are dealt with in appropriate sections in this book.

The Coriolis force is dominant in the blade-to-blade direction for centrifugal machinery and in the radial direction for axial machinery. The rotation affects the stability of the boundary layer in centrifugal machinery. In axial flow turbomachinery, the Coriolis force introduces radial flows and three-dimensionality and the boundary layer growth is thus indirectly affected. In both cases, the turbulence structure is altered due to rotation. The direct effect of rotation and curvature on the mean velocity is included in Eq. 5.1 and 5.9–5.12 through the rotation and curvature terms, whereas its influence on mean velocity through structural changes in turbulence will have to come through modeling of the turbulence terms in Eq. 1.88. It should be noted here that the Reynolds stress and algebraic Reynolds stress equations include rotation terms explicitly. Most of the influence of rotation on mean velocity via turbulence structure changes is included, if an RSM or ARSM is employed.

The influence of profile curvature is best characterized by δ/R_c, where δ is the boundary layer thickness and R_c is the radius of the blade camber. The curvature effect due to hub and annulus walls of radius r is similarly characterized by δ/r, where δ is the annulus or hub wall boundary layer thickness. The rotation effect is best characterized by the rotation number (R_o) given by

$$R_o = \frac{\text{Coriolis force}}{\text{Inertial force}} = \frac{\Omega L}{W}$$

where L is a characteristic length and W is a characteristic velocity. The rotation number is the inverse of the Rossby number used by meteorologists (Batchelor, 1967, p. 557). The rotation number is the ratio of the Coriolis force (ΩW) to the inertial force (W^2/L) and is a measure of the importance of rotation on the flow field. For centrifugal machinery, appropriate variables for L is boundary layer thickness or chord length, and the characteristic velocity is the total free-stream relative velocity. For axial turbomachinery, since the Coriolis force is in the radial direction, the appropriate choices would be $L = \delta$ and $W = (W_r)_{max}$ (maximum radial velocity). Lakshminarayana et al. (1972) proved that the most appropriate parameter characterizing the boundary layer growth for an axial flow turbomachinery blade is given by

$$\frac{2\Omega W_r}{W(\partial W/\partial s)} \sim \frac{2\Omega C \epsilon_w}{W} = R_o$$

where ϵ_w is the limiting streamline angle shown in Fig. 5.8.

5.5.3 Momentum Integral Equations for Blade Boundary Layers

In many design and analysis applications, the gross properties of the boundary layer (e.g., momentum thickness, displacement thickness, and skin friction stress) are of interest. For example, the displacement thickness provides an estimate for the blockage coefficient due to the blade boundary layers. This can then be incorporated into S_1 and S_2 surface equations (Section 4.2.3) to determine, approximately, the effects of boundary layer growth on the inviscid flow field through variables b and B. Similarly, the displacement thickness can be incorporated into Euler or inviscid codes to compute the global effects of boundary layers. Examples of such attempts are given by Thompkins and Usab (1982) and by Singh (1982). They computed the boundary layer displacement thickness and allowed for the flow blockage due to the blade thickness and the boundary layer displacement thickness to compute viscous effects on the pressure and velocity field in a blade row.

In addition, it is shown in Chapter 6 that the aerodynamic loss coefficient can be related to the momentum thickness for two-dimensional flow. Therefore, the calculation of momentum thickness (θ) and of displacement thickness (δ^*) becomes crucial in predicting performance and efficiency. Hence the momentum integral technique is attractive for preliminary design, and analysis of turbomachinery performance.

The derivation of momentum integral relations for two-dimensional flow can be found in any fluid mechanics textbook (e.g., Schlichting, 1979; White, 1991). The procedure is similar for a three-dimensional boundary layer. The momentum integral equations are obtained by integrating the boundary layer equations [s and r-momentum equations (Eqs. 5.10 and 5.11)] in the normal direction, and the normal velocity is eliminated using the continuity equation (Eq. 5.9). These equations can be simplified by suitable approximations for the streamwise (W_s) and cross (radial)-flow velocity (W_r) profiles. Anand and Lakshminarayana (1975, 1978) and Lakshminarayana and Govindan (1981) made the following assumptions for the streamwise and cross-flow velocity profiles (Fig. 5.8):

$$\frac{W_s}{W_{se}} = \left(\frac{n}{\delta}\right)^{(H-1)/2} \tag{5.13}$$

$$\frac{W_r}{W_s} = \epsilon_w\left(1 - \frac{n}{\delta}\right)^2 \tag{5.14}$$

where ϵ_w is the limiting streamline angle–that is, the angle of the streamline as the blade surface is approached, H is the shape factor (δ^*/θ) and n is the normal distance from the surface. The cross-flow profile is based on the model proposed by Mager (1952). Their analysis was for incompressible, three-dimensional, and turbulent flow, and the compressibility effects can be incorporated using Smith's (1982) analysis.

The momentum integral equations are written in terms of four parameters: θ_{ss}, ϵ_w, H, and C_{fs} (Lakshminarayana and Govindan, 1981).

$$\frac{1}{h_1}\frac{\partial\theta_{ss}}{\partial s} + \left(2 + H - M_e^2\right)\frac{1}{h_1}\theta_{ss}\frac{1}{W_{se}}\frac{\partial W_{se}}{\partial s}$$

$$+ \left\{\frac{\epsilon_w\theta_{ss}}{h_3}(2 - M_e^2)(J + 0.5L)\right\}\frac{1}{W_{se}}\frac{\partial W_{se}}{\partial r}$$

$$+ \frac{1}{h_3}\frac{\partial}{\partial r}(J\epsilon_w\theta_{ss}) + \frac{1}{h_1 h_3}\frac{\partial h_3}{\partial s}\left(\theta_{ss} - P\epsilon_w^2\theta_{ss}\right)$$

$$+ \frac{1}{h_1 h_3}\frac{\partial h_1}{\partial r}(J\epsilon_w\theta_{ss} + N\epsilon_w\theta_{ss}) - 2\frac{\Omega_n}{W_{se}}(L\epsilon_w\theta_{ss}) = \frac{C_{fs}}{2} \tag{5.15}$$

and

$$\frac{1}{h_1}\frac{\partial(N\epsilon_w\theta_{ss})}{\partial s} + \frac{1}{h_1}\frac{(2 - M_e^2)}{W_{se}}\frac{\partial W_{se}}{\partial s}(N\epsilon_w\theta_{ss})$$

$$+ \frac{1}{h_3}\frac{\partial}{\partial r}(P\epsilon_w^2\theta_{ss}) + \frac{(2 - M_e^2)}{W_{se}^2}\frac{1}{h_3}\frac{\partial W_{se}}{\partial r}(P\epsilon_w^2\theta_{ss})$$

$$+ \frac{1}{h_1 h_3}\frac{\partial h_3}{\partial s}(N\epsilon_w\theta_{ss} + J\epsilon_w\theta_{ss} + L\epsilon_w\theta_{ss})$$

$$+ \frac{1}{h_1 h_3}\frac{\partial h_1}{\partial r}\left(P\epsilon_w^2\theta_{ss} - \theta_{ss} - \theta_{ss}H\right) + 2\frac{\Omega_n}{W_{se}}H\theta_{ss} = \frac{\epsilon_w C_{fs}}{2} \tag{5.16}$$

where M_e is the freestream Mach number and,

$$\theta_{ss} = \int_0^{n > \delta} \frac{\rho}{\rho_e} \frac{W_s}{W_{se}} \left(1 - \frac{W_s}{W_{se}} \right) dn, \qquad \delta_s^* = \int_0^{n > \delta} \left(1 - \frac{\rho}{\rho_e} \frac{W_s}{W_{se}} \right) dn,$$

$$H = \frac{\delta_s^*}{\theta_{ss}}, \qquad C_{fs} = 2\tau_{ws}/\rho W_{se}^2$$

$$J = \frac{(30.0 + 14H)}{(H + 2)(H + 3)(H + 5)}$$

$$L = \frac{16H}{(H - 1)(H + 3)(H + 5)}$$

$$P = \frac{-24}{(H - 1)(H + 2)(H + 3)(H + 1)}$$

$$N = -\frac{2}{(H - 1)(H + 2)}$$

These two equations contain four unknowns (θ_{ss}, ϵ_w, H, and C_{fs}); therefore two additional closure equations are necessary to obtain a solution for θ_{ss} and ϵ_w. The closure for skin friction coefficient (C_{fs}) is usually based on empiricism. The second closure equations is either the entrainment equation or the energy integral equation described later. The most widely used skin friction relationship for two-dimensional flows is due to Ludwieg and Tillman (1949). This was modified by Anand and Lakshminarayana (1975) for rotation effects. The expression for the skin friction coefficient is given by (Anand and Lakshminarayana, 1975)

$$C_{fs} = 0.172 R_{\theta_{ss}}^{-0.268} 10^{-0.678 H} \left(1 + B_1\sqrt{\epsilon_w(s - s_t)/C} \right) \qquad (5.17)$$

where B_1 is a constant (a value of 0.52 was suggested by Anand and Lakshminarayana); $R_{\theta_{ss}}$ is the Reynolds number based on the streamwise velocity at the edge of the boundary layer and the streamwise momentum thickness θ_{ss}; and s_t is the distance between the leading edge and the transition point along the s direction.

5.5.3.1 Closure Model: Entrainment Equation

In turbomachinery blade passages, the flow experiences large pressure gradients. These pressure gradients cause substantial changes in velocity profiles from that encountered with zero pressure gradient and, consequently, the shape parameter H. The variation of H cannot be neglected, and an additional equation is required. Out of the available auxiliary equations, only the "energy integral equation" and the "entrainment equation" have been found to be suitable for the turbulent boundary layers. The former involves more empiricism in calculating the dissipation integral term, and it assumes the shear stress distribution in the turbulent boundary layer. Because the shear stress variation differs largely from flow to flow, especially those under the influence of rotation and curvature, the use of the entrainment equation rather than an energy integral equation is suggested. The entrainment equation is derived from the concept that a turbulent boundary layer grows by a process of "entrainment" of the inviscid fluid at the edge of the boundary layer into the turbulent region, and it has been found to give good predictions in a variety of flows (Head, 1958). The entrainment equation by Head (1958) has been extended by Nash and Patel (1972) for three-dimensional flow. The entrainment equation is obtained from the integration of the continuity equation from the blade surface $(n = 0)$ to the edge of the boundary layer $(n = \delta)$. The continuity equation is the same in both the rotating and stationary coordinate systems (s, n, r).

The entrainment equation for the rotor boundary layer in the coordinate system used in this analysis (Fig. 5.8) can be shown to be (Lakshminarayana and Govindan, 1981)

$$
\frac{1}{h_1} \frac{\partial}{\partial s}\left(\theta_{ss} H_{\delta-\delta_s^*}\right) + \theta_{ss} H_{\delta-\delta_s^*} \left\{ \frac{1}{h_1} \frac{1}{W_{se}} \frac{\partial W_{se}}{\partial s} + \frac{1}{h_1 h_3} \frac{\partial h_3}{\partial s} \right\}
$$

$$
- \frac{1}{h_3} \frac{\partial}{\partial r}\left(L\epsilon_w \theta_{ss}\right) - L\epsilon_w \theta_{ss}\left\{ \frac{1}{h_3} \frac{1}{W_{se}} \frac{\partial W_{se}}{\partial r} + \frac{1}{h_1 h_3} \frac{\partial h_1}{\partial r} \right\}
$$

$$
= F\left(H_{\delta-\delta_s^*}\right) \tag{5.18}
$$

where $H_{\delta-\delta_s^*} = (\delta - \delta_s^*)/\theta_{ss}$. The function F on the right-hand side of Eq. 5.18 represents the volume flow rate per unit area through the surface $\delta_s^*(s, r)$, and is the rate of entrainment of inviscid external flow into the boundary layer.

The entrainment process is a highly complex and inherently unsteady phenomena, and its direct measurement is difficult. The entrainment rate depends on the mean flow parameters (such as the streamwise velocity defect, the rate of boundary-layer growth, free-stream velocity, displacement thickness, etc.) as well as on the turbulent qualities (such as turbulence intensities and stresses in the outer layer). The streamwise velocity defect, in turn, is directly related to the free-stream velocity, shape factor, magnitude of cross flow, and blockage effects due to confinement of external inviscid flow in the channel. An empirical correlation for the entrainment function F for three-dimensional flow is not yet available. Therefore, the entrainment function F due to Head (1958) for two-dimensional flow is used. The function F is given by

$$
F\left(H_{\delta-\delta_s^*}\right) = 0.0306\left(H_{\delta-\delta_s^*} - 3.0\right)^{-0.653} \tag{5.19}
$$

It is assumed that the variation of the entrainment rate with $H_{\delta - \delta_s^*}$ follows the same relationship for three-dimensional flows.

Equations 5.15–5.19 are to be solved for θ_{ss}, ϵ_w, and $H(H_{\delta - \delta_s^*}$ is related to H) with the prescribed boundary conditions. Initial conditions are pre-scribed for θ_{ss}, ϵ_w, and H at the leading edge, θ_{ss} and ϵ_w are assumed to be zero at the leading edge, and an initial value of 1.4 is assumed for H. The boundary condition prescribed at the endwalls is that ϵ_w is zero there.

For two-dimensional cascades with camber $(h_1 = h_2 = h_3 = 1$, $W_r = 0$, $\epsilon_w = 0)$, Eqs. 5.15 and 5.18 reduce to

$$\frac{\partial \theta_{ss}}{\partial s} + (2 + H - M_e^2)\theta_{ss}\frac{1}{W_{se}}\frac{\partial W_{se}}{\partial s} = \frac{C_{fs}}{2} \tag{5.20}$$

and

$$\frac{\partial}{\partial s}(\theta_{ss}H_{\delta - \delta_s^*}) + (\theta_{ss}H_{\delta - \delta_s^*})\frac{1}{W_{se}}\frac{\partial W_{se}}{\partial s} = F(H_{\delta - \delta_s^*}) \tag{5.21}$$

The three-dimensional set of equations (Eqs. 5.15–5.18) and the two-dimensional set of equations (Eqs. 5.20, 5.21) are parabolic equations and can be solved by the techniques described in Appendix B. Lakshminarayana and Govindan (1981) solved these equations using the following steps:

1. The inviscid blade pressure distribution or inviscid velocity distribution is determined using the computer code and technique described in Section 4.2.5.1. (Katsanis, 1968; Katsanis and McNally, 1977)
2. A body-fitted boundary layer coordinate is generated using techniques described in Appendix B.
3. An initial value of the shape factor is assumed at the streamwise station at which the calculation is carried out.
4. Equations 5.15 and 5.16 are solved using the Runge–Kutta method described in Appendix B. A centered difference approximation was used in the radial direction. The method is explicit, because the radial derivatives are lagged in the calculation. Such an approach is similar to the "method of lines" approach used for partial differential equations, (PDE). The four-step Runge–Kutta method allows for the radial derivative to be corrected at each half-step of the calculation.

This technique and the computer code has been successfully used to predict the two-dimensional boundary layer in compressor and turbine cas-cades and three-dimensional boundary layers on compressor and rocket pump inducer rotor blades.

The measurement by Bammert and Sandstede (1980) in a turbine cascade with rough blade and large turning, $\lambda = 56.50°$, $S = 11.98$ cm, $C_L = 4.83$, $C = 17.50$ cm, $R_e = 5.6 \times 10^5$, $W_1 = 20$ m/s, $W_2 = 50$ m/s, and inlet turbulence intensity of 2% are compared with predictions in Fig. 5.9. Excellent agreement is obtained between the data and the prediction for the momentum thickness. Similar good agreement is obtained for the momentum thickness and the limiting streamline angle for a rocket pump inducer and a compressor rotor. The data and the predictions are compared for the Penn State compressor in Fig. 5.10 (Pouagare et al., 1985; Lakshminarayana and Popovski, 1987). The rotor is 0.932-m in diameter and $r_h/r_t = 0.5$, with 21 blades. The midradius values are: $C = 14$ cm, $S = 10.5$ cm, $C_L = 0.6$, $Re = 3.0 \times 10^5$, $N = 1060$ rpm $\lambda = 30°$. See Section 2.5.5 and Fig. 2.17 for additional details on this compressor. Excellent agreement between the data and analysis is obtained on both surfaces of the blade, with the exception of the trailing edge location at the blade tip. This indicates the usefulness of such techniques in evaluating the overall performance of turbomachinery, especially profile losses and blockage effect. The Profile losses predicted from the analysis agree well with the experimental data (Pouagare et al., 1985; Lakshminarayana and Popovski, 1987). Because this analysis does not include interaction between the wall and the blade boundary layer near the hub and annulus, this region is best handled by the full Navier–Stokes equation.

Hence an inviscid computation technique (Chapters 3 and 4) combined with the momentum integral technique can resolve most of the global flow features (pressure rise/drop, blockage effect, and losses) away from the endwalls and is adequate for subsonic axial rotors with thin blade boundary layers. As indicated earlier, centrifugal compressor/pumps are dominated by viscous and secondary flows, and this technique is not useful for such flows.

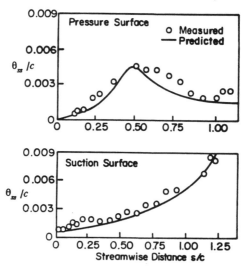

Figure 5.9 Chordwise distribution of momentum thickness for a rough turbine cascade blade. *s* is the streamwise distance from the leading edge. (Lakshminarayana and Govindan, 1981; reprinted with permission of AIAA.)

The shock–boundary-layer interaction in high-speed rotors results in flow separation and large boundary layer growth. The momentum integral analysis cannot resolve this effect. Thus, the utility of momentum integral technique is limited to unseparated, uncooled blade rows, with relatively thin boundary layers (say about 10–15% of blade spacing).

5.5.4 Annulus Wall Boundary Layers

The importance of the annulus wall boundary layer and its effect on the overall performance (e.g., stage characteristics, efficiency, stall, and surge) of multistage turbomachinery was recognized as early as 1942 (Howell, 1942). There is a lack of theoretical investigation, which is hampered by the complexity of the flow, some aspects of which are listed below:

1. The flow is highly three-dimensional.
2. The presence of the tip clearance flow, and the secondary flow caused by the inlet shear or density gradients in the radial direction results in appreciable radial and transverse flows.

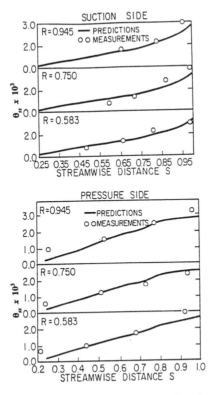

Figure 5.10 Comparison between the measured and predicted momentum thickness on Penn State Axial Flow compressor rotor blade. θ_{ss} is normalized by chord length. (Copyright © Pouagare et al., 1985; reprinted with permission of ASME.)

3. The presence of pressure gradients in all directions. The blade forces induce asymmetry to the flow. These, in addition to blade wakes, make the flow inside the boundary layer highly three-dimensional, with gradients in velocity in all coordinate directions.

4. The flow is unsteady. Because of the relative motion between the rotor and the stator and the disturbed rotational symmetry due to wakes, the flow to the successive blade row is unsteady.

5. Sudden perturbation is caused by rotating blade row, if the tip clearance is smaller than the annulus-wall boundary layer thickness. None of the investigations include this scraping effect.

6. Interaction between the blade boundary layers (which are three-dimensional in nature) and the annulus-wall boundary layer results in a complex flow field.

7. The flow is subjected to centrifugal and Coriolis forces due to wall curvature and rotation, respectively, and this results in complex flow.

The effects of annulus-wall boundary layer growth on performance are many:

1. Approximately one-third to one-half of the total losses in single-stage and multistage turbomachinery are due to endwall boundary layers (EWBLs), including tip clearance and secondary flow effects.

2. The EWBL introduces blockage to the flow, amounting to as much as 5–20%, depending on the type of blade row, blade height, and so on.

3. In a multistage compressor the EWBL has a direct influence on stall and surge phenomena. The stall inception usually occurs near the tip and propagates downward.

4. EWBL introduces off-design conditions near the tip. For example, the reduction of axial velocity in the EWBL results in larger incidence to the succeeding blade row.

5. The EWBL introduces unsteadiness in the next blade, resulting in vibrations and noise.

There have been several attempts made to calculate the growth of the annulus wall boundary layer. All of these methods are based on integral techniques. The equations of motion are averaged over the blade passage first, and the averaged equations are integrated over the EWBL thickness. The essential difference in various methods lies in the assumptions made for the second order terms and in the velocity profile assumed for the tangential component. The analysis given here is for incompressible flow. This can be easily modified to include the compressibility effect (see Eqs. 5.15 and 5.16). All the analyses neglect the effects of curvature. The boundary layer equa-

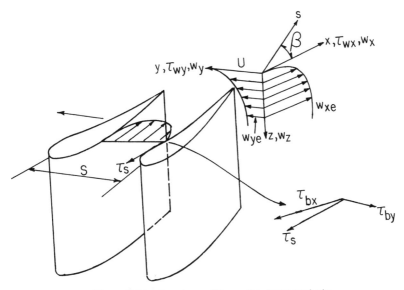

Figure 5.11   Annulus wall boundary layer analysis.

tions can be written as (see Fig. 5.11 for notations; curvature terms are neglected)

$$\frac{\partial W_x}{\partial x} + \frac{\partial W_y}{\partial y} + \frac{\partial W_z}{\partial z} = 0 \tag{5.22}$$

$$\frac{\partial}{\partial x}\left(W_x^2\right) + \frac{\partial}{\partial y}\left(W_x W_y\right) + \frac{\partial}{\partial z}\left(W_x W_z\right) = -\frac{1}{\rho}\frac{\partial p}{\partial x} + \frac{1}{\rho}\left(\frac{\partial \tau_{xx}}{\partial x} + \frac{\partial \tau_{yx}}{\partial y} + \frac{\partial \tau_{zx}}{\partial z}\right) \tag{5.23}$$

$$\frac{\partial}{\partial x}\left(W_y W_x\right) + \frac{\partial}{\partial y}\left(W_y^2\right) + \frac{\partial}{\partial z}\left(W_y W_z\right) = -\frac{1}{\rho}\frac{\partial p}{\partial y} + \frac{1}{\rho}\left(\frac{\partial \tau_{xy}}{\partial x} + \frac{\partial \tau_{yy}}{\partial y} + \frac{\partial \tau_{zy}}{\partial z}\right) \tag{5.24}$$

$$\frac{\partial}{\partial x}\left(W_z W_x\right) + \frac{\partial}{\partial y}\left(W_z W_y\right) + \frac{\partial}{\partial z}\left(W_z^2\right) = -\frac{1}{\rho}\frac{\partial p}{\partial z} + \frac{1}{\rho}\left(\frac{\partial \tau_{xz}}{\partial x} + \frac{\partial \tau_{yz}}{\partial y} + \frac{\partial \tau_{zz}}{\partial z}\right) \tag{5.25}$$

Mellor and Wood (1971) averaged the above equations in y (blade-to-blade) direction (similar to the procedure outlined in Section 4.2.3), invoked the usual boundary layer approximations, and then integrated Eqs. 5.23 and 5.24 across the annulus wall boundary layer (z direction) to derive the following momentum integral equations (see Fig. 5.11 for notations):

$$\frac{d}{dx}\left(\overline{W}_{xe}^2 \theta_x\right) + H\theta_x \overline{W}_{xe}\frac{d\overline{W}_{xe}}{dx} = \frac{\overline{W}_e^2}{2}\frac{dF_x}{dx} + \frac{\overline{\tau}_{wx}}{\rho} \tag{5.26}$$

$$\frac{d}{dx}\left(\overline{W}_{xe}\overline{W}_{ye} \theta_y\right) + H\theta_x \overline{W}_{xe}\frac{d\overline{W}_{ye}}{dx} = \frac{\overline{W}_e^2}{2}\frac{dF_y}{dx} + \frac{\overline{\tau}_{wy}}{\rho} \tag{5.27}$$

These equations can also be derived directly from generalized momentum integral equations (Eqs. 5.15 and 5.16). x and y are axial and tangential directions, whose velocity components are W_x and W_y, respectively, and $W_e = \sqrt{W_{xe}^2 + W_{ye}^2}$. In Eqs. 5.26 and 5.27, the velocities are passage-averaged values (Eq. 4.54). θ_x and θ_y are momentum thicknesses in axial (x) and tangential (y) directions, respectively, based on passage-averaged velocities. These are similar to the definition of θ_{ss} introduced earlier. τ_{wx} is the passage-averaged shear stress at the wall in x direction. The appearance of forces F_x and F_y is due to the blade-to-blade averaging procedure adopted by Mellor and Wood. They are defined as

$$\frac{\overline{W}_e^2 F_x}{2} = \int_0^\delta \frac{(f_e - f)_x}{\rho}\, dz \tag{5.28}$$

$$\frac{\overline{W}_e^2 F_y}{2} = \int_0^\delta \frac{(f_e - f)_y}{\rho}\, dz \tag{5.29}$$

F_x and F_y are called "Force defect thickness" arising from the wall boundary layers. f_x and f_y are blade forces in the boundary layer in x and y directions, respectively, defined by

$$f_x = \int_1^2 \left[\frac{\Delta p}{S}\tan\beta - \frac{2\tau_{bx}}{S} + \frac{\partial}{\partial x}\left(-\rho\overline{(W_x'')^2} + \overline{\tau}_{xx}\right)\right]dx$$

$$f_y = \int_1^2 \left[-\frac{\Delta p}{S} + \frac{2\tau_{bx}}{S}\tan\beta + \frac{\partial}{\partial x}\left(-\rho\overline{W_x''W_y''} + \overline{\tau}_{xy}\right)\right]dx \tag{5.30}$$

The first term inside the integration in the above equations represents the component of local blade force (Δp) in the x and y directions, (similar to Eq. 4.70), the second term represents the component of blade shear stresses (due to blade boundary layers), and the last term has two components $\rho\overline{(W_x'')^2}$ and $\rho\overline{(W_x''W_y'')}$ arising from averaging of the equation in blade-to-blade direction, W_x'' being the deviation from the passage averaged values ($W_x = \overline{W}_x + W_x''$). This averaging is similar to those illustrated in Fig. 4.10 ($A = W_x$ or W_y) and is represented by Eq. 4.53 and G_i in Eq. 4.74. $\overline{\tau}_{xx}$ represents the turbulent normal stress in the axial direction. The integration in Eq. 5.30 is carried out inlet to the exit or blade row. Four additional relationships are required for the complete solution of Eqs. 5.26 and 5.27, which have six unknowns: θ_x, θ_y, $\bar{\tau}_{wx}$, $\bar{\tau}_{wy}$, F_x and F_y.

The forces and velocities are related by

$$\frac{dF_x}{dx}\overline{W}_{xe} + \frac{dF_y}{dx}\overline{W}_{ye} = 0 \tag{5.31}$$

The derivation of this equation is based on the assumption that the blade force angle does not vary significantly through the boundary layer.

The assumptions for shear stress are

$$\frac{\bar{\tau}_{wx}}{\rho} = \overline{W}_{xe}\sqrt{\overline{W}_{xe}^2 + \left(\overline{W}_{ye} - \Omega r\right)^2}\,\frac{C_f}{2} \tag{5.32}$$

$$\frac{\bar{\tau}_{wy}}{\rho} = \left(\overline{W}_{ye} - \Omega r\right)\sqrt{\overline{W}_{xe}^2 + \left(\overline{W}_{ye} - \Omega r\right)^2}\,\frac{C_f}{2} \tag{5.33}$$

where Ωr = blade speed at the tip. These expressions allow for the relative motion between the blade and the annulus wall. C_f is the wall shear stress coefficient.

The fourth additional equation needed is derived from the secondary flow and tip clearance flow models. A collateral boundary layer (in which the flow direction is constant through the layer) is assumed in the absence of tip clearance. The spanwise flow due to secondary (Section 4.2.6) and leakage flow (Section 4.2.7) is assumed to vary linearly from pressure to suction surface, and the flow in the tangential direction is assumed to vary parabolically. With these assumptions, Mellor and Wood derive an equation for the blade exit conditions given (denoted by subscript 2) by

$$\overline{W}_{ye_2}\left(\theta_{y_2} - \theta_{x_2}\right) = K\delta_c\left[\frac{S}{C}\overline{W}_e\left(W_{ye_2} - W_{ye_1}\right)\right]^{0.5}\,\text{sgn}\left(W_{ye_2} - W_{ye_1}\right) \tag{5.34}$$

where δ_c is the blade tip clearance and K is a leakage coefficient.

This completes the theoretical formulation. Six constitutive equations (Eqs. 5.26, 5.27, and 5.30–5.34) can be solved for six unknowns. The empirical parameters involved in this solution are C_f and K. The boundary layer growth should be strongly dependent on C_f and K. Mellor and Wood's calculation provides an estimate of not only the stagnation pressure rise, but also the efficiency from the expression

$$\eta = \frac{\displaystyle\int_{r_h + \delta_h^*}^{r_t - \delta_t^*} r(\Omega r)f_y\,dr}{\displaystyle\int_{r_h}^{r_t} r(\Omega r)f_y\,dr} \tag{5.35}$$

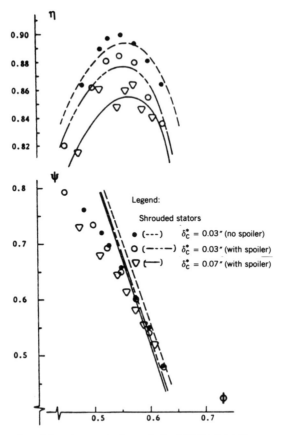

Figure 5.12 Experimental and numerical results for MGCR Axial Flow compressor. Re = 3 × 10⁵, M = 0.21, H = 1.4, C_ft = 0.011, C_th = 0.003, K = 0.55 (δ_c* is the clearance height). (Copyright © Mellor and Balsa, 1972; The original version of this material was first published by the AGARD / NATO in AG 164, "Boundary Layer Effects in Turbomachines.")

where δ_h^* and δ_t^* are displacement thicknesses of hub- and annulus-wall boundary layers, respectively. This includes only the losses due to annulus- and hub-wall boundary layers. Mellor and Balsa's (1972) calculation of η and ψ for a Westinghouse low-speed four-stage compressor (tip speed = 300 ft/s) is shown in Fig. 5.12. The agreement between the theory and the experiment for assumed values of K and C_f seems to be reasonable. One of the important features of this theory is that the stall limit can be predicted.

Some of the drawbacks of Mellor and Wood's analysis are the arbitrary assumptions for C_f and K and the assumption of collateral boundary layer (i.e., β does not vary inside the EWBL). The analysis can be easily modified to include (1) a more realistic assumption for C_f (e.g., Eq. 5.17) and (2) better representation of streamwise (Eq. 5.13) and cross-flow velocity profiles

(Eq. 5.14). Equations 5.13, 5.14, and 5.17 are valid for flows without endwall separation and tip clearance flow and should provide a reasonable estimate of EWBL growth on a stator or a rotor hub. It is not valid when horseshoe vortices are present (e.g., turbine endwall flows). Numerous investigators have incorporated these features to improve EWBL predictions (e.g., Hirsch, 1974, 1976; DeRuyck et al., 1979; DeRuyck and Hirsch, 1981; Horlock and Perkins, 1974; Adkins and Smith, 1982). A review of these techniques is covered in Leboeuf (1984). The comparison between predictions and data, shown in Fig. 4.31, is a good example of combining secondary flow theory, leakage flow analysis, and annulus wall boundary layer calculation in predicting the complex interaction effects in the endwall regions of turbomachinery.

Hirsch and his group (Hirsch, 1976; DeRuyck et al., 1979; DeRuyck and Hisch, 1981) have modified Mellor and Wood's analysis to include cross-flows within the boundary layers and the variation of edge velocity W_{se} across the passage due to blade boundary layers. Their modifications are listed below:

1. The momentum integral equations are written in the s, n, r system similar to Eqs. 5.15 and 5.16, neglecting all curvature terms.
2. A power law profile is used to represent variation of W_s within the EWBL. Mager's (1952) profile (Eq. 5.14) is used to represent cross flows (W_n) within the EWBL.
3. A skin friction correlation (similar to Eq. 5.17 without the rotation effect) is used to model C_f.
4. The density is allowed to vary across the EWBL using an empirical equation.
5. The shape factor is modeled using Head's entrainment function (Eq. 5.18).

These modifications result in a better and more realistic approach to modeling the EWBL. It should be recognized here that the boundary layers in the endwall region, resulting from the interaction of the tip clearance flow, secondary flow, and horseshoe vortex, are extremely complex and that the assumptions for velocity profiles are not representative of those that exist. Nevertheless, this approach should provide a reasonable estimate of momentum and displacement thicknesses.

Hirsch and his group have computed a large number of cases. The most successful one involves boundary layers in a stationary frame (e.g., endwall flow in cascades and stators). They also tried to compute the EWBL at the exit of a rotor measured by Hunter and Cumpsty (1982), with partial success. The assumed and measured velocity profiles deviate considerably, but the momentum thickness growth is predicted reasonably well. However, the use of arbitrary constants makes the method less universal.

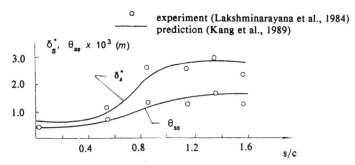

Figure 5.13 End Wall Boundary layer growth through Inlet guide vanes. (*s* is the distance from leading edge) (Copyright © Kang et al., 1989; reprinted with permission of ASME.)

Kang et al. (1989) provided a systematic approach through the use of S_2 surface equations and modified them (Section 4.2.3.2) to allow for warping of the surfaces near endwalls. Their momentum integral equations are based on the Mellor and Wood approach, but allow for the shape factor variation through the use of Head's entrainment equation (Eq. 5.18). They computed the EWBL of an inlet guide vane of a single-stage compressor (Lakshminarayana and Sitaram, 1984). The agreement between the data and predictions, shown in Fig. 5.13, is excellent.

It should be emphasized here that the closure equations employed for velocity profiles and shape factor are based on mildly three-dimensional boundary layer data, whereas the practical EWBL is unconventional. Additional complications are the modifications due to tip clearance and wall motion. The data from Lakshminarayana et al. (1983) clearly reveal the inadequacy of these models in the annulus wall region of rotors. Therefore, the applicability of the analyses presented in this section is restricted to EWBL in the absence of tip clearance, horseshoe vortex, and flow separation, rotation, and curvature. If these effects are present, it is better to adopt the numerical solution of the full Navier–Stokes equation.

5.5.5   Numerical Solution of Boundary Layer Equations

The boundary layer equations for incompressible, laminar, three-dimensional flow are given by Eqs. 1.44–1.48; the equations for compressible turbulent flow in generalized coordinates are given by Eqs. 5.9–5.12. In many turbomachines, the boundary layers are thin and it is not necessary to solve the full Navier–Stokes equation in all regions. The boundary layer techniques, with prescribed pressure or edge velocity distribution from inviscid codes, are about two orders of magnitude more efficient (computationally) than the corresponding full Navier–Stokes codes. In view of this, the viscous layers can be resolved quite accurately using large numbers of grid points within a boundary layer. These techniques are especially attractive in design systems.

The two-dimensional form of boundary layer equations is parabolic. Therefore, the numerical techniques and computer codes for these flows have reached a levels of sophistication leading to their routine use in design and analysis systems. The character of the equations changes from parabolic to parabolic–hyperbolic in three-dimensional flows, and the technique should be able to capture the reversal of radial flows. It is only recently that attempts have been made to predict the three-dimensional boundary layer on turbomachinery configurations.

5.5.5.1 *Numerical Solution for Two-Dimensional Boundary Layers*
There are many excellent textbooks and reviews dealing with the numerical solution of two-dimensional boundary layers (e.g., Bradshaw et al., 1981). Even though a large class of external flow problems have been solved, the application of these numerical solutions to turbomachinery flows has not found widespread use. The boundary layer equations are parabolic or hyperbolic. Some of the widely used schemes for the numerical solution of the boundary layer equations are the Crank–Nicholson scheme, the Keller box scheme, the shifted box scheme, the double-shifted box scheme, the zig-zag difference scheme, the characteristic difference scheme, and the two-step method covered in Appendix B. (For more details, see Smith, 1982; Keller, 1978; Wang, 1971; Cebeci et al., 1977; and Bradshaw et al., 1981.) One of the most widely used codes in turbomachinery for the computation of two-dimensional boundary layers (with heat transfer) is due to Crawford and Kays (1976).

Before we proceed to describe the numerical procedure, it is necessary to develop equations suitable for numerical solution. The boundary layer equations are always expressed in nondimensional form using a length scale (usually chord length of the blade) and reference velocity (usually the upstream velocity). In addition, it is essential to transform the equations to overcome the following undesirable features from a computational viewpoint: (1) The growth of the boundary layer necessitates the growth of the computational domain; (2) the leading edge is singular and $\delta = 0$ at $\xi = 0$; (3) rapid variation in properties occur near the walls, and it is essential to use stretched boundary layer coordinates.

Two of the successful techniques used to overcome these problems are as follows: one of the transformations consists of scaling the y (normal) coordinate by $\sqrt{Re}$ proportional to boundary layer growth in the laminar flow, the velocity is scaled by the edge velocity. A second and more widely used transformation (for compressible flow) is due to Levy and Lees (1954) and has been modified by Carter et al. (1980). This technique produces a computational domain in which the scaled boundary layer thickness is essentially constant, and it provides a simple procedure for specifying the grid point distribution across the boundary layer that is "adapted" to the velocity gradients. In the computational domain, the streamwise (ξ) and normal (η) coordinates are straight lines.

The transformation of the coordinates (which includes Blasius scaling) for a two-dimensional boundary layer is given by

$$\xi(x) = \int \rho_e W_e \mu_e \, dx \qquad (5.36)$$

$$\eta(x, y) = \frac{\rho_e W_e}{\sqrt{2\xi}} \int_0^y \frac{\rho}{\rho_e} \, dy \qquad (5.37)$$

where $\mu = \mu_l + \mu_t$, subscript e stands for conditions at the edge of the boundary layer.

If μ_t, the turbulent eddy viscosity, is chosen properly, this procedure should work for any turbulence model. By defining stream function ($\psi_y = \rho W_x$, $\psi_x = -\rho W_y$) and introducing a Blausius type of transformation, we obtain

$$\psi(x, y) = \sqrt{2\xi} F(\xi, \eta)$$

The two-dimensional form of boundary layer equation (compressible form of Eqs. 1.44–1.48) can be manipulated to give

$$\frac{\partial}{\partial \eta} \left(\frac{\rho \mu}{\rho_e \mu_e} \frac{\partial^2 F}{\partial \eta^2} \right) + (1 + \beta) F \frac{\partial^2 F}{\partial \eta^2} + \beta \left[\frac{\rho_e}{\rho} - \left(\frac{\partial F}{\partial \eta} \right)^2 \right]$$

$$= 2\xi \left(\frac{\partial F}{\partial \eta} \frac{\partial^2 F}{\partial \xi \partial \eta} - \frac{\partial F}{\partial \xi} \frac{\partial^2 F}{\partial \eta^2} \right) \qquad (5.38)$$

The value of μ_e is chosen such that the ratio $\rho \mu / \rho_e \mu_e$ is unity at the outer edge. The above equation contains the continuity equation. β is the pressure gradient parameter given by

$$\beta = \frac{2\xi}{W_e} \frac{\partial W_e}{\partial \xi} \qquad (5.39)$$

An equation similar to Eq. 5.38 can be derived for the energy equation (see White, 1991; Schlichting, 1979). A more generalized form of Eq. 5.38, valid for Three-dimensional compressible flow with heat transfer, applicable to turbomachinery, is given in Vatsa (1985).

The set of equations (Eqs. 5.36–5.39) can be solved by any of the numerical techniques described in Appendix B. The Keller Box (KB) scheme is also described in Appendix B. The shifted box scheme and the double shifted box scheme are variations of the KB scheme (Drela, 1983). The KB scheme (Keller, 1978) is widely used. To apply the KB scheme, Eq. 5.38 is written as a first-order system, as follows:

$$\frac{\partial F}{\partial \eta} = T, \qquad \frac{\partial T}{\partial \eta} = K \tag{5.40}$$

$$\frac{\partial}{\partial \eta}(bK) = 2\xi\left[T\frac{\partial T}{\partial \xi} - K\frac{\partial F}{\partial \xi}\right] - (\beta + 1)FK - \beta(\rho_e/\rho - T^2) \tag{5.41}$$

where $b = \rho\mu/\rho_e\,\mu_e$. The boundary conditions are

$$F(\xi,0) = 0, \qquad T(\xi,0) = 0, \qquad T(\xi,\eta_e(\xi)) = 1$$

Based on KB scheme illustrated in Fig. B.3 and Section B.2.2, let us use a grid (ξ_i, η_j) with spacing (k_i, h_j). The finite difference representation of Eq. 5.38 is given by

$$\left(\frac{\partial F}{\partial \eta}\right)_{i,j-1/2} = T_{i,j-1/2}$$

$$\left(\frac{\partial T}{\partial \eta}\right)_{i,j-1/2} = (K)_{i,j-1/2}$$

$$\left[\frac{\partial(bK)}{\partial \eta}\right]_{i-1/2,\,j-1/2} \tag{5.42}$$

$$= 2\xi_{i-1/2}\left[T_{i-1/2,j-1/2}\left(\frac{\partial T}{\partial \xi}\right)_{i-1/2,j-1/2} - K_{i-1/2,j-1/2}\left(\frac{\partial F}{\partial \xi}\right)_{i-1/2,j-1/2}\right]$$

$$-\left[(\beta + 1)FK + \beta(\rho_e/\rho - T^2)\right]_{i-1/2,j-1/2}$$

Newton's method can be used to solve FD equation (Eq. 5.42) using a block tridiagonal solver described in Appendix B.

5.5.5.2 Numerical Solution of Three-Dimensional Boundary Layer Equations
As mentioned earlier, the three-dimensional boundary layer equations present some unique problems, because as they are parabolic–hyperbolic in nature. A survey of techniques for three-dimensional boundary layers can be found in Smith (1982), Cousteix (1986), Hirschel and Kordulla (1981), and Murthy and Brebbia (1990). Let us examine the mathematical characteristics of these equations before presenting a solution methodology. Consider the incompressible, laminar form of Eqs. 5.9–5.12 in the Cartesian system (with $\Omega = 0$, $s \to x$, $n \to y$, $r \to z$). The characteristics determining these equations are given by (Wang, 1971; and Cousteix, 1986)

$$\lambda_y(W_x\lambda_x + W_y\lambda_y + W_z\lambda_z)^2 = 0$$

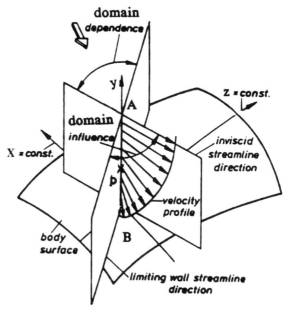

Figure 5.14 Regions of dependence and influence of a velocity profile within the boundary layer. (Copyright © Hirschel and Kordulla, 1981. "Notes on Fluid Mechanics")

where λ_i are components of the vector normal to the characteristics surfaces. Here, $\lambda_y = 0$ is related to the continuity equation and $W_x \lambda_x + W_y \lambda_y + W_z \lambda_z = 0$ is related to the convective terms in the momentum equation. This shows that the surfaces normal to the wall and the stream surfaces are subcharacteristic surfaces as shown in Fig. 5.14, where the spatial wedge of influence and the dependence of the flow profile along a wall-normal line is shown. For example, a perturbation at p is carried along the y axis by the diffusion process and along the boundary layer streamline by convection in the downstream direction. In view of this, the perturbation affects the stream surface normal to the wall. The domain of influence of point p is a wedge. In a similar way, a domain of dependence can be identified, as shown in Fig. 5.14. The disturbance at p is instantly felt along the normal AB (A is at the edge, B is on the surface). This disturbance is then swept downstream on all stream surfaces crossing AB.

The region of influence is given by the streamline surface, bounded by the limiting streamline on the wall to the inviscid streamline at the edge. For stable marching, the integration scheme must be oriented so as to follow the streamline, and the numerical domain of dependence should always include the physical domain of dependence of the boundary layer equation. This is the CFL condition illustrated in Appendix B (Fig. B.4). This explains the parabolic–hyperbolic nature of three-dimensional boundary layer equations.

Several investigators have solved various forms of the boundary equations (Eqs. 5.9–5.12) by the finite-difference method. Recent surveys of numerical schemes are given by Smith (1982) and Cousteix (1986). Because the primary characteristics of the equations governing the three-dimensional boundary layer are parabolic in nature, a step-by-step spatial marching technique can be used to obtain the solution starting from the leading edge. Most of the techniques employed are second-order accurate schemes; however, many of these techniques are the Crank–Nicholson type, which can result in an oscillatory solution when large perturbations in W_e, boundary conditions, and transition occur. In the regions of reverse cross-flow, a first-order accurate scheme must be employed to avoid instability. Vatsa (1985) has adopted a consistent first-order formulation for spatial derivatives.

Cebeci et al. (1977, 1979) employ boundary layer equations in a nonlinear curvilinear coordinate system, and the energy equation is expressed in terms of total enthalpy. The numerical integration is carried out by the KB scheme (Appendix B, Fig. 5.15). The two momentum equations are solved simultaneously, while the energy equation is solved retrospectively, and the solution is marched in the streamwise direction. The edge conditions (pressure gradient and velocity) are specified. The equations are centered inside the box [say $p(x_{i-1/2}, y_{j-1/2}, z_{k-1/2})$] and derivatives are written. This box technique can be used only if W_r or W_z is positive. Cebeci et al. (1979) proposed a new scheme to overcome this shortcoming. This comprised of a combination of zig-zag and box methods, which are illustrated in Fig. 5.15. Iyer and Harris (1989) have proposed a fourth-order accurate solution employing a two-point scheme in the η direction and a second-order zig-zag scheme in the cross-flow direction, and they computed the three-dimensional boundary layer on a wing. Their scheme is also shown in Fig. 5.15.

All the techniques described above were developed for external flows. There has been no comparison of various techniques, but the KB scheme is widely used. This technique is unconditionally stable. The application of these techniques to turbomachinery flows has been carried out by Thompkins and Usab (1982), Anderson (1987), and Zhang and Lakshminarayana (1990). Thompkins and Usab (1982) employed the KB scheme and solved the boundary layer equations in an uncoupled fashion, assuming that the cross-flow gradient is small ($\partial W_r / \partial r$). This is a quasi-three dimensional technique and is not applicable to flows with appreciable cross-flows and radial pressure gradient.

Most of the schemes are valid for unidirectional flow. It is not uncommon in turbomachinery applications to encounter reversal in radial component of velocity within a three-dimensional boundary layer (Fig. 5.8). The method due to Vatsa (1985) is probably the most suitable technique available for turbomachinery flows. He employs a first-order formulation for spatial derivatives and uses an upwind scheme in the radial direction (to capture flow reversal in the cross-flow direction only). The resulting linear algebraic equations are solved using a block tridiagonal solver. Good predictions were

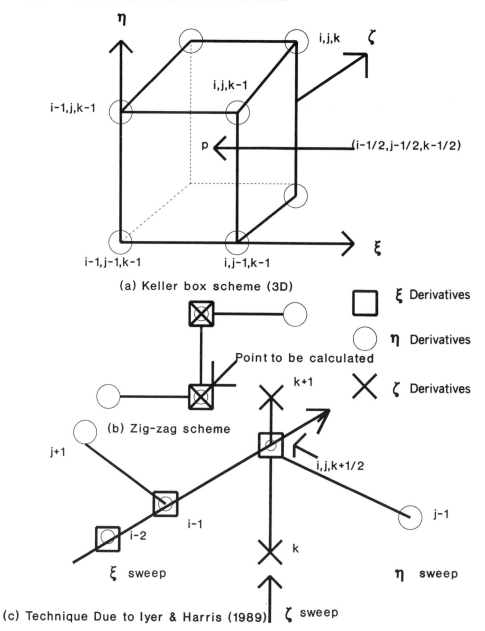

Figure 5.15 Numerical techniques for three-dimensional boundary layer. (a) Keller box scheme. (b) Zig-zag scheme. (c) Technique due to Iyer and Harris (1989).

obtained for the three-dimensional boundary layer on a rotor blade, as shown in Fig. 5.5. This illustrates the ability of boundary layer codes to predict complex three-dimensional boundary layers. Thus an inviscid solution of the flow field coupled with a boundary layer solution can resolve most features of the three-dimensional flow, especially in subsonic rotors.

The technique and code by Vatsa (1985) and Anderson (1987) are by far the most comprehensive. The technique includes the following:

1. The generalized boundary layer equations (Eqs. 5.9–5.12) are transformed into boundary layer coordinates using the Levy–Lees transformation (similar to Eqs. 5.36–5.39).

2. A general nonorthogonal surface coordinate is developed, and the equations are transformed into the system. ξ and ζ lie on the blade surface, ξ is in the general streamwise direction, ζ is in the radial direction, and η is normal to the wall.

3. An algebraic eddy viscosity model (Eqs. 1.76–1.79) and an algebraic heat flux equation (Chapter 7) are employed.

4. They use the experimental pressure distribution (C_P) and compute the velocities W_{se} and W_{re} by solving the Euler equation on the blade surface. This procedure is not very efficient because the edge velocities can be determined from analyses (e.g., panel method, S_1 and S_2 surface solutions) outlined in Chapters 3 and 4.

5. The boundary conditions are $W_s = W_r = 0$ (for $\eta = 0$), $W_s = W_{se}$, $W_r = W_{re}$, and $I = I_e$ at $\eta = \eta_e$.

6. Only the normal diffusion terms ($\partial^2 W_s / \partial \eta^2$, $\partial^2 W_r / \partial \eta^2$) are retained; therefore, the equations are parabolic in both directions. The streamwise derivatives are forward-differenced and the cross-flow derivatives are forward-differenced, with direction-dependent features.

This technique and code have been modified to include the following: (Zhang and Lakshminarayana (1990))

1. The inviscid or edge velocities in Eqs. 5.9–5.12 are determined from a Euler code (Katsanis, 1968; Katsanis and McNally, 1977).

2. The two-equation model (k–ϵ) is incorporated (Eqs. 1.84–1.86) in the code. These equations are cast in a generalized nonorthogonal coordinate system. To be consistent with the boundary layer equations, the Levy–Lees transformation is used to transform these into boundary layer coordinates.

3. An algebraic Reynolds stress model (Eq. 1.89) is used to resolve the components of normal and shear stresses.

These changes provided major improvements in boundary layer predictions. The simulated endwall boundary layer (Muller, 1982) was predicted

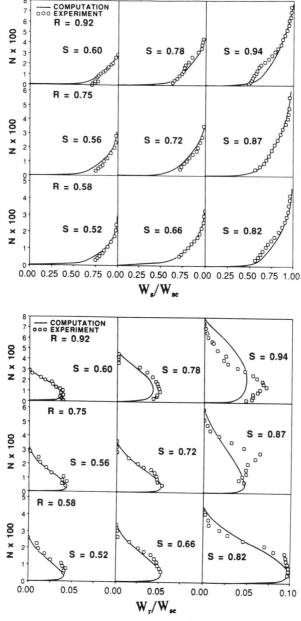

Figure 5.16 Streamwise (W_s) and radial velocity (W_r) profiles on suction surface of Axial Flow compressor rotor blade at $R = 0.58$, $R = 0.75$, and $R = 0.92$ (data from Lakshminarayana and Popovski, 1987). N is the distance normal to the blade, normalized by blade chord, S is the distance from leading edge normalized by blade chord. (Copyright © Zhang and Lakshminarayana, 1990; reprinted with permission of AIAA.)

accurately, including good predictions for mean velocities and all six Reynolds stress components inside the boundary layer.

The suction surface boundary layer profiles at the peak pressure rise coefficient for the Penn State compressor is shown in Fig. 5.16. This is a low-speed compressor with a hub/tip ratio of 0.5, an outer diameter of 0.936 m, and a ϕ value of 0.5 (peak pressure rise), and it is operated at 1080 rpm. Details of this compressor are given in Section 2.5.5 and in Fig. 2.17. The boundary layers are thin near midchord all the way from $R = 0.58$ (16% of the span from the hub) to $R = 0.92$ (16% of span from the tip) and grow rapidly as the trailing edge is approached. The profile near the trailing edge of the tip shows a tendency to separate, and this is not predicted well. The maximum radial velocities range from 5 to 10% of the free-stream velocity. Both W_s and W_r profiles are captured very well from the boundary layer code, except W_r profiles in the trailing edge region at $R = 0.75$ and 0.92. The velocity profiles at midradius are shown compared with all three turbulence models in Fig. 5.5. The ARSM shows the best performance.

The prediction of boundary layer on a turbine rotor blade due to Anderson (1987) is compared with data due to Dring et al. (1986) in Fig. 5.17. The computed and measured limiting streamline angle (ϵ_w), which is as high as 30° at some locations, is captured accurately by the code.

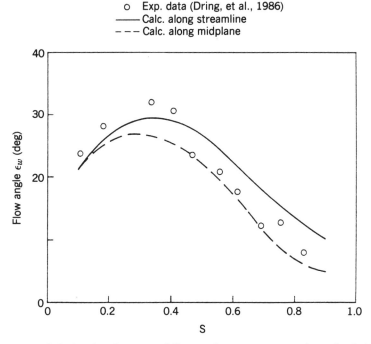

Figure 5.17 Calculated and measured flow angle on pressure surface of a turbine rotor blade. (Copyright © O. L., Anderson, 1987; reprinted with permission of ASME.)

This clearly reveals the ability of boundary layer codes to predict complex three-dimensional boundary layers. Thus an inviscid solution of the flow field coupled with a boundary layer solution can resolve most features of three-dimensional flow, especially in subsonic rotors. It is essential to introduce proper physics through turbulence transport equations (k, ϵ, and ARSM).

5.5.6 Assessment of Boundary Layer Solutions

1. The technique is not valid when boundary layers are thick and normal pressure gradients as well as streamwise and radial diffusion (viscous), are substantial. This limits the application to nonseparated, shock-free flows. Boundary layer solutions are not valid for transonic and supersonic rotors, where appreciable shock–boundary-layer interactions and three-dimensionality are present.

2. These techniques and codes are restricted to axial flow turbines and compressors where the boundary layers are relatively thin. The boundary layers in centrifugal compressors are thick and the entire passage is affected by the viscous flow; the boundary layer codes are not valid for such flows.

3. The resolution of the leading edge flow is a problem because the boundary layer thickness is very small there. It is best to interface the boundary layer code with an analytical solution in this region.

4. Boundary layer solutions are not valid when the flow is separated, as in a corner between the endwall and the blade. It is not valid in the shroud wall region where the blade boundary layer, endwall boundary layer, horseshoe vortex, and leakage flow can interact to produce a complex flow field. Caution should be exercised in attempting to capture both the radial inward and outward flow.

Despite these limitations, boundary layer techniques offer some unique advantages over Navier–Stokes codes. They include the following:

1. The grid resolution and computational time restrict the accurate solution of wall layers using Navier–Stokes codes. Because boundary layer codes are computationally more efficient, the wall layers, including skin friction, heat transfer, and profile losses, can be resolved accurately through usage of large number of grid points in the normal direction.

2. An efficient way to resolve both the viscous and the inviscid effects in turbomachinery, with three-dimensional boundary layers, is to couple the Euler code with a three-dimensional boundary layer code. This is all that is needed for subsonic rotors with thin boundary layers. The blade pressure distribution is predicted well, and the performance predictions are reasonable.

Description and assessment of coupled inviscid/boundary layer solutions are given in Section 5.8. These methods have been used successfully to predict the flow field in cascades and rotors (e.g., Barnett et al., 1990; Calvert, 1982; Singh, 1982). Fan and Lakshminarayana (1994) have successfully computed the unsteady inviscid flow and the unsteady boundary layer in cascades using an unsteady Euler/unsteady boundary layer code.

5.6 SINGLE-PASS SPACE MARCHING METHODS

In many instances, the boundary layer approximations are not valid, but the flow is not complex enough to necessitate a full Navier–Stokes solution. The situations in which the boundary layer approximations are not valid were listed earlier. The common feature of all of these cases is the presence of a significant pressure gradient component normal to the wall. This does not necessarily mean that one must resort to the full Navier–Stokes equation. If the streamwise pressure gradient is large and the streamwise diffusion is substantial, then a full Navier–Stokes solution is the only approach to take. In many instances, the flow can be resolved with some approximations to the Navier–Stokes equations, resulting in substantial savings in computational time and storage. The Navier–Stokes equation can be "parabolized" and "space marched" to resolve the flow field. Parabolization of the equations results in a substantial improvement in CPU time required over the time marching technique or the relaxation technique for a class of flows in which streamwise pressure gradients are small, or known from the inviscid solution. Fewer terms in the approximated equations result in a substantial reduction in the amount of computer storage. A computationally efficient solution is achieved without sacrificing much on accuracy. This technique has been successfully used for many internal flows such as ducts, cascades, and rotor/stator passages. Unlike boundary layer techniques, this technique retains the normal (to the wall) pressure gradient and normal (to the wall) momentum equations. This is an essential requirement in solving thick or three-dimensional boundary layers.

One of the approximations made is to parabolize the Navier–Stokes equation in one of the coordinate directions so that a marching type of computational scheme can be implemented in that direction. This approximation is well suited to internal flows, which have a dominant flow direction. The viscous diffusion terms in the streamwise direction are small in comparison to the diffusion terms in the transverse directions, and they can be dropped from the equations. However, the pressure terms have an elliptic influence in subsonic flows, and special care needs to be taken to handle them in a marching solution algorithm. Methods of treating the pressure terms differ among the many parabolic marching methods available in the literature.

These approximations are not without problems. In addition to error due to discretization, error due to approximation in flow physics is introduced.

Furthermore, stability problems are encountered, especially in situations for which the approximation is not valid. As indicated earlier, all parabolic and space marching techniques neglect the streamwise diffusion term ($\partial T/\partial x$ in Eq. 1.18).

There are two types of approximations to the inviscid term, which result in a parabolic system, as indicated in Tables 5.2 and 1.1. In the first technique, the streamwise pressure gradient term is assumed to be known from an inviscid analysis, and the transverse gradient in the cross-flow direction is resolved through an elliptic sweep. The assumed pressure gradient is then updated to satisfy the mass flow or continuity equation.

The streamwise pressure gradient in this technique is expressed as

$$\frac{\partial p}{\partial x} = \frac{\partial}{\partial x} p_i(x, y, z) + \frac{\partial}{\partial x} p_m(x)$$

where p_i is the assumed (or inviscid) pressure, and p_m is the bulk pressure correction due to viscous blockage effects and is assumed to be uniform over the entire cross section. The pressure gradient in the cross-flow momentum equation is expressed as

$$\frac{\partial p}{\partial z} = \frac{\partial}{\partial z} p_i(x, y, z) + \frac{\partial}{\partial z} p_c(y, z)$$

The cross-flow momentum equations are solved as a set of elliptic equations, thus resolving the pressure and the secondary velocity field accurately with only the approximation for the streamwise pressure gradient. The cross-flow pressure gradient retains a separate correction $p_c(y, z)$, which is allowed to vary in the cross section. The streamwise momentum equation is parabolic, and the cross-flow equations are elliptic. Patankar and Spalding (1972) derived a two-dimensional elliptic equation for pressure correction using a transverse momentum equation, while Briley (1974) utilized two-elliptic equations in the cross plane, one for potential function and the other for p_c, as explained later. This technique solves equations in an uncoupled fashion, while the technique due to Govindan and Lakshminarayana (1988) solves all of the equations simultaneously in a coupled fashion, resolving $\partial p_c/\partial x$ through global mass conservation. The former technique will be referred to as the *SM pressure correction (uncoupled) method*, and the latter will be referred to as the *SM pressure correction (coupled) method*, to draw a clear distinction between the two.

In the second approach due to Briley and McDonald (1984) and Govindan et al. (1991), a small scalar potential approximation is made, and the velocity is resolved into primary and secondary velocities. The secondary velocity potential is resolved into a scalar potential and a vector potential. The vector potential defines the streamwise vorticity and the secondary velocity component associated with streamwise vorticity. No approximation is introduced for the pressure. These decoupled equations are solved sequentially using an implicit numerical algorithm. Govindan et al. (private communication) have successfully solved a large class of flows including (a) tip vortex flows and (b) flow over marine vehicles, and claim space marching codes achieved the same accuracy as Navier–Stokes for some of the complex cases in about 1/30 of CPU time required for a full Navier–Stokes solution. This technique will be referred to as the *SM velocity decomposition technique*.

The space-marching/parabolized Navier–Stokes solutions are widely used to predict complex, three-dimensional, steady supersonic viscous flow fields. In this case, the outer inviscid region is supersonic. The Navier–Stokes equation is parabolized by neglecting the streamwise diffusion of momentum and manipulating the streamwise flux vector such that the product of the inverse of its Jacobian and the Jacobian of the transverse flux vector have real eigenvalues. In supersonic flow, the eigenvalues are all real and positive if the streamwise velocity is non-negative. In the boundary layer, one eigenvalue becomes imaginary and the system becomes unstable for space marching. In the boundary layer, retaining the streamwise pressure gradient in the implicit part of the streamwise flux vector yields an unstable system as the grid system is refined. Such departure solutions are well-documented. Keeping the streamwise pressure gradient term explicit removes this instability, but cannot account for any global interaction between the viscous and inviscid flows. Schiff and Steger (1980), following boundary layer theory, replaced the pressure in the viscous layer with the pressure in the outer supersonic flow.

Govindan and Lakshminarayana (1988) modified Schiff and Steger's method for internal subsonic flows. A special bulk pressure correction was required to conserve global mass flow. The method proved to be very sensitive to the value of the assumed (explicit) pressure gradient and had difficulty in low Mach number flows. The pressure gradient is thus specified in the subsonic region, and the entire flow is solved by a marching technique. Therefore, the space marching/parabolic methods are efficient numerical techniques for supersonic flows. Details of various techniques, stability, and convergency analysis can be found in Anderson et al. (1984).

The second category of flows where the parabolic/space marching method is widely used is subsonic viscous internal flow. Because this is a very common case occurring in turbomachinery and the technique has been widely used, a detailed treatment follows.

5.6.1 Pressure Correction (Uncoupled) Method

The basic assumptions made in this technique are as follows. Equation 1.18 is used with the following modifications (assuming x is the predominant flow direction, or the marching direction):

1. The streamwise pressure gradient $(\partial p/\partial x)$ in Eq. 1.18 or 5.1 is assumed to be a function of x only. Its value is initially specified from an inviscid solver. The value of this is updated in the marching process, using global mass constraint. The pressure is written as

$$p = p_i + p_m(x)$$

   where p_i is the inviscid pressure, p_m is the correction to the mean pressure across the entire channel cross section, and p_i is prescribed. Therefore, the streamwise solution is parabolically marched. Determination of p_m requires no information downstream; therefore, the technique is called "parabolic" or "space marching."
2. The flow in the cross plane (y, z) or (r, θ) is treated elliptically, solving for $\partial p/\partial y$ and $\partial p/\partial z$ in the cross-flow direction.

Thus, the pressure variation in the cross plane is neglected in the streamwise momentum equation. No restriction is placed on p in the y and z momentum equation, and p is written as

$$p = p_i + p_c(y, z)$$

where p_i is allowed to vary in the cross plane (r, θ) or (y, z) to satisfy the global mass flow. The procedure is, therefore, parabolic in the x direction (predominant flow direction) and elliptic in the cross-flow direction.

One of the first successful applications of the parabolic marching (uncoupled) method is due to Patankar and Spalding (1972). Patankar and Spalding assumed that the downstream pressure had no influence on the upstream flow, so that the flow was truly parabolic. They introduced approximate relations between the velocities and the pressure corrections from the transverse momentum equations. A more rigorous scheme for coupling the velocity and pressure corrections has been developed by Briley (1974). Recent extension and modification to these schemes is given by Mazher and Giddens (1991). New variables (called pressure–velocity variables) are introduced to couple the momentum and pressure equations. All of the above methods can be used to solve flows with moderate streamwise pressure gradients. One of the drawbacks of these methods is the difficulty in satisfying the local continuity over the cross section so as to obtain an accurate representation of the secondary flow field. Extensions of the parabolic marching methods to include the effects of strong streamwise pressure gradients require an additional equation to compute the pressure in the three-dimensional flow field. These methods are called "partially parabolic" methods or multipass space-marching methods, and they will be covered later.

The generalized procedure (Briley, 1974) will be illustrated employing incompressible viscous flow through a curved duct as a test case (Fig. 5.1c). The equations governing the flow field in a duct (see Fig. 5.1c for notations) can be written as (neglecting the streamwise diffusion term)

$$u\frac{\partial u}{\partial x} = -w\frac{\partial u}{\partial z} - v\frac{\partial u}{\partial y} + v\left(\frac{\partial^2 u}{\partial y^2} + \frac{\partial^2 u}{\partial z^2}\right) - \frac{1}{\rho}\frac{\partial}{\partial x}(p_i + p_m) \quad (5.43)$$

$$u\frac{\partial v}{\partial x} = -w\frac{\partial v}{\partial z} - v\frac{\partial v}{\partial y} + v\left(\frac{\partial^2 v}{\partial y^2} + \frac{\partial^2 v}{\partial z^2}\right) - \frac{1}{\rho}\frac{\partial}{\partial y}(p_i + p_c) \quad (5.44)$$

$$u\frac{\partial w}{\partial x} = -w\frac{\partial w}{\partial z} - v\frac{\partial w}{\partial y} + v\left(\frac{\partial^2 w}{\partial y^2} + \frac{\partial^2 w}{\partial z^2}\right) - \frac{1}{\rho}\frac{\partial}{\partial z}(p_i + p_c) \quad (5.45)$$

$$\frac{\partial u}{\partial x} + \frac{\partial v}{\partial y} + \frac{\partial w}{\partial z} = 0 \quad (5.46)$$

where u, v, and w are velocities in streamwise (x), normal (y) and cross flow (z) directions, respectively. The boundary conditions are $u = v = w = 0$ on walls and u, v, w specified at inlet.

The global mass flow conservation is given by

$$\int_0^H \int_{-L/2}^{+L/2} \rho u \, dy \, dz = \dot{m} \quad (5.47)$$

where H and L are the height and width of the duct, respectively. A Poisson equation is derived for the pressure correction using y and z momentum equations:

$$\frac{\partial^2 p_c}{\partial y^2} + \frac{\partial^2 p_c}{\partial z^2} = f(x, y, z) \quad (5.48)$$

where $f(x, y, z)$ contains velocity gradients.

The basic method consists of the following steps:

1. Express Eqs. 5.43–5.46 in finite difference form using central differencing.
2. Specify the initial condition $(u, v, w, p)^n$ at the nth streamwise station, p_i at all stations. The initial value of p_i could be based on assumed values or values derived from an inviscid procedure (e.g., panel method or technique as described in Chapter 4). In the cross plane, velocity is written as $v = v_i + v_c$ and $w = w_i + w_c$, where subscript i stands for the initial value and c stands for the correction.

3. Solve for u^{n+1} from Eq. 5.43, and update this using Eq. 5.47 to satisfy the global mass flow constraint.

4. Using the new u^{n+1}, solve the streamwise (x) momentum equation (Eq. 5.43) for pressure correction p_m.

5. Solve y and z momentum equations (Eqs. 5.44 and 5.45) to get approximate values for v_i^{n+1} and w_i^{n+1}.

6. Solve the continuity equation using u^{n+1}, v_i^{n+1} and w_i^{n+1} to derive velocity correction v_c and w_c from $v = v_i + v_c$ and $w = w_i + w_c$.

7. Solve Eq. 5.48 (with all terms known in the source term $f(x, y, z)$) to solve the pressure correction (p_c) in the cross plane.

8. Knowing $(u, v, w, p_m, p_c, p_i)^{n+1}$ at the $n + 1$ streamwise step, repeat steps 3–7 until the last streamwise station.

Various modifications have been suggested for improving this technique. Because the momentum and continuity equations have not been satisfied simultaneously, several iterations could be carried out at each plane before advancing the solution to the next streamwise station. Furthermore, a fraction of the corrections $(v_c$ and $w_c)$ could be added to the provisional values. In addition, only a fraction of the pressure corrections can be used at the nth step to update the values at the $(n + 1)$th step.

As indicated earlier in this section, these approximations are not without problems. In a flow field with strong secondary flow and transverse pressure gradients, the physical coupling between the pressure field and the transverse velocity would slow the convergence of the numerical solution. These disadvantages led Govindan and Lakshminarayana (1988) and Pouagare and Lakshminarayana (1986) to develop a coupled single-pass space marching technique. This is an extension of the method developed for parabolic regions in supersonic flow developed by Schiff and Steger (1980).

5.6.2 Pressure Correction (Coupled) Method

Internal flows of the type found in turbomachinery blade rows are marked by three-dimensionality, pressure gradient, and turbulence effects. Because the transverse velocities are strongly coupled to the streamwise velocities and flow field, it is best resolved using a "coupled" approach whereby all of the equations are solved simultaneously. The techniques which are based on simultaneous solution of all of the equations are dealt with in this section.

Here again, the dominant flow direction in most internal flows affords the possibility of "parabolizing" the steady state Navier–Stokes equation. The initial value problem, when well posed, is amenable to solution by marching procedure in the streamwise direction. The advantage gained by such a formulation is that an implicit noniterative numerical algorithm can be used to solve the system of equations. These techniques are suitable for both single-pass or multipass solutions. Only the single-pass solutions are described here; multipass techniques are described later.

The method due to Govindan and Lakshminarayana (1988) is valid for subsonic flows, whereas that due to Pouagare and Lakshminarayana (1986) is restricted to incompressible flow. Their methods will be illustrated using equations in Appendix A.

In the space marching and parabolic marching methods, the term $\partial T/\partial \xi$ is neglected in equation A.4 (Appendix A). If the flow is assumed to be steady, the governing equations reduce to the following form:

$$\frac{\partial \hat{E}}{\partial \xi} + \frac{\partial \hat{F}}{\partial \eta} + \frac{\partial \hat{G}}{\partial \zeta} = \frac{1}{\text{Re}} \left(\frac{\partial \hat{P}}{\partial \eta} + \frac{\partial \hat{Q}}{\partial \zeta} \right) \tag{5.49}$$

The presence of the streamwise pressure gradient in the streamwise momentum equation makes this equation elliptic for subsonic flows, even when the streamwise diffusion term is neglected. For Eq. 5.49 to be a well-posed initial value problem, the dependence of $\hat{E}$ vector on pressure must be removed. Govindan and Lakshminarayana (1988) separated the streamwise flux vector into two parts, one containing the convective terms and the other containing the pressure gradient terms, as follows:

$$\frac{\partial}{\partial \xi} \hat{E}(q) = \frac{\partial}{\partial \xi} \hat{E}_s(q) + \hat{E}_p(p_s) + \hat{E}_{pq}(p_s, q) \tag{5.50}$$

where

$$\hat{E}_s(q) = \frac{1}{J} \left[\rho U, \rho u U, \rho v U, \rho w U, e_0 U \right]^T$$

$$\hat{E}_p(p_s) = \frac{1}{J} \left[0, \xi_x \frac{\partial p_s}{\partial \xi}, \xi_y \frac{\partial p_s}{\partial \xi}, \xi_z \frac{\partial p_s}{\partial \xi}, U \frac{\partial p_s}{\partial \xi} \right]^T$$

where ξ is in the dominant flow direction, with contravariant velocity given by Eq. A.4 in Appendix A.

$$\hat{E}_{pq}(p_s, q) = \frac{1}{J} \left[0, 0, 0, 0, p_s \frac{\partial U}{\partial \xi} \right]^T$$

where p_s is the assumed pressure, and $\hat{E}_p(p_s)$ is treated as a known source term in the equation and computed from the prescribed or assumed pressure field.

It can be proven through an eigenvalue analysis that this formulation results in a well-posed problem for marching in the streamwise direction (ξ), provided that the streamwise velocity is positive. Therefore, the method is restricted to attached flows. The choice of the assumed pressure gradient depends on the flow being computed. Typically, it is derived from an inviscid or potential code. The assumed pressure is updated at each streamwise step, as described later.

A numerical marching procedure can now be formulated. Govindan and Lakshminarayana utilized the implicit technique due to Beam and Warming (1978) and Briley and McDonald (1977), which is described in Appendix B (Eq. B.47).

Equation 5.49 can be written as

$$\frac{\partial \hat{E}_s(q)}{\partial \xi} = D(q) - \hat{E}_p(p_s) \tag{5.51}$$

where

$$D(q) = \frac{1}{\text{Re}} \left[\frac{\partial \hat{P}(q)}{\partial \eta} + \frac{\partial \hat{Q}(q)}{\partial \zeta} \right] - \frac{\partial \hat{F}(q)}{\partial \eta} - \frac{\partial \hat{G}(q)}{\partial \zeta}$$

and the vector $\hat{E}_{pq}(p_s, q)$ has been absorbed into the vector $\hat{E}_s$. A finite-difference equivalent to Eq. 5.51 can be written in the form (similar to Eq. B.47)

$$\left[A + \Delta \xi L_\eta \right] \psi^* = \Delta \xi \left[\left(D_\eta + D_\zeta \right)^n - \hat{E}_p^{n+1/2} \right]$$
$$\left[A + \Delta \xi L_\zeta \right] \psi^{**} = A \psi^* \tag{5.52}$$
$$\psi^{n+1} = \psi^{**} = q^{n+1} - q^n$$

where

$$A = \left[\frac{\partial \hat{E}_s}{\partial q} \right]^n, \qquad L = \frac{1}{2} \left[\frac{\partial D}{\partial q} \right]^n, \qquad \Delta \xi \text{ is the streamwise step.}$$

The time step in Eq. B.47 (Appendix B) has been replaced by spatial step $\Delta \xi$. Equation 5.52 represents a two-step factorized scheme that is fully implicit and unconditionally stable. Each of the two steps in Eq. 5.52 involves the inversion of a block (5×5) tridiagonal matrix system. Any one step involves only one coordinate direction, typical of scalar ADI schemes. The scheme is direct, except for the global pressure iteration to update the assumed pressure p_s. The vector $\hat{E}_p$ is known from assumed p_s and is evaluated at the $n + 1/2$ spatial step. Simple second-order accurate finite-difference operators are used for discretizing the derivatives in Eq. 5.52.

To conserve global mass flow in the computations, the assumed streamwise pressure gradient needs to be updated by a mean streamwise pressure gradient correction. For this purpose, the streamwise pressure gradient in Eq. 5.50 is modified and written as

$$\frac{\partial p_s}{\partial \xi}(\xi, \eta, \zeta) = \frac{\partial p'_s}{\partial \xi}(\xi, \eta, \zeta) + \frac{dp''_s(\xi)}{d\xi} \qquad (5.53)$$

where $\partial p'_s/\partial \xi$ is now the assumed known streamwise pressure gradient and $dp''_s/d\xi$ is a mean streamwise pressure gradient correction, constant over the cross section (η, ζ), computed at each streamwise step to conserve global mass flow. The mean streamwise pressure gradient correction $dp''_s/d\xi$ is computed from the difference between the computed mass flux at a streamwise station and the inlet mass flow:

$$m' = -\int_\eta \int_\zeta \frac{(\xi_x^2 + \xi_y^2)}{J} \frac{1}{(\xi_x u + \xi_y v)} \frac{dp''_s(\xi)}{d\xi} \, d\eta \, d\zeta \qquad (5.54)$$

where m' is the difference between the computed mass flux at a streamwise station and the inlet mass flow. Equation 5.54 is easily integrated if we know the velocity field at a steamwise station, and the mean streamwise pressure gradient correction $dp''_s/d\xi$ is computed for use at the next streamwise station.

The coupled technique can be summarized as follows:

1. From initial conditions for the dependent variables (q) at the first streamwise station and a known streamwise pressure gradient derived from the assumed pressure field, the governing equation is solved numerically as a coupled system through one streamwise step using a noniterative algorithm to obtain the dependent variables at the second streamwise station.

2. The equation of state is used to compute the pressure field from the computed values of the dependent variables at the second streamwise station.

3. The computed mass flow along with the inlet mass flow is used to compute the mean streamwise pressure gradient correction from Eq. 5.54, which is used at the next streamwise station.

4. Computations for subsequent streamwise stations follow in a similar manner until the entire flow field has been computed. For internal flow computations, the mean streamwise pressure gradient correction is lagged one streamwise step.

Many complex flows were predicted using this technique. The flow in the endwall region of a cascade, measured by Flot and Papailiou (1975), is compared with predictions by Kirtley et al. (1986) in Fig. 5.18. The cascade consists of NACA 65-12-A10 airfoils, with a space chord ratio of 0.8, an

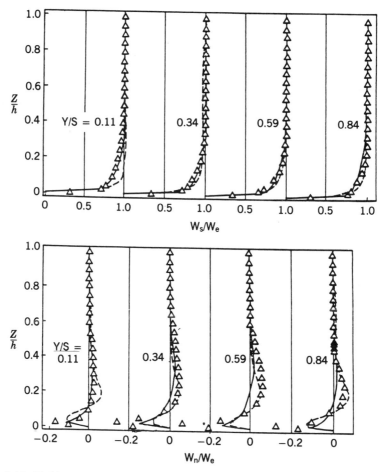

Figure 5.18 Turbine cascade endwall flow field near the trailing edge (data from Flot and Papailoiu, 1975; computation from Kirtley et al. 1986). Z is the spanwise distance measured from the endwall; Y is the tangential distance measured from the suction surface. W_s and W_n are streamwise and transverse velocities (blade-to-blade); subscript e represents the local edge conditions; solid curve represents space marching (parabolic); and dashed curve represents time marching (pseudo-compressibility code). (Kirtley et al., 1986 reprinted with permission of AIAA.)

aspect ratio of 2.1, a stagger angle of 15°, and a 41.2° flow turning. The prediction of both the mainstream velocity and the secondary velocity is good. A comparison between the solution obtained from a single-pass space-marching code and the results from a pseudocompressibility time-marching code (described in Section 5.7.7) are shown compared with the data for Flot and Papailiou's cascade. The grids and computer were identical for both computations. The space-marching algorithm requires a pressure field which was obtained from a panel code. The time-marching code computes the pressure field. The overall CPU time was a very small fraction of the time required for the time-marching code. The streamwise velocity profiles are almost identical, but there is some discrepancy between the predicted cross-flow profiles. The space-marching code seems to capture the wall region better than the time-marching code, while the outer region (the inviscid effect due to pressure gradient) is more accurately predicted by the time-marching code. An attempt to predict the flow through the Langston et al. (1977) cascade (Fig. 4.23), which has a much higher turning angle, was not successful, because elliptic effects due to the horseshoe vortex were substantial and its effect could not be captured.

It should be emphasized here that this technique is most valuable for duct flows, flow through an annulus, and flows in which the perturbation of the steamwise pressure gradient due to viscous and other effects is not large. For example, the technique can be used to predict subsonic flow through a rotor with thin boundary layers and for which the approximate pressure field is known from an inviscid code. But the effect of endwall flow cannot be captured, because elliptic effects in these regions are substantial.

Extensions of this computational technique have been carried out by Pouagare and Lakshminarayana (1986), who developed a technique for incompressible flow. They have shown that the single-pass solution of PNS equations can be achieved in approximately one order of magnitude less time than a full Navier–Stokes solution. Excellent agreement is achieved with the benchmark data for curved and S ducts.

5.6.3 Velocity Split Method

The velocity split method was developed by Briley and McDonald (1984) and by Govindan et al. (1991). The principle of velocity decomposition is illustrated in Fig. 5.19. The flow regime can be split into a potential core, rotational inviscid region (inviscid secondary flow region; see Section 4.2.6), and inner viscous region. The local primary direction is not normal to the cross section, an assumption implied in all inviscid secondary flow theories described in Section 4.2.6. The velocity is split into a primary velocity component (U_p) and secondary velocity components (v and w). The secondary velocity is further split into scalar potential and vector potential components as follows (Fig. 5.19):

$$v = v_\phi + v_\psi$$
$$w = w_\phi + w_\psi$$

The vector potential ϕ defines the streamwise vorticity and the large secondary velocity component associated with the streamwise vorticity, and the stream function ψ governs the flow in the rotational inviscid region. No approximation is introduced for the pressure. Govindan et al. (private communication) have successfully solved a large class of flows including tip vortex

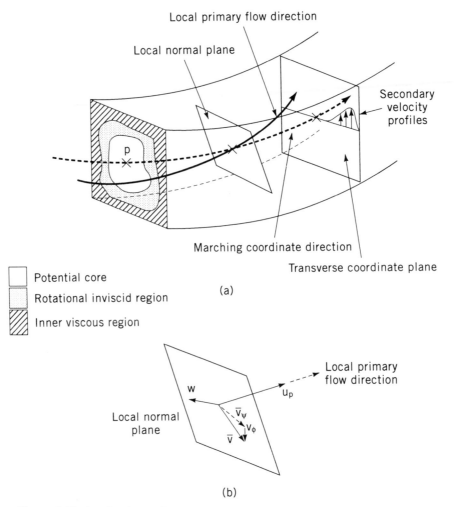

Figure 5.19 Local primary flow direction and velocity decomposition. (Copyright © Govindan et al., 1991; reprinted with permission of AIAA.)

flows, flow over marine vehicles, and so on, and claim that space-marching codes achieved the same accuracy as Navier–Stokes solution for some of the complex cases in a fraction of the time required from full Navier–Stokes solution.

The following equations are developed and used:

1. A scalar potential equation for ϕ.
2. A streamwise momentum equation for U_p.
3. A secondary vorticity (ω_s) transport equation (Eq. 4.121).

4. A second-order equation for p, which is derived by taking the divergence of transverse momentum equations (v and w) to derive an equation of the form

$$\nabla^2 p = f(y, z)$$

It should be noted here that the curl of the v and w momentum equations provide item 4 above. Therefore, this equation replaces the transverse momentum equations.

5. Equation of state for p.
6. Energy equation for e.

Thus, the dependent variables are ω_s, ϕ, ψ, ρ, e, and p. There are six equations for these six unknowns. Govindan et al. (1991) suggested the following sequence for the solution:

1. An artificial time derivative is added to the scalar potential equation (ϕ) and the streamwise vorticity equation. An iterative block implicit scheme is used to solve these equations at any streamwise step n.
2. The pressure equation is solved for the $(n + 1)$th step using a scalar ADI scheme (Appendix B). The values v_ψ and w_ψ are based on the $(n + 1)$th iteration, while other variables in this equation are evaluated at the nth level.
3. The energy equation is solved for e^{n+1} using the ADI scheme.
4. In solving the streamwise momentum equation, an assumed pressure ($p_i(x)$) given by $p = p_i(x) + p_v(x, y, z)$ is used. An assumed value of $p_i^{n+1}(x)$ is used to solve for U_p^{n+1} from the streamwise momentum equation. $p_v(x, y, z)$ is the perturbation due to secondary flow and viscous flow. The density is updated using the equation of state.
5. The initial guess for p_i^{n+1} is not exact, the global mass flow is usually not conserved. Step 4 is repeated iteratively to find a value of p_i^{n+1}, which leads to U_p^{n+1} and ρ^{n+1}, satisfying the integral mass flow in the cross plane. This step is similar to that outlined in the last section.
6. Finally, the scalar potential equation is solved for ϕ^{n+1} using the ADI scheme.

Steps 1–6 are repeated at the next streamwise step, and the solution is marched.

It should be emphasized here that this is a single-pass method, where the flow in the cross-flow plane is solved elliptically with an assumed streamwise pressure gradient, which is updated after each iteration. Even though the formulation is complex, it is computationally very efficient. The computed results from this technique for a ship propeller (including the rotation effects) in the tip region are shown in Fig. 5.20. The formation and roll-up of the vortex as well as the spanwise flow on both the pressure and suction sides

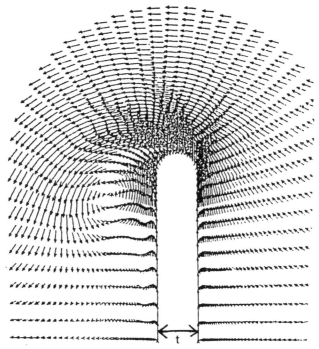

Figure 5.20   Computed flow field around a ship propeller blade ($C = 20t$, $Re = Ut / \nu = 1000$, $i = 6°$, advance ratio = 1.0; $z / t = 20.0$). (Copyright © Govindan et al., 1984.)

are captured. In addition, the predicted and the measured chordwise distribution of pressure were in good agreement (Govindan et al., 1984).

Even though the basic principle on which these techniques are based is similar, the numerical solution of algebraic equations differs. Patankar and Spalding use an ADI marching procedure in the streamwise direction, and they use ADI sweeps in the cross plane. Briley (1974) solves the Poisson equation for the velocity and pressure in the cross plane by the SOR technique.

5.6.4 Assessment of Space-Marching Techniques

1. It should be emphasized that all of the methods described in this section are single-pass where the corrected pressure is not utilized to update the velocities. These techniques are useful for ducts and cascades where an estimate for the pressure field is available or easily computed. Because only a single sweep is employed, space-marching methods are typically about one order of magnitude faster than solutions described in the next section.

2. The technique can be used to predict subsonic flow through a rotor with thin boundary layers if the approximate pressure field is known from an inviscid code. The endwall flow can be captured, provided that the turning is not large or there is no corner flow separation. The requirement of positive streamwise velocity for stable marching limits the applicability of the equations to unseparated flows.

3. The space-marching techniques are unsuitable for an endwall region in high-turning cascades with thick leading edges and with corner flow separation, or in the presence of a horseshoe vortex (e.g., turbine). Furthermore, because the space-marching technique requires an H grid, hence high stagger angles, high-turning (or high camber) cascades with thick leading edges are unsuitable for space-marching codes due to grid restrictions. The omission of the streamwise diffusion terms limits the application to regions away from the vicinity of a leading edge.

4. The flow field can be resolved as accurately as a full Navier–Stokes solution in a small fraction of the time. This technique is also most suitable in the design mode in view of its computational efficiency.

5. The space-marching code can be integrated with a Euler or Navier–Stokes code to develop an efficient zonal technique as described in Section 5.8.

5.7 EULER AND NAVIER–STOKES SOLUTIONS

As indicated earlier, the techniques typically used for the solution of the Euler and Navier–Stokes equations are essentially the same, even though there are some major differences in the behavior of the techniques. Therefore, techniques used for the solution of Euler and Navier–Stokes equations are treated together. The classification of numerical techniques is shown in Table 5.2. The approximations that can be made for the Euler equation and the resulting solution technique are explained in earlier chapters (e.g., 2D inviscid flow, irrotational flow in Chapter 3, numerical solution of 2D equations in Section 5.4). The approximations to Navier–Stokes equations are many. One must clearly distinguish between the approximations to the equations (which neglect some of the physical phenomena) and approximations used in discretization.

The approximations to the Navier–Stokes equations can be classified as thin-layer Navier–Stokes (TLNS), partially parabolic Navier–Stokes (PPNS), and reduced Navier–Stokes (RNS),[*] as explained in Chapter 1. These approximations are based on the nature and physics of the flow. For example, TLNS neglects streamwise diffusion (μu_{xx}) in the equation, and is valid for viscous flow over a body, with the viscous layer being small

---

[*]The acronym RNS is also used to denote Reynolds averaged Navier–Stokes equation. In this book, Navier–Stokes equation for turbulent flow always implies Reynolds averaged equation (Section 1.2.6, Eq. 1.68).

compared to a streamwise characteristic length (say chord length of a blade). The RNS equations are derived from an order-of-magnitude analysis or perturbation analysis of the NS equations applicable to high Reynolds number flows with a dominant direction. The RNS model neglects the viscous term in the normal momentum equation (normal to the body). Therefore, both TLNS and RNS are valid strictly for thin shear layers near the surface with a dominant flow direction. The equations are accurate for most flows; exceptions are separated flows and near wake regions where the neglected effects may be important. The RNS model can be utilized to solve the full Navier–Stokes equations where the neglected terms can be retained and solved by a deferred corrector approach. For the PPNS, the pressure is assumed initially and updated continuously through a pressure equation. The final converged solution in this case does represent the solution to the full Navier–Stokes equations, even though the solution was achieved through initial approximations. These are usually called the partially parabolic (or partially elliptic) marching methods. The classification of various computational techniques was given earlier (Section 5.3) and a more precise classification of Euler/Navier–Stokes solvers is given below.

Pressure-Based Methods. This is an extension of the single-pass space-marching technique described earlier. In this technique, the assumed pressure is updated using an auxiliary equation for pressure, and a multipass procedure is adopted to converge both the pressure field and the velocity field. In this technique, the domain is computed several times (equivalent to relaxation or time steps) using the previously computed pressure field. The procedure is repeated until the pressure correction p_c is zero. The steady form of Navier–Stokes equations can be solved by elliptic methods which allow, simultaneously, for all the elliptic effects including (a) streamwise and cross-flow pressure gradients and (b) streamwise and transverse diffusion. But the convergence of the techniques (which utilize the SOR, SLOR, Gauss–Seidel technique, etc.) for solving the discretized equations are slow and time-consuming.

Time-Dependent (Marching) Methods. In these techniques, which may be either explicit or implicit, the unsteady terms in the Navier–Stokes equations are retained, even though only the steady solution may be of interest. The unsteady form of the compressible (subsonic, transonic, and supersonic flow) system of continuity, Navier–Stokes, and energy equations are mixed type; therefore, efficient techniques available for these systems are employed to obtain the steady state solution.

Pseudocompressibility Technique for Incompressible Flow. One of the more efficient means of solving the Navier–Stokes equations for incompressible flow is to introduce artificial time derivatives in the continuity equation, thus transforming the elliptic equations into a hyperbolic system. Any of the time-dependent methods can then be used to solve the set of equations.

Zonal Techniques. The principal feature of zonal techniques is to utilize the most appropriate equations in regions of interest and iterate between zones to derive an accurate solution. The motivation is to reduce computer time and storage. For example, a Euler solution is used in the inviscid region and boundary layer or single-pass marching solution in the viscous region to resolve the entire flow, instead of employing the full Navier–Stokes equations in all regions, including regions where there are no viscous effects.

Four of the most apparent differences between the time-marching and pressure-based algorithms are (a) the utilization of the physical time variable, (b) the implementation of the continuity equation, (c) the handling of the coupling between the governing equations, and (d) the differencing schemes used for the convective terms. The typical solution procedure in a pressure-based algorithm is as follows:

1. Estimate the velocities and pressures.
2. Calculate the convection and diffusion coefficients.
3. Solve each momentum equation implicitly, but separately.
4. Calculate the coefficients of the pressure correction equation.
5. Solve the pressure correction equation implicitly.
6. Correct the velocities and pressures.
7. Solve any further transport equations (e.g., thermal energy, turbulent kinetic energy, etc.).
8. Repeat steps 2–7 until convergence is reached.

In contrast to this, the solution procedure for an implicit, compressible, time-marching formulation is as follows:

1. Estimate the densities and velocities.
2. Calculate the pressure from an equation of state.
3. Calculate the flux Jacobians and the residuals.
4. Calculate the change in the densities and velocities for one time step by solving the continuity and momentum and energy equations simultaneously and implicitly.
5. Update the values of densities and velocities.
6. Repeat steps 2–5 until convergence is reached.

In time-marching schemes, two approaches used to difference the convective terms are central differencing with directly added artificial dissipation terms (Hirsch, 1990; Jameson et al., 1981) and upwind differencing. In common pressure-based methods, the artificial dissipation is inherent in the scheme. Basson and Lakshminarayana (1994) suggested a procedure for controlling the artificial dissipation through explicit addition of artificial dissipation terms.

5.7.1 Pressure-Based Methods (Uncoupled Solution)

In the pressure-based methods, the momentum equations are solved initially with an assumed pressure field, which is continuously updated using an auxiliary pressure equation. One of the widely used pressure-based methods is the pressure correction method (PCM), a technique similar to that outlined in Section 5.6.1, modified to include a multipass or a feedback mechanism coupling the velocity and pressure fields. This technique was originally suggested by Chorin (1968). A review of these techniques is given by Patankar (1988), and a step-by-step numerical approach is described by Patankar (1980) and Moore (1985). The technique has been widely used in incorporating and testing new turbulence models.

In the PCM technique, a separate equation for pressure correction is developed for both $p_m(x)$ and $p_c(y, z)$ (e.g., Eqs. 5.43–5.48), relating the source term in the Poisson equation to the divergence of the velocity field through the continuity equation. This pressure correction required a staggered grid (pressure is solved on a different grid point than the velocity field) to eliminate oscillations in the solution of the pressure correction equation. This was one of the main disadvantages of the earlier techniques, which were restricted to Cartesian coordinates. Recent modifications include extension of the technique to curvilinear, body-fitted coordinate systems and nonstaggered grids. One of the earlier PCM techniques is due to Caretto et al. (1972) and Pratap and Spalding (1976). Both of these employ a pressure correction equation for p_m, which is partially converged through a multipass or several sweeps of the entire flow field. This technique was improved by Spalding (1972), who used an explicit treatment of the convection and implicit treatment of the diffusion terms in the momentum equations.

The detailed procedure will be illustrated for two-dimensional turbulent flow utilizing the two equation (k–ϵ) model (Eqs. 1.84 and 1.86) and the Cartesian coordinate system. Extension to three-dimensional flows is straightforward. The continuity equation, transport equations for momentum, k and ϵ, are written in the following form. [Detailed derivation of the procedure for the pressure correction method can be found in Patankar (1980)].

$$\frac{\partial}{\partial x}(\rho u) + \frac{\partial}{\partial y}(\rho v) = 0 \tag{5.55}$$

$$\frac{\partial}{\partial x}(\rho u \phi) + \frac{\partial}{\partial y}(\rho v \phi) = \frac{\partial}{\partial x}\left[\Gamma^\phi \frac{\partial \phi}{\partial x}\right] + \frac{\partial}{\partial y}\left[\Gamma^\phi \frac{\partial \phi}{\partial y}\right] + S^\phi \tag{5.56}$$

Table 5.4   Governing equations

Conservation Equation	ϕ	Γ^ϕ	S^ϕ
x-Momentum	u	$\mu + \mu_t$	$-\dfrac{\partial p}{\partial x}$
y-Momentum	v	$\mu + \mu_t$	$-\dfrac{\partial p}{\partial y}$
Turbulent kinetic energy	k	$\mu + \dfrac{\mu_t}{C_k}$	$P - \rho\varepsilon$
Turbulent dissipation rate	ε	$\mu + \dfrac{\mu_t}{C_\varepsilon}$	$\dfrac{\varepsilon}{k}[C_{\varepsilon 1}P - C_{\varepsilon 2}\,\rho\varepsilon]$

where u and v are velocities in x and y directions, respectively. The variables ϕ, Γ^ϕ, and source function S^ϕ are tabulated in Table 5.4 for both the laminar and the turbulent flow (k–ϵ) equations. Only the dominant viscous terms are retained.

The left-hand side of Eq. 5.56 is the convective term, the first two terms on the right-hand side are diffusion terms, and the last term is the source term, where P is the turbulence production term. Equations 5.55 and 5.56 are integrated over the control volume as shown in Fig. 5.21 and are written in the following finite volume form:

$$[\,\rho u\,]_w^e + [\,\rho v\,]_s^n = 0 \tag{5.57}$$

$$\left\{(\rho u\phi) - \Gamma^\phi\left[\frac{\partial\phi}{\partial x}\right]\right\}_w^e + \left\{(\rho v\phi) - \Gamma^\phi\left[\frac{\partial\phi}{\partial y}\right]\right\}_s^n = S^\phi \tag{5.58}$$

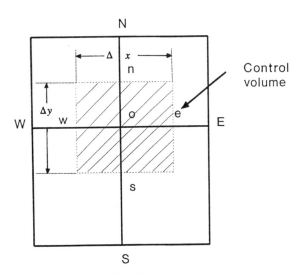

Figure 5.21   Schematic representation of a control volume for pressure-based method.

where n, s, e, and w are the locations of the intersection between the control volume faces and the grid lines.

The approximation of the transport equation is as follows. In the nonstaggered or "collocated" approach, all properties are defined at the nodes O, N, S, E, and W [see Patankar (1980) for a discussion of the "staggered" solution].

$$[\rho u \phi]_e \approx [\rho u]_e \tfrac{1}{2}[\phi_O + \phi_E] \qquad (5.59)$$

$$\left\{ \Gamma^\phi \left[\frac{\partial \phi}{\partial x} \right] \right\}_e \approx [\Gamma^\phi]_e \frac{(\phi_O - \phi_E)}{\Delta x} \qquad (5.60)$$

Diffusion coefficients $[\Gamma^\phi]_e$ are obtained by linear interpolation in the physical plane. These equations are substituted in Eq. 5.58, resulting in a relationship between ϕ_O and the neighboring values:

$$A_O \phi_O = A_N \phi_N + A_S \phi_S + A_E \phi_E + A_W \phi_W + S_P^\phi$$

or

$$A_O \phi_O = \sum A_i \phi_i + S_P^\phi \qquad (5.61)$$

where A_O, A_N, and so on are discretization coefficients. For example,

$$A_E = \left\{ \frac{\Gamma_e^\phi}{\Delta x} - \frac{(\rho u)_e}{2} \right\} \Delta y, \qquad A_W = \left[\frac{\Gamma_w^\phi}{\Delta x} + \frac{(\rho u)_w}{2} \right] \Delta y$$

$$A_N = \left\{ \frac{\Gamma_n^\phi}{\Delta y} - \frac{(\rho u)_n}{2} \right\} \Delta x, \qquad A_S = \left[\frac{\Gamma_s^\phi}{\Delta y} + \frac{(\rho u)_s}{2} \right] \Delta x$$

$$A_O = A_E + A_W + A_S + A_N$$

$$S_P^\phi = -\frac{(p_E - p_W)}{2\Delta x} \Delta x \, \Delta y \qquad \text{for} \quad \phi = u$$

$$= -\frac{(p_N - p_S)}{2\Delta y} \Delta x \, \Delta y \qquad \text{for} \quad \phi = v \qquad (5.62)$$

The central differencing scheme utilized above in the discretization of the convection terms gives rise to odd–even decoupling of the velocity field when local grid Reynolds numbers are high. Typically, the coefficients are modified according to alternative discretizations such as the hybrid differencing scheme which evaluates the convective term by first-order upwind differencing whenever grid Reynolds number exceeds 2 (Spalding, 1972). Equation 5.61 is solved at each point in the cross-stream plane with known values of S.

These are the basic equations for the pressure-based method. The method of updating the pressure differs. These can be broadly classified as the *pressure substitution method*, developed by Hobson and Lakshminarayana (1991), and the *pressure correction method*, developed by Pratap and Spalding (1976) and modified by Moore and Moore (1981), Rhie (1985), Hah (1986), and Basson and Lakshminarayana (1994). Both of these techniques will be described. To keep the equations simple, the method is illustrating neglecting such scalar quantities as temperature and turbulence and utilizing a simple Cartesian grid for two-dimensional flow.

In the pressure correction method, the pressure and velocities are expressed as

$$p = p^* + p', \qquad u = u^* + u', \qquad v = v^* + v' \qquad (5.63)$$

where the asterisk refers to initial or guessed or approximate values and the prime refers to values corrected to simultaneously satisfy the momentum and continuity equations. In the SIMPLE (semi-implicit method for pressure linked equations) scheme due to Patankar and Spalding (1972), the momentum equations are first solved using a guessed value for pressure (p^*). The pressure corrector equation is then solved, and the pressure and velocities are then updated to satisfy the continuity and the momentum equations. The procedure is continued until convergency is reached. The basic method is illustrated here.

The discretized momentum equation (Eq. 5.61) can be written as

$$u_O^* = \sum \frac{A_i^k u_i^*}{A_O^k} - B_O^k \frac{\partial p^k}{\partial x}\Big|_O$$

$$v_O^* = \sum \frac{A_i^k v_i^*}{A_O^k} - C_O^k \frac{\partial p^k}{\partial y}\Big|_O \qquad (5.64)$$

where $B_O^k = C_O^k = [\Delta x \, \Delta y / A_O^k]$. The superscript k denotes the initial guess, and the asterisk denotes the intermediate implicit solution of the relevant variable. In general, u^* and v^* will not satisfy the continuity equation; instead, a net source is produced. To remove this inequality, an equation for pressure correction can be derived from Eqs. 5.63 and 5.64, as indicated below. The momentum equation can be written for the $(k + 1)$th iteration as follows:

$$u_O^{k+1} = \sum \frac{A_i^{k+1} u_i^{k+1}}{A_O^k} - B_O^{k+1} \frac{\partial p^{k+1}}{\partial x}\Big|_O$$

$$v_O^{k+1} = \sum \frac{A_i^{k+1} v_i^{k+1}}{A_O^k} - C_O^{k+1} \frac{\partial p^{k+1}}{\partial y}\Big|_O \qquad (5.65)$$

Similar equations can be derived for u_E^{k+1} and u_e^{k+1}, and so on. u_O^{k+1} is the corrected velocity at $k+1$ iteration and the pressure correction (p') is related to the pressure p^{k+1} through the following equation:

$$p^{k+1} = p^k + p' \tag{5.66}$$

Equations 5.64–5.66 can be used to derive the following equation for the pressure correction assuming $A_i^{k+1} = A_i^k$ and neglecting the influence of velocities in the neighboring points:

$$
\begin{aligned}
u_O^{k+1} &= u_O^* - B_O^k \frac{\partial p'}{\partial x}\Big|_O \\
v_O^{k+1} &= v_O^* - C_O^k \frac{\partial p'}{\partial y}\Big|_O
\end{aligned}
\tag{5.67}
$$

The integral form of the continuity equation is used to update the pressure. Using a procedure similar to that used in deriving Eq. 5.61, the implicit pressure equation, in discretized form, can be written as

$$A_O^k p_O' = A_N^k p_N' + A_S^k p_S' + A_E^k p_E' + A_W^k p_W' + m_O^* \tag{5.68}$$

where

$$
\begin{aligned}
A_N^k &= \rho C_N^k, & A_S^k &= \rho C_S^k, & A_E^k &= \rho B_E^k, \\
A_W^k &= \rho B_W^k, & A_O^k &= A_N^k + A_S^k + A_E^k + A_W^k
\end{aligned}
$$

The source term of Eq. 5.68 is

$$m_O^* = [\rho u^*]_w^e + [\rho v^*]_s^n \tag{5.69}$$

This represents the degree to which the guessed velocity field u^*, v^* does not satisfy continuity. Equations 5.64–5.67 and 5.68 constitute the discretized equations for the pressure correction method. For turbomachinery application, it is more convenient and accurate to transform Eqs. 5.55 and 5.56 into a body-fitted coordinate system, as described in Appendix A. All the coefficients in Eqs. 5.61, 5.64, and 5.68 will then depend on metric terms similar to those in Appendix A. Derivation of the equations for pressure correction method in transformed coordinates is given by Basson (1992).

The procedure for the pressure correction algorithm can be summarized as follows:

1. Determine the intermediate velocity field (denoted by an asterisk) from an assumed pressure field at iteration level k, by solving Eq. 5.61, where $\phi = u^*, v^*$.
2. Solve Eq. 5.68 for the pressure correction, which is dependent on the intermediate velocity field.
3. Correct the velocity field, explicitly, by solving Eq. 5.67.

4. Correct the pressure field by solving Eq. 5.66.

5. Solve any additional scalar transport equations (e.g., energy, $k-\epsilon$, etc.)

6. Check for convergence. If necessary, repeat the above from step 1 by replacing values at k with those at $k + 1$.

This is the procedure used by Caretto et al. (1972), Rhie and Chow (1983), and Hah (1984), and it involves solution for the pressure correction (p'). As indicated in steps 3 and 4 above, the intermediate velocity (u^*) field and the previous pressure field (p^k) are updated by the gradient of pressure correction ($\partial p'/\partial x, \partial p'/\partial y$) (Eq. 5.67) and the actual pressure correction, respectively. This explicit pressure correction is neglected in the pressure substitution method (Hobson and Lakshminarayana, 1991) with no detrimental effect. In the pressure substitution method (PSM), steps 3 and 4 above are eliminated. Even though these are not the most computationally intensive steps, their omission improves the convergence rate and overall computational times. In the pressure substitution method, equations are manipulated to solve directly for p^{k+1}.

A detailed comparison of PSM and PCM methods and stability analysis of the PSM method are given by Hobson and Lakshminarayana (1991). A stability analysis of the PCM method is given by Shaw and Sivaloganathan (1988). When considering the PCM algorithm, it is evident that it is of the predictor type, with no corrector step. A logical extension of this would be to solve the transport equations and the pressure equation in a coupled fashion. A block solution of the equations would then be required. It should be remarked here that PCM and PSM performed equally well in simple flows (e.g., cavity flows), but PSM showed improved convergence in more complex cases such as a cascade. Further research is necessary to evaluate the relative performance of PCM and PSM methods in complex flows.

Various methods of solving the elliptic finite difference equations are described in Appendix B. Hobson and Lakshminarayana (1991) employed the Gauss-Seidel technique for the periodic tridiagonal systems which arise in two-dimensional cascade flows. Both finite difference and finite volume techniques can be used in solving the equations governing the fluid flow. In the finite volume method the equations are integrated over a finite volume and transformed into surface integrals. The dependent variables on control volume faces are approximated by finite difference methods. Many investigators (Hah, 1984; Chan and Sheedy, 1990) use a second-order upwind scheme to discretize the convective terms. The fluxes are interpolated from the two nearest neighboring nodes. Chan et al. (1988) use a seven-point finite difference scheme for three-dimensional flows.

Some of the recent advances are as follows:

Patankar (1980) introduced the SIMPLER method, in which an extra equation was solved for the evaluation of the pressure. The SIMPLEST procedure was developed by Spalding (1980), who recommended an explicit treatment of the convection and implicit treatment of diffusion terms in the momentum equations. Another variant is the SIMPLEC procedure developed by Von Doormal and Raithby (1984), which uses a consistent under-relaxation of the momentum and pressure corrections. Connell and Stow (1986) used a higher-order approximation to develop procedures and showed improvement over the SIMPLE scheme. Leonard (1979) proposed a quadratic upwind differencing (QUICK) to reduce numerical diffusion. Issa (1985) proposed a method called *pressure implicit split operator* (PISO), which, like SIMPLER, also solves an additional equation, but the implicit solution of the pressure is split between the double correction steps. Multi-grid methods have also been implemented in PCM algorithms.

The coupling between the velocities and pressure can be enhanced by solving them simultaneously at a grid point. For a point-by-point explicit scheme the convergence is very slow, especially for fine grids. Vanka (1986) has used the multigrid method with a coupled solution. This procedure was used to solve the discretized equations on a staggered grid. An explicit smoothing technique called *symmetrical coupled Gauss–Seidel* (SCGS) was proposed to update all four of the velocities at the faces of the computational control volume. An improvement on both convergence and CPU time was quoted.

Most earlier techniques utilized a staggered grid system. For a curvilinear coordinate system, which is most suited for turbomachinery flows, this staggered grid arrangement is extremely sensitive to grid smoothness. Chan and Sheedy (1990) overcame this problem by storing all dependent variables at the centroid of the control volume. The velocity components are linearly integrated from the discretized momentum equations in the neighboring node prints. Chan and Sheedy (1990) claim that this nonstaggered scheme provides sufficient ellipticity to maintain stability.

It is well known that excessive amounts of artificial dissipation will affect the accuracy of the solution, while convergence and stability of these solutions cannot be achieved without some dissipation. Therefore, it is necessary to use an optimum value for the artificial dissipation to maintain good accuracy, convergence, and stability characteristics. Basson and Lakshminarayana (1994) investigated the artificial dissipation inherent in many of the semi-implicit formulations (e.g., upwind differencing, power-law, QUICK, and pressure weighting) used in the pressure-based method. This can affect the accuracy of the solution in regions (e.g., inviscid flow) where physical dissipation is small. In view of this, second and fourth order difference artificial dissipation terms were introduced into the momentum and pressure equations (right-hand side of transformed equation 5.56) to control the amount of artificial dissipation, as follows:

$$\frac{1}{2}\epsilon_2\left[\frac{\partial}{\partial\xi}\left(|\rho JU|\frac{\partial\phi}{\partial\xi}\right) + \frac{\partial}{\partial\eta}\left(|\rho JV|\frac{\partial\phi}{\partial\eta}\right)\right]$$
$$-\frac{1}{8}\epsilon_4\left[\frac{\partial}{\partial\xi}\left(|\rho JU|\frac{\partial^3\phi}{\partial\xi^3}\right) + \frac{\partial}{\partial\eta}\left(|\rho JV|\frac{\partial^3\phi}{\partial\eta^3}\right)\right] \qquad (5.70)$$

where U and V are the contravariant velocities (Eq. A.3, Appendix A). A similar (fourth order difference term) is introduced into the pressure correction equation. The artificial dissipation can be controlled by varying values of ϵ_2 and ϵ_4. The effect of the amount of dissipation on the accuracy of the solution and the convergence rate was quantitatively demonstrated by Basson and Lakshminarayana (1994) for two-dimensional inviscid flow in a mildly curved duct, three-dimensional laminar flow in a square cross section elbow with strong secondary flows, and two-dimensional turbulent flow through a turbine nozzle. The adverse effects of artificial dissipation, particularly second-order dissipation, inherent in some commonly used algorithms, were clearly shown. The effect of artificial dissipation on the convergence rate was also demonstrated. The main conclusion drawn from their results was that the minimum amount of artificial dissipation that gives the required accuracy, but also an adequate convergence rate for a particular case, has to be used. This amount of dissipation is case-dependent. The direct inclusion of artificial dissipation terms provides control over the amount of dissipation used.

5.7.1.1 *Application of PCM and PSM Methods to Turbomachinery*
Application of these techniques to turbomachinery flows and subsequent modifications to improve accuracy and efficiency have been carried out by Moore (1985), Moore and Moore (1981, 1985), Moore et al. (1990), Hah (1984), Hah and Krain (1990), Rhie (1985, 1986), Hobson and Lakshminarayana (1991), Stow (1989), Basson and Lakshminarayana (1993), Ho and Lakshminarayana (1993), and others.

Moore's (1985) technique is similar to that of Pratap and Spalding (1976). They used a two-dimensional correction procedure on the cross-stream plane, which was not coupled to the longitudinal pressure gradient. A three-dimensional correction was employed to remove the local force residues after each sweep. A control volume with pressure corrections located at the center of the control volume was used. The turbulence was represented by an algebraic eddy viscosity model. They have computed a wide variety of flows, including flow through a varying area curved duct, centrifugal compressor, turbine cascade, and rocket pump inducer. Stow (1989) successfully used this code in the computational design of compressors and turbines. One typical computation by Moore et al. (1990) is the three-dimensional viscous flow in a rocket pump inducer (Fig. 1.7), shown in Fig. 5.22. Large radial outflow near blade surfaces as well as very high flow losses are predicted well. Similar comparisons between the experimental data and Navier–Stokes computations for a rocket pump inducer are shown by Bois et al. (1994). It should be noted here that the inducer operating at low flow coefficient has one of the most complex flow fields, including very high losses as measured by Lakshminarayana and Gorton (1977). It is encouraging that the present-day codes are able to capture such flows.

Hah (1984) solved the uncoupled equations on a staggered grid. A general coordinate transformation was used to represent the turbomachinery geome-

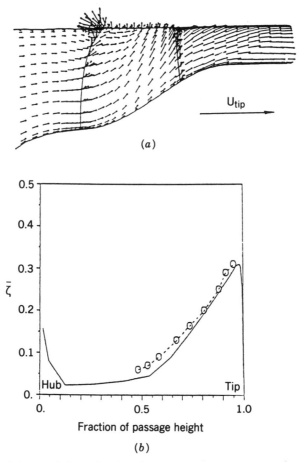

Figure 5.22 Rocket pump inducer flow field (noncavitating performance with air flow) (data from Bario et al., 1989; computation from Moore et al., 1990; reprinted with permission of AIAA.) (a) Velocity vectors near pressure side, 1% of pitch. (b) Pitch averaged loss coefficient at blade exit. Mass flow = 0.2109 kg / s, N = 10,000 rpm, ratio of axial velocity / blade tip speed = 0.084, $Re = \rho U_t D / \mu = 1.13 \times 10^6$, four-bladed, $\bar{\zeta} = 2(\bar{P}_{01})_{rel} - (\bar{P}_{02})_{rel}) / \rho U_t^2$.

try, with skewed upwind differencing. He used higher-order discretization for the convection term to reduce numerical diffusion; the quadratic upstream differencing by Leonard (1979) and skew upwind differencing by Raithby (1976) were extended and incorporated. The skew upwinding scheme is aimed primarily at reducing the error due to the streamline-to-grid skewness. Hah utilized one semi-implicit predictor step and two semi-implicit corrector steps and achieved third-order spatial accuracy. The code is capable of capturing both low-speed and high-speed shocked flows. Hah has computed a wide variety of turbomachinery flows (e.g., Hah, 1984; Hah and Krain, 1990; Copenhaver et al., 1993) such as diffuser flows, turbine endwall flows, centrifugal compressor flow fields, tip leakage flows, and transonic compressor flow. Hah's (1984) procedure, along with Chien's (1982) low-Reynolds-number turbulence model (Eqs. 1.84 and 1.86 modified to include the wall vicinity effect), was used by Copenhaver et al. (1993) to compute the flow field in a high-through-flow, single-stage transonic compressor flow field measured by Law and Wadia (1993). The characteristics were: (r_h/r_t) inlet $= 0.311$, tip diameter $= 43$ cm, speed $= 20,222$ rpm, pressure ratio $= 1.92$, $\eta_c = 85.4$, $\dot{m} = 210.5$ kg/s/m^2, and $(M_R)_{\text{tip}} = 1.629$. The measured and predicted performance (efficiency) and the casing pressure distribution are in good agreement, as shown in Fig. 5.23. They also note that Chien's model is the more appropriate model to use to capture the flow behind shock waves or separated flow regions. Leylek and Wisler (1991) used Hah's code to simulate spanwise and cross-flow mixing phenomena in compressors. They were able to identify several sources causing spanwise mixing and confirm the observations made through flow visualization.

Rhie and Chow (1983) used a pressure-weighted method to suppress pressure oscillations and developed a differencing scheme for a curvilinear coordinate system on a nonstaggered grid. The technique still employs PCM to couple the momentum and continuity equations. Rhie (1986) recently extended his earlier scheme to employ the *pressure-implicit split-operator* (PISO) concept with a multi-grid procedure to enhance convergence. During the pressure correction iteration, the density was treated implicitly for the mass flow balance. Numerical diffusion was introduced to damp instabilities associated with central differencing of convective terms. Rhie and his group (Rhie and Chow, 1983; Rhie, 1985) and Bansod and Rhie (1990) utilized the pressure-based method and computed a wide class of turbomachinery flows. This is a finite volume formulation, with a collocated grid and a central difference scheme for the basic equations. The pressure oscillations due to a nonstaggered grid are suppressed by the use of a fourth-order pressure dissipation term. They state that this term acts as a mass flux term and does not alter the second-order accuracy of the continuity equation; they computed the flow through a three-dimensional duct (Rhie, 1985), as well as through centrifugal compressors, with and without tip clearance (Bansod and Rhie, 1990). One of the complex cases computed by Bansod and Rhie (1990) is the centrifugal compressor tip loss mechanism as well as the entire passage flow. The blade tip was cusped and the Navier–Stokes code was run for different tip clearances. The results for the decrement in efficiency agree well with the measured data even with very coarse grid (within the tip clearance region), but the details of the flow field in the clearance region are predicted only qualitatively. It should be remarked here that all these computations were done with the k–ϵ turbulence model, and hence the complex features of tip vortex formation and jet–wake structure cannot be resolved as indicated earlier (e.g., see Fig. 5.4). However, the inviscid features of the flow, such as wall static pressure and meridional relative Mach numbers away from blade surfaces, are captured well.

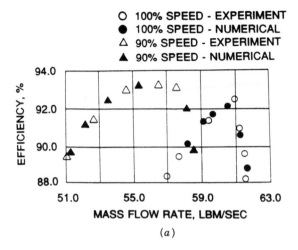

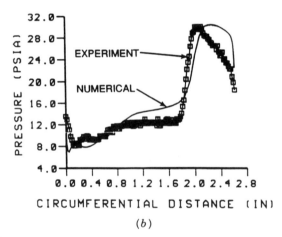

Figure 5.23 Prediction of efficiency and casing static pressure in an axial flow transonic compressor. (Copyright © Copenhaver et al, 1993; reprinted with permission of ASME.) (a) Comparison of rotor efficiency map. (b) Comparison of circumferential casing static pressure distribution 0.25 in. (6.4 mm) downstream of leading edge at stall condition (100% rotor speed).

Chan's (Chan et al., 1988; Chan and Sheedy, 1990) technique is similar to that of Rhie (1985) and utilizes (1) a second-order upwind scheme, (2) a preconditioned conjugate gradient method for acceleration, and (3) no user-defined smoothing function. They have computed the flow field in a turbine endwall region, the space shuttle main engine return duct, and rocket pumps. The code is used for both the design and the analysis.

Basson and Lakshminarayana (1993) developed an efficient grid generation scheme particularly suited to three-dimensional turbomachinery flows, especially the tip leakage flows. The resolution of tip clearance flow requires specialized grids. The abrupt transition from the blade passage to the tip gap, with sharp corners at the blade tip and large gradients in flow properties, make the grid generation a critical element in the computation. A commonly used approach, called "thin blade approximations" or "pinched tip approximations" has been used by earlier investigators to cusp the blade tips. (Bansod and Rhie, 1990; Kunz et al., 1993). This approach is only suitable for modeling the gross effects of the tip clearance flow because the tip geometry is not modeled correctly. Better modeling of the tip gap has been obtained by an embedded H grid (Basson and Lakshminarayana, 1993) which retains the actual geometry and provides a smooth transition from the blade tip to the gap.

In all the earlier schemes based on SIMPLE-type algorithms (Rhie, 1985; Hah, 1984; Hobson and Lakshminarayana, 1991), artificial dissipation is used indirectly (called *numerical dissipation*) as demonstrated by Basson and Lakshminarayana (1994). The direct inclusion of artificial dissipation is more common in methods which can be termed "time marching" (e.g., Pulliam, 1986). The numerical scheme developed by Basson and Lakshminarayana (1993, 1994), particularly suited to computational grids for the analysis of turbulent turbomachinery flows, is a semi-implicit, pressure-based scheme that directly includes artificial dissipation (Eq. 5.70) and is applicable to both viscous and inviscid flows. The values of these artificial dissipations are optimized to achieve accuracy and convergency in the solution. The numerical model is used to investigate the structure of tip clearance flows in a turbine nozzle. It is demonstrated, through optimization of grid size and artificial dissipation, that the tip clearance flow field can be captured accurately. The advantage of the inclusion of these terms directly, rather than indirectly through one-sided differencing schemes, is that the amount of artificial dissipation is clear and can be controlled. Basson and Lakshminarayana (1993) showed that the control provided by the direct inclusion of these terms is essential for the accurate modeling of complex flows on current practical grids. A low-Reynolds-number $k-\epsilon$ model (similar to Eqs. 1.84 and 1.86) has been incorporated into the code. They have computed the tip clearance flow field in a turbine cascade (Bindon, 1987) with $C = 0.186$ m, $S/C = 0.7$, tip gap $= 2.5\%$ chord, Re $= 4.7 \times 10^5$, $\alpha_1 = 0$, and $\alpha_2 = 68°$. The turning angle is therefore representative of typical turbine nozzles. The predicted and measured yaw angles (α_2) at 90% chord from the leading edge is shown in Fig. 5.24. The pitchwise location is measured from the pressure side. The spanwise location of the vortex (the spanwise location where the yaw angle crosses the midspan value, with underturning and overturning on either side) as well as the magnitude of under- and overturning are captured very well, except near the core of the vortex.

This procedure was also used to capture the secondary flow in a high-turning turbine cascade, measured by Gregory-Smith and Cleak (1992) in a

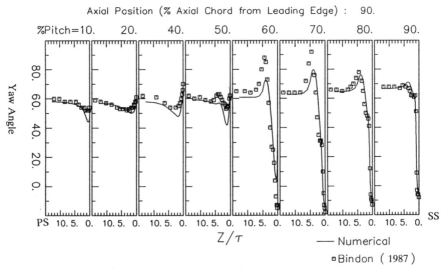

Figure 5.24 Yaw angle (α_2 in degrees) distribution at 90% of chord length from the leading edge. Z is the spanwise distance measured from end wall; τ is the tip clearance height. (Copyright © Basson and Lakshminarayana, 1993; reprinted with permission of ASME.)

cascade with $\alpha_1 = 42.75°$, $\alpha_2 = -68.7°$, $C = 224$ mm, $S = 191$ mm, and Re $= 4.0 \times 10^5$. The grid was clustered in the secondary flow region. The predicted and measured loss coefficient $\zeta [= (P_{01} - P_{02})/\frac{1}{2}\rho V_1^2]$ is plotted in Fig. 5.25. The major features of secondary flow vortex and the losses associated are captured well. Two vortex cores, which correspond to the passage vortex (with a maximum loss coefficient of 1.3) and the vortex resulting from the interaction between the passage vortex and the wake (see Fig. 4.25), are captured well. The maximum losses are slightly overpredicted. The computations were done with a fine grid ($165 \times 61 \times 45$). The computation with a coarser grid ($121 \times 51 \times 31$) was incapable of capturing the passage vortex and the loss cores accurately. Higher loss near midspan (150–200 mm) is caused by insufficient grids because most of the grids were clustered in the wake and secondary flow regions.

The code by Basson and Lakshminarayana (1993) has been extended to carry out time accurate solution, including rotor/stator interaction by Ho and Lakshminarayana (1993). An inner loop iteration scheme is used at each time step to account for the nonlinear effects. The unsteady flow field through the second-stage stator of a multistage compressor measured with a Laser Doppler velocimeter at United Technologies Research Center (UTRC) was computed. The experimental data in the second stage were acquired by Stauter et al. (1991). The UTRC multistage compressor is operated at $\phi = 0.51$, Re $= 2.5 \times 10^5$, and $U_m = 46.68$ m/s, which represents a reduced frequency of 8.48 based on the blade passing frequency.

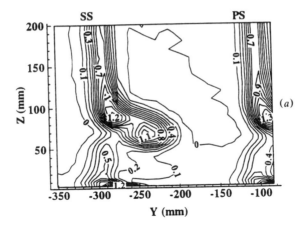

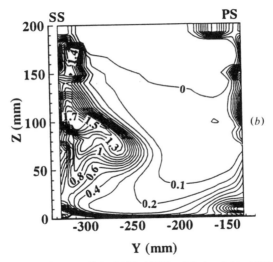

Figure 5.25 Total pressure loss coefficient (ζ) at 51% axial chord downstream of the trailing edge for a turbine cascade. The numbers on the contour denote values of ζ. (a) Experimental data. (b) Numerical prediction. (Copyright © Ho and Lakshminarayana, 1994; reprinted with permission of ASME.)

The time histogram of velocity profiles at 20% chord upstream of the stator (includes potential interaction between stator flow field and upstream wake) are shown in Fig. 5.26. The corresponding experimental data are placed side by side for comparison. The measured and the predicted profiles agree well, including distortions in the free stream. As the rotor wake approaches the stator leading edge, the fluctuating velocity increases significantly due to rotor/stator inviscid interaction. The rotor wake causes changes in incidence near the leading edge of stator causing unsteady pressure on the stator blade to increase substantially.

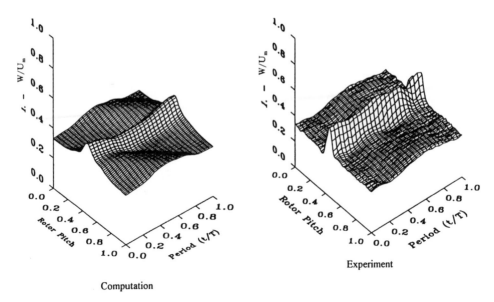

Computation

Experiment

Figure 5.26 Time-dependent rotor wake profile at 20% stator axial chord upstream of stator blade of a multistage axial flow compressor, T is period, W is total velocity in rotor frame. (Copyright © Ho and Lakshminarayana, 1993; reprinted with permission of ASME.)

5.7.1.2 Assessment of the Pressure-Based Methods (Uncoupled) The pressure-based methods have been in existence longer than any other scheme, and thus considerable development effort has been expended on this technique. Hence these methods have reached a high level of maturity and are widely used in industry and academia. They have also been used to incorporate and test a wide variety of turbulence models (algebraic eddy viscosity to full Reynolds stress) and have been utilized to predict a wide variety of mostly internal flows. Recent developments include (a) nonstaggered grids, (b) highly skewed and highly nonuniform grids, (c) multi-grid, (d) coupled-solution, (e) upwind differences, (f) introduction of artificial viscosity, and (g) extension to compressible and shocked flows. The method has found widespread use in the industrial design and analysis systems.

There are not many published attempts to evaluate various schemes, discretization, and grid sensitivity of pressure-based methods for turbomachinery. Chan et al. (1988) carried out such a systematic analysis, and some of their results are presented here. First- and second-order discretization schemes for convective terms were tried, and the results showed substantial influence on the pressure distribution. The convergence history for several solution schemes is shown in Fig. 5.27. A comparison between ADI, SIPSOL [Stone's (1968) strongly implicit procedure], and preconditioned conjugate gradient scheme [CGS (Kershaw, 1978)] shows the improved performance with CGS. The convergence rate for both ADI and SIPSOL was very rapid

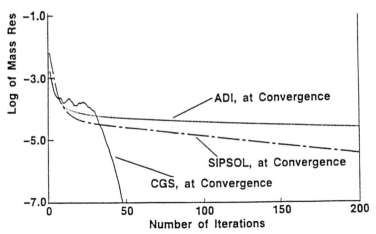

Figure 5.27 Convergence history of various techniques for Langston's turbine cascade (Langston et al., 1977). See Fig. 5.28 caption for details. (Copyright © Chan et al., 1988; reprinted with permission of the Pacific Center for Thermal Engineering.)

for the initial 20 iterations and slowed considerably after this. Chan et al. (1988) claim that CGS is the more suitable scheme in a highly stretched grid system (as in a turbine case) which results in an ill-conditioned matrix system. Likewise, Hadid et al. (1988) showed that the PISO scheme provides a much faster convergence than does a SIMPLE algorithm.

One of the dramatic differences between various computations using the pressure-based methods is the loss distribution for Langston's cascade as shown in Fig. 5.28 (see Fig. 4.23 for outlet angle predictions for this cascade). This leads to an apparent dilemma as to the ability of various codes to predict losses and sensitivity to grid and turbulence models. The loss computations with finer grids (Chan and Sheedy, 1990) resulted in improved prediction, but Moore's computations with an algebraic eddy viscosity model apparently show the best agreement with the data.

Basson and Lakshminarayana (1994) carried out a systematic study of the effects of artificial dissipation (ϵ_2 and ϵ_4 in Eq. 5.70 and a similar term in the pressure correction equation) on the flow field in a turbine cascade (Dring et al., 1986) and in a 90° curved duct (Taylor et al., 1981) for which accurate data are available. For the turbine cascade, a physically unrealistic total pressure loss was predicted even in the inviscid region. The higher values for second-order dissipation ($\epsilon_2 = 1$, $\epsilon_4 = 0$ in Eq. 5.70) provided unrealistic solution in the high-velocity gradient region near the leading edge, which resulted in stagnation pressure losses in the inviscid region. Lower values ($\epsilon_2 = \epsilon_4 = 0.4$) provided near-perfect agreement with the data in the inviscid region and the wake. Likewise, optimum values of $\epsilon_2 = 0.1$, $\epsilon_4 = 0.4$ were found to provide good prediction of the streamwise and the cross-flow velocity profiles in the 90° duct. Higher values of ϵ_2 and ϵ_4 did not provide

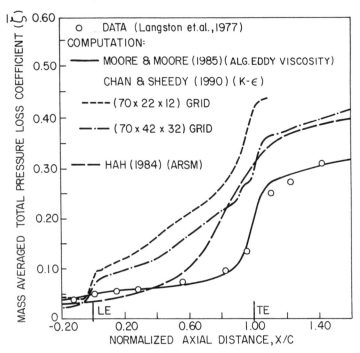

Figure 5.28 Comparison of mass averaged loss distribution from various codes for Langston's turbine cascade. $C = 281.3$ mm, $A = 1$, $S/C_x = 0.95$, turning angle = 110°, $\alpha_1 = 45.3°$. $\bar{\zeta}$ is the loss coefficient averaged over the entire passage, X is the axial distance from the leading edge. (Copyright © Lakshminarayana, 1991 reprinted with permission of ASME.)

good agreement with the data, especially in the high-gradient (vortex) regions.

It is thus clear that extreme caution should be exercised in using the Navier–Stokes code. Careful attention should be paid to the selection of artificial viscosity (dissipation), grid, discretization scheme, and turbulence models. The algorithm and the codes have to be validated, calibrated, and optimized for the class of geometry and flows for which they are to be used.

As indicated earlier, very efficient procedures are available for computation of the inviscid flow field for incompressible and subsonic flows, especially in 2D cascades. Hence, the pressure field can be resolved to good accuracy (except in the regions where viscous effects are large (e.g., suction surface near trailing-edge regions)). Availability of such codes makes PCM and PSM very attractive, and it provides a quick and accurate estimate of the input pressure field for these Navier–Stokes codes. In a coupled time-marching scheme, the pressure is resolved as part of the solution, and convergence of pressure is slower than the velocity field. Because the equations are solved sequentially in the pressure-based method, only the inversion of a scalar matrix is required. This can result in significant reduction in memory and storage as compared to coupled system (pseudocompressibility and time-marching solutions). Because convective velocity components are frozen at each iteration, and coefficients of the discretized equations are the same for all three momentum equations, the calculations can be distributed on three separate central processing units, thus taking advantage of parallel processors. This can provide a significant reduction in computational time.

The pressure-based method has been applied successfully to incompressible, compressible, laminar, turbulent, and low- and high-Reynolds-number flows for a large class of turbomachinery geometries. The grid system may be orthogonal or nonorthogonal. The method is flexible; either the finite-difference or finite-volume formulation can be employed. The technique has not been widely used for the computation of flows with shocks, even though many recent attempts (Rhie and Stowers, 1987; Copenhaver et al., 1993) indicate that shocked flows can be successfully computed; this method is as attractive as the time-marching technique described later.

5.7.2 Pressure-Based Method (Coupled Solution)

In the techniques described earlier, the momentum, continuity, and pressure equations are solved in an uncoupled manner and the solutions are updated. This may result in convergence problems, especially in situations where the gradients of flow variables are large (e.g., leading and trailing edges, etc.). Kirtley and Lakshminarayana (1988) developed a coupled pressure-based method, where all the equations were solved simultaneously. The continuity equation was modified with the introduction of a pressure correction term, analogous to the pseudocompressibility technique described later. The modified continuity equation is coupled to the momentum equation and solved using a spatial integration procedure.

The method differs from those of others in the manner in which the streamwise pressure gradient is handled. Various methods that have proved to be successful in computing internal incompressible flows have relaxed some conditions during the iterative process. In the techniques described in Section 5.7.1, the procedure involves relaxing the continuity constraint by lagging the solution of the continuity equation. In the pseudocompressibility technique, which utilizes a time-marching method, an extra term is introduced in the continuity equation. Kirtley and Lakshminarayana (1988) developed a new technique by introducing an additional term in the continuity equation and solving the entire set of equations as a coupled system. The additional term in the converged solution goes to zero. This is an implicit technique and utilizes the approximate factorization scheme due to Briley and McDonald (1977) and Beam and Warming (1978) described in Appendix B (Section B.5.3). The solution is advanced in the streamwise direction instead of time.

Because a space-marching algorithm is under consideration, the continuity equation for incompressible flow is written as

$$\alpha(p^n - p^{n-1}) + \nabla \cdot \mathbf{V} = 0 \tag{5.71}$$

where n is a global iteration index, and α is a relaxation parameter. This will introduce artificial mass through a source term, which goes to zero as the multiple pass procedure converges. The solution scheme employed was very similar to the approximate factorization linear block implicit scheme given in Appendix B (Eq. B.47). The equations can be cast into the form given by Eq. A.4 in Appendix A (with $\partial \hat{q}/\partial t = \partial \hat{T}/\partial \xi = 0$) and solved by an approximate factorization method. The Euler implicit scheme is used, which can be written as

$$\left(L_\xi \cdot + L_\eta \cdot + L_\zeta \cdot \right) \Delta \hat{q}_i = R \tag{5.72}$$

where: L_ξ = convection terms in ξ direction; L_η = convection plus diffusion terms in η direction; L_ζ = convection plus diffusion terms in ζ direction; $\Delta \hat{q}_i = \hat{q}_i - \hat{q}_{i-1} = (\Delta p, \Delta u, \Delta v, \Delta w)^T$; i = steamwise location; $\Delta u, \Delta v, \Delta w$, and Δp are change in variable from station $i-1$ to i; R = terms from Euler implicit differencing, cross-derivative diffusion terms, and the source term in the continuity equation.

If Eq. 5.72 is solved using the linear block implicit scheme for time-marching procedures, developed by Briley and McDonald (1977), the resulting factorization error significantly reduces the overall accuracy of the method. Unlike that of time-marching methods, the error is multiplied by the streamwise change in Δq, which does not go to zero. This error will accumulate in each streamwise step.

The following scheme was used by Kirtley and Lakshminarayana (1988) to reduce this factorization error. We can rewrite Eq. 5.72 as

$$\left(L_\xi \cdot + L_\eta \cdot + L_\zeta \cdot \right) \Delta s^n = R' = R - \left(L_\xi \cdot + L_\eta \cdot + L_\zeta \cdot \right) \Delta \hat{q}_i^{n-1} \tag{5.73}$$

where $\Delta s^n = \Delta \hat{q}_i^n - \Delta \hat{q}_i^{n-1}$ and n is the iteration level.

The operator on the right-hand side of Eq. 5.73 is split, as follows:

$$\left(L_\xi \cdot + L_\eta \cdot \right) \Delta s^* = R'$$

$$\left(L_\xi \cdot + L_\zeta \cdot \right) \Delta s^n = L_\xi \Delta s^*$$

This is repeated until R is very small. The convergence can be improved by including a relaxation parameter

$$\Delta \hat{q}^n = \Delta \hat{q}^{n-1} + \omega \Delta s^n$$

The optimum value of α was found to be 0–5 for laminar flows and 10–30 for turbulent flows. Artificial dissipation is intrinsically introduced through the one-sided differencing of the streamwise convection term. The method has good convergence characteristics, but the computational efficiency has not been compared with other techniques. A wide variety of flows have been computed using this code, including flow in a 90° duct; flow in an S duct; flow in cascade wakes; endwall flows, blade boundary layers and compressor rotor passage flow (Kirtley and Lakshminarayana, 1988; Kirtley, 1987).

5.7.3 Time-Marching Techniques

The most suitable technique for compressible flow is the time-marching technique. In this technique, the time derivatives in the Euler/Navier–Stokes equations are retained. For compressible flow, the Euler equation is hyperbolic and the Navier–Stokes is considered parabolic–hyperbolic. The time-dependent Navier–Stokes equation is essentially parabolic in time and space, although the continuity equation has a hyperbolic structure (Hirsch, 1990, Vol. 1, p. 135). Very efficient techniques have been developed to solve these equations for external and internal flows. The advantages of these time-iterative (or time-dependent or time-marching) techniques are as follows:

1. The same code/algorithm can be used for the solution of all flow regimes (low subsonic to hypersonic) and of viscous and inviscid equations (steady and unsteady flows).
2. The technique is easy to implement and vectorize to take advantage of supercomputing capabilities.
3. The equations are solved as a coupled system. The pressure, velocities, density, enthalpy, entropy, and temperature are solved simultaneously.
4. The code/technique is flexible and the same code can be employed (with modifications to boundary conditions) for external as well as a variety of internal flows: 2D cascade, 3D turbomachinery blade row and multistage turbomachinery flows, 3D unsteady rotor and stator interaction problems.
5. Extensive development work has been spent on these techniques by internal and external aerodynamicists, and hence the techniques are mature with considerable user experience in handling all flow situations and geometrics.
6. For complex flow geometries and flow fields, the technique is comparable in computational efficiency with pressure-based methods.

The major disadvantages are as follows:

1. Because the technique is time-iterative and often requires very small time steps, the computational efficiency may be low. This has been overcome, to some extent, by resorting to acceleration schemes such as local time stepping, multi-grid, and vectorization of the code.
2. The technique is not suitable (without modification) for incompressible or very low Mach number flows. (The pseudocompressibility technique described later overcomes this problem by introduction of a time derivative in the continuity equation.)

There are two classes of methods of solving the time-dependent hyperbolic equations: *explicit and implicit*. Explicit methods, where spatial derivatives are evaluated using the known conditions at the old time level, are simpler and more vectorizable. The implementation of the boundary conditions is also easier. The major disadvantage lie in the conditional stability dictated by a CFL limit (Appendix B), which, for most explicit schemes, states that the numerical domain of dependence must contain the physical domain of dependence for hyperbolic equation. Unless the interest lies only in the steady-state solution, the convergence is usually slower and requires more computational time than does an implicit method. Implementation of local time-stepping, residual smoothing, and multi-grid can enhance convergence rates.

Implicit methods, where the unknown variables are obtained from a simultaneous solution of a set of equations, usually allows for larger time steps and faster convergence and are attractive for both steady and unsteady flows. Direct solutions involve inversion of a block pentadiagonal matrix (2D) or block septadiagonal matrix (3D). This is computationally prohibitive, so approximate factorization is often employed to convert this into a block tridiagonal system. This in turn introduces factorization error and may lead to stability problems.

Nevertheless, both explicit and implicit schemes are widely employed in the computation of turbomachinery flows. Both of these techniques are undergoing development, and some of the disadvantages of both of these techniques have been overcome. Some of the earlier methods utilized the explicit techniques exclusively. Later developments involved the implicit schemes, and recently the explicit schemes have been revived for both steady and unsteady flow computation. A summary of the time-iterative techniques used and codes developed for turbomachinery are given in Table 5.2.

5.7.3.1 Boundary Conditions for Time-Iterative Solution of Flows for Turbomachinery Blade Rows
Specification and enforcement of proper boundary conditions are essential to accurately capture the physics of the flow. The specification should ensure that the boundary conditions lead to well-posed systems, which implies uniqueness of the solution. This depends on the boundary conditions, based on either the value of the variables or their derivatives. The important variables are $\mathbf{W}$, p, ρ, T, k, ϵ, P_o, and h_o. There are some subtle differences in the boundary conditions used for finite-difference, finite-volume (cell), explicit, and implicit techniques. Many of these boundary conditions are similar for all computational techniques. The general boundary conditions and method of implementing them are described below. The types of boundary conditions can be classified as follows:

Solid Surfaces (Blade Surfaces, Annulus and Hub Walls). If there is no injection, no slip boundary condition is imposed for $\mathbf{W}$ in the Navier–Stokes equation. For inviscid equations (without injection) the velocity vectors are parallel to the surface. For temperature, either the wall temperature or the normal (temperature) gradient is specified. For the nonadiabatic wall condition, it is necessary to carry out a heat transfer analysis as indicated in Chapter 7 to derive the wall temperature. If the wall is adiabatic, $\partial T / \partial n = 0$ is imposed on the surface. For inviscid and adiabatic flows, the pressure and velocities are related by energy equation; hence the pressure can be derived at the boundary from a prescribed or derived velocity. For the Navier–Stokes equation, typically, either $\partial p / \partial n = 0$ is prescribed or $\partial p / \partial n$ is evaluated by solving the normal momentum equation at the surface. For example, for two-dimensional viscous flow on a straight surface, the normal momentum equation can be approximated by

$$-\frac{1}{\rho}\frac{\partial p}{\partial n} = u\frac{\partial v}{\partial s} + v\frac{\partial v}{\partial n} - \mu\frac{\partial^2 v}{\partial n^2} \tag{5.74}$$

If we know the velocity field from a previous iteration or time step, the normal pressure gradient on the blade surface can be derived as a compatibility condition.

Inflow and Outflow Boundary Conditions. The prescription of inflow and outflow boundary conditions is one of the most sensitive tasks. Strictly speaking, these surfaces should be located far upstream and far downstream, where the influence of the blade row under consideration is negligible. Current computer capability and sometimes grid topology makes it prohibitive to employ the ideal conditions. Hence, most investigators locate them usually about one to one-half chord upstream and downstream. The prescription of the boundary conditions at inlet and exit boundaries depends on the flow regime (subsonic or supersonic). Most authors prescribe T_o (or h_o), P_o, and either the flow angles or the relative velocity components at inflow boundaries when the axial velocity is subsonic. The static pressure distribution is usually prescribed at the outlet for subsonic flow, and the inlet static pressure is resolved from the equations. For supersonic flow, properties T_o, P_o, and tangential velocity are prescribed upstream, and the downstream should have no influence on the upstream flow. If the equations are solved in the rotating coordinate system, inlet relative stagnation pressure (P_{oR}), temperature (T_{oR}), and the blade speed are typically prescribed. The static pressure is usually prescribed at the hub at the exit, and the radial distribution of exit pressure is determined from a simple radial equilibrium equation ($dp/dr = \rho V_\theta^2/r$) using the computed flow outlet angle. If the k–ϵ model is employed for turbulence closure, prescription of the upstream values for k and ϵ are typically derived from reasonable estimates for inlet turbulence intensity and length scales. The downstream condition has to be iterated or a gradient boundary condition is prescribed ($\partial k/\partial x = \partial \epsilon/\partial x = 0$).

One of the alternate approaches, if the flow is inviscid upstream and downstream, is to prescribe characteristic variables at these boundaries (Scott, 1985; Thompkins, 1982) given below:

$$s = \rho - \frac{p}{a_\infty^2}, \qquad m = \frac{p}{\rho_\infty a_\infty} + u, \qquad n = \frac{p}{\rho_\infty a_\infty} - u$$

Here s is the entropy, and m and n are the right and left running characteristics. Prescribing physical values instead of characteristic values often results in nonphysical behavior.

Application of these conditions to three-dimensional viscous flow in a stage is cumbersome. If only spatial nonuniformity is present, these can be easily prescribed at the inflow boundary. If the inflow is unsteady (for example, spatial distortions in a stator are sensed by the downstream rotor as temporal distortion) as in rotor/stator interaction, then time-dependent nonreflecting boundary conditions should be specified (Giles, 1990a, b).

Periodic Boundary Conditions. To simulate infinite blade-row conditions, it is essential to enforce periodic boundary conditions upstream and downstream of the blade row. This is accomplished by setting the dependent variables equal at the two periodic boundaries, using the adjoining periodic surfaces. In implicit techniques (elliptic and hyperbolic) a tridiagonal matrix of length N results for the solution of the flow between boundaries. For the enforcement of periodic boundary conditions, an additional grid point is incorporated into the system, resulting in a cyclic $(N + 1)$ system of simultaneous equations; this results in a cyclic tridiagonal matrix system.

5.7.4   Explicit Time-Marching Method

In the explicit schemes, the spatial derivatives are evaluated using known conditions at the old time level. The explicit schemes used widely for the computation of turbomachinery flows are as follows:

1. The Lax–Wendroff scheme (1964) is second-order accurate in time and space. This scheme has the advantage of simplicity and robustness, and the flow variables need to be stored only at a single time level.
2. The predictor–corrector method due to MacCormack (1969) is a modified version of the Lax–Wendroff scheme and has been widely used by external aerodynamicists. Many early turbomachinery computations were performed with this scheme, which is also second-order accurate in time and space and involves a two-step procedure. Hence, the variables have to be stored in two time steps. The method is less

complex than the Lax–Wendroff scheme and requires reasonable storage. The CFL limit is one for both of these schemes.

3. The Runge–Kutta-type schemes (Jameson et al., 1981) have found wide applications in both the internal and external flows. The four-stage scheme permits a maximum CFL number of $2\sqrt{2}$ and is a considerable improvement over the two methods listed above. This technique is likely to be explored widely for both the time-accurate (unsteady) and the steady-state solutions.

The advantage and disadvantage of the explicit versus implicit technique depends on the computational efficiency—that is, the computer time required to achieve a converged and accurate solution. Let us examine this aspect. The total stability criteria for convection, pressure, and diffusion terms in a two-dimensional viscous flow are given by (similar to the CFL condition given in Section B.5.2.1; see Anderson et al., 1984, p. 484, for details).

$$\Delta t \left\{ \left[\frac{|u|}{\Delta x} \left(1 + \frac{2}{(R_c)_x} \right) + \frac{|v|}{\Delta y} \left(1 + \frac{2}{(R_c)_y} \right) \right] + a \sqrt{\frac{1}{(\Delta x)^2} + \frac{1}{(\Delta y)^2}} \right\} \leq 1$$

$$(5.75)$$

where $(R_c)_x$ is the cell Reynolds number given by $(R_c)_x = \rho |u| \Delta x / \mu$.

Outside the viscous region, viscous diffusion is not significant, and grid spacings $(\Delta x, \Delta y)$ are chosen to resolve only the convection and pressure terms $(uu_x, \text{etc.})$, because moderate time steps can be taken. Inside the viscous layer and near the solid surface, diffusion terms dominate and the viscous layer can only be resolved with a fine mesh; small time steps are required in this region for turbulent flow, where the cell Reynolds number could be very small. The terms such as $2/(R_c)_x$ may dominate the criterion for the time step in the viscous and turbulent regions. This is one of the major drawbacks of the explicit methods, which require small time steps to resolve the viscous layer as well as the inviscid shock layers. In earlier computations, a constant time step was used in the entire region, but this is not necessary. If the interest lies only in steady-state solution, the scheme can be accelerated by using a time step consistent with the local CFL number for each grid point. Hence, $\Delta t_{i,j,k}$ can be varied to keep the CFL number constant over the grid. This procedure is called a *"variable or local time step"* and significantly increases the convergence rate on highly stretched grids.

There is very little information available on the comparison between finite difference and finite volume methods. Because the integration of conservation variables is carried over a control volume in the finite-volume technique, the error in conserving momentum fluxes should be lower than that in the finite-difference techniques. Hirsch (1990) lists other advantages such as direct discretization in physical space, utilization of arbitrary mesh configuration, and elimination of explicit computation of metric coefficients.

5.7.4.1 Lax–Wendroff Scheme and Its Variants

One of the most widely used class of schemes is the Lax–Wendroff (1960, 1964) scheme. Even though these techniques are developed for linear equation, many modifications have been made to make them applicable to nonlinear, viscous flow in complex situations. An excellent treatment of these methods, including various modifications, stability and convergence characteristics, and boundary treatment, is given by Hirsch (1990, Vol. 2).

The basic Lax–Friedrichs scheme can be written as (for two-dimensional inviscid form of Eq. 1.18, in Cartesian coordinates) ($\partial q / \partial t = -\partial E / \partial x - \partial F / \partial y$)

$$q_{i,j}^{n+1} = \frac{1}{4}\left(q_{i+1,j}^n + q_{i-1,j}^n + q_{i,j+1}^n + q_{i,j-1}^n\right)$$

$$-\frac{\Delta t}{2\Delta x}\left(E_{i+1,j}^n - E_{i-1,j}^n\right) - \frac{\Delta t}{2\Delta y}\left(F_{i,j+1}^n - F_{i,j-1}^n\right) \quad (5.76)$$

This scheme is mildly dissipative (which may improve robustness), but does not introduce enough numerical dissipation to capture shocks correctly or even to prevent odd–even uncoupling on moderately stretched grids.

One of the earliest applications of this explicit scheme to turbomachinery is due to McDonald (1971), who computed inviscid flows through transonic cascades. He employed a finite-volume technique, writing the continuity, momentum, and energy equations in time-dependent integral form, for a hexagonal grid (finite area) system. He modified the scheme by automatically adjusting the damping (artificial dissipation) to provide a proper balance between accuracy and stability, and the value of artificial dissipation decreased as the steady state was approached. He showed good agreement with the measured pressure distribution for a turbine cascade. This technique was extended to 3D Euler flows by Denton (1975, 1982). The earlier technique developed by Denton in 1975 has undergone many changes. These efforts were mainly directed at the solution of 2D and 3D Euler equations for turbomachinery.

McDonald's (1971) and Denton's (1975) methods employ a finite-volume approach (see Section B.1, Appendix B, for basic technique). The grid is set up and the equations are solved in the physical plane. If the flow field is divided into a finite number of control volumes Δv, then continuity, axial, tangential, and radial momentum equations and energy equations (Eqs. 5.1,

1.2–1.5, and 4.44–4.47) in the conservative form for inviscid flow are given, respectively, by Denton (1975):

$$\Delta t \sum (\rho W_z dA_z + \rho W_\theta dA_\theta + \rho W_r dA_r) = \Delta \nu \cdot \Delta \rho \tag{5.77}$$

$$\Delta t \sum \left[(p + \rho W_z^2) dA_z + \rho W_\theta W_z \, dA_\theta + \rho W_z W_r \, dA_r \right] = \Delta \nu \cdot \Delta (\rho W_z) \tag{5.78}$$

$$\Delta t \sum \left[(\rho W_z r W_\theta \, dA_z + (p + \rho W_\theta^2) r dA_\theta + \rho W_r r W_\theta \, dA_r) \right]$$
$$+ \Delta t (2 \Omega r \rho W_r) \Delta \nu = \Delta \nu \cdot \Delta (\rho r W_\theta) \tag{5.79}$$

$$\Delta t \sum \left[\rho W_r W_z \, dA_z + \rho W_\theta W_r \, dA_\theta + (p + \rho W_r^2) \, dA_r \right]$$
$$- (\Delta t) p \cdot \frac{\Delta \nu}{r} - (\Delta t) \rho \frac{V_\theta^2}{r} \cdot \Delta \nu = \Delta \nu \cdot \Delta (\rho W_r) \tag{5.80}$$

$$\Delta t \sum (\rho W_z I dA_z + \rho W_\theta I dA_\theta + \rho W_r I dA_r) = \Delta \nu \cdot (\rho e_o) \tag{5.81}$$

$$I = h_o - \Omega r(V_\theta) = h_{oR} - U^2/2 = c_p T_{oR} - U^2/2$$

In the above equations W_r, W_θ, and W_z are relative velocities, $\Delta \nu$ is the volume of the element and dA_r, dA_θ, and dA_z are the projected areas of the control volume of the element in r, θ, and z coordinate directions, respectively. Refer to Eq. 4.33 and its interpretation for limitations of energy Eq. 5.81. Denton (1975) employed cuboids for the finite volume, with the node at each center. The fluxes through all sides of the control volume are calculated using the average flow properties at the four corners of the face. For stability reasons, it is necessary to provide underrelaxation in the scheme. Details are given in Denton (1982).

A large number of computations have been carried out using this technique. In many instances the component codes are used in quasi-three-dimensional and quasi-viscous iterations. McDonald (1981) computed the flow in a transonic compressor (Dunker et al., 1977) using an axisymmetric streamline curvature calculation (Section 4.2.3.6 and 4.2.4) followed by blade-to-blade solutions using the technique just described (McDonald, 1971). The blade-to-blade flow calculation uses input data from a through-flow calculation in the form of inlet Mach number and angle, exit angle, static pressure ratio, meridional streamline angle, and stream tube height distribution. The profile loss is modeled by using a skin friction correlation. This is a prime example of how an inviscid, two-dimensional time-marching code can be integrated with existing quasi-three-dimensional iteration techniques to predict and design the flow path. The results indicate excellent agreement with the data from 18% to 68% span of a transonic compressor, but the presence of complex shock structures near the tip (89% span) is not captured well. This region may need a full Navier–Stokes solver to accurately predict the interaction between the end wall and the blade-to-blade flow field. Povinelli

(1984) carried out a detailed assessment of Denton's inviscid code and its ability to predict turbine rotor flows. The code gave reasonable agreement with outlet temperature distribution but poor agreement with outlet angle distribution; it did not predict the significant underturning at midspan rotor position, due to strong secondary flows present in the low-aspect-ratio blades. He concluded that "development of a three-dimensional viscous analysis for the rotor flow fields appears necessary."

Similar computations have been carried out by Denton (1982). The viscous corrections were applied using the displacement thickness calculated from a momentum integral technique (Section 5.5.3 and Singh, 1982). This involves a fully three-dimensional Euler calculation with viscous corrections applied through boundary layer global parameters. Hence, this technique has less empiricism than the McDonald (1981) code. The results were compared with LDV data for a transonic fan by Pierzga and Wood (1985).

The local time step approach can be used to accelerate the scheme. Because most of the applications of this technique are for the Euler equations, the improvements are not substantial and depend on grid stretching. The most successful acceleration scheme for the Euler equation is the multi-grid scheme. In the multi-grid scheme, a sequence of coarser grids is used to more effectively damp the low-frequency error component of the solution. In the classical multi-grid technique (e.g., Brandt, 1979; Jameson and Baker, 1984), a forcing function is added to ensure that the coarse-grid solutions maintain fine-grid accuracy. In Ni's (1982) multi-grid scheme, the coarse-grid equations are written in delta form, so that if the scheme converges, the coarse grid does not affect the fine-grid accuracy. Denton (1982) was able to accelerate the computation by a factor of three through the use of multi-grid.

Ni (1982) used a one-step Lax–Wendroff technique to predict the inviscid flow field in cascades and turbomachinery rotors. This is a finite-volume, explicit technique with the cell-centered solution algorithm and with multi-gridding to improve convergence. The scheme is second-order accurate in time and space. Ni has extended this technique to a time-accurate solution (Ni and Bogoian, 1989; Ni and Sharma, 1990). The finite-volume equations are similar to Eqs. 5.77–5.81, and the procedure consists of two steps. In the first step, a finite-volume spatial approximation of the governing equation is applied to nonoverlapping flow cells. The corrections to each cell node are determined by the distribution formulae, to provide the effect of upwind differencing to ensure the proper domain of dependence. As explained earlier, the computational efficiency of explicit methods is dictated by CFL numbers, and hence these techniques tend to be expensive. The computation can be accelerated by using local time-stepping and multi-gridding, both of which were incorporated by Ni. In the multi-grid technique developed by Ni (1982), the solution on the fine grid is obtained by cycling the numerical procedure between the fine and coarse grids. The coarse grid is used to propagate the corrections on the fine grid rapidly throughout the flow field, thus improving the convergence rate. The multi-grid technique can have several levels, as many levels as one can divide the grid approximately in half. Ni employs two levels, one coarse grid and the other fine grid. The coarse grid is obtained by removing every other fine grid point. In Ni's scheme, one iteration cycle consists of solution on the fine grid, and one application of coarse-grid procedure to each level of progressively coarser meshes. Details of the transfer formulation and the technique can be found in Ni (1982), and a brief summary is given in Appendix B (Section B.5.2.1). Ni and Bogoian (1989) utilized these techniques, developed a three-dimensional Euler code, and obtained good predictions for a turbine blade. They showed excellent agreement between the data and the predictions for blade pressure distributions. A savings of 70% CPU time was achieved with the multi-grid procedure as compared to an all-fine-grid solution. Huber and Ni (1989) later used the code for an advanced turbine design.

The above technique has been extended to solve the Navier–Stokes equation by Chima (1985) and Davis et al. (1988). The viscous stress terms and conduction terms are calculated by Davis et al. using central differences and stored at the computational nodes and, as such, are included in the integration step to calculate first-order changes in the flow variable at the cell center. Viscous effects are included by integrating the Navier–Stokes equation around a secondary control volume, consisting of cell centers of adjoining primary cells. The flow variables are updated according to Ni's distribution formulae and the multi-grid scheme. The Baldwin–Lomax (1978) algebraic turbulence model (Eqs. 1.77–1.79) is used for closure. Barnett et al. (1990) used Davis et al.'s (1988) code to predict flow field in a transonic compressor cascade (Design incidence angle of 7°, $M_1 \sim 0.73$, $M_2 = 0.49$, $\alpha_1 = 40°$, $\alpha_2 = 91°$, Re $= 9 \times 10^5$). The results shown are compared with data in Fig. 5.29. The Navier–Stokes solver and the zonal method (described later) provide good agreement with the data. The scheme successfully predicted the losses, as well as onset of separation. The prediction marked IVICAS is based on coupled Euler-boundary layer solution (described in Section 5.8.1). The IVICAS-total loss includes shock and viscous losses (boundary layer).

Vuillez and Veulliot (1990) used a two step Lax–Wendroff scheme (explicit, cell-centered, finite volume). This is second-order accurate in the inviscid zone and only first-order accurate in the viscous regions, because dissipative terms are computed only once per iteration. The stability is ensured by second-order nonlinear and fourth-order linear artificial viscosity terms. Both local time-stepping and multi-grid techniques are utilized for accelerating the convergence to steady state. Good agreement with data is shown for a supersonic cascade and wide chord supersonic fan.

Another modified version of the Lax–Wendroff technique is the hopscotch method, which combines features of both the explicit and the implicit schemes. The explicit and implicit schemes are combined at alternate points in a computational mesh to obtain improvement over previous explicit

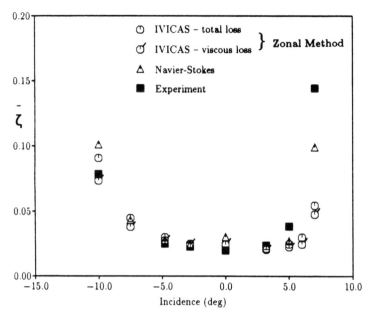

Figure 5.29 Loss distribution for a supercritical transonic compressor cascade (data from Stephen and Hobbs, 1979). (Copyright © Barnett et al., 1990; reprinted with permission of ASME.)

schemes. Gourlay and Morris (1972) were the first to apply the hopscotch technique to hyperbolic systems. This is a two-step scheme employing central differences for the spatial derivatives and backward differences for the time derivatives. The first sweep is equivalent to a single, linear extrapolation in time for the dependent variable. This scheme has been modified by Delaney (1983) and Sheih and Delaney (1987) and has been used to develop a 3D Euler code for turbomachinery, with the first sweep ($i + j + n$ = even) the same as Eq. 5.76 and the second sweep ($i + j + n$ = odd) given by

$$q_{i,j}^{n+1} = q_{i,j}^{n} - \frac{\Delta t}{2\Delta x}\left[E_{i+1,j}^{n+1} - E_{i-1,j}^{n+1}\right] - \frac{\Delta t}{2\Delta y}\left[F_{i,j+1}^{n+1} - F_{i,j-1}^{n+1}\right] \quad (5.82)$$

where $q_{i,j}^{n}$ is the first term on the right side of Eq. 5.76. For a repetitive application of this two-sweep process, the first sweep is equivalent to a simple linear extrapolation in time of the dependent variable:

$$q_{i,j}^{n+1} = 2q_{i,j}^{n} - q_{i,j}^{n-1}$$

This simplification, they observe, results in a twofold increase in the speed of the algorithm. The three-time-level process indicated in the above equation requires only two-time-level computer storage. The hopscotch method is first-order-accurate in time, which is sufficient for the steady-state solution.

Delaney (1983) and Sheih and Delaney (1987) used this technique with the boundary conditions and the CFL conditions described earlier, as well as an "O" grid to predict 2D and 3D inviscid flows. The 3D predictions are for a flared-wall turbine cascade. Excellent agreement between the measured and predicted pressure distribution, the critical velocity ratio, and the flow angle is shown. The efficiency of the hopscotch method as compared to other explicit techniques and implicit techniques is not yet known.

The results from these Euler codes (which employ the Lax–Wendroff scheme) for compressor flows indicate very good shock-capturing ability at both the peak efficiency and the peak pressure rise conditions. The 3D Euler code with a simple boundary-layer correction scheme (Singh, 1982) and a 2D Euler scheme combined with axisymmetric solution (McDonald, 1981) were able to predict the complex flow features in a transonic compressor; this demonstrates the utility of these codes in the analysis and design of turbomachinery. Some of the deviations between the computed and the measured flow fields in the tip and trailing-edge regions and in the shock–boundary-layer region indicate the need for a full Navier–Stokes solution in these regions.

5.7.4.2 Predictor–Corrector Method

The predictor–corrector method proposed by MacCormack (1969) is a two-step procedure based on the Lax–Wendroff scheme, and it is widely used for both internal and external flows. The method is second-order-accurate in both time and space. It can be used for both steady and unsteady compressible flows, as well as for viscous and inviscid flows. His method is briefly described in Appendix B (Section B.5.3.1, Eq. B.41) for 2D inviscid flows. Inadequate mesh size in regions of large gradients (e.g., shocks) results in numerical oscillations giving rise to instabilities in the code. MacCormack and Baldwin (1975) added an artificial viscosity or dissipation term (source term) of the form

$$\epsilon(\Delta x)^4 \frac{\partial}{\partial x}\left[\frac{|u| + a}{4p}\left|\frac{\partial^2 p}{\partial x^2}\right|\frac{\partial u}{\partial x}\right]$$

in the Navier–Stokes equation. This provides the necessary stability to the code. It affects the accuracy of the solution near the shock where $\partial^2 p/\partial x^2$ is large; its effect elsewhere is small. It should be emphasized that the technique is slow without the modifications (local time step, multi-grid, etc.) mentioned earlier, because the time steps are limited by the CFL number of one. For high-Reynolds-number turbulent flows, the viscous regions (which require small time steps) are small and a locally variable time step has to be adapted to make the technique attractive. The efficiency of the scheme can be enhanced by vectorization.

Most of the early Euler and Navier–Stokes codes for the prediction of turbomachinery flows were based on the MacCormack scheme; both the finite-difference and the finite-volume formulations have been employed using this technique. Thompkins (1982) used the MacCormack scheme for the Euler equation in the cylindrical coordinate system and showed qualitative comparison with the measured shock structure. Chima (1985) employed the multigrid technique, variable time step, and a vectorized code to solve the viscous flow through cascades. The two acceleration schemes reduced the CPU time for this cascade by a factor of seven. A comprehensive validation of explicit codes was carried out by Chima and Strasizar (1983). Laser doppler velocimeter measurements in a transonic compressor rotor (tip radius 254 mm, 52 blades, tip chord 44.5 mm, tip speed 16,100 rpm) were compared with the predictions from a 3D Euler code developed by Thompkins (BLADE 3D, 1982) and the quasi-3D code (MERIDL/TSONIC) developed by Katsanis (Katsanis, 1968; Katsanis and McNally, 1977) described in Section 4.2.5.1. The results shown in Fig. 5.30 indicate that the flow predictions are reasonably good, considering that the viscous effects are neglected. The Mach number, flow angle, and shock structure as well as the average flow properties from hub to tip and shock location are predicted reasonably well. The authors had considerable difficulty in specifying the downstream boundary condition, which is based on the specification of static pressure on the hub surface. The specification of measured static pressure produced erroneous mass flows and losses because the measured static pressure included the viscous effects not accommodated in the code. The authors suggested use of an axisymmetric through-flow code (e.g., see Katsanis and McNally, 1977) as a consistent and computationally efficient tool for choosing the downstream boundary.

Fourmaux and LeMeur (1987) and LeMeur (1988) used a predictor–corrector method to compute unsteady flows in turbomachinery, but showed no comparison with measured data. A novel approach was used to compute the flow field in a stage where the blade numbers were unequal. The computational domain contained k_1 and k_2 channels, where k_1/k_2 was as close as possible to the ratio of number of blades in the stator to those in the rotor. The computation of the exact geometry involves solution of flow field in almost all the passages, which is prohibitively expensive. Giles (1990a) utilized Ni's (1982) formulation of the Lax–Wendroff scheme and incorporated "time-inclined" computational planes to compute cases for which the ratio of stator and rotor pitches was not a simple integer. He predicted 40% variation in lift due to unsteady shocks. The code and technique have been validated mostly against analytical solutions.

5.7.4.3 *Runge–Kutta Technique*

One of the more recent advances is the use of Runge–Kutta, multistage methods, similar to those used for the numerical solution of ordinary differential equations. Jameson et al. (1981)

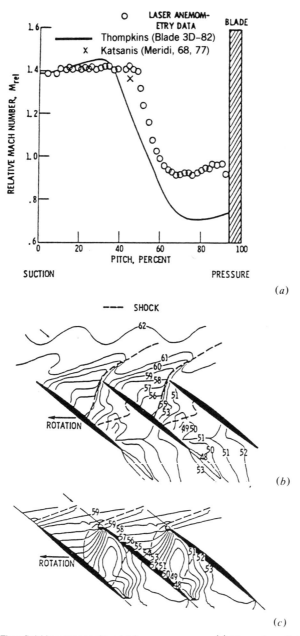

Figure 5.30 Flow field in a transonic axial flow compressor. (a) Circumferential distribution of relative Mach number at 15% of span from tip at 23% chord (through passage shock) at near-stall point (b) measured contours of relative flow angle in degrees at midspan, maximum flow point (c) predicted relative flow angle contours (BLADE 3D) (Copyright © Chima and Strazisar, 1983; reprinted with permission of ASME.)

developed and implemented this technique for the Euler equations. This technique has received widespread acceptance for both external and internal flows including turbomachinery flows.

The method for two-dimensional inviscid flow (Eq. 1.18) is illustrated in Section B.5.2.2 and Eq. B.39 (for 2D Euler, replace ϕ with q and replace R with $\partial E/\partial x + \partial F/\partial y$). Martinelli (1987) carried out a stability analysis to include both convection and diffusion terms in the numerical scheme. Kunz and Lakshminarayana (1992a) carried out a stability analysis of the discrete, coupled system of equations (including the $k-\epsilon$ equations) in rotating coordinates to ascertain the relative importance of grid stretching, rotation and turbulence terms, and the effective diffusivity on the stability of the scheme. As with other explicit schemes, Runge–Kutta schemes are amenable to convergence acceleration and can be easily extended to unsteady-flow computations in a straightforward manner.

Various groups have used Runge–Kutta procedures to develop both 2D and 3D viscous codes, as well as time-accurate viscous codes, to predict the flow field in cascades and rotors. Chima (1987) and Subramanian (1989) have successfully employed a multi-grid technique. Some of the salient features of the technique are as follows:

1. The standard four-stage scheme is fourth-order accurate in time for a linear equation and second-order accurate for nonlinear equations. The second-order central differences are used for the spatial derivatives.

2. Viscous and source terms are often evaluated prior to the first stage for computational efficiency. This renders the scheme first-order-accurate in time. Convective terms are computed at every stage. Typically, flux vectors, E, F, and G (inviscid part) (Eq. 5.1), are computed at midpoints between nodes in a finite-volume method. For viscous fluxes, this scheme incorporates information from all nine points in the differencing molecule.

3. To accelerate the solution to a steady state, locally varying time steps are computed based on a linear stability analysis of the discretized Navier–Stokes equations. The four-stage Runge–Kutta method allows a maximum CFL number of $2\sqrt{2}$, based on such an analysis.

4. Central difference schemes that do not inherently damp high-wave-number disturbances require the addition of artificial dissipation when applied to a hyperbolic system. Even for viscous flow calculations, artificial dissipation must be introduced into the scheme because the physical viscous terms are only effective in damping frequencies at higher wave numbers than can be resolved on practical grids. The fourth-order operators are included to damp high-wave-number errors, and the second-order operators are included to improve shock capturing (Jameson et al., 1981).

5. Implicit residual smoothing can be used to introduce some implicit character, as well as for increasing the maximum operational CFL number. This has the effect of smoothing the high-frequency variations of the residual. The smoothing can be applied at alternate stages. Detailed discussion can be found in Jameson and Baker (1984) and Hirsch (1990, Vol. 2, p. 336).

The technique has been used to predict a wide variety of flows, including turbine rotor and endwall flows, supersonic and subsonic compressor cascade flows, single and multistage compressor flows, centrifugal compressor flows, mixed flow pump flow field and hydraulic turbine flows. The technique has also been extended to compute rotor–stator interactions.

The following list contains the applications and assessment of the method:

1. The method has been used successfully to predict flow inside a Francis water turbine (Thibaud et al., 1989) and a mixed flow pump (Goto, 1992).
2. The method has been used successfully to predict two-dimensional viscous flows in cascades, quasi-three-dimensional flows in turbomachinery, and endwall flows in turbines (Chima et al., 1987; Chima and Yakota, 1990; Liu et al., 1989; Kunz and Lakshminarayana, 1992c; Zimmerman, 1990; Luo and Lakshminarayana, 1995a).
3. The technique has been used to predict compressor and turbine rotor flows (Subramanian, 1989; Kunz and Lakshminarayana, 1992b, d; Erickson and Billdal, 1989; Subramanian and Bozzola, 1987; Adamczyk et al., 1991; Crook et al., 1993).
4. The technique has been extended to include time-accurate unsteady flows and flows in multistage turbomachinery (Jorgenson and Chima, 1989; Kirtley et al., 1990; Hall and Delaney, 1991; Fan and Lakshminarayana, 1994).

Kunz and Lakshminarayana (1992c) computed the flow in a cascade tested at DFVLR by Schreiber (1988) at an inlet Mach number of 1.5. This precompression blade was especially designed to investigate shock–boundary-layer interaction with separation. At the test free-stream Mach number, a standoff leading-edge shock forms, which gives rise to a separated shock–boundary-layer interaction aft of midchord on the suction surface of the adjacent passage. This cascade operates at a "unique incidence" condition. This phenomenon along with the complex wave interaction field within the passage and shock–boundary-layer interaction, provides a challenging test case for both the numerical scheme and the turbulence model. The predicted isentropic blade surface Mach number is plotted against the experimental values in Fig. 5.31. The calculation and experiment show fairly

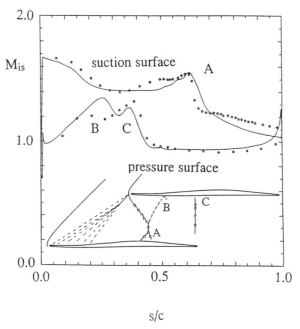

Figure 5.31 Isentropic blade surface mach numbers for the supersonic compressor cascade (data from Schreiber, 1988; computation from Kunz and Lakshminarayana, 1992c). The inset shows shock wave pattern deduced from flow visualization and laser measurements. (Copyright © Kunz and Lakshminarayana, 1992c; originally appearing in the AIAA Journal Vol. 30.)

good agreement. The features labeled A, B, and C in Fig. 5.31 correspond to local compression regions where the bow shock impinges on the suction surface and where the Mach reflection and the passage shock impinge on the pressure surface. Similar good agreement is also obtained for (a) the flow downstream including the wake and (b) the shock–boundary-layer interaction regions. In addition, good agreement is shown with the measured and computed profile losses (including shock losses at various Mach numbers).

Kunz et al. (1993) utilized the three-dimensional Euler and Navier–Stokes computational procedures to simulate the flow field in an axial compressor cascade with tip clearance. An embedded H-grid topology was utilized to directly resolve the flow physics in the tip gap region. Additionally, both Euler and Navier–Stokes computations were performed to investigate the relative performance of these approaches in reconciling the physical phenomena considered. Results indicate that the Navier–Stokes procedure, which utilizes a low-Reynolds-number k–ϵ model, captures a variety of important physical phenomena associated with tip clearance flows with good accuracy. These include tip vortex strength and trajectory, loading near the blade tip, the interaction of the tip clearance flow with passage secondary flow, and the effects of relative endwall motion. The Euler computation provides good but somewhat diminished accuracy in resolution of some of these clearance phenomena. It was concluded that the level of modeling embodied in the present approach is sufficient to extract much of the tip region flow field information useful to designers of turbomachinery. The configuration chosen for this computational study is the "Liverpool cascade." This configuration was the subject of a variety of experimental tip clearance flow investigations undertaken by the author in earlier work (see Lakshminarayana and Horlock, 1967b; Lakshminarayana, 1970). The cascade had nine "split" blades of 10(C4)30C50 profile with $C = 0.152$ m, $\lambda = 36°$, $A = 4.8$, $S/C = 1.0$, $\alpha_1 = 51°$, $\alpha_2 = 31°$, and Re $= 2.0 \times 10^5$. In Experiment A, the blades were separated by a distance 2τ and the upstream velocity profile was uniform. This served to isolate tip clearance physics without the influence of an endwall or the interaction between passage secondary motion and the tip gap flow. Contours of the dynamic head predicted by the Euler solution using the embedded H grid is shown in Fig. 5.32. In the Euler calculation, the leakage velocities are larger, and hence leakage flow tends to move somewhat further away from the suction surface before interacting with the main flow and rolling up into a vortex. The location of the tip vortex at this outlet location is predicted to be close to midpassage by the Navier–Stokes solution, in good agreement with experimental observation. The outlet location of the tip vortex, predicted by the Euler solution, is seen to be in poorer agreement with experiments.

Kunz and Lakshminarayana's (1992a,b,d) procedure also includes eigenvalue and velocity scaling of artificial dissipation to improve accuracy, especially in the viscous layers. This, along with the algebraic Reynolds stress model (ARSM) was used to predict flow in axial and centrifugal compressor rotors. The flow field in a centrifugal compressor tested by Krain (1988) was predicted accurately, including the tip clearance region and jet–wake features at the exit. The flow field in the Penn State compressor (Kunz and Lakshminarayana, 1992d) was resolved accurately including blade-to-blade profiles, passage-averaged properties at exit and the blade boundary layers and wakes, as shown in Fig. 5.33. See Section 2.5.5 and Fig. 2.17 for details of this compressor. Accurate prediction of flow properties far downstream indicate that the spanwise mixing effects are also captured accurately. This provides considerable confidence in the ability of the numerical procedure to capture important flow physics in these rotor passages.

Subramanian (1989) and Subramanian and Bozzola (1987) computed the three-dimensional viscous flow through turbine cascades, turbine stators, and transonic compressors. They were able to capture the endwall flow phenomena, including the presence of a horseshoe vortex, saddle point, and secondary flow. They also computed the flow field for the NASA transonic compressor (designated as NASA Rotor 67) for which detailed flow data (LDV) are available (Pierzga and Wood, 1985). The design pressure ratio was 1.629, the tip relative Mach number was 1.38, and the aspect ratio was 1.56. The computed Mach numbers at 30% of span from the tip are shown

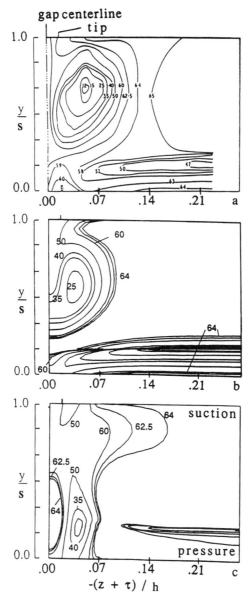

Figure 5.32 Contours of dynamic pressure ($\rho V_2^2 / 2$, in percent of inlet dynamic head) for compressor cascade at outlet measurement plane. Y is the pitchwise distance; Z is the spanwise distance. (a) Experimental. (b) Navier\Stokes. (c) Euler. (Copyright © Kunz et al., 1993; reprinted with permission of ASME.)

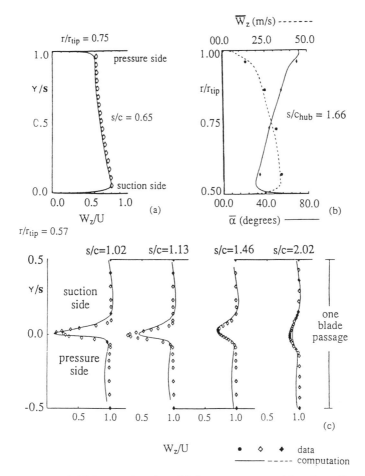

Figure 5.33 Flow field in a subsonic axial flow compressor (data from Popovski and Lakshminarayana, 1986; Prato and Lakshminarayana, 1993; computation from Kunz and Lakshminarayana, 1992d). See Fig. 5.5 for details of the compressor; s is the chordwise distance from the leading edge, y is the pitchwise distance (a) Blade-to-blade axial velocity distribution at midspan region, $s/c = 0.65$, (b) Radial distribution of passage averaged flow distribution downstream of rotor, (c) Axial velocity profiles inside the wake at $r/r_t = 0.57$. (Copyright © Kunz and Lakshminarayana, 1992d; reprinted with permission of ASME.)

compared with the measured values in Fig. 5.34. Also shown in this figure is the Mach number distribution of midpassage at 30% span from the tip, compared with Denton's (1982) explicit finite-volume inviscid procedure. The shock location as well as Mach number distributions are predicted well by the 3D viscous code; the inviscid code predicted the shock location further downstream. The inviscid analysis predicts a sudden jump in properties, while the viscous code captures the physics of the flow near the shock more accurately. The inviscid codes are fairly accurate in the shock-free region

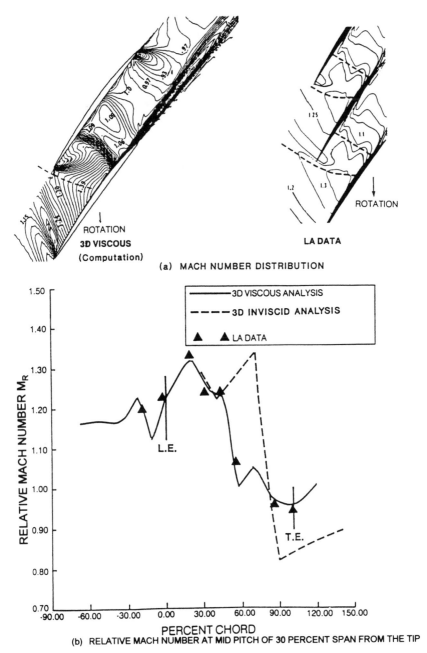

Figure 5.34   Flow field in a transonic axial flow compressor. (a) Mach number contours at 30% span from the tip (b) Relative Mach number at mid pitch. (Copyright © Subramanian and Bozzola, 1987; reprinted with permission of AIAA.)

upstream, but it is essential to employ viscous solutions to capture shock location, shock smearing, and shock–boundary-layer interaction effects. Similar conclusions have been drawn by Erickson and Billdal (1989), who computed the flow field in a high-speed centrifugal compressor (Eckardt, 1976) using both the Euler and Navier–Stokes version of their 3D code based on the Runge–Kutta method. A dramatically different secondary flow pattern was observed when viscous effects were neglected in the computation. The Euler solution does not capture the exit flow for a centrifugal impeller, where the "jet–wake" flow field makes it one of the most complex flow fields encountered in turbomachinery. Hah and Krain (1990) seem to have resolved this flow more accurately using the pressure-based method (viscous).

Dawes (1987) developed an algorithm consisting of a two-step explicit and one-step implicit scheme derived as a pre-processed simplification of the Beam and Warming (1978) algorithm. It is called "pre-processed" as it involves Beam and Warming's classical implicit method and premultiplying the implicit algorithm gives it an explicit character. The scheme is similar in implementation to a two-step Runge–Kutta method with implicit residual smoothing. This method is hybrid in character, with both implicit and explicit schemes, and is very similar to the schemes described in this section. Dawes (1991) has recently developed a four-stage Runge–Kutta method with a solution-adaptive mesh methodology and predicted the secondary flow and the corner stall in a transonic compressor cascade. Goto (1992) has provided a comprehensive validation [Dawes (1987) code] for a 5 bladed mixed-flow impeller. The Euler head rise was predicted well at four different tip clearance values. The actual head rise prediction was reasonable. The head loss coefficient (Euler head rise − actual head rise), defined by,

$$\zeta = \frac{U_2 V_{\theta 2} - U_1 V_{\theta 1}}{U_{2m}^2} - \frac{g(H_2 - H_1)}{U_{2m}^2}$$

is shown in Fig. 5.35. There is general agreement with the measured values. The details of the tip vortex (location, diameter) are not predicted well. As indicated earlier (see Section 5.2), it is essential to include higher-order turbulence models to capture the physics of tip vortex. In the present method, the authors use an algebraic eddy viscosity. Nevertheless, the ability of these Navier–Stokes codes to capture the overall features of the flow field is encouraging. The jet–wake flow pattern at the exit has been captured qualitatively.

Hathaway et al. (1993) utilized Dawes' code to complement and interpret their experimental data in a low-speed centrifugal compressor (described in Section 4.2.6.3 and in Fig. 4.22). The meridional velocity distribution, including the jet–wake structure, is qualitatively captured, but the secondary flow and tip clearance effects are not predicted well. As indicated earlier (Section 5.2), rotation and curvature introduce major structural changes in the turbulence and eddy viscosity models cannot capture these features.

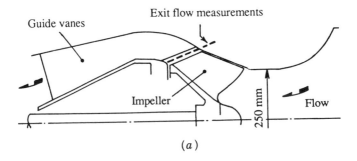

(a)

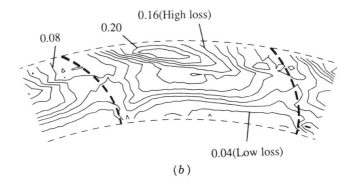

(b)

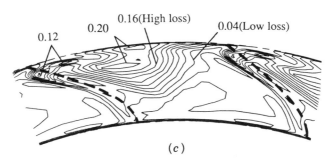

(c)

Figure 5.35 Predicted and measured loss coefficient (ζ) contours at the exit of a mixed-flow pump. Design flow rate = 8.7 m³ / min, head = 7.0 m H₂O, speed = 800 rpm, and tip clearance = 1.5 mm. (a) Pump Geometry (b) Experimental data (c) prediction (Copyright © Goto, 1992; reprinted with permission of ASME.)

Zimmerman (1990) employed the four-stage Runge–Kutta scheme and carried out a detailed simulation of turbine endwall flow, including the formation of a corner vortex, horseshoe vortex, and secondary vortex and their eventual roll-up. The loss distribution for high-speed flow ($M = 0.7$, 0.9, and 1.1) shows major differences in flow losses as the Mach number is increased. The agreement between the data and experiment is not good, and the authors attribute this to the turbulence model used (Baldwin–Lomax).

The shock–boundary-layer interaction and its effect on the flow field is not captured well.

There are instances where viscous effects may be small and a Euler code should be able to predict the flow accurately. Thibaud et al. (1989) used a Runge–Kutta method and computed the flow field inside and at the exit of a Francis turbine (Fig. 2.77c) using a Euler code. The flow properties at the exit are predicted accurately, except near the outer wall.

Hall and Delaney (1991) used a four-stage Runge–Kutta method and implicit residual smoothing and developed a time-accurate 3D viscous code. The prediction of the suction surface static/total pressure ratio for a two-bladed SR7 prop fan (Section 2.6, Figs. 1.3, 2.33) operating in steady flow is shown in Fig. 5.36. The predictions agree well with the experimental data. The leading edge vortex path is indicated by the region of low-pressure bending across the suction surface from approximately 30% span toward the blade tip. The experimental data suggest that the leading edge vortex remains attached to the leading edge (indicated by very low suction pressures), while both the Euler and the Navier–Stokes solutions suggest that the vortex drifts further off axially of the blade. Hall and Delaney (1991) suggest grid resolution as a potential source of error. The plot also indicates that the viscous effects are small (with the exception of leading edge vortex region), and a Euler code should resolve the blade pressures fairly accurately for this particular configuration. However, the performance prediction (efficiency, losses, state) would need a Navier–Stokes solver.

5.7.4.4 *Assessment Summary for Explicit Schemes*   The preceding discussion indicates that the explicit techniques are highly successful in predicting complex flow fields in turbomachinery. The choice between a Euler solution and a Navier–Stokes solution depends on the type of flows and the machinery. Centrifugal machinery and high-speed compressor flows are best resolved with a Navier–Stokes solution. The hydraulic turbine flow field can be captured with a Euler code. This should be combined with boundary layer codes to predict the efficiency and losses.

Among the explicit methods, all the techniques seem to perform well for viscous flow through turbomachinery. The two-step formulation of Ni, the Lax–Wendroff scheme, and the Runge–Kutta method all perform well. The Runge–Kutta method allows for a larger time step than that of other schemes. All of these schemes, except the Runge–Kutta procedure, provide inherent artificial dissipation due to truncation error. The Runge–Kutta schemes using central differencing for the convective terms are less dissipative at high wave numbers than other formulations when applied to convective problems. For this reason, artificial dissipation must be purposefully added to these schemes, usually after each stage in the update procedure. This may provide the user with better control over levels of artificial dissipation present in a solution.

The Runge–Kutta method has been widely used and is currently undergoing major development. The Runge–Kutta method is especially attractive if

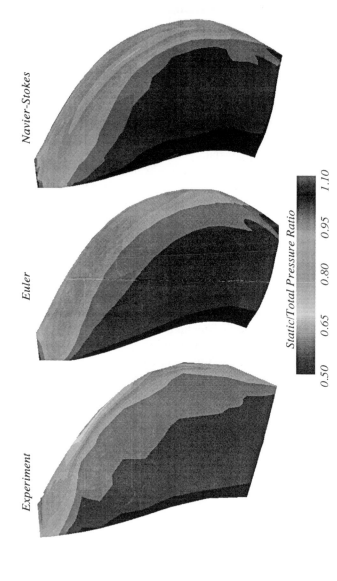

Figure 5.36 Static / total pressure ratio on an SR7 prop fan (data from Bushell, 1988; computation from Hall and Delaney, 1991). $M_\infty = 0.2$, diameter = 2.75 m, advance ratio = 0.881, two blades; SR7 is designed for $M_\infty = 0.8$, operating with eight blades. (Copyright © Hall and Delaney, 1991; reprinted with permission of ASME.)

implemented with implicit residual smoothing, local time-stepping, multi-gridding, and the finite-volume formulation. These and other techniques can be vectorized, and accuracy can be improved with velocity scaling for artificial dissipation. Moreover, all these techniques especially the Runge–Kutta method, are attractive for unsteady flow.

5.7.5 Implicit Time-Marching Methods

One of the major disadvantages of explicit methods is that the size of the time step is limited by CFL conditions. This led to the development of implicit techniques which are unconditionally stable for all time steps. In the implicit method, the calculation of an unknown variable requires the solution of a set of simultaneous equations. The basic techniques are described in Appendix B, Section B.5.3. A review of various implicit techniques used in computational fluid dynamics can be found in Anderson et al. (1984) and Hirsch (1990). Only the techniques that are widely used in turbomachinery flow computation are discussed here.

5.7.5.1 *Approximate Factorization* One of the most widely used implicit techniques is that due to Briley and McDonald (1977) and Beam and Warming (1978). These techniques are used, with modifications, to predict viscous flow fields in cascades, stators, and rotors. Operational CFL numbers which are of the order of 10 appear to provide the best error damping properties. As in the explicit techniques, both finite-volume and finite-difference formulations are employed; these are briefly summarized in this section.

From the point of numerical analysis, it is necessary to express viscous terms in the Navier–Stokes equation in a different manner. Let us consider the two-dimensional Navier–Stokes equation and develop the algorithm based on Beam and Warming (1978). The Navier–Stokes equation can be expressed as (Eq. 1.18 in dimensional form)

$$
\frac{\partial q}{\partial t} + \frac{\partial E(q)}{\partial x} + \frac{\partial F(q)}{\partial y}
$$
$$
= \left(\frac{\partial T_1(q, q_x)}{\partial x} + \frac{\partial T_2(q, q_y)}{\partial x} + \frac{\partial P_1(q, q_x)}{\partial y} + \frac{\partial P_2(q, q_y)}{\partial y} \right) \quad (5.83)
$$

In this formulation, the vector T (Eq. 1.20) is split into T_1 and T_2 which are functions of (q and q_x) and (q and q_y), respectively. Similarly, vector P (Eq. 1.21) is split into P_1 and P_2.

Following the procedure leading to Eq. B.47 in Appendix B (Eqs. B.42–B.46), the discretization of Eq. 5.83 can be written as (two-step ADI scheme)

$$\left\{ I + \frac{\Delta t}{2} \left[\frac{\partial}{\partial x} (A - N + R_x)^n - \frac{\partial^2}{\partial x^2} (R)^n \right] \right\} \Delta q^*$$

$$= \frac{\Delta t}{2} \left[\frac{\partial}{\partial x} (-E + T_1 + T_2)^n + \frac{\partial}{\partial y} (-F + P_1 + P_2)^n \right]$$

$$+ \frac{\Delta t}{2} \left[\frac{\partial}{\partial x} (\Delta T_2)^{n-1} + \frac{\partial}{\partial y} (\Delta P_1)^{n-1} \right] \qquad (5.84)$$

and

$$\left\{ I + \frac{\Delta t}{2} \left[\frac{\partial}{\partial y} (B - Q + S_y)^n - \frac{\partial^2}{\partial y^2} (S)^n \right] \right\} \Delta q^n = \Delta q^* \qquad (5.85)$$

$$q^{n+1} = q^n + \Delta q^n \qquad (5.86)$$

where I is the identity matrix, N is a Jacobian $\partial T_1 / \partial q$, Q is the Jacobian of $\partial P_2 / \partial q$, R is the Jacobian of $\partial T_1 / \partial q_x$ and $R_x = \partial R / \partial x$.

The Jacobian matrices for one-dimensional flow are given by Eq. B.28 in Appendix B. For two-dimensional Navier–Stokes in the Cartesian system, these are given by Beam and Warming (1978):

$$\frac{\partial E}{\partial q} = A$$

$$= - \begin{bmatrix} 0 & -1 & 0 & 0 \\ \dfrac{3-\gamma}{2}u^2 + \dfrac{1-\gamma}{2}v^2 & (\gamma-3)u & (\gamma-1)v & 1-\gamma \\ uv & -v & -u & 0 \\ \gamma e_o u + (1-\gamma)u(u^2+v^2) & -\gamma e_o + \dfrac{\gamma-1}{2}(3u^2+v^2) & (\gamma-1)uv & -\gamma u \end{bmatrix}$$

$$(5.87)$$

$$\frac{\partial F}{\partial q} = B$$

$$= - \begin{bmatrix} 0 & 0 & -1 & 0 \\ uv & -v & -u & 0 \\ \dfrac{3-\gamma}{2}v^2 + \dfrac{1-\gamma}{2}u^2 & (\gamma-1)u & (\gamma-3)v & 1-\gamma \\ \gamma e_o v + (1-\gamma)v(u^2+v^2) & (\gamma-1)uv & -\gamma e_o + \dfrac{\gamma-1}{2}(3v^2+u^2) & -\gamma v \end{bmatrix}$$

$$(5.88)$$

$$\frac{\partial T_1}{\partial q_x} = R$$

$$= \rho^{-1} \begin{bmatrix} 0 & \vline & 0 & \vline & 0 & \vline & 0 \\ -(\lambda + 2\mu)u & \vline & (\lambda + 2\mu) & \vline & 0 & \vline & 0 \\ -\mu v & \vline & 0 & \vline & \mu & \vline & 0 \\ -(\lambda + 2\mu - k/c_v)u^2 - (\mu - k/c_v)v^2 - (k/c_v)(e) & \vline & (\lambda + 2\mu - k/c_v)u & \vline & (\mu - k/c_v)v & \vline & k/c_v \end{bmatrix}$$

$$(5.89)$$

$$\frac{\partial P_2}{\partial q_y} = S$$

$$= \rho^{-1} \begin{bmatrix} 0 & \vline & 0 & \vline & 0 & \vline & 0 \\ -\mu u & \vline & \mu & \vline & 0 & \vline & 0 \\ -(\lambda + 2\mu)v & \vline & 0 & \vline & (\lambda + 2\mu) & \vline & 0 \\ -(\lambda + 2\mu - k/c_v)v^2 - (\mu - k/c_v)u^2 - (k/c_v)(e) & \vline & (\mu - k/c_v)u & \vline & (\lambda + 2\mu - k/c_v)v & \vline & k/c_v \end{bmatrix}$$

$$(5.90)$$

The temperature and the pressure (p) in Eq. 1.18 have been replaced by

$$T = \frac{1}{c_v}\left[e_o - \frac{(u^2 + v^2)}{2} \right]$$

$$p = \rho(\gamma - 1)\left[e_o - \tfrac{1}{2}(u^2 + v^2) \right] \text{ and } \lambda = -\tfrac{2}{3}\mu$$

These equations are similar to Euler 2D equations (Eq. B.47 in Appendix B), with the exception of additional viscous terms P, T, N, R, Q, S, R_x, and S_y. This is the basic algorithm used by most authors employing the approximate factorization method. Implementation of the difference formulation, turbulence models, and grid generations differs in various schemes.

The following procedure can be used to solve these equations:

1. The spatial derivatives in Eqs. 5.84 and 5.85 can be approximated by second-order central differencing (Eqs. B.3 in Appendix B). This results in a block tridiagonal system of equations with each block having dimensions of q (4 and 5 for 2D and 3D, respectively). These can be solved efficiently using the solution methodology described in Appendix B.

2. These equations can be used for laminar and turbulent flows. For turbulent flows, μ is replaced by $\mu + \mu_t$, with μ_t being specified by one of the models described earlier. There is no need to update μ_t at every time step. The Jacobians A, B, R, and S could be simplified, even though λ, μ, and μ_t are all functions of temperature and hence q. These coefficients can be assumed to be constant (for low subsonic flow) or updated at every 10–30 time steps.

3. Higher-order schemes can be implemented by using the scheme similar to Eq. B.48 in Appendix B. Beam and Warming (1978) have shown that this scheme is unconditionally stable for all values of $\zeta \geq 0.385$.

4. The numerical scheme described above is temporally dissipative except for the longest and shortest wavelengths. Because the phase error of the short waves is large, it is necessary to add dissipative terms to damp the short wavelengths. Beam and Warming (1978) suggest adding the following artificial dissipation terms to the right-hand side of Eq. 5.84 and 5.85, respectively:

$$-\frac{\Delta x^4}{2}\frac{\epsilon_x}{8}\frac{\partial^4}{\partial x^4}q^n, \qquad -\frac{\Delta y^4}{2}\frac{\epsilon_y}{8}\frac{\partial^4}{\partial y^4}(q^n + \Delta q^n) \qquad (5.91)$$

where ϵ_x and ϵ_y are dissipative coefficients (≤ 1). These are higher-order terms and do not affect the physics or accuracy of the method.

5. Although Eqs. 5.84 and 5.85 contain three time levels of data ($n + 1, n, n - 1$), only two levels of data q and Δq need to be stored.

6. The methods for accelerating the schemes include local time-stepping and multi-grid techniques described earlier.

The major computational effort is in inverting the block tridiagonal matrix system that arises. Even though large time steps are allowed (for steady-state problems), the additional effort involved in the inversion of these matrices and large amounts of storage required make the two methods (explicit and implicit) comparable in computational effort. Hence, there is no clear advantage of one method over the other. The increased stability bounds for the implicit schemes offer some advantages. Various improvements have been made to the original Beam and Warming technique. These include: diagonalization of the blocks in implicit operators, pressure–velocity split, grid refinement, improved boundary conditions, implicit treatment of artificial viscosity terms, variable time-stepping, and multi-gridding. These can be found in Pulliam (1986).

Most investigators use the ADI scheme, with each factorized step requiring inversion of a block tridiagonal matrix. For three dimensions, one of the ways to improve the computational efficiency is to implement LU factoring (lower and upper diagonal, Appendix B). In this method, the matrix coefficients in the discretized algebraic equations are split into two matrices. The lower and upper diagonals are derived from the original matrix through the Gauss elimination procedure, and this scheme is called *LU decomposition*. LU decomposition reduces storage requirements. Yakota (1990) has developed an implicit LU method with a multi-grid scheme for turbomachinery and computed turbine stator and endwall flow.

Knight and Choi (1989) employed the implicit approximate factorization method described in this section, finite-volume formulation, second-order upwind differencing, and a two-equation turbulence model. A cell-based finite-volume formulation was used for spatial discretization. Local time stepping was employed to accelerate the convergence. Knight and Choi

(1989) computed many turbomachinery flows including the flow field in an energy-efficient engine (E^3) turbine vane (see Fig. 2.64 and the corresponding text for details). Their results are shown compared with data in Fig. 5.37. The flow field computation indicates the presence of a supersonic region at the exit. The losses and outlet angles are predicted reasonably well. The location of passage vortices (the location of peak values away from the wall) are also predicted well. The underturning near the wall and a slight overturning away from the wall is also captured. (See Fig. 4.23 for similar results based on secondary flow theory.)

One interesting application of CFD is the computation of flow field in automotive torque converters. The passages in these devices are narrow, with very close pump–turbine–stator spacings all enclosed in a casing. The flow field is very complex. Fujitani et al. (1988) discretized the Navier–Stokes equation using a first-order Euler implicit scheme for the convection term. Because no detailed flow data are available in these torque converters, it is difficult to assess the code's capability to capture details of the local flow field. But the major interest in automotive torque converters is the torque ratio and efficiency, and these were predicted reasonably well (Fig. 5.38). The efficiency prediction under cruise conditions is not good. Here again, it is doubtful whether a laminar flow solution with coarse grid can provide the detailed flow resolution needed to improve torque converter performance and design. By et al. (1993) have provided detailed flow predictions in the pump using the pseudocompressibility technique described later, and Marathe et al. (1994) predicted the stator flow using a pressure-based method.

Nakahashi (1989) developed an LU implicit scheme and predicted turbine endwall flow. This code has been used to design an efficient bowed turbine vane. Weber and Delaney (1991) have developed a 3D code based on Beam and Warming's (1978) scheme and have predicted flow field for a NASA 67 compressor with both C and H grids and concluded that C grids provide better resolution of flow field.

5.7.5.2 Upwind Schemes The conventional techniques described earlier, both explicit and implicit, use artificial viscosity terms, which must be specifically chosen and tuned for a particular problem. If strong shocks are present, employment of large numerical dissipation may obscure the real physics. One of the recent developments to overcome this problem is the development of upwind schemes (Osher and Solomon, 1982). This has been applied to both explicit (Chakravarthy and Osher, 1983) and implicit techniques (Rai and Chakravarthy, 1986). Upwind schemes are attractive because they provide a rational mathematical framework for adding dissipation to a scheme. They may be more robust than schemes with artificial viscosity. They do not usually have operator prescribed constants.

Most of the techniques described earlier use conservative central difference schemes to capture shocks, requiring arbitrary smoothing parameters (artificial dissipation) to stabilize the calculation. The upwind difference

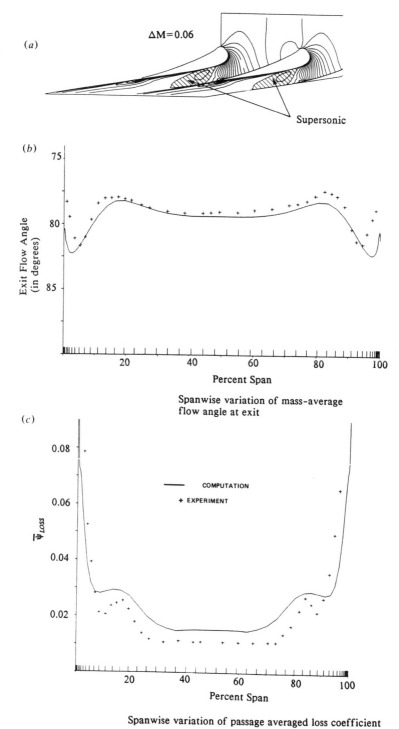

Figure 5.37 Flow field in energy-efficient-engine turbine nozzle (data from Kooper et al., 1981a, b; Knight and Choi, 1989). $C = 9$ cm, $S = 8.46$ cm, $A = 0.513$, $\alpha_1 = 0°$, $M_1 = 0.109$, $M_2 = 0.85$. (a) predicted Mach number contours. (b) Spanwise variation of mass-averaged flow angle at exit. (c) Spanwise variation of passage-averaged loss coefficient ($\bar{\psi}_{loss} = (P_{01} - P_{02})/P_{01}$) (reprinted with permission of AIAA.)

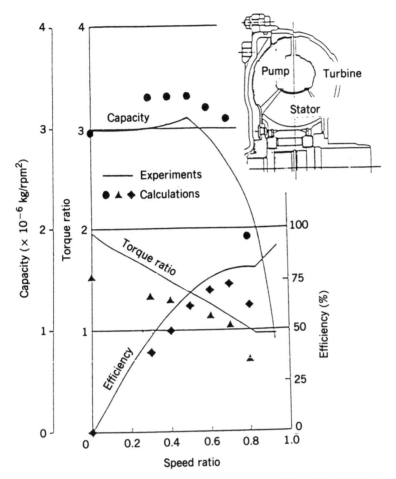

Figure 5.38 Performance of an automotive torque converter. (Copyright © Fujitani et al., 1988; reprinted with permission of the Society of Automotive Engineers, Inc.)

schemes are used to capture strong shocks without requiring arbitrary parameters, which may alter the physics of the problem. The flux evaluation is numerically more intensive in upwind schemes; the scheme is more complicated to code and requires more computational time.

The upwind scheme for an implicit algorithm for a one-dimensional Euler equation ($q_t + E_x = 0$) can be written as (Rai and Chakravarthy, 1986)

$$\left[I + \frac{\Delta t}{\Delta x}\left[\Delta_x A^- + \nabla_x A^+\right]^n\right]\Delta q_i = -\frac{\Delta t}{\Delta x}\left[\hat{E}_{i+1/2}^n - \hat{E}_{i-1/2}^n\right] \quad (5.92)$$

$$\hat{E}_{i+1/2} = \frac{1}{2}\left[E_i + E_{i+1} - \int_{q_i}^{q_{i+1}}\left\{\left(\frac{\partial E}{\partial q}\right)^+ - \left(\frac{\partial E}{\partial q}\right)^-\right\}dq\right] \quad (5.93)$$

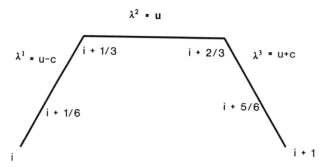

Figure 5.39 Subpaths in phase space for upwind scheme.

and $\Delta q_i = q_i^{n+1} - q_i^n$. Δ and ∇ are forward and backward difference operators (unlike central differences used in Eqs. 5.84 and 5.85), respectively, and

$$A^+ = \left\{ \frac{\partial E}{\partial q} \right\}^+, \qquad A^- = \left\{ \frac{\partial E}{\partial q} \right\}^- \qquad (5.94)$$

where $+$ and $-$ depend on the path of integration shown in Fig. 5.39. For example,

$$\Delta E^+ = \int_{q_i}^{q_{i+1}} \left(\frac{\partial E}{\partial q} \right)^+ dq = K_{i,i+1/3} \left[E_{i+1/3} - E_i \right]$$

$$+ K_{i+1/3,i+2/3} \left[E_{i+2/3} - E_{i+1/3} \right] + K_{i+2/3,i+1} \left[E_{i+1} - E_{i+2/3} \right]$$

$$K_{i+(m-1)/3,\,i+m\,/3} = \begin{cases} 1, & \text{if } \lambda_{i+(m-1)/3}^m, \ \lambda_{i+m\,/3}^m \geq 0 \\ 0, & \text{if } \lambda_{i+(m-1)/3}^m, \ \lambda_{i+m\,/3}^m < 0 \end{cases} \qquad (5.95)$$

$m = 1, 2, 3$; $\lambda^1, \lambda^2, \lambda^3$ are eigenvalues corresponding to three subpaths shown in Fig. 5.39.

This scheme has been extended to two- and three-dimensional Navier–Stokes equations (Rai and Chakravarthy, 1986; Rai, 1989). An approximate factorization scheme similar to Eqs. 5.84 and 5.85 are used.

This technique has also been extended to compute time accurate flows, and the major feature of the unsteady effects (unsteady pressures) has been captured (Rai, 1989). Rai was one of the first ones to predict viscous rotor–stator interaction in 2D and 3D flows. His work has prompted considerable interest in the field. Gundy-Burlet et al. (1990) have used Rai's code to predict the quasi-steady and unsteady flows in a low-speed multistage compressor using a zonal grid system (alternatively moving and stationary grids).

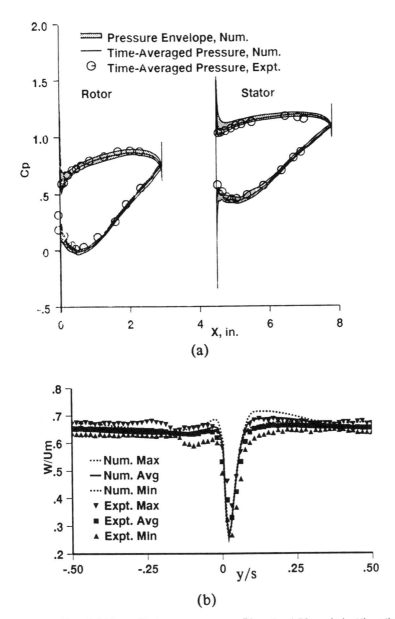

Figure 5.40 Flow field in multi-stage compressor. Diameter 1.52 m, hub / tip ratio = 0.8, U_m = 46.7 m / s, number of blades = 44 in all rows, axial velocity / mean blade speed ratio = 0.510. (a) Blade surface-pressure on second stage. (b) Wake profiles 2% aft of the second-stage rotor (y is the pitchwise distance). (Data from Joslyn et al., 1989; computation from Gundy-Burlet et al., 1990.) (Reprinted with permission of ASME)

The time-averaged pressure as well as the time-averaged wakes are predicted well for Joslyn et al. (1989) multistage compressor (Fig. 5.40). The qualitative features of unsteady pressure distribution and unsteadiness in the wake are also captured. This is an extremely complex case to compute, and recent advances made in this direction are noteworthy. What is needed now are improvements in transition and turbulence modeling in unsteady flow. Very little is known about the applicability of the turbulence models developed for steady flow to unsteady flows. The author's experience is that many of the important physical features are not captured by algebraic eddy viscosity models.

5.7.5.3 *Assessment of Implicit Schemes*
Implicit techniques for steady 3D Euler and 3D Navier–Stokes have reached a high level of maturity and are widely used in turbomachinery flow field computation. The techniques enable large time steps, but may introduce factorization error. Combined with local time-stepping and multi-gridding, they are as attractive as explicit codes. A wide variety of turbomachinery flows have been computed, including incompressible and compressible turbomachinery flow fields. Future directions will involves incorporation of higher-order turbulence models and parallel processor implementation.

A comprehensive assessment of Rai's code for unsteady flow in turbomachinery has been carried out by Griffin and McConnaughey (1989). The aerodynamic and thermal data from Dring et al. (1986) were compared against Rai's (1989) code, as well as against Katsanis and McNally's (1968, 1977) code (Section 4.2.5.1) combined with a boundary layer code (STAN 5; see Crawford and Kays, 1975). The predictions from both codes for Dring's stator at midspan (time-averaged heat transfer) are shown in Fig. 5.41. Both approaches yield similar agreement with the heat transfer data. Rai's code required six CPU hours, whereas Katsanis and McNally's code, including the boundary layer calculation, required eight CPU minutes on a Cray XMP. Griffin and McConnaughey conclude that if the interest lies only in the time-averaged quantities, a conventional method is the better approach, because it requires much less computational time. If the interest lies in characterization of unsteady flow (e.g., viscous wakes and unsteady boundary layers), then the Navier–Stokes approach may be better. Even here, one has to qualify this statement. If the interest lies only in unsteady pressure on the blade, the linearized theories [see Platzer and Carta (1987) for a review] can provide satisfactory results. For example, Chen and Eastland (1990) used Lakshminarayana and Davino's (1980) wake correlation and a linearized potential flow analysis to predict accurately the unsteady blade pressures measured by Dring et al. (1986) in a small fraction of computer time required by the unsteady Navier–Stokes code. The usefulness of the unsteady Navier–Stokes code lies in resolving the passage of wakes, unsteady flow field including boundary layers, and off-design performance in a multistage blade row.

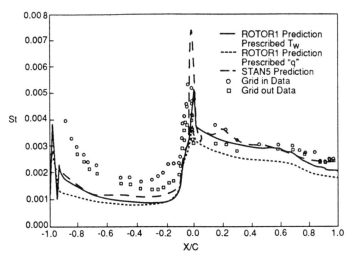

Figure 5.41 Heat transfer prediction for a turbine stator blade. Diameter = 1.52 m, hub / tip ratio = 0.8, aspect ratio = 1, 22 stator vanes, 28 rotor blades. (computation by Griffin and McConnaughey, 1989, data from Dring et al., 1986). St is the Stanton number (see Chap 7 for definition) ROTOR 1—Rai's code; STAN5—boundary layer code. (Copyright © Griffin and McConnaughey, 1989; reprinted with permission of ASME.)

5.7.6 Explicit Versus Implicit Time-Iterative Techniques

Both explicit and implicit techniques are widely used, with explicit techniques more commonly used in industry. Both techniques can incorporate local time stepping. The maximum time step (CFL) for a four-stage Runge–Kutta method is 2.8, and typical CFL numbers for the implicit LU scheme and ADI schemes are 5 and 10 respectively (Hall, 1989). But ADI schemes have more operations per iteration, and are harder to vectorize. Hence, there is no great advantage in ADI over Runge–Kutta, may be a factor of 2 at best. Rai (1989) uses much higher CFL numbers in his upwind scheme. Multi-gridding is extensively and successfully used in explicit schemes, with some success in implicit schemes. The implicit residual smoothing is an attractive feature in the Runge–Kutta method. Both techniques are sensitive to (a) grid-stretching and skewness and (b) turbulence models. Explicit methods are easier to code, vectorize, and adapt to massively parallel computers.

One of the extensive comparisons of various explicit and implicit techniques is due to Hall (1989) and Hall et al. (1990). They have made a systematic (same grid, turbulence model, and computer) comparison of the explicit hopscotch and Runge–Kutta scheme, the implicit Beam and Warming's approximate factorization linear block implicit scheme (AFLBI), and the upwind generalized conjugate gradient total variation diminishing (GCGTVD) scheme for the following cases: (1) inviscid flow over a cylinder, (2) steady viscous/inviscid flow through a turbine cascade, and (3) unsteady viscous flow over an impulsively started cylinder.

Table 5.5   Comparison of Various Techniques (Hall, 1989)

Method	CFL Number	Iterations for Convergence	CPU Seconds for Convergence	CPU Seconds per Grid Point/Iteration
Viscous High-Turning Turbine Cascade CPU Time Comparison				
Hopscotch	0.7	4,043	224.70	1.18×10^{-5}
Runge–Kutta	1.5	3,100	365.03	2.50×10^{-5}
AFLBI	3.0	1,745	616.42	7.50×10^{-5}
GCGTVD	5.0	1,497	706.50	10.02×10^{-5}
Impulsively Started Cylinder Flow CPU Time Comparison, Re = 3000				
Hopscotch	5.0×10^{-6}	25,000	1759.00	1.968×10^{-6}
Runge–Kutta	1.0×10^{-5}	12,500	1799.00	5.033×10^{-6}
AFLBI	2.5×10^{-5}	5,000	2500.00	2.798×10^{-5}
GCGTVD	2.5×10^{-5}	5,000	3600.00	4.029×10^{-5}

The results for a turbine cascade indicate that all techniques/codes provide nearly identical pressure distribution; major differences lie in the CFL number and CPU seconds for convergence, as shown in Table 5.5. It is evident that for viscous flows, explicit methods require appreciably less computer time, even though this is not significant in terms of new developments that may occur in computer architecture. Hall (1989) concludes "that the choice of explicit versus implicit technique and the time accuracy of a given algorithm may have less to do with the accuracy of the predicted results than the nature of numerical damping and application grid." All of the schemes can predict most of the complex flow fields in turbomachinery as indicated in earlier sections. But the Runge–Kutta method shows somewhat superior performance as the artificial viscosity is introduced and controlled by the user; it is easy to code and vectorize and has received wide acceptance by the turbomachinery community.

Liu et al. (1991) have provided a comparison of the explicit (Subramanian and Bozzola, 1987) and implicit (Dawes, 1987) codes described earlier for the three-dimensional viscous flow in a high-work turbine stator using identical grids and turbulence models. The flow field, losses, and outlet angles predicted are nearly identical. The authors have not provided any details on the computational efficiency.

5.7.7   Pseudocompressibility Technique for Incompressible Flow

Because a large number of turbomachines (especially liquid-handling pumps and turbines) operate in the low-speed range, incompressible flow prediction capability is important to the efforts of researchers to improve analysis and design techniques for these turbomachines.

The pressure-based methods solve a Poisson equation for the pressure and the momentum equations in an uncoupled manner, iterating until a divergence-free flow field ($\nabla \cdot \mathbf{V} = 0$) is satisfied. On the other hand, it is evident from the material presented in earlier sections, that very efficient and accurate algorithms have been developed for solving hyperbolic systems governing compressible flow using time-iterative methods. Prediction of incompressible flow (very low M) using these techniques is not efficient and is not recommended. Furthermore, the energy equation in Eq. 5.1 has to be decoupled from the momentum equation for isothermal flows. Hence there is a need for efficient techniques for solving incompressible flow. One approach involves recasting the incompressible flow equations (which are elliptic in steady state) as hyperbolic systems. One of the techniques which is finding wide use is the pseudocompressibilty technique due to Chorin (1968). To make the governing equations behave like a hyperbolic system, Chorin manipulated the continuity equation by adding a time derivative of a pressure term. In the steady state, these equations and solutions are exactly those of steady incompressible flow, even though the temporal solutions are not accurate. The continuity equation in this formulation is written as

$$\frac{1}{\beta}\frac{\partial p}{\partial t} + (\nabla \cdot \mathbf{V}) = 0 \tag{5.96}$$

where β is a pseudocompressibility parameter (because $(1/\beta)(\partial p/\partial t)$ behaves like the $\partial \rho/\partial t$ term in the compressible formulation). This modification results in a hyperbolic system of equations. For example, in Eq. 5.1, the modifications include

$$q = [p/\beta, W_r, W_\theta, W_z]^T$$
$$E = \left[W_r, (W_r^2 + p/\rho - \tau_{rr}/\rho), (W_r W_\theta - \tau_{r\theta}/\rho), (W_r W_z - \tau_{rz}/\rho)\right]^T \tag{5.97}$$

and similarly the vectors F, G, and S are modified.

The formulation given by Eq. 5.96 introduces spurious pressure waves of finite speed into the fluid through the pseudocompressibility parameter β. When $\beta \gg 1$, the transient effect on the solution is minimized. The artificial term drops out at steady state regardless of the size of β. A detailed discussion of the pseudocompressibility technique, including a stability analysis, can be found in Kwak et al. (1986). Typically, the pseudocompressibility parameter is selected to satisfy

$$\beta^* = \frac{\beta}{V_{\text{ref}}^2} \approx 5\text{--}10$$

The basic laminar pseudocompressibility code developed at NASA Ames Research Center (Kwak et al., 1986) was modified by Warfield and Lakshminarayana (1989) for turbomachinery configuration to include rotation and turbulence models to compute the flow in a low-speed compressor rotor. The pseudocompressibility formulation is linearized, using a truncated Taylor series and discretized using a two-point central difference in the spatial directions. The solution technique is the standard implicit approximate factorization scheme described earlier. Constant explicit and implicit smoothing was utilized to facilitate stability and convergence of the solution. Convergence acceleration was implemented in the form of a variable time step (constant local CFL number).

It should be remarked here that any of the explicit and implicit techniques can be used to solve Eq. 5.1 with pseudocompressibility formulation. Kwak et al. (1986) and Warfield and Lakshminarayana (1989) used the implicit technique due to Beam and Warming (1978) described earlier. Merkle and Tsai (1986) have employed the Runge–Kutta method.

The pseudocompressibility technique has been utilized to predict the turbulent flow field in a 90° turning duct, a cascade, a rotating channel (Warfield and Lakshminarayana, 1987b), and the Penn State compressor (Warfield and Lakshminarayana, 1989; Suryavamshi and Lakshminarayana, 1992) at two different operating conditions, with good agreement with the measured data. Extensive data are available for the Penn State compressor rotor, including the entire inviscid flow field, blade boundary layer, hub wall boundary layers, and tip clearance flow field. This is a low-speed compressor with a 21-blade rotor and a hub-to-tip ratio of 0.5. The design conditions were a flow coefficient of 0.56 and a loading coefficient of 0.4884. Further

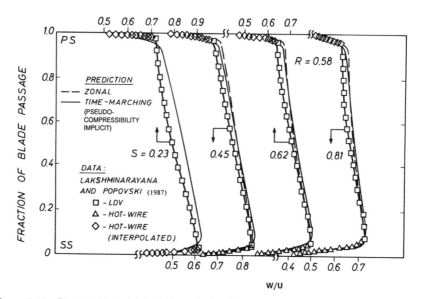

Figure 5.42 Blade-to-blade total relative velocity distribution in a subsonic single-stage axial compressor at $R = 0.58$; peak pressure rise condition: $\psi = 0.55$, $\phi = 0.5$. (S is the streamwise distance normalized by blade chord). (Copyright © Warfield and Lakshminarayana, 1989; reprinted with permission.)

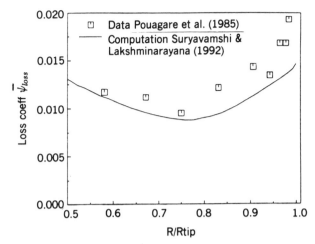

Figure 5.43   Radial distribution of passage averaged loss in a subsonic single-stage axial compressor at design operating condition at the trailing edge. (Copyright © Suryavamshi and Lakshminarayana, 1992; reprinted with permission of ASME.)

details on this compressor is given in Section 2.5.5 and in Fig. 2.17. The blade boundary layer and the complete flow field are predicted at the peak pressure rise coefficient, where the boundary layer growth is substantial. This is a severe test case for checking the validity of a three-dimensional Navier–Stokes code. The blade-to-blade flow, including the growth of the boundary layer along the blade, is predicted well (Fig. 5.42). Furthermore, the development of the wake, the spanwise mixing effects, and radial distribution of losses are also predicted well as shown in Fig. 5.43.

By et al. (1993) have utilized this technique to predict the flow field in an automotive torque converter pump. The measured (LDV) and predicted flow fields are in qualitative agreement. The reversal of secondary flow from midchord to trailing edge is captured as well as the pressure distribution on the blade.

5.7.7.1   *Assessment of Pseudocompressibility Technique and Comparison with Pressure-Based Methods*   The pseudocompressibility technique (PCT) utilizes well-developed time and iterative techniques, but the stability and convergence are sensitive to the value of the parameter β. This value varies with configuration and flow regime (laminar or turbulent). Once an optimum value for β is achieved, the technique is more efficient than a corresponding compressible-flow formulation. This technique is suitable for pumps, hydraulic turbines, and other low-speed turbomachinery. It is suitable for flows with separation, and the technique can be extended to unsteady flows (Rogers et al., 1989). A wide variety of geometries and flows (cascade endwall flow, low-speed compressor, S and 90° bends, space shuttle main engine hot gas manifold, rocket pump inducer, artificial heart pump) have been computed.

Recent developments on this technique include the use of a diagonal scheme for the inversion of the equations at each iteration (Rogers et al., 1987), extension to unsteady flow (Rogers et al., 1989), acceleration by introducing artificial bulk viscosity to dissipate the sound waves rapidly (Ramshaw and Mousseau, 1990), and preconditioned upwind method to accelerate the scheme to steady state (Hsu et al., 1990). These techniques are still in the developmental stage and have not been applied to turbomachinery flows. When these are incorporated, the technique and code may compete effectively with the pressure-based methods described earlier.

With regard to the pressure-based method (PBM) versus the pseudocompressibility technique (PCT), PBM is well established, widely used, and user friendly, and it has lower storage requirements than does PCT. It is easier to implement $k-\epsilon$ and ARSM models in the PBM. In situations where large pressure gradients exist (leading edge, large camber, etc.), PBM convergence slows down in an uncoupled approach. In PCT, the pressure is solved in a coupled fashion, and this may have some advantage. The PCT requires large storage and an experienced user to choose an optimum value of β. The PCT can be further developed to become more attractive. Incorporation of multigrid, an LU scheme (implicit), or Runge–Kutta scheme (explicit) could enhance its efficiency.

5.8 ZONAL TECHNIQUES

It is evident from the various techniques presented in this chapter that the solution of the Navier–Stokes equations by iterative techniques requires a large amount of computer storage and time for three-dimensional turbulent flows. Application of these techniques to the entire blade row and to single-stage and multistage turbomachinery flows will challenge the computing power of present and future computers. Utilization of these techniques on a day-to-day basis for design in industry requires efficiency and cost effectiveness. As indicated earlier, the Navier–Stokes equations can be reduced to Euler equations, potential equations, boundary layer equations, and parabolized Navier–Stokes equations. These are some of the approximations made to reduce the equations to forms amenable to efficient computation. Unfortunately, these equations do not provide solutions to the flow in the entire region. On the other hand, the solution of Navier–Stokes equations in the entire region is prohibitively expensive and challenges the computing power in terms of storage and cost. Zonal techniques overcome this problem. Instead of using one equation system in the entire region, several zones are identified and the solution is obtained by using the appropriate equations (Euler, PNS, BL) and patching and integrating these regions. For example, one widely used technique is a cascade solution consisting of potential equations in the inviscid region (Chap. 3) and boundary layer equations in the viscous region (Sect. 5.5), with interaction of these two regions to update the pressure and velocity fields due to boundary layer

growth. The efficiency, cost, accuracy, and robustness of a code are impor-
tant for repetitive runs for design and analysis of numerous blade geometry.
Duct flows can be often resolved through a combination of inviscid codes and
PNS codes.

Furthermore, the boundary layer codes or space-marching codes are
computationally very efficient, and finer grid could be used near the surface
to capture the viscous layers accurately. Examples where such zonal techniques are useful are ducts and interblade rows, cascades, and inviscid-
dominated subsonic rotors and multistage turbomachinery, where
Navier–Stokes solutions are expensive. Some examples of the zonal approach
are given in Fig. 5.44.

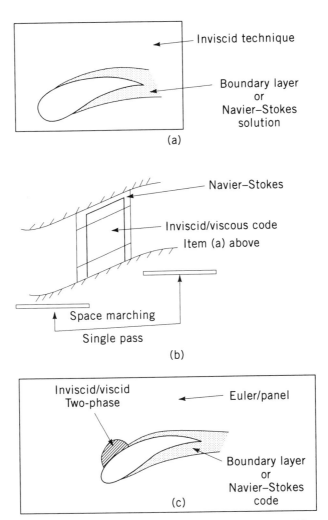

Figure 5.44 Examples of the zonal technique. (a) Cascade / rotor blade. (b) Rotor passage.
(c) Two-phase flow.

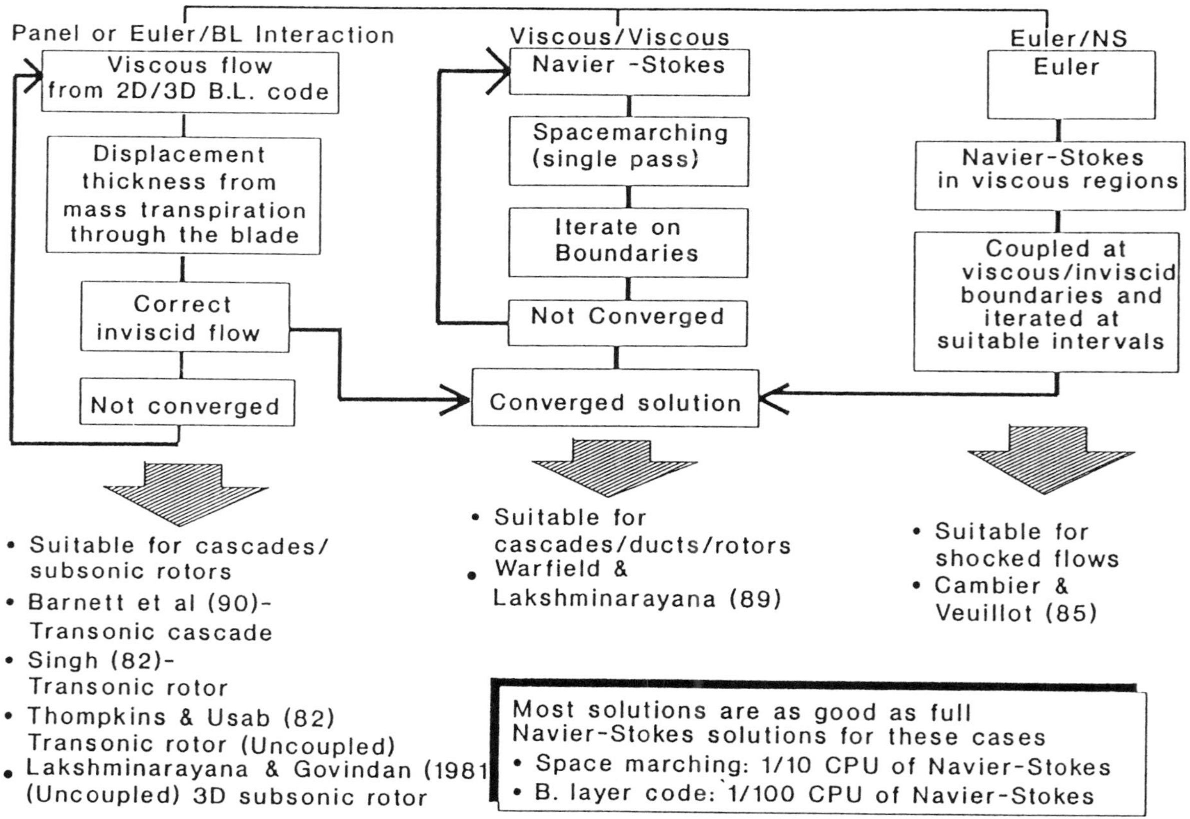
Panel or Euler/BL Interaction
Viscous flow from 2D/3D B.L. code
Displacement thickness from mass transpiration through the blade
Correct inviscid flow
Not converged
Viscous/Viscous
Navier -Stokes
Spacemarching (single pass)
Iterate on Boundaries
Not Converged
Converged solution
Euler/NS
Euler
Navier-Stokes in viscous regions
Coupled at viscous/inviscid boundaries and iterated at suitable intervals
• Suitable for cascades/ subsonic rotors
• Barnett et al (90)- Transonic cascade
• Singh (82)- Transonic rotor
• Thompkins & Usab (82) Transonic rotor (Uncoupled)
• Lakshminarayana & Govindan (1981 (Uncoupled) 3D subsonic rotor
• Suitable for cascades/ducts/rotors
• Warfield & Lakshminarayana (89)
• Suitable for shocked flows
• Cambier & Veuillot (85)
Most solutions are as good as full Navier-Stokes solutions for these cases
• Space marching: 1/10 CPU of Navier-Stokes
• B. layer code: 1/100 CPU of Navier-Stokes

Figure 5.45 Classification of zonal techniques.

An additional point is that even though the Navier–Stokes equations are universally valid, for certain classes of flows, reduced forms are actually more appropriate. For example, many boundary layer and duct flows are parabolic flows where downstream boundary conditions (which are necessary for Navier–Stokes solutions) can be difficult to determine. These situations actually make the Navier–Stokes system difficult to apply. The principal motivation behind zonal equation methods is to decrease the computation time and computer storage necessary to compute the solution for flows which exhibit only localized regions of ellipticity. These regions inhibit the use of parabolized methods but are small enough to make a full iterative NS solution seem inefficient. The zonal techniques can be interactive or noninteractive depending on the requirement, and they use the most efficient techniques in each region.

A brief review of various zonal techniques is given in Warfield and Lakshminarayana (1987a). These can be classified as (1) viscous/inviscid techniques, (2) supersonic/subsonic techniques, and (3) viscous/viscous techniques. Some of the current zonal techniques are shown schematically in Fig. 5.45.

5.8.1 Coupled Inviscid / Boundary Layer Solutions and Assessment

Normally this is an interactive procedure; that is, the inviscid solution feeds the boundary layer edge information to the viscous solver, while the viscous solution feeds displacement thickness information to the inviscid solver. The interaction is complete when a converged boundary layer solution and the inviscid solution are reached. The boundary layer equations, being the appropriate equation set within the boundary layer, are normally solved either in differential or integral form for the viscous effects. The inviscid solution is achieved through Euler or potential solvers.

A general procedure can be outlined as follows (Calvert et al., 1982; Barnett et al., 1990):

1. The inviscid flow field is computed from any of the techniques described in this and earlier chapters, depending on the flow regime and geometry. For example, 2D incompressible flow is best handled by a panel method, supersonic and transonic flows are best resolved by an Euler equation, and subsonic flows are best handled by either a potential or a stream function solver.

2. A boundary layer code is used with the prescribed pressure (from item 1) to predict the local properties or the global properties (integral technique).

3. If viscous effects are small (very thin boundary layer), the computation can be terminated here. If the viscous effects are substantial, the displacement thickness is calculated and the blade profile is modified to

recompute the inviscid flow field. The viscous displacement effect is represented by a mass transpiration through the surfaces of the blades and by a jump in the normal velocity component along the wake streamlines. These are specified so that the inviscid streamlines are displaced to conform to the effective displacement, which includes the blade profile plus the displacement thickness. The transpiration velocity is given by

$$\rho v|_s = \frac{d}{ds}(\rho_e U_e \delta^*) \qquad (5.98)$$

where δ^* is the displacement thickness, U_e is the boundary layer edge total velocity at the streamwise location s.

4. The inviscid and viscous equations are coupled through the appropriate boundary conditions on the blade surface and the wake edge. The solution of the complete flow is obtained by "global" iteration between the viscous and the inviscid solutions, varying the displacement thickness parameter until the viscous-layer edge velocity matches with the corresponding inviscid velocity. Inverse or semi-inverse coupling or fully coupled techniques are needed to ensure convergence.

Barnett et al. (1990) utilized Ni's (1982) multi-grid Euler solver described earlier to resolve the inviscid flow. The boundary layer code utilized is that due to Carter (1978), which is a finite-difference formulation of a compressible stream function approach. They computed the viscous flow field in a transonic cascade (Stephen and Hobbs, 1979) and a subsonic cascade and found good agreement with experimental data. The turbulence model used in this analysis is due to Baldwin and Lomax (1978). A comparison between Navier–Stokes analysis (Davis et al., 1988), and the results based on interaction of the inviscid (Ni, 1982) and boundary layer code (marked IVICAS) are shown in Fig. 5.29. The prediction from the zonal and Navier–Stokes codes are almost identical except in the stall regime. The incorporation of viscous effects through the transpiration model need not be carried out at every time step because $\partial \delta^*/\partial t$ is not large. Hence, the inviscid solution can be updated every 30–50 time steps depending on the geometry and the problem. The inviscid-flow–boundary-layer interaction code can provide very accurate results (at least for cascades) in a fraction of the computer time needed for a full Navier–Stokes solution. The shock–boundary-layer interaction regions may need more frequent updating than the unseparated boundary layers. None of the codes were able to predict the losses in stall regions with positive incidence.

Singh (1982) and Calvert (1982) incorporated a boundary layer integral technique for compressible flow in a finite-volume explicit procedure due to Denton (1975) and computed the shock-boundary–layer interaction effects on inviscid flow. Singh showed considerable enhancement in the quality of flow computations for a transonic compressor rotor. This is an injected mass technique (similar to the one described above), which allows fluid to pass through the walls in proportion to the rate of growth of the boundary layer thickness. The velocity normal to the wall is given by Eq. 5.98. Pierzga and Wood (1985) employed this code to predict the flow field in a NASA transonic rotor for which laser-doppler velocimeter data are available. The static pressure rise and exit tangential velocities were predicted more accurately with the inclusion of blockage in an Euler code, but no improvement was observed in detailed flow field predictions. Thompkins and Usab (1982) adopted a similar procedure and incorporated a differential 3D boundary layer code in their explicit time-marching finite-difference inviscid code described earlier. The two solutions were not coupled; hence, the effect of the boundary layer on the inviscid flow was not computed, but the effect of the inviscid flow on the boundary layer growth was resolved.

It should be remarked here that all the inviscid-flow/boundary-layer interaction techniques described above resolve the global effects of viscous flow on the inviscid pressure, but the local effects are not computed accurately. Furthermore, convergence may be slow in separated flows, including shock–boundary-layer interaction regions. An assessment of a coupled inviscid flow/boundary-layer interaction technique is similar to assessment of boundary layer solutions described in Section 5.5.6.

There have been techniques developed to resolve the interaction effects more accurately through the use of a Euler equation in the inviscid region and the use of Navier–Stokes equations in the viscous regions and couple these two techniques through an overlapping grid, where both equations are solved. The procedure used by Cambier and Veuillot (1985) can be summarized as follows:

1. About five zones are incorporated in a cascade passage: three inviscid zones in the core regions of the flow (one above the suction surface, one above the pressure surface, and one upstream) and two viscous zones (one near the suction surface and one near the pressure surface, including downstream wake).
2. The explicit technique due to MacCormack is used.
3. A compatibility condition (all variables have to be identical) is enforced in the overlapping zone between the inviscid and viscous regions.

Using this procedure, Cambier and Veuillot (1985) were able to resolve the inviscid pressure flow field as well as the boundary layers, wakes, and shock–boundary-layer regions in a transonic cascade. This procedure showed substantial improvement in CPU time required for computing the flow in a cascade over the full Navier–Stokes solutions in the entire domain.

The inviscid/viscous techniques are also used in supersonic/subsonic flows. For example, the supersonic external flow may have subsonic viscous layers. The solution at supersonic speeds can be accomplished by a space-marching technique, whereas an iterative procedure is necessary in subsonic viscous regions. If these regions are thin, they can be treated by a single-pass space-marching Navier–Stokes procedure described earlier. Such a procedure for supersonic flows was developed by Schiff and Steger (1980). The supersonic/subsonic zonal approach has been used for both potential and Navier–Stokes formulations in predominantly supersonic flows with embedded subsonic zones. These methods are employed primarily in external flows at the present time. They are attractive for turbomachinery analysis and design due to large savings in computer time. The use of the space-marching technique in supersonic regions results in decreased computer storage.

5.8.2 Viscous / Viscous Techniques

Unlike the supersonic/subsonic zonal method, the streamwise pressure gradient in the incompressible flow equations needs special treatment to create a fully parabolized formulation. The change in the character of the elliptic or time–hyperbolic Navier–Stokes equations to parabolic form creates an opportunity for significant improvement in computation time. A zonal technique for turbomachinery was developed by Warfield and Lakshminarayana (1989). The incompressible flow equations in rotating coordinates are solved for a turbomachinery rotor using the pseudocompressibility technique in the elliptic regions of the flow and a space-marching technique in the parabolic or nearly parabolic regions. The zonal technique developed is flexible, is valid for three-dimensional flows, allows for zonal interaction, and arbitrary zone location, and has potential for adaptive zone location. The techniques has been applied to Penn State compressor rotor flow with a 69% saving in computation time over the full Navier–Stokes solution.

The procedure developed by Warfield and Lakshminarayana (1987a, 1989) consists of the following steps:

1. The entire flow field is solved elliptically to partial convergence. The purpose of the partial convergence (approximately 10% of computer time) of the time-marching code over the entire field is to establish an initially assumed pressure for the space-marching code and also to establish the location and extent of the elliptic boundaries. Two ellipticity parameters derived (Warfield and Lakshminarayana, 1987a) are utilized to facilitate the placement of the boundaries between the elliptic and parabolic zones.

2. The parabolic zones are solved by the space-marching method.

3. The elliptic zone is solved using initial and boundary conditions from a partially converged solution and from the parabolic zone solution.

4. The parabolic zones are solved again utilizing initial and boundary conditions from the solution of the elliptic zone.

5. Steps 3 and 4 are repeated until convergence.

The zonal method for a compressor rotor application consisted of an elliptic zone extending from the inlet plane to the 25% chord location. The parabolic zone continued from the 25% chord location to the outflow plane. Both zones extended from blade-to-blade and from hub-to-tip. Only one sweep of the zonal method was utilized; the combination of zones with the elliptic zone upstream and the parabolic zone downstream reduced the need for multiple zonal sweeps due to the inability of the parabolic solution to influence the upstream elliptic solution. A comparison between the zonal technique and the full Navier–Stokes solution for the Penn State compressor is shown in Fig. 5.42. The overall features of the flow were predicted very well by both the Navier–Stokes and the zonal technique. This means that both techniques capture these features. The solution for the entire rotor (including upstream and downstream of the blade row) was obtained in 31% of the time needed for the full Navier–Stokes solution.

The zonal technique combines the efficiency of the space-marching technique with the accuracy of the time-iterative methods. The application of these techniques to the multistage compressor flow field is promising. The zonal techniques could be improved with an adaptive buffer zone, and could possibly include analytical solutions or the passage-averaged equations (Adamcyzk, 1985) in some regions of the flow field. The improvement in zonal techniques also depends on improvements in component codes. As new and efficient techniques become available, they can be incorporated to provide improvements in zonal techniques.

5.8.3 Assessment of Zonal Techniques

1. The zonal techniques, especially the inviscid/viscous techniques, are widely used in industry for design/analysis. The Navier–Stokes code is mainly used in the final iteration/research mode.

2. Zonal techniques with embedded analytical solutions (i.e., tip clearance flow, leading-edge laminar solutions), loss correlations (secondary and tip clearance losses), and quasi-viscous representation (shear forces) are most attractive for turbomachinery.

3. Zonal techniques save substantial computer time, but require experienced personnel and are difficultto code and operate.

4. The technique can accurately resolve overall performance, efficiency, blade pressure distribution, boundary layer and wake profiles, losses, and spanwise mixing.

5. Because space-marching and boundary layer codes are very efficient,a very fine grid can be used in the viscous/near-wall regions to resolve these flows more accurately than Navier–Stokes codes, particularly where boundary layers are thin (e.g., near the leading-edge stagnation point).

6. Choice of component codes is critical. Overall efficiencyand accuracy of zonal techniques depend on the efficiencyand accuracy of component codes.

7. Interaction between component zones is arbitrary at the present time. Systematic investigation is needed.

8. Zonal techniques are not valid in endwall and tip clearance regions.

5.9 COMPUTATIONAL DESIGN

Major advances will take place in the use of powerful computational techniques in the design of turbomachinery flow systems. Many industries have already updated their preliminary design codes with a computational design/analysis system. In a recent AGARD (1989) lecture series, some authors described comprehensive state-of-the-art design methods for blading: compressors (Stow, Meauze), turbines (Bry, Hourmouziadis), 2D blading (Starken), and so on. Howard et al. (1987) presented a description of the design system used for the aerodynamic and the thermal design of turbines. Nojima (1988) provided a similar description for industrial centrifugal compressors. The methods solely based on empiricism, analytical, or iterative analytical methods at arriving blade shape are not covered in this section. The design methodology depends on the application, geometry, and type of industry, and there is no unified approach.

The inverse design technique, where the blade profile is derived from a prescribed velocity or pressure in the flow path, is restricted to two dimensions at the present time. If the flow is irrotational, a hodograph method (Section 3.4.3; Figs. 3.21 and 2.65; Sanz, 1987; Leonard, 1990) can be used to transform the equations into the hodograph plane, and the potential equation is solved using computational techniques to derive the blade geometry. A good example of design methodology is illustrated in Fig. 5.46, taken from Stow's paper in AGARD (1989). Stow employed mixed-mode blading codes. He specified the initial geometry and calculated the blade pressure distribution. This was the analysis mode. The pressure distribution was then specified over part of the chord to eliminate the shock, and an inverse design technique was used to derive a new (or design) blade shape. It should be remarked here that these designs are strictly valid for the design condition, with performance deterioration occurring at off-design conditions.

One of the interesting applications of inverse design techniques was carried out by Favre et al. (1987). They developed an inverse design method, based on the panel technique, to reduce suction peaks at the suction side near the leading edge of a Francis turbine. This reduced the inlet and leading-edge cavitation over an extended range, thus improving the performance. Furthermore, inverse design techniques can be made to interact with the boundary layer calculations to optimize for boundary layer growth and to minimize flow separation.

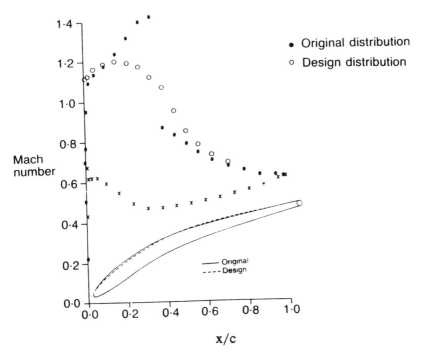

Figure 5.46 Shock-free supercritical compression blade derived from inverse design with prescribed surface Mach number. (Copyright © Stow, 1989, AGARD LS 167. Reprinted with permission.)

If the flow is rotational, a Euler technique has to be adopted; a first guess of the geometry is essential for this calculation. The initial geometry is modified during the flow calculation (Zannetti and Larocca, 1990). The most commonly used technique is the iterative inverse method, where the inverse and analysis methods are used in a suitable combination to derive an optimum blade shape and an optimum geometry of the annulus. These techniques are reviewed by Leonard (1990).

Most industries now have some iterative schemes for the design of blading and flow path (e.g., AGARD, 1989; Worth and Plehn, 1990; Nojima et al., 1988; Howard et al., 1987; Tyner and Sullivan, 1990). Most aircraft engine companies are beginning to iterate the design with Euler and Navier–Stokes codes. A typical system used successfully by Nojima et al. (1988) to design high-performance industrial compressors include the specification of the pressure distribution and the desired overall performance. In the preliminary design system, the main dimensions of the impeller and diffuser are generated. A first approximation of the impeller and diffuser shape are generated by the program for succeeding detailed design. Optimization of the geometry is then carried out to achieve maximum efficiency. The analysis system consists of blade-to-blade calculation (Chap. 3), boundary layer computation

(Sect. 5.5) and three-dimensional time-marching (Denton, 1982) techniques which were described earlier. The geometry and flow path are modified to achieve the desired velocity/pressure distribution along the flow path. The database (loss correlations, etc.) is then used to carry out performance prediction. This step could be replaced by a Navier–Stokes solver. Nojima et al. (1988) were thus able to achieve very high efficiency of a family of centrifugal compressors, and this was verified by measurements.

Hourmouziadis (AGARD, 1989) describes a similar system for low-pressure turbines, with optimization to achieve minimum boundary layer growth, minimum trailing-edge thickness, no flow separation, highest blade pitch, and high acceleration on the suction side and delayed transition. Tyner and Sullivan (1990) designed a small high-performance, low-aspect-ratio,

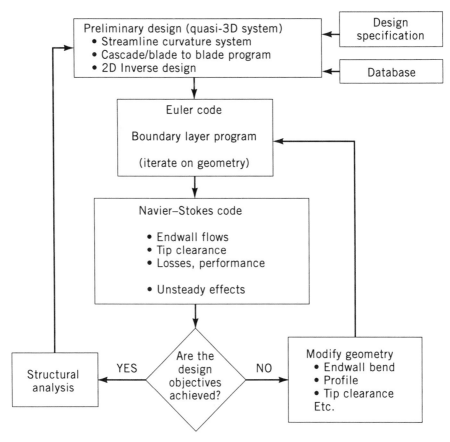

• Three-dimensional viscous codes should be employed in early stages of design to locate problem areas and to check validity of loss correlations.

Figure 5.47 Turbomachinery design: a suggested approach.

high-reaction compressor using a quasi-3D and 3D design system. The high performance achieved by this compressor is attributed partly to a 3D design system based on Ni's (1982) explicit Euler technique described earlier. Rhie et al. (1994) have successfully employed Euler and Navier–Stokes procedure to design advanced turbofan and achieved 0.5% improvement in fan efficiency, 1% improvement in compressor efficiency, and 3% improvement in transonic stall flutter boundary relative to the earlier conventional design.

A suggested approach for turbomachinery design is shown in Fig. 5.47. The design specifications include velocity and pressure field, desired efficiency, and losses. The database includes the profile loss correlations, tip clearance model, endwall loss correlations, deviation angle, spanwise mixing model, and other correlations developed for that particular family of turbomachinery. For axial flow turbomachinery, the cascade inverse design technique (such as that of Sanz, 1987) or Favre et al.'s (1987) inverse panel code could be used to generate blade profiles from hub to tip. A Euler code is then used to iterate on the geometry in a coupled fashion. The desired hub-to-tip, blade-to-blade, and leading-to-trailing edge flow is achieved through iteration. The final step involves analysis of the blade row to detect high loss, separation, and shock–boundary-layer interaction regions. If the design objectives are not met, the procedure is repeated until the desired performance and blading are achieved. The choice of technique and code for the analysis is based on the assessment carried out earlier. For example, the liquid-handling machinery is best handled with a pressure-based method or pseudocompressibility technique.

5.10   CONCLUDING REMARKS

The technique to be employed for the computation of a flow field in turbomachinery depends on the geometry and flow regime. The full Navier–Stokes code is most suitable for the final stage of design and analysis and is only now being adapted in the design system. This situation will change as advances are made in the algorithm development, turbulence modeling, and computer architecture. The recommendation made here refers to the cost-effective computation to capture the flow field accurately. These recommendations are summarized in Fig. 5.48. If the interest lies only in the pressure distribution, the inviscid codes are most appropriate, provided that the viscous regions are small. If, for example, the flow field in a constant-area duct has large viscous regions, a space-marching code can capture most of the flow physics.

For incompressible flow and low-Mach-number cascades, the most suitable approach would be to use a panel code, followed by a Keller-box-type finite-difference scheme for the boundary layer. The high subsonic flow case can be best resolved using a potential code integrated with a boundary layer code. The transonic and supersonic case, with shock–boundary-layer separa-

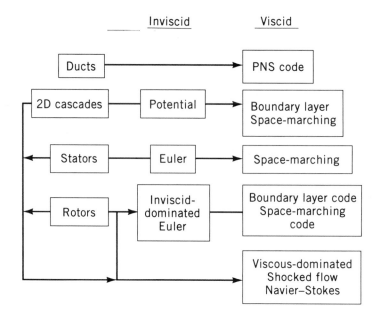

Euler / Navier–Stokes codes

- No definitive conclusion; pressure-based methods, RK explicit widely used

- Advances in LU scheme may make implicit attractive

- None of the Navier–Stokes codes has resolved endwall flow, tip clearance flow field, shock–boundary-layer iteration regions satisfactorily

Figure 5.48   Summary recommendations for turbomachinery computation.

tion and interaction, is best handled by a Euler/boundary layer code or a Navier–Stokes code depending on the extent of flow separation. Likewise, incompressible and subsonic rotor flows can be resolved with a quasi-three-dimensional techniques and a boundary layer code, with empirical correlations or secondary flow analysis in the endwall regions (Chap. 4). Transonic and supersonic rotor flows are best resolved by a Navier–Stokes code. The viscous dominated regions such as the endwall, shock–boundary-layer interaction regions are most accurately resolved through the solution of the full Navier–Stokes equation. If the majority of the flow field is viscous or dominated by shocks, the full Navier–Stokes solution would be the most appropriate route to take.

The flow in a hydraulic turbine, some pumps, and other subsonic turbomachinery, where inviscid flow dominates, is best resolved with a Euler code in the inviscid region and a boundary layer code or a single-pass space-marching code in the viscous regions. On the other hand, low-aspect-ratio turbomachinery is dominated by endwall effects with a large viscous loss core. This, as well as transonic axial compressor and centrifugal compressor flow fields, is better resolved with a full Navier–Stokes code.

Techniques for turbomachinery flow computation have reached a high level of maturity. A wide variety of flows have been computed: the Francis runner, mixed-flow pumps, automotive torque converter, transonic and supersonic compressors and turbines, propellers and prop fans, steam and gas turbines, and rocket pumps, including inducers. However, the design is still based on classical techniques such as the through-flow method and correlation, even though some companies are now beginning to do the final analysis and design iteration using 3D Navier–Stokes codes. The potential for resolving the complex flow physics in turbomachinery is recognized. Improvements in physical approximations and models (e.g., turbulence, transition), numerical accuracy, stability, convergence rate, and computational efficiency are needed before the Navier–Stokes method can be used extensively in the analysis and design phase. Ideally, the codes should be independent of user-specified constants (e.g., pseudocompressibility factor β, artificial dissipation, or constants in various turbulence models).

Furthermore, there is a need for calibration and validation of these codes before they are adapted for production runs. The calibration should be based on an exact analytical solution, and validation should be based on benchmark-quality data for complex geometries and flow fields. For example, it is necessary to validate turbulence quantities predicted against data to check the accuracy of turbulence modeling. Benchmark quality data are scarce in the area of turbomachinery. This issue should be addressed before complete confidence in these codes can be achieved.

It should be remarked here that potential codes and panel codes are limited to cascade flows. The technique is not applicable for turbomachinery rotor flows. The Euler codes have matured and are routinely used in the design and the analysis of turbomachinery flows. The choice of Euler code/technique depends on the flow regime. The pressure-based methods and the pseudocompressibility techniques are attractive for low-speed flows, whereas the explicit Runge–Kutta method and the pressure-based methods are widely used for steady high-speed shocked flows. Similar conclusions can be made for Navier–Stokes. One of the most efficient computational techniques is the zonal method, especially in the design mode. The inviscid-flow/boundary layer code is efficient, accurate, and widely used. The other zonal techniques (Euler/Navier–Stokes, Euler/space-marching, Navier–Stokes/space-marching) are very attractive, but have not received wide acceptance.

The Navier–Stokes analyses are most useful for applications where viscous flows dominate (e.g., centrifugal compressor and low-aspect-ratio turbomachinery). Even though great progress has been made in this area, the codes have not been able to accurately capture some of the complex flow features, including annulus–wall and hub–wall boundary layers in rotors.

There is a tendency among the new generation of turbomachinery computors to ignore the early advances made in the analysis of turbomachinery flows (e.g., boundary layer solutions, secondary and leakage flow theories, etc.). This has led to widespread misuse of Navier–Stokes solvers. In some cases, for example, a panel code combined with boundary layer and secondary flow or other linearized solutions can provide predictions which are as good as Navier–Stokes solutions, but carried out in a fraction of the time it takes to compute with a full Navier–Stokes equation. The computors should be aware of these and incorporate these analyses in their codes for cost-effective computation. Navier–Stokes is an overkill for some flows which may be as much as 95% inviscid. In addition, the designers should be willing to accept Navier–Stokes solutions and eliminate or update design correlations (e.g., blockage coefficient, loss correlations, deviation, etc.) using these solutions.

6

TWO- AND
THREE-DIMENSIONAL
VISCOUS EFFECTS
AND LOSSES

6.1 INTRODUCTION

Turbomachinery flows are among the most complex flows encountered in fluid dynamic practice. Both the geometric description of the fluid flow domain and the physical processes present are extremely complicated.

Turbomachinery flows are always three-dimensional, viscous, and unsteady. The working fluid may be single-phase liquid or gas, or two phase (liquid–solid or gas–solid or liquid–gas). Real gas effects may be important, such as gas mixture effects, non-perfect gas effects, evaporation, and sublimation. The flow may be incompressible or compressible, with subsonic, transonic, and supersonic regimes which may be present simultaneously in different regions. The viscous flow usually has high free-stream turbulence which may include multiple length and time scales. Regions of laminar, transitional, and turbulent flows, separated flows, and fully developed viscous profiles may all be present simultaneously due to the multiplicity of length scales introduced by the complicated geometry of the flow field. The viscous and turbulent regions encounter complex stress and strain due to the presence of three-dimensionality, large pressure gradients in all directions, rotation, curvature, shock waves, shock wave–boundary layer interaction, interacting boundary layers and wakes, heat transfer and cavitation. The flow field may be dominated by vortical flows: secondary flow, tip leakage vortices, shed vortices, leading-edge horseshoe vortices and scraping vortices. The equations describing the flow physics are strongly coupled, and complex boundary conditions are often present (transpiration, heat transfer, periodicity, moving walls, etc.). Schematic representation of the flow field in two typical turbomachinery configurations are shown in Figs. 1.15 and 1.16, respectively.

The geometric description of the flow regions includes many parameters having different length scales. Associated with the blade profile are stagger angle, blade camber, chord, blade spacing, maximum thickness, camber distribution, thickness distribution, leading and trailing edge radii, surface roughness, and cooling hole distribution. Associated with the flowpath are radial distributions of stagger angle, camber and thickness, lean, twist, dihedral, sweep, skew, flare, aspect ratio, hub/tip ratio, tip clearance, end-wall curvature (hub and tip), flow path area change, axial spacing between blade and vane rows, radial distribution of cooling holes, and the presence of part-span dampers (which changes the domain from periodic simply con-nected to periodic multiply connected). Also associated with the flow path can be recirculating passages at hub or tip, surface roughness, platform cooling holes, and offtake bleeds.

Both the flow and geometrical variables dictate the nature of the govern-ing equations and the flow solvers to be used. A large number of flow and thermal parameters are encountered in turbomachinery fluid dynamics, in-cluding Reynolds number, Mach number, rotation number, Richardson num-ber, Prandtl number, Eckert's number, and Thoma coefficient.

This chapter is devoted to viscous effects in turbomachinery. These viscous effects include: laminar, transitional, turbulent, and separated boundary layers and wakes due to blade profile (both 2D and 3D effects); annulus- and hub-wall boundary layers; mixing of tip clearance leakage flow, secondary flow and horseshoe vortex, and annulus-wall boundary layers in the endwall; spanwise mixing downstream of the blade row; shock–boundary-layer inter-action; shock losses.

6.2 NATURE OF REAL FLUID EFFECTS

The viscous and turbulence effects play a major role in affecting the perfor-mance of turbomachinery, and they have been alluded to in Chapters 2, 4, and 5. Some of the major effects are as follows:

1. The losses associated with these viscous effects decrease the efficiency of turbomachinery through stagnation pressure loss and increase in entropy. This also decreases the pressure rise in compressors and pumps, and it decreases the pressure drop in turbines.

2. The viscous effects affect the local and global flow properties inside the passage and downstream of the blade row.

3. The viscous effects not only introduce three-dimensionality in the flow, but also affect the flow properties (which vary in both the spanwise and in blade-to-blade directions), thus introducing off-design conditions in subsequent blade rows.

4. Viscous effects have a major effect on the cooling and heat transfer in turbines. These are described in Chapter 7.

5. The viscous layers introduce blockage to the flow, thereby limiting the performance through decreased effective area, pressure rise (or drop) in turbomachinery. In a compressor, as the mass flow is reduced at constant rotational speed, pressure rise tends to increase due to increased angle of incidence to the blade. However, the viscous effects also increase, increasing loss and blockage as the mass flow is reduced. This limits the increase in pressure rise, creating a characteristic pressure rise/flow curve which peaks as the mass flow is reduced (Figs. 2.17, 2.18, and 2.41). This compression system characteristic leads to a system instability called rotating stall (Section 2.5.7); if the mass flow is reduced beyond the peak pressure rise, rotating stall quickly degenerates into a more severe system instability called surge (Section 2.5.8).

6. In single-stage or multistage turbomachinery, with alternating stationary and rotating blade rows, the viscous layers (and wakes) introduce unsteadiness in the subsequent blade row. The viscous-induced unsteadiness causes vibration of the blade row and noise generation; thus, the overall performance (aerodynamic, thermal, mechanical, and acoustics), reliability, and life of these turbomachinery are influenced by viscous effects.

With an increased emphasis on compact, efficient, reliable, and high-performance turbomachinery, major advances have to come from a thorough and basic understanding, evaluation, and minimization of these effects.

In Chapter 2, the global effects of viscosity and turbulence on overall performance were described. In Chapter 5, the predictive capability of Navier–Stokes codes were discussed. This chapter is devoted to a detailed physical understanding of these effects. The empirical or semitheoretical correlations available for estimating these effects will also be covered. Some of the general features of the viscous effects and three-dimensional flow common to all turbomachinery will be covered first, followed by features and effects specific to configuration and types of turbomachinery. For example, compressors and pumps operate with an overall pressure rise (adverse gradient) thereby the tendency exists for boundary layers to separate. This effect is not as critical in turbine flow, where there is an overall pressure drop. Nevertheless, local adverse pressure gradients occur on both the turbine and compressor (or pump) blades. The extent of viscous effects vary from one type of machinery to another. The axial flow inducers used in rocket pumps are at one end of the spectrum, where there might not be a distinct inviscid region, especially at very low flow coefficient. The centrifugal compressor/pump is also dominated by viscous and turbulence effects in most regions. Axial flow compressors (subsonic) and pumps, along with axial flow turbines, are at the other end of the spectrum in regard to the real fluid effects, where distinct boundary layer and 'inviscid' core flow regions can be distinguished.

The general features of viscous flow in turbomachinery are shown in Figs. 1.15 and 1.16. Some of the physical aspects of real fluid effects are covered in Chapter 5. The major features, in addition to those covered earlier, can be classified as follows:

1. The blade boundary layer may be laminar, transitional [as in front stages, inlet guide vane (IGV), and nozzle], and turbulent. It is unsteady due to relative motion between the blade rows. The nature of two- and three-dimensional boundary layers on blade profiles and endwalls, along with the various parameters controlling them, was described in Section 5.5.1 and shown in Figs. 5.1, 5.8, 5.11, and 5.16. The presence of strong pressure gradients (chordwise, radial, and pitchwise), combined with high Mach number, curvature, rotation, free-stream turbulence, and heat transfer effects make them very complex to understand and analyze. Not all of these effects are present in any given turbomachinery blade row. The presence of pressure gradients (normal to the surface) and three-dimensionality introduce features that are beyond the scope of the boundary layer theory. In addition, the presence of shocks and/or strong pressure gradients cause flow separation on the blade profile, and this increases the losses.

2. The presence of corners (especially the corner formed by the suction surface and the wall), introduces corner flow separation (e.g., Fig. 6.1). This is caused mainly by the strong streamwise pressure gradient,

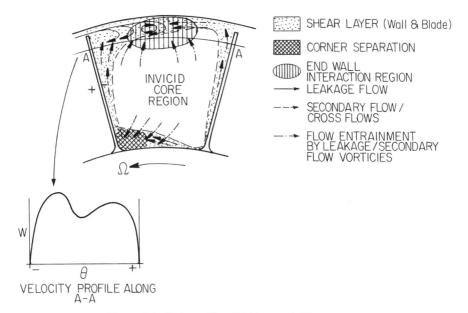

Figure 6.1 Nature of flow field in an axial flow compressor.

presence of secondary flow (Section 4.2.6), and merging of the wall and blade boundary layers. These are commonly referred to as *corner stall* or *wall stall*. The presence of horseshoe vortices in the case of a thick leading edge introduces corner separation, three-dimensionality, and additional losses. The effect of horseshoe vortices is more severe in a turbine than in a compressor.

3. There is a strong interaction between the inviscid core flow and the viscous layers, both in the boundary layer region and in the endwall regions. In the endwall region, the mixing of leakage flow, secondary flow, and annulus wall boundary layer with the inviscid region generates a very unusual and complex flow, resulting in "wake" and "jet" types of unconventional velocity profile (Fig. 6.1).

4. The flow is nonuniform in both the spanwise and blade-to-blade directions (Fig. 1.15). This introduces problems in identifying the "free stream" for the particular viscous layer. This "free-stream velocity" has to be derived from interaction of each of the viscous layers, the pressure and suction surfaces, and the hub-wall and annulus-wall boundary layers. In a rotor or stator preceded by another blade row, the boundary layers are inherently unsteady. This is especially true for closely packed blade rows. The cause of this unsteadiness is partly potential interaction (relative motion between two pressure fields) and partly viscous (viscous wakes impinging on the subsequent blade row).

There are many experimental data showing some of the features described above. These will be presented in the next few sections to extract the physics of the flow field. In view of these complexities, the evolution of turbomachinery design has relied heavily (by necessity) on empirical correlation.

The ideal or inviscid flows can be predicted accurately using the techniques described in Chapters 3, 4, and 5. The viscous and real fluid effects can be predicted using loss correlations in approximate governing equations (Chapter 4) or by solving the full Navier–Stokes equations (Chapter 5). The two-dimensional profile and mixing losses can be adequately resolved using the Navier–Stokes procedure or inviscid-viscid interaction techniques described in Chapter 5. The losses due to separation and shock–boundary-layer interaction have not yet been accurately captured.

One of the most attractive approaches widely employed in industry is to modify idealized or inviscid solutions to allow for real fluid effects through loss correlations. These loss correlations are developed on the basis of sound physical principles and the empirical constants derived from systematic testing. In this respect, cascade as well as low-speed testing are extremely useful tools. These loss correlations are similar to pipe loss correlations and induced drag coefficient for aircraft wings widely used in industry. As indicated earlier, the losses are not yet predictable to the same engineering accuracy as is idealized or inviscid flow. To appreciate the role played by loss correlations, it is sufficient to say that all turbomachinery designed and operational today is based to a greater or lesser degree on the use of loss correlations in quasi-inviscid models described in Chapter 4. These loss correlations are an integral part of computer codes and design procedures. They have certain limitations and are usually based on certain turbomachinery or specific flow regime or zonal losses. Hence, it is essential to know the limitations and assumptions made in deriving them.

The approach used in this chapter is to present only those correlations that are based on sound physical principles and the ones that are widely accepted by designers. The loss correlations are usually expressed in terms of the variables on which they depend (e.g., Mach number, Reynolds number, lift coefficient or turning angle, blade geometry) and empirical constants and/or exponents. For example, it is well known that the skin friction coefficient in a laminar flow varies as $(Re^{-1/2})$ and in turbulent flow it varies as $(Re^{-1/5})$. Hence, depending on the loss mechanism, such functions appear in the loss correlations.

6.2.1 Three-Dimensional Flows in Axial Flow Compressors / Pumps

This section will be devoted to a physical understanding of the complex flow field that exists in axial flow compressors and pumps. Experimental data will be presented and interpreted to show these features. As indicated in Section 5.5, the effects of rotation, radial pressure gradient, and curvature introduce three-dimensionality into the blade boundary layer. Depending on the geometry, the radial flows inside these blade boundary layers can be substantial. The presence of three-dimensional shocks introduce large radial flows, as described in the beginning of Chapter 4 (see Figs. 4.1 and 1.15). The nature of the flow fields in an axial compressor and pump rotor are shown in Figs. 6.1 and 1.15. Some of the salient features of this flow are as follows:

1. Rotation, curvature, and radial pressure gradients introduce three-dimensionality inside rotor blade boundary layers; hence the profile boundary layers, and the associated losses, may be substantially different from the cascade profile boundary layers. These interact in the tip region to provide unconventional velocity profiles across the passage as shown in Fig. 6.1. The blade boundary layers are usually thin at the hub and thick at the endwall (near the tip). Likewise, the wakes are also three-dimensional and give rise to appreciable radial mixing as the flow proceeds downstream to the next blade row.

2. As explained in Section 4.2.6, there are appreciable secondary flows both in the hub and in the endwall region. The hub-wall secondary flow, combined with merging of the wall and blade boundary layers in the presence of adverse pressure gradient, results in "corner stall" (or wall separation) as shown in Fig. 6.1. These separated flows reattach

and mix with the mainstream in the hub region downstream, giving rise to losses, flow redistribution, and spanwise mixing. This may result in potentially severe off-design conditions in the subsequent blade rows.

3. The most complex flow features occur in the annulus wall regions. As explained in Section 4.2.7, the leakage flow develops into a vortex, which tends to mix with the annulus-wall boundary layer, blade boundary layer, and secondary flow, resulting in extremely complex flow. Large turbulence production, due to mixing in this region, causes the wakes to decay rapidly. The losses in the annulus wall region usually accounts for 30–50% inefficiency in these blade rows.

4. In a transonic rotor, the three-dimensional shock waves present are due to radial variation of relative inlet Mach number. The variation in chordwise location and height of these shock sheets along the passage introduce three-dimensional flow in both the inviscid and viscid regions as shown in Fig. 4.1. Across a shock wave, there will be a jump in radial pressure gradient, resulting in substantial increase in radial flow and variation in boundary layer growth in the spanwise direction (e.g., at Q in Fig. 4.1).

6.2.1.1 Subsonic Single-Stage / Rotor Flow Field The Pennsylvania State University (PSU) group has carried out comprehensive measurement of the flow field in a subsonic axial flow compressor rotor, using a rotating miniature five-hole probe, rotating three-sensor hot-wire probe, and a two-component laser system. The performance of this compressor is shown in Fig. 2.17, and the details of the compressor are given in Section 2.5.5. The spacing between the IGV and the rotor, and that between the rotor and the stator, was deliberately kept large (about three to four chord lengths) to understand the rotor flow in the absence of rotor/stator interaction.

The following measurements were acquired to understand the three-dimensional nature of flow field in isolated rotors: (1) the passage blade-to-blade flow field from hub-to-tip, including wall and blade boundary layers, (2) flow inside and near the tip clearance region, (3) the nature and decay of the rotor wake from hub to tip, and (4) loss measurement near the trailing edge and far downstream from blade to blade and from hub to tip.

Some of the data were presented in earlier chapters. For example, the inviscid flow field measured from a rotating five-hole probe is presented and compared with predictions in Fig. 4.17. The flow field near the blade tip region is shown in Fig. 4.26. The blade boundary layer profiles are shown in Figs. 5.5, 5.10, and 5.16, hub-wall boundary layer characteristics are shown in Fig. 5.13, and the rotor wake characteristics are shown in Fig. 5.33. The profile loss distribution is shown in Fig. 5.43 and distribution of composite velocity profile (blade-to-blade flow distribution) in Fig. 5.42. Some of these data are revisited and additional data presented in this section to offer an integrated assessment of the flow field.

The blade boundary layer data for the PSU compressor at the peak pressure rise coefficient ($\phi = 0.50$, Fig. 2.17) are shown in Fig. 5.16. The boundary layer profile at three typical radial locations ($R = 1$ corresponds to the blade tip, and $R = 0.5$ corresponds to the hub) and at several chordwise location are shown. As expected, the boundary layer thickness increases from the leading to the trailing edge. The radial velocities are substantial close to the trailing edge ($S > 0.8$). Even though radial velocities at most locations are small, their influence in transporting mass, momentum, and energy along the span and in redistributing flow properties (spanwise mixing) is substantial. These radial flows, migrating toward the tip, cause boundary layer growth to be substantial toward the tip. The composite velocity profile near the hub ($R = 0.58$) and near midspan ($R = 0.75$) (Warfield and Lakshminarayana, 1989), for the same rotor at $\phi = 0.50$ (peak pressure rise condition), indicates that the boundary layer growth influences the inviscid region, and it is essential to couple these solutions for accuracy when computing the flow. The boundary layer momentum thickness for this rotor is shown in Fig. 5.10. The momentum thickness, which is a measure of the profile loss, increases substantially from midradius to the tip (Fig. 5.43) and is lower on the pressure side of the blade.

The structure of the rotor wake measured from a rotating five-hole probe is shown in Fig. 5.33, and the data from a rotating three-sensor hot wire are given in Ravindranath and Lakshminarayana (1980, 1981) and in Reynolds and Lakshminarayana (1979). These investigations clearly reveal strong three-dimensional structure of the wake, with appreciable radial outward velocity inside the wake near the trailing edge. The rate of wake decay is a function of the blade loading, turbulence levels, radial pressure gradient, Reynolds and Mach numbers, and the trailing-edge shape. Three-dimensionality in the wake, along with intense mixing due to this and the turbulence, causes the rotor wakes to decay faster than the wakes of corresponding (similar loading) isolated and cascade airfoils. The data on wake decay due to Prato and Lakshminarayana (1993) are shown in Fig. 6.2. Three distinct regions can be recognized in this plot. In the trailing-edge region, with intense mixing of two merging boundary layers (Fig. 5.7), the wakes decay very rapidly from trailing edge to 10% of the chordwise distance downstream, the maximum wake defect decreases by nearly 60% near the midspan. The wake decay is slower near the hub region. The next region is the near-wake region (covering approximately 10–50% of the chord downstream of the trailing edge) where the wake decay rate is slower, followed by the far-wake region ($Z/\cos \beta_o > 0.5$) where the wake decay is much slower. The maximum defect in total velocity (W_c) reaches 10% of W_0 at about one chord length. The radial velocities decay much slower. Furthermore, the static pressure inside the wake (not shown) is not uniform, as is usually assumed in classical analyses, even though the maximum difference in static pressure across the wake decays more rapidly than other flow properties. These wakes introduce additional mixing losses and also cause unsteady flow and off-de-

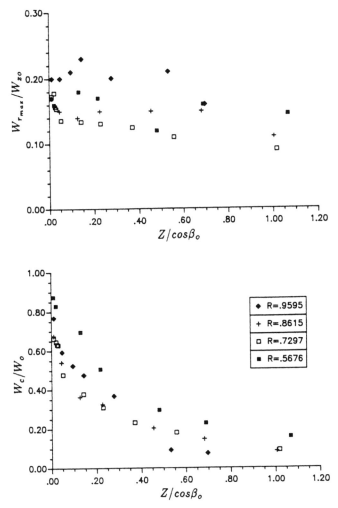

Figure 6.2 Decay of maximum radial velocity $(W_r)_{max}$ and decay in maximum velocity defect in relative total velocity (W_c) downstream of the Penn State axial flow compressor rotor. $Z / \cos \beta_o$ is the downstream distance in the streamwise direction, Z is measured from trailing edge, and W_o is the free-stream relative velocity at the trailing edge, $R = 0.5$ at hub, $R = 1.0$ at the tip. (Prato and Lakshminarayana, 1993; reprinted with permission of ASME.)

sign conditions in subsequent blade rows, resulting in noise, inefficiency, and vibration. Several correlations have been developed to estimate the wake profile and the decay characteristics and these can be found in the references quoted.

As mentioned earlier (Figs. 6.1, 4.24, 4.26, and 4.27), one of the most complex flow features occurs near the tip region. The leakage flow reduces blade loading, with an appreciable decrease in the pressure rise across the compressor and pump. The interaction of this flow with other flow features in the endwall region produces complex flow field as shown in Figs. 6.3 and 4.26. The data shown in Fig. 6.3 were required with a rotating three-sensor hot wire inside the rotor of the Penn State compressor described earlier. At 25% chord from the leading edge, where the leakage flow has not yet developed, the flow is predominantly inviscid. At 50% chord from the leading edge, a "wake" type of profile is observed near the midpassage up to 5% of the span from the tip. This is the interaction region shown in Fig. 6.1. The leakage jet is augmented by the relative motion between the wall and the blade. Hence the interaction region for a compressor rotor is located closer to the pressure surface. As the flow proceeds downstream, the velocity deficiency (or wake type of profile) increases substantially and this interaction region extends to the lower radii. The interaction region extends to about 8% of span near the trailing edge ($Z = 0.979$). The composite blade-to-blade velocity profile, with blade boundary layers followed by a wake type of profile near midpassage is common to many types of turbomachinery tip flows. The intense mixing in this region results in high turbulence intensities as shown in Fig. 6.4. The highest turbulence intensities for this configuration occur near the midpassage and midchord (where the leakage-main flow interaction is highest). Furthermore, the turbulence is highly anisotropic; maximum $\tau_r\left(\sqrt{\overline{(w_r')^2}}\,/W_s\right)$ is 35%, compared to maximum τ_n of 25% and maximum τ_s of 15%. This region is within the annulus-wall boundary layer. Turbulence intensity near the leading edge is nearly isotropic at this radial location, with a value ranging from 5% to 7.5%. The presence of highly anisotropic turbulence in this region reveals the need for higher-order (RSM, ARSM) turbulence models for the prediction of this flow.

The blade boundary layers and wakes in the tip regions are very thick, as shown in Fig. 4.26b. The wake width in the tip region is large, and this is caused by the radial migration of low-momentum boundary layer flow toward the tip. It is also observed that these wakes decay more rapidly (than do wakes in other regions) downstream due to high turbulence levels in these regions (Davino and Lakshminarayana, 1982). Dominant influence of leakage flow is evident from Fig. 4.26b. The influence of leakage flow is felt up to about 15–20% of the span in the downstream region. The interaction region is located closer to the pressure surface, indicating the influence of blade rotation. The flow distribution from 4% chord downstream ($Z = 1.04$) to 64% of chord downstream is found to be very rapid and dramatic (Lakshminarayana et al., 1995). The wake in this region ($R > 0.90$) decays rapidly due to high turbulence levels and mixing associated with the leakage flow. The leakage flow interaction region occupies nearly middle third of the passage. The passage-averaged outlet angles indicate that beyond $R = 0.95$ the flow is predominantly underturned.

The iso-contour of loss coefficient (ζ) in the tip region near the trailing edge and at 64% chord downstream, shown in Fig. 6.5, indicates the major

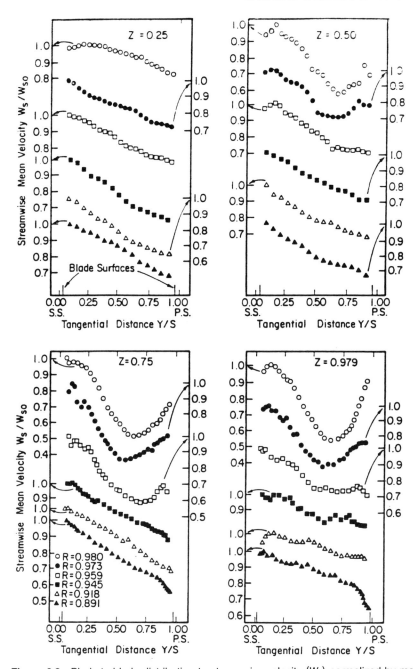

Figure 6.3 Blade-to-blade distribution to streamwise velocity (W_s) normalized by maximum value (W_{so}) at a given radial and axial location for Penn State axial flow compressor rotor. Z is the axial distance from the leading edge, normalized by axial chord length. (Lakshminarayana et al., 1982; reprinted with permission of ASME.)

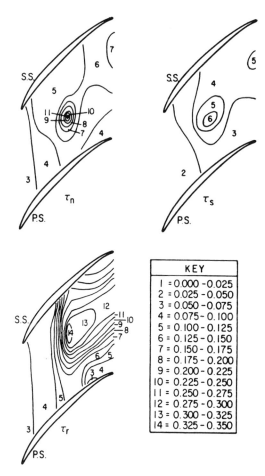

Figure 6.4 Contours of streamwise (τ_s), normal (τ_n), and radial (τ_r) turbulence intensities (normalized by W_s) at 4% span from the tip ($R = 0.98$) for Penn State compressor rotor. (Lakshminarayana et al., 1982; reprinted with permission of ASME.)

effects of leakage flow on losses and inefficiency. The corresponding velocity field is shown in Fig. 4.26b. The high-loss region covers almost the entire passage up to about 18% of the span, with the highest values occurring at the same location where the minimum total velocity is observed (compare Figs. 4.26b and 6.5). The flow is mixed out rapidly downstream, achieving nearly uniform flow (Lakshminarayana et al., 1995) and loss distribution at 64% chord downstream of the trailing edge. The development of the three-dimensional annulus-wall boundary layer (mean) at various axial locations is shown in Fig. 6.6 for the Penn State compressor. The velocity profile is well behaved from upstream to the leading edge and shows marked change beyond $Z = 0.25$; this is caused by the scraping effect of the blade and by the leakage

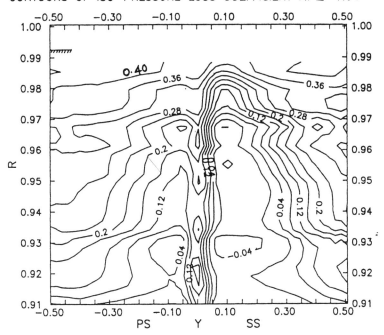

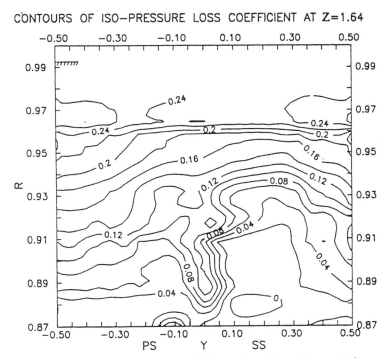

Figure 6.5 Contours of iso-pressure loss coefficient at $Z = 1.04$ and $Z = 1.64$ (the numbers denote relative stagnation pressure loss coefficient, $\zeta = 2[(P_{01})_{rel} - (P_{02})_{rel}] / \rho U_t^2$. Z is the distance from rotor leading edge, normalized by tip axial chord; Y is tangential distance normalized by blade spacing. (Lakshminarayana et al., 1995; reprinted with permission of ASME.)

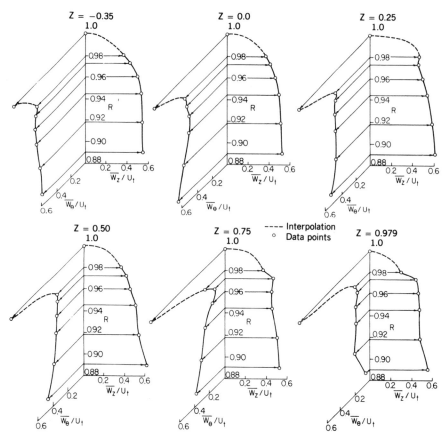

Figure 6.6 Annulus-wall boundary layer development inside the passage of an isolated rotor $\overline{W}_z$ and $\overline{W}_\theta$ are passage-averaged axial and tangential velocity, Z is the distance from leading edge normalized by blade chord. (Lakshminarayana et al., 1982; reprinted with permission of ASME.)

flow. A wake type of profile existing in this region near the midpassage (Fig. 6.3) results in the observed complex velocity profiles shown in Fig. 6.6. This result along with those of Pandya and Lakshminarayana (1983) indicate a very complex, perturbed, and nonuniform flow field influenced by viscous and turbulence mixing effects in the tip clearance region. The boundary layer is no longer of the conventional type, and the terminology "annulus-wall boundary layer" is misleading. A more appropriate terminology would be "complex viscous layers" or simply "endwall flows." The periodic perturbations and scraping caused by the rotor blade, as well as large gradients in velocity (especially W_θ), indicate the need for dense computational grids and higher-order (RSM, ARSM) turbulence models in this region. Hunter and Cumpsty (1982) carried out similar measurements downstream of a compressor rotor at different tip clearance heights and loading and concluded that the annulus-wall boundary layer thickens as the blade loading and the tip clearances are increased.

The wall or corner stall (or separation) in the corner formed by the endwall and the suction surface has been observed in both cascades and stators, and it is known to exist in rotors also. A review of corner stall in cascades is given by Horlock et al. (1966). Schulz and Gallus (1988) and Schulz et al. (1990) carried out extensive measurement in a stator (24 blades, $r_h/r_t = 0.75$, $D_t = 428$ mm, $A = 0.86$, $\sigma = 0.78$, $\alpha_1' = 44°$, $\alpha_2' = 15°$) preceded by a simulated rotor (consisting of rotating rods). Based on extensive flow visualization, data on the surface, and the flow field survey at the exit, they formulated a composite model for structure of this corner flow in the absence of an upstream rotor. The streamlines on the surfaces (Fig. 6.7a) show the presence of a vortex on the hub (marked a). In the core of these vortices the flow is transported out normal to the surface; points (a) and (b) seem to represent the saddle points. At the leading edge of the separated region, vortex (c) is formed by the main flow when a sudden obstruction due to flow separation is encountered. The back flow inside the separated region moves upstream and coils up into another vortex marked (d). The separated region is closed off from the main flow by limiting streamlines (the angle of limiting streamline as the wall is approached) at the hub and on the suction side. The vortex axis is normal to the hub on the hub wall and normal to the blade on the blade surface. Hence, it is anticipated that a ring vortex is formed as shown in 6.7b, covering part of the blade suction side and the hub wall. A possible structure of flow separation is shown in Fig. 6.7b. The limiting stream surfaces, bounded by the limiting streamlines on the hub and the suction side, are closed off by a ring vortex near the trailing edge. The downward motion of the flow field due to this vortex, along with a strong flow deflection downstream of the separation bubble, has been observed by Schulz and Gallus (1988) and Schulz et al. (1990) (Fig. 6.7c). The separated region cannot be sustained downstream of the cascade, and it reattaches before the exit measuring station (Fig. 6.7c). The flow perceives the corner stall region as a solid body obstruction. Substantial deflection of the main flow is found to occur due to the blockage caused by this flow separation. Dring et al. (1982a) observed that the hub corner stall on the blade suction surface at lower blade loading developed into a full-span stall as the blade loading was increased. Dring et al. (1982a) and Wagner et al. (1985) provide additional details on three-dimensional flow in axial flow compressors.

The flow field in the hub region of a compressor rotor will be different from those of stators and cascades. The major additional effect here is the centrifugal and Coriolis forces associated with rotation. As described in Section 5.5.1, the radial flow in a compressor rotor boundary layer is outward, unlike that of a stator (see Fig. 6.7). The effect of rotation in centrifuging of blade boundary layer, secondary flow, and (possibly) the corner separation zone would be beneficial in the case of rotor hub flow. The

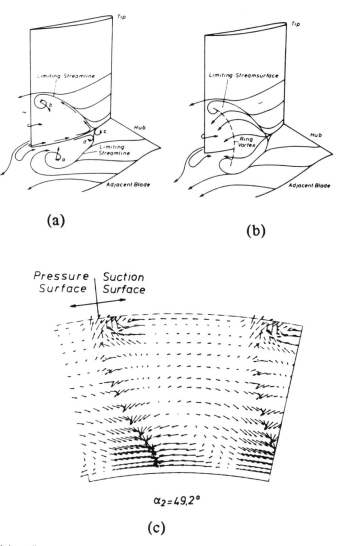

(a)

(b)

(c)

Figure 6.7 Hub wall corner separation phenomena. (a) Composite flow structure near the hub-based flow visualization. (b) Proposed topology of the hub corner stall. (c) Secondary flow vectors at exit. (Copyright © Schulz et al., 1990; reprinted with permission of ASME.)

flow field data in the hub region of the Penn State compressor rotor, described earlier in this section and shown in Fig. 6.8, reveals no evidence of corner separation even though the hub is heavily loaded with turning angle in excess of 30°. The relative velocity (W, normalized by blade tip speed) and loss (ζ) contours reveal no major corner separation or loss regions usually observed in stator and cascade endwall flows. The secondary flow accumulation has been swept from suction surface toward a higher radius ($R = 0.52$),

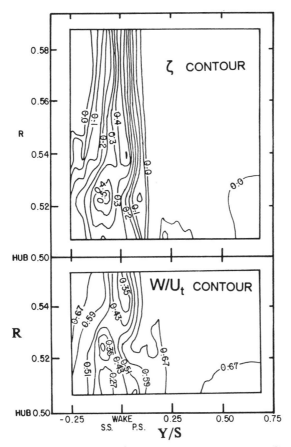

Figure 6.8 Contours of loss coefficient $\zeta (= 2[(P_{01})_{rel} - (P_{02})_{rel}] / \rho U_t^2)$ and W / U_t in the Hub Region at the Exit (17% chord from trailing edge) at $\phi = 0.56$ for the Penn State Compressor Rotor. (Lakshminarayana et al., 1986; reprinted with permission of ASME.)

providing a slightly thicker wake. These differences may invalidate the use of cascade data in the analysis or the design of rotor hub flow field.

The flow in noncavitating axial flow pumps will be similar to those described for the subsonic compressors. The inducers used in rocket pump (Fig. 1.7) feed systems, water jet propulsion, and other applications have the most complex flow fields among all turbomachinery. These devices are inefficient and serve mainly to pressurize the liquid and avoid cavitation in subsequent impellers. The Penn State group (Lakshminarayana, 1982; Lakshminarayana and Gorton, 1977; Anand and Lakshminarayana, 1977) measured the flow in two-, three-, and four-bladed inducers using rotating hot-wire and conventional probes. The hub/tip ratio of the 3-bladed inducer used varied from 0.25 at inlet to 0.5 at exit; in addition, $\phi = 0.065$, annulus wall diameter $= 0.932$ m, $C(\text{tip}) = 2.10$ m, $\sigma(\text{tip}) = 2.15$, Re $= 7 \times 10^5$ and

$N = 450$ rpm. The test medium was air, hence, the simulation was restricted to noncavitating flows. A composite of the flow field based on extensive data is shown in Fig. 6.9. Because the inducer passages are long and narrow (sometimes extending to about 300° as shown in Fig. 1.7), the viscous effects are large. The boundary layers are nearly fully developed at most locations. Furthermore, these operate at very low flow coefficient (0.05–0.1). Hence, the effects of rotation are substantial. This makes the flow highly three-dimensional, viscous, and turbulent, with complex mixing leading up to high losses. In some respects, the inducer behaves like a shear pump, where the head rise comes from viscous stresses. The radial variation of head-rise coefficient (ψ), shown in Fig. 6.9, indicates large flow departures and deviations from the design condition. This inducer was designed to have free vortex distribution (radially constant ψ), yet the actual head-rise coefficient is more than two times the design value at the tip. The section from midradius to tip behaves like a shear pump (Lakshminarayana, 1978), where the pressure or head rise occurs mostly due to viscous or shear pumping effect (as in a rotating disk). In addition, the spanwise mixing and losses associated with this viscous phenomena are large. The inducer with cambered blades had better performance than did the flat-plate inducer presently used in liquid rockets.

Zierke et al. (1993) report very detailed laser and five-hole probe measurements in an axial flow pump, including detailed trajectory and velocity profiles across the rotor passage and in the tip clearance region (including visualization of tip vortex).

6.2.1.2 *Rotor Flow Field with Shocks*   As indicated earlier and shown in Fig. 4.1, the shock surfaces inside transonic rotors are three-dimensional and highly warped. The nature of flow fields and performance of transonic and supersonic compressors were discussed briefly in Sections 2.5.9, 2.5.10, and 3.5. The jump in the radial and chordwise pressure gradients (due to jump in properties across a shock) is substantial (Figs. 5.30 and 5.34). This, combined with shock-induced separation, introduces considerable three-dimensionality in the free stream as well as in blade boundary layers. Even the two-dimensional flow is complex as illustrated in Figs. 5.31 and 5.46. The presence of leading-edge and passage shocks make the flow extremely complex, with substantial influence on pressure distribution and blade-to-blade velocity profiles.

The performance of transonic compressors and fans is very sensitive to the strength and location of shock waves. The outer flow in a transonic compressor/fan is dominated by the shock wave, as indicated in Fig. 4.16. This results in lower efficiency and higher temperature rise. This is clear from the numerical and experimental results presented by Wadia and Law (1993). The results shown in Fig. 6.10 (also see Fig. 5.23) are for a low-aspect-ratio transonic fan ($\dot{m} = 27.8$ kg/s, $N = 20{,}222$ rpm, pressure rise ratio $= 1.92$, D(inlet) $= 0.43$ m, $M_2 = 0.592$, and $A = 1.31$; 20 blades). In this configura-

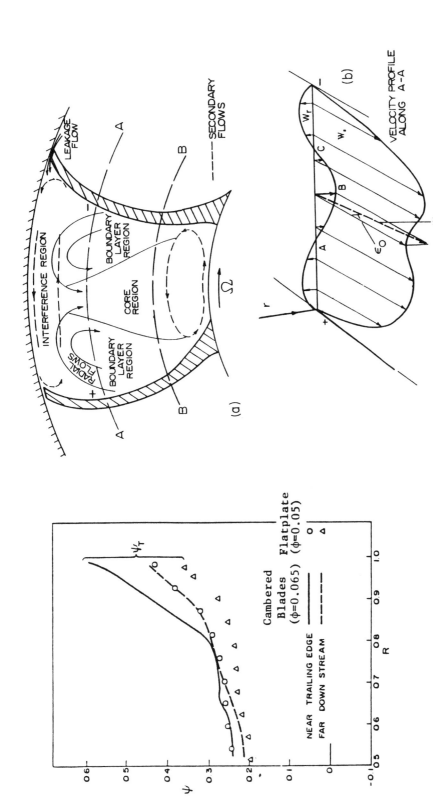

Figure 6.9 (a, b) Nature of flows in rocket pump inducers and (c) the radial variation of passage averaged pressure rise coefficient $(\psi)(2(P_{02} - P_{01}) / \rho U_t^2)$. (Copyright © Lakshmi-narayana, 1982; reprinted with permission of ASME.)

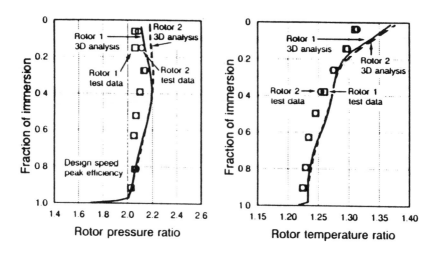

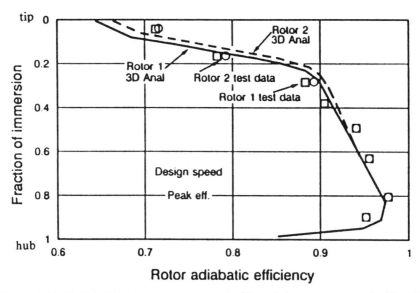

Figure 6.10 Radial distribution of pressure ratio (P_{03} / P_{01}), temperature ratio (T_{03} / T_{01}), and efficiency at the exit of a low-aspect–ratio axial flow transonic compressor. (Copyright © Wadia and Law, 1993; reprinted with permission of ASME.)

tion, with shock waves present, the temperature rise is much higher near the tip. Even though pressure rise is also higher, the losses are also high, giving rise to only moderate increase in pressure rise in the tip region. The efficiency in the outer 20% of span is very low due to shock-blade boundary layer and shock–annulus-wall–boundary-layer interactions. Good agreement between Navier–Stokes computations from an explicit Runge–Kutta code (see Section 5.7.4.3 and Jennions and Turner, 1993) and the measured data, shown in Fig. 6.10, provides confidence in the Navier–Stokes flow solver or code or algorithm for analysis, design modifications, and predicting trends. In addition, Wadia and Law undertook a study on the effect of the location of maximum thickness. Rotor 1 blade had a maximum thickness at 40% of chord, and this was moved to 55% of chord in the Rotor 2 configuration. Even this minor change (which affects shock location) had an measurable effect on radial distribution of flow property and 1% improvement in overall efficiency as shown in Fig. 6.10.

Comprehensive data available in a transonic compressor is due to Strazisar (1985) and Pierzga and Wood (1985). These data were taken in a transonic fan (NASA Rotor 67), with a design pressure ratio of 1.63, $\dot{m} = 33.25$ kg/s, tip Mach number of 1.38, and tip speed of 429 m/s. The rotor had 22 blades, $A = 1.50$, D_t(inlet) = 51.4 cm, and hub/tip ratio (inlet) = 0.375. The blade-to-blade distribution of relative Mach numbers at 30% of span from the tip is shown in Fig. 5.34. The leading-edge shock is detached. The Mach number distribution along the midpassage is substantially different from a smooth distribution expected in a subsonic flow. The presence of two shock systems and the associated reflections produce locally high gradients in the chordwise pressure gradient. This results in a substantial increase in the boundary layer growth at these locations. The three-dimensional shock structure at three operating conditions (peak pressure rise, midrange, and peak efficiency conditions) are shown in Fig. 6.11. Both the bow and the passage shocks are curved, revealing the presence of a highly twisted shock surface. The bow shock angle varies significantly across the span. The flow downstream of such a shock system is likely to be highly three-dimensional due to the jump in radial pressure gradient near the shock surface. This introduces large radial velocities in this region, but no LDV data on radial velocity are available to assess the magnitude of these radial velocities. It was further observed (Strazisar, 1985) that these shocks were not stationary and that the shock oscillated ±3 to 4% of the chord; similar oscillations (±4 to 6% of chord) were found in the wake. Hence the transonic flow in compressor (even in the isolated, single-stage rotors) is inherently unsteady in the relative frame of reference and highly three-dimensional downstream of the shock. This has led some to doubt the validity of cascade data in designing and analyzing flow in transonic rotors.

The relative Mach number contours presented by Strazisar (1985) show that the relative inlet Mach number is supersonic at 10% span. The inlet shock is nearly normal around the leading edge but is oblique inside the

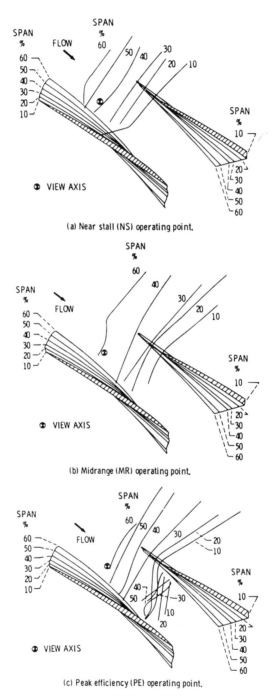

(a) Near stall (NS) operating point.

(b) Midrange (MR) operating point.

(c) Peak efficiency (PE) operating point.

Figure 6.11   Three-dimensional shock structure in NASA transonic axial flow fan (Rotor 67) derived from LDV measurements (spanwise location from tip, 0 at tip, 100% at the hub; the solid line within the passage denote the shock wave). (Copyright © Strazisar, 1985; reprinted with permission of ASME.)

passage (Fig. 6.11). The flow field at 30% span (blade-to-blade) shows the existence of the shocks (Fig. 5.34). The second shock on the pressure surface introduces boundary layer separation downstream of the shock and this is evident from the Mach number contours. The flow at 70% span from the tip is subsonic and, the datashow the presence of a very small supersonic region near the leading edge of the suction surface. The flow is well-behaved at this spanwise location. The predicted isentropic Mach number (Chima and Strazisar, 1983; Also see Fig. 5.30) contours from hub-to-shroud clearly reveal the presence of large gradients in relative Mach numbers from hub to tip, especially near the shock surface. The region of supersonic flow varies substantially from hub to tip as well as from blade to blade. Recent laser-doppler measurements and aspirating probe data (Norton et al., 1989) indicate that the spanwise variation in the flow field is not periodic, especially at the design operating point.

Recent measurements by Cherrett et al. (1994) in a single-stage transonic fan (with 25 rotor blades and 52 stator blades, $P_{o3}/P_{o1} = 1.807$, $\Delta h_o/U^2 = 1.21$, $T_{o3}/T_{o1} = 1.214$, $U_t = 442$ m/s, $D_t = 633$ mm, $r_h/r_t = 0.39$ (inlet) to 0.62 (exit), $\dot{m} = 53.3$ kg/s, $\sigma = 1.53$ at midradius) acquired at the exit of the stator (located downstream of the rotor) with a fast response yaw probe reveal the complex features of unsteady and steady flow at these locations. The rotor exit absolute ensemble averaged stagnation pressure is shown in Fig. 6.12a. Large variation in stagnation pressures across the wake are clear from this plot. The variations are higher on the suction side of the rotor wake than on the pressure side, and very large pressure variations occur in the outer 20–30% span from the tip, because the rotor inlet Mach numbers are supersonic in this region. The corner separation zone is located at the hub near the suction surface and is more pronounced at near-surge condition. The interesting feature is the unresolved unsteadiness (includes all unsteadiness, including random, not clocked with the shaft frequency, see Section 4.2.5.2) shown in Fig. 6.12b. Large unsteadiness can be seen in the wake region, with maximum values occurring near the shock locations (outer 20–30% of span). The core region, where the flow is inviscid and fluctuations are small, exists only near the hub region. The pressure fluctuations are large in the corner separation region near the hub, especially near surge condition. This indicates that the corner separation is a highly unsteady phenomenon in single- and multi-stage compressors. As mentioned earlier, the shock structure is not only three-dimensional but also unsteady. This introduces considerable spanwise mixing in the rotor passage in addition to unsteadiness.

6.2.1.3 *Multistage Compressor Flow Field* The flow field in a multistage compressor is three-dimensional, turbulent and viscous, and inherently unsteady. The blade and the annulus-wall boundary layer development and their mixing in inter- and intra-blade rows introduces large viscous losses and flow deviation from the design conditions. Even in a single-stage or rotor-alone configuration, the presence and interaction of secondary flow, leakage

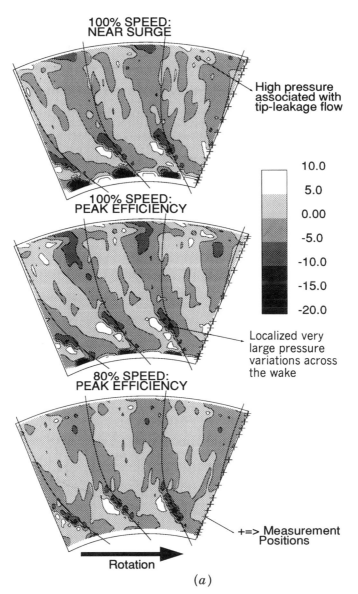

(a)

Figure 6.12a The absolute stagnation pressure variation (ensemble-averaged value), $\{[P_{o2}(r, \theta) - \overline{P_{o2}}(r)] \, / \, \overline{P_{o2}}(r)\} \times 100$, at the exit of a transonic axial flow fan rotor. ($\overline{P}_{o2}(r)$ is the passage averaged and time averaged value.) (Crown copyright © Cherrett et al., 1994; Defense Research Agency; reprinted with permission of the Controller, Her Britannic Majesty's Stationary Office.)

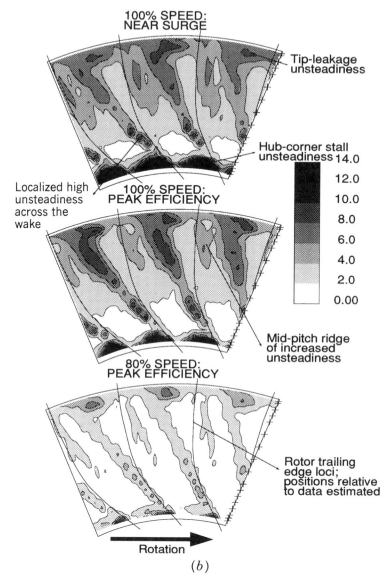

Figure 6.12b Unresolved unsteadiness (RMS value in percent) in absolute stagnation pressure at the exit of a transonic axial flow stator (due to rotor expressed as a percentage of local time-averaged stagnation pressure $\bar{P}_{o2}(r, \theta)$) (see Fig. 6.12a). Cherrett et al., 1994.

flow, blade boundary layer and wakes, and endwall boundary layer growth (all of which are three-dimensional) makes it truly a challenging flow to understand, measure, and compute. These complexities are compounded in a multistage compressor (especially aft or core stages) due to the presence of several closely packed blade rows and low hub-to-tip ratio. These complexities have eluded designers and researchers, and very few attempts have been made to gain a good understanding of the flow in a multistage environment. The design of these compressors is still based on empirical correlations derived from single-stage and cascade testing.

The present practice is to represent the nonaxisymmetric flow as blade blockage and include some of the gross effects in the design process. The nonaxisymmetric effects are included in the calculation through body force and correlation of spatial variation in blade-to-blade flow field as described in Sections 4.2.3 and 4.2.5.2. It is not known how the blade responds to nonaxisymmetric or unsteady inflow in a real machine. The major question to be answered is the nature of unsteady or nonaxisymmetric flow in a rotor or stator (including hub-to-tip and blade-to-blade distribution) and the response of the blade to such inflow. The time-average flow process inside a passage in a multistage compressor responds to a time-averaged flow field upstream, but this rarely provides a physical insight into the response of viscous layers and endwall flows (included the leakage flow) to unsteady flow field upstream.

The flow in a multistage compressor develops with substantial viscous effects, unsteady flow, losses, and off-design conditions. Most of the measurements in a multistage compressor have been aimed at an understanding of the overall performance. Detailed measurement to understand the small structure and detailed flow field is now possible due to development of miniature fast-response probes. Detailed data will be forthcoming before the end of this decade. One of the early data due to Smith (1970) in a 12-stage compressor is shown in Fig. 6.13. The wall boundary layer grows continuously through the machine, and the profiles undergo changes as the flow progresses downstream. If the stages are not identical in design, the tangential velocity profiles will undergo major changes. The stagnation temperature and pressure gradients are large, especially in the hub and tip regions; these are likely to lead to additional secondary flows as explained in Section 4.2.6. More recent data in a multistage compressor can be found in Behlke et al. (1979). Substantial improvements have been made in the design of embedded stages to improve efficiency and performance. Recent performance and interstage data by Marchant et al. (1989) in a 14:1 pressure-rise multistage compressor indicates the importance of matching the embedded stage properly. Three types of high-pressure compressor designed for the energy efficient engine (Fig. 2.18) featuring advanced controlled diffusion airfoils were tested to demonstrate that very high efficiencies can be achieved through accurate correlations.

Robinson (1992) provides a critical review of the recent advances in understanding the nature of the flow in multistage compressors. Recent aerodynamic studies include area traverses at the exit of stator 2 in a highly loaded compressor (Calvert et al., 1989) and radial traverses at the exit of each blade row of a high-speed multistage compressor (Falchetti, 1992). The measurements acquired by Falchetti (1992) with laser 2 focus anemometry and miniature probes show substantial spanwise mixing from blade row to blade row. Stauter et al. (1991) and Stauter (1993) provides details of

(a)

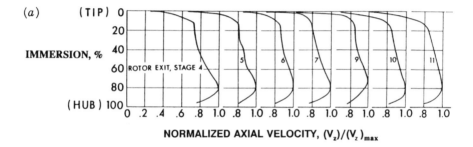

NORMALIZED AXIAL VELOCITY, $(V_z)/(V_z)_{max}$

(b)

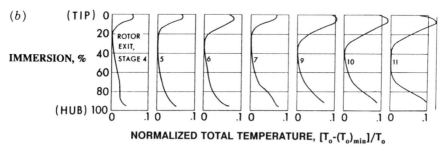

NORMALIZED TOTAL TEMPERATURE, $[T_o-(T_o)_{min}]/T_o$

Figure 6.13 Traverse measurements obtained from a 12-stage axial flow compressor (Smith, 1970). (a) Axial velocity profiles. (b) Total temperature profiles.

unsteady flow field acquired from laser-doppler velocimetry in a two-stage low-speed axial flow compressor. The data reveal the complex flow field associated with the wake transport and the leakage flow. Cherrett and Bryce (1992) conducted radial traverses of a high-frequency pressure transducer between the first three stages of a high-speed multistage compressor operating at three different throttle settings. The analysis of the data revealed the presence of both periodic and random fluctuations in the flow field properties. In spite of these advances, there is a lack of cohesive and detailed understanding of the flow features.

One of the phenomena in multistage compressors that has received major attention in recent years is spanwise mixing. Spanwise mixing refers to the mixing of the flow field in inter- and intra-blade rows caused by secondary flow field (including wakes and tip leakage flow) and unsteadiness, including random turbulence. Radial transport of mass, momentum, energy, and turbulence causes considerable nonaxisymmetry in the flow as well as large radial gradients in flow properties as the flow progresses downstream from one stage to the other. The radial velocity and transport are caused by secondary flows, leakage flows, radial pressure gradient, three-dimensional boundary layers and wakes, turbulence (random component of overall unsteadiness), and viscous diffusion. The blade boundary layer is centrifuged outward on the rotor blade, and the secondary flow is usually augmented by this on the suction side and opposed on the pressure side of a blade. The other effects which can cause spanwise mixing are the sweep and lean of the blades. The magnitude of the spanwise mixing depends on the rotation parameter (expressed as flow coefficient, based on the blade tip speed), flow turning, aspect ratio, tip clearance, blade solidity, radial pressure gradient, lean and sweep, flow unsteadiness, and turbulent fluctuations and the blade row spacing. If the flow is transonic with shockwaves, spanwise mixing is influenced by the location and strength of the three-dimensional shock. Adkins and Smith (1982) were the first to recognize the potential importance of mixing. They also developed a model for calculating the magnitude and the effect of mixing on the multistage compressor performance based on inviscid, small perturbation secondary flow theory (Section 4.2.6). The calculation from Adkins and Smith (Fig. 4.31; see Section 4.2.8 for details of these compressors), with and without allowing for mixing, shows that the mixing is strong; this substantially alters the radial distribution of T_o and α. Furthermore, the flow velocity and loss distribution are also altered substantially.

It is only recently that attempts are being made to gain a basic understanding of the flow field in multistage compressors. Calvert et al. (1989) measured the flow field at the exit of the second stage of a four-stage compressor (pressure ratio = 4.0, $\dot{m}$ = 33.2 kg/s, N = 6380 rpm, r_h/r_t = 0.912 at inlet, D_{inlet} = 0.892 m) using a cobra probe. They show that the passage-to-passage repeatability is maintained. The stator wakes extend to about 20% of the blade passage, and annulus-wall and hub boundary layers extend 5–10% of the blade height. Some of these features are very similar to those described earlier for a single-stage compressor. The overturning observed near the annulus wall reinforces the hypothesis from classical secondary flow theories (Section 4.2.6). The hub-wall region encounters a distorted wake and boundary layer due to rotating hub and clearance.

Similar measurements are carried out at Penn State's three-stage axial flow compressor facility (D_t = 0.6096 m, N = 5410 rpm, $\dot{m}$ = 8.609 kg/s, pressure ratio = 1.354, blade tip Mach number = 0.5, average r_h/r_t = 0.843). The overall performance and time-averaged data acquired at the exit of stator 2 from a miniature five-hole probe is presented in Lakshminarayana et al. (1994), and the time-accurate data acquired from a high response Kulite probe is presented in Suryavamshi et al. (1994). The contours of absolute stagnation pressure rise coefficient $[C_{pt} = (P_o - \bar{p}_1)/(\bar{P}_{o1} - \bar{p}_1)$, where $\bar{p}_1$ is the inlet average static pressure, and $\bar{P}_{o1}$ is the inlet average stagnation pressure) and the absolute yaw angle at the exit of stator 2 are shown in Figs. 6.14a and 6.14b respectively. The blade on the right-hand side had a leading-edge Kiel probe (stagnation pressure probe located in a channel to increase angular sensitivity) near the tip at 90% of the span from the hub. Hence, this region shows a much thicker wake. Lowest pressures are observed in the wake near the endwall region. The blade wakes are much thicker near the casing, and this region extends to about 20–30% of the span from the tip. There was no major corner separation, even though secondary flow effects and sweeping of low-energy regions toward the suction surface is

(a)   Cpt: 5.6% chord DS2; peak efficiency

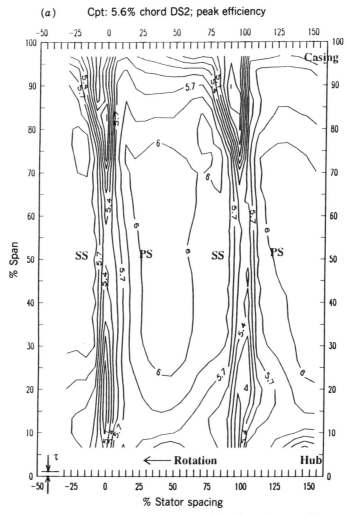

Figure 6.14a Contours of C_{pt} at the stator 2 exit of the Penn State multistage axial flow compressor at peak efficiency. (Lakshminarayana et al., 1994; reprinted with permission of ASME.) Interval = 0.15, min = 4.5, max = 6.1.

evident. The wakes in the inviscid core region, which extend from 20% to 80% of span from hub, is about 20% blade spacing. The stagnation pressures in the hub wall and endwall (about 20% span from each wall) are low. Because the hub is cantilevered, the leakage flow between the hub and stator and the "scraping" effect due to hub rotation generates a low-energy region as shown. The yaw (or swirl) angle contours shown in Fig 6.14b indicate that the flow is overturned on the suction side and underturned on the pressure side of the wake (design outlet angle varies from 31.63° at the tip to 23.21° at

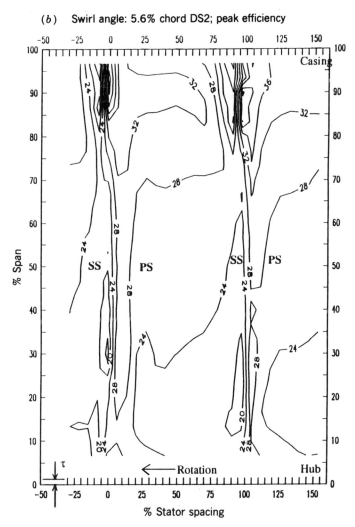

Figure 6.14b   Contours of swirl angle (α_2 in degrees) at peak η at stator 2 exit of Penn State multistage compressor; interval 4°, min = 12°, max = 48° (Lakshminarayana et al., 1994).

the hub), and this is consistent with the wake data of an isolated rotor (Prato and Lakshminarayana, 1993). The endwall region has overturning near the corner formed by the suction surface and the casing, caused by the secondary flow.

The unsteadiness downstream of stator 2 is caused by the rows preceding it and the potential effects of downstream rotor, with major contributions coming from rotor 2. This is evident from Fig. 6.15, where the contours of total RMS unsteadiness in C_{pt} are plotted. The data were acquired from a high response Kulite probe at the exit of stator 2 and ensemble-averaged to

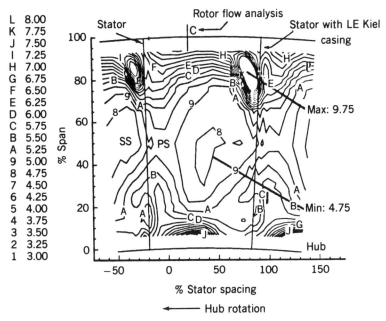

Figure 6.15 Contours of RMS of total unsteadiness in C_{pt} (as a percent of local time-averaged value) at the exit of stator 2 of the Penn State multistage axial flow compressor. (Suryavamshi et al., 1994; reprinted with permission of ASME.)

derive the magnitude of the total unsteadiness caused by rotor 2. The region where the tip vortex of the rotor and the stator corner flow is located has the highest unsteadiness. The unsteadiness in the midspan region is mainly caused by rotor wakes and the unsteady boundary layer on the stator blades. The magnitude of unsteadiness in this region is about 5%. The hub region where the rotor endwall flow and the stator leakage flow interact also has very high flow unsteadiness. It is evident from the plots that it is necessary to include unsteadiness (Adamczyk, 1985) in the prediction and the design of multistage turbomachinery flows.

6.2.2 Flow Field in Centrifugal Impellers

The flow phenomenon in centrifugal compressors and pumps is extremely complex. The flow phenomenon is schematically illustrated in Fig. 1.16. There are strong viscous effects, significant regions of flow with separation and secondary flow, and shock waves in high-speed machines. Further complications arise due to tip clearance, whose influence is more dominant in small centrifugal compressors than in axial compressors or pumps. The flow field is very complex, three-dimensional, and turbulent all under the influence of curvature (due to curved and swept blades) and rotation. It is also well known that the turbulence structure in these curved and rotating passages is affected by rotation and curvature, thus invalidating the models developed for stationary, noncurved 2D flows (Section 5.2, Fig. 5.3). In view of these complications, centrifugal compressor design and analysis have relied heavily on empirical correlations. This makes the improvement and design a slow process. The one-dimensional analysis of centrifugal compressors was covered in Section 2.7, the quasi-three-dimensional analysis was discussed in Section 4.2 (Fig. 4.7), the secondary flow analysis was covered in Section 4.2.6.3 (Figs. 4.21 and 4.22), and Navier–Stokes analysis was described in Section 5.7 (see Fig. 5.35).

The centrifugal compressor achieves part of the pressure rise from the centrifugal and Coriolis forces due to rotation and the change of radii. This is in addition to the pressure rise achieved through flow-turning as in an axial compressor, where the interchange is through the change of tangential momentum. In view of the large pressure rise and the presence of centrifugal and Coriolis force fields, the flow through the centrifugal compressor passages is affected adversely by large boundary layer growth, flow separation, and secondary flow. In view of these features, the centrifugal compressors usually have lower efficiency than do axials, even though some lean-back centrifugal impellers have been successfully designed to operate in excess of 94% polytropic efficiency. The boundary layers are highly three-dimensional in nature. Because hub and shroud boundary layers thicknesses are substantial, considerable secondary flows are likely to arise due to large turning of the shear flow associated with these boundary layers. Furthermore, the blade boundary layers are subjected to Coriolis and centrifugal forces, and these give rise to appreciable spanwise flows inside the blade boundary layer. The radial or spanwise transport of mass interact with the shroud boundary layers and tip leakage flow to produce a very complex flow field. Such a flow field is similar to that of an axial flow compressor and inducer flow fields shown in Figs. 6.1, 6.3 and 6.9, respectively. These effects are compounded in a small centrifugal compressor due to small blade height, narrow and curved passages, high rotational speed, and large tip clearance/blade height.

The quantitative nature of the flow field measured in a high-speed compressor (using laser anemometer) by Eckardt (1976, 1980) is shown in Fig. 6.16. Meridional velocity (W_m-resultant of axial and radial components) at various streamwise locations are shown in Figs. 6.16a, b, c. The performance details of this compressor can be found in Section 2.7.1 and Figs. 2.41 and 2.43, and the geometry is shown in Fig. 6.16. Substantial boundary layer growth is evident even at about midchord position (Fig. 6.16a). The inviscid core shows steeper profiles than estimated from the inviscid theory due to blade boundary layer blockage and flow separation. The low velocity region increases rapidly as the trailing edge is approached. The most striking feature of this flow is the "jet–wake" structure which occurs at the discharge, and significant losses occur as the jet and wake mix in the diffuser. In Fig. 6.16b and c, a region of slow-moving fluid is shown near the discharge suction

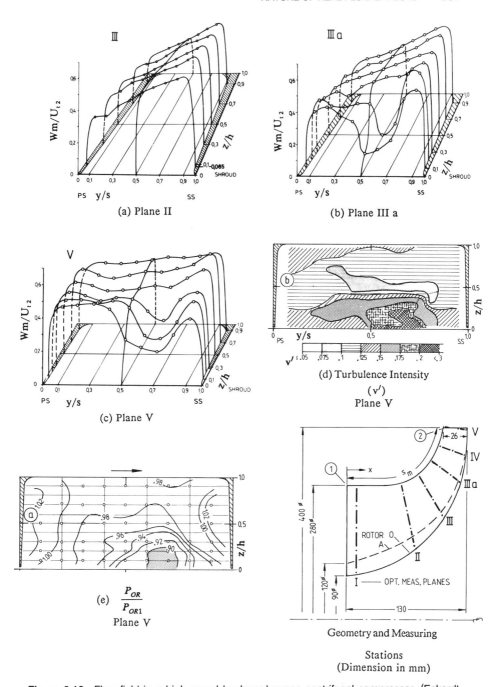

Figure 6.16 Flow field in a high-speed backward-sweep centrifugal compressor. (Eckardt, 1980; reprinted with permission of ASME.) $N = 14000$ rpm, $\dot{m} = 4.54$ kg / s, $P_{o2} / P_{o1} = 1.91$, rotor A at peak efficiencypoint in Fig. 2.41, z / h is the nondimensionalized blade height from Shroud, v' is the amplitude of the maximum absolute velocity fluctuation normalized by local mean velocity $(V_{max} - V_{min}) / 2\bar{V}$, and W_m is the meridional velocity. U_{t2} is the blade tip speed at the exit. (a) W_m at Station (plane) III (b) W_m at Station IIIa (c) W_m at Station V (d) Turbulence intensity (v') at Station V (e) relative stagnation pressure ratio (P_{oR} / P_{oR1}) at Station V.

surface at the shroud. Directly adjacent to it is a region of fairly uniform flow. These two flow patterns constitute the jet–wake structure. The wake typically fills approximately 20–30% of the flow passage. The extent of the "wake region" increases with the blade loading. The "jet–wake" structure is influenced substantially by secondary flows. Slow-moving fluid from the pressure surface and the hub boundary layer are both swept toward the suction surface/shroud region at the impeller discharge (Figs. 4.21 and 4.22). A measure of the stagnation pressure loss can be discerned from the ratio $P_{OR}/(P_{OR})_1 = \exp(-\Delta s/R)$, where relative stagnation pressure P_{OR} is based on the local rothalpy, $I = h + (W^2 - U^2)/2$. The ratio of $P_{OR}/(P_{OR})_1$ equal to 1 indicates no loss. A plot of this ratio for Eckardt's compressor is shown in Fig. 6.16e. It is evident that the "jet–wake" region has the maximum loss due to intense mixing in that region.

The turbulence structure in these passages is affected by rotation and curvature. The rotation effects make the leading (the rotation) or pressure surface boundary layers unstable and the trailing surface boundary layers stable. The turbulence is amplified by the rotation on the leading surface as described (and substantiated by model experiments) in Section 5.2. The secondary flow pattern measured by Eckardt (1976) at the exit shows clear evidence of large spanwise and cross flows (blade-to-blade) as well as separation zone. Overall fluctuations in most regions were found to be between 10% and 20% (Fig. 6.16d). In the separated region, velocity fluctuations were as high as 30%. Large stagnation pressure losses are also observed in this region (Figure 6.16e).

Krain (1988) repeated Eckardt's measurements in a newly designed 30° back swept impeller and found decreased jet–wake effects observed by Eckardt (Figs. 6.16b and 6.16c) and other investigators. Many design advances have been made to reduce or eliminate "jet-wake" phenomena through proper choice of streamwise/spanwise work distributions and blade and wall contouring. The design total pressure ratio was 4.0, $\dot{m} = 4.0$ kg/s, and $\eta_c = 95\%$. This flow is predicted reasonably well by Hah and Krain (1990), Kunz and Lakshminarayana (1992b). The flow fields away from the blade and inside the passage were similar to those of Eckardt (Fig. 6.16). The boundary layer interaction region near the tip produces a wake type of profile near the midpassage, similar to those observed in the axial flow compressor (Fig. 6.3). Recent laser-doppler velocimetry measurements by Fagan and Fleeter (1991) in a low-speed backward-swept impeller at both the design and the off-design conditions indicate that both the traditional jet–wake structure (Fig. 6.16) and opposing vortex secondary flow are present at the design conditions. At the stall operating point, the secondary flow in the aft 25% of the passage is found to be dominated by reverse flow.

Hathaway et al. (1993) recently reported detailed laser and hotwire measurements of the three-dimensional flow field in a low-speed centrifugal compressor designed to duplicate essential flow physics of a high-speed centrifugal compressor. The test compressor was a backward-swept impeller with a design tip speed of 153 m/s, , 20 blades, a back sweep of 55°, an exit diameter = 1.524 m, a blade height of 0.218 m at inlet and 0.141 m at exit and $\dot{m}$ = 30 kg/s at 1862 rpm. The meridional velocity profiles (W_m) measured are similar to those reported by Krain (1988) in a high-speed impeller. The laser data on secondary velocity vector at the trailing edge is shown in Fig. 4.22. The data at S/S_m = 0.644 (nondimensional shroud meridional distance = 0 at LE, 1 at TE of shroud) shown in Fig. 6.17 is compared with the Navier–Stokes computation (Dawes, 1987; also see Section 5.7.4.3). The measuring station corresponds (approximately) to the station in between Section III and IIIa in Fig. 6.16. Figure 6.17 clearly shows the development of tip vortex and flow migration toward the tip. The vortex is located near the midpassage, and the inset shows additional details on the leakage flow. The results show that the inward flow near the pressure surface/shroud corner of the passage is caused by the roll-up of the endwall fluid near the tip of the blade. The flow is outward in the tip region of the suction surface. The radial velocity is outward inside the blade boundary layer on both surfaces. The flow field undergoes rapid change as it progresses towards the trailing edge, as evidenced by the secondary velocity vectors at the exit of this compressor, shown in Fig. 4.22. Additional data in a centrifugal impeller (air-flow model of a two-stage regulating pump–turbine) reported by Ubaldi et al. (1993) are similar to those of Eckardt in a compressor (Fig. 6.16) and are described in detail in Section 4.2.6.3 and Fig. 4.22. Detailed measurement of turbulent kinetic energy and Reynolds stress provide additional evidence on the mixing of jet and wake flows. Johnson and Moore (1980) have reported three-dimensional flow measurement in a low-speed centrifugal impeller and have provided a detailed interpretation of the secondary flow induced by turning of the shear flow and by the rotation.

Hence the main features of a centrifugal flow field are large secondary flows within the entire passage and, in many instances, separated flow. Most importantly, the close coupling of the inviscid core, viscous regions, and the secondary flows makes this a truly challenging flow to compute. The "wake" region observed is similar to the corner separation measured in axial flow compressors.

Although most experimental work has been done using radial or backward-swept impellers, there has been some investigation of forward-swept impellers. One example is the study of Goulas and Mealing (1984). They measured velocities and turbulent kinetic energies in a forward-swept fan. Their data showed the presence of jet–wake pattern at the impeller discharge. However, the wake appeared on the hub rather than at the usual shroud location although it was still on the suction surface.

Engeda and Rautenberg (1987) and Senoo and Ishida (1986) found that the centrifugal impeller performance is very sensitive to tip clearance and loading. Engeda and Rautenberg measured an 8% decrease in efficiency when the tip clearance was increased from 0.5% to 2% of blade height at the highest loading.

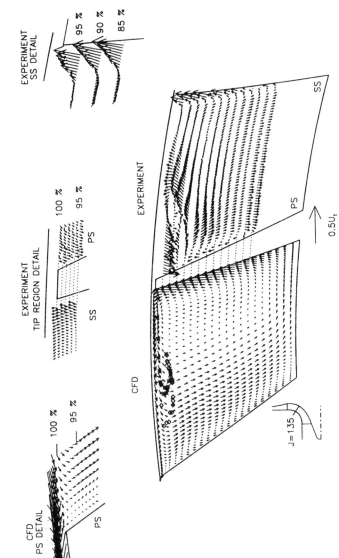

Figure 6.17 Secondary flow velocity vector plots at station $S / S_m = 0.644$ for centrifugal compressor. The symbols $\diamond$ and $*$ denote CFD tracers released along the blade leading edge on the pressure and suction surfaces, respectively. Insets show additional details on the tip region flow using the actual CFD grid and full laser measurement resolution. (Copyright © Hathaway et al, 1993; reprinted with permission of ASME.)

In the area of centrifugal pumps, the data of Goto (1992) at the exit of a mixed-flow pump (shown in Fig. 5.35) indicate substantial three-dimensional effects, wake and blade boundary layer growth, and tip clearance effects. An additional effect in a mixed-flow pump is the longitudinal curvature of hub and casing which influences the stability of boundary layer growth as illustrated in Fig. 5.2. The data acquired in an automotive torque converter (Fig. 5.38) show large regions of flow separation, wake, and substantial three-dimensionality in flow, with velocity fluctuations as high as 15–20% at the exit of the turbine rotor (Marathe et al., 1994).

6.2.3 Flow Field in Axial Turbines

The flow field in axial turbines is similar in some respects to the axial flow compressor flow field described earlier. But there are some important differences: (1) As the flow in these passages is accelerating, the flow may be transitional even on a rotor blade. (2) The flow in the endwall region is much more complex due to large flow-turning, the secondary flow dominates the endwall region and covers greater spanwise extent than that in a compressor. (3) Because the leading edge is thick, the presence of wall boundary layers in the corner formed by the blade and the wall results in a horseshoe vortex, which is split into a suction leg and a pressure leg and transported through the passage. (4) In cooled gas turbines, additional complications arise due to thermal effects. The injection coolant (film cooling) has a substantial effect on aerodynamics and losses. This topic is covered in the next chapter. The convective cooling involves change in blade thickness, which modifies aerodynamics of blading.

The three-dimensional flow field due to endwall secondary flows (Figs. 4.23, 4.25, 4.30, 5.18, and 5.25), tip clearance flow (Figs. 4.25 and 5.24), three-dimensional boundary layers (Fig. 5.17), and losses (Fig. 5.37) were covered in earlier chapters. Additional features will be covered in this section and integrated with the material covered in earlier chapters.

The turbine flow (both two- and three-dimensional) may be laminar (for substantial portions of chord length). Hence, the transitional flows (depicted in Fig. 5.7) are often encountered. An understanding of the transition and its effect on aerodynamics and heat transfer is important in turbines. The latter part is covered in Chapter 7. Laminar flows with laminar separation bubbles are often encountered in these accelerating flows. Turbine flow with a laminar separation bubble is illustrated in Fig. 6.18. The laminar flow may separate because of the local adverse pressure gradient present as shown in Fig. 2.64). This separation bubble will reattach, and this may in turn initiate transition (over the bubble) resulting in turbulent flow downstream. The presence of these separation bubbles result in increased losses, three-dimensionality, and change in heat transfer and pressure distribution.

Detailed surface conditions measured in a two-stage low-pressure turbine by Hourmouziadis et al. (1987) reveal some interesting features. At Reynolds

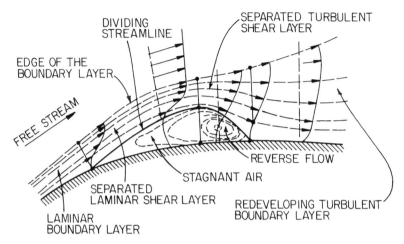

Figure 6.18   Nature of laminar separation bubble on a turbine blade.

number (based on chord) of 2×10^5, the data show substantial laminar boundary layer (beyond midchord), followed by a laminar instability region on the second guide vane of the two-stage turbine. The separation occurred near 75% of chord, resulting in transition and subsequent turbulent boundary layer. Similar features have been measured by others in low-speed tests. Whether such a transitional phenomena occurs in a real engine environment is still a controversial issue. Transitional flow depends on inlet turbulence levels, flow acceleration, roughness, and most importantly, Reynolds number. Other investigations in a transient facility at high Mach numbers reveal no such transitional flow. It is likely that such transitional flows do occur in the nozzle and first-stage rotor; but subsequent flow will encounter high free-stream turbulence (due to mixing of wakes, secondary flow, and leakage flow), and transitional flow is likely to occur very near the leading edge. This type of transitional phenomenon is rare in axial compressors, and it is mostly confined to cascades and the first blade row. The effect of Mach number on the transitional boundary layer was studied by Oldfield et al. (1981) in a turbine cascade with $\alpha_1 = 0$, $\alpha_2 = 65°$, $M_{2is} = 0.955$ and $\text{Re} = 5 \times 10^5$ $C = 8$ cm. They systematically investigated four different bladings with similar loading and their results are shown in Fig. 6.19. They used three methods to detect inception of transition; thin film guage, a flattened pitot probe, and the blade static pressure tap. The blades and the corresponding pressure distribution are shown in Fig. 2.76. By changing the leading-edge profile, the authors were able to delay transition, eliminate the separation bubble, and decrease profile losses for the profile T8 compared to profile T4, especially at higher Mach numbers. Even though there is discrepancy among the three methods of locating transition, the trend in delaying the transition is clear from Fig. 6.19. This indicates that improvements in blade design can provide reduced losses.

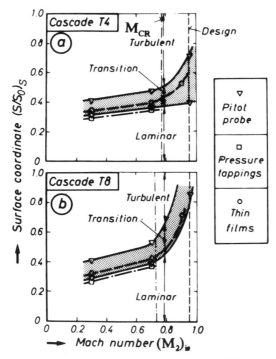

Figure 6.19 Effects of profile and exit Mach number on transition on suction surface of a turbine cascade. M_{CR} is the critical Mach number (Section 3.4), S is the surface distance normalized by total distance, M_{2is} is the exit isentropic Mach number. (Copyright © Oldfield et al., 1981; reprinted with permission of ASME.)

6.2.3.1 Endwall Flow Near the Leading Edge and in the Passage of a Cascade

The secondary flow phenomenon in a turbine is much more complex than that in compressors or pumps; this is because of large flow-turning and the presence of a thick leading edge. Let us first consider the formation of a horseshoe vortex. The phenomenon is shown schematically in Fig. 6.20 and 5.1f. Most of the data and interpretation are based on cascade tests (Langston et al., 1977; Langston, 1980; Klein, 1966; Sieverding, 1985; Sharma and Butler, 1987; Gaugler and Russell, 1980; Moore and Ransmayr, 1984; Yamamoto, 1987). A review of the endwall flow phenomenon in turbine cascades can be found in Sieverding (1985). The reconstruction of flow field, shown in Fig. 6.20, is based on measurements and flow visualization in a cascade.

The complex flow field induced by the horseshoe vortex, which combined with wall boundary layers and secondary flow, results in flow separation in corners. The inception of the horseshoe vortex can be explained on the basis of secondary flow theory (Section 4.2.6). Let us consider the endwall region of a cascade shown in Figs. 4.19. The wall boundary layers encounter large flow-turning upstream of the thick leading edge. The curvature of the streamline upstream of the leading edge (in the presence of normal vorticity due to the wall boundary layer) introduces pitchwise motions oriented away

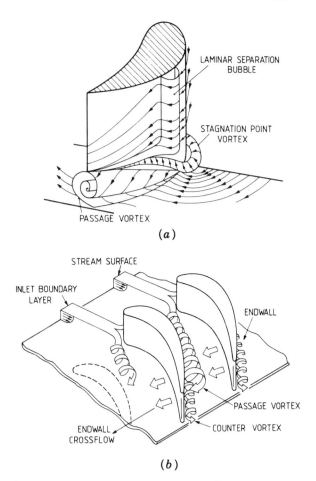

Figure 6.20 Endwall flow models for turbine cascades, [proposed by (a) Klein (1966) and (b) Langston (1980)]. (Copyright © Sieverding, 1985; reprinted with permission of ASME.)

from the blade, along with spanwise motion toward the wall along the blade surfaces resulting in vortex structure shown in Fig. 6.20 and Fig. 5.1f. This vortex is counterclockwise (looking from downstream) on the suction side. This vortex interacts with the corner boundary layer and entrains the fluid from these viscous layers to produce a strong vortex which may persist along the passage depending on Reynolds number, leading edge radius, turning angle, and initial boundary layer thickness. Even though the formation of a horseshoe vortex can be explained on the basis of inviscid secondary flow theory (Section 4.2.6), the subsequent roll-up, increased size, and flow separation associated with it is predominantly a viscous phenomenon and should be treated as such. A qualitative estimate of the strength of the horseshoe vortex can be made from Eqs. 4.125 or 4.121, but it should be emphasized that viscous effects have a dominant influence on its growth, size, effect, transport, and decay inside the passage. The horseshoe vortex formation occurs near the leading edge and forms a distinct core. In addition to this, a secondary vortex due to curvature (reversed) of streamline inside the passage develops classical secondary flow described in Section 4.2.6. Thus two distinct (but arising from the same physical phenomena) vortices are present on the suction side, as shown in Fig. 6.20b. They may merge, interact, or stay separate. The secondary or passage vortex (arising from the turning of the flow inside the passage) occurs near the suction surface due to transverse fluid motion from the pressure to the suction surface along the endwall (Fig. 4.19). Endwall flow models suggested by Klein (1966) and Langston (1980) are shown in Fig. 6.20. Klein's "stagnation point vortex" is the same as the terminology "counter vortex" by Langston. This is also called "leading-edge vortex" or "horseshoe vortex" by many authors. The "passage vortex" (secondary flow vortex) arises from secondary flow due to flow-turning inside the passage as explained in Section 4.2.6.

It should be mentioned here that the strength of horseshoe vortex depends on the leading-edge radius. Langston et al. (1977) observed that the pressure side of the horseshoe vortex merges with the passage vortex near the suction surface (both have the same sense of rotation). They observed that the suction leg of the horseshoe vortex (which is opposite in direction to passage vortex) continues on the suction side. Klein's (1966) results (Fig. 6.20) showed that the suction leg of the horseshoe vortex diffused and dissipated by the time it reached the trailing edge. This was confirmed by computational simulation by Ho and Lakshminarayana (1994). Their computational simulation indicates that the horseshoe vortex diffuses and dissipates quickly. The results for Gregory-Smith and Cleak's (1992) turbine cascade (e.g. Fig. 5.25) indicate that the horseshoe vortex grows in strength up to 40% chord and dissipates quickly with no trace of it at 80% chord. Whether this is true at high speeds and in a turbine rotor is yet to be established. The presence of horseshoe vortex and passage vortex as well as separation of the wall and the blade boundary layer and their interaction with the vortices makes the flow extremely complex. A word of caution is in order. Most of these interpretations are based on low-speed cascade data. Even though there is evidence to show that the cascade flow in a turbine is representative of flows in rotors, the structure of endwall flow may be substantially altered due to rotation as mentioned earlier.

Yamamoto's (1989) data, shown in Fig. 4.25, indicate the dominance of passage vortex (caused by passage flow-turning). It is also evident from these data that the secondary vortex interacts with the blade wake in the endwall flow region to generate an additional vortex, which hugs the secondary flow. This feature is clearly revealed in Gregory-Smith et al.'s data (1992) shown in Fig. 5.25. The top figure shows two loss cores. The one on the right side is due to passage secondary flow, and the one on the left is due to interaction between the blade wake and the secondary vortex. The numerical simulation and secondary velocity vectors shown by Ho and Lakshminarayana (1994) reveal that the corner or horseshoe vortex dissipates quickly and that the structure observed at the exit includes only the passage vortex and the wake–secondary-flow interaction vortex. Large underturning and overturning are measured across the secondary vortex (Gregory-Smith et al., 1988).

The secondary flow/vortex introduces large deviation in blade and end-wall static pressure as revealed by Langston et al.'s (1977) data. Details of the cascade are given in section 4.2.6.5, and some of the flow features are shown in Figs. 4.23 and 5.28. The endwall static pressure distribution, shown in Fig. 6.21a, indicate the trajectory of secondary vortex (as indicated by a low-pressure region) as well as the presence of nonuniformity in pressure near the leading edge of the pressure surface. This can be explained on the basis of flow velocity vectors predicted at first grid point from the endwall by Hah (1984) and shown in Fig. 6.21b. Two attachment lines a_1 and a_2 and two separation lines s_1 and s_2 can be identified from these velocity vector plots. The separation lines correspond to the two legs of the horseshoe vortex formed near the leading edge. The separation line (s_2) move around the leading edge and merge with the fluid coming from the free-stream direction to form the suction side leg of the horseshoe vortex (see Fig. 6.20). The second separation line s_1 is pushed away from the leading edge and merges into the passage or secondary vortex. The saddle point, marked A, corresponds to the highest static pressure measured on the endwall. The trajectory of horseshoe vortex and separation lines can also be identified in this plot. The numerical simulation of Gregory-Smith et al.'s (1992) cascade ($\alpha_1 = 42.75°$, $\alpha_2 = -68.7°$, C = 224 mm, S = 191 mm, Re = 4×10^5) endwall flow by Ho and Lakshminarayana (1994), shown in Fig. 6.22, reveals the trajectory of secondary vortex on the blade in the endwall region. The vortex occupies nearly 25% of the span. The suction peak for this case is near the 30% of chordwise location at the midspan, and the suction peak near the endwall has moved downstream of this location due to the presence of secondary vortex. The location of the suction peak ($C_p = -4.88$) indicates the position at which the vortex starts to move away from the blade surface. The secondary loss distribution at the exit of the cascade is presented in Fig. 5.25.

Because the blade loading is very high in a turbine, the tip clearance plays a major role and increases losses and three-dimensionality in a turbine. The tip clearance flow field measured by Yamamoto (1989) in a cascade is shown in Fig. 4.25. Very high tip leakage flow, its roll-up into a vortex, and interaction with the secondary flow are clearly evident in this figure. The secondary flow and leakage flow interaction produces a distinct interface; this can be seen in Fig. 4.25 and is confirmed by the numerical simulation results of Basson and Lakshminarayana (1993). The yaw angle distribution in the tip clearance region (Fig. 5.24) reveals underturning as high as 60° (near the wall) and overturning as high as 25° (at $Z/\tau = 5$) in the clearance region. The extent of leakage flow/vortex covers nearly 50% of the pitch. Even the passage-averaged flow angles show large variation in yaw angle in this region

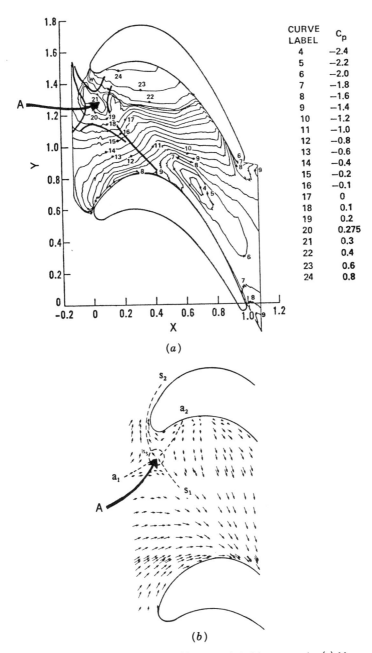

Figure 6.21 Flow field near the endwall of Langston's turbine cascade. (a) Measured static pressure coefficient. (b) Predicted velocity near the wall. (Copyright © H ah, 1984; reprinted with permission of ASME.)

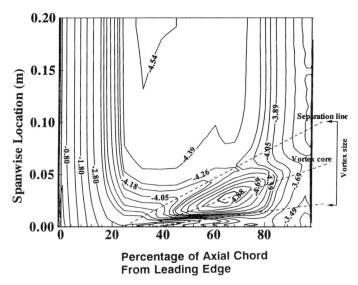

Figure 6.22 Blade static pressure coefficient C_p on suction surface for Gregory-Smith turbine cascade. (Ho and Lakshminarayana, 1994; reprinted with permission ASME.)

(Fig. 4.30). Sjolander and his group (e.g., Yaras and Sjolander, 1992, and Sjolander and Cao, 1994) have carried out extensive research to identify various flow features including the effect of rotation and tip gap geometry.

6.2.3.2   Flow Field in Turbine Stages   The information on the flow field in a turbine rotor passage is scarce. Dring and colleagues (Dring and Joslyn, 1981; Dring et al., 1982b, 1986) and Sharma et al. (1985) have acquired steady and unsteady flow data in a 1.5 stage (nozzle–rotor–stator), low-speed turbine [diameter = 1.52 m, $r_h/r_t = 0.8$, $A = 1.0$, $N = 405$ rpm, $V_z = 23$ m/s, $M_2 = 0.2$, flow-turning angle for nozzle = 67.5°, rotor (midspan) = 66°, stator (midspan) = 69°]. The exit pressure contours measured by a rotating Kiel probe shows features similar to that in a cascade (Fig. 5.25 and 4.25), but the casing and hub-wall secondary vortices have moved much closer to the midspan region and they virtually fill the entire span from the hub to the tip. The tip leakage vortex is located very close to the casing and the suction surface. The relative motion between the endwall and the rotor blade opposes the leakage flow; hence the leakage vortex is located closer to the suction surface, unlike that observed in a compressor rotor (Fig. 4.26). The unsteady blade pressure time history on the rotor blade (Dring et al., 1982b) at 15% (of axial chord) gap between the rotor and the stator is shown in Fig. 6.23. Very-high-amplitude pressure fluctuations (caused by nozzle wake) near the leading edge of the suction surface is evident from this plot, with the maximum amplitude as high as the time mean pressure. This results in unsteady boundary layer on rotor blades, and it has a major consequence on

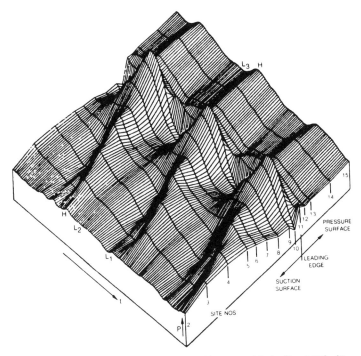

Figure 6.23 Time history of static pressure on turbine rotor blade (f = 0.78). (Copyright ©
Dring et al., 1982b; reprinted with permission of ASME.)

losses and heat transfer (Chapter 7). The Cambridge group (e.g., Hodson
et al., 1993) has acquired very detailed data on the flow properties on the
blade surface of the cascade, rotor, and multistage blades to investigate
rotor–stator interaction effects.

Detailed unsteady flow data at the midspan of a single-stage turbine rotor
with high turning angle (125° at the hub, 95° at the tip) including the nozzle
vane wake transport and unsteady rotor wake characteristics are reported by
Zaccaria and Lakshminarayana (1995b). The details of the stage are as
follows: r_h/r_t = 0.73, nozzle chord at midspan = 17.68 cm, nozzle turning
angle = 70°; rotor blade chord at midspan = 12.86 cm, ψ = 1.88. The development of nozzle passage flow field including wake and secondary flow in the
endwall region of the same stage is reported in Zaccaria and Lakshminarayana (1995a). The measured secondary flow in the nozzle was not as
severe as that reported in cascades. The nozzle wake was distorted considerably (due to large blade-to-blade variation in convective velocities) as it
passed through the rotor.

Leach (1983) tested one of the stages (nozzle-rotor) of the energy-efficient
engine high-pressure turbine and measured the three-dimensional distribution of pressure and temperature. The blading and the design parameters are

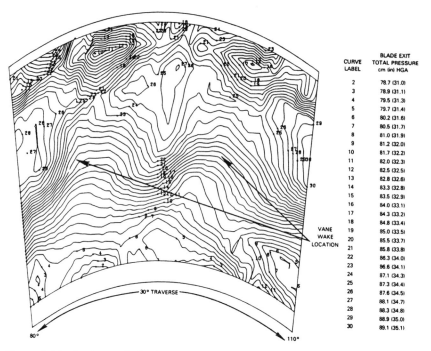

CURVE LABEL	BLADE EXIT TOTAL PRESSURE cm (in) HGA
2	78.7 (31.0)
3	78.9 (31.1)
4	79.5 (31.3)
5	79.7 (31.4)
6	80.2 (31.6)
7	80.5 (31.7)
8	81.0 (31.9)
9	81.2 (32.0)
10	81.7 (32.2)
11	82.0 (32.3)
12	82.5 (32.5)
13	82.8 (32.6)
14	83.3 (32.8)
15	83.5 (32.9)
16	84.0 (33.1)
17	84.3 (33.2)
18	84.8 (33.4)
19	85.0 (33.5)
20	85.5 (33.7)
21	85.8 (33.8)
22	86.3 (34.0)
23	86.6 (34.1)
24	87.1 (34.3)
25	87.3 (34.4)
26	87.6 (34.5)
27	88.1 (34.7)
28	88.3 (34.8)
29	88.9 (35.0)
30	89.1 (35.1)

Figure 6.24 Contours of absolute stagnation pressure at the exit of turbine stage (HGA stands for absolute pressure in terms of mercury height) (Leach, 1983; NASA CR 168189, PWA 5594-243).

shown in Fig. 2.64. The turbine test had the following operating conditions and geometric properties $\Delta h_o/U^2 = 1.61$, pressure ratio $= 4.01$, $M_2 = 1.24$, exit tip diameter $= 82$ cm, average $h = 5.81$ cm, number of rotor blades $= 54$, number of vanes $= 24$, $T_{o1} = 566$ K, mass flow rate of cooling air $= 12.36\%$ of primary flow. The contours of absolute stagnation pressure at the exit of the turbine, measured by Leach and shown in Fig. 6.24, indicate the complex nature of the flow field. The stagnation pressure is measured in a stationary frame downstream of the rotor; hence the circumferential survey provides absolute flow field at the exit (including the presence of nozzle wakes), but the rotor flow is averaged out in the process. The pressure inside the nozzle wake is lower (or pressure drop is higher). The pressure increases toward the tip, up to about 70% of the span, and then decreases to about 90%, beyond which the tip clearance effects tend to increase pressure again (as tip clearance decreases the pressure drop and the work done). The contours near the tip should not be mistaken for tip vortices, because the measurements were done with a stationary probe downstream of the rotor. the region near the tip reveals complex interaction between the nozzle secondary flow and the rotor tip clearance flow in the annulus wall region. The exit temperature contours (not shown) show a similar pattern. The flow in the annulus wall region has the highest temperature, which is clear evidence of the effects of rotor tip clearance flow and nozzle secondary flow. The clearance allows the hot gas to bypass the rotor and produce the highest temperatures. the temperatures were found to be lowest in the nozzle wake region; this was caused by dual effect, higher pressure drop, and coolant effect in this region.

Features similar to those observed in a multistage compressor (Fig. 6.13) have been measured in a multistage turbine by Brideman et al. (1983). the data at the exit of the first stage and at the exit of the fifth stage of a five-stage energy-efficient engine turbine (similar to that described earlier and in Figs. 2.62 and 2.64) are shown in Fig. 6.25. The full five-stage scale model representing the core-low-pressure turbine spool design that was tested by Brideman et al. had five stages with a rotational speed of 3208 rpm, $\dot{m} = 24.14$ kg/s, and a pressure ratio of 4.37. The testing was done using a building-block approach; nozzle, nozzle and first stage, and all five stages were tested separately in the same facility. The stagnation pressure distribution at the exit of the nozzle indicates the presence of the secondary flow region, along with the associated losses near the tip and the hub-wall regions. This loss core region is mixed very effectively by the rotor as seen by the survey at the exit of the first stage. The annulus and hub wall boundary layers are repeatedly perturbed as it passes consecutively through rotating and stationary blade rows, resulting in distribution shown at the exit of the fifth stage in Fig. 6.25. The distribution from 30% to 90% span follows the design distribution, deviating considerably near the hub and annulus walls. But the annulus- and hub-wall boundary layer growth is not as large as that observed

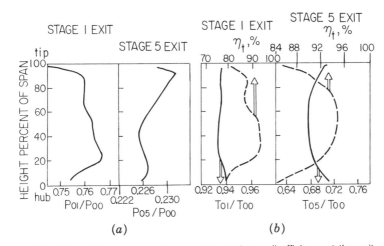

Figure 6.25 Stagnation pressure, temperature, and overall efficiency at the exit of stage 1 and stage 5 of a multistage axial flow turbine. (a) Pressure ratio. (b) Temperature ratio and efficiency. (Subscript o refers to upstream of first stage, 1 is downstream of stage 1, 5 is downstream of stage 5.) (Brideman et al., 1983; NASA CR 168290)

in multistage compressors due to the presence of favorable pressure gradient. The temperature profile (Fig. 6.25) is nearly uniform at the exit of the first stage, but shows the effect of wall boundary layers at the exit of the fifth stage. The temperature drop is fairly uniform in the middle third of the annulus. The local total-to-total efficiency (calculated from Eq. 2.22; the annulus area was divided into several streams, and η_t was calculated from $\eta_t = \Sigma_i m_i (\Delta h_o)_{\text{actual}} / \Sigma_i m_i (\Delta h_o)_{is}$, where m_i is the mass flow in individual stream) varies considerably along the span. Redistribution of flow temperature and pressure (as well as other flow properties) along the radial direction occurs as the flow progresses from stage to stage, and the efficiency decrement from midspan to wall at the exit of the fifth stage is as high as 10%.

Huntsman and Hodson (1994) report detailed flow visualization and measurement in a low-speed radial turbine, carried out in a rotating frame of reference. One of the main features observed is the presence of a scraping vortex (e.g., Fig. 1.16). The leakage flow, the secondary flow, and the scraping vortex accounted for major proportion of losses at off-design conditions. The profile or boundary layers losses were found to be small.

6.3   LOSS MECHANISMS AND CLASSIFICATION

The loss mechanisms in fluid mechanics and methods of evaluating them (either analytically or empirically) are covered in many textbooks (e.g., Schlichting, 1979, White, 1991). This section gives a general overview and classification of the mechanisms and sources specific to turbomachinery flows.

The overall efficiency can be defined on the basis of entropy changes as described in Section 2.3. Because the entropy cannot be measured directly, the approach is to relate entropy to other measurable thermodynamic or fluid mechanic quantities. In Section 2.3, the approach taken was to compare an isentropic turbomachinery ($\Delta s = 0$) to an actual turbomachinery, keeping the pressure ratio and the mass flow constant for both the actual and the ideal machine, and define efficiencies in terms of actual to isentropic expansion or compression process. Knowing the final and initial thermodynamic states, the efficiencies can be calculated using the expression given in Section 2.3.

To explore various sources of losses and their contribution to overall inefficiency, it is necessary to revert to basic equations of thermodynamics. Let us consider adiabatic flow through a stationary blade row (e.g., cascade, stator, nozzle). The loss manifests itself in the form of stagnation pressure loss (or entropy increase) through the equation (for perfect gas)

$$\frac{\Delta s}{R} = -\ln \frac{P_{o2}}{P_{o1}} = -\ln \left[1 - \frac{(\Delta P_o)_{\text{loss}}}{P_{o1}} \right] \qquad (6.1)$$

where Δs = entropy change, P_{o2} is the stagnation pressure at the exit, and $(\Delta P_o)_{\text{loss}}$ is the viscous or other losses (e.g., shockwave). For example, along a streamline in stationary components, $P_{o1} - P_{o2} = (\Delta p_o)_{\text{loss}}$. For example, in boundary layers, the viscous dissipation ϕ or Φ in Eqs. 1.12, 1.18 and 1.24 would increase the internal energy level and the entropy, and hence is not available for performing useful work. We thus call it a loss. Likewise, viscous and turbulent dissipative mechanisms of vortices and shocks result in an increase in the entropy and the internal energy, and hence they are also a source of losses. For shocks (with known upstream Mach numbers) the entropy increase can be calculated from the compressible gas dynamic relationships (e.g., Shapiro, 1953, Anderson, 1982).

For viscous layers and vortices, the interest usually lies in overall or integrated losses rather than local losses. An estimate of the local loss involves solution of the energy equation, Eq. 1.18 or 5.1, with a suitable turbulence model. In addition to viscous dissipation, the turbulence dissipation plays a significant, if not a major role, in turbomachinery flows. A detailed phenomenological description of turbulence is given in Tennekes & Lumley, "turbulent flows are always dissipative. viscous shear stresses perform deformation work which increases the internal energy of the fluid at the expense of kinetic energy of the turbulence. Turbulence needs a continuous supply of energy to make up for these viscous losses. If no energy is supplied, turbulence decays rapidly. Random motions, such as gravity waves in planetary atmospheres and random sound waves (acoustic noise), have insignificant viscous losses and, therefore, are not turbulent. In other words, the major distinction between random waves and turbulence is that waves are essentially nondissipative (though they often are dispersive), *while turbulence is essentially dissipative*." The dissipation of turbulent energy thus constitutes additional losses. Most boundary layers and vortices in turbomachinery (wall and blade boundary layers, wakes, leakages, and secondary vortices) are turbulent in nature. Hence the dissipative mechanism includes both the molecular and the turbulent part and is dependent on Reynolds number, pressure gradient, free-stream turbulence and length scale, turbulent Reynolds number, roughness, and other relevant parameters.

Equation 6.1 provides a basis for estimating losses. For shocks, Δs can be estimated directly through shock wave relationships. For viscous layers (where flow may be subsonic or incompressible), by expanding Eq. 6.1 in an infinite series we obtain

$$\frac{\Delta s}{R} = \frac{[(\Delta P_o)_{\text{loss}}]}{P_{o1}} + \frac{1}{2}\left(\frac{(\Delta P_o)_{\text{loss}}}{P_{o1}}\right)^2 + \cdots \tag{6.2}$$

For viscous layers, $(\Delta P_o)_{\text{loss}}/P_{o1}$ is very small ($\ll 1$) and the second term can be neglected. Thus,

$$\frac{\Delta s}{R} = \frac{(\Delta P_o)_{\text{loss}}}{P_{o1}} \tag{6.3}$$

In most turbomachinery practice, the loss coefficient is expressed as (for stationary components)

$$\zeta = \frac{(\Delta P_o)_{\text{loss}}}{\frac{1}{2}\rho V_1^2} = \frac{P_{o1} - P_{o2}}{\frac{1}{2}\rho V_1^2} = \frac{\Delta s}{R} \frac{P_{o1}}{\left(\frac{1}{2}\rho V_1^2\right)} \quad \text{or} \quad \frac{\Delta s}{R} = \left\{\frac{\zeta\rho V_1^2}{2P_{o1}}\right\} \quad (6.4)$$

In most instances, the mass-averaged (either the blade-to-blade or the entire passage) values are of interest. For example, the mass-averaged value at any radius (blade-to-blade average) is given by

$$\bar{\zeta} = \frac{2\int_{SS}^{PS}\left(P_{o1}\,\rho_1 V_{z_1} - P_{o2}\,\rho_2 V_{z_2}\right)dy}{\rho_1 V_{z_1} S \rho_1 V_1^2} \quad (6.5)$$

where subscript 2 refers to exit and 1 to inlet; S is the blade spacing and y is the tangential distance. A similar expression can be written for mass average value for the entire passage.

These quantities are useful in axial turbomachines, because the streamline path is nearly axial. In radial turbomachinery, the intense mixing of the flow and three-dimensionality makes it difficult to define a passage-averaged loss coefficient unless P_{o1} is constant (radially) at the inlet.

These loss coefficients (measured or correlated) can be used to estimate the efficiency from expressions given in Section 2.3. Referring to Fig. 2.4, the enthalpy change between an ideal compressor and an actual compressor, with the same pressure ratio (P_{o3}/P_{o1}) and mass flow, can be related to the entropy change through the equation

$$T_o ds = dh_o \quad (6.6)$$

This process is taken along $o3_s$ to $o3$ along the constant stagnation pressure line. Hence,

$$T_o ds = h_{o3} - h_{o3s} \quad (6.7)$$

Strictly speaking, in the shear layers and shocks, Eq. 6.6 should be integrated from $o3_s$ to $o3$ because T_o is changing along this line. As an approximation, T_{o3s} is assumed to be very close to T_{o3}. It should be emphasized once again that the definition of η_c involves two different compressors, operating with the same $\dot{m}$ and P_{o3}/P_{o1}. For the same compressor, an isentropic process will produce higher pressure rise than an actual adiabatic compressor, and the change in stagnation enthalphy (work input) will be the same for both the actual (adiabatic) and an ideal compressor. Hence the compressor efficiency can be written using Eqs. 2.13 and 6.7

$$\eta_c = \frac{h_{o3s} - h_{o1}}{h_{o3} - h_{o1}} = 1 - \frac{T_o ds}{h_{o3} - h_{o1}} \quad (6.8)$$

Equations 6.3–6.8 provide the necessary relationship between efficiency, entropy rise, and loss coefficient ζ.

The corresponding relationship for a turbine is given by (Eq. 2.22 and Fig. 2.7)

$$\eta_t = \cfrac{1}{1 + \cfrac{T_o\, ds}{h_{o1} - h_{o3}}} \tag{6.9}$$

Horlock (1973a) assumes, in deriving similar relationships between the efficiency and the loss coefficient for a turbine, that the exit kinetic energy at $o3$ and $o3_{ss}$ are the same. For improved accuracy, T_o in Eqs. 6.6–6.9 can be assumed to be $(T_{o3s} + T_{o3})/2$; T_{o3s} is derived from the isentropic relationship, and T_{o3} is derived from the estimated value of ds or η.

In practice, a large number of cascade, rotor, and stages are tested to derive loss and loss correlations (ζ). The losses from various sources are added to derive an overall (global) loss coefficient and are substituted in Eq. 6.8 or 6.9 to estimate efficiency. It should be emphasized here that the computations, based on Navier–Stokes equations (provided that the code is optimized for accurate turbulence model, grid resolution, and minimum artificial dissipation), would provide prediction of local dissipation (Eqs. 1.18 and 1.24), as well as local increase in the internal energy and enthalpy. Thus h_{o3} for a compressor or turbine is computed directly. Equations such as Eqs. 6.3 and 6.4 are employed only when loss correlations or loss analysis based on integral method (global losses) are utilized. For rotor flows, the loss coefficient is usually based on measured or estimated or computed values of inlet and exit relative stagnation pressure (normalized by $0.5\,\rho U_t^2$) or entropy.

The sources of loss (or increase in entropy) in turbomachinery can be classified as follows:

1. *Profile Loss.* This includes the losses due to blade boundary layers (including separated flow) and wakes (Figs. 5.7 and 6.26) through viscous and turbulent dissipation. The mechanical energy is dissipated into heat within the boundary layer. This increases the entropy (Eq. 6.3) and results in stagnation pressure loss, even though the stagnation enthalpy (relative for rotor) is constant for adiabatic flow. In addition, the nonuniform velocity profiles in both the boundary layer and the wake (bounded and shear flows) are smoothed out by viscous and turbulence effects. Furthermore, the trailing vortex systems in the blade wake and its eventual mixing and dissipation give rise to additional losses. This again can be budgeted under profile loss. The profile loss correlation should be based on mixed out conditions; otherwise considerable inaccuracy in the estimation of losses can occur. For example,

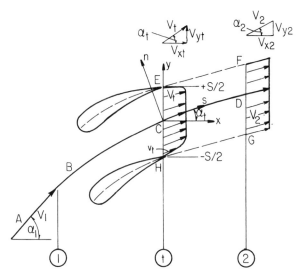

Figure 6.26 Profile and mixing loss analysis (v_t is the local velocity and V_t is the free stream velocity).

the losses measured at one-half chord downstream of the blade row are not the same as those based on data one chord downstream.

2. *Shock Losses*. Here again the losses arise due to viscous dissipation (in a very thin layer) across the shock. This could be estimated fairly accurately using Rankine–Hugoniot relations (e.g., Shapiro, 1953), but the estimate of indirect (additional) losses associated with boundary-layer–shock interaction has to be based on computation or correlations. Sudden jump in static pressure across the shock results in thickening of the boundary layer and/or flow separation. This loss could be a substantial portion of total profile losses, depending on Mach number and Reynolds number.

3. *Clearance Losses*. This loss arises from leakage flow in the tip clearance region of the rotor (Section 4.2.7) or tip clearance region of a can-tilevered stator or clearance losses due to axial gaps in the flow path. The leakage flow and the main flow in the tip clearance region are at differing angles, and mixing of these two dissimilar flows and resulting dissipation account for losses. In addition, the formation of a leakage vortex (Section 4.2.7) and its eventual dissipation (viscous and turbulent) along with its interaction with the main flow (e.g., entrainment of fluid by the vortex), result in losses. The loss due to leakage flow or vortex accounts for approximately 20–35% of total losses, depending on the machinery. For example, a centrifugal compressor with a long flow path along the shroud has higher leakage or tip clearance losses, as does a turbine with heavy blade loading.

4. *Secondary Flow Losses.* The secondary flow (Section 4.2.6) and formation of vortex result in mixing and dissipation of energy similar to those described under clearance losses. In addition, the indirect effect such as initiation of corner stall, interaction of secondary flow with the wall and the blade boundary layer, and wake (e.g., Fig. 5.25) results in additional losses. The losses due to this source are usually lumped together with the annulus/hub-wall boundary layer losses and budgeted under "endwall losses." Here again, the source of loss mechanism is viscous and turbulent mixing, and dissipation.

5. *Endwall Losses.* These usually include losses due to annulus- and hub-wall boundary layer (Section 5.5.4) which are almost invariably three-dimensional in nature. The three-dimensional (or skewed) boundary layer does account for the secondary flow. The growth of endwall boundary layer and mixing, along with the dissipative mechanisms associated with them, result in additional losses as explained earlier. The endwall losses (including tip clearance and secondary flow losses) may account for approximately 50–70% of total losses, depending on the type of turbomachinery.

6. *Cavitation Losses.* The formation of cavitation bubble, its eventual collapse, and the resulting shear layers would introduce (a) losses due to mixing and (b) viscous and turbulent dissipation. The information on this source of loss is scarce.

In many instances, a separate category called "trailing edge losses" is included to account for losses due to blunt trailing edge, which causes flow separation (mixing of jets) and shock–expansion-wave interactions due to sharp corners. This loss could be appreciable in transonic and supersonic turbines.

In many instances, it is difficult, if not impossible, to distinguish between various sources of losses. For example, in centrifugal compressors with narrow passages, the nonlinear effects, where the blade and the wall boundary layers interact to produce complex shear layers and separation, make it very difficult to distinguish between secondary flow losses and endwall losses or the profile losses. Even the tip leakage/secondary flow loss cannot be separated due to intense interaction/mixing of these two flows. In any case, it is useful to (a) evaluate the various sources of losses to estimate the performance of these machines and (b) minimize losses through improved design or flow control (active and passive devices).

There have been many papers and reviews published on sources of losses, loss correlations, and methods of predicting turbomachine performance. The papers by Sieverding (in Ucer et al., 1985), Kacker and Okapuu (1982), and Horlock (1973a) deal with axial flow turbines; Cetin et al. (1987) and Koch and Smith (1976) deal with transonic compressors, and Denton (1993) and Balje (1981) provide more general approach valid to all turbomachines.

6.4   TWO-DIMENSIONAL LOSSES AND CORRELATIONS

The two-dimensional losses are relevant only to axial flow turbomachines. These are mainly associated with blade boundary layers (Figs. 5.7 and 6.26), shock–boundary-layer interactions and separated flows, and wakes. The mixing of the wake downstream produces additional losses called *mixing losses*. The profile boundary layer may be laminar, transitional, or turbulent as shown in Fig. 5.7. The maximum losses occur near the blade surface, and the minimum losses occur at the edge of the boundary layer. This variation gets mixed out as the flow proceeds downstream. Consider a cascade (stationary) shown in Fig. 6.26. The stagnation pressure (relative for rotor), P_{o2} is uniform downstream as shown in Fig. 6.26, and thus $P_{o1} - P_{o2}$ represents the average loss for the entire cascade. If the stagnation pressure is measured at station t for example, or anywhere between t and 2, the mass-weighted (passage) average is given by Eq. 6.5 (where station 2 now represents station t or in between t and 2). For incompressible flow, the loss is given by $\zeta = 2g(\overline{H}_2 - \overline{H}_1)/V_1^2$ where $\overline{H}_2$ and $\overline{H}_1$ are average head (relative for rotor).

The two-dimensional losses can be classified as (1) profile losses due to boundary layer, including laminar and/or turbulent separation, (2) wake mixing losses, (3) shock losses, and (4) trailing-edge losses due to the blade. The two-dimensional profile losses depend on the (a) Reynolds number, Mach number, longitudinal curvature of the blade, inlet turbulence, free-stream unsteadiness and the resulting unsteady boundary layers, main-stream pressure gradient, and shock strength and (b) blade parameters such as thickness, camber, solidity, sweep, skewness of the blade, stagger angle, and blade roughness. The mixing losses arise due to mixing of the wake with the free stream, and this depends on, in addition to the parameters listed above, the distance downstream. The physical mechanism is the exchange of momentum and energy between the wake and the free stream. This transfer of energy results in the decay of the free shear layer, increased wake center line velocity, and increased wake width. At far downstream, the flow becomes uniform. Theoretically, the stagnation pressure loss far downstream and the trailing edge ($\overline{P}_{ot} - \overline{P}_{o2}$) represent the mixing losses. Most loss correlations are based on measurements downstream of the trailing edge (usually between 1/2 and 1 chord length) and thus do not include all the mixing losses. If there is flow separation, the losses would include the losses due to this zone and at its eventual mixing downstream.

The profile and mixing loss evaluation should take into consideration the boundary layer growth, as well as the static pressure variation along the flow path. The profile and mixing losses for a cascade can be related to the blade boundary layer properties through an analysis carried out by Speidel (1954). Detailed derivation of this analysis, which is briefly described here, can also be found in Csanady's book (1964, p. 271). The analysis is valid for 2D steady incompressible flow, both laminar and turbulent.

Referring to Fig. 6.26, the stagnation pressure loss along the streamline ABCD is given by $P_{o1} - P_{o2}$. Because ABC is an inviscid streamline, $P_{o1} = P_{ot}$, and hence the profile and wake mixing losses can be written as

$$\bar{\zeta}_p = \frac{2(P_{ot} - P_{o2})}{\rho V_1^2} \tag{6.10}$$

To determine $\bar{\zeta}_p$ and P_{o2}, it is necessary to relate the static pressure difference $(p_2 - p_t)$ and velocities V_t and V_2 to the displacement and momentum thickness of the blade boundary layer at the trailing edge. This can be carried out by (1) using continuity equation across stations t and 2 (Eq. 1.2) and (2) tangential (y) and axial momentum balance between stations t and 2 (Eq. 1.3).

By assuming that the flow is steady and one-dimensional and applying Eq. 1.2 to control surface EFGH (EF and HG are periodic surfaces), we can derive the following equation

$$V_t \cos \alpha_t = \frac{V_{x_2}}{1 - \Delta} \tag{6.11}$$

where the blockage Δ is

$$\delta^* = \Delta(S \cos \alpha_t) = \int_{-S'/2}^{+S'/2} \left(1 - \frac{v_t}{V_t}\right) dn, \qquad S' = S \cos \alpha_t \tag{6.12}$$

The displacement thickness δ^* is based on integration normal (n) to the blade surface, v_t is the velocity inside the blade boundary layer. The y momentum balance provides the following relationship:

$$V_t \sin \alpha_t = V_{y_2} \left\{ \frac{1 - \Delta}{1 - \Theta - \Delta} \right\} \tag{6.13}$$

where

$$\theta = \Theta(S \cos \alpha_t) = \int_{-S'/2}^{+S'/2} \frac{v_t}{V_t} \left(1 - \frac{v_t}{V_t}\right) dn \tag{6.14}$$

where θ is the momentum thickness in the n direction. The x momentum balance between t and 2 is given by

$$\frac{p_2 - p_t}{\rho} = \frac{\cos^2 \alpha_t}{S} \int_{-S/2}^{+S/2} v_t^2 dy - V_{x_2}^2 \tag{6.15}$$

The integration in Eq. 6.15 can be expressed in terms of Θ and Δ to prove

$$\frac{P_2 - P_t}{\rho} = V_{x2}^2 \left[\frac{1 - \Theta - \Delta}{(1 - \Delta)^2} - 1 \right] \tag{6.16}$$

Now along ABCD,

$$\bar{\zeta}_p = \frac{2(P_{ot} - P_{o2})}{\rho V_1^2} = \frac{2(p_t - p_2)}{\rho V_1^2} + \frac{V_t^2 - V_2^2}{V_1^2} \tag{6.17}$$

Equations 6.11, 6.13 and Eq. 6.16 can be substituted into Eq. 6.17 to prove

$$\bar{\zeta}_p \sec^2\alpha_1 = \left[\frac{2\Theta + \Delta^2}{(1 - \Delta)^2} + \tan^2\alpha_2 \left\{ \frac{(1 - \Delta)^2}{(1 - \Theta - \Delta)^2} - 1 \right\} \right] \tag{6.18}$$

If the boundary layer is thick, it is preferable to retain all the terms in Eq. 6.18 and compute the loss coefficient. If δ is small (one order of magnitude less than S), Eq. 6.18 can be simplified (neglecting terms of the order Δ^2, Θ^2, $\Delta\Theta$, etc.) to give

$$\bar{\zeta}_p \sec^2\alpha_1 = 2(\Theta + \Theta \tan^2\alpha_2) = \frac{2\theta}{S \cos^3\alpha_2} \tag{6.19}$$

Thus the blade profile loss can be estimated from Eq. 6.18 or 6.19, using the momentum thickness computed from the analysis given in Section 5.5. Extensive measurements carried out in the cascade tunnel at the University of Braunschweig (Speidel, 1954) show good agreement with the predictions from Eq. 6.19. Scholz (1965) carried out an analysis of the losses from 1 to t and t to 2 and estimated that the mixing losses due to wake (from t to 2) represent about 20% of the total losses for an unseparated cascade. The flows with separation and trailing shocks will have much higher mixing losses.

The loss correlation (Eq. 6.18) includes both the profile loss and the wake mixing loss. If flow separation occurs, an additional loss is caused by the fact that the pressure distribution, beyond the separation point, is substantially altered. The losses increase due to larger values of Δ and Θ. Speidel (195′) provided an estimation of the additional losses due to separation. This can be more accurately estimated from boundary layer/Navier–Stokes analysis that includes computation of separated flows. The effect of compressibility (shock-free) can be accounted for by allowing for the compressibility and expressing the density changes in terms of Mach numbers (Eq. 3.17). The analysis given above can be easily extended to include density changes.

In addition to the losses described above, boundary layer growth and the subsequent decay of the wake cause deviation in outlet angle (over and above those calculated from a potential flow). An estimate of this deviation can be obtained from Eqs. 6.11 and 6.13:

$$\tan \alpha_2 \approx (1 - \Theta - \Delta)\tan \alpha_t \tag{6.20}$$

Hence the viscous effect in turbomachinery always causes a decrease in turning angle. It is evident from the analysis that the profile losses depend on the momentum and the displacement thicknesses for a given inlet and outlet angle. The values of Θ and Δ in turn depend on the variation of free-stream velocity, Mach number, skin friction coefficient, free-stream pressure gradient, free-stream turbulence intensity, and Reynolds number as indicated in Section 5.5.

The loss calculation involves one of the following two steps:

1. Calculation of potential or inviscid flow using methods described in Chapter 3 or 5, and computing momentum and displacement thicknesses using the procedure outlined in Sections 5.5.3 or 5.5.5 (e.g., Figs. 5.9, 5.10, 5.16). The losses can then be calculated using Eq. 6.18 or 6.19.

2. If a Navier–Stokes code is employed, the local and integrated losses can be derived directly. As mentioned earlier, the predicted loss is very sensitive to the turbulence models employed. In many instances, step 1 described above provides much more accurate results because boundary layer codes are inexpensive to run, and a large number of grids can be employed within the boundary layer to predict Θ and Δ accurately.

The profile losses for the cascade tested [NACA 65(12)10] by Herrig et al. (1957) were calculated from these equations. The pressure distribution was determined from a panel code (Sect. 3.3.2) and the momentum thickness distribution was determined from the momentum integral technique described in Section 5.5.3 (Eqs. 5.20 and 5.21). The loss coefficient was computed from Eq. 6.18, and the drag C_D coefficient was computed from Eq. 3.19. The measured and predicted C_D are found to be 0.014 and 0.017, respectively. The discrepancy between the two may be due to the presence of transition in the experimental blade. Computation is based on turbulent flow from the leading edge. The location of the transition point should be known for accurate prediction of boundary layer growth and losses.

A good example of the predictions based on the coupled Euler/boundary layer code and Navier–Stokes solution is shown in Fig. 5.29. The predictions, based on boundary layer and Navier–Stokes codes, are good except at very low ($i = -10°$) and very high ($i = 7°$) incidences. The flow separation at these incidences makes it extremely difficult to predict the momentum thickness and losses accurately. The profile loss at midspan of the Penn State compressor is also predicted accurately by the Navier–Stokes code (Fig. 5.43). Most turbomachinery blades are designed to operate at incidences

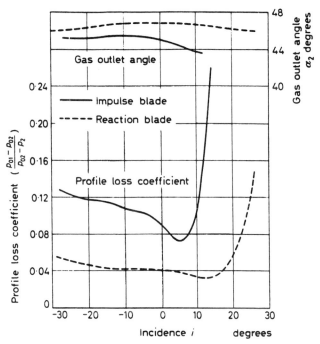

Figure 6.27 Variation in outlet angle and profile loss with incidence for typical turbine blade from Ainley (1948). Impulse blade; $\alpha_1 = 45.5°$, $\alpha_2 = -45.8°$, $S/C = 0.625$, $t_{max} = 0.22$ C. Reaction blade; $\alpha_1 = 18.9°$, $\alpha_2 = -47.1°$, $S/C = 0.58$, $t_{max} = 0.15$ C, Re = 1.5×10^5 Base profile T6; P_{02} is passage averaged stagnation pressure at the exit. (From Horlock, 1973a; reprinted with permission of Kreiger Publication Company.)

below the stall point, and thus inability to predict the separated flow in the cascade is not a major handicap.

The measured profile losses in a turbine cascade are shown in Fig. 6.27. The design exit Mach number is 0.5. As the incidence is increased, the losses decrease up to a minimum value after which a sudden increase is observed. This is due to the formation of a shock wave at the exit. The shock losses and the associated increase in boundary layer thickness and subsequent mixing increase the profile losses. The losses in an impulse blade, where turning angles are higher, are higher than those of a reaction turbine. The change in outlet angle, with increase in the incidence angle, is small. Additional data by Ainley and Mathieson (1951) indicate that there is an optimum value of S/C, where the losses are minimum. The losses increase with an increase in the blade turning angle. Turbine industry personnel have developed numerous correlations for profile loss (see Ainley and Mathieson, 1951; Kacker and Okapuu, 1982; and Horlock, 1973a). Profile losses can be computed quite accurately, using the methods indicated in this section; and the use of correlation, except in very complex cases, is no longer justified. The procedure for calculating losses in a subsonic turbine cascade is similar to that described for the compressor in this section. Sharma et al. (1982) have demonstrated the validity of this procedure. Alternately, the Navier–Stokes code can be used to predict the viscous losses, including mixing, for the subsonic, transonic, or supersonic cascades (see Fig. 5.37).

6.4.1   Effect of Mach Number and Shock Losses in Compressors

As indicated in Section 3.4 the static pressure rise in a compressor increases with an increase in the relative Mach number for a fixed value of C_p. Furthermore, the C_p values also increase with an increase in the Mach number. Thus the streamwise pressure gradient increases with an increase in the Mach number. Hence the boundary layer momentum thickness (θ) and the losses also increase with the Mach number. This explains reasons for the trend observed (Fig. 2.20a). For a subsonic cascade at a given incidence the off-design losses increase with increase in the Mach number, or at some critical Mach inlet number (which results in shock waves inside the passage) the losses increase drastically (see Fig. 3.19).

Shock losses increase with an increase in Mach number, and at transonic speeds, the losses and flow field are very sensitive to trailing-edge and leading-edge geometry. This is evident from the tests carried out at the NASA Lewis Research Center in a compressor cascade (Boldman et al., 1983; Schmidt et al., 1984). The tests were carried out using supercritical airfoil (designed for shock-free operation) similar in shape to the one shown in Figure 5.46, with $\alpha_1 = 35.7°$, $\alpha_2 = 0.5°$, $M_1 = 0.75$, $(M_2)_{\text{design}} = 0.564$, Re $= 1.4 \times 10^6$, and $\sigma = 0.91$. The results are shown in Fig. 6.28. The original design was intended to accelerate the flow rapidly near the suction surface leading edge followed by a constant velocity up to about 40% chord (which was never achieved experimentally). The experimental data indicate the presence of a laminar separation bubble in the original design. Subsequent reattachment of the bubble and the resulting adverse pressure gradient resulted in turbulent boundary layer separation near 60% of chord. This resulted in lower pressure rise as well as increased losses. When the blading was modified very slightly near the leading edge (the blade thickness was changed by less than 1% of chord similar to that shown in Fig. 5.46) the airfoil behaved very well, resulting in moderate acceleration near the leading edge, followed by a peak at about 35% and a well-behaved boundary layer. The modified isentropic Mach number distribution is also shown in Fig. 6.28a and is similar to those shown in Figs. 2.12 and 5.46. The losses associated with laminar separation in the original design were high as shown in Fig. 6.28b. The effect of Mach number on losses was similar to that shown in Fig. 3.19. The losses increased (Boldman et al., 1983) only moderately up to about $M_1 = 0.7$ (which had no passage shocks). Beyond this Mach number, the passage shocks appeared and the losses increased rapidly. As the Mach number was increased further, the adverse pressure gradient associated with

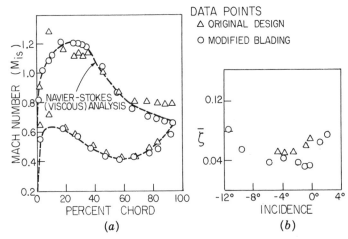

Figure 6.28 Effect of blade geometry on blade Mach number and losses for a compressor cascade. (a) Mach number distribution on the blade, (b) loss coefficient. (Copyright © Schmidt, 1984; reprinted with permission of AIAA).

the shocks resulted in boundary layer separation. For example, at $M_1 = 0.75$ the increase in the profile losses was only moderate compared to those at $M_1 = 0.3$, and the large gain in pressure rise more than offset the increased losses. The effect of trailing-edge shape on losses was also investigated by Schmidt et al. (1984) and Boldman et al. (1983). The losses with thin blunt trailing edges were found to be minimal, but substantial increases in losses were observed with thick, rounded, and blunt trailing edges.

An estimate of two-dimensional shock losses for a compressor should include the following:

1. The losses due to leading-edge bluntness with supersonic upstream Mach number.
2. The location of the passage shock can be determined from inviscid theories. If the shock strength is known, the losses due to passage shock can be determined.
3. The losses associated with the boundary growth and the shock–boundary-layer interaction are most difficult to estimate. The contribution is small for weak shocks.

Koch and Smith (1976) used empirical correlations for leading-edge shock losses. The passage shock loss model is based on the assumption that the loss is equivalent to the entropy rise of an oblique shock that reduces a representative passage inlet Mach number to unity. This applies for exit subsonic or sonic Mach numbers. The shock losses calculated by Koch and Smith for various high-speed fans are in reasonably good agreement (based on predicted efficiency) as shown in Fig. 6.29. The shock loss coefficient (ζ_{sh}, Eq.

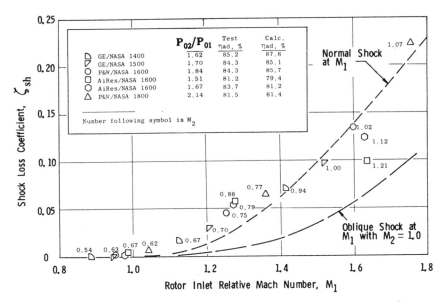

Figure 6.29 Shock loss coefficients calculated for several high-speed fan stages. (Copyright © Koch and Smith, 1976; reprinted with permission of ASME.)

6.4), including both the leading-edge bluntness and the passage shock losses, estimated by Koch and Smith (1976) is shown as symbols. There are three symbols plotted for each case listed; these correspond to rotor tip, midspan, and hub. The figure also shows the efficiency estimated from the loss coefficient, as well as the measured efficiency for each case. It should be remarked here that the overall efficiency estimate is based on Koch and Smith's complete loss model (including endwall loss described later, the profile loss, and the partspan shroud loss). Even though the agreement between the measured and the estimated efficiency is good, further improvement in predictive capability is necessary for some of the cases shown.

Freeman and Cumpsty (1989) used an approach similar to this and those outlined in Section 3.5.1.2 and Fig. 3.24 and derived the following expression for the shock loss, assuming that the incidence is zero [valid for a flat plate—a unique incidence condition discussed in Section 3.5.1.2 (Fig. 3.26)].

$$\zeta_{sh} = \frac{(\Delta P_o)_{\text{loss}}}{P_{o1} - p_1} = \left[\frac{(\Delta P_o)_{\text{loss}}}{P_{o1} - p_1}\right]_{\text{normal shock}}$$
$$+ [2.6 + 0.18(\alpha'_1 - 65°)]10^{-2}(\alpha_1 - \alpha'_1) \quad (6.21)$$

where α'_1 is the blade inlet angle. Equation 6.21 is valid for up to $5°$ incidence $(\alpha_1 - \alpha'_1)$. The loss is equal to that of a normal shock at the inlet relative Mach number. The authors showed good agreement with the measured data in a transonic compressor at various incidence angles and corrected speed.

It should be emphasized here that the shocks are almost invariably three-dimensional as shown in Fig. 4.1. Good agreement between predictions and data of Koch and Smith (1976) and Freeman and Cumpsty (1989) are in conflict with the three-dimensional data presented earlier (Fig. 6.11). Cumpsty (1989, p. 217) argues that their one-dimensional approach may be satisfactory due to the imposition of conservation of mass, momentum, and energy, and this is true if the radial fluxes are small.

A more satisfactory approach for predicting losses is based on the Navier–Stokes equations. If a proper turbulence model and sufficient grid points are used, it is possible to predict shock losses in a cascade. One such effort was by Kunz and Lakshminarayana (1992c) and described in Section 5.7.4.3. They predicted flow in a supersonic cascade (Schrieber, 1988), shown in Fig. 5.31. Their prediction of the downstream wake (inset in Fig. 6.30) and

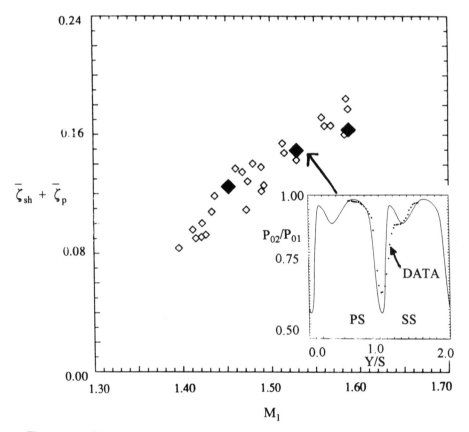

Figure 6.30 Total stagnation pressure loss coefficients at several cascade operating points for a supersonic cascade [calculated (solid symbols) and experimental (open symbols) values]. The predicted and measured wake profile at M_1 = 1.53 are shown in inset. Data due to Schreiber (1988). (Copyright © Kunz and Lakshminarayana, 1992c; reprinted with permission of AIAA.)

profile losses (both boundary layer and shock losses) at various Mach numbers are shown in Fig. 6.30. The losses are captured well with the k-ϵ turbulence model. This provides confidence in the ability of the Navier–Stokes code to predict losses.

6.4.2   Effect of Mach Number and Shock Losses in Turbines

The common blade configurations used in turbines are illustrated in Figs. 2.63–2.66. The operating boundaries are shown in Fig. 3.28. The procedures used for calculating 2D shock losses in a turbine are very similar to those for a compressor described earlier. Because the turbine flow is developing under a favorable pressure gradient, the boundary layers are usually thin and the shock–boundary-layer interaction and flow separation and the resulting complexity are not as severe as those in a compressor. Moreover, the passage shocks are located closer to the trailing edge. Hence the shocks and their effect are easier to predict. The profile loss in a transonic turbine consists of (1) the loss due to blade boundary layer and wakes (ζ_p), as explained earlier, and (2) the loss (ζ_{sh}) associated with the trailing-edge shock system.

The trailing-edge shock system (Figs. 2.63–2.66 and 3.28) and its interaction with the blade boundary layer and wake may overshadow the losses due to the blade boundary layer alone. The trailing-edge and mixing losses account for a large portion of the total losses. The loss associated with the trailing-edge shock, wake mixing, and separation bubble may account for nearly 30–70% of total losses measured, depending on blade geometry and Mach number. The trailing edge flow in a transonic turbine is extremely complex. The trailing edge losses are dependent on blade profile (including curvature) near the trailing edge and its thickness. Loss due to the trailing edge shock alone can be predicted adequately by the Euler code (Denton and Xu, 1990), but the mixing effects have to be resolved through either empirical correlations or Navier–Stokes analysis. Mee et al. (1992) carried out a systematic experimental program to identify contributions to the loss from various sources in a turbine blade row. The investigations were carried out in a blow-down wind tunnel. The cascade parameters were as follows: $\alpha_1 = 42.8°$, $M_1 = 0.31$, $T_{o1} = 280$–290 K, $T_u = 4.3\%$, $\alpha_2 = -68°$, $M_2 = 0.92$ (design), axial chord = 230.7 mm, and $S = 252.1$ mm. The contributions to the losses, shown in Fig. 6.31, were calculated from the following procedure:

1. The profile loss due to boundary layer alone ($\bar{\zeta}_p$). This was calculated from the wake–boundary-layer survey. The loss is estimated by subtracting the actual (integrated) kinetic energy from that which would be present in the absence of the boundary layer.

2. The shock losses (ζ_{sh}) are estimated from inviscid consideration. It should be remarked here that this expression includes only the shock

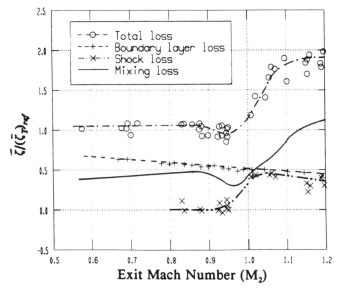

Figure 6.31 Individual components of loss in a turbine cascade. Exit Reynolds number $Re_{ex} = 1,000,000$. (Copyright © Mee et al., 1992; reprinted with permission of ASME.)

losses in the inviscid region and does not include the mixing or shock boundary layer interaction losses.

3. The mixing loss due to dissipation of the wake and losses due to shock–boundary-layer interaction (essentially all losses downstream) are called *mixing loss* and are given by $\bar{\zeta}_m = \bar{\zeta}_T - \bar{\zeta}_p - \bar{\zeta}_{sh}$.

The total losses $\bar{\zeta}_T$ are determined from a survey of the flow field downstream, from which values of $\bar{\zeta}_p$ and $\bar{\zeta}_{sh}$ are calculated. The data are shown in Fig. 6.31, where $(\bar{\zeta}_T)_{ref}$ is the loss at design condition (shock-free). It is evident that the boundary layer profile loss dominates at subsonic exit Mach numbers, and the downstream wake mixing is about 30% of the total loss. This is consistent with the observation made earlier that the wake mixing loss for a compressor cascade may represent about 20% of total losses. When the shock wave develops, the shock and the mixing losses dominate with nearly 100% increase in total losses at $M_{exit} = 1.2$. What is significant here are the total losses. One can argue about the budgeting procedure. The downstream flow is complex and includes shock–wake inter-action and wake dissipation regions. Part of the mixing losses can be attributed to either the shock (which brings about sudden increase in the thickness of the boundary layer) or the wake mixing loss. Mee et al. (1992) attribute most of the mixing losses to the wake decay downstream. The wake width measured downstream of the cascade increases rapidly with an increase in the Mach number (nearly 25% increase when M_2 is increased from 0.7 to 1.0), while the wake width at the trailing edge is found to be nearly identical. Mee's cascade data is in agreement with the predictions by Luo and Lakshminarayana (1995a).

At transonic exit flows, the shock system and its interaction with the blade boundary layer may overshadow the losses due to boundary layer alone. Martelli and Boretti (1985) developed a method for calculating the losses for a transonic turbine cascade. For subsonic turbine cascade, the method is similar to that described earlier for the compressor. For supersonic exit conditions, a shock structure with two oblique shocks is formed at the trailing edge, as shown in Fig. 6.32. Such shock configurations have been reported by many (e.g., AGARD 1989, article by Hourmouziadis). The pressure side leg of the trailing-edge shock (a) crosses the passage and is reflected back as shock (b). The location of the shock is very sensitive to local blade profile. Any small defect in manufacturing can result in a major change in shock structure and increased losses. The flow acceleration downstream of shock (b) results in a trailing-edge shock. The mixing of two dissimilar supersonic jets, along with the associated expansion, results in a reattachment shock (d) as shown in Fig. 6.32. A method of estimating (Martelli and Boretti, 1985) the

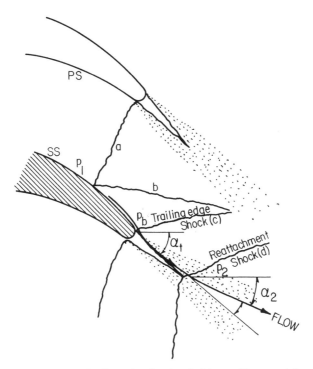

Figure 6.32 Trailing edge flow in a turbine and loss model.

losses due to this system is as follows:

1. The pressure distribution is calculated from an inviscid time-marching code (Section 5.7).
2. The trailing-edge flow (Fig. 6.32) is calculated from the procedure described in Section 3.5.1.4. This provides an initial estimate for α_2, p_2, and P_{o2}.
3. A boundary layer calculation (Section 5.5) is carried out to determine momentum and displacement thicknesses at the trailing edge.
4. A one-dimensional analysis is carried out inside a control volume (such as EFGH in Fig. 6.26) using equations similar to Eqs. 6.10–6.19. This includes all the losses, and solution of the equations provide updated values for P_{o2}, p_2, α_2, and the loss coefficient (ζ). The exit station 2 is located far downstream, where the flow is uniform.

Martelli and Boretti computed and compared with measured losses for various types of blading, two of which are plotted in Fig. 6.33. The inset shows the blade geometry and shock wave pattern at design conditions. The top plot is a typical steam turbine section, and the bottom part is a typical gas turbine blading with subsonic inlet and supersonic exit flow. The measured and predicted losses are in very good agreement for the gas turbine blade, and the trends are correctly predicted for the steam turbine profile. The losses are found to be very sensitive to cascade parameters (suction side blade curvature), and internal area increases in the supersonic part of the blade channel.

As indicated earlier, the losses increase as the flow is accelerated from low subsonic to high subsonic values (Fig. 6.31). This is caused by increased boundary layer growth. When the exit Mach number is increased further, shock wave formation occurs and this increases the shock losses and total losses substantially (see Fig. 6.31). In the case of Mee et al. (1992), the sudden increase occurs when M_2 is increased from 0.9 to 1.0, while for the data plotted in Fig. 6.33 the maximum losses occur beyond $M_2 = 1$. In the latter case, as the exit Mach number is increased further, the shock wave emanating from one vane swings downstream as the Mach number is increased, and this impinges further downstream on the suction surface. This results in lower losses as shown in Fig. 6.33. Similar distributions have been observed by Moustapha et al. (1993), who conclude that it is advantageous to operate the nozzle at higher than design pressure ratio, provided that the downstream rotor can be designed to accommodate the higher inlet Mach number.

The total losses (including shock losses) can also be computed from the Navier–Stokes code directly without employing any correlations and assumptions. The critical issue is again turbulence modeling. The computational results due to Knight and Choi (1989) for a supersonic exit turbine blade,

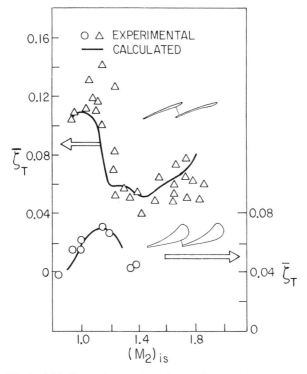

Figure 6.33 Effect of Mach number on total losses in a turbine cascade. (Martelli and Boretti, 1985; reprinted with permission of ASME.) $\bar{\zeta}_T$ is total profile loss coefficient.

shown in Fig. 5.37, indicate that the Navier–Stokes code is capable of capturing the total losses, including shock and trailing-edge losses.

6.5 THREE-DIMENSIONAL LOSSES AND CORRELATIONS

Three-dimensional losses in turbomachinery were described briefly in Section 6.3. These include:

1. *Profile Losses Due to Blade Boundary Layer and Wake.* These are due to the profile boundary layer as in two-dimensional flow, but involves three-dimensionality due to cross flows present (Fig. 5.1). In addition to the parameters described in Section 6.4, the losses depend on the following: radial distribution of variables including $\partial p / \partial r$; three-dimensional blade geometry including lean, dihedral, sweep, twist, aspect ratio, and meridional curvature; rotation parameter $2\Omega C / V$; and the flow coefficient ϕ. The profile losses depend on spanwise location.

2. *Shock Losses.* Losses arise due to leading-edge, passage, and trailing-edge shocks, which are almost invariably three-dimensional (both in strength and in geometry as shown in Figs. 4.1 and 6.11). The shock losses and associated boundary layer interaction are very difficult to estimate. In addition to the variables that control two-dimensional shock losses (Section 6.4.1 and 6.4.2), the three-dimensional blade geometry and radial variation of properties, including inlet Mach number, have a major influence on radial variation of shock losses.

3. *Secondary Flow and Endwall Losses.* The viscous and turbulence dissipation associated with endwall boundary layers (Section 5.5.4 and Fig. 6.6) introduce flow losses and decrement in efficiency. These are direct losses associated with endwall boundary layers. The endwall losses include both the secondary flow (or vortex) losses and the losses due to the wall boundary layer. It is difficult to separate these two sources of losses, and thus they are lumped together and called *endwall losses.* The endwall losses may also include losses due to the corner separation induced by adverse pressure gradient and secondary flow as shown in Fig. 6.7a for a compressor. The secondary flow losses depend (as indicated in Section 4.2.6) on ψ (or C_L), Re, M_1, δ_1, H, T_u, α_1, α_2, A, L_t, σ, and the blade camber shape. The boundary layer losses and the losses due to corner separation depend on ψ (or C_L), $\partial p / \partial s$, M_1, Re, δ_w, δ_b, A, H_w, H_b, $\Omega C / V$, σ, and other parameters.

4. *Clearance Losses.* The leakage flow and its eventual roll-up results in aerodynamic losses and decrement in efficiency. The leakage losses are associated with the loss of kinetic energy attributed to leakage velocity, the dissipation of leakage vortex, and the disturbance (perturbations) caused by the flow. These losses depend on blade static pressure distribution, ψ (or C_L), τ, σ, Re, M, C, A, α_1, α_2, δ_w, δ_b, tip shape, $t_{\max}$, T_u, $\Omega C / V$, and ϕ.

6.5.1 Three-Dimensional Profile Losses

The three-dimensional profile losses (in the absence of separation and shock waves) can be calculated from either a boundary layer code (Section 5.5) or Navier–Stokes code (Sections 5.6 and 5.7). Pouagare et al. (1985) used the momentum integral technique (Section 5.5.3) and the quasi-three-dimensional inviscid analysis (Fig. 4.17) to predict the boundary layer momentum thickness for the Penn State low-speed compressor. Assuming that the mixing losses are two-dimensional (neglecting the spanwise mixing), they estimated the losses from the predicted momentum thickness using Eq. 6.18. The predictions were nearly identical to those from the Navier–Stokes predictions shown in Fig. 5.43 (Suryavamshi and Lakshminarayana, 1992) for the region $R = 0.54$ and 0.94 (approximately 12% away from endwalls). In a rotor, one can expect large radial gradient ($\partial s / \partial r$) in losses from midspan to tip (increasing toward the tip), and this will have a substantial effect on the spanwise mixing and three-dimensionality in the flow. The losses measured at two downstream locations, one very close to trailing edge and one at 66% chord downstream of Penn State compressor at peak pressure ratio condition (Fig. 2.17), reveal some interesting phenomena (Prato and Lakshminarayana, 1993). High losses measured near the annulus wall and tip clearance region near the trailing edge decrease as the flow proceeds downstream; this is caused by flow redistribution toward the inner radius and flow mixing. The losses are predicted (Kunz and Lakshminarayana, 1992d) reasonably well except very near the endwalls. Most importantly, the radial redistribution of losses as the flow proceeds downstream is captured accurately. Thus the Navier–Stokes codes can capture the magnitude of losses as well as their variation downstream. It is evident that for subsonic, unseparated flows, either a boundary layer code or a Navier–Stokes code can provide a good estimate of the profile losses, and the approach based on correlations is no longer justified.

The radial distribution of total losses at the exit of the turbine vane of the energy-efficient engine (E^3), described earlier and shown in Fig. 2.64, is plotted (Leach, 1983) in Fig. 6.34. The profile loss is nearly constant (except for the cooled high-pressure turbine) in the middle 60% of the span, indicating that the three-dimensional effects are confined to hub and annulus walls. The coolant air increases the profile losses, and the platform cooling increases the endwall flow losses. This is covered in detail in Chapter 7. The three-dimensional effects in the turbine rotor of the above stage are large, and the radial gradient in losses is very high at the exit of this single-stage turbine (high-pressure E^3 engine turbine rotor blade profile shown in Fig. 2.64) tested by Leach (1983) and shown in Fig. 6.35. The absolute stagnation pressures at the exit of this stage are shown in Fig. 6.24 (see the corresponding text for details of this stage). The peak efficiency for this stage occurs at nearly 40% span, decreasing by almost 16 percentage points towards the tip. The tip clearance effect, combined with migration of the blade boundary layer toward the tip, results in an increase in losses and a decrease in efficiency. Likewise, the efficiency decreases as the hub wall is approached. The secondary flow and the hub wall boundary layer growth are responsible for this increase.

6.5.2   Three-Dimensional Shock Losses

The nature and structure of three-dimensional shocks for a compressor rotor are shown in Figs. 4.1 and 6.11. In both the transonic and supersonic rotors, the strength of the shock increases continuously toward the tip; and the chordwise location of the shock also varies, resulting in a three-dimensional warped shock sheet. This develops large spanwise gradient in loss distribution. Development of correlation or analytical techniques for predicting or calculating the losses due to profile boundary layer, shock losses, and

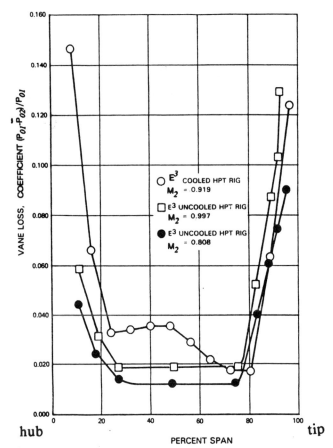

Figure 6.34 Vane loss profile showing the influence of cooling on performance for a turbine nozzle. $\bar{P}_{o2}$ is the passage averaged exit stagnation pressure. (Leach, 1983; NASA CR 168189, PWA-5594-243).

shock–boundary-layer interaction losses is extremely difficult. Furthermore, these three sources of losses, which together constitute total profile loss, cannot be separated or isolated for modeling purposes. Hence the practice in industry is to model or calculate these losses from the data acquired in a similar machine; examples of this are shown in Figs. 6.36 and 6.37. The data shown in these figures were acquired by Behlke et al. (1983) in a transonic compressor stage (IGV, rotor, and stator) at a corrected speed of 12,210 rpm and where $\dot{m} = 47.28$ kg/s, rotor total pressure ratio $= 1.845$, tip diameter $= 0.69$ m, $(\sigma_R)_{tip} = 1.26$, $A = 1.30$, $N_s = 27$, $N_R = 24$. The Mach number distribution at inlet to rotor (Fig. 6.36) indicates that the inlet flow is supersonic everywhere except for 15% span from the hub. The rotor exit Mach number is subsonic everywhere. The rotor total losses (Fig. 6.36) increase rapidly from midspan to tip, hence it is dominated by shock and shock–boundary-layer interaction losses in the outer half of the blade.

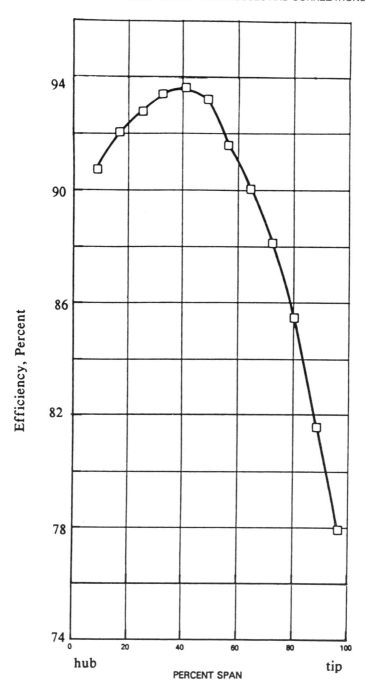

Figure 6.35 Spanwise variation of average efficiency at the exit of the rotor of a single-stage high-pressure turbine used in an E^3 engine. See Figs. 6.24 and 6.34 for related data (Leach, 1983).

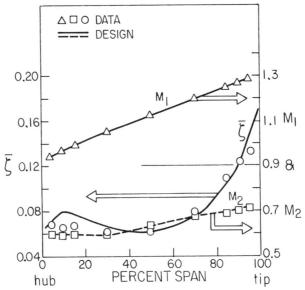

Figure 6.36 Total rotor loss coefficient and rotor inlet and exit Mach number distribution for a single-stage transonic axial flow compressor (Behlke et al., 1983).

Increased rotor losses near the tip are attributed to secondary flow losses and the annulus-wall boundary layer and the tip clearance effects. The losses in these high-speed rotors are substantially higher than those measured in low-speed rotors (Fig. 5.43). The design target (presumably based on correlation of losses in similar machines) is close to the measured values, except near the hub and tip. The stator downstream of the rotor (Fig. 6.37) had very high losses near the hub, up to about 20% of span. The turning angle in this region was also large. The stator profile losses from 20% to 80% span are small. The combined losses due to the rotor (which has the highest losses toward the tip, Fig. 6.36) and stator (which has the highest losses toward the hub, Fig. 6.37) result in large radial variations in stage efficiency as shown in Fig. 6.37. The measured stage efficiency peaks at 92%, between 20% and 50% span, decreasing to nearly 82% near the tip and 88% near the hub.

The spanwise distribution of pressure rise and efficiency for a transonic fan, measured by Norton (1989), shows a similar trend as indicated in Fig. 6.38. The single-stage fan blade had an aspect ratio of 2.8, hub/tip ratio of 0.316, a pressure ratio of 1.72, $\dot{m} = 90.99$ kg/s, and $N = 10,766$ rpm. Peak efficiency occurs from 30% to 50% span, decreasing gradually to 89% at 80% span, reducing rapidly to 68% at 100% span. The effect of shock losses is clearly evident in the loss distribution from 50% to 90% span. Because this is a shroudless fan blade, there is no tip clearance effect; only the secondary flow is present. This fan had a high overall efficiency, a 2% increase over the conventional shrouded configuration.

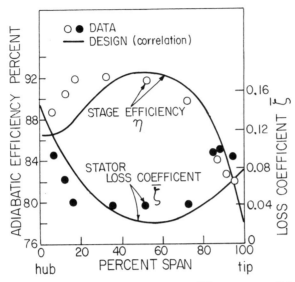

Figure 6.37 Radial distribution of stator loss coefficient ($\bar{\zeta}$) and stage adiabatic efficiency (η) for a single-stage transonic axial flow compressor (Behlke et al., 1983).

As indicated earlier, it is difficult to derive a correlation for the radial loss distribution because it is dependent on a large number of variables including blade element geometry and performance, blade speed and inlet Mach number, hub/tip ratio, aspect ratio, and, most importantly, strength and location of the shock sheet and radial distribution of geometrical and flow properties. Most industries use correlations based on geometrically similar (operational) turbomachinery for which performance data are available.

Navier–Stokes codes are now capable of predicting three-dimensional losses reasonably well. A good example of this is shown in Fig. 6.10. The results shown are for a low-aspect-ratio transonic fan (see Section 6.2.1.2 for details of the fan). The data clearly reveal the influence of shock losses from the tip to about 20% of span from the tip. The measured and predicted adiabatic efficiency of rotors are in reasonable agreement with the data. The endwall losses, near the tip in this case, include the tip leakage, secondary flow, and profile and shock losses. The shock losses (as stated by Wadia and Law, 1993) represent nearly 50% of the losses near the midspan (the rest is profile loss) and nearly 20% of the total losses near the tip.

6.5.3 Secondary Flow and Endwall Losses in Axial Turbomachinery

As explained in Section 4.2.6, the radial gradient in stagnation pressure develop secondary flow, and in many instances this flow rolls up into vortices. The secondary flow losses include some of the kinetic energy in secondary velocities as well as losses associated with the formation, development, diffusion, and dissipation of these vortices. The inviscid theories available for

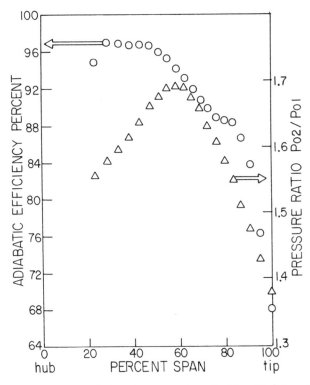

Figure 6.38 Adiabatic efficiency (η_c) and pressure rise ratio (P_{02}/P_{01}) for a transonic axial flow fan at 100% design speed, at peak efficiency operating condition (Norton et al., 1989; NASA CR 182220).

the prediction of secondary flows, including flow deviation angle, is described in Section 4.2.6. An attempt will be made to derive an expression for the functional dependency of these losses. Such an approach is valid only for axial flow turbomachinery. The secondary flow in radial and mixed-flow turbomachinery arises due to several causes as explained in Section 4.2.6. These introduce complex interaction, making it difficult to develop loss correlations. An attempt is made here to provide a theoretical basis for the loss correlations developed/used for estimating secondary flow losses. The kinetic energy in the secondary flow can be calculated using the procedure proposed by Hawthorne (1955b) and is given by (referring to Fig. 4.19)

$$\text{SKE} = \frac{\text{Kinetic energy in secondary flow}}{\text{Kinetic energy in inlet flow}} = \frac{\dfrac{1}{2}\displaystyle\int_{-S'/2}^{+S'/2}\int_{0}^{h/2}\rho(v^2 + w^2)\,db\,dn}{\dfrac{1}{2}\rho_1 V_1^2 \dfrac{h}{2} S \cos\alpha_1}$$

$$(6.22)$$

where V_1 is the inlet velocity, v and w are secondary velocities and $S' = S \cos \alpha_2$. The inlet kinetic energy is based on the area from hub to midspan or from midspan to tip. Not all the kinetic energy is lost; nevertheless, the loss is proportional to the above ratio. Hawthorne (1955b) has provided a solution to Eq. 4.131 [see Horlock and Lakshminarayana (1973) for these expressions]. Based on this analysis, Eq. 6.22 can be approximated by (for incompressible flow)

$$\bar{\zeta}_s \sim SKE \sim \frac{\epsilon^2 S^2 \cos^2 \alpha_2}{\delta_1 h/2} \frac{\cos \alpha_2}{\cos \alpha_1} f(\delta/S')$$

Where δ_1 is the inlet wall boundary layer thickness. Furthermore, for moderate turning (ϵ), Eq. 3.18 can be simplified to

$$C_L = 2\frac{S}{C}(\epsilon)\cos \alpha_m$$

Hence

$$\bar{\zeta}_s \sim SKE \sim \frac{C_L^2 \cos^3 \alpha_2}{\cos^2 \alpha_m} \frac{C^2}{\delta_1 h/2} \frac{1}{\cos \alpha_1} f\left(\frac{\delta}{S'}\right) \qquad (6.23)$$

Hence the passage-averaged (averaged over the entire passage, hub-to-tip and blade-to-blade) secondary flow losses depend on blade loading (C_L, ϵ), flow angles (α_1, α_2, α_m), inlet boundary layer thickness (δ_1/S' or δ_1/C), and the blade geometry (C, S, h). The additional parameters for the more generalized case include the shape factor (H), Mach number (M), rotation parameter $\Omega C/V$, and Reynolds number. Thus the secondary flow theory (Section 4.2.6) has provided valuable information on the nature of losses and their dependency on flow and blade parameters. This correlation does not include the roll-up of secondary flow to vortices and their eventual dissipation.

The empirical correlations proposed by various authors is a subset of this correlation (see reviews by Horlock and Lakshminarayana, 1973; Sitaram and Lakshminarayana, 1983). For example, Dunham (1970), based on a large amount of cascade data, proposed the following correlation for turbine cascades:

$$\bar{\zeta}_s = \left[0.0055 + 0.078\sqrt{\delta_1/C}\right]C_L^2 \frac{\cos^3 \alpha_2}{\cos^3 \alpha_m} \frac{C}{h}\left(\frac{C}{S}\right)^2 \frac{1}{\cos \alpha_1'} \qquad (6.24)$$

It should be emphasized here that $\bar{\zeta}_s$ is the average secondary flow loss coefficient for the entire passage, hub to midspan or tip to midspan. For a rotor, the absolute angles are replaced by relative angles. This correlation is valid for small-turning, incompressible flow through axial turbomachinery and does not include the rotation (e.g., rotating hubwall boundary layer or relative motion in the tip clearance region) and compressibility effects.

In most instances the "endwall" loss and secondary flow losses cannot be separated. The endwall loss includes all losses in the endwall region. In the region without tip clearance, it includes losses due to corner separation, horseshoe vortex, secondary flow and vortex, annulus-wall boundary layer, as well as the wake mixing losses. The losses in one of the endwalls (e.g., hub for a rotor) may account for approximately 20–40% of the losses depending on inlet conditions, type of turbomachinery, blade loading, Mach number, and Reynolds number. How much of this is due to secondary flow is an unresolved issue. For a large-flow-turning turbine passage, most of the "endwall" loss (in the absence of tip clearance) is due to secondary flow vortex and horseshoe vortex, while in a compressor the annulus-wall boundary layer growth and the corner flow separation account, in the author's opinion, for majority of the losses.

6.5.3.1 Endwall Losses in Axial Turbines

Most of the secondary and endwall loss data available is either for a cascade or for a nozzle or stator vane. The loss distribution differs considerably between turbine and compressor cascades and hence are dealt with in separation sections. The loss mechanism in a turbine cascade is summarized by Yamamoto and Yamasi (1985) (also see Figs. 4.25, 5.18, 5.25, and 5.28). The secondary flow development occurs downstream of the leading edge; and the wall boundary layer fluid is swept by secondary flow toward the suction side, resulting in mixing between the wall and the blade boundary layers in the corner. The loss core resulting from this process is then transported away from blade and wall surfaces due to convection by secondary flow. Hence the center of the loss core is located away from the walls. This is evident from Fig. 5.25. The loss core is located away from the wall and the wake, downstream of the blade passage. The loss contours and secondary velocity vectors for a large-turning cascade, measured by Yamamoto (1989), indicate the details of secondary flow roll-up and transport mechanism (Fig. 4.25). Similar measurements have been reported by Hodson and Dominy (1987), Langston et al. (1977), and Gregory-Smith et al. (1988). The loss data acquired by Kooper et al. (1981b) in a turbine nozzle, shown in Fig. 5.37, also indicates the presence of loss peaks away from the endwalls, and these are captured qualitatively by the Navier–Stokes code.

In a turbine cascade, the presence of large flow-turning, horseshoe vortex, and secondary flow results in an extremely complex flow field and loss mechanisms. The nature of the endwall flow field in a turbine cascade is shown in Figs. 4.25, 5.25, and 6.20. Thus the interaction effects result in the presence of more than one loss core. The investigations by Gregory-Smith et al. (1988) indicate that the total losses measured near the endwall represent only a fraction of secondary flow kinetic energy. Thus the secondary flow theories based on small flow-turning is not valid for such flows. The secondary flow associated with the high flow-turning induces large passage (toward the suction surface) and spanwise flows, resulting in sweeping of the blade and the endwall boundary layers toward the corner, eventual roll-up into a vortex, and its interaction with the horseshoe vortex. Hence the discrepancy between the kinetic energy in the secondary flow and losses should be viewed as the inability of the linear inviscid theories to predict the viscous and real fluid effects and its interaction with the horseshoe vortex. On the other hand, data in an annular nozzle, with modern blading, indicate that the secondary flows and losses are not as severe as reported in cascades (see Zaccaria and Lakshminarayana, 1995a).

Sharma and Butler (1987) have developed a method for estimating the endwall losses for an axial flow turbine blading. Their analysis of cascade data has demonstrated that the inlet boundary layer loss is convected through the passage without causing additional loss, and this can be distinguished from the passage loss. The passage loss is split into two-dimensional profile loss, which can be calculated using methods described earlier (Section 6.4) and the endwall secondary flow loss. The following semiempirical expression, developed by Sharma and Butler, seems to provide good agreement with data by Marchal and Sieverding (1977), Gregory-Smith and Graves (1983), Sharma et al. (1982), Langston et al. (1977), Graziani et al. (1980), and Kooper et al. (1981a).

$$\bar{\zeta} = \bar{\zeta}_p + \bar{\zeta}_{ew}$$

$$= \bar{\zeta}_p \left[1 + \left(1 + \frac{4\epsilon}{\sqrt{\rho_2 V_2 / \rho_1 V_1}} \right) \frac{S \cos \alpha_2 - t_{TE}}{h} \right] \qquad (6.25)$$

where ϵ is turning angle in radians. It should be noted here that $\bar{\zeta}$ is the total losses (integrated over the passage and span), including blade profile ($\bar{\zeta}_p$), endwall ($\bar{\zeta}_{ew}$), and inlet losses. The endwall loss was derived from momentum integral analysis of the endwall boundary, similar to those described in Section 5.5.4. Allowance was made for flow acceleration (V_2/V_1), blade boundary layer and secondary flow. There have been several attempts (e.g., Wang et al., 1992) to study the effect of lean and endwall contouring on turbine endwall losses. Properly designed lean and endwall contouring could reduce the endwall losses.

6.5.3.2 *Endwall Losses in Axial Flow Compressors*

In axial flow compressors, the secondary flows are milder than those in turbines, and in most instances the inviscid theories can provide a good estimate of the change in the turning angle and secondary flow field. The losses can be estimated from empirical correlations presented earlier (e.g., Eq. 6.24). The kinetic energy in the secondary flow is usually overestimated by the inviscid theories. Lakshminarayana and Horlock (1967a) allowed for viscous effects and the change in secondary vorticity due to Bernoulli surface rotation and improved prediction of kinetic energy in secondary flow and showed good

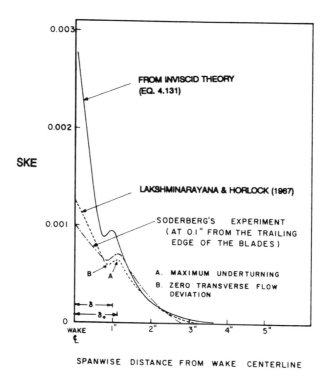

SPANWISE DISTANCE FROM WAKE CENTERLINE

Figure 6.39 Comparison between experimental (Soderberg, 1958) and predicted spanwise distribution of kinetic energy in secondary flow in a compressor cascade. SKE-kinetic energy in Secondary flow, Eq. 6.22. (Copyright © Lakshminarayana and Horlock, 1967a; reprinted with permission of ASME.)

comparison with Soderberg's (1958) data in a low-speed compressor cascade (Fig. 6.39). The secondary flow in this case was simulated with a wake from a perforated plate placed upstream at midspan. Kinetic energy in secondary flow represents only a fraction of total losses in the endwall regions.

The indirect effect of the secondary flow is to transport the wall boundary layer toward the corner formed by the blade suction surface and the wall and initiate "wall stall" and interact with the "blade stall" region in the corner. In some cases, with large secondary flow and high blade loading, the wall stall is initiated first, followed by "blade stall." The merging of these two stall regions makes it difficult to distinguish the source of losses. If the secondary flow is mild, the blade stall may occur first and accelerate the wall stall. The combination of these two results in what is known as "corner stall." Details of the physical mechanism and physical nature of these flows can be found in Horlock et al. (1966). The secondary flow migration to the corner would result in mixing and generate additional losses. These will all be lumped under "endwall" loss. It should be emphasized here that the designers in recent years have made successful attempts to prevent "corner stall" and thus reduce the endwall losses. In the absence of corner stall, the wall boundary layer and secondary flow losses dominate. In the presence of corner stall, the secondary flow is overshadowed by corner stall. The tests carried out by Schulz et al. (1990) indicate complex mechanisms in this region (Fig. 6.7). At low incidence, with mild pressure gradients, the losses due to the wall boundary layer and the secondary flow were found to dominate. At high incidence angles, the losses due to corner stall were found to overshadow all other losses. The secondary flow is swept into the corner region, and the peak passage-averaged losses increased nearly 100% when the inlet angle is increased from 40° to 49°. Furthermore, the loss core moved away from the wall.

The leakage flow is beneficial in reducing the effects of corner stall. The data from Lakshminarayana and Horlock (1967b) acquired in a compressor cascade (British 10C430C50 blading with $C = 0.152$ m, $\lambda = 36°$, $A = 4.8$, $S/C = 1.0$, $\alpha_1 = 51°$, $\alpha_2 = 31°$, $\mathrm{Re} = 2 \times 10^5$) at $\tau/h = 0$ and 0.016 indicate beneficial effects as shown in Fig. 6.40. The overall losses are reduced by the leakage flow, and the corner stall region is swept away from the endwall. The overall (area averaged) endwall losses (for blading with an aspect ratio of 4.83) was nearly 54% of the two-dimensional losses, and the presence of leakage flow reduced this to 12.7%. For low-aspect-ratio blades, endwall losses due to corner separation may represent a major portion of the total losses.

In a rotor, the secondary flow is modified by the rotation. The blade boundary layers are thin at the root of the rotor (due to radial transport of the blade boundary layers). The hub-wall boundary layer is swept toward the suction side and transported radially outward (or mixed in the downstream direction) (see Murthy and Lakshminarayana, 1987). The data by Lakshminarayana et al. (1986), plotted in Fig. 6.8, indicate that the secondary flow is not as severe as in a cascade and that the endwall losses are not very high. Furthermore, the data do not reveal the presence of separated flow. It is very likely that the endwall losses in the hub region of a rotor will be lower than those measured in a cascade. The rotation has a beneficial effect in minimizing the blade and hub wall boundary layer growth, corner separation, and secondary flow.

Koch and Smith (1976) have developed an endwall loss correlation scheme for axial flow compressors. Their approach was to relate the efficiency loss to two of the dominant endwall boundary layer parameters, the displacement thickness (δ^*) and the tangential force defect thickness (F_θ), similar to Eq. 5.35. The displacement thickness is a measure of the amount that the mass flow is reduced by the presence of the endwall boundary layer from what it would be if the free-stream flow profiles were extended to the wall. Similarly, the tangential force defect thickness represents the amount that the tangential component of blade force is reduced from its free-stream value by the

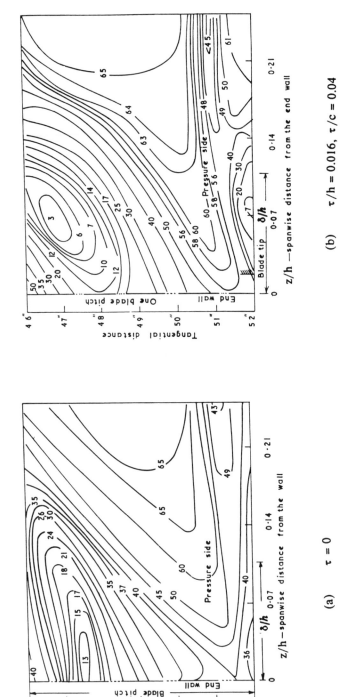

Figure 6.40 Contours of dynamic pressure at outlet $(\frac{1}{2}\rho V_2^2)$ near the endwall for a compressor cascade; (a) zero tip clearance (b) $\tau/h = 0.016$. The numbers on the contours denote the dynamic pressure at outlet as a percentage of reference dynamic pressure at inlet. (Copyright © Lakshminarayana and Horlock, 1967b; R & M No. 3483.)

(a) $\tau = 0$

(b) $\tau/h = 0.016,\ \tau/c = 0.04$

presence of the boundary layer. The tangential force defect F_θ is defined as (similar to Eq. 5.29)

$$F_\theta = \int_0^\delta \left(1 - \frac{f_\theta}{(f_\theta)_e}\right) dr \qquad (6.26)$$

The definition of f_θ is the same as Eq. 5.30.

For incompressible flow in high-radius-ratio multistage compressors, the following expression (similar to Eq. 5.35) provides a good estimate of the endwall loss or inefficiency for axial flow compressors (Koch and Smith, 1976):

$$\eta = \bar{\eta} \frac{1 - (\delta_h^* + \delta_t^*)/h}{1 - (F_{\theta h} + F_{\theta t})/h} \qquad (6.27)$$

where $\bar{\eta}$ is the efficiency in the absence of endwall boundary layers and h and t refer to values near the hub and tip, respectively. Empirical correlations can be used for the stress and other terms in Eq. 5.30, or, alternately, data for the force defect can be acquired directly and correlated. Koch and Smith (1976) deduced values of F_θ and δ^* from flow profiles measured in several low-speed multistage compressors and derived correlations for F_θ and δ^* and obtained good correlations for the displacement thicknesses, but scatter occurred in the data on tangential force defect thickness. However, by assuming that F_θ was a fixed fraction of δ^* and by using a representative profile loss formulation, they were able to construct fairly accurate pressure rise and efficiency characteristic curves for a series of stages with aspect ratios varying from 2 to 5 and with tip clearances varying from 0.8% to 3.6% of the annulus height. The displacement and force defect thicknesses can be calculated from the method described in Section 5.5.4. Alternatively, the Navier–Stokes procedure can be used to predict endwall flow losses and efficiency directly. This attempt is limited to rotor-alone or single-stage configurations at the present time (see Fig. 6.10). Multistage configuration involves time-accurate solution (which is very prohibitive) to capture inflow for each stage and the cumulative effects accurately.

There have been many attempts to reduce secondary flow and losses in the endwall region of an axial flow compressor. Some of the early attempts are described in Lakshminarayana and Horlock (1963). Recent attempts are due to Wisler (1985) and Behlke (1986). Referring to Fig. 2.10, the secondary flow in the endwall region of rotor 1 tends to decrease β_2 (overturning), which increases α_2 and incidence to stator 1. Likewise, the annulus wall boundary layer at inlet to rotor 1 increases β_1 at the inlet. It is desirable to twist the blade at the endwall to decrease the effects of the annulus wall boundary

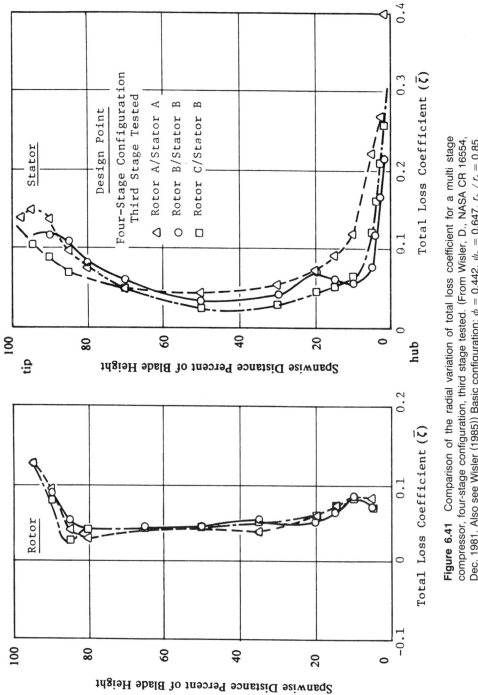

Figure 6.41 Comparison of the radial variation of total loss coefficient for a multi stage compressor, four-stage configuration, third stage tested. (From Wisler, D., NASA CR 16554, Dec. 1981. Also see Wisler (1985)) Basic configuration; $\phi = 0.442$, $\psi_p = 0.647$, $r_h / r_t = 0.85$, $N_R = 54$, $N_S = 74$, Low speed model of Stage 7 of E^3 engine. Also see Fig. 2.29 for improvement in profile losses for this configuration.

layer and the secondary flow. The flow at the exit of the rotor and stator in the endwall region has a higher turning, and it is desirable to twist the blade to accommodate this higher turning. An increase in the stagger angle and turning would result in lower losses. The early attempts to do this were conflicting, because the increased camber was likely to increase chances of flow separation. Very few investigators have tried to adjust the inlet incidence and stagger the angle. Wisler (1985) seem to have succeeded in modifying both the inlet incidence and the camber, thus achieving about a 10% reduction in losses. The results for an axial flow compressor are shown in Fig. 6.41. The rotor A/stator A configuration was a conventional design without any endwall modification. Rotor B/stator B and rotor C/stator B had "endwall bend" with increased stagger angle and turning. The modifications to tip sections are also shown. Rotors B and C had only the tip section modified, and stators C and B had both the hub and tip sections with "endwall bends." It is clear that appreciable reduction in endwall losses was achieved through modification of inlet incidence (through stagger angle) and flow-turning (through camber angle).

6.5.4 Tip Leakage Flow Losses

The nature of tip leakage flow and its effect on the flow field (velocity perturbations, flow angles, and blade pressure) were covered in Sections 4.2.7 and 6.2. This section is concerned with the aerodynamic losses and the inefficiency resulting from it. The leakage flow losses can be classified as follows:

1. *Direct Flow Through the Clearance Space.* The flow in the clearance space undergoes no change in angular momentum, and thus it is not available for doing work. This loss can be estimated from a knowledge of inlet boundary layer thickness and the clearance height. The fraction of mass flow through the gap does not participate in the pressure rise or drop.

2. *The Indirect Effect.* This is due to leakage flow caused by blade unloading; its interaction with main flow causes flow perturbation, along with vortex formation and its dissipation, resulting in mixing losses. This contribution to the leakage flow losses is the dominant one.

The leakage flow or tip clearance loss accounts for nearly 20–40% of the total losses, depending on the machine. An estimate of these losses can be made using Eqs. 4.141–4.148 and the approach outlined by Lakshminarayana (1970). This approach is very similar to the estimate of the secondary flow loss described earlier (Eq. 6.23). The losses should be proportional to kinetic energy in velocity perturbations caused by the leakage flow and the leakage vortex. An attempt is made here to provide a theoretical basis for the loss correlations developed/used for estimating losses due to leakage flow. For incompressible flow with constant axial velocity, this is given by (refer to Fig.

4.28 for notations)

$$\bar{\zeta}_L \sim \frac{\text{Kinetic energy in perturbations caused by leakage flow/vortex}}{\text{Inlet kinetic energy}}$$

$$\sim \frac{\int_0^h \int_0^S (w_L^2 + v_L^2)\, dy\, dz}{V_1^2 Sh} \tag{6.28}$$

where $\bar{\zeta}_L$ is the average loss coefficient due to tip clearance; v_L and w_L are velocity perturbations in y and z directions, respectively, caused by the leakage vortex. Based on Eqs. 4.147 and 4.148, Eq. 6.28 can be written as

$$\bar{\zeta}_L \sim \frac{(1 - K)^2 (C_L^2)\left(\dfrac{C}{S}\right)^2 SLV_m^2}{V_1^2 Sh} f\left(\frac{a + \tau}{S}\right) \tag{6.29}$$

Where L is the farthest spanwise distance at which influence of leakage flow is present. This is used to approximate the integration in Eq. 6.28. The cascade data due to Lakshminarayana and Horlock (1967b) and loss measurements by other investigators (see Section 4.2.7.3) indicate that $(1 - K)^2 Lf[(a + \tau)/S]$ is proportional to τ. Hence, Eq. 6.29 can be expressed as

$$\bar{\zeta}_L \sim \frac{C_L^2}{A} \frac{C}{S} \frac{\tau}{S} \frac{\cos^2\beta_1}{\cos^2\beta_m} \tag{6.30}$$

Hence the total kinetic energy in the leakage flow varies directly with C_L^2, tip clearance height, solidity, and the flow angles. The effect of the aspect ratio is obvious: the larger the aspect ratio, the smaller the fraction of total energy. This was recognized as early as 1951 by Ainley and Mathieson (1951), who proposed the following correlation for the axial flow turbines:

$$\zeta = 0.5 \frac{C_L^2}{A}\left(\frac{\tau}{S}\right) \tag{6.31}$$

Additional losses arise due to entrainment and mixing between the blade boundary layers as they are transported toward the tip due to low pressure caused by the tip vortex in the tip clearance region, and the endwall flow. Lakshminarayana (1970) made an estimate of these losses as

$$\bar{\zeta}_W \sim \frac{\delta_S^* + \delta_P^*}{S} \frac{1}{A} \frac{C_L^{3/2}(\tau/S)^{3/2}V_m^3}{V_2V_1^2} \tag{6.32}$$

where δ_S^* and δ_P^* are the blade boundary layer displacement thickness on the suction and pressure surfaces, respectively, near the tip and $\bar{\zeta}_w$ is the average indirect losses associated with entrainment and dissipation process.

The total losses due to clearance is given by Eqs. 6.30 and 6.32. Furthermore,

$$C_L = \frac{\psi_p}{\phi} \frac{S}{C} \cos \beta_m \quad \text{and} \quad \Delta\eta = \frac{(\Delta P_o)_{\text{loss}}}{(\Delta P_o)_{\text{isentropic}}} = \frac{2(\Delta P_o)_{\text{loss}}}{\rho U^2 \psi_p} \tag{6.33}$$

where $(\Delta P_o)_{\text{Loss}}$ is the stagnation pressure loss due to clearance flow (includes both the direct and indirect losses given by Eqs. 6.30 and 6.32), ψ_p = pressure rise or drop coefficient. Hence, the decrement in efficiency due to clearance can be written as

$$\Delta\eta = \frac{0.7\psi_p\tau/h}{\cos \beta_m}\left[1 + 10\sqrt{\frac{\phi}{\psi_p} \frac{\tau/C}{\cos \beta_m}}\right] \tag{6.34}$$

The constants in the above equation are evaluated from the experimental data. It is shown by Lakshminarayana (1970) that the decrement in efficiency due to clearance estimated by Eq. 6.34 agrees well with data from a variety of turbomachines (fans, compressors, pumps, and turbines). Some typical examples of the estimated and measured loss in efficiency are shown in Figs. 6.42 and 6.43. The corresponding flow parameters are listed in Table 6.1. The agreement is reasonably good for all types of turbomachinery. The trend is predicted well for a high-speed transonic fan rotor. The leakage flow also influences the pressure rise, efficiency, and stall margin of compressors. The data in a transonic fan rotor from Moore (1982) ($P_{o2}/P_{o1} = 1.75$, $m = 29.5$ kg/sec, $U_t = 423$ m/s, $r_h = 0.5$, $r_t = 25$ cm, C/S(tip) $= 1.5$, $N_R = 49$, $A = 2.4$, $N_S = 54$) clearly reveal this effect.

This approach has certain limitations as shown by Waterman (1986). The loss correlation given by Eq. 6.34 overpredicts the efficiency decrement for small turbines (which usually have very large clearance/height). Waterman evaluated data from nine sets of small turbines and concluded that the simple correlation due to Ainley (Eq. 6.31) provides the best correlation. For the turbine tested by Haas and Kofskey (1979), $T_{o1} = 288$K, $N = 31,400$ rpm, $P_{o1}/P_{o2} = 2.77$, $\psi_p = 3.34$, $\phi = 0.366$(tip), $r_t = 6.383$ cm, $A = 1$, and r_h/r_t

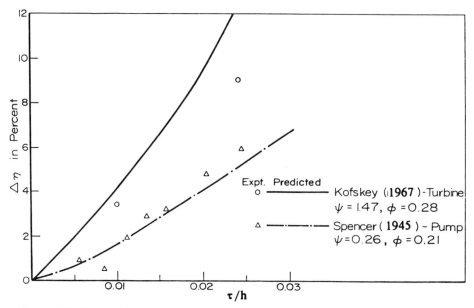

Figure 6.42 Decrease in stage efficiency ($\Delta\eta$) with clearance for axial flow pump and turbine (Copyright © Lakshminarayana, 1970; reprinted with permission of ASME.)

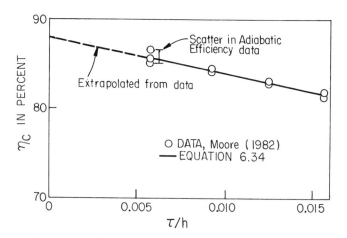

Figure 6.43   Tip clearance effects for a transonic fan rotor.

$= 0.835$, the measured value of $\Delta\eta/(\tau/h)$ was 1.8. The corresponding values from Eq. 6.34 and 6.31 are 3.4 and 1.8, respectively. Similar good agreement between Ainley's correlation and small turbine data indicate that the three-dimensional effects (Eq. 6.32) introduced into loss correlation by Lakshminarayana may not be important in very small turbomachinery (e.g., auxiliary power units). The second term in Eq. 6.34 represents the spanwise

Table 6.1. Characteristics of stages

Figure	Authors	Type of Machinery	ϕ	Ψ_p	$\cos \beta_m$	A
6.42	Kofskey and Nusbaum (1967)	Two-stage axial turbine	0.28	1.47	0.42	1
6.42	Spencer (1945)	Axial flow pump	0.21	0.26	0.246	0.47
6.43	Moore (1982)	Axial flow transonic fan	0.47	0.6	0.46	2.4
6.44	Senoo and Ishida (1986)	Radial (R) impeller	Varies (Fig. 6.44)	1.76	NA	NA
6.44	Senoo and Ishida (1986)	Backward vane (B) Impeller	Varies (Fig. 6.44)	1.07– 1.35	NA	NA

mixing effects in the tip region. Neglecting this term will result in an estimated value of $\Delta\eta/(\tau/h) = 2.71$ for Haas and Koskey turbine, which is closer to the experimental data than that given by Eq. 6.34.

The effect of tip clearance for a centrifugal blower has been investigated by Senoo and Ishida (1986, 1987). They have also developed a correlation for the loss in efficiency. The contribution to the losses is assumed to be arising from (1) leakage flow through the gap, (2) the stagnation region due to interaction of the secondary flow and the leakage flow, and (3) the effect of rotation near the shroud. Their prediction based on this hypothesis is compared with their data in Fig. 6.44. The impeller aerodynamic parameters are tabulated in Table 6.1. The loss in efficiency due to clearance varies almost linearly with τ/h_2 as in all other types of turbomachinery. The blade loading has a major influence on losses. It should be noted here that τ/h_2 is the ratio of clearance height to the blade height at the exit. This value is about 50% of τ/h_2 for the impeller, if it is based on the blade height at the throat.

Basson and Lakshminarayana (1993) utilized a modified pressure-based method (Section 5.7.1), a $k-\epsilon$ model, an embedded H mesh to capture the leakage flow and losses accurately. The predicted loss distribution for Bindon's (1987) cascade (see Fig. 5.24 for outlet flow angle distributions for this cascade) is shown in Fig. 6.45. Both the local and the averaged losses are predicted accurately except in the trailing-edge region. This clearly demonstrates that with properly calibrated/validated codes, sufficient grid resolution (at least 10–15 grid points within the clearance region, and similar fine grid up to about $Z = 5\tau$), and good turbulence models, the clearance and endwall losses can be predicted reasonably well.

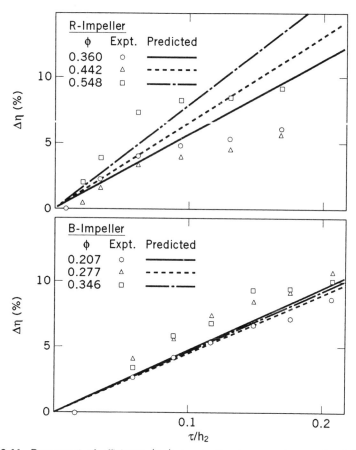

Figure 6.44 Decrement of efficiency ($\Delta\eta$) due to tip clearance in centrifugal impellers. (Copyright © Senoo and Ishida, 1986; reprinted with permission of ASME.)

6.5.5 Radial Distribution of Overall Losses and Estimation of Efficiency

As explained earlier, the nature and magnitude of radial distribution of total losses depends on the type of machinery and geometry. Hence, the approach taken in the industry has been to develop loss models for each class of machinery. The axial compressor has received more attention. The nature of the radial distribution of overall losses and/or efficiency is shown for a low-speed compressor in Fig. 5.43, for a transonic fan in Fig. 6.38, for transonic compressors in Figs. 6.10, 6.36, and 6.37, and for a high-speed turbine in Figs. 6.34 and 6.35.

There have been many attempts to derive correlations for the radial variation of losses for particular types of machines (e.g., Figs. 6.36 and 6.37).

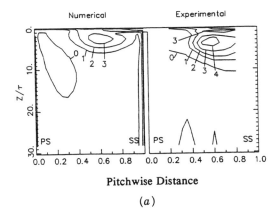

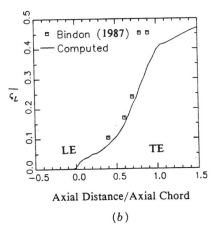

Figure 6.45 Measured and predicted leakage flow losses for Bindon's (1987) turbine cascade at the trailing edge. Z is the spanwise distance from the wall. (a) Loss distribution (ζ_L) in the passage. (b) Axial development of mass-averaged loss coefficient ($\bar{\zeta}_L$). $\zeta_L = 2(P_{01} - P_{02}) / \rho V_1^2$. (Bisson and Lakshminarayana, 1993; reprinted with permission of ASME)

In many instances these are based on sound physical principles, and in other instances, a trial-and-error approach is used to derive loss correlations that fit large amounts of data. Lakshminarayana (1974), while developing a correlation for a rocket pump inducer (tested in both air and water), argued that the flow in an inducer with long and narrow passages (Fig. 1.7) behaves like fully developed flow in a channel. Hence, the loss in relative stagnation pressure is given by

$$\frac{(\Delta P_o)_{\text{loss}}}{\rho} = \frac{\lambda}{2R_N^{0.25}} \frac{L}{D_h} \overline{W}^2$$

where λ is the Blausius friction factor ($= 0.316$ for stationary channels). $R_N = \overline{W} D_h / \nu$ is the Reynolds number based on the hydraulic mean diameter (D_h), L is the length of the channel, $\overline{W}$ is the mean or bulk velocity. He further argued that the losses depend inversely on the flow coefficient (caused mainly by strong rotation effects, because $\phi = V_z / \Omega r$ is very small for inducers) and that rotation causes radial outward flow inside rotor blade boundary layers as explained in Section 5.5.1. In addition, the losses should depend on hub/tip ratio, because this controls the radial distribution. He was thus able to collapse all the inducer loss data available (similar to data shown in Fig. 5.22) and develop a correlation for the passage averaged total loss coefficient:

$$\bar{\zeta}(r) = \frac{2(\Delta P_o)_{\text{loss}}}{\rho U_t^2} = \lambda_R \frac{R_{ht}}{\phi} \frac{1}{R_N^{1/4}} \frac{L}{D_h} \left(\frac{\overline{W}}{U_t} \right)^2 \tag{6.35}$$

where λ_R is the friction factor derived from experimental data and is a function of the radius and R_{ht} is the hub/tip radius ratio. The losses calculated from Eq. 6.35 are in good agreement with most of the rocket pump inducer data available.

Similar attempts have been made by Roberts et al. (1988) to derive correlations for axial flow compressors. They examined large amounts of axial flow compressor data from the middle stages of compressors (15 sets of data) and derived correlations that are nearly universal for such stages. The radial loss distribution proposed includes such variables as hub and casing wall boundary layer thicknesses, camber, solidity, aspect ratio, and rotor tip clearance. The three-dimensional loss distribution is superimposed on the two-dimensional variation calculated from the analysis. They showed good comparison between their correlations and the General Electric four-stage compressor data. It is thus evident that it is feasible to use the existing database and derive loss correlations that are reasonably accurate for use in the design system. Such correlations can be incorporated into inviscid codes or quasi-viscous methods described in Chapter 4.

The overall efficiency of axial flow turbomachinery can be estimated from Eqs. 6.4, 6.8, and 6.9 using various correlations proposed for loss coefficient ζ, which can be written as

$$\bar{\zeta} = \bar{\zeta}_p + \bar{\zeta}_{sh} + \bar{\zeta}_s + \bar{\zeta}_L + \bar{\zeta}_w \tag{6.36}$$

The profile and shock losses (for two-dimensional flow) can be predicted using analysis presented in Section 6.4 (Eqs. 6.19 and 6.21), and secondary or endwall losses can be estimated from Eqs. 6.24, 6.25, or 6.27 and leakage flow losses from Eqs. 6.30–6.34. There are no loss correlations available for three-dimensional shock losses. The three-dimensional profile losses can be predicted using the analysis presented in Section 5.5.

For liquid handling machinery (e.g., pumps), Eq. 2.16 for η_p (or the one below it for η_H) can be used to estimate efficiency. The actual head rise can be related to ideal head rise through the equation

$$(H_3 - H_1)_{actual} = (H_3 - H_1)_{ideal} - (\Delta H_0)_{loss}$$

where $(\Delta H_0)_{loss}$ is the head loss and is given by

$$(\Delta H_0)_{loss} = (\Delta P_0)_{loss}/\rho g$$

All the loss correlations given in this chapter (except ζ_{sh}) can be used to derive the head loss or head loss coefficient (usually defined as $\psi_{loss} = (\Delta H_0)_{loss}/(\rho U_t^2)$) and the efficiency estimated from Eqs. 2.16 and 6.36.

It is very likely that future analysis and design will be based on losses predicted directly from the Navier–Stokes procedure. The computation will provide the entropy increase or pressure losses or head losses directly, which can then be substituted in Eqs. 2.16, 6.8, or 6.9 to derive the efficiency. This approach is preferable to the correlation as the Navier–Stokes procedure includes nonlinear effects. The approach based on correlation assumes linear superposition, as indicated by Eq. 6.36, and this may not be valid when interaction effects are large. For example, secondary flow and leakage flow interact to produce complex nonlinear effects, and linear superposition of these flows and the resulting losses would be inaccurate.

6.5.5.1 *Overall Losses in Centrifugal Impellers*

The efficiency prediction of centrifugal machinery is best handled from a Navier–Stokes procedure, as profile, secondary, shock, and leakage flow losses are almost indistinguishable. For centrifugal machinery with thin wall and profile boundary layers, the tip clearance loss correlation, according to Senoo and Ishida (1986, 1987), shown in Fig. 6.44, may provide a reasonably good estimate of the inefficiency due to tip clearance.

A discussion of overall losses in centrifugal impellers (compressors and pumps) is given in this section. The major sources of losses are (1) tip clearance losses (Section 6.5.4, Fig. 6.44), (2) secondary flow and losses (Section 4.2.6.3 and Figs. 4.21 and 4.22), and (3) profile losses. The profile losses are influenced by rotation, curvature, and modifications of the structure of turbulence as described in Section 5.2.1. Nevertheless, isentropic efficiencies as high as 95% (Krain 1988), have been achieved. Even though losses are high in these impellers, the pressure rise is also high. The centrifugal effect is a major contributor to the pressure rise (as explained in Section 2.7), and this is an inviscid effect. Hence losses in relation to pressure rise are not very high, and the centrifugals are as good as axials in efficiency for high-pressure rise and low-mass flow application.

The loss distribution in a centrifugal impeller shows peak losses near the leakage flow region (shroud) and near the suction surface corner of the shroud (Fig. 6.16e), caused by secondary flow. Rodgers (1980) estimated the losses based on a procedure similar to that indicated by Eq. 6.35, using a correlation for the pressure loss in a 90° pipe bend with a mean hydraulic diameter (D_h) and a mean length (L). He found that this correlation provides a reasonably good estimate. Moore et al. (1984) carried a numerical simulation to understand various sources of losses and found that major sources of losses (entropy increase) are due to (1) leakage flow/secondary flow interaction near the shroud-suction surface corner, (2) losses due to shroud (presumably due to shroud boundary layer) and the shearing effect caused by the relative movement between the blade and the shroud, and (3) losses due to interaction of the suction surface and the wall boundary layers in the inducer. The investigation and relevant data on loss mechanisms for centrifugal impellers are very scarce.

7

TURBINE COOLING AND
HEAT TRANSFER

7.1 INTRODUCTION

Among various types of turbomachinery, high-temperature gas turbines are the only components which require cooling and heat transfer analysis. Because this is specific to only one type of turbomachinery, detailed treatment is given separately in this chapter; the computational techniques, however, are the same as described in Chapter 5, with heat transfer as additional constitutive equations.

The need for higher turbine inlet temperature and turbine cooling can best be illustrated by examining the performance relationships of a turbojet. Similar arguments are valid for turbo fan, turbo prop, and industrial and power gas turbines. Two important performance parameters of a turbojet are the thrust/unit mass flow rate $(F/\dot{m}_a)$ of air and thrust specific fuel consumption [TSFC (fuel consumption per unit thrust)]. The expressions for these for an ideal turbojet are given by (e.g., Kerrebrock, 1977)

$$\frac{F}{\dot{m}_a a_o} = M_o\left[\left\{\left[\frac{\theta_o}{\theta_o - 1}\right]\left[\frac{\theta_t}{\theta_o \tau_c} - 1\right][\tau_c - 1] + \frac{\theta_t}{\theta_o \tau_c}\right\}^{1/2} - 1\right] \quad (7.1)$$

$$\mathrm{TSFC} = \frac{\dot{m}_f}{F} = \frac{c_p(\theta_t - \theta_o \tau_c)T_1}{Q_R\left(\dfrac{F}{\dot{m}_a}\right)} \quad (7.2)$$

where $\theta_o = T_{o1}/T_1$, $\theta_t = T_{ot}/T_1$, $\tau_c = \pi_c^{1-(1/\gamma)}$, π_c = compressor stagnation pressure rise ratio, T_{ot} = turbine inlet stagnation temperature, $\dot{m}_a$ = rate of

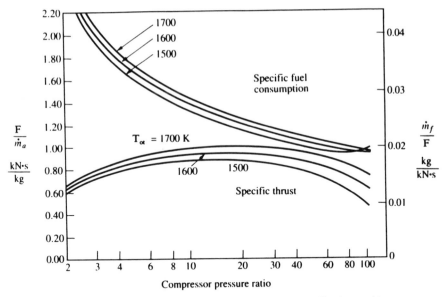

Figure 7.1 Variation of thrust specific fuel consumption and specific thrust with compressor pressure ratio for typical turbojet engine. M_o = 0.85, altitude = 12,200 m, η_c = 0.85, η_t = 0.90, η_N = 0.98, Q_R = 45,000 kJ / kg. (From Hill and Peterson, 1992; reprinted with permission of Addison-Wesley Publishing Co.)

air mass flow, $\dot{m}_f$ = rate of fuel mass flow, Q_R = heat value of the fuel, M_o = flight Mach number, and T_1 = inlet static temperature.

It should be emphasized here that the actual engine performance differs substantially from that of an ideal engine, but the concepts can be best illustrated (briefly) using equations for an ideal engine. Expressions similar to Eqs. 7.1 and 7.2 can be derived for an actual engine (Hill and Peterson, 1992). The power output is $F \cdot V_o$, where V_o is the flight speed. The specific thrust $(F/\dot{m}_a)$ and specific power $(F \cdot V_o/\dot{m}_a)$ for a typical core engine (compressor/combustor/turbine configuration) can be calculated from these equations. Performance of a typical turbojet is shown in Figs. 7.1 and 2.69. It is clear from Eq. 7.1 and Fig. 7.1 that the thrust power per unit mass flow rate increases continuously with an increase in the turbine inlet temperature, with all other variables held constant.

The interpretation of the variation of TSFC with turbine inlet temperature is more complicated. The dual effect of the compressor pressure ratio and the turbine inlet temperature can be clearly understood using an ideal turbojet optimized for maximum $F/\dot{m}_a$. By differentiating Eq. 7.1 with respect to compressor stagnation pressure ratio (π_c) (with M_o, θ_o and θ_t fixed), it can be proven (Kerrebrock, 1977) that the optimum compressor pressure ratio (for an ideal turbojet) for maximum thrust-to-mass ratio at a fixed Mach number M_o and T_{ot} is given by

$$(\tau_c)_{\text{optimum}} = \frac{\sqrt{\theta_t}}{\theta_o} \qquad (7.3)$$

Most compressors operate near the optimum; increasing the turbine inlet temperature increases $(\tau_c)_{\text{optimum}}$. Therefore, both θ_t and $\theta_o \tau_c$ increase in Eq. 7.2, thus reducing any increase in TSFC due to an increase in the θ_t alone. But an increase in $F/\dot{m}_a$ with an increase in T_{ot} reduces the TSFC. The numerator as well as the denominator in Eq. 7.2 increases with an increase in T_{ot} for an optimum cycle. Hence both performance parameters, $F/\dot{m}_a$ and TSFC, improve for an optimum engine. This will also result in lower thrust-to-weight ratio, resulting in a compact, efficient, and high-performance engine, unless the high temperature requires a heavy turbine. For a nonideal engine, actual cycle analysis has to be carried out to find an optimum τ_c shown in Fig. 7.1. It is clear that improved efficiencies and power can be achieved by an increase in the turbine inlet temperature. It is also clear from these equations that increasing the flight speed increases the thrust power. But the phenomena related to TSFC are complicated. It will suffice to say that there is an optimum value of M_o, at which TSFC is minimum. Ainley has carried out an analysis (see Andrews and Bradley, 1957) that shows that the optimum turbine inlet temperature increases with the flight Mach number; hence the turbine cooling becomes more critical for a high-speed aircraft.

These arguments are equally valid for a turbo fan. The turbine in this case drives both the fan and the compressor, so that the turbine becomes a much more critical component for a turbo fan. This evidently shows the desire to improve the turbine engine performance through higher inlet temperatures, larger pressure ratios, and higher bypass ratios. It is widely acknowledged that of all the improvements that can be made to component performance, the largest benefit will result from an increase in the turbine inlet temperature. This is also true for industrial and power gas turbines.

The materials that are presently available cannot stand temperatures in excess of 1300 K (Fig. 2.70), but the maximum temperature of the cycle can be increased by cooling the surfaces. Therefore, cooled components (combustion liner, blades and walls, exhaust nozzles, disks, etc.) are widely used in aerospace applications. An excellent summary of various techniques used for cooling and the method of analysis are presented in LeGrives (1986). Incorporation of cooling and improvements in materials have resulted in an increase in turbine inlet temperature in excess of 750 K for an aircraft gas turbine since 1960 (Fig. 2.72). It is acknowledged that cooling devices allow operation at a total mean temperature of at least 300–350 K higher. Present-day commercial aircraft engines operate with temperatures in excess of 1600 K, and the military engines operate at higher temperatures than this.

Associated with the gain in performance is the mechanical, aerodynamic, and thermodynamic complexities involved in the design and analysis of these cooling passages and holes. Furthermore, the bleeding of high-pressure air and the specific cooling method cause reduction in overall thermal efficiency. Still, the advantages of cooling outweigh the disadvantages, for both aerospace and power gas turbines.

The environment in which the turbine airfoil (both cooled and uncooled) operates is very hostile. Heiser (1978) has provided a very good description of such an environment. The nozzle and blade rows are subjected not only to high temperatures, but also to a large variation in temperature along the blade (10–25%). In addition, the rotor/stator thermal field is unsteady due to blade-row interactions. The inlet turbulence intensity is also high, resulting in random temperature fluctuations. The nozzle is subjected to the most severe conditions of all the blade rows. Cooling this component is very critical (Demuren, 1976; Back et al., 1970). It is also evident from the velocity triangles shown in Fig. 2.56 that

$$T_{o0} = T_{o1} = T_o\left[1 + \frac{\gamma - 1}{2}M_o^2\right] = T_1\left[1 + \frac{\gamma - 1}{2}M_1^2\right]$$

$$(T_{o1})_R = T_1\left[1 + \frac{\gamma - 1}{2}M_{1R}^2\right]$$

$$= T_{o0}\frac{1 + \frac{\gamma - 1}{2}M_{1R}^2}{1 + \frac{\gamma - 1}{2}M_1^2}$$

$$M_1 = \frac{W_1}{a_1} \tag{7.4}$$

Because $M_{1R}^2 < M_1^2$, the turbine rotor would perceive lower stagnation temperatures (about 200–300 K lower) than would the nozzle; the rotor blade environment is less hostile only from the point of temperature. The subsequent blade rows (first-stage stator, second-stage rotor and so on) operate at a much lower temperature, which makes cooling much simpler. Before we proceed with cooling techniques and the analysis of the cooled blade, it is desirable to understand some basic aspects of heat transfer. This is covered in the next section.

7.2   BASIC CONCEPTS IN HEAT TRANSFER

The field of heat transfer is a highly developed major subject area, much the same as fluid mechanics. There are a large number of books and papers devoted to heat transfer. The objective of this section is to provide only a brief background in this area. The reader is referred to excellent books such as Eckert and Drake (1972), Özisik (1985), Burmeister (1983), and Kays and Crawford (1993) for a thorough treatment of the subject.

Let us consider the case of an uncooled blade, as shown in Fig. 7.2. Three distinct heat transfer phenomena may be identified. These are as follows:

1. Heat transfer by conduction
2. Heat transfer by convection
3. Heat transfer by radiation

At low to moderate temperatures, convection and conduction play a major role in heat transfer. These topics are covered in this chapter. All gases and solids at high temperatures emit, reflect, and absorb energy through electromagnetic waves. This form of heat transfer is called *radiation*. At temperatures encountered in turbines, this form of heat transfer may not be significant, and therefore it is not addressed here. The reader is referred to many specialized books available on this topic. Radiative heat transfer is important in situations such as the primary zone of a combustion chamber, nuclear

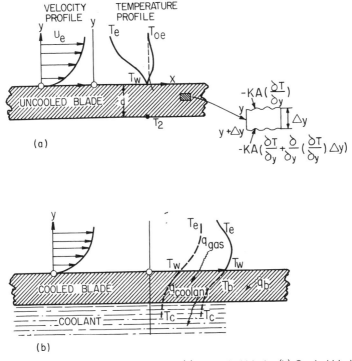

Figure 7.2 Heat transfer phenomena. (a) Uncooled blade. (b) Cooled blade.

reactors, and so on. Radiative heat transfer may be appreciable from the leading and trailing edges of nozzle vanes of an aircraft gas turbine. Because this chapter is primarily concerned with turbine cooling and heat transfer, only the basic aspects of conduction and convection heat transfer are covered.

7.2.1 Heat Transfer by Conduction

Conduction refers to the phenomenon whereby heat flows from one part of a medium (solid, liquid, and gas) at a higher temperature to another part at a lower temperature. An example of conduction is shown in the lower portion of Fig. 7.2a. The upper part of the plate is at a higher temperature (T_w) than the lower part, and the heat transfer by conduction will occur from the upper to the lower part of the wall. The rate of heat transfer can be written as (Fourier conduction law)

$$\frac{Q}{A} = q = -k\frac{dT}{dy} \tag{7.5}$$

where Q/A is the rate of heat transfer per unit area of the surface, and dT/dy is the temperature gradient (Fig. 7.2). Thermal conductivity, k, can be defined on the basis of Eq. 7.5 as the amount of heat conducted per unit time per unit area per unit negative temperature gradient. Conduction is an intermolecular diffusion process, and this can be predicted easily as long as the accurate value of k is known. This is a property of the metal and is governed by the kinetic theory of matter. Thermal conductivity depends on the material processing, impurities, structure, temperature, and composition. It varies widely from one material to another, differs from metallic to nonmetallic compounds, and varies between the liquid and solid phases.

The transport equation governing the heat transfer by conduction can easily be derived by referring to Fig. 7.2. Consider an infinitesimal strip of the slab of height Δy; the rate of heat flux into the slab is given by

$$-k\left(\frac{\partial T}{\partial y} - \frac{\partial T}{\partial y} - \Delta y\frac{\partial^2 T}{\partial y^2}\right) = \frac{\partial T}{\partial t}\left(\rho c_p \Delta y\right) \tag{7.6}$$

where c_p is the specific heat and ρ is the density of the material. Hence, the one-dimensional heat transfer equation is given by

$$\frac{k}{\rho c_p}\frac{\partial^2 T}{\partial y^2} = \frac{\partial T}{\partial t} \tag{7.7}$$

For a more generalized case, $T = T(x, y, z, t)$ and $q = q(x, y, z, t)$, the governing equation is a three-dimensional Poisson equation given by

$$\frac{k}{\rho c_p} \nabla^2 T = \frac{\partial T}{\partial t} \tag{7.8}$$

This is known as the Fourier equation. ∇^2 is the Laplace operator; parameter $k/\rho c_p$ is called *thermal diffusivity* and is a property of the conducting material.

Simplified forms of these equations have been solved by many investigators. One of the simplest solutions is for a one-dimensional slab of uniform thickness, uniform property, and steady flow. The governing equation for such a case (e.g., Fig. 7.2, with $\partial T/\partial x$ and $\partial T/\partial t = 0$) is given by

$$\frac{d^2 T}{dy^2} = 0 \tag{7.9}$$

The solution of Eq. 7.9 is given by

$$T = ay + b \tag{7.10}$$

or

$$\frac{T - T_w}{T_2 - T_w} = \frac{y}{d} \tag{7.11}$$

Hence

$$Q = -kA\frac{dT}{dy} = \frac{T_w - T_2}{d/kA} \tag{7.12}$$

The denominator in Eq. 7.12 is the thermal resistance.

The assumptions made in deriving Eqs. 7.9–7.12 are an oversimplification. In actual situations (say a turbine nozzle or rotor blade), there would be heat transfer and temperature gradients in all directions. Because of this, the Laplace equations have to be solved with the boundary conditions (Dirichlet boundary conditions) $T_w(x, y, z)$ and $T_2(x, y, z)$ specified on $y = 0$ and $y = d$, respectively. In many instances, the rate of heat transfer at the wall (q_w) is specified (Neumann type of boundary condition). There are very efficient numerical techniques to solve this equation. The finite difference method described in Chapter 5, or the finite element method (Baker, 1983) can be used.

The heat transfer analyses involve simultaneous calculations of conduction and convection of heat transfer from the hot gas to the blades, heat transfer across the blade from the external surface to the internal surface, and heat transfer from the inner surface of the blade to the coolant. The values of T_w

and T_2 are established from the convective heat transfer analysis described below. The temperatures within the blade can then be determined by the solution of Eq. 7.8.

7.2.2 Heat Transfer by Convection

Unlike that in a solid, the transfer of heat in a fluid can occur through conduction as well as convection through the movement of the fluid. The latter is the most common form of heat transport in turbomachinery flows and aerospace applications. The overall heat transfer in a moving media, including both conduction and convection, is usually referred to as *convective heat transfer*. There are many specialized books on this topic, because it represents a complex interplay between fluid mechanics and heat transfer. The two fields have evolved as two separate entities, without any major interaction between fluid dynamicists and heat transfer specialists. This situation must be rectified before the complex interactions and mechanisms and their dependency on parameters governing fluid mechanics (gas to blade temperature ratio, Re, M, T, structure of turbulence, body geometry, Ec) and heat transfer (k, c_p, PR, Nu, St, Gr, etc.) can be clearly understood.

In general, the temperature and velocity fields are coupled and interact strongly; thus the velocity field affects the temperature distribution, and vice versa. In many situations, when the velocity is high and temperature differences ($T_w - T_e$) are small, the temperature is influenced by the velocity field, but the velocity field is mildly affected by the temperature differences. The velocity is influenced by the density variations caused the temperature variations. When the flow field is created by external influences (e.g., blower, compressor, pump, fan, etc), the heat transfer process in such situations is called *forced convection*. In modern turbines, both the velocity and $T_w - T_e$ are high. In many situations, where the buoyancy is the dominant driving force, the velocities are small and the temperature distribution, both the absolute value and the differences ($T_w - T_e$), has a major influence on the flow. Heat transfer in such a situation is called *natural convection* or *free convection*. Good examples of this are jet engine combustion flows, convection in heated rooms, and atmospheric flows. Forced convection is the dominant phenomenon in turbine flows.

The heat transfer phenomena in an uncooled blade (Fig. 7.2) can be described as follows. The transfer of heat from the flowing media to the solid body (or vice versa) is of interest here. The critical elements in the process are the boundary layer developing on the surface and the free-stream total temperature. The boundary layers, which act as a buffer zone between the main stream and the solid, offer resistance to the heat transfer. Heat transfer occurs in this viscous layer between the solid and the fluid through both conduction and convection mechanisms. If the body is at a higher temperature, the heat transfer will occur from the solid to the fluid, and vice versa if the fluid total temperature is higher than the body. Once the heat has penetrated into the flow, the energy transport occurs mainly through the convection by the moving media. Therefore, the buffer region, or boundary layer, plays a very critical role in heat transfer. The condition and property of this layer determines the rate at which the heat is transferred.

A good physical understanding of the convective heat transfer can be obtained by examining the basic heat transfer/energy equation. In a laminar flow, the heat transfer is governed by Eqs. 1.18 or 1.34, and in a turbulent flow it is governed by Eqs. 1.72 and 5.1. The heat flux equation for two-dimensional turbulent boundary layer (derived from Eqs. 1.34 and 1.72) can be written as

$$\rho c_p \left[\frac{\partial T}{\partial t} + u \frac{\partial T}{\partial x} + v \frac{\partial T}{\partial y} \right] = \frac{\partial}{\partial y} T \left(k \frac{\partial T}{\partial y} \right) + u \frac{\partial p}{\partial x}$$

$$+ \frac{\partial u}{\partial y} \left(\mu \frac{\partial u}{\partial y} \right) - \rho \overline{u'v'} \frac{\partial u}{\partial y} - \rho c_p \frac{\partial}{\partial y} \left(\overline{v'T'} \right)$$

$$(7.13)$$

The quantities p, T, u, v, and ρ are time-averaged variables. The velocity v' is the fluctuation normal to the wall in the direction y, $\rho \overline{u'v'}$ is the turbulent shear stress and $-\rho \overline{v'T'}$ is the turbulent heat flux. This equation can be derived by using decomposition implied in Eq. 1.71.

In Eq. 7.13, the left-hand side denotes the convection of energy or heat transfer, the right-hand side denotes, respectively, the heat transfer by molecular conduction in the fluid, the work done by pressure forces, the work done by laminar stresses, the work done by turbulent stresses, and heat transfer by turbulent velocity and temperature fluctuations.

At least seven distinctly different regimes of blade heat transfer can be identified as follows (Fig. 7.3): stagnation point, laminar boundary layer, transitional boundary layer, turbulent boundary layer, shock–boundary layer interaction, separation with reattachment, and separation without reattachment. Because the heat transfer is closely coupled to fluid mechanics phenomena, each of these features involves a separate analysis valid for that particular region. In embedded turbine stages, transition may occur near the leading edge. In addition, the heat transfer is influenced by (a) different types of cooling and regions of heat transfer, (b) flow parameters such as Mach number, Reynolds number, free-stream turbulence, pressure gradient, free-stream-to-wall temperature, (c) blade curvature, roughness, and material, and (d) gas properties such as ρ, k, μ, c_p, c_v, and so on.

In heat transfer applications, the use of dimensional analysis is important, because the heat transfer depends on a large number of parameters. The dimensional analysis enables the reduction of a large number of variables to a manageable number of groups (similar to Section 2.4). This, in turn, enables experimentalists to design experiments to carry out a parametric study.

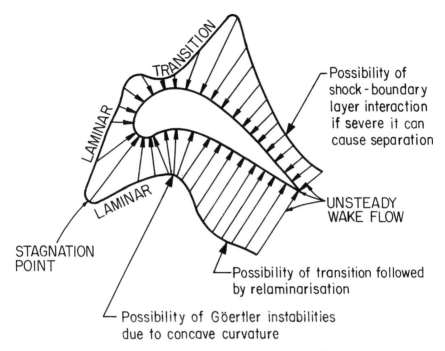

Figure 7.3   Variation of heat transfer rate around a turbine blade. (Adopted from Daniels and Schultz, 1982).

There is considerable empiricism in the analysis of convective heat transfer because of complicated geometry and also because of complex interaction between the flow and the heat transfer fields. In Chapter 1, the following groups were identified by a dimensional analysis of Eq. 1.18: Reynolds number (Re), Mach number (M), Prandtl number (PR), and Eckert's number (Ec). In heat transfer (internal as well as external) applications, many additional nondimensional groups are important. A knowledge of the heat transfer between a solid and a fluid is all that is required in many engineering applications, and a detailed knowledge of the flow and temperature is not essential. Overall heat transfer is related to the temperature difference between the fluid and the solid (T_w) through Newton's cooling law. The heat flux ($q_w(x)$) from the fluid to the wall is determined by (Fig. 7.2)

$$q_w(x) = h(x)(T_r - T_w) = k\left(\frac{\partial T}{\partial y}\right)_w \qquad (7.14)$$

where $h(x)$ is the heat transfer coefficient (kW/m$^{2\circ}$ K). For a turbine blade, T_r is the adiabatic recovery temperature (the temperature attained by an insulated surface in the flow). At low Mach numbers, this can be assumed to be equal to the stagnation temperature relative to the blade. At high speeds $T_r = T_e(1 + r((\gamma - 1)/2)M_e^2)$, $r = \sqrt{PR}$ for laminar flow, $= (PR)^{1/3}$ for turbulent flow. The heat transfer coefficient can be nondimensionalized by k, the thermal conductivity of the fluid/solid, and a characteristic length to define Nusselt's number:

$$\text{Nu}_x = \frac{h(x)L}{k} = \frac{L}{T_e - T_w}\left(\frac{\partial T}{\partial y}\right)_w \quad (7.15)$$

where $T_e - T_w = \Delta T_o$, the temperature difference between wall and free stream. It can be easily recognized that Nu is equal to the nondimensional value of temperature gradient at the wall:

$$[\partial(T/\Delta T_o)/\partial(y/L)]$$

Hence, according to Eqs. 1.28 and 1.34,

$$\text{Nu}_x = f(\text{PR}, \text{Re}, \text{Ec}, \mathbf{x})$$

where $\mathbf{x}$ is the local streamwise coordinate.

In addition to these groups, there are situations where gravitational, rotational, and centrifugal forces are present. These are represented, respectively, by Grashof, rotation, and Richardson numbers and are described in the section on free convection.

The prediction of convective heat transfer involves solution of transport equations for momentum and energy described in Chapter 1. For example, for laminar boundary layer, Eqs. 1.44–1.48 and 1.34 have to be solved simultaneously. For the most complex situation, the equations have to be solved numerically because the equations are highly nonlinear, but for simple situations (e.g., flat-plate boundary layer) it is possible to obtain analytical solutions which provide great insight into the physical phenomenon and its dependency on various heat transfer parameters.

There are variety of boundary conditions encountered; the temperature on the surface of the body may be constant or variable (and prescribed), the heat flux, q_w, may be constant (and prescribed); hence the temperature gradient at the wall may appear as a boundary condition.

7.2.3   Heat Transfer in Laminar Flows (Forced Convection)

Let us consider incompressible laminar flow over a flat plate at zero incidence. Equations 1.44–1.46 and 1.34 can be simplified considerably (including neglection of dissipation due to friction) to derive the following equation, where velocities are normalized by the free-stream velocity, U_e:

$$\frac{\partial(u\phi)}{\partial x} + \frac{\partial(v\phi)}{\partial y} = \alpha\frac{\partial^2\phi}{\partial y^2} \qquad (7.16)$$

where $\phi = u$ or θ, $\alpha = \mu/\rho$ or $k/\rho c_p$, and $\theta = (T - T_w)/(T_e - T_w)$. Boundary conditions are $y = 0$, $\phi = v = 0$; and $y \rightarrow \infty$, $\phi = u = \theta = 1$.

The velocity field has a direct effect on the temperature field (Eqs. 1.34, 1.44–1.46), while the temperature field influences the momentum equation through the density field. In turbine flows, where very high temperature and velocities are encountered, the classical practice of neglecting the coupling between the velocity and the temperature may not be valid. Nevertheless, the similarities between the two fields (velocity and temperature) are evident from Eq. 7.16. Transport equations are similar. The convection and diffusion terms are similar in form; hence the velocity and temperature profiles within the boundary layer can be expected to be similar in nature for PR = 1. The qualitative nature of the velocity field and temperature field are shown in Fig. 7.2.

It is clear from Eq. 7.16 that both u and θ are identical functions of y and α. This will result in a very important relationship between the heat transfer and the skin friction coefficient. The heat transfer from the fluid to the wall, or vice versa, depends on $(\partial T/\partial y)_w$. Let us relate this to the velocity gradient near the wall. For PR = 1, the gradients $(\partial\theta/\partial y)$ and $(\partial u/\partial y)$ are related due to similar solutions,

$$\frac{k}{\rho c_p}\left(\frac{\partial\theta}{\partial y}\right)_w = \frac{\mu}{\rho}\left(\frac{\partial u}{\partial y}\right)_w \qquad (7.17)$$

Hence,

$$St = \frac{q_w(x)}{\rho c_p U_e(T_e - T_w)} = \frac{k}{\rho c_p U_e}\frac{1}{T_e - T_w}\left(\frac{\partial T}{\partial y}\right)_w = \frac{C_f}{2} \qquad (7.18)$$

This clearly indicates that the rate of heat transfer depends on the wall skin friction coefficient (C_f). This result is not surprising. As indicated earlier, the boundary layer acts as a resistor, and this property is characterized by the shear stress. At the wall, the wall shear stress is the dominant factor. The nondimensional group on the left-hand side of Eq. 7.18 is called the *Stanton number*, because it represents the nondimensional heat flux. The important relationship St = $C_f/2$ has a major consequence. Most empirical relationships for a generalized case are based on this relationship.

The analytical solution of Eq. 7.16 can be found in most fluid mechanics books (e.g., Schlichting, 1979). The analytical solution for skin friction coefficient is $C_f = 0.664/\sqrt{Re_x}$. Similarly, a solution can be derived for the thermal boundary layer; $St(PR)^{2/3} = C_f/2$. It can be proven that the heat transfer is related to the Reynolds number and the Prandtl number through the Nusselt's number, given by

$$\mathrm{Nu}_x = \frac{q_w x}{k(T_e - T_w)} = (\mathrm{St})(\mathrm{Re}_x)\mathrm{PR}$$

$$= 0.332(\mathrm{Re}_x)^{1/2}(\mathrm{PR})^{1/3} = \frac{C_f}{2}(\mathrm{PR})^{1/3}\mathrm{Re}_x \qquad (7.19)$$

A relationship such as Eq. 7.19 is widely used by experimentalists to derive empirical heat transfer correlations for more complex cases. The Nusselt's number can be interpreted as the ratio of the characteristic length scale of the body and the thickness of the thermal boundary layer. Some of the conclusions based on the laminar boundary layer analysis can be summarized as follows:

1. Heat transfer is a function of the Reynolds number $(\mathrm{Re}_x)^{1/2}$ and Prandtl number $(\mathrm{PR})^{1/3}$ and is directly proportional to the skin friction stress. A thin boundary layer (smaller resistance) has a larger heat transfer. Thus the maximum heat transfer in a turbine usually occurs near the stagnation point and near the leading edge.
2. For $\mathrm{PR} = 1$, the thickness of the thermal and the velocity boundary layers is the same, and the velocity and the temperature $(T - T_w)$ profiles are the same for a laminar flat-plate boundary layer with zero pressure gradient.

7.2.4 Heat Transfer in Turbulent Boundary Layers (Forced Convection)

The solution for flow and thermal fields in turbulent flows depend on the flow and blade configuration. The Navier–Stokes equation is given by Eq. 5.1, combined with one of the turbulence closure models described in Section 1.2.7. The energy equation could be written in terms of either stagnation enthalpy (Eq. 5.1) or static temperature (Eq. 7.13). If boundary layer approximations (thin viscous layers) are valid, the constitutive equations are Eqs. 5.10–5.11. This is coupled with the energy equation (Eqs. 5.12 or 7.13). The turbulent heat flux, represented by the last term in Eq. 7.13 (or similar terms in Eq. 5.1 or 5.12), can be treated much the same way as Reynolds stress in the momentum equation. Velocity–temperature correlation is assumed to be proportional to the temperature gradient in the y direction (normal to wall) as in laminar flows $(q_l = -k\partial T/\partial y)$. Hence, the heat transfer due to turbulent fluctuations is written as

$$q_t = \rho c_p \overline{v'T'} = -c_p \epsilon_t \frac{\partial T}{\partial y} \tag{7.20}$$

where ϵ_t is the eddy diffusivity.

The similarities between the exchange of heat and momentum by turbulent fluctuations should be emphasized here. The fluctuating motion causes momentum exchange through the term $\rho \overline{u'v'}$ in the presence of velocity gradient. Similarly, the fluctuating motion also enables heat transfer between layers of different temperatures through the term $-\rho c_p \overline{v'T'}$ in Eq. 7.13 or $\rho \overline{w_i'e'}$ in Eq. 5.1 or $\rho \overline{h_{ORV}'v'}$ in Eq. 5.12. In both cases, exchange takes place in the presence of a gradient in velocity and temperature. There is a close coupling between heat transfer and momentum transfer, which also translates into a close coupling between heat flux and shearing stress. Because the major concern here is the heat transfer between fluid and solid, the shear stress on the wall plays a key role in this interaction. Because the momentum and heat transfer are closely coupled, the following dimensionless number is of relevance in heat transfer:

$$\text{Turbulent Prandtl number} = \frac{\mu_t}{\epsilon_t} = \text{PR}_t$$

Hence, the ratio of heat flux and momentum flux is given by

$$\frac{q_t}{\tau_t} = -\frac{c_p(\partial T/\partial y)}{\text{PR}_t(\partial u/\partial y)} \tag{7.21}$$

The total rate of heat transfer, due to both molecular and turbulent motions, can be written as

$$q = q_{\text{molecular}} + q_{\text{turbulent}} = -c_p \left(\frac{\mu}{\text{PR}} + \frac{\mu_t}{\text{PR}_t} \right) \frac{\partial T}{\partial y} \tag{7.22}$$

In a laminar flow, the heat transfer through the viscous layer to the wall is only by conduction given by the term $(\partial/\partial y)[k(\partial T/\partial y)]$. In turbulent flow, the last term in Eq. 7.13 will have dominant influence on heat transfer. The value of $(\partial T/\partial y)_w$ is different due to differing flow and thermal boundary layer profiles. There is a clear distinction between PR and PR_t. The Prandtl number (PR) is a physical property of the fluid. The turbulent Prandtl number (PR_t) is a property of the flow field.

Many assumptions are made in understanding the complex heat transfer phenomena associated with turbulent motion. One of the widely used assumption is that $\mu_t = \epsilon_t$, or $\text{PR}_t = 1$. This assumption implies that the eddy moving from one location to another (as in mixing length theory), where the velocity and temperature differ, achieves the new state (T and u) at the same time. This also implies the same mixing length for both the energy and the momentum transfer. If this concept or assumption is correct, then Eq. 7.16 is valid for turbulent motion on a flat plate as well, with $\alpha = (\mu + \mu_t)/\rho$ or $\{\mu/(\text{PR})\rho + \mu_t/(\text{PR})_t \rho\}$. This also implies that the thermal and velocity boundary layers are of the same thickness, and that the ratio of q_t/τ_t remains the same for all values of y.

It should be emphasized here that Eq. 7.22 represents one of the hypotheses, much the same as the hypothesis for Reynolds stress (Section 1.2.6). The constitutive relationship governing the turbulent heat flux ($\overline{u_i'T'}$ correlation) is much more complex. One probably needs to use the turbulent Reynolds stress and turbulent energy equation and a transport equation for $\overline{u_i T'}$ in the formulation. In many applications, the eddy diffusivity and eddy viscosity are expressed in algebraic form as functions of y. Both of these concepts, termed *algebraic eddy viscosity* and *algebraic eddy diffusivity*, respectively, need to be used with caution, but the concepts provide an excellent starting point for a qualitative understanding of this complex situation and a quantitative prediction of simple heat transfer flows in engineering applications.

Advanced closure models that involve transport equations for velocity and temperature correlation ($\overline{u_i T'}$) are given later in this chapter. Semiempirical expressions are available for simple geometries. For example, for the turbulent boundary layer on a flat plate, Reynolds analogy can be invoked (Eqs. 7.21 and 7.22) replacing μ by $\mu_l + \mu_t$ in Eq. 7.16. In this case, the temperature and velocity profiles are identical and Eqs. 7.17 and 7.18 are valid for turbulent flows as well. The skin friction coefficient for a flat-plat boundary layer is given by $C_f = 0.058(\text{Re}_x)^{-1/5}$, and the expression for $\text{Nu}(x)$ for a flat plate boundary is thus given by

$$\text{Nu}(x) = 0.029(\text{Re}_x)^{4/5}\text{PR}^{1/3} \tag{7.23}$$

The laminar and turbulent flow analysis presented in this section leads to a more general heat transfer relationship given by

$$\text{Nu}(x) = A\text{Re}_x^m \text{PR}^n \tag{7.24}$$

where A, m, and n are constants for a particular flow. This equation is called *Nusselt's equation* and is of great significance in the heat transfer field. It should be emphasized here that Nusselt's number is a function of x ($\text{Nu}(x)$), and $(\text{Re})_x$ is the local Reynolds number.

There have been many empirical/semitheoretical modifications made to this expression to account for laminar sublayer, change in free-stream velocity, laminar/transitional/turbulent boundary layers, strongly accelerated flow, surface curvature, surface roughness, and so on. These can be found in Kays and Crawford (1993) and Özisik (1985) and other textbooks.

7.2.5 Free- or Natural Convection and Other Relevant Parameters

As indicated earlier, free convection is a phenomenon whereby the heat is transferred naturally within the fluid. Excellent examples of this are (a) buoyancy-driven flows, combined with density or temperature gradients, and (b) flows subjected to rotational (Coriolis) and centrifugal forces as in a turbine. An example of the first case is the heat transfer in an enclosed room (room heating), where the buoyancy and gravitational forces are mainly responsible for the fluid motion as well as heat transfer. In turbine internal cooling passages, which are subjected to high rotation, the fluid motion as well as the heat transfer is affected by the rotation, both forced convection and natural convection are present. The additional parameters that are important in such flows are those related to the forces described below.

1. Gravitational Forces. In Eq. 1.16, the term responsible for natural convection is given by

$$\rho g = g\left[\rho_\infty + \left(\frac{\partial \rho}{\partial T}\right)\Delta T + \left(\frac{\partial \rho}{\partial p}\right)\Delta p + \cdots \right] \qquad (7.25)$$

It is generally recognized that the third term is small when the flow field is dominated by the gravitation effects and the forced convection effects are small. Hence

$$\rho g = g\left[\rho_\infty - \rho_\infty \beta(T - T_\infty)\right], \qquad \beta = -\frac{1}{\rho_\infty}\frac{\partial \rho}{\partial T} \qquad (7.26)$$

where β is the volumetric coefficient of thermal expansion.

The first term in Eq. 7.26 is the static field, and the second term is responsible for convection. This equation is substituted in Eq. 1.16 to derive the following nondimensionalized group (length normalized by L, velocity by U_∞, and others by the respective free-stream values).

$$\frac{g\,\rho_\infty \beta(T - T_\infty)L}{\rho_\infty U_\infty^2} = \frac{g\beta(\Delta T_o)L}{U_\infty^2} = \underbrace{\left[\frac{g\beta L^3(\Delta T_o)}{\nu_\infty^2}\right]}_{(1)}\underbrace{\left[\frac{\nu_\infty^2}{U_\infty^2 L^2}\right]}_{(2)} = \frac{\mathrm{Gr}}{\mathrm{Re}^2} \qquad (7.27)$$

The first term (nondimensional) in the above equation is the ratio of the buoyancy forces to the viscous forces and is called the *Grashof number* (*Gr*), given by,

$$\frac{\rho g}{\left(\mu_\infty U_\infty / L^2\right)} \sim \frac{\rho_\infty g\,\beta\,\Delta T_o\,L^3}{\mu_\infty U_\infty} \sim \frac{\rho_\infty g\,\beta\,\Delta T_o\,L^2 \rho_\infty L}{\mu_\infty^2(\mathrm{Re})} \sim \frac{g\,\beta L^3(\Delta T_o)}{\nu_\infty^2\,\mathrm{Re}}$$

where subscript '∞' refers to free-stream values. In the above equation, the free-stream velocity is replaced by the Reynolds number $U_\infty \cong (\mathrm{Re})\mu_\infty/\rho_\infty L$. In forced convection, the Reynolds number represents the ratio of inertial to viscous forces in the Navier–Stokes equation. In free convection, the Grashof number plays the same role as the Reynolds number.

Many investigators employ another nondimensional parameter, called the *Rayleigh number* (Ra), instead of the Grashof number (Eckert and Drake, 1972). This is defined by

$$\mathrm{Ra} = \mathrm{Gr}\,\mathrm{PR} = \frac{\beta g L^3\,\Delta T_o}{\nu_\infty^2}\frac{\mu_\infty c_p}{k} = \frac{\beta g L^3\,\Delta T_o\,\rho_\infty^2 c_p}{\mu_\infty k} \qquad (7.28)$$

For wall flows, $\Delta T_o = T_w - T_\infty$.

2. Coriolis Forces. Consider the momentum equation in relative frame of reference developed in Chapter 4 (Eqs. 4.28 and 4.32). The ratio of Coriolis to inertial force is given by

$$\Omega U D / U^2 = \frac{\Omega D}{U} = \mathrm{Ro} \qquad (7.29)$$

Ro is called the *rotation number*. The rotation effect depends on the rotation number $\Omega D/U$, where Ω is angular velocity, D is the characteristic dimension of the coolant passage (e.g., hydraulic diameter), and U is the coolant velocity in the internal cooling passages. In situations where the coolant velocity is very small and the coolant passages are very narrow, a more appropriate ratio would be (Coriolis/viscous) forces given by

$$J \sim \frac{\Omega U \rho}{\mu\, \partial^2 U/\partial y^2} \sim \frac{\Omega U \rho L^2}{\mu U} = \frac{\Omega \rho L^2}{\mu} \qquad (7.30)$$

This is sometimes referred to as the *rotational Reynolds number*. In turbomachinery external flows (forced convection), the rotation number is more relevant because the inertial forces and acceleration are large in the flow field (see Section 4.2.6.3). The rotational Reynolds number is more important for flows in viscous layers (Section 5.2.1 and Figs. 5.3 and 5.8) on rotating blades. In internal cooling passages, both the rotation number and the rotational Reynolds number may be important, depending on the geometry and the coolant velocity.

3. Centrifugal Forces. The curvature introduces centrifugal forces in the flow field causing additional convection. The ratio of the centrifugal force to the inertial force in a wall layer can be represented by the *gradient Richardson number* given by (Bradshaw, 1973),

$$\text{Ri} = \frac{U^2/R}{U(\partial U/\partial r)_w} \sim \frac{U/R}{(\partial U/\partial r)_w} \sim \frac{\delta}{R} \qquad (7.31)$$

It is evident that the Richardson number is proportional to δ/R; this effect becomes important when δ/R is substantial. Even values as small as 0.1 can have a significant effect on the wall shear stress (see Table 5.3).

In addition to the effects mentioned above, one can consider turbulence fluctuations also as causing natural convection. Assuming that the effects of free-stream turbulence can be represented by turbulence intensity (Tu) and length scale (l), the heat transfer due to natural convection can be represented by

$$h \text{ or Nu or St} = f(\text{Gr, Re, PR, Ro or J, Ri, Tu, } l)$$
$$= f(\text{Ra, PR, Ro or J, Ri, Tu, } l) \qquad (7.32)$$

In most cases, it is extremely difficult to identify the individual contributions by the free convection and the forced convection. Generally speaking, forced convection dominates the external heat transfer, while both are important in internal cooling passages. External blade and vane heat transfer depend on PR, PR_t, Re, Ri, Ro, Tu, l, ∇p, and T_e, and the internal heat transfer depends on Gr, J, PR, PR_t, Ro, Re, Ri, ∇p, T_e, and Tu. More specific correlations will be presented later.

Based on the analysis of heat transfer in laminar and turbulent motions presented in Section 7.2, several conclusions can be drawn:

1. Heat transfer is higher for a thin boundary layer (provided $(T_w - T_e)$ is the same) than for a thick one, because $\partial T/\partial y$ is higher for a thin boundary layer.
2. The value of μ_t/PR_t is much greater than that of μ/PR. Heat transfer in a turbulent boundary layer (for the same boundary layer thickness) is higher than that in a laminar boundary layer.
3. Heat transfer in very thin viscous layers, such as those that exist near the stagnation point and the leading edge, is very high. Because the velocity and the temperature gradients have very high local values near the stagnation point, the heat transfer rates are especially high in these regions.

In separated flows, the mechanism of heat transfer is much more complicated than indicated by the analysis presented in this section. The velocity and temperature gradients are small near the wall in the separated region. Hence, the heat transfer below the separated region is small due to low momentum and heat exchange normal to the wall. When flow reattachment occurs, the reattachment point is a singular point, and associated with this are high heat fluxes on the blade.

7.3 COOLING TECHNIQUES

Various techniques of cooling the blade are shown in Fig. 7.4. Many techniques of cooling have been proposed over the years: air cooling, water cooling, steam cooling, fuel cooling, liquid metal cooling, the heat pipe, thermosyphon, and so on. Air cooling is by far the most common method, especially in aerospace turbines. Only the air-cooling techniques are described here.

In air-cooled turbines of gas turbines, the air is usually drawn from the exit of the compressor. This involves loss in efficiency; however, a considerable gain in turbine performance offsets this loss. An example of an air-cooled

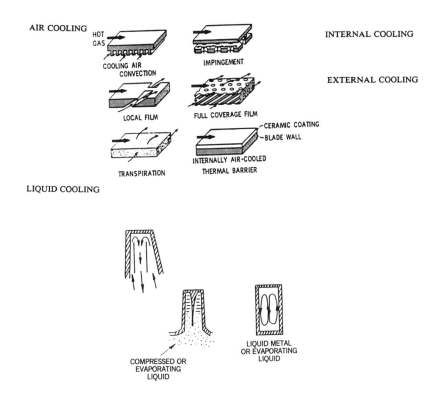

Figure 7.4   Blade and vane cooling techniques.

turbine for 6-MW industrial application is given in Hannis and Smith (1982). The General Electric Company and the U.S. Department of Energy (High-Temperature Turbine Technology Program) have demonstrated the use of water-cooled blades for an industrial gas turbine combusting a coal-derived low-BTU gas or coal-derived liquid fuel with a turbine inlet temperature range of 2600–3000°F. Examples of steam-cooled combined power plant are described in Rice (1979), and water-cooled turbine is discussed in Schilke and DeGeorge (1983). Devices based on the heat pipe or thermosyphon (Stuart-Mitchell and Andries, 1978) or liquid metal seem to be future potential candidates. However, many of these techniques have yet to be tested in a real engine environment.

Major advances have been made in incorporating cooling in aircraft gas turbines. A recent example is the energy–efficient engine (E^3) designed to operate in excess of 1660 K. Some of the aircraft engine technology is slowly being transferred to industrial gas turbines. The air-cooling technique can be classified as follows (LeGrives, 1986):

1. Internal cooling (for external stream temperatures of 1300–1600 K): convection cooling, impingement cooling, and internally air-cooled thermal barrier
2. External cooling (for external stream temperatures > 1600 K): local film cooling, full-coverage film cooling, and transpiration cooling

All of these techniques are illustrated in Fig. 7.4. Convection cooling is the simplest and one of the earliest techniques used. The coolant is passed through a multipass circuit from hub to tip and ejected at the trailing edge or the blade tip. In impingement cooling, the cold air from one row or many rows of small holes ejected from an insert within a blade impacts the blade wall and reduces its temperature. The mean blade temperature will be at a lower temperature than the gas, thus enabling a higher turbine inlet temperature to be used. Another method of keeping the blade cool is through a thermal barrier, which is a low-conductivity ceramic coating. Internal cooling is not as effective as external cooling, so its use is limited to a temperature range of 1300–1600 K. Modern gas turbines employ a combination of convective cooling and film cooling to achieve higher turbine inlet temperature. In film cooling, a thin layer of cool air insulates the blade from the hot gas stream, especially near the leading-edge regions where high temperatures are encountered. A more efficient technique is the full-coverage film cooling, which utilizes a large number of closely spaced holes as shown in Fig. 7.4. The temperature range where this technique is employed is 1560–1800 K. One of the most efficient techniques, which is only in the conceptual stage, is transpiration cooling, where a layer of cool air is deposited on the surface after passing through porous or woven material. This technique has not been tested in real engines. It is useful when inlet temperature is in excess of 1800 K.

The modern gas turbine employs a combination of these techniques to achieve a lower metal temperature, enabling higher operating gas temperature. The E^3 vane- and blade-cooling arrangement, reproduced from Thulin et al. (1982) (see Chapter 2, Figs. 2.62 and 2.64, and Fig. 5.37, for related data), is shown in Fig. 7.5. Internal surfaces are cooled by convection, whereas external surfaces are film-cooled. Because the nozzle vane experiences the highest temperature, a very elaborate arrangement of cooling is used (Fig. 7.5). The cooling air enters the vane from the tip and the root at 284 kPa and 850 K (at sea-level take-off conditions). The coolant is distributed within the internal structure of the vane, which is designed with three cavities. These cavities are convectively cooled through the use of sheet metal impingement tubes that fit into the three cavities. This probably represents the state of the art of the nozzle cooling as of 1986. The maximum turbine inlet temperature (sea-level take-off conditions) is 1981 K. The cooling air requirement is 6.4% of the core engine inlet flow. The calculated surface temperatures are also shown. It is evident that the leading edge,

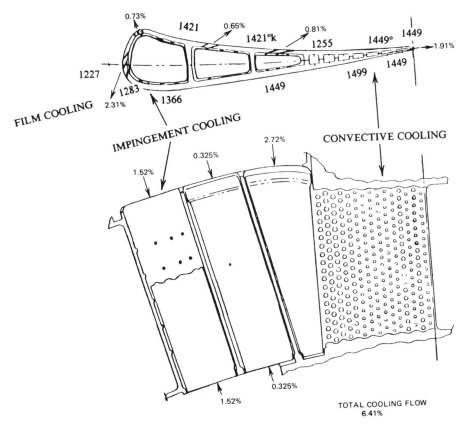

Figure 7.5 E^3 turbine vane cooling arrangement (Thulin et al., 1982; NASA CR 165608). Distribution of intake and exhaust coolant air, as a percent of total air flow is also shown.

where the cooling requirements are critical, is at a fairly low temperature. The inner and outer platforms are cooled by a combination of convective cooling, impingement cooling, and thermal barrier coatings. The ideal situation for vane, blade, and platform cooling is to maintain a uniform metal temperature to minimize differential thermal growth (and thus thermal fatigue).

The rotor blade cooling arrangement is shown in Fig. 7.6. It employs convective cooling inside the blade, and external surfaces are film-cooled from the leading-edge shower-head holes, tip-pressure side holes, and trailing-edge holes. Unlike the nozzle vane, there are no cooling holes on the pressure or suction surface (Fig. 7.5). The first stage of the rotor requires only 2.75% of the core-engine mass flow. The coolant is supplied at 1661 kPa and 556 K from the blade root. The predicted blade temperatures are also shown in Fig. 7.6. It is evident that the maximum temperature occurs on the suction side near the leading edge, in spite of leading-edge cooling. In the case of a rotor, pressure losses (due to injection from a stationary frame to a rotating frame) can be minimized by expanding the air through adjustable nozzles oriented tangentially, in order to impart a velocity to the coolant equal to the blade speed. One example of such an arrangement is given in LeGrives (1986).

There has been considerable research toward increasing the turbine inlet temperature of a radial turbine through cooling devices. One of the earliest cooling arrangements for a radial turbine is shown in Fig. 7.7. (Rohlik, 1983, based on Ewing, 1980) (see Section 2.10). The cooling arrangement is relatively simple, with a single two-pass channel in each blade discharging the coolant through a radial slot on the blade surface. Recent developments include a cooled inlet nozzle, as well as more effective cooling methods (AGARD, 1987; Ewing, 1980).

Typical temperature profiles for uncooled and cooled blades are shown in Fig. 7.2. Also shown are the stagnation temperature profiles. For an uncooled blade, the wall temperature is nearly equal to the stagnation temperature, because the Mach number is zero at the wall. The wall temperature attained when there is no heat transfer is called the *recovery temperature*, described by a recovery factor given by (see Fig. 7.2 for notations)

$$r = \frac{T_r - T_e}{T_{oe} - T_e} = \frac{T_r - T_e}{U_e^2/2c_p} \tag{7.33}$$

It has been found (Eckert and Drake, 1972) that $r = \sqrt{\text{PR}}$ and $(\text{PR})^{1/3}$ for laminar and turbulent flows, respectively. The recovery temperature is given by

$$T_r = T_e\left(1 + r\frac{\gamma - 1}{2}M_e^2\right) \tag{7.34}$$

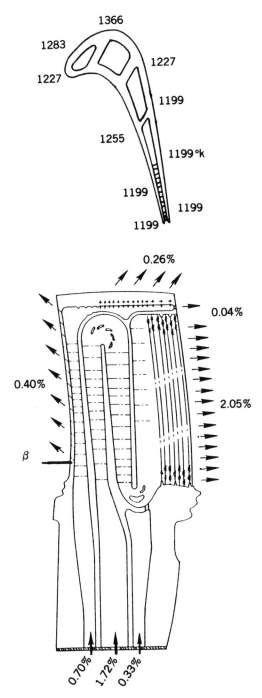

Figure 7.6 E[3] blade cooling system (Thulin et al., 1982; NASA CR 165608). Distribution of intake and exhaust coolant air, as a percent of total air flow is also shown.

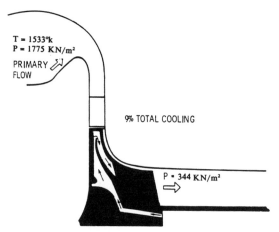

Figure 7.7 Cooled radial turbine (see Chapter 2, Fig. 2.78). Rohlik, 1983, NASA TM 83414.

Many designers use an empirical factor for r. The recovery factor, r, accounts for the fact that the deceleration of the flow as the wall is approached is not precisely adiabatic.

For a cooled blade (Fig. 7.2), the wall temperature could either be higher or lower than the static temperature in the free stream as shown in Fig. 7.2b. The interior temperature of the blade decreases continuously along its thickness due to the coolant. This is a complex heat transfer phenomenon. The heat is transferred (q_{gas}) from the hot gas (external) to the blade by conduction and convection; the blade heat transfer (q_b) occurs through conduction, resulting in a temperature gradient within the blade; and the blade heat is transferred ($q_{coolant}$) to a coolant, increasing its temperature to T_c. The coolant temperature $T_c(y)$ and the gas temperature $T_e(y)$ have to be determined from the solution of the flow and heat transfer equations. The blade temperature (for known T_c and T_w) can then be determined from Eq. 7.8 or Eq. 7.11.

7.4   CONVECTIVE COOLING

Convective cooling is one of the earliest forms of cooling used in practice, and it is employed in both industrial and aircraft gas turbines. The coolant air passes through very complex passages, which may include ribs and turbulence generators to increase their effectiveness. Some of this air is bled at the leading-edge or the blade tip or the trailing edge, or it goes through a 180° return channel to finally be vented near the trailing edge. Figure 7.6 shows the complexity of such passages.

This section is concerned only with convectively cooled and impingement cooled blades, where the coolant is not ejected on the blade surface, but instead stays internal during the heat transfer process.

The analysis—or, more precisely, the prediction—of the heat transfer coefficient and coolant and blade temperatures is extremely difficult. This is compounded by close coupling between the external and internal heat transfer and the need for precise boundary conditions (e.g., T_w and T_2 in Fig. 7.2). Analytical solutions have been obtained only for simple configurations, such as those described in the earlier section. The complexities of analyzing heat transfer with internal cooling are summarized by Heiser (1978). Some of the major complications are as follows:

1. The coolant passages are very complex and narrow, with large turning and obstructions as shown in Fig. 7.6. The fluid mechanics of coolant flows are extremely complex due to the effects of curvature and roughness. There is a great deal of turbulent mixing and separation associated with these coolant jets, cavities, and so on. Furthermore, these coolant passages are subject to rotation in a rotor.

2. The temperature of the cooling air changes continuously through its path, causing the density and property of the coolant to change as well.

3. The effectiveness of internal heat transfer is best expressed in terms of the heat transfer coefficient, which depends on the flow and fluid properties.

Hence, the analysis and design of internal heat transfer/cooling passages is very empirical and is based on the data collected from a large number of geometries, including geometries that have pins, baffles, inserts, roughness, tubes, and so on.

The analysis given here is approximate. Recent developments in numerical techniques for fluid flow and heat transfer, based on finite element and finite difference methods, have led to a decreased usage of empirical relationships, increased accuracy in prediction, and the ability to handle complex geometries (Chapter 5). These techniques will be dealt with in a later section on computational techniques. This section is mainly concerned with analytical techniques.

Some of the basic concepts and nondimensional groups used in heat transfer analysis were described earlier. Let us consider a convectively cooled blade as shown in Fig. 7.8. The approximate analysis given below follows the approach described in Horlock (1973a).

Heat transfer is related to the heat transfer coefficient (for that particular configuration and flow) as indicated by Eq. 7.14. Three distinct mechanisms of heat transfer can be recognized in the configuration shown in Figs. 7.2 and

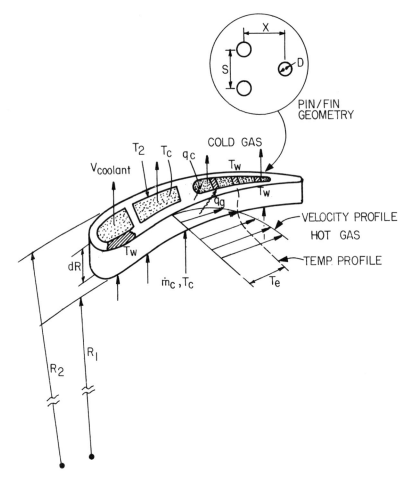

Figure 7.8   Analysis of convectively cooled blade.

7.8:

1. The first mode is the heat transfer from the hot gas to the blade. This
   heat transfer rate can be written as

$$q_{gas} = h_g A(T_r - T_w)\, dR \qquad (7.35)$$

where h_g is the gas-to-blade heat transfer coefficient, A is the outside
surface area exposed to heat transfer, and dR is the spanwise height. T_w
is the chordwise mean wall temperature and T_r is the average adiabatic

recovery temperature given by

$$A(dR)T_r = \int_{\text{blade surface}} \left[T_e \left(1 + r\frac{\gamma - 1}{2} M_e^2 \right) \right] ds \qquad (7.36)$$

2. Conductive heat transfer within the blade material given by

$$q_b = -k\frac{\partial T}{\partial y} \qquad (7.37)$$

where k is thermal conductivity of the blade material.

3. The heat transfer rate from the blade element to the coolant is given by

$$q_c = h_c A_c (T_2 - T_c) \, dR \qquad (7.38)$$

where T_c is the bulk temperature of the coolant, h_c is the blade to coolant heat transfer coefficient, and A_c is the area exposed to coolant.

Heat transfer variation in the spanwise direction of the blade is neglected; hence, the analysis is not valid close to the tip of the blade or low-aspect-ratio blading. Three-dimensional effects on the flow field have also been neglected. Furthermore, let us assume that the blade is thin and that the wall temperature is constant across the thickness; hence $T_w = T_2$ in Figs. 7.2 and 7.8. The heat flux equation for the blade is given by

$$Ah_g(T_r - T_w) + A_c h_c (T_c - T_w) = 0 \qquad (7.39)$$

In this equation, the coefficients h_g and h_c must be evaluated from empirical relationships or computational techniques. T_r is known from analysis of external flow, and T_w and T_c must be computed using Eq. 7.39—hence the need for an additional relationship. This relationship can be derived from the energy balance, equating the heat transfer q_c to the temperature rise of the coolant along the spanwise path. The increase in temperature of the coolant from radius R_1 to radius R_2 (Figs. 7.2 and 7.8) is given by

$$q_c = \dot{m}_c c_{pc} \frac{dT}{dR} \qquad (7.40)$$

where $\dot{m}_c$ is the mass flow rate of the coolant. Hence, equating the expressions in Eqs. 7.38 and 7.40, we get

$$\dot{m}_c c_{pc} \frac{dT_c}{dR} + h_c A_c (T_c - T_w) = 0 \qquad (7.41)$$

A general solution of Eqs. 7.39 and 7.40 can be obtained for cases where h_c, h_g, A_c, and A are all functions of radius (Horlock, 1973a). If the variation of these quantities with radius is small, then the solution is fairly simple and is given by

$$\frac{T_w - T_{c0}}{T_r - T_{c0}} = 1 - \frac{\exp M}{1 + h_g A / h_c A_c}$$

$$\frac{T_c - T_{c0}}{T_r - T_{c0}} = 1 - \exp M \tag{7.42}$$

where

$$M = -\frac{(R - R_h) h_g A}{\dot{m}_c c_{pc}} \frac{1}{1 + h_g A / h_c A_c} \tag{7.43}$$

The subscript zero refers to value of T_c at entry, $R = R_h$.

The determination of temperatures T_w, T_c, or $\dot{m}_c$ requires a knowledge of h_c and h_g for the particular flow medium. As indicated earlier, due to complexities of the equations (Eq. 1.18) as well as configuration, the analytical solutions available are for laminar flow in simple configurations. For internal coolant flows, the influence of centrifugal and Coriolis forces on the flow field makes it very difficult to obtain an analytical solution. A great deal of effort has been spent in deriving correlations that provide a reasonable estimate of these coefficients. The correlations for h_g and h_c are usually expressed in terms of a Nusselt number (e.g., $h_c D / k$ or $h_g C / k$ where D is the diameter of the coolant hole and C is the chord length). Detailed analysis and correlations for convective cooling are covered under the following three subsections: internal heat transfer (stationary blades), internal heat transfer (rotating blades), and external heat transfer.

7.4.1 Internal Heat Transfer (Stationary Blades)

There have been numerous experimental investigations conducted as well as correlations proposed for heat transfer in various-shaped ducts, fins, and banks of tubes. All of these configurations are simplified models of turbine nozzle and rotor cooling passages. In many instances, the geometry is simplified, and in most instances the experiment is carried out at low temperatures in a well-controlled environment. Nevertheless, these correlations show a reasonably good trend in determining the heat transfer coefficient. There has been a recent trend to carry out tests in a realistic geometry as well as with realistic flow conditions (high temperature and rotation, etc.) Clifford et al. (1985) used a multipass geometry (shown in Fig. 7.9), operated at ambient conditions with a cool, gaseous nitrogen. The model was also rotated to evaluate the effects of Coriolis and centrifugal forces. Complex

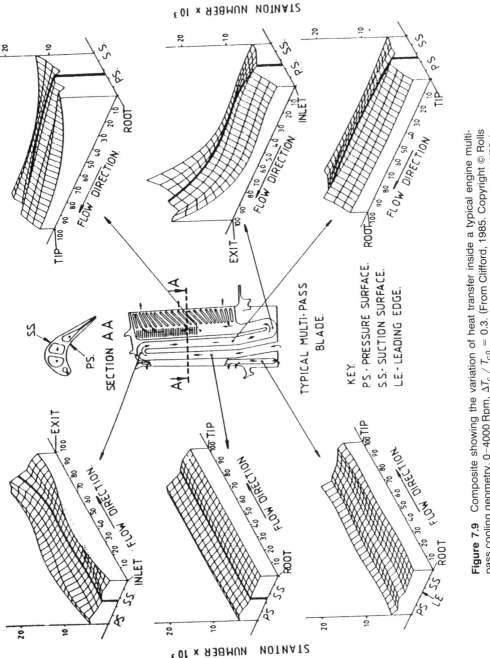

Figure 7.9 Composite showing the variation of heat transfer inside a typical engine multipass cooling geometry, 0–4000 Rpm, $\Delta T_c / T_{c0} = 0.3$. (From Clifford, 1985. Copyright © Rolls Royce, plc. The original version of this material was first published by the AGARD / NATO in CP 390, ''Heat Transfer and Cooling in Gas Turbines,'' September 1985.)

heat transfer characteristics and their variation from pressure to suction surface and from hub to tip are shown in Fig. 7.9. High heat transfer rates (Stanton number, Eq. 7.18) near bends as well as near fins are clearly evident. Systematic experimental investigation of heat transfer in a return channel with a sharp 180° bend has been carried out by various investigators. These investigations reveal that the heat transfer is substantially higher in bends, with the highest values measured immediately downstream of the bend. These are associated with secondary flows and flow separation in bends (Section 4.2.6).

As indicated in the previous section, the heat transfer coefficient for a nonrotating duct with negligible natural convection is a function of geometry, Reynolds and Prandtl numbers, and coolant-to-wall temperature ratio. The precise relationship depends on whether the flow is laminar or turbulent. The heat transfer in such a configuration is represented by Eq. 7.24, where $m = 0.8$ and $n = 0.33$ for turbulent flow and $m = 0.5$ and $n = 0.33$ for laminar flows. Most of the correlations follow the general correlation given by Eq. 7.24, and additional functions are very specific to that configuration. For example, in a pipe flow, the mean Nusselt number would depend on the length/diameter ratio of the coolant passage. For fin-type heat exchangers (Fig. 7.8), additional geometrical parameters include the diameter of the pin, pin/fin spacing, stagger, and so on. For impingement cooling, one has to include the mass ratio of the impingement jet/main stream, coolant hole diameter, spacing, and so on. It is clear that it is almost impossible to generalize heat transfer characteristics of various geometries used, much less predict the heat transfer coefficient in these passages. But basic equations and dependency on Prandtl and Reynolds numbers, as postulated by Eq. 7.24, are generally valid for most configurations. For the coolant flow through a pipe of general cross section, one can write the following equation (based on Humble et al., 1951):

$$\mathrm{Nu}_m = A(\mathrm{Re})^m (\mathrm{PR})^n \left(\frac{L}{D}\right)^E \left(\frac{T_c}{T_w}\right)^F \tag{7.44}$$

where L is the length of the pipe and D is the diameter (or hydraulic diameter), Nu_m ($= h_c L/k$, where L is characteristic length, D or d, and h_c is the average heat transfer coefficient for the coolant). The Reynolds number $\mathrm{Re} = \bar{\rho}_c \bar{V}_c D_c / \bar{\mu}_c$, where superscript bar refers to mean values inside the coolant channel. The constants proposed by Humble et al. and by other investigators are given in Table 7.1. Caution should be exercised when using these correlations. The precise definition of the Reynolds number, diameter, and Nusselt's number should be noted. These are defined in Table 7.1. For fully developed flow and long channels, the heat transfer rate should be constant and hence $(L/D)^E = 1$. The analysis and experimental work by Deissler (1959), in a fully developed flow, indicates that the constant, A, is a function of a friction factor which depends on such factors as roughness.

Table 7.1 Internal heat transfer coefficient (Nu_m)

Reference	Configuration	Equation	Re	Nu	A	m	n	E	F	D
Humble et al. (1951)	Smooth tubes of uniform cross section	7.44	Bulk values for coolant velocity	Mean values for h and k	0.034	0.8	0.4	−0.1	0.55	Hydraulic mean diameter
Deissler (1959)	Fully developed	7.44	Based on pipe diameter	—	0.07f f is a friction factor	1.0	0.25	NA[a]	NA	
Arora and Abdelmesseh (1985)	Circular and oblong pin fins (see Fig. 7.8)	7.44	Based on rod diameter	—	0.35	0.6	0.36	NA	NA	Rod diameter
Lokai and Limanski (1975)	Rotating pipe	7.53 with $f = 1 + 137 \dfrac{J^{0.7}}{Re^{1.08}}$	Duct flow Reynolds number	Mean value of h	0.023	0.8	0.33			Hydraulic mean diameter

Table 7.1 (Continued)

Reference	Configuration	Equation	Re	Nu	A	m	n	E	F	D
Morris and Ayhan (1979)	Circular tubes (rotating)	7.53	Duct Reynolds number	Mean value of h						Hydraulic mean diameter
	Radially outward flow	$f = \left(\dfrac{RRa}{Re^2}\right)^{-0.186} Ro^{0.33}$			0.022	0.8	NA			
	Radially inward flow	$f = \left(\dfrac{RRa}{Re^2}\right)^{0.112} Ro^{-0.083}$			0.036	0.8	NA			
Clifford (1985)	Complex 3D multipass geometry (Fig. 7.9)	7.53 $$f = \left(\frac{RRa}{Re^2}\right)^{B} Ro^{c}$$ constants A, B, and C depend on the location in Fig. 7.9 (See Table 1 in Clifford [1985])	Hydraulic smean diameter	Mean value of h	A	0.8				Hydraulic mean diameter

[a]NA, not available.

Deissler's constants are also tabulated in Table 7.1. It seems that correlations, such as Eq. 7.44, adequately model the heat transfer in coolant passages (straight passages without turbulence generators, fins and rotation).

In many cases, fins/pins are used inside these coolant ducts to augment heat transfer between coolant and blade as shown in Figs. 7.5, 7.6, 7.8, and 7.9. The heat transfer in a duct with fins/pins (Fig. 7.8) should depend on the dimensions (S, D, X) in addition to other parameters relevant to the channel. Considerable work has been done with this configuration (Zukauskas, 1972; Arora and Abdelmesseh, 1985; Metzger et al., 1981). Zukauskas (1972) proposed the following correlation for heat transfer due to pins/fins:

$$\mathrm{Nu}_m = 0.35\left(\frac{S}{X}\right)^{0.2} \mathrm{Re}^{0.6} \mathrm{PR}^{0.36} \qquad (7.45)$$

where $\mathrm{Re} = \rho(V_c)_{\max} D/\mu$. These correlations agree well with Arora and Abdelmesseh's (1985) data. It should be emphasized here that the heat transfer correlation given by Eq. 7.45 represents the augmentation due to pins/fins; for total heat transfer, one has to include the duct or pipe heat transfer given by the correlation such as Eq. 7.44. The heat transfer characteristics of different shapes of turbulence promoter ribs, shown in Fig. 7.10,

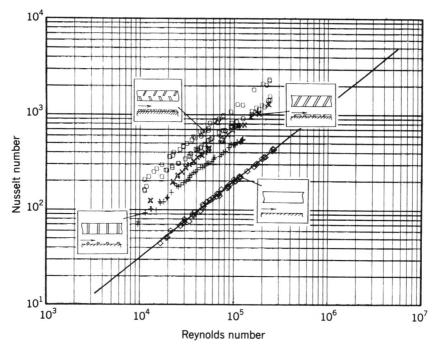

Figure 7.10 Heat transfer characteristics of different shapes of turbulence promotor ribs. Nusselt's and Reynolds numbers are based on mean values. (Courtesy of Hitachi Company, Japan.)

indicates that the most effective configuration is the one with a double row of zig-zag ribs.

Heat transfer with impingement jet cooling is also a complex configuration from the point of view of evaluating the heat transfer coefficient. The configuration is shown in Figs. 7.5, 7.6, and 7.11. The configurations and parameters in this case are numerous. Some attempts have been made by various groups (e.g., Florschuetz and Tseng, 1984; Metzger and Afgan, 1984) to understand the heat transfer coefficients in such a case. The jet flow, after impingement, is constrained to exit in a single direction or dual directions. There may be several cross flows interacting together. The flow field and heat transfer field are truly three-dimensional. It should be emphasized that impingement results in stagnation point in the flow (such as A in the inset of

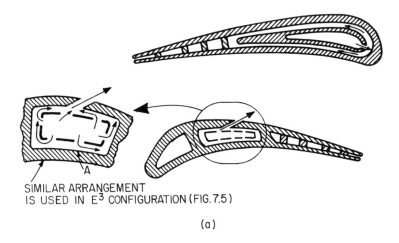

SIMILAR ARRANGEMENT
IS USED IN E^3 CONFIGURATION (FIG. 7.5)

(a)

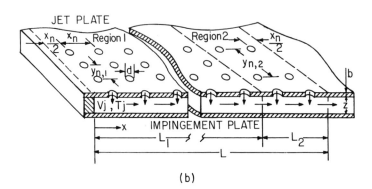

(b)

Figure 7.11 Impingement cooling—examples and laboratory models. (a) Examples of Impingement Cooling. (b) Models Used in Laboratory.

Fig. 7.11a), resulting in high heat transfer rate. One such geometry studied (Florschuetz and Tseng, 1984; Metzger and Afgan, 1984) (single-pass jet) has variables x_n/d, y_n/d, z/d, x/L_1, x/L_2, x/L and $\mathrm{Re}_j = \rho V_j d/\mu$, as shown in Fig. 7.11b, in addition to Re and PR. Even though it is a complex configuration, it is known to have very good heat transfer performance. Florschuetz and Metzger provide a correlation for this case in Metzger and Afgan (1984). In addition to geometric variables, there are at least three temperature fields to reckon with: T_{j1} of the initial cross flow, T_j of the jet, and T_m of the mixture. The heat transfer group at Arizona (Florschuetz and Tseng, 1984; Metzger and Afgan, 1984) have studied this three-temperature problem as well as initial cross-flow effects. Metzger (1985) provides a brief summary of these efforts. One can expect the correlation given by Eq. 7.24 to be valid, with A being a function of geometrical variables shown in Fig. 7.11. The most effective impingement cooling near the leading edge is one with two impingement holes separated by a rib, as shown in Fig. 7.12.

The configuration and geometry of the convective cooling passage, including the nature of impingement cooling, have an appreciable influence on the surface heat transfer. One such investigation by Kohler et al. (1977) reveals the importance of cooling geometry. The blade, with impingement cooling near the leading edge and film cooling near the pressure surface of the trailing edge, had the most desirable property as measured by cooling

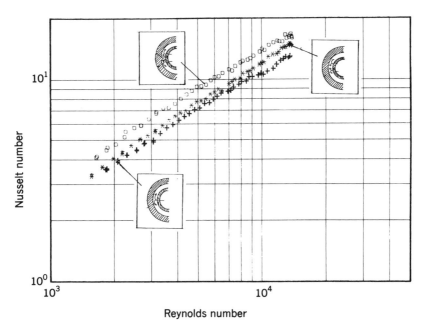

Figure 7.12   Heat transfer characteristics of different impingement cooling system. Reynolds and Nusselt's numbers are based on mean values. (courtesy of Hitachi company, Japan.)

effectiveness (defined later). The leading-edge cooling effectiveness was also maximum for this configuration. At the leading edge, the cooling is most effective with the insert blade, where the entire cooling air is first used to cool the leading edge.

An additional effect, which has not been properly explored, is the effect of centrifugal forces on internal heat transfer in a stationary blade. It is clear from Figs. 7.5, 7.6, and 7.9 that the internal passages contain bends and that the heat transfer in these bends is augmented by the centrifugal forces, which generate secondary flows, as explained in Section 4.2.6. The secondary flow or secondary vortices in these bends introduce intense mixing of the flow, increasing the heat transfer rate substantially. One can expect the constants A and m in Eq. 7.44 to be different. The coefficient A would probably be a function of the radius of curvature. Because δ/R approaches near-unity in these passages, this effect should be very important. The data shown in Fig. 7.9 indicate that this is indeed the case in inner and outer bends. Abuaf and Kercher (1992) recently took detailed flow and heat transfer measurements in a model (10 times the size of a typical turbine blade) with three-pass turbulated cooling circuit geometry. The heat transfer measurements were conducted on rib-roughened channels with staggered turbulators along the convex and concave surfaces using thin foil heater and liquid crystal temperature sensors. Very high turbulence intensities were measured at the exit, and the authors attributed the higher transfer rates measured to these high turbulence intensities.

7.4.2   Internal Heat Transfer (Rotating Blades)

As indicated in an earlier section, heat transfer in a rotating coolant passage includes both forced and free convection. The Coriolis, buoyancy, and centrifugal forces change the heat transfer rates substantially from those of straight nonrotating passages. These forces alter the flow characteristics, including the skin friction coefficient, which control the rate of heat transfer. As explained earlier, the effect of Coriolis force is to destabilize the boundary layer on the leading surface and stabilize the boundary layer on the trailing surface. The effect of rotation on the flow field in a rotating channel was described in Section 4.2.1 (Fig. 4.7), Section 4.2.6.3 (Fig. 4.21), and Section 5.2.1 (Fig. 5.3). As indicated in Fig. 5.3, the turbulence is amplified near the leading side* with an increase in $\overline{(v')}^2, \overline{u'v'}$ (Table 5.1) and larger gradient in $\partial u/\partial y$ and $\partial T/\partial y$ near the wall. The opposite effect occurs near the trailing side, where the flow may relaminarize due to the suppres-

---

*The terminology used in centrifugal turbomachinery field (Fig. 5.3) is opposite to those used by some investigators of heat transfer in coolant passages. To avoid confusion, the notations used in Section 5.2.1 and Section 4.2.6.3 are retained. The leading side (Fig. 7.13) referred to here is the pressure side and is designated by some references quoted in this section as "trailing surface". Likewise, the trailing side referred to here is designated as "leading surface" by some investigators.

sion of turbulence. The heat transfer augmentation (due to rotation) as high as 35% on the leading side and decrement as high as 40% on trailing side have been observed by many investigators.

The effect of rotation on an inward and an outward coolant passage is demonstrated in Fig. 7.13 (note the differences in notation used here and in Figs. 5.3 and 4.7). The velocity, rotation, and Coriolis force vectors $(-2\mathbf{\Omega} \times \mathbf{W})$, where $\mathbf{W}$ is the coolant velocity (radial), are shown for passages with coolant flow from hub to tip. The viscous flow on the leading side of this passage (designated BC in Fig. 7.13b) is destabilized, resulting in thinner boundary layer (with larger values of $\partial W/\partial x$, and $\partial T/\partial x$) and hence has larger heat transfer from blade to coolant. This phenomenon is designated as "+" to denote increased effectiveness. The blade pressure-side heat transfer for this coolant passage is augmented by rotation. The reverse effect exists on the trailing side, designated as AD in Fig. 7.13b, with diminished heat transfer from suction surface to coolant. These effects are reversed in an inward coolant passage (Fig. 7.13c), because the direction of Coriolis force $(-2\mathbf{\Omega} \times \mathbf{W})$ is also reversed. In Fig. 7.13c, the flow near the surface A'D' is destabilized. Hence, the heat transfer is augmented between the blade suction surface and the coolant passage, while it is diminished on the pressure side (B'C'). This effect has been experimentally confirmed by several authors. The data due to Clifford et al. (1985) in the configuration shown in Fig. 7.9 confirm this trend.

Many researchers have carried out investigation of flow and heat transfer in a rotating duct (Lokai and Limanski, 1975; Morris, 1977b; Clifford et al., 1985; Guidez, 1989; Han et al., 1994; Johnson et al., 1994; Kuo and Hwang, 1994). No attempt will be made here to present all of the analysis, but governing equations for a simplified case will be presented here to illustrate the major physical features of the flow and thermal field (see Fig. 7.13 for configuration and notations). The following assumptions are made: The flow is fully developed ($U = V = 0$), the radius R is large compared to D, only dominant viscous terms are retained, and the work done by pressure and shear forces is assumed to be small.

The constitutive equations (nondimensionalized) for this flow are (similar to equations developed in Chapter 4, Eqs. 4.32, 4.44–4.47) as follows:

$$R \text{ momentum:} \qquad \rho W \frac{\partial W}{\partial R} - \rho \Omega^2 R = \frac{1}{\text{Re}} \nabla^2 W + \frac{\rho g D}{W_o^2} - \frac{\partial p}{\partial R} \qquad (7.46)$$

$$X \text{ momentum:} \qquad \frac{2\Omega D}{W_o} W = -\frac{1}{\rho} \frac{\partial p}{\partial x} \qquad (7.47)$$

$$\text{Energy:} \qquad W \frac{\partial \theta}{\partial R} = \frac{1}{\text{Re PR}} \nabla^2 \theta,$$

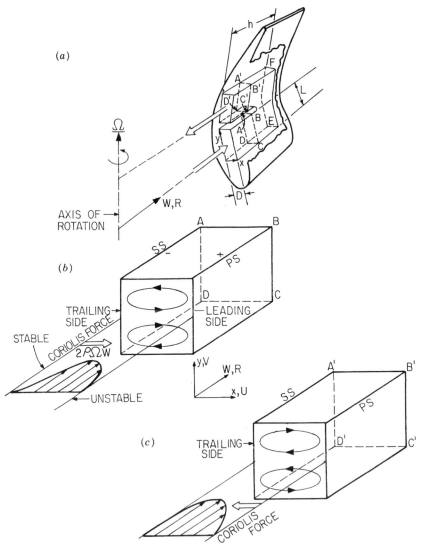

Figure 7.13 Effect of rotation on flow field and heat transfer in coolant passage. (a) Geometry. (b) Flow into ABCD (outflow). (c) Flow out of A'B'C'D' (inflow).

$$\text{where } \theta = \frac{T - T_o}{T_w - T_o}, \text{Re} = \frac{\rho_o W_o D}{\mu_c} \quad (7.48)$$

where all velocities are normalized by W_o (the coolant velocity at inlet), lengths by duct width D, ρ by ρ_o, p by $\rho_o W_o^2$ and Ω by W_o/D and so on, subscript o represents reference values (bulk mean or inlet). Equation 7.46 corresponds to Eq. 4.45 (with $W_r = W$, and $F_r = \rho g$), and Eq. 7.47 corresponds to Eq. 4.46 (with $r\theta = x$) and energy equation corresponds to Eq.

1.34. It is evident from these equations that heat transfer depends not only on the Reynolds and Prandtl numbers, but also on the rotation number and other parameters. Because radius R is assumed to be large compared to D, the curvature effect (W^2/R) is small. Both forced and natural convection are important in this flow. The relative importance of each depends on the Reynolds number, rotation number, and the gravitational effect. As indicated earlier (similar to Eq. 7.27), the buoyancy term in Eq. 7.46 can be expressed as

$$\frac{\rho g D}{\rho_o W_o^2} \sim \frac{Gr}{Re^2}\frac{T - T_o}{T_w - T_o} \sim \frac{Gr}{Re^2}\theta \tag{7.49}$$

In many instances, the rotation term $\Omega^2 R/W_o^2$ and the buoyancy term $\beta g D(T - T_o)/W_o^2$ can be combined to define a rotational Rayleigh number given by (manipulation of buoyancy term is very similar to that carried out earlier in Eqs. 7.27 and 7.28)

$$[\Omega^2 R]\left[\frac{gD}{W_o^2}\right]\beta(T - T_o)(PR) \sim R\,\Omega^2 g\frac{\beta D^3(T_w - T_o)}{\nu^2}\frac{(PR)\theta}{Re^2}$$

$$\sim \left\{\frac{R\,\Omega^2\beta D^3(T_w - T_o)(PR)}{\nu^2}\right\}\frac{\theta}{(Re)^2} \tag{7.50}$$

The term in curly brackets is called the *rotational Rayleigh number* (*RRa*) and characterizes the combined effect of the centrifugal force (caused by noninertial frame and not the streamline curvature) and the buoyancy force. The gravitational constant, g, is dropped, because it is constant in the flow field. Some investigators (Han et al., 1994; Johnson et al., 1994) prefer to use buoyancy parameter given by

$$\left(\frac{\Delta\rho}{\rho}\right)\left(\frac{\Omega R}{W}\right)\left(\frac{\Omega D}{W}\right)$$

which can be proven to be the same as Eq. 7.50 for PR = 1 because $[(T_o - T) \sim (\rho_o - \rho)]$

$$\left(\frac{\rho_o - \rho}{\rho}\right)\frac{gD}{W_o^2}\Omega^2 R = \left(\frac{\Delta\rho}{\rho}\right)\left(\frac{\Omega R}{W_o}\right)\left(\frac{\Omega D}{W_o}\right)g$$

Many conclusions and interpretations can be drawn from these equations and others presented earlier.

1. The Coriolis force sets up pressure gradient in the x direction and modifies the velocity profiles of the coolant air as explained earlier and shown in Fig. 7.13.

2. The pressure gradient necessary to force the flow through the coolant passages should be sufficient to overcome the frictional losses, the centripetal acceleration, and the generation of forced convection of coolant flow.

3. There will be secondary flow (velocity in x and y directions) induced in the duct due to both the Coriolis force and the density gradients as indicated in Section 4.2.6. Consider Eq. 4.133 in Chapter 4, with s', n', b' representing R, x, and y directions, respectively, in Fig. 7.13. Let $\Omega_y = \Omega$, $\Omega_x = \Omega_z = 0$. Then

$$\frac{\partial}{\partial R}\left(\frac{\omega_R}{\rho W}\right) = \frac{2\Omega \omega_x}{\rho W^2} - \frac{1}{\rho^3}\frac{1}{W^2}\left(\frac{\partial p}{\partial x}\frac{\partial \rho}{\partial y}\right) \qquad (7.51)$$

where $\omega_x = \partial W/\partial y - \partial V/\partial R$, and ω_R is the streamwise vorticity in the coolant passage.

The secondary flow and vorticity is induced by rotation, as shown in Fig. 7.13. ω_R is positive on the bottom wall and negative on the top wall of the outflow channel. This is reversed for the inflow channel. The density stratification occurs in the y direction and this combined with Coriolis force induced pressure gradient in the x direction induces secondary flow and vorticity. Thus, the cross coupling between the buoyancy terms and the Coriolis force terms $[(\partial p/\partial x)(\partial \rho/\partial y) \sim 2\rho\Omega W(\partial \rho/\partial y)]$ introduces additional secondary flow and has a major effect on the heat transfer. The secondary flow enhances the heat transfer. In addition to these two types of secondary flow, another source of secondary flow is due to turning of the flow in a bend (first term in Eq. 4.133). As explained in Section 4.2.6, this secondary flow arises due to the turning of the flow with vorticity. Additional secondary flows would arise in return bends such as EF (Fig. 7.13). The effect of buoyancy and rotation is already included in the rotational Rayleigh number, and this should be adequate to take into account the secondary-flow-induced heat transfer augmentation. Therefore, the heat transfer in a rotating duct, such as the one shown in Fig. 7.13, can be represented as follows:

$$\text{Nu} = f(\text{Re}, \text{PR}, \text{RRa}, J), \qquad \text{Re} = \frac{\overline{W}D}{\nu}, \qquad \text{Ro} = \frac{\Omega D}{\overline{W}}, \qquad \text{Nu} = h_c D/k$$

$$\text{RRa} = \frac{\Omega^2 R D^3 \beta \,\Delta T\, \text{PR}}{\nu^2}, \qquad J = \text{Re}\,\text{Ro} = \frac{\Omega D^2}{\nu} \qquad (7.52)$$

The rotational Rayleigh number (RRa, defined in Eq. 7.50) accounts for buoyancy and centrifugal force effects, both direct and indirect. The rotational Reynolds number, J (Eq. 7.30), includes the effect of rotation in altering the flow and thermal field. From Eq. 7.47, it is clear that the rotation number, Ro, would be the more appropriate parameter, but almost all the investigators in this field prefer to correlate the flow with the rotational Reynolds number J.

There have been many basic experiments carried out to determine the precise relationship between heat transfer and the nondimensional variables in Eq. 7.52. Some of the representative data can be found in Lokai and Limanski (1975), Morris (1977b), Clifford et al. (1985), Guidez (1989), Han et al. (1994), Johnson et al. (1994), Mochizuki et al. (1994), El-Husayne et al. (1994), Wagner et al. (1991), and Kuo and Hwang (1994). Representative data by Guidez (1989) are shown in Fig. 7.14. A horseshoe type of return duct, shown in Fig. 7.13, was tested in an electric furnace (temperature range 700–1400 K) with the test section heated by radiative flux, with coolant air flow as shown in Fig. 7.13 (with $D = 8$ mm, $L = 16$ mm, $R = 286$ mm, $h = 122$ mm, $\dot{m}_c = 4$–10 g/s, $N = 0$–5000 rpm, $\overline{W} = 30$–50 m/sec, $T_w = 270$–$320°C$, $T_{blade} = 600°C$). Very detailed flow field and heat transfer data, including velocity profiles and secondary flow patterns, have been acquired. The nondimensional data at various Reynolds numbers are shown plotted against the Rotation (Ro) number in Fig. 7.14. It is clear that the heat transfer increases on the leading, or the pressure side, and decreases on the trailing, or suction side. This is consistent with the rotation effects described earlier. The secondary flow augments heat transfer on both the inward and the outward flow duct, while the direct effect of rotation is to augment heat transfer on the leading or pressure side (e.g., BC in Fig. 7.13b) and reduce heat transfer on the trailing side for outflow of the coolant passage. This rotation effect probably dominates, thereby causing the observed effect,

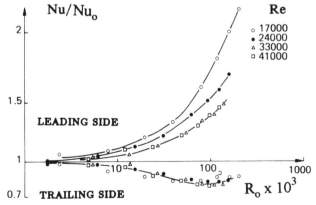

Figure 7.14 Variation of Nusselt number with rotation number. Reynolds number and rotation numbers are based on the hydraulic diameter, D_H, of the coolant passage. Nu_0 is the Nusselt's number for $\Omega = 0$, $\overline{W}$ is the mean velocity, $Ro = \Omega D_H / \overline{W}$. (Copyright © Guidez, 1989; reprinted with permission of ASME.)

namely, an enhanced heat transfer on the leading side and decreased heat transfer on the trailing side. The decrease in the heat transfer due to rotation on the trailing side (AD) is substantially reduced due to an increased heat transfer from the secondary flow. The ratio of Nu/Nu_0 (where Nu_0 is the heat transfer without rotation) varies with both the rotation number and the Reynolds number. These nondimensional parameters can be combined into one parameter, rotational Reynolds number J. The Nusselt's number on the leading (pressure) side increases as much as 50–100% at Ro = 0.1.

The experiment data due to Guidez (1989), Wagner et al. (1991), and Mochizuki et al. (1994), analyzed by Mochizuki et al. (1994), are shown in Fig. 7.15. The results clearly confirm the trends postulated earlier, namely, that the heat transfer is augmented on the surfaces leading the rotation. For example, surface BC in Fig. 7.13 with positive rotation encounters augmentation; when the speed is reversed, these surfaces encounter a decreased heat transfer rate. The trailing surfaces encounter opposite effects. The skin friction coefficient measurement carried out in a rotating duct by Johnston et al. (1972) clearly reveals that C_f increases almost linearly on the leading side with rotation number and decreases almost linearly on the trailing side for low rotation numbers. This clearly confirms the validity of Eq. 7.18, relating the skin friction to heat transfer coefficient. In addition, the effect of buoyancy force is also revealed in this plot (Fig. 7.15). The dependency of Nu on the parameter, Gr/Re^2, as revealed by Eq. 7.49, is also confirmed by this plot. The experiments by Guidez and by Mochizuki et al. were with uniform wall heat flux, while Wagner et al.'s data utilized uniform wall temperature. All the experiments used a square cross section. The three sets of data (Fig. 7.15) agree fairly well on the Nusselt's number distribution on the leading surface over the entire range of Gr/Re^2, as well as on the trailing surface up to $Gr/Re^2 > 0.1$. In summary, the heat transfer in coolant passages can be

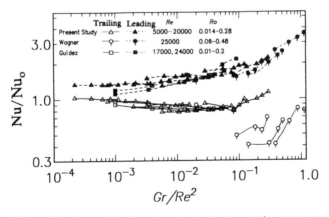

Figure 7.15   Effect of Ro, Gr, and Re (based on hydraulic diameter) on internal heat transfer in a rotating channel. (Copyright © Mochizuki, 1994; reprinted with permission of ASME.)

represented by

$$\mathrm{Nu}_m = f(\mathrm{Re}, \mathrm{PR}, \mathrm{RRa}, J) = A\,\mathrm{Re}^m\,\mathrm{PR}^n f(J, \mathrm{RRa}, \mathrm{Ro}) \qquad (7.53)$$

The effects in a stationary duct are separated from the rotation effects through the function f in Eq. 7.53. Correlations proposed by various investigators are tabulated in Table 7.1. Morris and Ayhan (1979) and Clifford et al. (1985) use H instead of R in rotational Rayleigh number (Eq. 7.50), where H is the eccentricity of mid-cross-sectional plane of the cooling duct. Clifford's (1985) data acquired in a multipass passage (Fig. 7.9) show large scatter in the correlation for function f in Eq. 7.53. Clifford found that rotation has a significant effect on the heat transfer distribution in cooling passages, with reduction and enhancement as much as 30% of the value for a stationary duct.

It is evident that for the practical configuration, such as the one shown in Fig. 7.9, there is no satisfactory correlation. However, most of the correlations based on rotating pipe flow (Table 7.1) provide the correct trend for the effects of rotation, even though the magnitudes are not predicted well.

In summary, rotation and buoyancy have a major effect on internal heat transfer. The leading-side (radial outward flow) heat transfer is augmented [as much as 3.5 times (Wagner et al., 1991)] and attenuated on the trailing surface (as much as 40% from the stationary case measured by Wagner et al., 1991). The trend is reversed in the return (inward) passage. In an actual turbine, the leading and trailing surfaces are at differing temperatures. Han et al. (1994) studied this effect experimentally and found that unequal buoyancy forces caused by uneven wall temperatures appreciably alters the heat transfer characteristics.

7.4.3 External Heat Transfer

This section is concerned with heat transfer between the external gas and the blade, both with and without convective cooling. External heat transfer is very complex and is affected by several fundamental mechanisms, as explained earlier and as shown in Fig. 7.3. Among the important mechanisms and flow variables are: stagnation point heat transfer; laminar, transition, and turbulent flow heat transfer; pressure gradient; shock–boundary-layer interaction and shock oscillations; flow unsteadiness (due to wakes, etc.); separation and reattachment; free-stream turbulence intensity and structure; compressibility, coolant jet/free-stream and coolant jet/blade boundary layer interaction; secondary flow and other three-dimensional effects; and leakage flow. The last two mechanisms are essentially three-dimensional phenomena, and all others are present in both two-dimensional and three-dimensional flows. The blade row and blade variables that control the heat transfer are camber (curvature), solidity, aspect ratio, incidence, rotation, roughness, leading edge radius, and profile shape and so on.

These mechanisms are in addition to the dependency of heat transfer on the temperature ratio, Reynolds number, Prandtl number, and Mach number as described earlier. The fundamental features in all of these are the viscous layer near the surface and the wall shear stress, which are affected by the mechanisms mentioned earlier. A detailed understanding and prediction of external heat transfer necessarily involves our ability to predict the boundary layer structure, including the effects mentioned above.

As indicated earlier, heat transfer is very sensitive to transition. Transitional behavior (laminar, transition, turbulent) has a major influence on the magnitude of the heat transfer. The free-stream turbulence, roughness, and curvature not only influence the transitional behavior, but also affect the growth of laminar and turbulent boundary layers. Concave curvature may produce Görtler vortices, which influence the heat transfer. As indicated in Section 5.2.1, convex curvature suppresses the generation of turbulent kinetic energy and concave curvature augments the generation of turbulent energy. This should have a corresponding effect on the heat transfer. Heat transfer reduces on a convex (suction) surface and increases on a concave (pressure) surface. The influence of the pressure gradient on boundary layer growth is well known, and this has a direct influence on the $(\partial u/\partial y)_{\text{wall}}$, $(\partial T/\partial y)_{\text{wall}}$, separation, and reattachment. A sharp increase in local heat transfer is caused by reattachment just after the separated zone. A separated flow zone is normally a cold spot. This effect is also present in shock–boundary-layer-interaction induced separation. The unsteadiness in the free stream controls not only the transition (being time-dependent), but also its location, which may vary widely (from leading to trailing edge) depending on the Reynolds number and the magnitude of unsteadiness. The secondary flow and leakage flow are three-dimensional effects which augment heat transfer from the gas to the blade. Figure 7.3 clearly depicts the realistic heat transfer phenomenon on turbine blades.

The internal heat transfer correlations presented earlier have been successful in predicting heat transfer from blade to coolant. Simple correlations, presented earlier, have been used very effectively in predicting the heat transfer or cooling requirements. On the other hand, there is no satisfactory correlation for the prediction of external heat transfer, which is complicated by the various mechanisms mentioned earlier. This is compounded by our inability to predict the precise location of the transition, separation, and effect of free-stream turbulence. An additional problem is the lack of adequate and accurate data that reveal the dependency of heat transfer on parameters that characterize the condition of the boundary layer. Earlier data on external heat transfer (e.g., Hodge, 1958; Turner, 1971), obtained at restricted variation of parameters, did not provide the necessary information to generate either the empirical correlations or analytical or computational techniques. Heat transfer in general can be represented as a function of the following variables: $\text{Nu} = f(\text{Re}, M, \text{Tu}, \text{PR}, C_p(x), T_g/T_w, X_t, X, R, \Omega, \omega, k, \text{Ro}, \text{3D effects})$. The review by Simoneau and Simon (1993) provides an excellent overview of the data available on external heat transfer and an assessment of heat transfer computational techniques.

Several early investigators (Hodge, 1958; Turner, 1971) provided data and correlations at low Reynolds number in a cascade. In recent years, there has been an increased effort in understanding the various mechanisms and in generating good-quality data to enable the development of the high-pressure and high-performance turbine. Measurements have been carried out in cascades and rotors. Some of these and other data will be examined here to present an integrated treatment of various mechanisms responsible for heat transfer. The methods available for the prediction of these effects will be presented in a later section.

Nealy et al. (1984) carried out a systematic study of heat transfer on a highly loaded, low-solidity contemporary turbine blading at an inlet temperature of 811 K. The measured local heat transfer rates over the highly loaded nozzle cascade airfoil (blade spacing = 11.77 cm, C = 14.49 cm, $\alpha_2 = 72.38°$, Re_2 in Fig. 7.16 is based on exit velocity, S is the non-dimensional surface distance from leading edge) and experimental parameters are shown in Fig. 7.16. High heat transfer in the stagnation region (where velocities are extremely small and boundary layers are very thin) is clearly evident in Fig. 7.16. As the laminar boundary layer grows, the heat transfer decreases due to decreased values of $\partial u / \partial y$ and $\partial T / \partial y$, reaching the lowest rate near the transition inception point. The heat transfer increases again in the

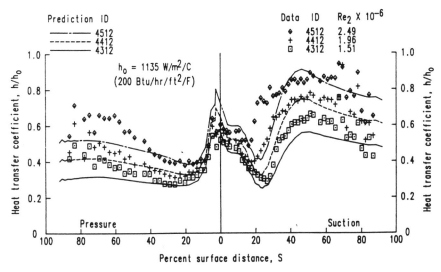

Figure 7.16 Predicted and measured blade heat transfer coefficient on a cascade blade. S = 0 is the leading edge ($T_w / T_g = 0.73$, $T_u = 6.5\%$, $M_1 = 0.18$, $M_2 = 0.90$) (Nealy et al., 1984; NASA CR 168015).

transitional region. The reasons for higher heat transfer in the turbulent regions have been explained earlier. The transition on the suction surface occurs between 20% and 30%, and the heat transfer decreases in the fully turbulent region (beyond 50% chord).

It should be noted that heat transfer is very sensitive to the streamwise pressure gradient. The influence of the Reynolds number on transition as well as on heat transfer is substantial. The transition region becomes smaller and occurs earlier with an increase in the Reynolds number. Abrupt changes in heat transfer on the suction side with changes in the Reynolds number are largely controlled by changes in the inception of transition. The pressure surface heat transfer distributions are largely dependent on the Reynolds number and show a tendency toward transitional behavior. The predictions are from the boundary layer code (Crawford and Kays, 1976) mentioned in Section 5.5.5.1. An interpretation of this will be covered Section 7.8.3. Nealy et al., also successfully calculated the internal heat transfer rate and coolant temperatures using the correlations presented earlier, and they used this and the measured external heat transfer rate as boundary conditions to solve the temperature distribution within the blade using Eq. 7.9 and the finite element technique. Data similar to those shown in Fig. 7.16 have also been obtained by Arts et al. (1992). In some cases, a laminar separation bubble, followed by transition and reattachment, may exist. It has been observed that downstream of separation point or transition point, the heat transfer varies as $(Re)^{0.8}$ as indicated by Eq. 7.23. A summary of all cascade heat transfer experiments and data can be found in Simoneau and Simon (1993).

The experimental data available in three-dimensional rotating rig configuration were carried out either in short duration (30–3000 ms) facilities or low speed facilities. In the former category, most comprehensive data have come from Dunn and co-workers (Dunn, 1985, 1986; Dunn et al., 1994), Epstein and co-workers (Abhari et al., 1992; Abhari and Epstein, 1994), and Ashworth et al. (1985). Most comprehensive data in a low-speed facility are due to Blair (1994).

Dunn and his group at Calspan (Dunn, 1985, 1986; Dunn et al., 1994) have built a short-duration shock-tunnel facility to duplicate the conditions that exist in an actual engine. The mass flow parameter, temperature ratio T_w/T_g, and pressure ratios are duplicated. Heat transfer measurements were carried out in a space shuttle main engine turbine stage (shown in Fig. 1.6 with two stages). The measurements were taken at various spanwise locations on the first- and second-stage vanes and the second-stage rotor with thin-film resistance thermometers. The operating conditions were: speed 9075–9690 rpm, Re ~ 1.38–3.00×10^5, temperature ~ 970–1112 K, stage pressure ratio 1.38–1.65. The heat transfer at 50% of span for the first stage rotor (63 blades) is shown in Fig. 7.17. The data on the suction surface shows a rapid decrease in Stanton number from stagnation point to about 10% of surface length from the leading edge, followed by a sharp increase with a peak at 15% of the surface length from the leading edge; beyond this point

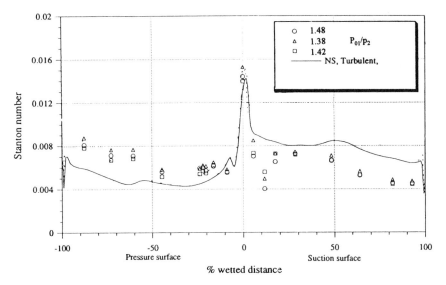

Figure 7.17 Stanton number distribution (Eq. 7.18) on first SSME rotor blade, 50% span, Re ~ 250,000. (Copyright © Dunn et al., 1994; reprinted with permission of ASME.)

there is a decrease in Stanton number. On the pressure side, the heat transfer falls sharply from the stagnation point, reaching a minimum value at about 10% surface distance, and then increasing steadily toward the trailing edge. This is similar to the heat transfer data measured in a cascade (Fig. 7.16) and interpreted earlier. The data reported for an SSME vane and blade (Fig. 7.17) are similar to the data from other engine (aircraft) configurations (Dunn 1985, 1986). The predictions shown in Fig. 7.17 were carried out with a Navier–Stokes code and will be covered in Section 7.8.3.

The data of Dunn et al. (1994) and Dunn (1985, 1986) and Abhari et al. (1992) indicate that the rotor flow is probably fully turbulent, and hence the low-speed and cascade data should be used with caution. Some striking differences can be noticed between the rotor data (Fig. 7.17) and the cascade data (Fig. 7.16), especially on the suction surface. Dunn et al.'s data indicate that the laminar length on the suction side is relatively short. The Stanton number drops rapidly from the stagnation point to about 10% of the surface length, and this corresponds to the laminar region; unlike the cascade vane data, the heat transfer is nearly constant from 15 to 50% of surface length. The transition is probably abrupt (one can only make an intelligent guess with this data because the corresponding aerodynamic data was not acquired by either one of the authors). The pressure side also shows differences between the cascade and rotor blade data. The transition region is very short and the turbulent part shows a nearly constant heat transfer rate for 50% of the surface from the leading edge. This is probably caused by early transition brought about by impinging wakes (where turbulence is high) from a nozzle vane preceding the rotor.

The transition, laminar region, and the boundary layer growth on a rotor blade is likely to be substantially different from a cascade, as explained in Sections 5.2, 5.5.1, and 6.2. The heat transfer on nozzle vanes is likely to be closer to that of stationary cascades, provided that inlet flows are simulated properly in a cascade. Because the flow upstream of the rotor is likely to be highly turbulent due to nozzle wakes, early transition can be expected. Blair (1994) has provided comprehensive heat transfer data on a low-speed turbine rotor, including the endwall regions. The effects of surface roughness, off-design operating conditions, and Reynolds number were investigated. In addition, the loading (variation of inlet angles) had a major effect on the heat transfer near the leading edge. The heat transfer in trailing-edge regions were unchanged. The incidence effect manifests itself in the form of change in local pressure gradient, change in local Reynolds number, and change in transition location. Increased surface roughness greatly increased the heat transfer rates for all locations.

There has been considerable interest in recent years to measure and evaluate the effect of upstream flow unsteadiness on the heat transfer. This is caused by upstream wakes of nozzle vanes, the potential interaction due to the relative motion between a nozzle and a rotor blade row, and high turbulence intensities at the exit of the combustor. In addition, the presence of shock waves in these passages induce large unsteadiness, because any small movement of shock wave location causes large fluctuations in the static pressure and in the boundary layer behavior. Shock locations as well as transition will be time-dependent, causing an extremely complex and unsteady flow and thermal field.

The Oxford University group (Ashworth et al., 1985; Doorly, 1987) has simulated the unsteady heat transfer in a linear cascade ($S = 3.43$ cm, $C = 4.18$ cm, $h = 5$ cm, Re $= 9 \times 10^5$, $M_2 = 1.18$, $T_o = 432$ K, turning angle 125°) using an isentropic, light piston facility (0.3–1.0 s) and a moving rod upstream. The experiment provides a basic understanding of unsteady phenomena associated with the wake passing. Heat transfer data (Nusselt's number Eq. 7.15 with $L = C$), with and without the passing of a moving wake upstream, are shown in Fig. 7.18. There is an enhancement of heat transfer due to upstream unsteadiness (due to wake) from the leading edge to about 50–60% of chord on both surfaces. The mean turbulence level is likely to be higher with an upstream wake, and this alters the boundary layer character. Both mechanisms are present in the data shown in Fig. 7.18. The instantaneous traces of heat transfer shown in the insets indicate that fluctuations as high as 100% are observed on the pressure surface at about 10% of the chord near the leading edge. This is clearly caused by a substantial movement of transition location with time. The oscillations represent the transient passing of laminar, transitional, and turbulent regions at these locations. On the other hand, instantaneous values on the suction side

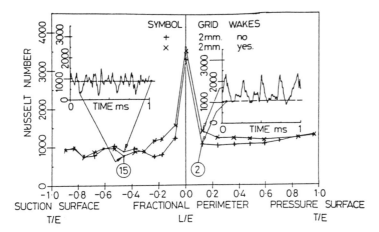

Figure 7.18 Effect of wake interaction on mean heat transfer to blade with free-stream turbulence of ~ 4% and normal "design" operating conditions. The insets show typical transient recorder signals at two chordwise locations showing a 1-ms interval of the record together with the time-averaged value in the absence of wake interaction. (Copyright © Ashworth et al., 1985; reprinted with permission of ASME.)

at 45% of chord show diminished fluctuations, indicating lower unsteadiness at these locations. The fluctuations in heat transfer can have a significant effect on cooling air requirements, thermal fatigue, and the life of the blade.

Recent measurements by Liu and Rodi (1992) in a simulated unsteady flow in a low-speed cascade indicate that the boundary layer transition on the suction surface moves upstream at increased wake passing frequency. In addition, the increased frequency increases the free-stream turbulence, causing increased heat transfer on both the sides.

As indicated earlier [e.g., Fig. 7.16 and Blair (1994)], the Reynolds number has a major influence on local and global heat transfer through direct effect on such phenomena as laminar, transitional, and turbulent boundary layer, but also through indirect effect on various mechanisms mentioned in the beginning of this section. Furthermore, local heat transfer is very sensitive to additional effects such as unsteadiness, curvature, and 3D effects. It is worthwhile examining the global heat transfer (mean for the entire surface) to derive a reasonable correlation to predict the trend. Such an attempt was made by Louis (1977), who brought together all of the earlier data (1958–1977) and plotted it against the Reynolds number. At high Reynolds number, the data (slope) follow the trend predicted by Eq. 7.24, even though the constant, A, differs widely from one configuration to another. The constant should depend on such effects as transition location, curvature, turbulence intensity, and so on. Such correlations provide valuable information to the designer, who can develop correlations and determine the external heat transfer coefficient and use it to determine the coolant requirements and wall temperatures.

7.4.3.1 *Effect of Free-Stream Turbulence and Pressure Gradient*   Turbulence in the free stream has a major influence on the laminar, transitional, and turbulent boundary layers, which, in turn, affects heat transfer. The pressure gradient effects on laminar and turbulent boundary layer growth are well known (Schlichting, 1979) and have a major influence on laminar, transitional, and turbulent boundary layer growth and heat transfer. It is well known that the favorable pressure gradient (as in the case of a turbine) stabilizes the laminar boundary layer, delays transition, and increases transition length. The flow acceleration also has a damping effect on turbulence intensities inside the boundary layer. In almost all practical situations, both the free-stream turbulence and pressure gradient effects are present, and there is a close coupling between these two mechanisms. In addition, the wall cooling (or heating) has a strong influence on the boundary layer properties, and here again the effects and mechanism associated with free-stream turbulence, streamwise pressure gradient, and wall cooling (or heating) are closely coupled. Many investigators have attempted a basic study of the effects of each of these phenomena when the other effects are absent. For example, Schlichting (1979) has listed large numbers of references associated with pressure gradient effects alone, carried out in the presence of very low turbulence intensities simulated in well-controlled flow facilities, such as a wind tunnel. Several references (e.g., Krishnamoorthy, 1982; Blair, 1983; Rued and Wittig, 1985) deal with the effects of turbulence on heat transfer. Krishnamoorthy (1982) observed significant changes in heat transfer beyond 5% turbulence intensity. Rued and Wittig (1985, 1986) have reported extensive measurement with a systematic variation of all of these conditions. Some of their results will be presented here.

Let us first consider the effect of free-stream turbulence on heat transfer. As indicated earlier, free-stream turbulence affects transition. This is clear from Fig. 7.19. The laminar and turbulent parts of the heat transfer distribution follow the relationships for laminar flat-plate boundary layer (Eq. 7.19) and turbulent flat-plate boundary layer (Eq. 7.23), respectively. But none of the techniques are satisfactory for the prediction of transition or heat transfer in the transitional boundary layer, much less the prediction of the onset of transition at different free-stream turbulence levels. It is clear from this plot that the transition point moves continuously toward upstream with an increase in the free-stream turbulence intensity, and at very high turbulence intensities, transition occurs very near the leading edge. This is also clear from the comparison of well-controlled cascade experimental results and highly turbulent rotor results presented in Figs. 7.16 and 7.17, respectively. In actual turbines, the turbulence encountered at the nozzle could be very high (10–15%) due to intense mixing in the combustion chamber. Therefore, the effect of free-stream turbulence as well as pressure gradient

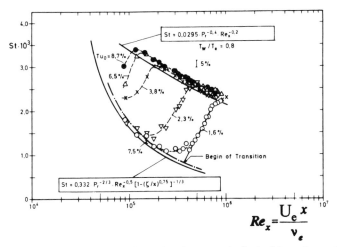

Figure 7.19 Laminar and transitional boundary layers and effect of free stream turbulence on a flat-plate boundary layer heat transfer (St Eq. 7.18), (Tu$_0$ is the free stream turbulence intensity, U_e is the free stream velocity and ζ is the uncooled starting length of the plate). (Copyright © Rued and Wittig, 1985; reprinted with permission of ASME.)

should be taken into account in predicting blade temperatures and air-cooling requirements.

Blair (1983) observed that the combination of unsteadiness, high Reynolds number, and concave surface can produce high heat transfer distributions well in excess of fully turbulent levels. In addition, laminar separation bubble and its reattachment also caused enhanced heat transfer in the reattachment location.

Moss and Oldfield (1992) reported appreciable effect of the turbulent length scale on heat transfer. A systematic experimental program was designed to vary inlet turbulence as well as the length scales to a nozzle cascade using grids of varying geometries. The data based on a large number of tests were correlated to derive the following empirical relationship [based on Blair's (1983) earlier investigation]:

$$\frac{\text{Nu}}{\text{Nu}_0} = 1.185 \left\{ \frac{\text{Tu}}{\left(\dfrac{1.5 L_x}{\delta} + 2 \right) \left[1 + 3 \exp\left(\dfrac{-\text{Re}_\theta}{400} \right) \right]} \right\}^{0.222} \tag{7.54}$$

Where L_x is the turbulent integral length scale (longitudinal), δ is the boundary layer thickness, Tu is the turbulence intensity, Re_θ is the Reynolds number based on momentum thickness, and Nu_0 is the Nusselt number without grid. Similar correlations have been developed by Dullenkopf and Mayle (1994), based on the data by Priddy and Bayley (1988) and others.

The effect of pressure gradient on heat transfer was investigated by Rued and Wittig (1985, 1986) in a well-controlled experiment on a flat plate. These data were acquired at the same turbulence level (7% and 10%) at several pressure gradient parameters (accelerated flow) given by

$$K = \frac{\nu}{U_e^2} \frac{\partial U_e}{\partial x} \tag{7.55}$$

The values of acceleration factor K were varied from 3.2×10^{-6} to 5.6×10^{-6}. The pressure gradient has appreciable influence on the laminar and turbulent boundary layers, because the profile and thicknesses are modified by the pressure gradient. Decreased heat transfer in turbulent and laminar boundary layers at large values of K can also be inferred from these data. The acceleration has a damping effect on turbulence and strongly increases the viscous sublayer thickness, and this in turn reduces heat transfer. Rued and Wittig (1985) compared their data with the correlation given by Kays and Crawford (1993)

$$\text{St} = 0.0295(\text{PR})^{-0.4}\text{Re}^{-0.2}\left(1 - F_k\frac{K(x)}{\text{St}}\right)F(\text{Tu}) \tag{7.56}$$

$F(\text{Tu})$ is a function of turbulent length scale and intensity (e.g., Eq. 7.54). F_k accounts for the effects of pressure gradient (its value varies from 90 to 165), and Re is an integral Reynolds number given by

$$\text{Re} = \int_0^x \frac{U_e(x)\, dx}{\nu_e(x)}$$

The correlation is good for low values of K at most Reynolds numbers. However, it underpredicts the heat transfer in the initial part of the plate, even at low turbulence intensities. The result is not surprising, because Kays and Crawford correlation does not include the influence of free-stream turbulence. When the parameter F_k was reduced to 90, the correlation provided better agreement with the data. For turbine flows, one may generalize the correlation as follows:

$$\text{Nu} = A(\text{Re})^m(\text{PR})^n\left(\frac{T_e}{T_w}\right)^c F(\text{Tu})F(k) \tag{7.57}$$

In view of recent developments in Navier–Stokes solvers, boundary layer codes, and thermal and turbulence models, these correlations may have applications only in initial and preliminary estimates.

7.4.3.2 *Stagnation Point Heat Transfer*   The cooling schemes used in a turbine nozzle and rotor require coolant holes/film cooling near the leading edge. This invariably results in a thick leading edge, leading to large acceleration and curvature effects. Such sudden acceleration can bring about considerable changes in heat transfer rates. In many cases, the flow near the leading edge may separate. There have been a large number of analytical/ experimental studies related to laminar boundary layer behavior near the stagnation point (Eckert and Drake, 1972; Schlichting, 1979). Furthermore, the stagnation point flow is very sensitive to free-stream turbulence. For example, it has been found that even low turbulence levels can cause the formation of streamwise vortices, which augment heat transfer. Kestin's data (Schlichting, 1979; Back et al., 1970) indicate that the heat transfer near the leading edge of a cylinder is doubled when the turbulence is increased from zero to 2.2%. Even though the cylinder case does not represent the leading edge of a turbine blade, the qualitative effects are similar, and one can expect the trend to be the same in turbine leading-edge heat transfer.

As indicated earlier, cooled turbine blades usually have thick leading edges and many bladings may have a laminar separation bubble near the leading edge. Separation and reattachment can result in a substantial change in the heat transfer. Bellows and Mayle (1986) undertook a systematic study of this phenomenon using a two-dimensional blunt body. They have also measured the boundary layer and the skin friction characteristics in this region. The data near the stagnation point agree quite well with the correlation for the laminar flat-plate boundary layer (Eq. 7.19), but caution should be exercised; the situation may be different in turbine blading where the free-stream turbulence is high (compared to 0.4% in Bellows and Mayle's experiment), and much higher heat transfer rates can be anticipated. At about $Re_\theta = 100$, the Stanton number decreased rapidly to half its value and then increased rapidly in a short distance. A laminar separation bubble gave a minimal heat transfer rate. Subsequent reattachment resulted in high heat transfer rates with peak values. Downstream of the reattachment, the data follow the turbulent heat transfer relationship given by Eq. 7.23. The location and extent of the separation bubble would depend on free-stream turbulence as well as free-stream temperatures, and these areas need to be addressed.

Because a turbine blade has a cylindrical leading edge, the data and correlations based on cylinder stagnation point heat transfer are useful in turbine applications. Lowery and Vachon (1975) carried out measurement and correlation for the stagnation point heat transfer and provided the following correlation [see Daniels and Schultz's article in VKI Lecture Series (1982) for other correlations and data]:

$$Nu_d = Re_d^{0.5}\left[1.01 + 2.624\left(\frac{Tu\,Re_d^{1/2}}{100}\right) - 3.07\left(\frac{Tu\,Re_d^{1/2}}{100}\right)^2\right] \quad (7.58)$$

The Nusselt and Reynolds numbers are based on the diameter of the cylinder (d). This correlation is valid for $Tu\,Re_d^{1/2}$ from 0 to 40, and it underpredicts Nu_d after Nu_d reaches a maximum value. The last two terms incorporate the effect of free-stream turbulence. Daniels and Schultz (1982) show that their leading edge (HP blade) data agrees to within 10% of the value predicted from the above correlation. The physical mechanism by which the free-stream turbulence affects the leading-edge flow field and heat transfer is not well understood. Further complications arise due to curvature and large deceleration or acceleration of the flow. The breaking down of the streamwise vortices originating near the stagnation point has a significant effect on the heat transfer. This breakdown occurs where the curvature changes on the pressure surface. Vanfossen and Simoneau (1994) provide correlation similar to Eq. 7.58 and include the effect of turbulence integral length scale. The correlation, based on a systematic experiment with leading edges of different shapes, indicates that the heat transfer augmentation, due to turbulence, is independent of the body shape.

7.4.3.3 Additional Effects   In addition to the effects dealt with earlier, rotation, curvature, roughness, and three-dimensionality do have a substantial effect on the external heat transfer. Because these areas have not yet been thoroughly investigated and understood, they are all included in this section with a brief introduction to the subject.

The surface curvature of turbine blading has substantial influence on boundary layer growth, velocity profile, wall shear stress, turbulence structure, and heat transfer rate. The effect of curvature on turbulence and boundary layer growth was dealt with in Section 5.2.1 (Figs. 5.2 and 5.6). It is well known that the surface curvature, even with small values of δ/R (as low as 0.01), a parameter controlling the curvature effect, can have an appreciable influence on the wall shear stress and the heat transfer. Gillis and Johnston's (1983) aerodynamic data on a convex wall clearly indicate that C_f decreases rapidly in the curved region, and curvature should have a corresponding effect on the heat transfer. The C_f values decrease as much as 30% for $\delta/R = 0.1$. The conclusions from this and other studies summarized by Kim and Simon (1988) for a turbulent boundary layer on a convex wall are as follows:

1. A rapid decrease in Stanton number and C_f were observed at the beginning of the curved portion, followed by a slower decrease beyond the initial part. Decreases of 20% to 10% in C_f and St from the flat-plate values were observed for $\delta/R = 0.13$ and 0.03, respectively. Recovery was very slow beyond the curved section, with Stanton number recovering faster than C_f.

2. The turbulent heat flux $\overline{v'T'}$ (Eq. 7.13), which controls heat transfer, was affected by curvature more strongly than the the turbulent shear stress $\overline{u'v'}$.

The correlation $(\overline{v'T'})$ at $y/\delta = 0.4$ (shown in Fig. 7.20a) for $\delta/R = 0.03$ decreased by nearly 50% compared to the value for a flat plate, with a much

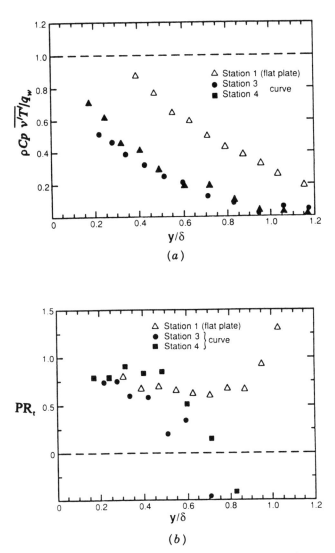

Figure 7.20 Effect of convex curvature on heat transfer (Tu = 2.0%). y is the distance normal to wall. (a) Effect of curvature on $\overline{v'T'}$ normalized by wall heat flux. (b) Effect of curvature on PR_t. (Copyright © Kim and Simon, 1988; reprinted with permission of ASME.)

more substantial reduction in outer layer of the boundary layer. It is also clear that there is a substantial variation in turbulent Prandtl number across the boundary layer. The effects of curvature on concave (pressure) surface is substantially different from that on the convex surface (see Section 5.2.1, Figs. 5.2 and 5.6). The turbulence intensities and shear stresses increase, and this should lead to increased heat transfer rates. The effect of curvature can be captured accurately using the algebraic Reynolds stress model described in Sections 1.2.7 and 5.2.2. Several empirical correlations have been suggested (Lakshminarayana, 1986) to include curvature effect on C_f, which can then be used in Eq. 7.23 (with modified coefficients) to calculate the heat transfer. No such empiricism is required for the laminar boundary layer as the curvature terms in the momentum equation (in curvilinear coordinates) account for all of the curvature effect.

The effect of rotation on the boundary layer growth in axial and centrifugal machines is described in Sections 5.2.1, and 5.5.1. There has been no corresponding work to isolate its effects on heat transfer. Based on the physical phenomena described in Section 5.2.1, it can be concluded that the effect of rotation on heat transfer is more dominant in radial turbines than in axial turbines. The effect of rotation on heat transfer in a radial turbine is similar to the effect of rotation on internal rotating passage flows described in Section 7.4.2. For an inward radial turbine, the situation is very similar to that depicted by passage $A'B'C'D'$ in Fig. 7.13. The surface $A'D'$, trailing the rotation, will have enhanced heat transfer, and there will be decreased heat transfer on the surface leading the rotation. The effect of rotation on an axial turbine is mainly in the form of three-dimensional effects. The boundary layer is three-dimensional, and thus the mass and heat transport in the radial direction becomes important. The additional gradients $\partial T/\partial r$, $\partial W_r/\partial r$, and $\partial W_s/\partial r$ (where W_s is the streamwise and W_r is the radial velocity), make both the flow equations and heat transfer equations truly three-dimensional; hence, Eq. 7.13 will have an additional heat transport term, $\rho c_p(\partial/\partial r)(\overline{w_r'T'})$. This is likely to introduce nonuniform heat transfer in the radial direction. The data by Blair (1994, UTRC Report) clearly show the effect of rotation. There were substantial differences in the heat transfer distribution measured with identical flow and geometry in a cascade and on a rotor blade.

An additional effect, which is important in turbine cooling, is the roughness. The surface roughness adversely affects heat transfer, efficiency, and the life of a turbine blade (Tabakoff, 1984). There are two types of roughness encountered in practice. The first is the result of the manufacturing process, which is present even in a new turbine. The second is due to the surface erosion and pitting caused by dirt, dust, sand, rust, or carbon particles in the real engine environment. This is associated with either the combustion preceding the turbine or the operation of a turbine in a desert or dusty environment. The latter effect has been investigated by Turner et al. (1985). The experiments were conducted in an HP turbine stage, with $T_w = 60°C$, $T_g = 90°C$, $M_2 = 0.8$, and Re $= 1.2 \times 10^6$. The roughness was varied using an abrasive powder of aluminum oxide and silicon carbide, which adhered uniformly over the blades. The roughness grades, 1, 3, 4, and 9 (Fig. 7.21) correspond to particle sizes of 54, 76, 105, and 250 μm. The test, with roughness 41, was conducted with the cooling only on the pressure surface. It should be noted here that the finest roughness (54 μm) is about two orders of magnitude larger than the roughness of a polished blade. Their results, shown in Fig. 7.21, indicate that even such small roughness could cause a

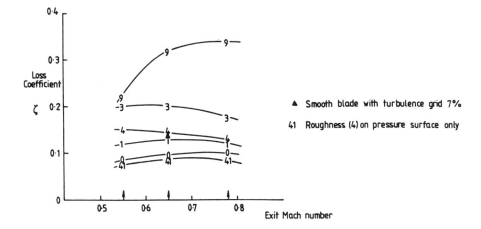

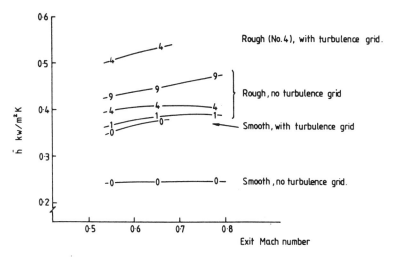

Figure 7.21 Effect of roughness and free-stream turbulence (7%) on overall loss and mean heat transfer coefficients (Turner et al., 1985). (The original version of the material was first published by the AGARD / NATO in CP 390, "Heat Transfer and Cooling in Gas Turbines, September 1985.)

substantial increase (as much as a factor 2–3 near the leading edge) in heat transfer rates on both surfaces (especially at high free-stream turbulence), an undesirable effect. The aerodynamic effects result in lower efficiency. The surface roughness influences transition, leading-edge separation, and the streamwise vorticity development near the stagnation point as well as the turbulent boundary layer growth. These phenomena need further investigation. The losses increase substantially with an increase in roughness, as shown in Fig. 7.21.

The three-dimensional effects on heat transfer are important, especially in the hub and tip regions as well as in a low-aspect-ratio blading. There are secondary and leakage flows in the hub and endwall regions. The three-dimensionality (due to rotation, curvature, and radial pressure gradient) affects transition, laminar, and turbulent boundary layer growth. A detailed treatment of secondary and leakage flow is given in Sections 4.2.6 and 4.2.7. These flows are likely to enhance the heat transfer in these regions. Heat transfer in the tip region is known to be very high, and it has the same order as those present near the leading edge. The tip leakage flow is in the direction normal to the blade. Its eventual roll-up causes intense mixing and turbulence in this region (Lakshminarayana et al., 1982). This causes a very high heat transfer rate to occur. Large near-gap heat transfer enhancement has been observed by Kumada et al. (1994). The heat transfer enhancement depends on rotational speed, tip clearance height, and the inlet flow angle. Kumada et al. found that the average Nusselt number becomes smaller with an increase in the tip clearance. This is probably caused by downward movement of the leakage vortex (Section 4.2.7).

The secondary flows that exist in the endwall region are likely to cause enhanced heat transfer due to cross flows as well as high turbulence intensities present in this region. The low-speed cascade results of Graziani et al. (1980), plotted in Fig. 7.22, reveal the dramatic influence of these secondary flows. The experiments were conducted in a large-scale, low-speed turbine cascade endwall region: $C = 34.3$ cm, $A = 1.0$, $S/C = 0.79$, $\alpha_1 = 44.6°$, and $\alpha_2 = -26°$. The gas temperature was ambient. The steady-state heat transfer boundary conditions on the airfoil and on the endwall surfaces was one of constant heat flux. The Stanton number (St) definition used is given by

$$\text{St} = q_w / \rho c_p U_e (T_w - T_{aw})$$

where T_w is the endwall temperature with simultaneous heating of the endwall and the airfoil, and T_{aw} is the temperature without endwall heating and U_e is the free-stream total velocity. The presence of a horseshoe vortex near the leading edge increases the heat transfer on the suction side near the leading edge by as much as 60–80% at some locations. This was a major problem in early cooled turbines, causing burnout of the airfoils. Likewise, the presence of passage secondary vortices on the suction side beyond about midchord increases the heat transfer by a similar magnitude. This case has been computed by Choi and Knight (1988) using the technique described in Section 5.7.5. More recent data in the endwall region is due to Blair (1994), carried out in a similar large-scale, low-speed, rotating turbine rotor. Some of the conclusions drawn by these authors are as follows:

1. Passage secondary flows greatly influence the heat transfer on the suction surface and endwall of the cascade; the pressure surface is not affected.

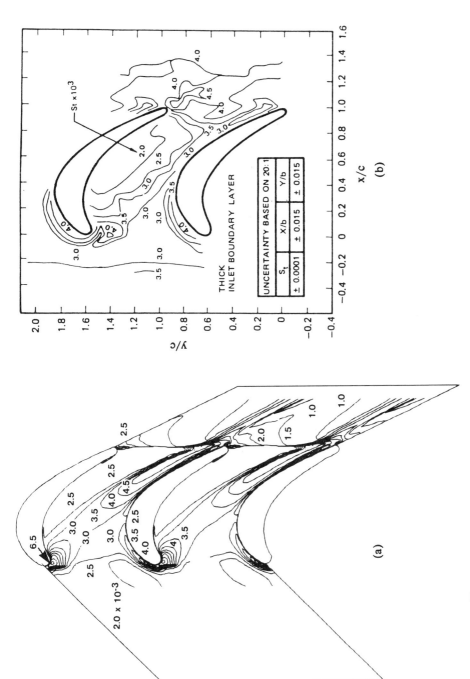

Figure 7.22 Predicted and measured Stanton number distribution on the endwall of a turbine cascade. Numbers on the contours are Stanton numbers. (Simoneau and Simon, 1993). (a) Computation: Choi and Knight, 1988. (b) Experiment: Graziani et al., 1980.

2. The thickness of the inlet boundary layer has a significant effect on the heat transfer in the endwall regions.

3. The conventional boundary layer approximations, and thus the conventional approach to heat transfer prediction, is not applicable to such flows.

Additional information on the effects of secondary flow on the endwall heat transfer can be found in Hylton et al. (1981), Boyle and Hoose (1989), and Sieverding and Wilputte (1981). Enhanced effects of leakage and secondary flow on heat transfer have also been observed by Kumada et al. (1994).

7.4.3.4 *Approximate Analysis of Convective Cooling and Performance*
Various correlations available for calculating the heat transfer coefficients h_g and h_c were presented in this section. There are several approaches for predicting the heat transfer performance of convectively cooled blades. The least empirical approach is the numerical solution of all governing equations. The accuracy of this method depends on the numerical technique, turbulence and transition modeling, heat flux model, grid resolution, and the effects of artificial dissipation. These techniques will be described later.

In analytical techniques, two levels of approximation can be employed. In the empirical category, one uses the best correlations available for h_g and h_c solving equations such as Eq. 7.42 or the more exact form, Eqs. 7.39–7.41. This should provide values for coolant temperature (T_c) and wall temperature (T_w), or T_c and $\dot{m}$ for a specified T_w. In the second method, experimentally measured h_g and h_c are used in equations such as Eq. 7.41. An approximation method of predicting the spanwise variation of temperatures is given in Horlock (1973a). The use of boundary layer codes to determine the heat transfer coefficient has been successful (see Fig. 7.16), but the high speed, high temperature flow with shocks and separation is best resolved by the Navier–Stokes procedure described in Section 7.8.

7.5 FILM COOLING

The mechanism by which the film cooling produces a lower blade temperature is different from that of convective cooling. In film and transpiration cooling, the surface is covered with a thin film of cool air, which acts as a buffer between the high-temperature gas and the blade. The film and convective cooling are almost invariably used in combination as shown in Figs. 7.4–7.6 and 7.11. The film cooling configuration may consist of a slot, a single row of holes (local film cooling), or a multiple row of holes as needed.

The topic of film cooling is one of the most complex in the turbine cooling area and is still not well understood. Unlike convection cooling, there is a direct interaction between the coolant, the mainflow, and the mixing of the neighboring jets. This will not only alter the inviscid character of the flow field, but will substantially modify transition, turbulence, and boundary layer characteristics. If proper care is not taken, this may result in increased aerodynamic losses and decreased efficiency.

The three-dimensional nature of a coolant jet interacting with the main flow is illustrated schematically in Fig. 7.23. The well-behaved boundary layer upstream may undergo dramatic changes, including separation. Far downstream, the reattachment may occur, resulting in a normal boundary layer as shown. If there are multiple rows, the boundary layers may never reach a state of equilibrium, being perturbed consecutively by a series of film-cooling holes in the downstream region. The resulting three-dimensional flow field (including spanwise flows) and thermal field are very complex.

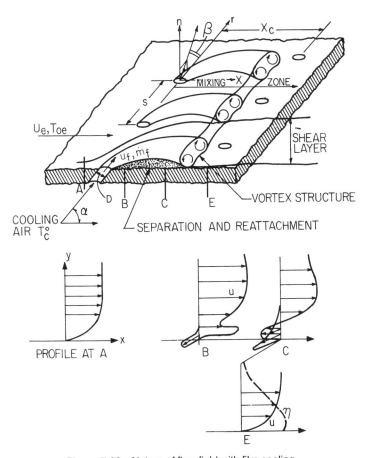

Figure 7.23   Nature of flow field with film cooling.

Some of the additional complexities are as follows:

1. The coolant is at a different temperature, density, and velocity than the mainstream; therefore, all of the flow and fluid properties of both the mainstream and the coolant jet enter into the governing equations.

2. The cooling configuration as well as blade configuration are complex. For example, hole diameter and spacing, angle, location, and finish, as well as the surface curvature and roughness alter the aerothermal flow field, including pressure distribution and the viscous nature of the flow.

3. Turbulence properties are equally complex, because mixing increases production and dissipation of turbulent kinetic energy. The distribution of turbulence properties as well as mean properties are irregular in the jet mixing region.

4. In many cases (e.g., rotor and stator), the mainstream is unsteady and the coolant air is influenced by the Coriolis and centrifugal forces. The mixing is also unsteady. The three-dimensional instantaneous velocity contains three components of mean velocity, three components of periodic unsteady component, and three components of random unsteady component.

5. The flow near the surface may be supersonic or transonic with shock waves. The coolant jets cause a major disturbance to the flow field, shock structure, and location. The separation caused by subsonic mixing and the shock–boundary–layer interaction is a nightmare for fluid dynamicists. Needless to say, these effects have a major influence on the temperature and the heat transfer field, as well as on the temperature fluctuations.

6. The injection causes early transition of the flow, and this may eventually relaminarize under some circumstances.

This section will be mainly concerned with an understanding of the physical principles, simple analyses, effect of slot, and jet and flow geometry on heat transfer. It is only recently that attempts have been made to predict the aerothermal field due to film cooling. These attempts will be described in a later section.

The analysis presented in this section relies heavily on experimental data. A large number of investigators have acquired data in a variety of configurations; some are relevant, whereas others are only of academic interest. The field is confusing and complex, and there has been no systematic effort to generalize the behavior and provide systematic correlations. This area as mentioned earlier, suffers from the same lack of interplay between fluid dynamicists and heat transfer researchers. A comprehensive survey of various aspects of cooling can be found in LeGrives (1986), VKI lecture series (1982, 1986), Boyle and Hoose (1989), and Metzger and Afgan (1984).

Heat transfer and flow losses associated with film cooling depend on many of the nondimensional parameters introduced earlier: Reynolds number, Prandtl number, Eckert's number, Richardson number, rotational Reynolds number, and Grashof or Rayleigh number. The additional variables introduced in film cooling are: (1) cooling geometry such as diameter, slot width, angle, shape, hole spacing, and row spacing, (2) number of rows of slots or holes, (3) surface roughness, and (4) properties of the coolant jet: density, velocity, mass flow, temperature, pressure, turbulence properties, c_{pc}, c_{vc}, γ_c, and R_c. In view of this complex dependency on a large number of parameters, designers and researchers have relied heavily on experimental and empirical correlations. Even here, one encounters large amounts of data on a wide variety of configurations. For example, Goldstein (1971) tabulated 42 simple film cooling configurations for which data are available. All of these are usually for one flow variable (e.g., the Reynolds number).

The basic slot configuration is shown in Fig. 7.23. It is characterized by the following: diameter D; injection angle α with respect to chordwise axis; injection angle β with respect to radial direction (Figs. 7.23 and 7.6); the tip radii; location with respect to the leading edge and the hub or the tip; spacing in the chordwise direction X_c; spacing in the radial direction s; and number of rows, n. The injection angles α and β are usually around 30°.

The cooling arrangement usually consists of one or more of the following (LeGrives, 1986):

1. Series of spanwise rows of holes centered around the leading edge stagnation line in order to cool the leading edge region (Fig. 7.6). Cooling is very critical in this region; convection cooling alone is not sufficient. These holes are referred to as *shower head*, because the individual jets penetrate into the inviscid region and are in a region of strong curvature.

2. Groups of two or three rows of holes are drilled along the pressure or suction side as shown in Fig. 7.5.

3. The spent air is usually ejected through holes at the trailing edge, as shown in Fig. 7.6.

4. For very-high-temperature turbines, full-coverage film cooling, consisting of a large array of holes from the leading to the trailing edge, is used. This is expensive and difficult to maintain. Aerodynamic penalties associated with this are also high.

The cooling effectiveness and aerothermal performance of film cooling depends on various flow parameters as follows: (1) the density ratio (DR): ρ_f/ρ_e; (2) the blowing or mass ratio (m_R): $\rho_f u_f/\rho_e u_e$; (3) the coolant mass flow ratio (MR): $\dot{m}_f/\dot{m}_p$; (4) dilution mass ratio (E): $\dot{m}_e/\dot{m}_f$; and (5) momentum flux ratio (I) $\rho_f u_f^2/\rho_e u_e^2$, where f refers to the coolant flow and p to primary external gas flow and e to conditions in the free stream.

These are in addition to the variables dealt with for convectively cooled blades (Section 7.4). It should be noted here that the turbine cooling performance is evaluated using primary airflow $\dot{m}_p$ which is the primary air through one passage, while $\dot{m}_f$ is the coolant air through one whole blade (both pressure and suction surfaces). On the other hand, $\dot{m}_e$, used in the dilution ratio, is the mass of the external flow (drawn into the film) associated with the coolant jet defined later.

A comprehensive set of flow data for a single row of film cooling holes, relevant to gas turbine application, was acquired by Pietrzyk et al. (1990), with hole inclination of 35°, hole spacing (S) of 3D, and mass ratio (m_R) varying from 0.25 to 1. The corresponding heat transfer data were acquired by Sinha et al. (1991a) for this case with $T_e = 302$ K and $T_f = 150$ K. The experiments were carried out on a flat plate in a wind tunnel, and a laser-doppler velocimeter was used to acquire the three-dimensional velocity data. The flow data include three-dimensional velocity and turbulent stresses at several cross sections from $X/D = -0.5$ to 30 (where X is the downstream distance from the coolant hole center line). They found that the jet causes major disturbance to the boundary layer, and its effect persists even after 10 diameters from the injection point. This case has been simulated numerically by Leylek and Zerkle (1994) and the result shows general agreement with the qualitative nature of flow and the temperature data. The computation was carried out with a pressure-based method (Section 5.7.1) and a standard $k-\varepsilon$ model (Eqs. 1.84 and 1.86). The computed results for $m_R = 2.0$ and density ratio (DR) = 2.0 are shown in Fig. 7.24. In Fig. 7.24a, the secondary velocity vectors in the lateral plane are superimposed on temperature contours. The formation of two counter-rotating vortices, depicted in Fig. 7.23, is clearly visible. The experimental data due to Pietrzyk et al. (1990) show similar features. The temperature distribution follows a pattern similar to those of secondary flow, being smallest near the core and largest away from the core. Hence, the physical mechanism is the mixing of the coolant jet with the main stream, through the formation and diffusion of the vortex. The computed temperature contours reveal the mixing process as the flow/film progresses from $X/D = 2$ to 25. The data at $X/D = 25$ show that the mixing is still continuing at this location. The strength of the secondary flow/vortex depends on the various parameters mentioned earlier. These secondary flows lift the coolant film vertically away from the surface at high m_R. The vortical motion also has an undesirable effect of hot cross flow being forced down under the film layer.

7.5.1 Adiabatic Film-Cooling Effectiveness

The heat transfer phenomenon associated with film cooling is very complex. We must reckon with three modes of heat transfer; from hot gas to film, from film to the surface, and the conduction through the blade material to an internal coolant. It is further complicated by the fact that film cooling is used

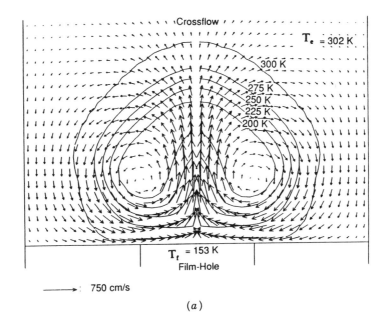

$T_e = 302$ K

300 K

275 K
250 K
225 K

200 K

$T_f = 153$ K
Film-Hole

⟶ : 750 cm/s

(a)

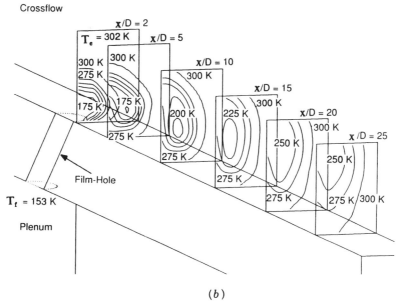

Crossflow

x/D = 2

$T_e = 302$ K   x/D = 5

300 K   300 K   x/D = 10

275 K   300 K   x/D = 15

300 K   x/D = 20

175 K   175 K   200 K   225 K   300 K   x/D = 25

0

275 K   250 K   300 K

275 K   250 K

Film-Hole   275 K   275 K

$T_f = 153$ K   275 K   275 K   300 K

Plenum

(b)

Figure 7.24 Computed velocity and temperature contours for a single row of coolant jets ($m_R = 2.0$, DR = 2.0). (a) Secondary velocity vectors and temperature contours. (b) isometric view of computed temperature contours on many cross-flow planes shows coolant jet trajectories and lateral diffusion. (Copyright © Leylek and Zerkle, 1994; reprinted with permission of ASME).

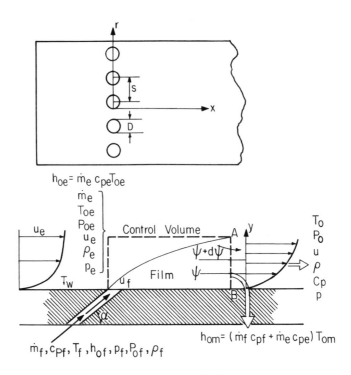

Figure 7.25 Geometry and notation for film-cooling analysis.

in combination with convective cooling. There have been a large number of basic studies on film cooling in the laboratory. These can be found VKI Lecture Series (1982, 1986), Metzger and Afgan (1984), and Goldstein (1971), as well as in numerous references quoted in these reviews and lectures. In film cooling, two analysis problems are encountered (see Fig. 7.25). First, for a given main flow, coolant mass flow ($\dot{m}_f$), T_f, the geometry of the hole or slot, it is required to find the wall temperatures, and the heat transfer. Alternatively, for a given main flow, T_w, and q_w, it is required to find the mass flow needed to maintain the prescribed wall temperature.

Consider the configuration shown in Fig. 7.25, with only external cooling applied through a film. The external stream static temperature is T_e, the film temperature is T_f, and the local wall temperature is T_w. To define the heat transfer, a datum temperature is needed. The datum temperature is called the *adiabatic wall temperature* (T_{aw}) and represents the surface temperature of a perfectly insulated surface. The rate of heat transfer can be then written (similar to Eq. 7.14)

$$q = h_f(T_w - T_{aw}) \tag{7.59}$$

where h_f is the film heat transfer coefficient. Without film cooling, $T_{aw} = T_r$, the recovery temperature is given by

$$T_r = T_e\left(1 + r\frac{\gamma - 1}{2}M_e^2\right) = T_{oe}\left[\frac{1 + r\dfrac{\gamma - 1}{2}M_e^2}{1 + \dfrac{\gamma - 1}{2}M_e^2}\right] \qquad (7.60)$$

For incompressible flow ($M_e \sim 0$), T_r, T_e, and T_{oe} are nearly identical. The performance of cooling can be expressed as *adiabatic film cooling effectiveness* given by

$$\eta_f = \frac{T_{aw} - T_r}{T_{of} - T_r} \qquad (7.61)$$

where T_{of} is the stagnation temperature of the coolant near entry (Fig. 7.25). This definition satisfies all of the boundary conditions, namely, $\eta_f = 0$ with no film cooling ($T_{aw} = T_r$), $\eta_f = 1$ near the injection point ($T_{aw} = T_{of}$) and $\eta_f \to 0$ far downstream. For very low speeds, T_r is replaced by T_e, T_{of} by T_f. The adiabatic film effectiveness is primarily a function of the blowing (mass flux) ratio $m_R = \rho_f u_f / \rho_e u_e$, the density ratio (DR) = ρ_f / ρ_e, the width and height of the injection hole, the distance X, and the injection angles α and β. T_f is the temperature of the coolant near entry.

Eckert (1983) has provided formal proof for the definitions used for adiabatic cooling effectiveness (Eq. 7.61) and the film heat transfer coefficient (Eq. 7.59). He has made the following assumptions in deriving Eqs. 7.59–7.61.

1. The coolant has the same property as the main flow, ejected through one or more holes at the exit temperature of T_{of} (Fig. 7.25).
2. The flow is assumed to be steady with constant properties. The aerodynamic heating effects are assumed to be negligible.
3. The assumption of constant fluid property implies that velocities are independent of the thermal fields and are assumed known, or prescribed.

The analytical/numerical solution for the temperature and velocity fields with film cooling involves solution for the velocity field (momentum equation Eq. 1.18 or the reduced form for the boundary layer) and the energy equation Eq. 7.13 with a suitable closure model for turbulent heat flux, the simplest being the eddy diffusivity model (Eq. 7.22).

As mentioned earlier, the analytical solution of Eq. 7.13 is extremely difficult; therefore, an empirical relationship for h_f (Eq. 7.59) is sought. The adiabatic wall temperature T_{aw} can be expressed in the form of a performance parameter given by Eq. 7.61. If η_f, T_{of}, T_e, M_e, and r are known, T_{aw} can be calculated. In addition, if h_f and q_w are known, T_w can be calculated, or vice versa. The analytical procedure used generally involves developing a correlation for η_f from the measured data. This correlation is then used in the design to determine T_{aw} and T_w for a given coolant mass flow, or T_{aw} and $\dot{m}_f$ are computed for a prescribed wall temperature (T_w).

The adiabatic film cooling effectiveness is closely related to the velocity and temperature profiles as well as velocity and thermal boundary layer thicknesses. For example, the turbulent boundary layer velocity and temperature profiles are represented by (Mayle, 1976) (see Fig. 7.25 for notations).

$$\frac{\rho u}{\rho_e u_e} = \left(\frac{y}{\delta}\right)^n \tag{7.62}$$

$$\frac{T_e - T}{T_e - T_w} = \exp\left[-\Gamma\left(\frac{3 + n}{2 + n}\right)\frac{y}{\delta_t}\right]^{2+n} \tag{7.63}$$

where the exponent n depends on the pressure gradient and the change in velocity profile due to blowing, Γ is the gamma function, δ is the boundary layer thickness, y is the coordinate normal to the wall, and δ_t is the thermal boundary layer thickness given by

$$\delta_t = \int_0^\infty \left(\frac{T_e - T}{T_e - T_w}\right) dy \tag{7.64}$$

Using these equations, Mayle (1976) proved that

$$\eta_f = \frac{m_R}{F(n)} \left(\frac{\delta}{\delta_t}\right)^n \frac{D}{\delta_t} \tag{7.65}$$

where D is the slot width for boundary layer blowing with slot and $F(n)$ is a function of n (e.g., $F = 0.877$ for $n = 1/6$).

It is clear from Eq. 7.65 that η_f is a function of δ/δ_t. The value of δ/δ_t is high (~ 3.0) near the injection location and decreases continuously to a value of about 1.5 far downstream. The adiabatic cooling effectiveness increases with the cooling mass flow rate up to a limit, beyond which substantial changes in boundary layer thickness/perturbation tend to diminish the values of η_f. Let us consider a case where δ/δ_t is nearly constant, and then the effectiveness depends only on δ_t. This suggests that the effectiveness increases with a decrease in the thermal boundary layer thickness.

As mentioned earlier, film cooling is used in actual practice in combination with internal convective cooling. There would be heat transfer from the wall to the internal coolant. This could be calculated separately to determine the wall temperatures. Lee and Han (1983) suggest a modified definition for nonadiabatic effectiveness as given by

$$\eta_{fn} = \frac{T_r - T_f(x)}{T_r - T_f} \tag{7.66}$$

where $T_f(x)$ is the local film temperature, and they have provided a method of relating η_f to η_{fn}.

7.5.2 Analysis and Correlation

The aerodynamic configurations vary in practice. Among the cooling configurations, one encounters (1) a single row of holes, (2) several rows of holes (Fig. 7.23), (3) full coverage, and (4) porous walls.

Among the aerodynamic interactions, one encounters:

1. Inviscid–Inviscid Interaction. In locations where the boundary layer is thin, as near the leading edge region, the jet penetrates into the inviscid stream, and this can be analyzed by inviscid theories.
2. Inviscid–Viscid Interaction. If the jet is injected, say, beyond about midchord where the boundary layers are thick, the jet interacts with the blade boundary layers, resulting in complex interactions, including substantial growth near the injection point.
3. Inviscid–Inviscid/Viscid Interaction. In many instances, the jet may partly penetrate into the inviscid region in addition to interacting with the viscous layer.

It is therefore impossible to develop a unified theory that encompasses all aerodynamic configurations and aerodynamic interactions. There are several theories available for the first case (inviscid–inviscid interaction) as well as for the second case (inviscid–viscid interaction). These will be described.

7.5.2.1 *Single-Row Film Cooling*   Goldstein (1971) provides a review of various simplified analyses available for the prediction of η_f or T_{aw}. One of the simplest techniques utilizes the overall energy and mass conservation based on the control volume concept, shown in Fig. 7.25. The mass flow due to primary flow ($\dot{m}_e$) includes both the inlet flow and the entrained flow (from external stream) at a temperature of T_e. The coolant flow has properties $\dot{m}_f$, T_f, and (c_{pf}). The Gibbs–Dalton's law for mixture (at the exit) is given by

$$c_p = \frac{\dot{m}_f c_{pf} + \dot{m}_e c_{pe}}{\dot{m}_f + \dot{m}_e} \tag{7.67}$$

and the energy equation is given by

$$\left(\dot{m}_f + \dot{m}_e\right)c_p T = \dot{m}_f c_{pf} T_f + \dot{m}_e c_{pe} T_e \tag{7.68}$$

where $\dot{m}$, c_p, and T are properties of the mixed gas at the exit station (AB in Fig. 7.25). Equations 7.61, 7.67, and 7.68 can be used to prove (assuming $T_r = T_e$, $T = T_{aw}$, and $T_{of} = T_f$)

$$\eta_f = \frac{T_{aw} - T_e}{T_f - T_e} = \left\{ \frac{1}{1 + \dfrac{\dot{m}_e c_{pe}}{\dot{m}_f c_{pf}}} \right\} \tag{7.69}$$

The ratio $\dot{m}_e / \dot{m}_f$ is called the *dilution mass ratio* (E), and it is the ratio of the mass of external stream drawn into the film and the mass of the injected coolant (Fig. 7.25).

Various methods available for the analysis of film cooling differ in the way $\dot{m}_e$ is calculated. This involves a knowledge of the boundary layer profile as well as its growth. None of them are satisfactory, and many assumptions made are arbitrary. The assumption that the average mixture temperature $T = T_{aw}$ is incorrect, because the temperature varies across the boundary layer. In practice, $T = \lambda T_{aw}$, where λ is a parameter representing the velocity and temperature profiles. Furthermore, the boundary layer thickness increases significantly near the injection location, as does the entrained mass flow. Goldstein (1971), Goldstein et al. (1985), and Jabbari and Goldstein (1978a) tried to rectify this by using empirical data for λ as well as for $\dot{m}_e$, and they provided the following correlation for the film-cooling effectiveness for a single row of discrete holes of diameter D (Fig. 7.25):

$$\eta_f = \frac{1.9 \, \text{PR}^{2/3}}{1 + 0.329 \left(\dfrac{c_{pe}}{c_{pf}} \right) \xi^{0.8} \beta}$$

$$\xi = \frac{X}{m_R L} \left[\left(\frac{\mu_f}{\mu_e} \right) \text{Re}_f \right]^{-0.25} \tag{7.70}$$

$$\beta = 1 + 1.5 \times 10^{-4} \, \text{Re}_f \left(\frac{\mu_f w_e}{\mu_e w_f} \right) \sin \alpha$$

where w is the molecular weight, $\text{Re}_f = \rho_f u_f L / \mu_f$, and L is the equivalent slot width $(= \pi D^2 / 4S)$.

Goldstein (1971) showed excellent agreement between this correlation (Eq. 7.70) and the experimental data. It is thus obvious that cooling effectiveness is a function of injection location, geometry, and coolant property as well as the mass flow entrained. Additional parameters that are important are density ratio (DR) = ρ_f/ρ_e, momentum flux ratio $(I) = \rho_f u_f^2/\rho_e u_e^2$, blowing or mass flux ratio $(m_R) = \rho_f u_f/\rho_e u_e$.

Inviscid–Inviscid Interaction and Viscous Diffusion Due to a Single Row of Holes. LeGrives (1978, 1986) has identified two mechanisms controlling film cooling effectiveness. Let us consider a case where $c_{pf} = c_{pe}$. Equation 7.69 reduces to

$$\eta_f = \frac{1}{1 + E}, \qquad E = \frac{\dot{m}_e}{\dot{m}_f} \qquad (7.71)$$

Analysis of the problem involves prediction of E.

The dilution process includes not only the mixing of the film jet by turbulent diffusion, but also the entrainment of the external flow by the vortices. The process is illustrated in Fig. 7.26. The component of velocity, $u_e \cos \alpha$ (of the external stream), and its interaction with the film jet can be viewed as turbulent diffusion of the jet with a velocity of $\pm (u_f - u_e \cos \alpha)$. Let us denote the entrainment rate due to this mechanism as E_T. In addition, the vortices entrain fluid from the external stream. Even though viscous effects play a role, this can be analyzed by an inviscid analysis. The normal component of velocity $u_e \sin \alpha$ interacting with the jet can be viewed as a flow over a cylinder D in a uniform stream of $u_e \sin \alpha$ (Fig. 7.26c). Let us denote the entrainment rate due to this mechanism as E_V. Hence, the adiabatic effectiveness can be written as

$$\eta_f = \frac{1}{1 + E_V + E_T}$$

LeGrives (1978, 1986) has provided an analysis for the jet penetration and cooling effectiveness considering the formation of vortices (and the entrainment associated with it) as well as turbulent diffusion of the jet. Let us consider methods available for evaluating E_T and E_V. LeGrives (1978, 1986) utilizes the force balance (centrifugal force CF, drag force D, and the inviscid momentum (M) equation in the streamwise direction, shown in Fig. 7.26b) to obtain an expression for u_f and α in terms of initial values u_{fo} and α_o. His expression is given by (see Fig. 7.26 for notations)

$$\frac{u_f}{u_{fo}} = \frac{\cos(\alpha_o - \phi)}{\cos(\alpha - \phi)}, \qquad \text{where } \phi = \tan^{-1}\left(\frac{u_e - u_{fo}\cos\alpha_o}{u_{fo}\sin\alpha_o}\right) \qquad (7.72)$$

The subscript o refers to values at the injection point.

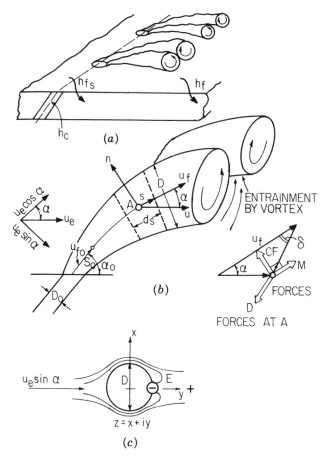

Figure 7.26 Analysis of film cooling (modified and adopted from LeGrives, 1986. Reprinted with permission of Concepts ETI Inc.).

Let us now consider the first mechanism (E_V). LeGrives (1978, 1986) has carried out extensive measurement, flow visualization to model this flow. The measurement (based on injection of O_2 and N_2) indicates that the formation of vortices and the flow field resembles the flow field around a circular cylinder of diameter D with a sink of strength E at $z = D/2$, shown in Fig. 7.26c. Using the complex potential for such a configuration (similar to Eq. 3.32) and $C_D = 2$, he derives an expression for the mass entrainment ratio E_V given by (Fig. 7.26b)

$$E_V = \frac{\dot{m}_e}{\dot{m}_f} = \frac{\int_{So}^{S}(\rho u_e \sin \alpha) D\, ds}{\rho u_{fo}(\pi D_o^2/4)} = \log \frac{\sin \alpha_0}{\sin \alpha} \frac{\cos(\alpha - \phi)}{\cos(\alpha_o - \phi)} \quad (7.73)$$

For moderate values of α_o, these simple expressions lead to fairly accurate predictions of η_f up to 10 jet diameters along the wall.

The second process is the turbulent diffusion and entrainment due to spreading of the jet (E_T). The rate of entrainment due to this can be derived from the empirical expressions given by Keffer and Bains (1963). The resultant velocity for the jet is $\pm(u_f - u_e \cos \alpha)$ as shown in Fig. 7.26b. As indicated earlier, the component of free-stream velocity $u_e \sin \alpha$ is responsible for the vortex generation and mass entrainment, whereas the component $u_e \cos \alpha$ is responsible for turbulent diffusion and entrainment. The entrainment due to this phenomenon can be analyzed using Prandtl's mixing-length hypothesis. The expression derived by LeGrives (1978) is given by

$$(1 + E_T)^{1/2} = 1 + 0.1226 \frac{\pi}{2} \left[\frac{E_V}{\cos \phi} - \frac{1}{\sin \phi} \frac{\sin (\alpha_o - \alpha)}{\cos (\phi - \alpha) \cos (\phi - \alpha_o)} \right]$$

$$(7.74)$$

The mass entrainment ratios are averaged values (averaged in the lateral direction), so the lateral average film cooling effectiveness can be written as

$$\overline{\eta}_f = \frac{1}{1 + E_V + E_T} \tag{7.75}$$

where E_V and E_T are given by Eqs. 7.73 and 7.74, respectively.

It should be emphasized here that all of these expressions are valid when the jet penetrates into the inviscid stream and the wall boundary-layer–jet interaction is negligibly small. If the boundary layers are very thin (for instance, from leading edge to about half a chord on both blade surfaces), Eq. 7.75 should be employed. In cases where there is boundary-layer–jet interaction, Eq. 7.70 may be more suitable.

The analysis given in this section is approximate. As indicated earlier, the jet mixing process and its interaction with the main flow as well as boundary layer flow are too complex and they are best handled by computational techniques described later. But the analysis presented in this section is adequate for many engineering design applications, if properly combined with realistic empirical correlations for the Stanton number.

A comparison between the data (Liess, 1973) and the above analysis (LeGrives, 1986) is shown in Fig. 7.27. The agreement is very good. As indicated earlier, the mass entrainment due to vortex dominates up to $X = 10D_o$. Very high values of η in this region are clear evidence of this. The comparison is for boundary layer thickness of about 10% of the jet diameter, and thus the assumptions made in the analysis are clearly valid for this data.

The data from Goldstein et al. (1985) for a single hole shows large scatter and does not correlate well with Eq. 7.70. The three-dimensionality and the resultant vortex formation may be better handled by the analysis of LeGrives.

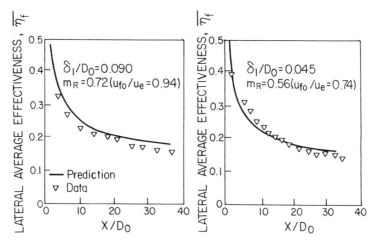

Figure 7.27 Comparison of measured and predicted average film cooling effectiveness with injection from one row of holes. X is the downstream distance measured from the hole ($S / D_o = 3$, $\alpha_o = 35°$, $M_e = 0.61$). (Copyright © LeGrives, 1986; reprinted with permission of Concepts ETI, Inc.).

7.5.2.2 *Multi-row Film Cooling*

In many applications, multi-row and full-coverage film cooling are employed. As indicated earlier, it is necessary to have multi-row cooling (shower head) near the leading edge of first stage nozzle and rotor, where high temperatures are encountered. Furthermore, small changes in incidence would make a single-row ineffective. Furthermore, if blades are small, convective cooling increases the blade thickness considerably, and multi-hole film cooling/full-coverage film cooling is desirable.

Sinha et al. (1991b) have reported detailed flow measurement carried out behind a second row of holes located 40 hole diameters (D) downstream of the first row and staggered with it as shown in Fig. 7.28. The facility configuration is the same as that investigated by Pietrzyk et al. (1990). Results indicate that the dominant structures identified in the flow field downstream of the first row (Figs. 7.23 and 7.26) were also present behind the second row. Because of a thicker boundary layer (caused by the first row) the mean flow field and the turbulence structure were altered. They also observed that thicker boundary layer enabled the second row of jets to penetrate deeper into the main stream.

LeGrives (1986) has developed an approximate analysis for multiple hole cooling, using the following assumptions (Fig. 7.28):

1. The interval between rows is larger than the hole diameter; hence, the successive films can be regarded as forming stratified layers.
2. The input of mass flow into one layer is contributed to by diffusion from the external layer.

Therefore, the total dilution ratio for n rows of injection holes is given by (Seller's superposition hypothesis)

$$1 - \eta_{fn} = \prod_{i=1}^{i=n} (1 - \eta_{fi}) \tag{7.76}$$

where η_{fn} is the total adiabatic film cooling effectiveness for n row of holes.

The results from this expression, based on $E(=E_V + E_T)$ calculated from Eqs. 7.75 and 7.76, are shown compared with the data from LeGrives and Nicolas (1977) in Fig. 7.28. The agreement is excellent, giving confidence in the simple analysis presented in the previous section. The advantage of multihole cooling is obvious. The several layers formed (Fig. 7.28) provide a very efficient shield for the blade from the high-temperature gas. The effectiveness is increased by nearly a factor of two at most locations.

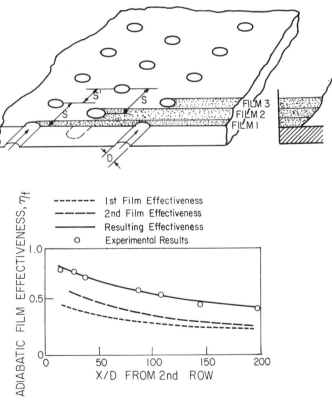

Figure 7.28 Effectiveness with two rows computed from $1 - \eta_f = (1 - \eta_{f1})(1 - \eta_{f2})$ (LeGrives and Nicolas, 1977) (The original version of the material was first published by AGARD / NATO in CP 229.)

Far downstream, where the jets have mixed with the wall layers, the flow is found to be nearly two-dimensional and well-behaved. The mass entrainment can be calculated from the turbulent boundary layer growth using one of several numerical procedures described in Chapter 5 and later in this chapter. It is better to utilize these rather than the approximate analysis/correlation in the calculation of heat transfer and film effectiveness in this region.

Goldstein et al. (1985) and Jabbari and Goldstein (1978a, b) have carried out extensive measurements with both rough and smooth walls, and single and multi-row configurations to derive correlations for η_f and h (or St). The experiments were carried out at various values of m_R and DR with $D = 6.5$ mm, $\alpha_o = 35°$, $S = 3D_o$, and spacing of rows $= 2.6D$. The holes were staggered for the two-row case. The adiabatic film cooling effectiveness is correlated against the parameter ξ introduced earlier (Eq. 7.70). For the two-row configuration, Goldstein et al. (1985) modified the parameter ξ as follows:

$$\xi = \left(\frac{X + 1.909D}{m_R L} \right) \left(\frac{\mu_f}{\mu_e} \mathrm{Re}_f \right)^{-0.25} \tag{7.77}$$

Jabbari and Goldstein (1978a, b) defined an effective slot width (L) for injection through holes, requiring that the coolant flow per unit span have the same value when leaving the slot with effective width as the one leaving the holes, given by

$$L = \frac{n\pi D^2}{4S}$$

where n is the number of rows of holes.

7.5.2.3 Full-Coverage Film Cooling

The experimental data acquired by various groups demonstrate the advantages of full-coverage film cooling. The full-coverage film cooling provides better protection of the blade from high temperature. This is especially true for blades in low-aspect-ratio turbines with a high hub/tip ratio. Full-coverage film cooling is very effective in small turbines, as well as in situations where very high temperatures are encountered. It also involves higher aerodynamic losses and decreased aerodynamic efficiency. An optimization study is essential before one resorts to single, multi-row or full-coverage film cooling. For example, the E^3 engine blade and vane (Figs. 7.5 and 7.6) does not have full-coverage film cooling due to the large size of the blading. The small high-performance gas turbines usually have full coverage.

Eckert (1983) used the comprehensive data available from the Stanford group (Crawford et al., 1980) as well as the Minnesota group (Goldstein et al., 1985; Jabbari and Goldstein, 1978a, b) to derive correlations for the

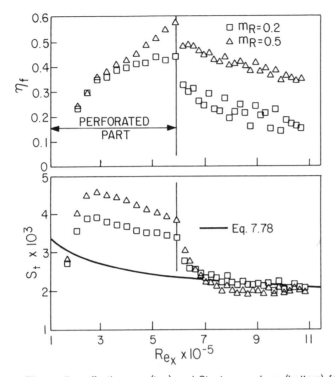

Figure 7.29 Film-cooling effectiveness (top) and Stanton numbers (bottom) for $S/D = 5$ and normal injection through 11 rows of holes for two blowing parameters m_R. (Copyright © Eckert, 1983; reprinted with permission of ASME.)

film-cooling effectiveness and Stanton number. The Stanford data are for full-coverage film cooling of an isothermal flat plate at zero pressure gradient with a $u_e = 10–15\,\text{m/s}$. The perforated part had 11 rows of holes, followed by a solid wall. The plate was electrically heated so that $T_w - T_e = 15°\text{C}$. This case represents full-coverage film cooling on the perforated part and film cooling downstream. The data for $\alpha = 90°$ and $30°$ injection with differing hole geometry are reported. The data, replotted by Eckert to conform to the definitions in his analysis (Eqs. 7.59, 7.61), are shown in Fig. 7.29. It is evident that the effectiveness (η_f) is a strong function of the Reynolds number (Re_x) as well as m_R. An increase in the value of m_R increases η_f as expected. As in a single-hole/multihole covering, the effectiveness decreases downstream. The maximum values occur at the last injection point. The effectiveness increases continously from the first row to the last row, a result consistent with the observations made earlier.

The Stanton number, representing heat transfer, is high in the perforated region and is a strong function of m_R and Re_x, falling off rapidly in the solid wall region. The dependency of the Stanton number on m_R is observed only for the perforated part. The downstream region, where the dependency on m_R is weak, behaves as a conventional turbulent boundary layer. All of the correlations developed for this case are valid for the downstream region of full-coverage film cooling.

For example, the solid line shown in Fig. 7.29 is based on the correlation given by

$$\text{St} = 0.0295\text{Re}_x^{-0.2}\text{PR}^{-0.4} \qquad (7.78)$$

The Reynolds number dependency is similar to that of a turbulent boundary layer (derived from Eq. 7.23), and the agreement with Eq. 7.78 is excellent in the downstream region. No such correlation exists for the covered part, where the correlation should include m_R as well as Re_x.

A word of caution is in order. The boundary layer development in the presence of pressure gradient (adverse or favorable) and film cooling is different from those with zero pressure gradient for the data shown in Fig. 7.29. An additional drawback is that most of these tests did not simulate D/δ (where δ is local boundary layer thickness) that exists in real engines. These values may be much different than those employed in these experiments. Caution should be exercised in using these data and correlations when pressure gradient is present or in calculating the heat transfer in the covered region. The engine manufacturers may have correlations which represent practical situations (high temperature, high turbulence, curvature, pressure gradient, compressibility effects, etc.), but these are not available in the open literature.

An approximation analysis for the prediction of film-cooling effectiveness, wall temperature, or mass flow ratio (m_R) for a prescribed wall temperature is as follows. For the downstream region, one can use a correlation such as Eq. 7.78 for Stanton number, and use analysis and correlation for the effectiveness developed in this section to derive values of h_f and η_f. These can then be used to calculate T_{aw} and T_w (Eqs. 7.59–7.61). Although such a procedure provides qualitative predictions, a final analysis should include data/correlations from realistic configurations simulating the engine conditions. This should involve most of the variables, including m_R, M, Re_f, Re_x, PR, DR, slot geometry (actual), number of slots, number of rows, pressure gradient, curvature, rotation, free-stream turbulence effects, and so on.

The present trend is toward computation using the full Navier–Stokes and the energy equations governing the aerothermodynamic field. Those attempts will be described in Section 7.8.4.

7.5.3 Heat Transfer Coefficient

For a complete analysis of film cooling, prediction for adiabatic film-cooling effectiveness (Eq. 7.61) and the heat transfer coefficient (Eq. 7.59) are needed. Most analyses available for the calculation of h_f or the Stanton number are empirical in nature due to the complexities of configuration, flow, and the aerothermal field involved. The heat transfer phenomenon consists of three mechanisms, as shown in Fig. 7.26a:

1. Convective heat transfer from blade to coolant in the injection slot (h_c).
2. Heat transfer between the main stream and the blade in the region of coolant injection (h_{fs}). This is probably the most difficult to calculate because the flow field and thermal field are influenced by D, α, u_f/u_e, M, DR, m_R, and other parameters described earlier, as well as by the resulting formation of vortices, mixing, and so on.
3. Heat transfer in the downstream region (h_f).

Heat transfer in the cooling hole can be calculated from one of the internal heat transfer correlations given in an earlier section (Table 7.1). The following correlation for h_c has been successfully used by Wadia and Nealy (1985) for predicting the film-cooling effects near the leading edges of a blade:

$$h_c = \left[1 + \left(\frac{D}{l}\right)^{0.7}\right] 0.023 \frac{k_f}{D} (\text{Re}_D)^{0.8} (\text{PR}_f)^{0.33} \qquad (7.79)$$

The term in brackets is the entrance effect, where l is the length of the coolant hole. k_f is the coolant thermal conductivity. LeGrives (1986) provides three different expressions: one each for $\text{Re}_D \leq 2300$, $3000 < \text{Re}_D < 15,000$, and $\text{Re}_D > 15,000$. The Stanton number is expressed as a function of Re_D, PR_f, and $1/D$, as in Eq. 7.44.

The heat transfer correlation near the injection hole should include the presence of vortices as well as mixing within the boundary layer. No universal correlation can be derived because it depends, in addition to various parameters listed earlier, on the location of the hole and the local boundary layer thickness. Nevertheless, LeGrives (1986) provides a correlation based on vortex-induced dilution effect. He modified Eq. 7.23 to allow for film-cooling effects, and the modified expression includes Re_x, PR, η_f, u_e/u_f, T_{of}, and T_{oe}. The predictions from these correlations agree well with the flat-plate film cooling data.

Finally, the heat transfer coefficient for the downstream region, where the flow has reached an equilibrium state (Fig. 7.29), is similar to that of a turbulent boundary layer, given by equations such as Eq. 7.23. Because this heat transfer coefficient is the same for full-coverage film cooling, Eq. 7.78 should also be valid.

The overall procedure is to calculate T_{aw} from one of the analyses given above (depending on the application) and utilize one of the heat transfer correlations (again depending on the location) to determine T_w, or the heat transfer and the mass flow of the coolant required if T_w is prescribed.

7.5.4   Effect of the Geometry of the Slot Configuration and the Cooling Mass Flow

Even though the injection hole configuration shown in Figs. 7.24 and 7.25 is one of the most widely used by laboratory investigators, several other configurations have been tested. These are mainly aimed at increasing the cooling effectiveness. One successful design involves diffusing the flow (similar to a conical diffuser) before it exits the blade. This prevents penetration of the jet into the mainstream, thus increasing the cooling effectiveness. Many different configurations have been tested by various groups. Goldstein (1971) had counted as many as 42 configurations as of 1971. However, transferring the laboratory configuration to the practical turbine is seriously limited by the difficulties in manufacture, durability, maintenance, clogging problems, sensitivity to solid particles, and so on. The turbine manufacturer is therefore limited in its choice of the slot geometry, but it is made easier by laser and ECM hole drilling. There are wide variations in the internal structure and the connection to the main coolant reservoir. This depends on the type of engine, the size of the blading, and so on.

Injecting from a slot increases the η_f as compared to a corresponding injection from a row of holes (LeGrives, 1986). This is a more effective, but less practical way of increasing η_f.

Some of the conclusions by LeGrives and Nicolas (1977) and LeGrives (1986) based on their data are summarized below:

1. Film-cooling effectiveness increases with the number of rows of holes. A staggered row configuration provides higher cooling effectiveness than does a single row, resulting in more uniform effectiveness in the spanwise direction.
2. Compound angle injection (α, β shown in Fig. 7.23) was tried. Injection through holes drilled along the streamwise (X) direction leads to higher effectiveness than does injection with a spanwise component. LeGrives suggests only a moderate value of β. These conclusions may not be valid near the leading edge (see Fig. 7.6).
3. Cooling effectiveness increases with a decrease in the hole spacing, but the mechanical constraint requires $S = 6{-}8D$.

Kruse (1985) has also carried out experiments with a systematic variation of the slot geometry, angle, pressure gradient, and so on, and has drawn similar conclusions.

A systematic work was carried out by Pederson et al. (1977) with various values of u_f/u_e and ρ_f/ρ_e with a single row of holes, $S = 3D$, $\alpha = 35°$. Their data clearly reveal that η_f is a strong function of u_f/u_e as well as ρ_f/ρ_e. But these data collapse when η_f/m_R is plotted against I (momentum flux ratio $\rho_f u_f^2/\rho_e u_e^2$). For a fixed ρ_f/ρ_e, η_f increases with m_R in the

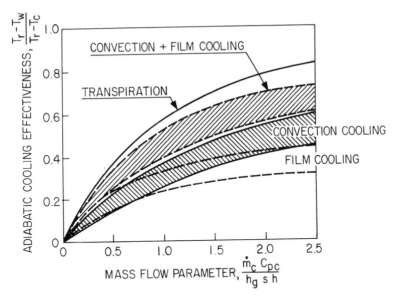

Figure 7.30 Plot of cooling effectiveness versus mass flow parameter from Fullagar (1974). (Subscript c stands for coolant or film.)

practical range. One has to pay attention to practical values of ρ_f/ρ_e and u_f/u_e and then draw conclusions.

The effect of coolant mass flow on adiabatic efficiency was also investigated by Fullagar (1974) in a turbine configuration, shown in Fig. 7.30. The adiabatic effectiveness is plotted against the mass flow parameter given by

$$m^* = \frac{\dot{m}_c c_{Pc}}{h_g Sh} \tag{7.80}$$

where $\dot{m}_c$ is the coolant mass flow, h_g is the gas heat transfer coefficient, S is the blade perimeter, and h is the blade height. The adiabatic film-cooling effectiveness increases rapidly up to $m^* = 1.0\text{--}1.2$, beyond which the increase is much slower. It is interesting to note from this plot that film cooling requires more coolant flow than do other types of cooling for the same T_w. Film cooling is used only if it is essential to achieve lower temperature.

Adiabatic film cooling effectiveness is also sensitive to the injection location. Controlling parameters here are the local pressure gradient and the structure and thickness of the boundary layer at the injection point. The effect of the pressure gradient will be dealt with in the next section.

7.5.5   Film Cooling Near the Stagnation Point

Leading-edge film cooling has been investigated by Kruse (1977), Wadia and Nealy (1985), and others. The former investigation was carried out with an airfoil representative of gas turbine blades, and the latter was carried out with a cylinder. Kruse found that the film effectiveness is very sensitive to α, slot location, ρ_f/ρ_e, u_f/u_e, and m_R. The high blowing rate provided no improvement in η_f downstream (Fig. 7.31). The jets in both cases penetrated deeply into the free stream, and thus the cooling was ineffective. Inclination of the jet in the direction of the free stream had a similar effect. The adiabatic film effectiveness is found to be very sensitive to the relative distance between the hole and the stagnation point. It is clear from Fig. 7.31 that a shower head is more effective in forming a film. The mean adiabatic film effectiveness of a row of holes, slightly increased η_f downstream of the stagnation point, but far downstream $(X > 10D)$ the effectiveness was the same as a single hole. Such a shower head configuration is essential, at least

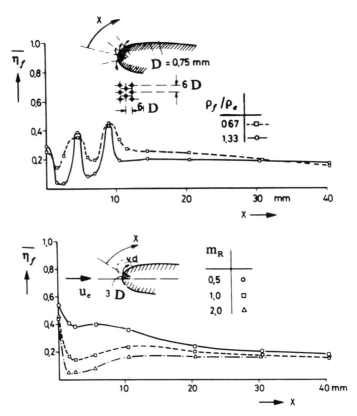

Figure 7.31 Influence of blowing rate on cooling effectiveness at the leading edge of an airfoil $(T_e = 400\,\text{K}, \; T_f/T_e = 0.8)$. (Copyright © Kruse, 1977; the original version of this material was first published by AGARD / NATO in CP 229.)

in the first-stage nozzle, to avoid excessive hot spots and melt down near the leading edge. Wadia and Nealy's results indicate that effectiveness is primarily influenced by α and is not adversely affected by the variation of pressure ratio (p_f/p_e), M_e, T_f/T_e, and Re. Camci (1989) reports extensive measurements in the stagnation region. He varied the incidence angle and blowing rate and found the heat transfer to be very sensitive to these as well as T_f/T_e.

7.5.6 Effects of Pressure Gradient, Free-Stream Turbulence, Roughness, Curvature, and Secondary Flow

LeGrives (1986) and Jabbari and Goldstein (1978b) have carried out a systematic study on the effects of free-stream pressure gradient on the film-cooling effectiveness. The results from LeGrives were derived from a real engine environment and are shown in Fig. 7.32. The adiabatic film cooling effectiveness is higher on the suction side in the flow acceleration region, which conforms with the results from Jabbari and Goldstein carried out on a flat plate with an imposed pressure gradient. Kruse (1977) reported very little change in η_f with either the adverse or the favorable pressure gradient. The values of η_f on the suction side (accelerating part) are generally higher than those on the pressure side. LeGrives (1986) postulates that the negative pressure gradient region leads to a blow off, separation, and a deceleration following the bubble as illustrated in Fig. 7.32. This may induce abrupt transition. Injection near the trailing edge, where the boundary layer is thick, may induce turbulent separation. The major effect of the pressure gradient is on the transition and formation of the separation bubble, accelerated by injection. It may not have any major direct influence on the film-cooling effectiveness,

Free-stream turbulence and roughness have an indirect effect as well. The free-stream turbulence in the nozzle following a combustion chamber is high (10–20%). High free-stream turbulence tends to increase the turbulent diffusion of the jet, especially if it penetrates into the free stream. Otherwise, high free-stream turbulence results in higher turbulence intensities inside the wall boundary layer. This should decrease film effectiveness. Brown and Saluja (1979) varied the free-stream turbulence intensity from 2% to 9% and measured decreased effectiveness, but Camci and Arts (1986a, b) indicate no significant change in heat transfer at increased turbulence intensity. There is very little data available on the effect of high free-stream turbulence on film cooling.

The roughness also leads to an increase in the boundary growth, as well as an early transition. These are likely to have an influence on the cooling effectiveness. Goldstein et al. (1985) found that for a rough wall, at low values of m_R, the effectiveness decreased by 10–20% as compared to a smooth wall. However, at a high blowing rate when the jets tended to lift off from the surface, there was a significant improvement in effectiveness. This improvement, they argue, follows from the greater turbulence and mixing associated with the rough surface, which tends to dissipate the injected flow

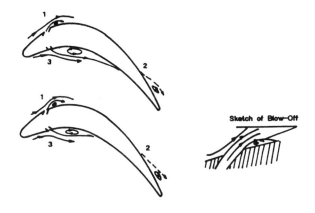

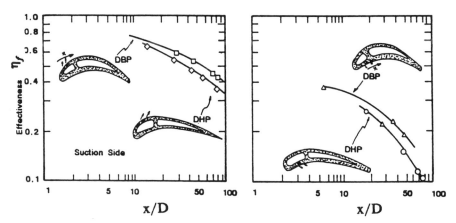

Figure 7.32 Typical data illustrating the influence of gas turbine environment on film effectiveness with coolant injection on the suction side or pressure side of a first stage NGV (DHP) and second-stage NGV (DBP). The top inset shows the effect of hole location on film / boundary layer interaction. (Copyright © LeGrives, 1986; reprinted with permission of Concepts ETI, Inc. from Advance Topics in Turbomachinery Technology).

and prevent it from penetrating far into the mainstream. Vanfossen and Simoneau (1985) reported no effect of roughness on the stagnation point heat transfer, but they measured increased heat transfer in regions where the turbulence fluctuations are small.

The effect of curvature is very similar to that described earlier for convective cooling (Section 7.4.3.3) and boundary layer growth (Section 5.2.1). Mayle et al. (1976) have carried out a systematic experiment for concave, flat, and convex surfaces, measuring the boundary layer and the heat transfer properties. They found that δ_t/D (where δ_t is the thermal boundary layer thickness given by Eq. 7.64) increases with X in all cases, more slowly for the convex case. On the suction side, where the curvature is convex, η_f is found to increase (from that of a flat plate); the opposite effect occurs on the pressure side. Kruse (1977) has drawn similar conclusions on the curvature effect for low blowing rate. The convex surface had increased effectiveness, while the concave surface had decreased effectiveness. Kruse found that the curvature effect was sensitive to hole diameter, spacing, and the blowing rate.

The streamline curvature, coolant blowing rate (u_f/u_e) and α are some of the major parameters influencing the coolant film trajectory and the resulting adiabatic film cooling effectiveness. In addition, the curvature (convex or concave), in the presence of streamwise pressure gradient (as in a turbine), can have major interaction and nonlinear effects. The combined effect may be different from isolated effects of curvature or pressure gradient alone. Hence, caution should be exercised in drawing conclusions based on isolated effects of curvature alone. The adiabatic film effectiveness is a function of the pressure gradient, curvature, thermal boundary layer thickness, the Reynolds number, blowing ratio, m_R, and I. It is essential to study the film effectiveness in the pressure of all these effects and then isolate them systematically.

In addition, the behavior of film cooling in the endwall region is very complex, brought about by the interaction between the coolant jet and the passage vortices. Some fundamental study and simulation by Ligrani and Mitchell (1992) suggest that the longitudinal vortices have a strong effect on local heat transfer in a film cooled blade. When the secondary flow in the endwall region is in the same direction as the injection velocity, a decrease in heat transfer is observed because the film is readily swept beneath the vortex core and into the upwash regions, making them ineffective. The effects are beneficial when the secondary flow and the injection velocity are in opposite directions.

Harasgama and Burton (1992) provide comprehensive data and analysis of film-cooling effects in the endwall of a turbine nozzle guide vane, carried out in simulated engine conditions (Re $= 2.55 \times 10^6$, M $= 0.93$, $T_e/T_w = 1.3$, $\rho_f/\rho_e = 1.8$). The placing of film-cooling holes along an iso-Mach line resulted in a very effective method of cooling the endwall. The secondary flow in the nozzle guide vane convects the coolant flow toward the suction side (due to secondary flow described in Section 4.2.6) and results in relatively high temperatures at the pressure side near the trailing edge. The cooling at this location reduced Nusselt number marginally (20%) compared to 50–75% in other areas of the endwall. The authors also found that the heat transfer to the endwall reduces with an increase in m_R.

7.5.7 Film-Cooling Effects on High-Speed Turbine Blading

Most of the data presented earlier (except as indicated otherwise) were taken in a well-controlled laboratory, where all of the parameters governing the

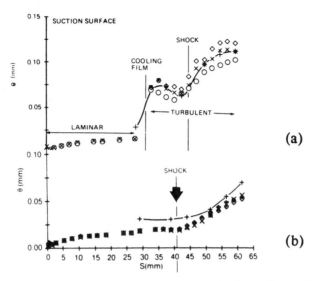

Figure 7.33 Momentum thickness distribution with and without film-cooling jet on the suction side surface. Turning angle 110°, $M_2 = 1.17$, $m_R = 0.76$, solid line represents data, and symbols represent predictions from integral and differential boundary layer techniques described in Section 5.5. (a) with film cooling (b) without film cooling (Copyright © Haller and Camus, 1984; reprinted with permission of ASME.)

flow and thermal field can be varied systematically. This has led to a basic understanding of the effects of α, Re, M, ∇p, m_R, Tu, DR, curvature, slot geometry, spacing, roughness, and so on, on film-cooling effectiveness. To carry out such an investigation in a high-speed/high-temperature blading, where all of these effects are present simultaneously, is very expensive and difficult. Many attempts have been made recently to understand the film cooling effectiveness, heat transfer, and aerodynamic effects in practical configurations. Some of the results are different from those obtained on a flat plate. The data in a high-temperature (1600 K), high-Reynolds-number (2.5×10^6, based on chord), high-pressure (18 atm) turbine (Gladden et. al., 1982) indicate that the cooling effectiveness data taken at low temperature is somewhat more optimistic when used to predict cooling performance at engine conditions.

The dramatic effect of coolant injection in a high-speed flow is illustrated in Fig. 7.33. Haller and Camus (1984) carried out aerodynamic measurements in a transonic cascade ($C = 4.17$ cm, $h = 10.16$ cm, $A = 2.4$, $S/C = 0.842$, turning angle $= 120°$) and found that injection dramatically changes transition and boundary layer growth. The presence of an abrupt transition, shock movement downstream, and momentum thickness (downstream of slot) which is two to three times that of the configuration without coolant injection is evident from Fig. 7.33. Caution should be exercised in extrapolating the low-speed data to the high-speed range, especially in the presence of shock waves. The authors did not report any heat transfer measurement. Camci

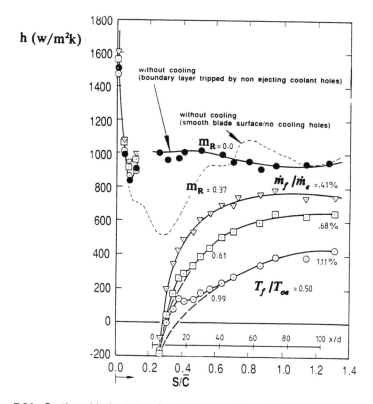

Figure 7.34 Suction side heat transfer with film cooling, effect of blowing rate. S is the distance from the leading edge, Tu = 5.2%, Re = 8.4 × 10^5, T_w / T_e = 0.72, T_{oe} = 408 K, M_1 = 0.25, M_2 = 0.9, and T_f / T_e = 0.51. (Copyright © Camci 1989, reprinted with permission of ASME.)

(1989), on the other hand, reported extensive heat transfer data but did not acquire boundary layer data. His experiment was done in a light piston compression tunnel in which engine conditions (M_2 = 0.9) were simulated. The cascade used had C = 8 cm, h = 10 cm, S/C = 0.67, turning angle = 99.5°. The heat transfer rate ($h = (q_w/(T_{oe} - T_w))$) as a function of distance for m_R = 0, 0.37, 0.61, and 0.99 is shown in Fig. 7.34. At m_R = 0, early transition due to the presence of a cooling hole can be observed. Heat transfer decreases substantially, even at low coolant mass flow due to the formation of the film, but the cooling effect diminishes substantially beyond about midchord for m_R = 0.37. A higher blowing rate increases the film thickness and decreases the heat transfer downstream. Nirmalan and Hylton (1990) incorporated internal cooling and film-cooling holes on the cascade they had tested earlier (Fig. 7.16), and they observed a large reduction in the Stanton number. They systematically varied the Mach number, Reynolds number, wall-to-gas temperature ratio, coolant-to-gas temperature ratio, and coolant-to-gas pressure ratio. Considerable cooling benefits were achieved by using downstream film cooling. The downstream cooling process was shown to be a complex interaction of two competing mechanisms: the thermal dilution effects due to cold coolant decrease heat transfer, and the turbulence augmentation due to jet–main-stream mixing results in an increased heat transfer to the blade. It is also observed that the favorable cooling effect is diminished when the coolant-to-gas temperature ratio is varied.

Full-coverage film cooling data on an entire blade in a high-temperature cascade facility acquired by Yoshida (1982), shown in Fig. 7.35, indicates that the cooling mass flow rate up to 6% of primary flow has a major effect on cooling effectiveness, especially in the leading-edge regions. The cooling holes had curved passages. High cooling effectiveness (almost uniform across

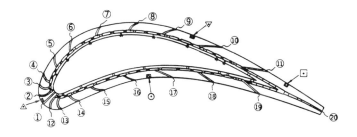

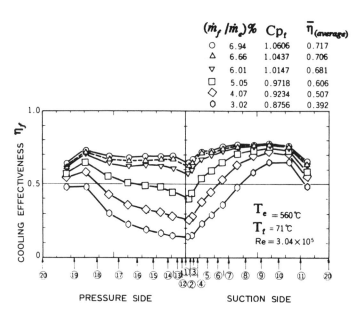

Figure 7.35 Chordwise cooling effectiveness distribution from cascade test. (Copyright © Yoshida, et al., 1982; reprinted with permission of ASME.)

the entire blade surface) has been achieved at the higher cooling rate. It is also clear that the transition probably occurs at the leading edge. Perhaps one major difference between the low-speed, low-turbulence, low-temperature data and the high-turbulence, high-temperature data is the early transition. This is also consistent with Dunn's data (Fig. 7.17) for convective cooling presented earlier. Gaugler (1985) emphasizes the need for a knowledge of transition (if any) before any attempt can be made to predict the adiabatic cooling effectiveness, heat transfer, and aerodynamic effects of film cooling.

Furthermore, at high speed and high Reynolds number (engine conditions), the boundary layer is thinner than the corresponding case at low speed and low Reynolds number (wind tunnel simulation).

The effect of injection near the trailing edge influences the base pressure, heat transfer, and wake structure. These are covered in Sieverding (1983) and Kost and Holmes (1985).

Abhari and Epstein (1994) reported heat transfer measurements in their blow-down facility. The turbine was tested in Ar-Fr 12, with $T_{o1} = 478$ K, Re $= 2.7 \times 10^6$, $P_{o1} = 4.3$ Atm, $P_{o2} = 10$ Atm, $T_{o2} = 343$ K, $N = 6190$ rpm, $\dot{m} = 16.6$ kg/s, $T_w/T_e = 0.63$. They found that film cooling reduces the time-averaged heat transfer by nearly 60% on the suction side, as compared to an uncooled rotor. Heat transfer on a rotor blade was found to be lower than that measured in a corresponding cascade, and the blowing was less effective on the pressure side. High blowing ratios provided much less effective cooling.

7.6 TRANSPIRATION COOLING

Transpiration cooling is probably the most efficient and effective film-cooling technique from the point of view of heat transfer and cooling. It is the most difficult one from the manufacturing and maintenance point of view, and is the least efficient (at the present time) from the aerodynamic efficiency viewpoint. It is still in the development stage.

The transpiration cooling concept is shown in Fig. 7.4. It consists of a porous skin connected to a coolant/plenum chamber. Each plenum chamber is fed by a separate orifice near the hub, so that the coolant mass flow into each chamber can be individually regulated. The present-day manufacturing technique allows for pore size on the order of 0.01–0.05 mm. The low Reynolds number within the pores result in a high Stanton number (and heat transfer coefficient). Therefore, convection heat transfer plays an important role. These pores have a tendency to be clogged by dirt, carbon, and other particles. These problems, as well as oxidation problems, will decrease its effectiveness during the life cycle.

Transpiration cooling has been successfully attempted in a nozzle guide vane by Morris (1977a), operating at $T_e = 1650$ K, $T_f = 817$ K, and $T_w = 1150$ K. It has also been incorporated by Wolf and Moskowitz (1985) in an

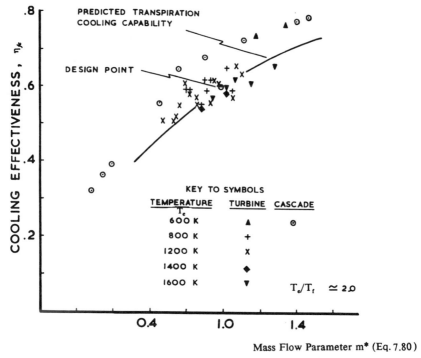

Figure 7.36 Cooling effectiveness of transpiration cooling in a nozzle (Morris, 1977a). (The original version of this material was first published by AGARD / NATO in CP 229.)

experimental high temperature turbine operating at 1700 K. They report a successful test for 900 hours without any major deterioration (oxidation, mechanical or thermal failure, erosion or corrosion, particle deposition).

The mean cooling effectiveness $(\eta_{fe}) = (T_e - T_w)/(T_e - T_f)$ measured by Morris (1977a) is shown in Fig. 7.36. It is plotted against a mass flow parameter m^* (Eq. 7.80). The results show high effectiveness, even at low coolant mass flow parameter. The superior cooling performance in terms of efficient use of coolant can be discerned by comparing this with a conventional film cooling system where $m^* = 4$ and $\eta_{fe} = 0.7$ at $T_e = 1600$ K. For the same effectiveness, transpiration cooling required about 25% of the air required for film cooling. This is also evident from Fig. 7.30, where Fullagar (1974) has compared transpiration cooling against film and convection cooling. Even though transpiration cooling is the most efficient from the point of view of the coolant air requirement, a suitable combination of convective and film cooling can achieve performance close to that of transpiration cooling and is much more practical. Morris (1977a) concludes that the aerodynamic penalty of transpiration cooling is significant, and will preclude its application in engines until such time as current cooling methods become untenable. The aerodynamic losses associated with transpiration cooling will be covered in a later section.

The major problem in implementing transpiration cooling in engines is a decrease in aerodynamic efficiency. A 4% decrease in aerodynamic efficiency was measured by Morris (1977a) when the solid vane was replaced by a porous vane, without cooling. This indicates the major effects of a porous surface on boundary layer characteristics. A word of caution is in order with regard to these data. The aerodynamic efficiency should be measured with coolant injection. Optimum coolant mass flow (from the point of aerodynamic efficiency and coolant effectiveness) may decrease skin friction coefficient compared to the one without it. This may result in lower than measured reduction in aerodynamic efficiency.

The effect of a porous surface on aerodynamic loading (e.g., blade pressure distribution), as well as its influence on transition, separation, and boundary layer growth, has not been well understood. It is possible to treat this as roughness, and calculate the boundary layer growth using the techniques described in Section 5.5.

Bayley (1977) has carried out basic research on transpiration cooling with various materials, porosity, and thickness. He has correlated all of the data and has provided the following expression for the heat transfer coefficient:

$$\frac{St}{St_o} = \left(1 - \frac{b}{b_{cR}}\right)^{1.6} \left[\frac{2}{1 + \sqrt{St/St_o}}\right]^2 \tag{7.81}$$

where

$$b = \frac{2\dot{m}_f}{A\rho_e u_e (C_f)_0}$$

where the subscript o refers to values for a solid blade, and b_{cR} is the critical blowing rate at which St/St_o is zero and $\dot{m}_f/A$ is coolant mass flow/unit area. It should be remarked here that $St/St_o = C_f/C_{fo}$; thus large blowing would lift the boundary layer from the surface, giving rise to $C_f = 0$, and causing low heat transfer. Bayley has carried out a boundary layer analysis to prove that $b_{cR} = 6$.

If the coolant reaches the metal temperature, the following equation can be used to determine the coolant rate (for a prescribed T_w) or T_w for a given $\dot{m}_f$:

$$h_g(T_g - T_w) = \dot{m}_f(c_p)_f(T_w - T_f) \tag{7.82}$$

where g refers to gas and f refers to film.

The right-hand side represents the enthalpy rise of the coolant, which should be equal to the heat transfer from the gas, neglecting the radiative heat transfer. The equation provides a preliminary estimate for $\dot{m}_f$ or T_w.

7.7 AERODYNAMICS, LOSSES, AND EFFICIENCY

All of the cooling techniques involve increased aerodynamic losses and decreased efficiency. The losses and inefficiency associated with turbine cooling include: the lost work or power of the coolant flow; increased profile losses due to thicker blade profiles arising from coolant holes; interaction of coolant film with the blade boundary layer; mixing losses; and the modification of transition, boundary layer, and wakes. The end-wall losses are also affected through both convective and film cooling. Many of these effects were covered in earlier sections. Anderson and Heiser (1969) cautioned and encouraged additional research and methodology to investigate sources of losses and assessment of aerodynamic performance, in addition to the heat transfer performance. In view of the complexity of the geometry and numerous parameters involved, a systematic method is yet to evolve for the prediction of cooled turbine efficiency. At the present time, testing is the only reliable method of determining the aerodynamic performance. It is necessary to optimize the thermal gain (increased turbine inlet temperature, cycle efficiency, etc.) with aerodynamic losses and decreased efficiency. An understanding of the losses is very important in comparing various cooling techniques. Loss analysis plays a significant role in the selection of the technique, location of the coolant holes, geometry of the coolant hole, coolant mass flow rate, and so on. A physical understanding of the sources of losses, as well as methods of estimating or predicting them, is covered in this section.

7.7.1 Efficiency of Cooled Turbines

In a cooled turbine, several streams of air are undergoing a similar thermodynamic cycle. The main flow, expanding through a nozzle and rotor (Fig. 2.7), has loss associated with frictional effects as described in Chapters 2 and 6. The coolant air consists of several streams: the coolant air from the nozzle and the rotor blades, platforms, and shrouds as well as coolant air from the turbine disks. There are several definitions of efficiency in the open literature, but the definition used by McDonel and Eiswerth (1977) is probably the most appropriate for a modern turbine. This definition will be adopted here, with some modifications. The thermal efficiency is defined as follows:

$$
\eta_t = \frac{\text{Net turbine power output}}{\text{Isentropic power available}} = \frac{\dot{m}_p (\Delta h_o)_p + \sum_j \dot{m}_j (\Delta h_o)_j - (\Delta h_o)_{cp}}{\dot{m}_p (\Delta h_o)_{ip} + \sum_j \dot{m}_j (\Delta h_o)_{ji}}
$$

$$(7.83)$$

where $\dot{m}_p$ is the primary air of the mainstream flow, $\dot{m}_j$ is the mass flow of the jth cooling stream, $(\Delta h_o)_p$ is the actual enthalpy drop of the primary air, and $(\Delta h_o)_j$ is the actual enthalpy drop of the jth cooling stream. $(\Delta h_o)_{cp}$ is the power required to pump the coolant. This is the work done to accelerate the tangential velocity of the rotor coolant at rotor entry to the blade meanline velocity. The denominator represents the ideal power developed by the primary air $[(\Delta h_o)_{ip}]$ and the secondary air $[(\Delta h_o)_{ji}]$, respectively.

In most cases, the isentropic work (the denominator of Eq. 7.83) includes the expansion of the primary gas and all of the coolant gases from the turbine inlet total pressure to the turbine exit total pressure. Let us assume that the mixing is complete, and the coolant air and primary air leave at the same total temperature. Because a comparison between an uncooled and a cooled turbine is sought, the coolant pressure is assumed to be the same as the primary air. For comparison with an uncooled turbine, the coolant is assumed to be mixed with the combustion products and expanded through a turbine. The enthalpy changes are given by (see Fig. 2.7 for notations)

$$(\Delta h_o)_{ji} = h_{o1c}\left[1 - \left(\frac{P_{o3}}{P_{o1}}\right)^{(\gamma-1)/\gamma}\right] \tag{7.84}$$

and $(\Delta h_o)_j = h_{o1c} - h_{o3}$, where $(h_{o1})_c$ represents the coolant inlet stagnation enthalpy, and h_{o3} is the exit stagnation enthalpy of both the coolant and the primary air. The coolant power $(\Delta h_o)_{cp}$ is given by

$$(\Delta h_o)_{cp} = \dot{m}_{cR}\left(U_m^2 - U_c V_{\theta_c}\right) \tag{7.85}$$

where U_m and U_c = blade velocity at the mean radius and coolant inducer radius, respectively; V_{θ_c} = tangential component of rotor coolant velocity at inducer discharge; and $\dot{m}_{cR}$ = mass of the rotor coolant air. Equation 7.85 is the Euler work done on the rotor blade coolant mass given by Eq. 2.7, with

$$U_2 = V_{\theta_2} = U_m, \quad U_1 = U_c, \quad \text{and} \quad V_{\theta_1} = V_{\theta_c}.$$

Using Eqs. 7.84 and 7.85, the efficiency defined by Eq. 7.83 can be written as

$$\eta_t = \frac{\dot{m}_p(h_{o1} - h_{o3}) + \sum_j \dot{m}_j(h_{o1c} - h_{o3}) - \dot{m}_{cR}\left(U_m^2 - U_c V_{\theta_c}\right)}{\left(\dot{m}_p h_{o1} + \sum_j \dot{m}_j h_{o1c}\right)\left[1 - \left(\frac{P_{o3}}{P_{o1}}\right)^{(\gamma-1)/\gamma}\right]} \tag{7.86}$$

Each of the expansion processes, for both primary and secondary air (say $j = 1$ to n), can be represented on an enthalpy–entropy diagram as in Fig. 2.7. The entropy change of each of these streams is due to viscous losses, heat removal, or heat addition. For the main flow, the viscous losses (end-wall, shock-boundary layer interaction, blade boundary layer, etc.), heat removal, or addition all contribute to the change in entropy.

The coolants undergo different thermodynamic paths. This should be recognized in calculating the actual and available work from different coolants. h_{o1c} for the nozzle coolant is the entry stagnation enthalpy, while for the rotor, it is the stagnation enthalpy at the entry to the rotor. Likewise, P_{o1c} and P_o for the nozzle coolant is the same as the primary air, but for the rotor coolant, it is again the value at entry to the rotor.

The efficiency of a cooled turbine can be expressed in terms of entropy change and the actual pressure ratio using the thermodynamic relationship, such as

$$\frac{T_{o3}}{T_{o1}} = \left(\frac{P_{o3}}{P_{o1}}\right)^{(\gamma-1)/\gamma} \exp\left(\frac{\Delta s}{c_p}\right)$$

where Δs is the increase in entropy. Therefore, Eq. 7.86 can be written as (assuming c_p is constant in the temperature range T_{o1} to T_{o3})

$$\eta_t = \frac{\dot{m}_p T_{o1}\left[1 - \left(\frac{P_{o3}}{P_{o1}}\right)^{(\gamma-1)/\gamma} \exp\left(\frac{\Delta s}{c_p}\right)\right] + \sum_j \dot{m}_j T_{o1c}\left[1 - \left(\frac{P_{o3}}{P_{o1}}\right)^{(\gamma-1)/\gamma} \exp\left(\frac{\Delta s}{c_p}\right)_j\right] - \dot{m}_{cR}\left[U_m^2 - U_c V_{\theta_c}\right]}{\left(\dot{m}_p T_{o1} + \sum_j \dot{m}_j T_{o1c}\right)\left[1 - \left(\frac{P_{o3}}{P_{o1}}\right)^{(\gamma-1)/\gamma}\right]}$$

$$(7.87)$$

There is still controversy in the profession as to the precise definition of efficiency of a cooled turbine. The compression work needed to bring the coolant to turbine entry pressure should be included as loss. Furthermore, the Euler work on the coolant is superfluous if the highest reservoir state is used for the coolant jet. Nevertheless, the definition implied by Eq. 7.87 is widely employed by practitioners, and this will be dealt with in the remainder of this section.

The entropy change or losses are calculated by considering each of these gases in a frame of reference in which the flow is steady. For example, in a nozzle, the frame of reference is absolute and $T_{o1} = T_{o2}$. In a rotor, the relative frame of reference is used and $T_{o2R} = T_{o3R}$. In an uncooled turbine, the entropy change is caused only by the viscous losses, and it is given by

$$\frac{\Delta s}{c_p} = -\frac{\gamma - 1}{\gamma} \ln\left(\frac{P_{o1} - \Delta P_o}{P_{o1}}\right) \qquad (7.88)$$

where ΔP_o and P_{o1} are stagnation pressure loss and inlet stagnation pressure respectively, in a corresponding frame of reference. In a cooled turbine, there is a transfer of heat from the gas to the coolant and vice versa, and thus the flow is nonadiabatic. In such a case,

$$\frac{\Delta s_i}{c_p} = \ln\left(1 + \frac{\Delta T_{oi}}{T_{oli}}\right) - \frac{\gamma - 1}{\gamma}\ln\left(1 - \frac{\Delta P_{oi}}{P_{oli}}\right) \qquad (7.89)$$

where i refers to the individual stream, and ΔT_{oi} is the stagnation temperature change due to heat transfer.

A knowledge of the fundamentals of heat transfer described in an earlier section provides an estimate for ΔT_{oi}; and a knowledge of losses to be described later provides an estimate for Δp_{oi} or stagnation pressure loss coefficient. These are then substituted in Eqs. 7.89 and 7.87 to estimate the efficiency of a cooled turbine.

In many cases, it is essential to examine the loss coefficient defined for the main flow only, coolant only, and the total air flow to understand the various loss mechanisms and to make a proper comparison of various cooling techniques for aerodynamic performance. The coolant has both a direct effect (pressure loss, mixing loss) and an indirect effect (e.g., change in transition and boundary layer growth) on the main flow. All of the basic research on loss mechanisms pertains to a stationary nozzle or cascade. Therefore, the following loss coefficients pertain to a nozzle or to a rotor in a relative frame of reference.

In addition to the overall efficiency defined by Eqs. 7.83, 7.86, or 7.87, several other efficiency expressions are used to evaluate the performance of cooled turbines. These will be described below.

In the stationary system of a nozzle and a cascade and in the rotating system for a rotor, the kinetic energy recovery is the main process. Performance of the system can be based on how efficiently the kinetic energy is being recovered. In the isentropic case, $T_{o1} = T_{o2}$, and $T_{o2R} = T_{o3R}$. The losses based on ideal kinetic energy of the total air flow are more appropriate and are given by

$$\zeta_{\text{KE}} = 1 - \frac{\text{Actual exit kinetic energy in total air flow}}{\text{Ideal exit kinetic energy in total air flow}}$$

$$= 1 - \frac{(T_{o2} - T_2)\left(\dot{m}_p + \sum_j \dot{m}_j\right)}{(T_{o1} - T_{2s})\dot{m}_p + \sum_j \dot{m}_j(T_{o1j} - T_{2sj})} \qquad (7.90)$$

The summation term in the denominator is the ideal kinetic energy in the coolant flow, T_{o1j} is the inlet stagnation temperature of the jth stream, and T_{2sj} is the isentropic static temperature of the jth coolant flow (corresponding to T_{2s} for primary flow), all in either a relative frame of reference for a rotor or an absolute frame of reference for a stator or nozzle. $T_{o1j} - T_{2sj}$ is an isentropic process (Fig. 2.7). The efficiency based on this will be termed *thermal efficiency*, given by

$$\eta_{th} = 1 - \zeta_{KE}$$

It should be recognized here that η_t and η_{th} defined in this section are different. η_{th} is related to η_{ts} as defined in Eq. 2.21. All the efficiencies introduced here are used in practice. For an uncooled turbine, the most commonly used efficiencies are η_t and η_{ts}.

7.7.2 Aerodynamic Losses

The aerodynamic losses, both direct and indirect, associated with cooling can be classified as follows: internal coolant loss; profile loss for convective cooling; film cooling losses; transpiration cooling losses, and three-dimensional losses.

Most of the investigations on cooling have concentrated mainly on heat transfer performance. Only scattered information exists on the losses classified above. Losses due to film cooling has received more attention than others due to their magnitude and importance. As indicated in an earlier chapter, the loss estimate is still very much based on empirical correlations. There has been no systematic investigation to isolate the three-dimensional (endwall and leakage flow) losses from those due to cooling. Hence, no detailed treatment on these losses is given here.

7.7.2.1 *Internal Coolant Loss* The losses associated with internal cooling arise from the flow of coolant in narrow passages, in pins and fins, and through ejection holes. In most cases the coolant holes are small, and the flow can usually be considered fully developed. If the flow is not fully developed, the boundary layer on the walls can be calculated using the techniques described in Section 5.5 (if the pressure along the centerline is known). A knowledge of the momentum thickness allows an estimation of the pressure loss coefficient (Chapter 6).

Let us consider the case of fully developed flows. Information on the losses in these ducts can be found in most engineering books on viscous flow (e.g., Eckert and Drake, 1972; Schlichting, 1979, p. 79). For example for fully developed laminar flow in a pipe, the pressure gradient is given by

$$-\frac{\partial p}{\partial x} = \frac{\lambda}{d} \frac{\rho}{2} (\overline{U})^2 \qquad (7.91)$$

where $\lambda = 64/\text{Re}$, for circular pipe of diameter d, $\overline{U}$ is the mean velocity. For triangle, trapezoidal, and rectangular pipes, an equivalent diameter is used in the definition of the Reynolds number.

For turbulent flow at high Reynolds number (Schlichting, 1979, p. 584) we have

$$\frac{1}{\sqrt{\lambda}} = 1.74 - 2\log\left[\frac{k_s}{r} + \frac{18.7}{\mathrm{Re}\sqrt{\lambda}}\right] \qquad (7.92)$$

where k_s is the roughness height and r is the radius of the pipe.

Because the coolant passage has bends, the friction factor increases due to secondary flow and associated mixing in these bends. The increase in the friction factor can be estimated from the relationship (Schlichting, 1979, p. 590)

$$\frac{\lambda}{\lambda_o} = 1 + 0.075(\mathrm{Re})^{1/4}\left(\frac{r}{R}\right) \qquad (7.93)$$

where λ_o is the friction factor for straight pipe or channel and R is the radius of curvature of the bend.

For a rotating coolant channel, there would be an additional increase in frictional losses. The friction coefficient would then depend on the rotation number as well as the Reynolds number. In this case, one would expect λ to be a function of the rotation number ($2\Omega D/\overline{W}$) or the rotational Reynolds number as indicated by Eqs. 7.46–7.48 and 7.52. There are no correlations available at this time to include the effect of rotation on pressure losses.

Arora and Abdelmesseh (1985) have carried out extensive measurements of frictional losses in a pin/fin configuration (Figs. 7.5, 7.6, and 7.8) and correlated the results with Eq. 7.91. The value of λ is found to be a strong function of the pin/fin diameter, shape, stagger distance, axial orientation, and so on (Fig. 7.8). But the Reynolds number dependency was similar in all cases; that is, λ decreases with the increasing Reynolds number.

Similarly, the pressure losses through film ejection holes are important in evaluating the overall performance. In addition to the loss correlation presented here, it is necessary to include the effect of sharp edges at the ejection point. These are likely to increase the losses. Tillman and Jen (1984) have provided an experimental correlation for estimating the coolant losses through film-cooling holes in the presence and absence of external flow.

Once the static pressure is calculated from Eq. 7.91, using the appropriate correlation for λ, the stagnation pressure loss can be computed knowing the inlet and exit velocities and densities. This can then be used in Eqs. 7.89 and 7.87 to determine the change in efficiency due to these losses. When the coolant holes are choked (or nearly choked), the total pressure can be found directly from gas dynamic equations knowing inlet and exit Mach numbers.

7.7.2.2 *Profile Loss Due to Convective Cooling and the Change in Geometry* Convective cooling alters the profile loss from that of an uncooled blade by changing the blade boundary layer characteristics, including

the laminar length, separation bubble, transition, and the turbulent boundary layer growth. One can distinguish two sources. The first source is the change in boundary layer properties due to a change in geometry, and the second source is the heat transfer from the gas to the coolant, which brings about the change in the boundary layer property. One additional source of loss is due to the ejection of spent coolant to the main stream. The last source will be dealt with under film cooling and trailing ejection losses in the next section.

The cooled blades must be thicker than uncooled blades to accommodate convective cooling and film-cooling passages. Because the boundary layer property is very sensitive to the profile, this results in increased profile (boundary layer and wake mixing) losses. The cooled blade has greater surface velocities, greater suction surface compression, and larger trailing-edge thickness. Furthermore, the coolant holes on the surface for film cooling or transpiration cooling, even in the absence of coolant, would introduce a substantial change in the roughness and pressure gradient. This would change the boundary layer properties and the profile losses.

A dramatic example of this effect is shown in Fig. 7.37, where the data from Morris (1977a) for a transpiration-cooled nozzle vane is shown ($T_e = 1650\,\mathrm{K}$, $T_f = 817\,\mathrm{K}$, $P_{oe} = 407\,\mathrm{KN/m^2}$, $P_{of} = 414\,\mathrm{KN/m^2}$, $\Delta h_o/U^2 = 2.00$). The difference in efficiency between a solid and a perforated blade (without any coolant mass flow) is as much as 3.5%, a substantial increase in the profile losses. This is a clear indication that direct effects such as those

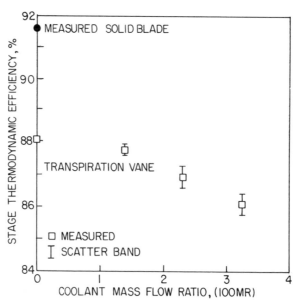

Figure 7.37 The effect of transpiration cooling on turbine stage thermodynamic efficiency. η_t is from Eq. 7.83, MR = $\dot{m}_f / \dot{m}_p$. See Fig. 7.36 for η_f for this blading. (Morris 1977a. The original version of this material was first published by AGARD / NATO in CP 229.)

arising from profile change and surface conditions (e.g., roughness) have an appreciable influence on the profile losses. It should be cautioned here that the losses due to perforations would be substantially higher than those due to change in the profile due to internal coolant holes. At low or zero $\dot{m}_f$ there can be recirculation through the surface. The data in Fig. 7.37 are presented to illustrate the sensitivity of losses to profile and surface changes alone.

In the same way, indirect effects due to heat transfer will affect the transition, laminar, and turbulence boundary layer growth. Very few attempts are made to isolate these effects, but they can be predicted using a boundary layer code (Section 5.5) that includes higher-order turbulence models. The prediction techniques will be covered in Section 7.8. Once the momentum thickness is accurately predicted, the profile losses in the presence of heat transfer can be calculated using the method described in Chapter 6.

7.7.2.3 Aerodynamics of Film Cooling and Losses

Aerodynamics and losses due to film cooling have been investigated by many researchers. Complex phenomena associated with film cooling are depicted in Figs. 7.23–7.26. The coolant, depending on its location and velocity, will induce flow separation and a reattachment zone with the formation of a separation bubble. This will eventually mix with the boundary layer, causing additional losses. The boundary layer characteristics depend on many parameters in addition to those detailed in Section 5.5. These include coolant hole geometry (shape, size, spacing, angle, etc.), location, coolant mass flow, velocity and pressure, Reynolds number, and temperature. The physical phenomena associated with these effects and the variation of flow field and losses will be described first before presenting analyses to predict these effects.

The location of the coolant hole is critical and can be optimized through a knowledge of inviscid pressure distribution, transition, and boundary layer characteristics. The location is very crucial from the aerodynamic point of view, which is often in conflict with the heat transfer requirements. This is clearly evident from Fig. 7.38, due to Heiser (1978). The shower head and 90° angle injection usually experience the highest losses. The additional losses due to film cooling, as shown in Fig. 6.34 for the E^3 engine vane, can be substantial. The losses are nearly doubled from hub to mid-span. This stage had platform cooling, hence the increased losses (due to film cooling) observed near the hub.

A comprehensive investigation of the effects of various cooling configurations on aerodynamic performance is reported by Kiock et al. (1985). A summary of their results is given in Fig. 7.39. The cascade parameters were: turning angle = 100°, $M_2 = 0.985$, $T_f = 303$ K, $T_e = 310$ K, $Re_2 = 8.4 \times 10^5$, and $C = 100$ mm. Design values for MR are 3.2%, 5.2%, and 1.2%, respectively, for configurations A, B, and E. The boundary layer profile at 2% chord upstream of the trailing edge is also shown in Fig. 7.39. Case A is characterized by shock-laminar boundary layer interaction. With injection, the boundary layer at the trailing edge shows a strong tendency to separate.

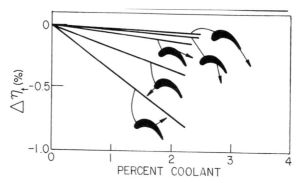

Figure 7.38 Effect of film cooling upon turbine efficiency ($\Delta\eta_t$ is the change in turbine efficiency) (Heiser, 1978).

The trailing edge jets influence the onset of early separation, which is substantiated by a large increase in the boundary layer shape factor (Kiock et al., 1985). The loss coefficient and momentum thickness all increase with an increase in the value of mass ratio MR. The ejection at the throat section (case B) caused an early transition and an appreciable increase in the momentum thickness. The loss coefficient is appreciably higher than case A, as well as the solid blade. Similar increases are observed for case E, which has only shower head. Configuration A, with low momentum thickness and loss, is an optimum one from an aerodynamic point of view. Because no heat transfer data are taken, it is difficult to assess its thermal performance.

Kollen and Koschel (1985) have made an attempt to study the effect of location of the injection holes on the aerodynamic profile losses in a turbine nozzle ($\alpha_1 = 88°$, $\alpha_2 = -19.5°$, $C = 48.1$ mm, $S/C = 0.693$, $h = 44$ mm, mean diameter = 350 mm). No heat transfer data were taken by the authors. The blade profile, experimental data, injection hole location, and the exit Mach numbers are shown in Fig. 7.40. Interpretation of predictions will be covered in a later section. The data indicate that the loss coefficient ζ_{KE} increases with an increase in coolant mass flow rate for the stagnation region cooling. The various sources causing the increased losses are not clear from the investigation. It could be caused by premature laminar separation or blowing of the boundary layer in this region. The blowing on the suction side, shown in Fig. 7.40, seems to have a beneficial effect on the aerodynamic losses. One can speculate that this is caused by early transition. It is also likely that the boundary layer at this location is blown away, giving rise to a decreased boundary layer growth and the wake. It should be cautioned here that the boundary layer behavior is very sensitive to the location of the injection hole. A slight movement of this location on either side could have a pronounced effect.

In many instances, the spent coolant in a convectively cooled blade is ejected at the trailing edge, or film ejection is located there to provide

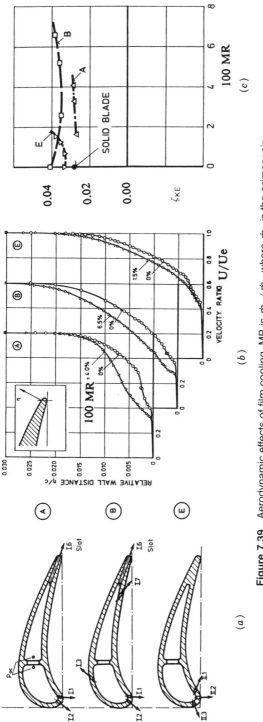

Figure 7.39 Aerodynamic effects of film cooling. MR is $\dot{m}_f / \dot{m}_p$ where $\dot{m}_p$ is the primary air through one passage, ζ_{KE} is from Eq. 7.90, U_e is the free stream velocity, and n is the distance normal to the surface. (a) Geometry. (b) Velocity profiles at 98% chord on suction surface. (c) Aerodynamic losses. (From Kiock et al., 1985; the original version of this material was first published by AGARD / NATO in CP 390, "Heat Transfer and Cooling in Gas turbines," September 1985.)

697

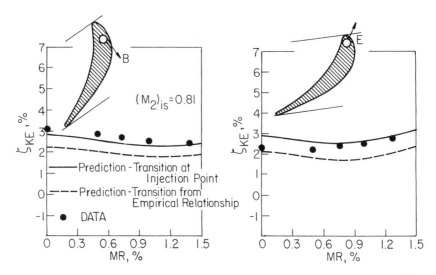

Figure 7.40 Energy loss coefficient for suction side and leading-edge blowing (MR = $\dot{m}_f / \dot{m}_p$; ζ_{KE} is from Eqs. 7.94 and 7.94a) (Kollen and Koschel, 1985). (The original version of this material was first published by AGARD / NATO in CP 390.)

additional cooling. This is especially critical in an HP turbine blade due to its cooling requirement. Therefore, the trailing edges are thick, increasing the blockage, wake thickness, base drag, and shock wave interaction with the adjacent blading. The trailing-edge ejection has a beneficial effect, and it tends to prevent separation in this region as well as increase wake diffusion and decreased shock interactions. Results from Kost and Holmes (1985) in a turbine rotor, shown in Fig. 7.41, reveal the beneficial effect of trailing edge ejection. The loss coefficient ζ_{KE} decreases as much as 20–25% up to MR = 4% at $M_2 = 0.77$, beyond which additional blowing tends to increase the losses. Once again, it is essential to study the effect of overall efficiency η_t to determine its beneficial effect.

The data by Sieverding (1983) indicate that the ejection of coolant flow near the trailing edge affects the base pressure considerably for a transonic cascade. The increase in the base pressure was as high as 15% of the downstream dynamic head. The influence of the trailing-edge bleed on the pressure distribution is found to be of minor importance. Shock waves are shifted slightly upstream without influencing the shock strength. In relatively low Mach number blading with a thin trailing edge, the ejection actually increases the wake thickness and depth and decreases the temperature defect in the wake (Hempel and Friedrich, 1978). Only at a higher mass flow does the wake defect begin to decrease. This configuration will encounter adverse aerodynamic effects due to blowing. In summary, the trailing-edge ejection for a thick trailing edge at high Mach numbers improves aerodynamic

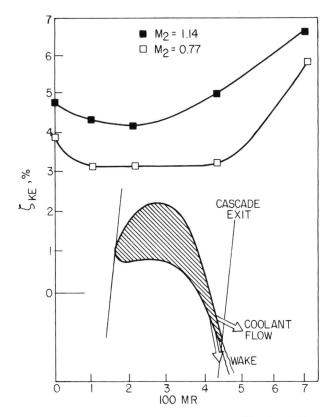

Figure 7.41 Aerodynamic effects of trailing-edge ejection ($S/C = 0.76$, $\lambda = 32.8°$, $\alpha_1 = 60°$, $\alpha_2 = -69.4°$, Re $= 8.8 \times 10^5$, $M_1 = 0.45$, $M_2 = 1.15$) (Kost and Holmes, AGARD CP 390 1985).

performance, and for a trailing edge at low Mach numbers (subsonic), there is an increase in aerodynamic losses.

The aerodynamics of full-coverage film cooling has been investigated by McDonel and Eiswerth (1977) and Prust (1978). The data reported by Prust indicate that the losses increase continuously and almost linearly with an increase in the coolant mass ratio MR. The data reported by McDonel and Eiswerth (1977) are some of the very few available in the open literature on the effect of full coverage film cooling on a full-scale high-temperature turbine. The data were taken in a single-stage turbine with $T_{o1} = 783$ K, $P_{o1} = 3.9 \times 10^5$ N/m², $\dot{m}_1 = 11.2$ kg/s, $N = 10{,}146$ rpm, and $D = 0.508$ m. The measured efficiencies are shown in Fig. 7.42. The "discrete film" configurations had discrete film cooling and an impingement system. The "full film" configuration also had impingement cooling. In all cases, the spent air was discharged along the flow path. It should be noted that MR = 4% for a

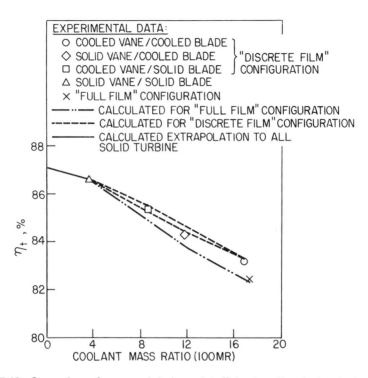

Figure 7.42 Comparison of measured design point efficiencies with calculated values for a single stage turbine (η_t is from Eq. 7.83) (McDonel and Eiswerth, 1977. The original version of this material was first published by AGARD / NATO in CP 229.)

solid blade represents the cooling air for endwalls, disk cooling, and so on. The cooled vane/cooled blade, which is the most practical case, has the largest decrease in efficiency. As much as a 3.5% decrease in efficiency for 12% blade coolant air (16% total coolant air) as compared to a solid blade (86.5% efficiency) is observed. McDonel and Eiswerth have also provided an estimate of the component losses. The losses associated with the rotor cooling were approximately 60%, and nozzle cooling were 40% of the total vane/blade losses due to cooling.

As in film cooling, the transpiration cooling modifies the transition, boundary layer growth, and separation. The efficiencies measured by Morris (1977a) for a transpiration-cooled nozzle vane in a turbine facility with $T_{o1} = 1650\,\mathrm{K}$, $T_{of} = 817\,\mathrm{K}$, $P_{o1} = 407\,\mathrm{kN/m^2}$, $\psi = 2.0$, and $P_{o1}/P_{o2} = 1.28$ are shown in Fig. 7.37. It is evident that the losses are substantial. The measured loss in efficiency is as high as 6% for a 3% coolant mass ratio. Not all of this is attributable to coolant ejection and its mixing with the main stream. Nearly half of it can be attributable to geometry change and its effect on boundary layer growth.

There has been very little work done to quantify the three-dimensional effects of film cooling. Some of the major effects are as follows:

1. Film cooling on endwalls would modify hub and annulus wall boundary layers, endwall losses, and secondary flow.
2. Spanwise temperature gradients, discrete cooling holes in the spanwise direction, and spanwise mixing of coolant will introduce three-dimensionality.
3. If there is tip blowing, the structure of the tip vortex will be modified.
4. The trailing-edge blowing in a rotor introduces three-dimensionality due to radial transport of the coolant near the trailing edge. This is likely to alter the wake characteristics from those of two-dimensional flows.
5. Compound angle (α and β in Figs. 7.23 and 7.6) injection introduces severe three-dimensionality.

7.7.2.4 *Analysis of Aerodynamic Losses Due to Film Cooling* The total profile losses in a cooled blade row consist of the following:

1. The profile loss due to the blade boundary layer and the wake in the absence of surface cooling jets, but with trailing edge ejection cooling. This includes base pressure drag, shock-boundary layer interaction losses, wake dissipation losses, and so on. The losses are estimated according to the procedure described in Chapters 5 and 6. Let us denote this loss as ζ_p, to indicate that it is the profile loss in the absence of surface coolant.
2. The jet, which emerges from the blade, as shown in Fig. 7.43, mixes with the boundary layer and the main stream. The losses due to the mixing of the coolant jet with the main stream from the location of the jet to far downstream is an *additional* loss. Let us denote this by ζ_{jm}. The velocity profile shown in Fig. 7.43, at station 1, is assumed to be uniform for the jet mixing analysis, because the profile loss (due to blade boundary layer) is already included in ζ_p. The flow is assumed to be one-dimensional across the mixing layer and far downstream.

This is only a hypothesis to enable the development of a calculation method for ζ_{jm}. In practice, these two processes are coupled. This is the approach taken by Crawford and Kays (1976). In this case, the jet is assumed to stay within the boundary layer, and the process is treated as one of mass addition. A similar approach is taken by Goldman and Gaugler (1980). As indicated earlier, the film-cooling process involves both effects. If the coolant holes are located near the leading edge or up to about midchord, the jet is likely to penetrate into the main flow. This approach then involves the

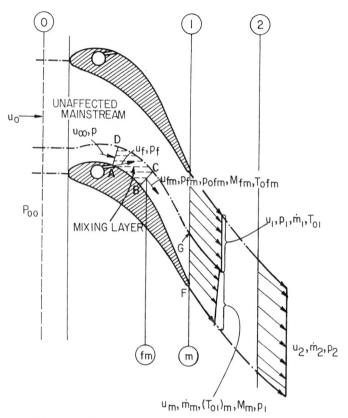

Figure 7.43 Schematic of film-cooling mixing process for loss estimate for a cascade blade.

solution of not only the boundary layer equation, but almost the entire flow field, including coolant injection. If the jet is confined within the thick boundary layer (say from the midchord to the trailing edge) then the boundary layer approach described in Crawford and Kays (1976) and Goldman and Gaugler (1980) is valid. This section is devoted to simple and approximate analyses. Numerical and boundary layer calculations methods will be described in the next section.

The calculation method for ζ_p is indicated in Chapter 6. Total losses can be computed from the equation

$$\zeta_{KE} = \zeta_p + \zeta_{jm} \tag{7.94}$$

The analysis for predicting the film-cooling effectiveness were presented earlier. The aerodynamic analysis for the prediction of film cooling losses (ζ_{jm}) will be presented in this section. The loss is usually expressed in the form of loss in the kinetic energy given by Eq. 7.90. This can be recast in the following form (see Fig. 7.43 for notations):

$$\zeta_{jm} = 1 - \frac{(1 + MR)u_2^2}{u_1^2 + MRu_{fi}^2}$$ (7.94a)

Once again, it should be emphasized here that the profile losses in the above equation (ζ_{jm}) include only the additional losses due to jet mixing. In the absence of boundary layer, the ideal kinetic energy at station 1 is $0.5\dot{m}_o u_1^2$ (where $\dot{m}_o$ is the total mass flow of the primary air at both stations 0 and 1). The velocity components u_1 and u_2 are defined in Fig. 7.43, and MR = $\dot{m}_f/m_p$. u_{fi} is the ideal coolant velocity at the same pressure as the mixed gases. These are the velocity components in the absence of viscous layers. The purpose of the analyses given below is to predict u_2, u_{fm}, and u_1 using a control volume approach. These quantities can also be predicted using the computational techniques described in the next section.

There are several methods available for predicting the aerodynamic mixing resulting from the coolant injection. Many investigators have attempted to model the mixing of the coolant jet and the main flow through one-dimensional analysis. Some of the earlier attempts are due to Tabakoff and Hamed (1975), Harstel (1972), and Prust (1972). All of these methods assume that the coolant and primary flow are mixed one-dimensionally at constant pressure. Only the component of coolant flow parallel to the main-stream flow at the coolant hole contributes to the blade-row output. Harstel (1972) used the same technique on both sides of the airfoil, while Prust (1978) used a different approach. In the diffusion region, Prust assumes that the component of coolant momentum flux that is parallel to the main-stream flow at the ejection location is maintained at the blade-row exit, where the coolant mixes with the main flow. His analysis neglects the compression energy required to raise the pressure of the coolant fluid p_f to p_2.

Ito et al. (1980) and Kollen and Koschel (1985) have modified Harstel's analysis for a more generalized case. Because Kollen and Koschel include compressibility, their analysis is given here. Most of the authors assume that the mixing process occurs in a constant pressure field. Even though this is true in the vicinity of the jet, it is not valid for the accelerating/decelerating regions from the vicinity of the jet to the trailing edge. It is easy to incorporate the pressure gradient effect. This modification is included in the analysis presented here. The mixing process is illustrated in Fig. 7.43. The velocity upstream of cascade is u_o and the main stream velocity in the vicinity of the coolant jet is u_∞. The following processes are identified.

1. The jet mixes with the main stream and is assumed to penetrate into the main stream. The velocity at the exit of the "mixing layer" is u_{fm} which is assumed to be one-dimensional.

2. The unaffected main stream and the mixing layer continue toward the trailing edge, resulting in a velocity profile shown in Fig. 7.43. Most authors assume that the jet continues to mix in a constant pressure field with a resultant velocity u_m. This assumption is not valid and is unnecessary. The diffusion of velocity u_{fm} to u_m can be calculated from the known pressure and density field, assuming that the mass flow is conserved along the stream tube ABF and DCG.

3. The "mixing layer" and the unaffected main stream, both of which are inviscid, mix into a uniform stream from station 1 to station 2. It is required to find the stagnation pressure loss $(P_{o0} - P_{o2})$ due to the mixing, and this involves analyses of properties at stations BC, station 1, and station 2, shown in Fig. 7.43.

In carrying out these analyses, the inlet stagnation pressure P_{o0} and the exit static pressure p_2 are assumed not to be influenced by the jet; that is, these values are the same as those calculated without the jet and with an identical upstream flow field. The main-stream flow near the ejection point is assumed to be unaffected by the jet; that is, the value of u_∞ is unaffected by u_f. That is a valid assumption, because the injection velocity and coolant mass flow are usually small. The analysis is based on control volume approach (Eqs. 1.2–1.5), similar to that carried out for the analysis of profile loss coefficient (ζ_p) in Section 6.4. The viscous effects of the boundary layer and the wake are included in the estimate for ζ_p. The main and the coolant fluid are assumed to be perfect gases.

Mixing Layer Analysis: Property Change from 0 to 1. As mentioned earlier, the region inside the "mixing layer" in Fig 7.43 is assumed to be a constant pressure mixing; therefore, the continuity, momentum, and energy equations for this region (ABCD) are given by

$$\underset{(\dot{m}_\infty)}{\rho_\infty u_\infty A_\infty} + \underset{(\dot{m}_f)}{\rho_f u_f A_f} = \underset{(\dot{m}_{fm})}{\rho_{fm} u_{fm} A_{fm}}$$

$$\dot{m}_\infty u_\infty + \dot{m}_f u_f \cos \alpha_f = \dot{m}_{fm} u_{fm}$$

$$\dot{m}_\infty h_{o\infty} + \dot{m}_f h_{of} = \dot{m}_{fm} h_{ofm} \qquad (7.95)$$

where A_∞ and A_{fm} are the areas of stream tube at AD (denoted by ∞) and BC (denoted by fm), respectively; A_f is the coolant hole area; and u_∞ is the inviscid velocity calculated from a panel or a Euler code. The right side with subscript fm denotes conditions at the exit of the mixing layers (BC). Solution of these equations for the exit Mach number (M_{fm}) results in the following expression (Kollen and Koschel, 1985):

$$M_{fm}^2 = \cfrac{1}{\cfrac{(1+\chi)(h_{o\infty}+\chi h_{of})\left(1+\cfrac{\gamma-1}{2}M_\infty^2\right)}{\left(\chi\cfrac{u_f}{u_\infty}\cos\alpha_f+1\right)^2 M_\infty^2 h_{o\infty}} - \left(\cfrac{\gamma-1}{2}\right)} \qquad (7.96)$$

where $\chi = \dot{m}_f/\dot{m}_\infty$ is the inverse of dilution ratio E. α_f is the injection angle. The total pressure loss due to mixing between AD and BC is then given by (assuming $p_\infty = p_{fm}$)

$$\frac{P_{ofm}}{P_{o\infty}} = \frac{\left(1+\cfrac{\gamma-1}{2}M_{fm}^2\right)^{\gamma/(\gamma-1)}}{\left(1+\cfrac{\gamma-1}{2}M_\infty^2\right)^{\gamma/(\gamma-1)}} \qquad (7.97)$$

The assumption $p_\infty = p_{fm}$ implies that the injection of the jet does not alter the local pressure distribution on the blade at the location of injection.

Mixing Between Stations 1 and 2. Let us now consider the downstream mixing process (stations 1 and m, and 2) in Fig. 7.43:

$$\dot{m}_2 = \dot{m}_1 + \dot{m}_m = \dot{m}_o + \dot{m}_f$$
$$\dot{m}_2 u_2 - \dot{m}_1 u_1 - \dot{m}_m u_m = (p_1 - p_2)A_2$$
$$\dot{m}_m(T_{o1})_m + \dot{m}_1 T_{o1} = \dot{m}_2 T_{o2}$$

where $\dot{m}_1$ is the mass flow in the unaffected main stream at station 1 and $\dot{m}_m(=\dot{m}_{fm})$ is the mass flow in the mixing region at station 1 and m as shown in Fig. 7.43. In these equations, c_p and areas are assumed to be the same at stations 1 and 2. A_2 is the area normal to $u_2(S/\cos\alpha_2)$. These equations can be solved to derive expressions for u_2, T_{o2}, and p_2 (Kollen and Koschel, 1985):

$$u_2 = \frac{A}{2} - \sqrt{\frac{A^2}{4} - \frac{2\gamma}{\gamma+1}\cdot RT_{o2}} \qquad (7.98)$$

$$A = \frac{2\gamma}{\gamma+1}\left[\frac{1-\sigma}{1+\chi}u_1 + \frac{\sigma+\chi}{1+\chi}u_m + \frac{p_1 A_2}{\dot{m}_o(1+\chi)}\right] \qquad (7.99)$$

$$\sigma = \frac{\dot{m}_m - \dot{m}_f}{\dot{m}_o}$$

$$T_{o2} = \frac{1-\sigma}{1+\chi}T_{o0} + \frac{\sigma+\chi}{1+\chi}(T_{o1})_m \qquad (7.100)$$

$(T_{o1})_m = (T_o)_{fm}$, because the flow is one-dimensional and u_m is calculated from a knowledge of u_{fm} (previous analysis) and the pressure gradient in the stream tube from B to m. If the properties at station BC are known, the properties at station 2 can be calculated. The values of u_1 and p_1 are derived from an inviscid analysis of the cascade flow.

This analysis provides all the quantities in Eq. 7.94a. The ideal coolant velocity u_{fi} is derived by assuming an isentropic expansion of the coolant flow from the injection point to station 2. This is given by

$$u_{fi} = \sqrt{\frac{2\gamma}{\gamma - 1} RT_{of}\left[1 - \left(\frac{p_2}{p_{of}}\right)^{(\gamma-1)/\gamma}\right]} \qquad (7.101)$$

The following steps enable us to calculate ζ_{jm}:

1. Compute u_1, p_1, A_2, T_{o1}, and u_∞ from an inviscid analysis.
2. Calculate M_{fm}, P_{ofm}, and u_{fm} from Eqs. 7.95–7.97.
3. Using the inviscid blade pressure distribution, calculate the quantities u_m, $(T_{o1})_m$, and so on.
4. Calculate u_2 and u_{fi} from Eqs. 7.98 and 7.101, respectively, to derive losses from Eqs. 7.94 and 7.94a. It should be remarked here that none of the earlier analyses (Kollen and Koschel, 1985; Kost and Holmes, 1985; Hempel and Friedrich, 1978; Prust, 1972, 1978; Crawford and Kays, 1976; Goldman and Gaugler, 1980; Tabakoff and Hamed, 1975; Harstel, 1972; Ito et al., 1980) include step 3; all the comparisons shown neglect this step.

Kollen and Koschel (1985) have compared their data with predictions from their analysis described in this section. These are shown in Fig. 7.40. ζ_p was calculated from a boundary layer code (without injection) and includes wake and boundary layer losses as well as base drag. The prediction from the above analysis, which assumes transition at the injection point, provides the best agreement. This seems to confirm that with suction surface blowing, the transition inception is likely to occur at the injection point.

The method of Harstel (1972) mixes the individual coolant flow and a portion of the main flow sequentially at each ejection location along the blade suction and pressure surfaces. The resultant flow is then mixed with the remainder of the main-stream flow at the blade-row exit. Prust's (1972) method expands the individual coolant flow from the ejection location to the blade exit and then mixes the coolant flow with the main-stream flow. The two analyses are very similar to the analysis presented here. The agreement between Prust's analysis and his data is very good.

The analytical prediction of efficiency from Eqs. 7.87 and 7.89 for a complete turbine stage is very complex, and a consistent system should yield the same result for η_t. An energy budget has to be carefully drawn. For example, $(\Delta s/c_p)_j$ in Eq. 7.87 involves several streams: the nozzle vane coolant, nozzle endwall coolant, rotor blade and endwall coolant, and the disk coolant. Each of these are again subdivided into finer elements. For example, for the nozzle vane we have

$$\left(\frac{\Delta s}{c_p}\right) = \left(\frac{\Delta s}{c_p}\right)_{FC} + \left(\frac{\Delta s}{c_p}\right)_p$$

where FC stands for film-cooling losses and p stands for profile losses due to boundary layer growth, both of which were treated earlier. Most of the cases presented hitherto involve a cascade or a single blade row. McDonel and Eiswerth (1977) carried out a careful analysis of all losses, including film-cooling and profile losses from Hartsel's analysis, and predicted the efficiency η_t. Their data and the estimated values of η_t for the turbine stage, are shown in Fig. 7.42. The agreement is good, thus providing confidence in the film-cooling analyses presented earlier.

It should be remarked here that the analysis presented in this section concerns only losses. For the prediction of local properties (e.g., transition, local velocity, and pressure field), one must resort to computational techniques described in the next section.

7.8   AERODYNAMICS AND HEAT TRANSFER COMPUTATION

In most cases, the flow field over a turbine blade consists of laminar, transitional, and turbulent regions as shown in Fig. 7.3. If the flow is laminar, Eq. 1.18 is relevant. The boundary layer approximations can be made in most cases, because the viscous layer in these regions is thin. The turbulent flow, depending on the approximations made, is given by either the Navier–Stokes equation (Eq. 5.1 or 1.18) or the boundary layer equations (Eqs. 5.9–5.12 and 7.13). Additional complications arise from the specification of boundary conditions and in formulating the closure equations. The equations must be solved numerically, because they are highly nonlinear with large number of variables and difficult boundary conditions.

The following ingredients are essential for the prediction of aerodynamic and heat transfer fields and losses:

1. Computation of pressure distribution, if a boundary layer code is employed.
2. Prediction of laminar boundary layer growth and heat transfer.
3. Prediction of transition length and heat transfer.
4. Turbulence and heat flux closure model for $\overline{u_i' u_j'}$, $\overline{\rho' u_i'}$, $\overline{T' u_i'}$, and so on.

5. Prescription of coolant velocity for the film/transpiration cooling.
6. Numerical solution of the system of equations using techniques described in Chapter 5 and Appendix B.
7. Prediction of the temperature within the blade material using Eqs. 7.8 or 7.9. These equations can be solved numerically, or an approximate analytical solution can be employed (e.g., Eq. 7.11).

In addition, the codes must be calibrated for accuracy and validated against well-documented data. The methods for solving the boundary layer equations and Navier–Stokes equations are described in Chapter 5. In this section, the additional features needed to calculate the heat transfer and aerodynamic interactions will be described. The differences in various approaches lie in the choice of technique/model for items 3, 4, and 6. All of the early attempts utilized boundary layer assumptions (e.g., Crawford and Kays, 1976; Crawford et al., 1980; Schonung and Rodi, 1987; Daniels, 1979; Wang et al., 1985a) and parabolic marching techniques (Bergeles et al., 1980, 1981). The boundary layer techniques utilized compressible flow boundary layer equations with eddy viscosity models. These techniques have been continually updated to include higher-order models (Chapter 5). A partially parabolic technique developed by Patankar and Spalding (1972), or modifications described earlier (Chapter 5), are also used with an extensive menu on turbulence modeling. Recent attempts (Birch, 1987; Griffen and McConnaughey, 1989; Choi and Knight, 1991; Boyle, 1991; Dunn et al., 1994; Luo and Lakshminarayana, 1995a; Tekriwal, 1994; Fourgeres and Heider, 1994; Leylek and Zerkle, 1993; Ameri et al., 1992) have mainly concentrated on numerical solution of full Navier–Stokes equations. Here again, a wide variety of turbulence models (AEVM, k-ϵ, ARSM) as well as transition models are employed. The list is by no means exhaustive, but includes sample papers on each of the techniques that are related to turbine cooling and heat transfer prediction. A comprehensive survey of two-dimensional boundary layer computations (with heat transfer) is given by Daniels (1979) and Crawford and Kays (VKI Lecture Series, 1986) and a review of Navier–Stokes and boundary layer computations is given by Simoneau and Simon (1993).

7.8.1 Heat Flux Models

Before we proceed with the computation of viscous layers and heat transfer, there is a need to develop constitutive equations for the temperature and velocity correlations $\overline{u_j'T'}$ appearing in Eq. 1.72, Eq. 7.13, or Eq. 5.1 ($\overline{w_i'e'}$) and for the $\overline{h_{oR}'w_n'}$ correlation appearing in Eq. 5.12. A review of various constitutive equations and/or closure models available is given by Launder (1988) and Jones (Kollmann, 1980).

Similar to models used for Reynolds stresses $\overline{(u_i'u_j')}$ described in Section 1.2.7, the heat flux models can be classified as (a) algebraic eddy diffusivity model, (b) two-equation models, (c) algebraic heat flux model (AHFM), and (d) transport equations for heat flux.

These models are briefly described below. The detailed derivation can be found in the references quoted.

The simplest model is the model based on Reynolds analogy, given by Eq. 7.22. The turbulence heat transport term in Eq. 7.13 is given by

$$-\rho\overline{(u_i'T')} = \frac{\mu_t}{PR_t}\frac{\partial T}{\partial x_i} \tag{7.102}$$

where T is the time mean temperature and i is the coordinate direction.

As mentioned in Section 7.2.4, this concept is based on the assumption that turbulent heat transport is proportional to the local temperature gradient. The value of PR_t is assumed to be constant.

Some authors allow for a variation of PR_t. The turbulent Prandtl number $PR_t(=\mu_t/\epsilon_t)$ is a function of the Reynolds number, molecular Prandtl number $(c_p\mu/k)$ and distance from the wall. The value of PR_t seems to increase as the wall is approached. Kays and Crawford (1993, p. 267) suggested a modification to the turbulent Prandtl number to improve the accuracy near the wall. Their expression is given by

$$\frac{1}{PR_t} = \frac{1}{2PR_{t\infty}} + C_h\frac{c_p\mu_t}{k}PR_{t\infty}^{-0.5}$$

$$-\left(C_h\frac{c_p\mu_t}{k}\right)^2\left[1 - \exp\left(\frac{-1}{C_h\frac{c_p\mu_t}{k}(PR_{t\infty})^{0.5}}\right)\right] \tag{7.103}$$

where C_h is a constant. Kays and Crawford showed good agreement between Eq. 7.103 and the near-wall data for air with $PR_{t\infty} = 0.85$, $C_h = 0.3$, and $PR = 0.7$.

The use of the algebraic eddy diffusivity model suffers from the same criticism as the algebraic eddy viscosity model. Neither is based on physical process or governing equations. This is precisely the reason that the $k-\epsilon$ model and the ARSM/RSM models are more realistic and should be more accurate for dynamic simulation. The higher-order turbulent heat flux models are based on transport equations for turbulent heat flux.

An approach similar to the $k-\epsilon$ model can be adopted by defining eddy diffusivity in Eq. 7.102 as a function of velocity and temperature scales for fluctuations. One of the anisotropic heat flux models is the generalized gradient diffusion hypothesis (GGDH) (Launder, 1988), given by

$$\overline{u_i' T'} = -C_\theta \frac{k}{\epsilon} \overline{u_i' u_k'} \frac{\partial T}{\partial x_k} \qquad (7.104)$$

where $C_\theta = 0.3$ (empirical constant). The above equation can be used in combination with k and ε equations (Eqs. 1.84 and 1.86 with near-wall corrections) to solve for $\overline{u_i' u_k'}$, k, ϵ, and $\overline{u_i' T'}$.

Two equation models for turbulent thermal diffusivity have been developed. These are similar to the dynamic modeling of turbulence (k–ϵ equation). Nagano and Kim (1988) expressed eddy diffusivity (μ_t / PR_t) in Eq. 7.102 in terms of dynamic time scale (k/ϵ) and a scalar time scale $(\overline{(T')^2}/2\epsilon_T)$. They suggested the following model:

$$\alpha_t = \frac{\mu_t}{\rho \mathrm{PR}_t} = C_\lambda f_\lambda k \sqrt{\frac{k}{\epsilon} \frac{\overline{(T')^2}}{\epsilon_T}} \ , \qquad -\overline{u_i' T'} = \alpha_t \frac{\partial T}{\partial x_i} \qquad (7.105)$$

where C_λ is a model constant, and f_λ is a wall damping function similar to f_μ in Eq. 1.80. ϵ_T is the dissipation rate of $\overline{(T')^2}/2$, which is analogous to the ϵ for dissipation of dynamic fluctuations (Eq. 1.86). This procedure then involves development of scalar transport equations for $\overline{(T')^2}$ (which is similar to the k equation, Eq. 1.84) and ϵ_T (which is similar to the ϵ equation, Eq. 1.86). Details of the two-equation model $[\overline{(T')^2}, \epsilon_T]$ can be found in Nagano and Kim (1988). Sommer et al. (1993) have provided improved models for ϵ_T and $\overline{(T')^2}$ transport equations and showed good agreement between the measured and the predicted thermal boundary layer properties on a flat plate with and without cooling. The use of the two-equation model for turbulence diffusivity $[\epsilon_T, \overline{(T')^2}]$ relaxes the asssumption of dynamic similarity between momentum and heat transport, and allows for variation of the turbulent Prandtl number through the thermal boundary layer. The use of the two-equation model provides a physically realistic approach to account for thermal effects and heat transfer.

A more accurate approach would be to develop a transport equation for $\overline{u_i' T'}$, and this is described below. Following the procedure used in modeling Reynolds stress, a transport equation for $\overline{u_i' T'}$ can be developed and approximations made to arrive at physically realistic models for $\overline{u_i' T'}$. This correlation represents the transport of heat or temperature by velocity fluctuation. An exact transport equation for $\overline{u_i' T'}$ can be derived much the same way as a Reynolds stress equation (Eq. 1.87). The instantaneous energy or heat equation (for example, Eq. 1.11 with $T = \overline{T} + T'$) is multiplied by u_i', and the momentum equation (e.g., Eq.1.10) is multiplied by T'. The two resulting equations are then added and ensemble-averaged to derive the following transport equation for $\overline{u'_i T'}$. For clarity, all mean temperatures are now denoted by T instead of $\overline{T}$.

$$
\left(\bar{u}_j \overline{u'_i T'}\right)_{,j} = -\left[\underbrace{\left(\frac{\overline{p' T'}}{\rho}\right)\delta_{ij}}_{(2a)} + \underbrace{\overline{u'_i u'_j T'}}_{(2b)} - \underbrace{\nu \overline{T' u'_{i,j}}}_{(2c)} - \underbrace{\gamma \overline{u'_i T'_{,j}}}_{}\right]_{,j} + \underbrace{\frac{\overline{p'}}{\rho} T'_{,i}}_{(3)}
$$

$$
\underbrace{-\left(\overline{u'_i u'_j} T_{,j} + \overline{u'_j T'} \bar{u}_{i,j}\right)}_{(4)} - \underbrace{(\nu + \gamma)\overline{T'_{,j} u'_{i,j}}}_{(5)} \tag{7.106}
$$

where γ is the molecular thermal diffusivity $(k/\rho c_p)$ and $\bar{u}_j$ is the mean velocity.

The gravitational effects are neglected in these equations. The significance of the terms are similar to those of the Reynolds stress transport terms (Eq. 1.87). The corresponding terms are marked by the same number. For example, term 2a in Eq. 1.87 is the diffusion of $\overline{u'_i u'_j}$ by the pressure fluctuations, and term 2a in Eq. 7.106 is the diffusion of $\overline{u'_i T'}$ by the pressure fluctuations. The other terms in Eq. 7.106 are as follows:

2b. Transport or diffusion due to velocity fluctuations
2c. Viscous diffusion
3. Fluctuating pressure–temperature gradient correlation
4. Production due to combined actions of the mean temperature (or density) and the mean velocity gradients. The first term tends to increase the velocity fluctuations, and the second term increases temperature fluctuations.
5. Dissipation.

Detailed interpretation of Eq. 7.106 and modeling of various terms are given by Launder et al. (1984) and Launder (1988). The dissipation term is small if the flow is isotropic or the Reynolds number is high. The terms that need to be modeled are 2a, 2b, 2c, 3, and 5. Shih et al. (1987) provide comprehensive (and complicated) models for various terms in Eq. 7.106.

For heat transfer application, near-wall physics is very important; unfortunately, none of the full Reynolds stress equations (Eq. 1.87 and 1.88) or the modeled heat flux equations accurately capture the dynamics of the flow and thermal field near the wall. A preliminary attempt to derive a low-Reynolds number heat flux model is described in Lai and So (1990). This is a very rapidly developing area.

A compromise between the two-equation model and the full transport equation model for heat flux is the algebraic heat flux model (AHFM). In this approach the transport terms (convection + diffusion) in Eq. 7.106 are assumed to be proportional to the transport terms in the $(T')^2$ equation, much the same way as the ARSM model development (Section 1.2.7.3). Gibson and

Launder (1976) and Gibson (1978) derived an AHEM. Because this model is presently limited to a high Reynolds number region, it is not suitable for capturing the thermal field near the wall.

At the present time, the engineers engaged in heat transfer calculation have to be content with either the assumption of constant turbulent Prandtl number, with modifications to include its variation near the wall, or the two-equation model for $\overline{(T')}^2$ and ϵ_T. The full transport equation in its present form is not suitable for capturing the wall heat transfer accurately. Future development may result in an accurate low-Reynolds number form of Reynolds stress and heat flux transport equations. The prediction of transition is very critical in many situations. The low-Reynolds number form of $k-\epsilon$ model may adequately capture the transition. More sophisticated models using direct numerical simulation data are under development. These will eventually be used for the prediction of flow field, including transition, near-wall flow and thermal physics, and heat transfer.

7.8.2   Governing Equations and Boundary Conditions

There are various approaches to a coupled solution of aerodynamics and heat transfer. The most complicated one is the complete system of Reynolds averaged equations (Eq. 5.1) coupled with a suitable model for transition, turbulence and heat flux. Because the equations are complex, most authors prefer simple models of turbulence (AEVM, Eq. 1.77) and the algebraic eddy diffusivity model (Eq. 7.102). On the other hand, one can employ simple and computationally more efficient techniques (e.g., boundary layer or parabolic marching method) in combination with more accurate transport equations for turbulence and heat flux. But the full Reynolds-averaged equation and the complete heat flux and Reynolds stress transport equations have not been employed due to lack of adequate computer resources and numerical stability problems. For three-dimensional flows, a complete set of equations may consist of 15 PDE's, consisting of the following: one continuity equation, three momentum equations, one mean energy equation, six Reynolds stress equations, one dissipation equation, and three turbulent transport equations for heat flux.

The uncooled and convective heat transfer computation can be carried out with a boundary layer code or Navier–Stokes procedure described in Chapter 5. The computation with film cooling , in view of the elliptic nature of interactions, should use an elliptic code or other approximate methods to include the elliptic region in an otherwise parabolic flow. A brief review of computational efforts for heat transfer prediction is covered in this section. Future efforts will involve the solution of the complete set of exact equations, including direct numerical and large eddy (sub-grid model) simulations.

The boundary conditions are an important aspect of the computation. The boundary conditions for flow variables were covered in Section 5.7.3.1. When dealing with film cooling with a boundary layer code, the boundary conditions in the injection region have to be modified as follows:

$$u(x,0) = u_f, \qquad v(x,0) = v_f, \qquad w(x,0) = w_f$$
$$u(x,h) = u_e, \qquad v(x,h) = v_e, \qquad w(x,h) = w_e, \qquad h > \delta$$

where u_f, v_f, and w_f are streamwise, normal, and radial components of coolant velocity, respectively, for film-cooled or transpiration-cooled blades at the location of slot or holes. Everywhere else, it is zero.

The thermal wall boundary condition can be prescribed with the wall heat flux or the wall temperature as follows:

$$c_p \frac{\partial T(x,0)}{\partial y} = -q_w(x)\frac{\text{PR}}{\mu}$$

$$T(x,0) = T_w(x)$$

The inlet conditions are also specified, either from the analysis or experimental data:

$$u(0,y) = u(y), \quad v(0,y) = v(y), \quad w(0,y) = w(y), \quad T(0,y) = T(y)$$

If the $k-\epsilon$ model is used, the free-stream turbulence intensity is specified, and low-Reynolds number $k-\epsilon$ models (Eq. 1.84 and 1.86 with near-wall terms) are used to capture the location of transition. The boundary conditions for $k-\epsilon$ equations are (for two-dimensional boundary layer flow with free-stream velocity u_e, Eqs. 1.84 and 1.86) as follows:

$$x = x, \quad y = 0, \quad k_w = \left(\frac{\partial \epsilon}{\partial y}\right)_w = 0$$

$$x = x, \quad y > \delta, \quad u_e\frac{\partial k_e}{\partial x} = -\epsilon_e; \qquad u_e\frac{\partial \epsilon_e}{\partial x} = -C_{\epsilon\epsilon}\frac{\epsilon_e^2}{k_e}$$

Other boundary conditions are also used for ϵ_w (see Patel et al. 1985). Furthermore to start the solution, it is necessary to prescribe an initial condition, including free-stream turbulence intensity (or the solution can be started from the free-stream conditions through the leading edge). In such a case, the prescription of u, k, and ϵ profiles becomes necessary. The velocity profile is prescribed either from experimental data or from a known analytical solution (e.g., law of the wall). The low-Reynolds-number form of two-equation models can capture both steady and unsteady transition. A discussion of this can be found in Fan and Lakshminarayana (1994).

A word of caution is in order when computing the flow field in the vicinity of blade leading edge. The free-stream turbulence intensity changes substantially in the vicinity of the stagnation point. This is mainly brought about by the large velocity gradient due to flow deceleration, an inviscid effect. The isotropic $k-\epsilon$ model is incapable of capturing turbulence production controlled by normal components of Reynold stress. Taulbee et al. (1989) employed both the full Reynolds stress equations (Eq. 1.88) and the $k-\epsilon$ equation to predict the large change in turbulence intensities from upstream to the leading edge region. The prediction of k profile from the $k-\epsilon$ model was poor. The predictions from the full Reynolds stress equations showed good agreement with the data from a cylinder. But downstream of the leading edge and in the boundary layer region, the $k-\epsilon$ model performed well.

The difference in the various approaches usually lies in the numerical method, transition, and turbulence models, but the basic equations employed are similar for all the techniques. A comprehensive review of various efforts, assessment of turbulence and transition models, and computational techniques for heat transfer calculation can be found in Simoneau and Simon (1993) and aerodynamic computation in Lakshminarayana (1991).

7.8.3 Computation for Uncooled / Convectively Cooled Blades

Most of the early attempts to predict viscous flows and heat transfer employed boundary layer equations. The general techniques for the solution of these equations are described in Section 5.5 and will not be repeated here. The boundary layer procedure involves the following steps:

1. The boundary layer equations are solved by one of the techniques described in Section 5.5 (Keller box, zig-zag). The equations may be transformed into a body-fitted coordinate system or through a Levy–Lees transformation for compressible flow. The blade static pressure is derived from a Euler code (compressible) or one of the techniques described in Chapter 3 for incompressible flow. If the boundary layer growth is large or separation is present, it is better to employ the full Navier–Stokes equation.

2. Transition is still a thorny issue. There are no accurate methods to predict it. The models described in Mayle (1991) can be used. The low-Reynolds-number $k-\epsilon$ model can predict the transition reasonably well.

3. The choice of turbulence and heat flux models depends on the particular case. For many two-dimensional flows, AEVM or $k-\epsilon$/ARSM may give good predictions, and the eddy diffusivity model (Eqs. 7.102–7.105) may be adequate. For three-dimensional flows, it may be better to employ $k-\epsilon$/ARSM and the two-equation model for turbulent thermal diffusivity.

4. The use of wall functions in the boundary layer code should be avoided. The near wall flow can be captured using a low-Reynolds-number $k-\epsilon$ model and a large number of grid points near the wall.

Among the boundary layer codes, the one developed by Crawford and Kays (1976) and Crawford and Stephens (1988) seem to provide fair predictions. However, all of these codes have various constants, transition models, and turbulence models. If the transition location is not predicted, then heat transfer predictions will be poor. The predictions from Crawford and Kays' (1976) STAN5 code are compared with the data by Nealy et al. (1984) and shown in Fig. 7.16. The inviscid pressure distribution (input to the boundary layer code) was derived from a Euler code, utilizing the Hopscotch method (Section 5.7.4.1 and Delaney, 1983). The transition on the suction surface (about 20% of S) is predicted well by the code, but the prediction in the turbulent region and on the pressure surface is only qualitative. In addition, the agreement is reasonably good at low Reynolds number, but deteriorates progressively at higher Reynolds numbers. The degree to which the transition is advanced with an increase in Reynolds number is underestimated. Blair (1994) has shown similar comparisons with his midspan rotor data with various roughness. Heat transfer computations by various other investigators show reasonable agreement with the data. Taulbee et al. (1989) computed the heat transfer using (a) the full Reynolds stress equation near the stagnation region and (b) the $k-\epsilon$ equation downstream of the stagnation region. A boundary layer code was employed. The effect of free-stream turbulence was also included.

The Navier–Stokes procedure involves solution of Eq. 5.1 using several techniques described in Chapter 5 (e.g., pressure-based method, time marching explicit and implicit techniques). Most authors employ algebraic eddy viscosity models and algebraic eddy diffusivity (Eq. 7.102) models. Some examples of computation using Navier–Stokes procedure can be found in Boyle (1991), Dunn et al. (1994), Luo and Lakshminarayana (1995a), and Choi and Knight (1988). Boyle (1991) utilized Chima's explicit Navier–Stokes code (Section 5.7.4.3; Chima, 1987) and the Baldwin–Lomax (1978) model (Eqs. 1.76–1.79) and computed the heat transfer on various low- and high-speed flow configurations and demonstrated the ability of Navier–Stokes code to capture the heat transfer reasonably well, including the region downstream of shock-induced separation. This code was also used by Dunn et al. (1994) to capture the heat transfer on the space shuttle main engine turbine rotor blade, and the results are shown in Fig. 7.17. The predictions are qualitative, and the transition location on the suction surface is not captured. It should be emphasized here that boundary layers in accelerating flows are very thin; hence, accuracy of the Navier–Stokes solver is limited by the small number of grid points within the boundary layer. Typically, the first grid point from the surface should be located at $y^+ \sim 1$ with low-Reynolds-number $k-\epsilon$ model. On the other hand, boundary layer codes can capture the near-wall properties well (because large number of grid points can be used within the boundary layer due to the parabolic nature of the equations) but suffers accuracy if proper inviscid–viscous coupling is not enforced.

Luo and Lakshminarayana (1995a), used an explicit Runge–Kutta method (Section 5.7.4.3; Kunz and Lakshminarayana, 1992d) and several different

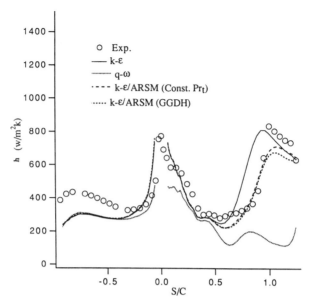

Figure 7.44   Heat transfer prediction for a turbine cascade M_{is2} = 0.93, Re_{is2} = 1.15 × 10⁶, Tu_∞ = 6%, T_e = 410 K, T_w = 300 K, α_1 = 0°, α_2 = 74°, C = 6.75 cm, S/C = 0.85, $h = q_w/(T_{oe} - T_w)$). (Luo and Lakshminarayana, 1995a; reproduced with permission of AIAA.)

turbulence and heat-flux models to compute the flow and heat transfer fields in a high-speed turbine cascade (Arts et al., 1992). The $k-\epsilon$ model is due to Chien (1982)(Eqs. 1.84 and 1.86), and the $q-\omega$ ($q = \sqrt{k}$, $\omega = \epsilon/k$) model is due to Coakley (1983). The ARSM model (Eq. 1.89) is interfaced with the low-Reynolds-number $k-\epsilon$ model away from the wall ($y^+ = \sim 200$). The GGDH model used is due to Launder (1988), given by Eq. 7.104. The heat transfer predictions are shown in Fig. 7.44. The blade pressure distribution and downstream wakes are predicted well by all models. Both the $k-\epsilon$/ARSM and $k-\epsilon$ models provide good predictions of the blade heat transfer from the leading edge to the transition location on the suction surface. Beyond this location, $k-\epsilon$/ARSM provides better prediction of the heat transfer, except very near the trailing edge. The predictions on pressure surface are good only up to the transition point and qualitative beyond this location. It should be remarked here, once again, that the pressure surface boundary layers are extremely thin and the boundary layer code would do better in this region.

Choi and Knight (1988) utilized an implicit technique (Section 5.7.5.1) and the two-equation $q-\omega$ model (Coakley, 1983; also see Wilcox and Rubesin, 1980) and computed the three-dimensional flow field and heat transfer in the endwall region of a turbine cascade, where the influence of secondary flow introduces large spatial variation in heat transfer rates. Their predictions are compared with Graziani et al.'s (1980) data in Fig. 7.22. All qualitative features of the flow and the heat transfer fields due to horseshoe vortex and secondary flows are captured, but the magnitudes are not. The turbulence is highly anisotropic in this region, and properties of the endwall boundary layers have to be captured accurately. Here again, the limitations are in the grid resolution and the turbulence model. Because of upstream boundary layers and the intense mixing near the leading edge in the endwall region, the flow is turbulent from the leading edge, and transition plays no role in this prediction.

The problem areas to be addressed in the compuation of heat transfer are the following (see Simoneau and Simon, 1993, for detailed discussion of current problems):

1. The coupling of pressure field, flow field, and thermal field near the leading edge should be addressed. This is a very complex flow region with high-velocity gradients and large changes in turbulent structure.

2. The prediction of transition and transition length is critical. None of the transition models available address such issues as turbine geometry, high temperature, curvature, compressibility effects and roughness, and coolant injection.

3. The improvement in turbulence and heat transfer models should come through the use of (a) nonisotropic models for the turbulence and (b) transport equations for velocity-temperature correlations.

4. The turbulence models and heat flux models applicable to the separated zone and the shock-boundary-layer interaction regions need to be addressed.

5. It is only recently that attempts have been made to compute coolant channel flows (Tekriwal, 1994). An integrated approach involving coupled coolant flow and thermal field, blade thermal field, and gas flow and thermal field predictions needs to be pursued.

7.8.4 Computation for Film-Cooled Blades

A schematic of the physical processes associated with film cooling is sketched in Figs. 7.23–7.26 and 7.43. Film cooling has a major influence on the boundary layer characteristics, heat transfer, and mixing losses. Specifically:

1. The boundary conditions at the location of the injection hole are modified as indicated in Section 7.8.2.

2. The mixing of the jet and the free stream occurs along the streamwise path. Lack of an adequate turbulence model to account for such flows is a major handicap.

3. Large transport of mass, momentum, and heat transfer in a small region makes it a difficult computational problem.

There are three approaches to the computation of flow and heat transfer with film cooling. In the first approach, coolant jet flow is modeled on the basis of a control volume or integral approach, embedded in a code for primary flow. In the second approach, an elliptic region is used for the jet flow, which is then patched into a code for primary flow (either boundary layer or parabolic marching techniques as described in Section 5.8). In the last approach, the entire flow field is solved using the Navier–Stokes code.

The injection process can be either modeled or computed. Considering the fact that flow through each hole has to be resolved accurately, the computational techniques become extremely complicated. The injection model involves modeling the injection flow via an integral or control volume approach and incorporating this model (e.g., analyses described in Section 7.5) in the marching calculation of the boundary layer equations. One of the successful models is due to Crawford et al. (1980). The injection, entrainment, and diffusion processes are modeled together. The equations that describe the model are obtained from a one-dimensional mass, momentum, and thermal energy balance on the volume bounded by adjacent stream surfaces (ψ and $\psi + d\psi$) as shown in Fig. 7.25. The three-dimensional lateral mixing was accounted for by an augmentation of turbulence mixing length (by varying the coefficients in equations similar to Eqs. 1.76–1.79). The jet interaction model was incorporated into a boundary layer code and showed limited success in predicting the data (Crawford et al., 1980). Because the model is essentially two-dimensional and three-dimensional effects are very important in jet-mainstream mixing, subsequent investigators (e.g., Harasgama and Burton, 1992) report the inability of this technique to capture heat transfer in turbine endwalls with film cooling.

Some of the limitations of this model are as follows:

1. The model assumes that the jet stays within the boundary layer and the perturbations are small. The method is therefore not suitable in regions where jets penetrate into the main flow (e.g., leading-edge or near-leading-edge injection).

2. This model cannot account for the fact that at higher blowing rate and large injection angles, the velocity above the jet can be higher than that in the free stream (see Fig. 7.23).

3. The method is untested for flows with pressure gradient and with large temperature ratios.

4. The flow with discrete film cooling is inherently three-dimensional, and this should be accounted for in the computational procedure.

Some of these limitations are overcome by Schonung and Rodi (1987), who have proposed two models that can be incorporated into a boundary layer code. These are as follows:

1. *Injection Model.* In the parabolic marching procedure, the initial conditions are varied by leaping over the region near the injection and by prescribing profiles behind the blowing region. The prescribed profile takes into account the characteristics of the incoming boundary layer, as well as the injection jets.

2. *Dispersion Model.* The lateral mixing is taken into account by using a spanwise averaged equation (averaged over the semi-width of the jet), solving the averaged equation in this region, and integrating this into a two-dimensional boundary layer program.

The calculations which incorporate these models show improvement, but the improvement is not substantial.

Kulisa et al. (1992) have included the three-dimensional effects through a quasi-three-dimensional analysis. The gross properties of the film-cooling jet (e.g., maximum velocity, transverse area, flow angle, trajectory of the maximum velocity, and temperatures) are derived from an integral analysis of the jet. These properties are then used in spanwise averaged (similar to the procedure described in Section 4.2.3) boundary layer equations. This procedure eliminates derivatives in spanwise (z or r) direction (Fig. 7.25), but introduces additional source terms dependent on properties of the jet, similar to G_i functions in Eq. 4.75. The spanwise averaged two-dimensional boundary layer equations, with the source term due to the jet, is solved using the Keller box method (Appendix B and Section 5.5.5.1). Detailed derivations of the jet flow model equations and of spanwise averaged equations, along with a large number of cases validated, can be found in Kulisa (1989). Their predictions compare well with the data in a flat-plate boundary layer with injection ($D = 0.005$ m, $S = 0.015$ m, $\alpha_f = 45°$, $T_w = 310$ K, $T_f = 340$ K, T_m (downstream gas temperature) $= 301$ K, $\delta = 0.015$ m, $u_e = 30$ m/s).

The results from this technique show considerable promise. Such techniques or modifications (e.g., zonal jet model is introduced in a Navier–Stokes code) will be used in the future for practical applications. The resolution of film-cooling effects through the use of full-Reynolds-averaged equations is computationally intensive and does not (at the present time) predict the important features of the flow. A physically based flow model, embedded in a Navier–Stokes solver, may be more attractive.

Some of the shortcomings of two-dimensional or quasi-three-dimensional or zonal approaches can be overcome by employing the complete three-dimensional Navier–Stokes equations and solving the coupled system by numerical techniques described in Chapter 5. This is the approach taken by many authors. Recent efforts in this area are due to Vogel (1991), Garg and Gaugler (1993), Leylek and Zerkle (1993), and Hall et al. (1994).

These are two approaches used. In one of the approaches (e.g., Vogel, 1991; Garg and Gaugler, 1993), the cooling holes are modeled using boundary conditions. For example, Vogel (1991) assumes $p_f = p_e$ at the injection point on the blade (Fig. 7.25), and the Mach number and velocity (u_f) of the jet are derived using one-dimensional gas dynamic equations and from known values of T_{oe}, T_{of}, P_{of}, P_{oe}, and p_e. Vogel (1991) solves the external flow using Dawes' (1987) technique described in Section 5.7.4.3 and utilizing the Baldwin–Lomax turbulence model (Section 1.2.7.1). He has also coupled this code with a Laplace equation for blade temperature distribution (Eq. 7.9). The flow and thermal field solution (Navier–Stokes equation) and temperature field solution (Laplace equation) are handled on two separate processors. Good agreement is shown for adiabatic cooling effectiveness (similar to that presented in Fig. 7.27 for the data from the same author) for a flat plate with $\alpha_0 = 35°$, $S/D_0 = 3.0$, $T_{of} = 350\,\text{K}$, $T_{oe} = 276\,\text{K}$. Garg and Gaugler (1993) utilized an explicit technique (Chima and Yakota, 1990) described in Section 5.7.4.3 and an algebraic eddy viscosity model (similar to Eqs 1.76–1.78) and showed relatively good agreement between the data and the prediction of heat transfer on the suction surface and up to midchord of the pressure surface of a turbine blade of Nirmalan and Hylton (1990). This technique of solving only the external flow with film cooling represented by the boundary conditions may not be accurate in capturing the aerothermal mixing of jets. The temperature and velocity profiles across the jet at the injection point are not uniform, and this region should be resolved through either analysis or computation. The assumption that $p_e = p_f$ at the injection point may not be valid.

In the second approach, attempts are made to solve the flow field in the coolant holes (plenum and film holes), mixing region (Fig. 7.25) and the external flow simultaneously. Leylek and Zerkle employed a pressure-based method (Section 5.7.1) and a high-Reynolds-number k–ϵ model (Eqs. 1.84 and 1.86 with wall functions) and predicted the flow and thermal fields measured by Pietrzyk et al. (1990). The predictions are shown in Fig. 7.24. The predictions capture all salient features of flow and thermal fields. One interesting conclusion from this investigation is that the flow at the coolant ejection location is highly nonuniform, with high turbulence intensity (14%) at the center of the jet and very low momentum region on the side walls. This seems to indicate that specification of the jet as the boundary conditions derived from one-dimensional analysis may not be valid. The flow in the plenum and coolant holes should be analytically modeled or resolved computationally. The adiabatic cooling effectiveness predicted by Leylek and Zerkle show only qualitative agreement with the experimental data. The predictions are poor up to 5 diameters downstream of the coolant jet. Hall et al. (1994) utilized the Runge–Kutta technique (Hall and Delaney, 1991) described in Section 5.7.4.3, as well as the Baldwin–Lomax turbulence model (Section 1.2.7) and constant turbulent Prandtl number ($\text{PR}_t = 0.9$), and computed the heat transfer field in convectively cooled and film-cooled blades showing qualitative agreement with the data (Hylton et al., 1988). This investigation shows great promise of capturing the film–main-flow interactions and effectiveness of various design changes such as hole size, shape, and placement.

In summary, the convective heat transfer can be resolved reasonably well with a Navier–Stokes code for shock-free attached flows or from a boundary layer code coupled with a Euler code. Many are beginning to incorporate and use k–ϵ models. The need to develop better transition and heat flux models cannot be overemphasized. The computation of heat transfer with film cooling is in a less satisfactory stage. It is doubtful whether the Navier–Stokes solver alone can resolve the entire flow (plenum to downstream of blade). The zonal approach with physically based models may prove to be more effective in capturing the aerothermal properties with film cooling.

TRANSFORMATION OF EQUATION 1.18 INTO A CURVILINEAR (BODY-FITTED) COORDINATE SYSTEM

The transformation of the coordinate system is of the form

$$\xi = \xi(x, y, z)$$

$$\eta = \eta(x, y, z)$$

$$\zeta = \zeta(x, y, z)$$

The partial derivatives in the original coordinate, using the chain rule of differentiation, are given by

$$\frac{\partial}{\partial x} = \xi_x \frac{\partial}{\partial \xi} + \eta_x \frac{\partial}{\partial \eta} + \zeta_x \frac{\partial}{\partial \zeta} \qquad (A.1)$$

and, similarly, $\partial/\partial y$ and $\partial/\partial z$.

The Jacobian of transformation can be written as

$$J = \frac{\partial(\xi, \eta, \zeta)}{\partial(x, y, z)} = \left[\frac{\partial(x, y, z)}{\partial(\xi, \eta, \zeta)} \right]^{-1} = \begin{bmatrix} \xi_x & \xi_y & \xi_z \\ \eta_x & \eta_y & \eta_z \\ \zeta_x & \zeta_y & \zeta_z \end{bmatrix} = \begin{bmatrix} x_\xi & x_\eta & x_\zeta \\ y_\xi & y_\eta & y_\zeta \\ z_\xi & z_\eta & z_\zeta \end{bmatrix}^{-1} = (\bar{J})^{-1}$$

where $\bar{J}$ is the Jacobian of inverse transformation. Standard matrix manipulation yields (Hoffmann, 1989, p. 308)

$$
\begin{bmatrix}
\xi_x, \xi_y, \xi_z \\
\eta_x, \eta_y, \eta_z \\
\zeta_x, \zeta_y, \zeta_z
\end{bmatrix}
$$

$$
= \frac{1}{\det |\bar{J}|}
\begin{bmatrix}
y_\eta z_\zeta - y_\zeta z_\eta & -(x_\eta z_\zeta - x_\zeta z_\eta) & x_\eta y_\zeta - x_\zeta y_\eta \\
-(y_\xi z_\zeta - y_\zeta z_\xi) & x_\xi z_\zeta - x_\zeta z_\xi & -(x_\xi y_\zeta - x_\zeta y_\xi) \\
y_\xi z_\eta - y_\eta z_\xi & -(x_\xi z_\eta - x_\eta z_\xi) & (x_\xi y_\eta - x_\eta y_\xi)
\end{bmatrix}
$$

$$(A.2)$$

To illustrate the procedure, let us apply this to the continuity equation (first row in Eq. 1.19):

$$
\frac{\partial \rho}{\partial t} + \xi_x(\rho u)_\xi + \eta_x(\rho u)_\eta + \zeta_x(\rho u)_\zeta
$$

$$
+ \xi_y(\rho v)_\xi + \eta_y(\rho v)_\eta + \zeta_y(\rho v)_\zeta
$$

$$
+ \xi_z(\rho w)_\xi + \eta_z(\rho w)_\eta + \zeta_z(\rho w)_\zeta = 0
$$

or

$$
\frac{\partial \rho J^{-1}}{\partial t} + \frac{\partial(\rho U J^{-1})}{\partial \xi} + \frac{\partial(\rho V J^{-1})}{\partial \eta} + \frac{\partial(\rho W J^{-1})}{\partial \zeta} = 0
$$

where

$$
U = \xi_x u + \xi_y v + \xi_z w
$$

$$
V = \eta_x u + \eta_y v + \eta_z w
$$

$$
W = \zeta_x u + \zeta_y v + \zeta_z w \tag{A.3}
$$

These are called *contravariant velocities*. Using this procedure for all the equations, Eq. 1.18 in the curvilinear system can be written as (Pulliam and Steger, 1978)

$$
\frac{\partial \hat{q}}{\partial t} + \frac{\partial \hat{E}}{\partial \xi} + \frac{\partial \hat{F}}{\partial \eta} + \frac{\partial \hat{G}}{\partial \zeta} = \frac{1}{Re}\left(\frac{\partial \hat{T}}{\partial \xi} + \frac{\partial \hat{P}}{\partial \eta} + \frac{\partial \hat{Q}}{\partial \zeta}\right) \tag{A.4}
$$

where

$$
\hat{q} = J^{-1}\begin{bmatrix} \rho \\ \rho u \\ \rho v \\ \rho w \\ \rho e_o \end{bmatrix}, \qquad
\hat{E} = J^{-1}\begin{bmatrix} \rho U \\ \rho u U + \xi_x p \\ \rho v U + \xi_y p \\ \rho w U + \xi_z p \\ (\rho h_o)U \end{bmatrix}
$$

$$
\hat{F} = J^{-1}\begin{bmatrix} \rho V \\ \rho u V + \eta_x p \\ \rho v V + \eta_y p \\ \rho w V + \eta_z p \\ (\rho h_o)V \end{bmatrix}, \qquad
\hat{G} = J^{-1}\begin{bmatrix} \rho W \\ \rho u W + \zeta_x p \\ \rho v W + \zeta_y p \\ \rho w W + \zeta_z p \\ (\rho h_o)W \end{bmatrix} \qquad (A.5)
$$

The viscous flux terms are given by

$$
\hat{T} = J^{-1}\begin{bmatrix} 0 \\ \xi_x \tau_{xx} + \xi_y \tau_{xy} + \xi_z \tau_{xz} \\ \xi_x \tau_{yx} + \xi_y \tau_{yy} + \xi_z \tau_{yz} \\ \xi_x \tau_{zx} + \xi_y \tau_{zy} + \xi_z \tau_{zz} \\ \xi_x \beta_x + \xi_y \beta_y + \xi_z \beta_z \end{bmatrix}, \qquad
\hat{P} = J^{-1}\begin{bmatrix} 0 \\ \eta_x \tau_{xx} + \eta_y \tau_{xy} + \eta_z \tau_{xz} \\ \eta_x \tau_{yx} + \eta_y \tau_{yy} + \eta_z \tau_{yz} \\ \eta_x \tau_{zx} + \eta_y \tau_{zy} + \eta_z \tau_{zz} \\ \eta_x \beta_x + \eta_y \beta_y + \eta_z \beta_z \end{bmatrix}
$$

$$
\hat{Q} = J^{-1}\begin{bmatrix} 0 \\ \zeta_x \tau_{xx} + \zeta_y \tau_{xy} + \zeta_z \tau_{xz} \\ \zeta_x \tau_{yx} + \zeta_y \tau_{yy} + \zeta_z \tau_{yz} \\ \zeta_x \tau_{zx} + \zeta_y \tau_{zy} + \zeta_z \tau_{zz} \\ \zeta_x \beta_x + \zeta_y \beta_y + \zeta_z \beta_z \end{bmatrix} \qquad (A.6)
$$

The expression for sheer stress τ_{ij} is given by Eq. 1.26. The β_i terms in Eq. A.6 are

$$
\beta_x = -Q_x + \Phi_1, \qquad \beta_y = -Q_y + \Phi_2, \qquad \beta_z = -Q_z + \Phi_3 \quad (A.7)
$$

where viscous dissipation terms Φ_i are given by Eq. 1.24, and the heat flux terms q_i are given by Eqs. 1.29–1.31. The pressure is obtained using a perfect gas equation of state

$$
p = \rho(\gamma - 1)\big[e_0 - 0.5(u^2 + v^2 + w^2)\big]
$$

APPENDIX B

BASIC COMPUTATIONAL TECHNIQUES

The purpose of this section is to provide brief exposure to basic numerical techniques widely used in turbomachinery application. For details, the reader is referred to textbooks in the field (e.g., Smith, 1978, Anderson et al. 1984; Hoffmann, 1989; Hirsch, 1990; Shyy, 1994; Anderson, 1995; Chapra and Canale, 1988). The finite element technique is not included for the sake of brevity. Baker's textbook (1983) should be useful to those interested in the finite element method.

B.1 CLASSIFICATION, FINITE-DIFFERENCE REPRESENTATION, AND FINITE-VOLUME REPRESENTATION

Let us consider the stream function equation governing compressible cascade flows (Eq. 1.60). This can be represented by

$$A\psi_{xx} + B\psi_{xy} + C\psi_{yy} = 0 \tag{B.1}$$

where

$$A = 1 - M_x^2, \qquad B = -2M_x M_y, \qquad C = 1 - M_y^2$$

The slopes of the characteristic lines of the above equation are given by

$$\frac{dy}{dx} = \frac{-B \pm \sqrt{B^2 - 4AC}}{2A}$$

725

Let us consider three possibilities. If $(B^2 - 4AC) > 0$, there exist two real characteristics at a point. This case corresponds to $(M_x^2 + M_y^2) > 1$ (i.e, supersonic flow), and this type of equation is called the *hyperbolic equation*. If $(B^2 - 4AC) < 0$, no real characteristics exist and $(M_x^2 + M_y^2) < 1$ (i.e., subsonic flow). This type of equation is called an *elliptic equation*. If $(B^2 - 4AC) = 0$, characteristics are real and only one independent family exists and $M_x^2 + M_y^2 = 1$. This type of equation is called a *parabolic equation*.

Generally speaking, boundary layer equations are parabolic, and steady Euler and Navier–Stokes equations are elliptic for subsonic flows and hyperbolic for supersonic flows. If time derivatives are included, Euler and Navier–Stokes equations become hyperbolic for compressible flows.

The basic concept behind finite difference representation of Eq. B.1 is the Taylor series. The derivative, ψ_{xx}, in Eq. B.1 can be expressed in finite difference form as

$$(\psi_{xx})_{i,j} = \frac{\psi_{i-1,j} - 2\psi_{i,j} + \psi_{i+1,j}}{(\Delta x)^2} + 0(\Delta x)^2$$

$0(\Delta x)^2$ indicates that terms of order $(\Delta x)^2$ and higher are neglected. This is an indication of the approximation made in expressing derivatives in terms of finite differences. The neglected terms will decrease proportionally to $0(\Delta x)^2$ when the distance between (i, j) and its neighbors is reduced. This is exact when $\Delta x \to 0$; similarly,

$$(\psi_{yy})_{i,j} = \frac{\psi_{i,j-1} - 2\psi_{i,j} + \psi_{i,j+1}}{(\Delta y)^2} + 0(\Delta y)^2 \tag{B.2}$$

$$\underset{(1)}{(\psi_x)_{i,j} = \frac{\psi_{i+1,j} - \psi_{i-1,j}}{2\Delta x} + 0(\Delta x)^2}, \quad \underset{(2)}{(\psi_x)_{i,j} = \frac{\psi_{i+1,j} - \psi_{i,j}}{\Delta x} + 0(\Delta x)},$$

$$\underset{(3)}{(\psi_x)_{i,j} = \frac{\psi_{i,j} - \psi_{i-1,j}}{\Delta x} + 0(\Delta x)} \tag{B.3}$$

Finite difference equations represented by Eq. B.2 and the first term in Eq. B.3 are called "central differencing," and terms 2 and 3 in Eq. B.3 represent "forward" and "backward differencing," respectively. The accuracy can be improved by either a higher-order finite difference equation or by reducing the step size.

In the finite area or volume technique, the differential equations are approximated by integrating them over a finite volume. Because the equations encountered in fluid dynamics are based on conservation laws, the conservation character is preserved by integration over a finite volume. The finite volume approach has been widely used in fluid dynamics and turboma-

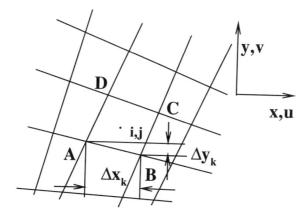

Figure B.1   Finite area / finite volume technique.

chinery and can be explained with reference to Fig. B.1. Let us consider two-dimensional, inviscid, compressible flow in a cascade with the flow field divided into cells as shown in Fig. B.1. The governing equations are given by Eq. 1.18 with $e_0 =$ constant, $h_0 =$ constant, Re $\rightarrow \infty$, $w = 0$, and $S = 0$. The equation for a cell ABCD, with the center located at i, j, can be written as follows using the divergence theorem:

$$\frac{\partial}{\partial t} \iint_{ABCD} q\, dx\, dy + \iint_{ABCD} (E\, dy - F\, dx) = 0 \qquad (B.4)$$

The above equation is applied to each cell, resulting in ordinary differential equations such as

$$\frac{d}{dt}(Aq) + Pq = 0 \qquad (B.5)$$

where A is the cell area, and the operation P represents an approximation to the second integral in Eq. B.4. This is defined as follows (Jameson et al., 1981): Let Δx_k and Δy_k be the increments of x and y along side k of this cell (e.g., AB, shown in Fig. B.1, with $k = 1, 2, 3, 4$ respectively for AB, BC, CD, and DA respectively). The flux balance for the continuity equation is given by

$$\frac{d}{dt}(A\rho) + \sum_{k=1}^{4} (\Delta y_k u_k - \Delta x_k v_k)\rho = 0 \qquad (B.6)$$

Similarly, for the x momentum equation we have

$$\frac{d}{dt}(A\rho u_k) + \sum_{k=1}^{4} \left[(\Delta y_k u_k - \Delta x_k v_k)\rho u_k + \Delta y_k p_k\right] = 0 \qquad \text{(B.7)}$$

A similar equation can be written for y momentum equation. The term $(\Delta y_k u_k - \Delta x_k v_k)$ in Eq. B.6 represents the operator P.

A method of solving equations such as B.6 or B.7 is as follows:

1. ρ, ρu and ρv are known at time t.
2. A linear variation of mass and momentum flux is assumed between cell centers, and the mass and momentum are computed for each side.
3. The solution of Eqs. B.6, B.7, and others are marched by advancing in time.

The algebraic equations for the unknowns are solved by methods indicated later. Kallinderis and Baron (see Murthy and Brebbia, 1990) have provided a comprehensive treatment of the finite volume method as applied to Navier–Stokes equation, including the discretization procedure, spatial accuracy, and effects of artificial dissipation. Details on finite volume and finite area methods can be found in Hirsch (1990, vol. 1, p. 241).

The Taylor series provides finite difference approximations to derivatives, and the governing equations are replaced by a system of these finite difference equations. In the control volume approach on the other hand, the conservation law is satisfied in integral form over a cell surrounding the point. The conservation law is satisfied over a finite region rather than at a point. The cell does not have to be four-sided or rectangular. Any arbitrary geometry can be used. Furthermore, in three-dimensional flows the cells are finite volumes surrounded by six surfaces. In an *explicit* scheme, the unknown values at the $(n + 1)$th time iteration step are expressed in terms of known values at the nth step. In an *implicit* scheme, the calculation of an unknown variable requires the solution of a set of simultaneous equations.

B.2   BASIC METHODS FOR PARABOLIC FLOWS

There are a large number of techniques available for the solution of parabolic equations. The parabolic equations encountered in fluid mechanics are (1) boundary layer equations (Section 5.5) and (2) parabolized Navier–Stokes equations (Section 5.6) where the streamwise diffusion term is neglected and the streamwise pressure gradient is small or known or assumed initially.

The Crank–Nicholson method (1947) is second-order accurate in time and space, unconditionally stable, and there is no upper limit on the time step. The technique is efficient and widely used in the solution of flow and heat transfer equations. The methods widely used for three-dimensional flows are the alternating-direction implicit (ADI) method and the Keller box scheme, and these are described briefly here.

B.2.1 Alternating-Direction Implicit (ADI) Method

The ADI method belongs to a general and powerful class of techniques called splitting or approximate factorization methods. These are used in elliptic, hyperbolic, and parabolic equations (Peaceman and Rachford, 1959; Douglas and Gunn, 1964). The idea is to split the three-dimensional problem (x, y, z, or t) into two two-dimensional problems. These are known as "splitting" techniques.

Let us consider the following equation governing the two-dimensional unsteady heat conduction problem (Eq. 7.8) or two-dimensional unsteady, fully developed viscous flow (neglecting pressure gradient) (Eq. 1.43):

$$\frac{\partial \phi}{\partial t} = \tau \left(\frac{\partial^2 \phi}{\partial x^2} + \frac{\partial^2 \phi}{\partial y^2} \right) \tag{B.8}$$

Peaceman and Rachford (1959) replaced one of the second derivatives above by the implicit finite difference approximation and replaced the other by the explicit finite difference approximation. Their solution procedure is as follows (Fig. B.2):

1. Choose implicit difference approximation for $\partial^2 \phi / \partial x^2$ and explicit difference for $\partial^2 \phi / \partial y^2$ and advance the solution to the $(k + 1)$ time step. The boundary conditions are known at $i = 1, I$; $j = 1, J$.

$$\frac{\phi_{i,j,k+1} - \phi_{i,j,k}}{\Delta t}$$
$$= \tau \left[\left(\frac{\phi_{i-1,j,k+1} - 2\phi_{i,j,k+1} + \phi_{i+1,j,k+1}}{(\Delta x)^2} \right) + \left(\frac{\phi_{i,j-1,k} - 2\phi_{i,j,k} + \phi_{i,j+1,k}}{(\Delta y)^2} \right) \right]$$

$$\tag{B.9}$$

Apply this equation to each of the rows parallel to the x axis (Fig. B.2), $i = 1, \ldots, I$. Application of Eq. B.9 at fixed j gives $(I - 2)$ algebraic equations for $I - 2$ unknowns ($\phi_{i=2, \ldots, I-1; j, k+1}$). Because there are J columns, the advancement of solution to $(K + 1)$th time step involves solution of the $(J - 2)$-independent set of equations, each set consisting of $(I - 2)$ unknowns. This is where the advantage of the ADI scheme lies. If a fully implicit method was chosen for both terms ($\partial^2 \phi / \partial x^2$ and $\partial^2 \phi / \partial y^2$), the number of equations to be solved would

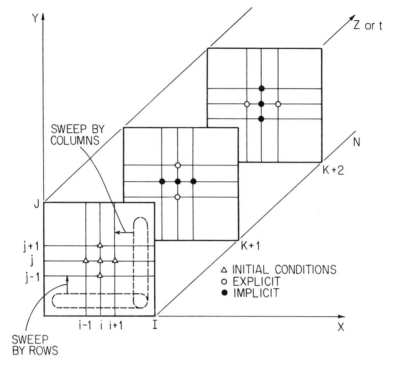

Figure B.2   ADI technique.

have been $(I - 2)(J - 2)$. This is much more computationally intensive to solve than the $(J - 2)$-independent set of equations with $(I - 2)$ unknowns in each set. The ADI method involves a line-by-line solution of a small set of simultaneous equations. These can be solved by noniterative methods. A method for solving these equations will be described later. The nodal points used are shown in Fig. B.2, planes K and $K + 1$. This provides solution of the equation in the rectangle at time step $(K + 1)$.

2. Advance to the next time step by using the explicit difference approximation for $\partial^2\phi/\partial x^2$ and the implicit difference approximation for $\partial^2\phi/\partial y^2$ given by

$$\frac{\phi_{i,j,k+2} - \phi_{i,j,k+1}}{\Delta t} = \tau\left[\left(\frac{\phi_{i-1,j,k+1} - 2\phi_{i,j,k+1} + \phi_{i+1,j,k+1}}{(\Delta x)^2}\right)\right.$$

$$\left.+ \left(\frac{\phi_{i,j-1,k+2} - 2\phi_{i,j,k+2} + \phi_{i,j+1,k+2}}{(\Delta y)^2}\right)\right] \quad \text{(B.10)}$$

Eq. B.10 provides an $(I - 2)$-independent set of equations for $(J - 2)$

unknowns in plane $K + 2$ for $(\phi_{i, j=2, \ldots, J-1, k+1})$. This completes the solution for $(K + 2)$th time step. This procedure is repeated until the last station or steady state is reached.

The set of equations from the ADI scheme result in a tridiagonal matrix, which can be solved very efficiently. The method is second-order accurate in time $(\Delta t)^2$ and space $((\Delta x)^2, (\Delta y)^2)$, and the method is unconditionally stable. Douglas and Gunn (1964) extended this method to three-dimensional heat conduction equations (t, x, y, z) and provided an unconditionally stable scheme, retaining the second-order accuracy. A description of other splitting methods for parabolic equations (fractional step method, alternate direction explicit method) can be found in Anderson et al. (1984).

B.2.2 Keller Box Method

One of the widely used methods for parabolic flows, especially the boundary layer equations, is the Keller box scheme (Keller, 1970; Keller and Cebeci, 1972). See Sections 5.5.5.1 and 5.5.5.2 for detailed application to turbomachinery. This technique is second-order accurate in space and time and allows for rapid variation in x or y or t. It is more complicated than the methods described and involves more computation per grid point. The steps involved in this scheme are as follows: (1) reduce the second-order partial differential equations to a set of first-order partial differential equations; (2) represent the first-order equations by central differences; (3) linearize the resulting algebraic equations and represent them in matrix form; and (4) solve the resulting matrix by a tridiagonal matrix solver described later. The technique is illustrated in Fig. B.3.

Let us apply this technique to

$$\frac{\partial \phi}{\partial t} = \tau \frac{\partial^2 \phi}{\partial y^2} \tag{B.11}$$

Let us write,

$$\psi = \frac{\partial \phi}{\partial y} \tag{B.12}$$

Hence Eq. B.11 can be written as

$$\frac{\partial \phi}{\partial t} = \tau \frac{\partial \psi}{\partial y} \tag{B.13}$$

Equations B.12 and B.13 have to be solved simultaneously for variables ϕ

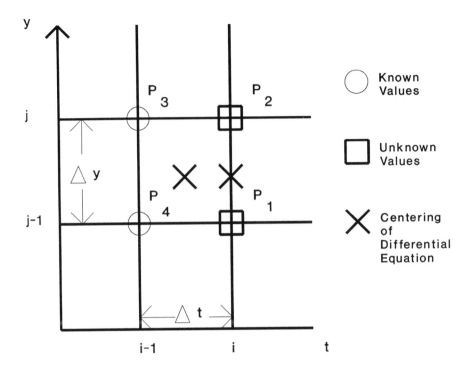

Figure B.3   Keller box scheme.

and ψ. The finite difference form of Eq. B.12 is given by

$$\psi_{i,j-1/2} = \left(\frac{\psi_{i,j} + \psi_{i,j-1}}{2}\right) = \left(\frac{\phi_{i,j} - \phi_{i,j-1}}{\Delta y}\right) \tag{B.14}$$

and finite difference form of Eq. B.13 is given by

$$\frac{1}{k_i}\left[\left(\frac{\phi_{i,j} + \phi_{i,j-1}}{2} - \frac{\phi_{i-1,j} + \phi_{i-1,j-1}}{2}\right)\right]$$
$$= \frac{\tau}{h_i}\left[\left(\frac{\psi_{i-1,j} - \psi_{i-1,j-1}}{2}\right) + \left(\frac{\psi_{i,j} - \psi_{i,j-1}}{2}\right)\right] \tag{B.15}$$

where $k_i = (\Delta t)_i$ and $h_i = (\Delta y)_i$. In Eq. B.14, the average value of ψ at i is taken. In Eq. B.15, $\partial\psi/\partial y$ and $\partial\phi/\partial t$ are represented by difference equations based on centered points, as shown in Fig. B.3.

Equations B.14 and B.15 can be manipulated to express the unknown (i) in terms of known quantities ($i - 1$). In each row, there are $2(I - 1)$ unknowns ($\phi_{i,j=1,\ldots,J-1}$), ($\psi_{i,j=1,\ldots,J-1}$) and $2(J - 1)$ equations, and these can be solved efficiently by a tridiagonal matrix solver described later. This method is similar to the Crank–Nicholson method, and it requires more computer time per gridpoint and fewer gridpoints but is more accurate.

B.3   METHOD OF SOLUTION OF ALGEBRAIC EQUATIONS

There are various methods of solving the algebraic equations resulting from the finite difference approximation of partial differential equations. It is beyond the scope of this book to describe these techniques. The methods can be classified as follows:

1. Direct methods for a system of linear equations
    a. Gauss elimination method
    b. Matrix inversion (Gauss–Jordan) method
    c. LU decomposition
2. Iterative methods for systems of linear equations
    a. Gauss–Seidel elimination (GS)
    b. Successive overrelaxation (SOR)
    c. Successive line overrelaxation method (SLOR)
    d. ADI methods

Chapra and Canale (1988) provide a detailed description and computer programs for solving the algebraic equations using each of these methods. Gauss elimination method is one of the earliest methods of solving a linear set of algebraic equations and is attractive for its simplicity and ease of programming. The algebraic equations resulting from the finite difference equation of PDE or ODE can be expressed in a matrix form as follows:

$$[A]\{U\} = \{e\} \tag{B.16}$$

where matrix A is a known matrix, and U and e are column vectors. The matrix inversion method involves writing Eq. B.16 in the form

$$\{U\} = [A]^{-1}\{e\} \tag{B.17}$$

where $[A]^{-1}$ is the inverse of matrix A. Instead of solving each equation individually, the matrix $[A]$ is inverted and the vector on the right-hand side is multiplied by this inverse to solve for U. The matrix multiplication with $[A]^{-1}$ is more efficient than matrix inversion. Hence $[A]^{-1}$ for the entire problem set can be stored to solve for various column vectors $\{e\}$. One widely used technique for computing the inverse of a matrix is the LU decomposition. The LU decomposition scheme involves only the operations on the matrix coefficients A in Eq. B.16. The technique is very efficient and is especially suited for situations where many right-hand vectors e are to be evaluated for a fixed value of A. In this technique, the matrix A is split into two matrices:

$$[L][U] = [A]$$

The lower and upper diagonals are derived from the original matrix through the Gauss elimination procedure, and this LU scheme is called *Crout's decomposition*. Detailed techniques, derivation, and computer programs can be found in basic textbooks on numerical analysis (Chapra and Canale, 1988). This decomposition can also be used to determine the inverse of a matrix.

B.3.1 Tridiagonal Matrix Solver

Many finite difference representations of partial differential equations result in a tridiagonal matrix. Consider the simple implicit scheme for Eq. B.11. This can be written as

$$a\phi_{i-1,j+1} + b\phi_{i,j+1} + c\phi_{i+1,j+1} = e \tag{B.18}$$

where

$$a = c = -\tau\frac{\Delta t}{(\Delta y)^2}, \qquad e = \phi_{i,j}, \qquad b = 1 + 2\tau\frac{\Delta t}{(\Delta y)^2} \tag{B.19}$$

If the boundary conditions at $j = 2, \ldots, N$ are specified, then $\phi_{1,j+1}$ and $\phi_{N,j+1}$ are known. Hence the algebraic equation with $i = 1, \ldots, N$ at the $(j + 1)$th time step can be written as

$$
\begin{bmatrix}
b_1 & c_1 & \cdot & \cdot & \cdot & \cdot & 0 \\
a_2 & b_2 & c_2 & 0 & \cdot & 0 & 0 \\
0 & a_3 & b_3 & c_3 & \cdot & \cdot & 0 \\
0 & 0 & a_4 & b_4 & c_4 & \cdot & 0 \\
0 & 0 & 0 & a_5 & b_5 & \cdot & \cdot \\
\cdot & \cdot & \cdot & \cdot & \cdot & a_N & b_N
\end{bmatrix}
\begin{bmatrix}
\phi_{1,j+1} \\
\phi_{2,j+1} \\
\phi_{3,j+1} \\
\vdots \\
\phi_{N,j+1}
\end{bmatrix}
=
\begin{bmatrix}
e_1 \\
e_2 \\
e_3 \\
\vdots \\
e_N
\end{bmatrix}
\tag{B.20}
$$

This is a tridiagonal system, which is a banded matrix where all elements are zero, except for a band along the main diagonal. An efficient method of solving this system is the Thomas algorithm. As with the conventional LU decomposition, this algorithm consists of decomposition and forward and back substitution. Details of this method, the method of handling other boundary conditions, the method of solution for $\phi_{i+1,j+1}$, and an algorithm to solve the matrix are given in Anderson et al. (1984) and Chapra and Canale (1988).

In many numerical schemes the coefficient matrix in Eq. B.20 is in the form of a block tridiagonal matrix. Details on the Thomas algorithm and the solving of block tridiagonal system are given in Anderson et al. (1984) and Hoffmann (1989).

B.4 BASIC METHODS FOR ELLIPTIC EQUATIONS

A large class of problems in fluid dynamics, especially the steady, viscid and inviscid subsonic flows, are governed by elliptic equations. The Navier–Stokes equation (Eqs. 1.9–1.11), the stream function equation (Eqs. 1.60–1.63), and the potential function equations (Eqs. 1.52–1.55) are good examples of elliptic equations. The best known examples are the Laplace equation and the Poisson equation

$$\nabla^2 \phi = 0, \qquad \nabla^2 \phi = f(x, y) \tag{B.21}$$

Elliptic equations are more difficult to solve than parabolic equations because the entire flow field has to be solved simultaneously; it is computationally more expensive and requires large computer storage. The Poisson equation arises in rotational flows and in quasi-three-dimensional flows (e.g., secondary flows, S_1 and S_2 surface equations, and passage-averaged equations, covered in Chapter 4).

The five-point scheme, which includes four adjoining corners, with the central point located at (i, j), is by far the most common finite difference formula used in the solution of elliptic equations. The Laplace equation is approximated by

$$\frac{\phi_{i+1,j} - 2\phi_{i,j} + \phi_{i-1,j}}{(\Delta x)^2} + \frac{\phi_{i,j+1} - 2\phi_{i,j} + \phi_{i,j-1}}{(\Delta y)^2} = 0 \tag{B.22}$$

The truncation error is $0(\Delta x)^2$ and $0(\Delta y)^2$.

The finite difference approximation introduces an additional term (truncation error) in the equation. The equation that is actually solved (called *modified equation*) can be derived by substituting the Taylor expansion for $\phi_{i+1,j}$ in Eq. B.22. The following modified equation can be derived for the grid with $\Delta x =$ const. and $\Delta y =$ const.

$$\nabla^2 \phi = -\frac{1}{12}\left[\phi_{xxxx}(\Delta x)^2 + \phi_{yyyy}(\Delta y)^2 \right] + \cdots \tag{B.23}$$

This is the actual equation solved, with a truncation of error of $(\Delta x)^2$ and $(\Delta y)^2$. In most cases, higher derivatives ϕ_{xxxx} are small, and thus the correction is negligible. Accuracy can be improved by using small values of Δx and Δy. In iterative techniques, initial guesses are employed and then iterated upon to obtain improved accuracy. These techniques are sometimes referred to as "relaxation methods" and can be broadly classified as Gauss–Seidel, SOR, SLOR, and ADI methods (indicated earlier). They can be further classified as point (or explicit) iterative methods and block (or implicit) iterative methods. In the former, the unknown function at each grid point is solved by successive iterative steps, while the latter subgroups are solved by the iterative method. In many elliptic problems, with large domain, it is better to resort to block iterative schemes, where blocks of unknowns are iterated to convergence by the method of elimination. These are usually expressed in terms of block tridiagonal matrix and solved by efficient algorithms. A good description of this technique can be found in Anderson et al. (1984) and Hirsch (1990). The convergence rate can be improved by sweeping rows and columns alternately as in the ADI scheme for the parabolic equations (Section B.2). This provides a more efficient method for including the influence of the boundary in the iteration process. Boonkkamp and Ten (1988) have proposed a new method of applying the residual smoothing to an ADI scheme. They employ a smoothing matrix for each of the ADI steps and show improved convergence, where both high-frequency and low-frequency oscillations are damped.

B.5   BASIC METHODS FOR HYPERBOLIC EQUATIONS

The methods available for the solution of hyperbolic equations can be classified as follows: (1) method of characteristics; (2) explicit techniques; (3) implicit techniques, and (4) splitting techniques. Some of the most commonly encountered hyperbolic equations in turbomachinery and fluid dynamics are time-dependent Euler/Navier–Stokes equations governing the flow (Eq. 5.1). The techniques available for hyperbolic systems will be demonstrated using simple equation governing practical flows. The emphasis is on presenting some typical techniques that are widely used. A survey of all techniques can be found in Anderson et al. (1984) and Hirsch (1990).

B.5.1   Method of Characteristics

Let us consider two-dimensional potential or stream function given by Eq. B.1. This represents two-dimensional inviscid flow over a cascade or blade row and is hyperbolic when the flow is supersonic. Furthermore, the coefficients A, B, and C are functions of x, y, ϕ_x, and ϕ_y only. It can be shown that at every point in the x–y plane, there are two directions along which the integration of partial differentials reduces to the integration of an equation involving total differential only. The characteristic lines are curves on which $\psi(x, y)$ is constant. For two-dimensional flow, the characteristic equation is an ODE and it is a PDE for three-dimensional flow. Details on the application of the method of characteristics to compressible flow can be found in Schreier (1982) and Anderson et al. (1984).

The method of characteristics was one of the earliest methods used for the calculation of supersonic flows and has been adopted to numerical application in recent years. One of the major advantages is that the important physics are retained in this formulation and the discontinuity and shocks occur along these characteristics. For simple two-dimensional flows, the method is attractive, but for three-dimensional flows, storage of the coordinates of all the characteristics becomes prohibitive and difficult.

B.5.2 Explicit Method

As mentioned earlier, the method of characteristics has some disadvantages for three-dimensional flow. Furthermore, when the flow is of mixed type (subsonic, transonic, supersonic in the same passage), use of finite difference equations in one part and the method of characteristics in the supersonic regions is very cumbersome. Hence, for such flows (mixed two- and three-dimensional flows), the finite difference solutions of the entire set of finite difference equations become very attractive.

B.5.2.1 *Lax–Wendroff Scheme*   One of the very popular explicit schemes is due to Lax and Wendroff (1960, 1964).

Consider a simple hyperbolic equation given by (first-order convection equation)

$$\frac{\partial \phi}{\partial t} + a \frac{\partial \phi}{\partial x} = 0 \tag{B.24}$$

Representing the time step by j and the spatial step by i, the Taylor expansion for Eq. B.24 can be written as

$$\phi_{i,j+1} = \phi_{i,j} + (\Delta t)\left(\frac{\partial \phi}{\partial t}\right)_{i,j} + \frac{(\Delta t)^2}{2}\left(\frac{\partial^2 \phi}{\partial t^2}\right)_{i,j} + 0(\Delta t)^3 \tag{B.25}$$

Using Eq. B.25, $\partial \phi / \partial t$ in Eq. B.24 can be eliminated to give

$$\phi_{i,j+1} = \phi_{i,j} - a\,\Delta t \left(\frac{\partial \phi}{\partial x}\right)_{i,j} + \frac{(a\,\Delta t)^2}{2}\left(\frac{\partial^2 \phi}{\partial x^2}\right)_{i,j}$$

The derivatives in the above equation can be replaced by central difference approximations to give

$$\phi_{i,j+1} = \phi_{i,j} - \frac{a(\Delta t)}{2\,\Delta x}\left(\phi_{i+1,j} - \phi_{i-1,j}\right)$$

$$+ \frac{a^2(\Delta t)^2}{2(\Delta x)^2}\left(\phi_{i+1,j} - 2\phi_{i,j} + \phi_{i-1,j}\right) \tag{B.26}$$

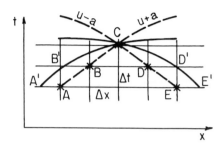

Figure B.4   CFL condition.

The unknown at the $(j + 1)$th time step can be calculated and marched with time. For stability, the domain of dependence of the difference equation must include the domain of dependence of the differential equation. This condition is illustrated in Fig. B.4. Point C is the calculation point and A'B'C and E'D'C are characteristics of the differential equation through point C given by $dx/dt = a$, and A'B'CD'E' is the domain of dependence of point C. ABCDE is the numerical domain of dependence. If the condition along AA' or EE' is changed, the solution at C from the finite difference equations will not be affected, even though the solution of the differential equation should change. Thus the numerical domain of dependence of this difference equation (ABCDE) should include domain of dependence of the differential equation (A'B'CD'E'). Hence A' and E' should be within AE; in other words, $a \Delta t/\Delta x \leq 1$ and the term is called the *Courant number*, and the condition is known as a *CFL condition* (Courant-Friedrichs and Lewy). This criterion can also be derived from a stability analysis and is valid for second-order hyperbolic systems as well (Anderson, 1995).

The Lax–Wendroff method can be extended to a set of equations [e.g., Euler and Navier–Stokes equations (Eq. 1.18)]. This technique is used widely in the solution of equations governing flow field. Let us apply this technique to one-dimensional Euler equations given by Eq. 1.18 with $v = w = F = G = S = 0$, Re $\rightarrow \infty$, and h_0, e_0, p = const. The equation can be expressed as follows, by writing $E(x, t) = E(q)$:

$$\frac{\partial q}{\partial t} + \frac{\partial E(q)}{\partial x} = \frac{\partial q}{\partial t} + \frac{\partial E}{\partial q}\frac{\partial q}{\partial x} = \frac{\partial q}{\partial t} + A\frac{\partial q}{\partial x} = 0 \qquad \text{(B.27)}$$

where A is a Jacobian matrix given by a simplified form of Eq. 5.87

$$A(q) = \begin{vmatrix} \dfrac{\partial E_1}{\partial q_1} & \dfrac{\partial E_1}{\partial q_2} \\[2ex] \dfrac{\partial E_2}{\partial q_1} & \dfrac{\partial E_2}{\partial q_2} \end{vmatrix} = \begin{vmatrix} 0 & 1 \\ u^2 & 2u \end{vmatrix} \qquad \text{(B.28)}$$

The Lax–Wendroff approximation to Eq. B.27 can be written by replacing x derivatives by central differences:

$$q_{i,j+1} = q_{i,j} - \Delta t \left(\frac{\partial E}{\partial x} \right)_{i,j} - \frac{(\Delta t)^2}{2} \frac{\partial}{\partial t} \left(\frac{\partial E}{\partial x} \right)_{i,j}$$

$$\frac{\partial}{\partial t} \left(\frac{\partial E}{\partial x} \right) = \frac{\partial}{\partial x} \left(\frac{\partial E}{\partial q} \frac{\partial q}{\partial t} \right) = -\frac{\partial}{\partial x} \left(\frac{\partial E}{\partial q} \frac{\partial E}{\partial x} \right)$$

Hence

$$q_{i,j+1} = q_{i,j} - \frac{\Delta t}{2 \Delta x} \left[E_{i+1,j} - E_{i-1,j} \right]$$

$$+ \frac{(\Delta t)^2}{2(\Delta x)^2} \left[A_{i+1/2,j} (E_{i+1,j} - E_{i,j}) - A_{i-1/2,j} (E_{i,j} - E_{i-1,j}) \right]$$

$$(\text{B.29})$$

The midpoint values can be averaged values—for example,

$$A_{i+1/2,j} = \frac{A_{i,j} + A_{i+1,j}}{2}$$

This is a one-step nonlinear version of the Lax–Wendroff scheme. The CFL condition for the equation is $\Delta t(|u| + a)/\Delta x \leq 1$. The Lax–Wendroff scheme is explicit, second-order accurate in time and space, and stable for CFL numbers less than 1. This technique, with variations, is widely used in solving compressible flow equations for internal and external flows. The shock structure is captured as a part of the solution.

Similarly, for two-dimensional flows with velocity components u and v in x and y directions, respectively, the CFL condition can be proven to be

$$\Delta t \left(\frac{|u|}{\Delta x} + \frac{|v|}{\Delta y} + a \sqrt{\frac{1}{(\Delta x)^2} + \frac{1}{(\Delta y)^2}} \right) \leq 1$$

An interpretation of Eq. B.29 for a control volume is given by Ni (1982). Equation B.29 can be expressed as

$$(\delta q)_i = q_{i,j+1} - q_{i,j} = \frac{1}{2} \left\{ \left[\Delta q_B + \frac{\Delta t}{\Delta x} \Delta E_B \right] + \left[\Delta q_C - \frac{\Delta t}{\Delta x} \Delta E_C \right] \right\}$$

$$(\text{B.30})$$

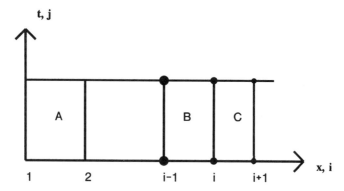

Figure B.5   Illustration of Ni's (1982) scheme.

where

$$\Delta q_{\rm B} = \left[E_{i-1,j} - E_{i,j} \right] \frac{\Delta t}{\Delta x}, \qquad \Delta q_{\rm C} = \left[E_{i,j} - E_{i+1,j} \right] \frac{\Delta t}{\Delta x}, \qquad \Delta E = \left(\frac{\partial E}{\partial q} \right) \Delta q$$

where B refers to the control area (volume for 3D flows) lying between $i - 1$ and i, and C is the control area lying between i and $i + 1$ as shown in Fig. B.5. This implies that the correction to point C consists of two parts. The terms in the first square bracket in Eq. B.30 are the correction $(\delta q_i)_{\rm B}$ due to change in $\Delta q_{\rm B}$ in the control area just upstream of i, and the other term is the correction $(\delta q_i)_{\rm C}$ associated with change of $(\Delta q)_{\rm C}$ occurring downstream of the control area. Similarly,

$$(\delta q_{i+1})_{\rm C} = \frac{1}{2} \left[\Delta q_{\rm C} + \frac{\Delta t}{\Delta x} \Delta E_{\rm C} \right] \tag{B.31}$$

Hence the change taking place in control area C is given by

$$\Delta q_{\rm C} \equiv (\delta q_i)_{\rm C} + (\delta q_{i+1})_{\rm C} \tag{B.32}$$

Based on this concept, Ni (1982) proposed the following scheme consisting of three operations for Eq. B.27:

1   Approximate the governing equation by

$$(\Delta q)_{\rm C} = \frac{\Delta t}{\Delta x} (E_{i,j} - E_{i+1,j}) \tag{B.33}$$

2   Determine the corrections to point i and $i + 1$ through the distribution formula

$$(\delta q_i)_C = \frac{1}{2}\left[\Delta q_C - \frac{\Delta t}{\Delta x}\Delta E_C\right] \tag{B.34}$$

$$(\delta q_{i+1})_C = \frac{1}{2}\left[\Delta q_C + \frac{\Delta t}{\Delta x}\Delta E_C\right] \tag{B.35}$$

where $\Delta E_C = (\partial E/\partial q)_C(\Delta q)_C$ and $\partial E/\partial q$ is the Jacobian of E.

3   Update the dependent variable by

$$q_{i,j+1} = q_{i,j} + (\delta q)_i, \qquad \delta q_i = (\delta q_i)_B + (\delta q_i)_C \tag{B.36}$$

where the right-hand side is known.

The above integration is second-order accurate in space and time. The distribution formulae proposed by Ni can be viewed as a systematic way for determining corrections to the surrounding grid points, $(\delta q_i)_C$ and $(\delta q_{i+1})_C$, due to change taking place in the control area (q_C). Application of this technique to turbomachinery flow is given in Section 5.7.4.1. Detailed treatment of finite volume implementation of Lax–Wendroff's scheme is given by Kallinderis and Baron (see Murthy and Brebbia, 1990).

B.5.2.2   *Runge–Kutta Scheme*   One of the more recent developments is the use of the multistage Runge–Kutta scheme, similar to the ones used for the numerical solution of ordinary differential equations. Jameson et al. (1981) have made extensive use of this technique. Runge–Kutta methods have been widely used for hyperbolic systems, and they achieve the accuracy of Taylor's series without calculating the higher-order derivatives.

In the explicit method, ϕ at the $(n + 1)$th step is calculated using its value at the nth step. The Taylor expansion for ϕ in Eq. B.24 can be written as (fourth order accuracy)

$$\phi^{n+1} = \phi^n + \Delta t(\phi_t)^n + \frac{(\Delta t)^2}{2}(\phi_{tt})^n + \frac{(\Delta t)^3}{6}(\phi_{ttt})^n$$

$$+ \frac{(\Delta t)^4}{24}(\phi_{tttt})^n + \dots \tag{B.37}$$

$$= \phi^n - \Delta t(a\phi_x) + \frac{a^2\,\Delta t^2}{2}(\phi_{xx}) - \frac{a^3\,\Delta t^3}{6}(\phi_{xxx})$$

$$+ \frac{a^4\,\Delta t^4}{24}(\phi_{xxxx}) + \dots \tag{B.38}$$

In the Runge–Kutta scheme, ϕ^{n+1} is evaluated at several intermediate steps between $n\,\Delta t$ and $(n + 1)\,\Delta t$ and then combine them to achieve higher-order approximations for ϕ^{n+1}. This considerably simplifies the difficulty involved in computing the higher-order derivative in Eq. B.38.

One widely used scheme is the four-stage Runge–Kutta method given by [for the equation $\phi_t = -R(\phi, t)$]

$$\phi^1 = \phi^n - \tfrac{1}{4}\,\Delta t\, R^{(0)}, \qquad \phi^2 = \phi^n - \tfrac{1}{3}\,\Delta t\, R^{(1)}$$
$$\phi^3 = \phi^n - \tfrac{1}{2}\,\Delta t\, R^{(2)}, \qquad \phi^{n+1} = \phi^n - \Delta t\, R^{(3)} \tag{B.39}$$

The values $\phi^1 \cdots \phi^3$ are provisional values at the new step. The spatial derivatives are usually approximated by the central difference scheme. Equation B.39 (through elimination of ϕ^1, ϕ^2, ϕ^3, etc.) can be shown to be the same as Eq. B.37. It can be shown that there is a close relationship between this and the Lax–Wendroff scheme described earlier. This technique has been widely used for 2D and 3D, inviscid and viscous, internal and external flows. This is covered in Section 5.7.4.3. Similar schemes can be written for two, three, and five stages. For a linear equation, the four-stage scheme is fourth-order accurate in time. For shock capturing and stability, artificial dissipation terms can be added to the term R. The CFL limit for the four-stage scheme can be proved to be (Jameson et al., 1981)

$$a\frac{\Delta t}{\Delta x} \le 2\sqrt{2}$$

For the two and three stages, CFL limits are 2 and 2.51, respectively. It is clear that the CFL limit for this scheme is higher than those presented earlier. Hence this scheme has found wide application in internal and external flows. The advantages of this method over the other schemes for Euler and Navier–Stokes equation are described in Sections 5.7.4 and 5.7.6.

B.5.3 Implicit Method

One major disadvantage of explicit methods is that the time step size is limited by the CFL condition. This has led to increased use of implicit techniques, which are unconditionally stable for all time steps. The simplest scheme is the Crank–Nicholson scheme. The choice of central difference is arbitrary, and other schemes can also be used (Anderson et al., 1984; Hoffman, 1989; Hirsch, 1990). This results in a tridiagonal matrix for a one-dimensional case, which can be solved by methods described in Section B.3. The simple implicit methods are not attractive for nonlinear problems. Moreover, in most practical applications, the fluid dynamic equations are either two- or three-dimensional, requiring large computer storage and time. For these situations, the splitting techniques or approximate factorization, described below, become attractive.

B.5.3.1 *Splitting and Approximate Factorization*   As mentioned earlier, the simple implicit techniques are difficult to apply to nonlinear problems and require large computer time. Even the explicit techniques suffer from similar problems when dealing with nonlinear, two- and three-dimensional problems. Ferziger (1981) provides a list of problems associated with the simple explicit and implicit techniques and the reasons for resorting to splitting techniques. In splitting techniques, the multidimensional problem is recast into a set of one-dimensional problems. One of the most successful concepts of splitting is originally due to Peaceman and Rachford (1959). A method for *explicit* nonlinear 2D/3D aerodynamic problems is due to Mac-Cormack (1969). Let us consider a two-dimensional Euler equation (Eq. 1.18):

$$\frac{\partial q}{\partial t} + \frac{\partial E}{\partial x} + \frac{\partial F}{\partial y} = 0 \tag{B.40}$$

where q, E, and F are given by Eq. 1.18 with $w = S = G = 0$, Re $\rightarrow \infty$, with h_o and e_o constant. MacCormack's two-step procedure is given by

$$\overline{q_{i,j}^{n+1}} = q_{i,j}^n - \frac{\Delta t}{\Delta x}\left(E_{i+1,j}^n - E_{i,j}^n\right) - \frac{\Delta t}{\Delta y}\left(F_{i,j+1}^n - F_{i,j}^n\right)$$

$$q_{i,j}^{n+1} = \frac{1}{2}\left[q_{i,j}^n + \overline{q_{i,j}^{n+1}} - \frac{\Delta t}{\Delta x}\left(\overline{E_{i,j}^{n+1}} - \overline{E_{i-1,j}^{n+1}}\right) - \frac{\Delta t}{\Delta y}\left(\overline{F_{i,j}^{n+1}} - \overline{F_{i,j-1}^{n+1}}\right)\right]$$

$$\tag{B.41}$$

where n refers to time step ($t = n\,\Delta t$), and i, j refers to calculation point x_i, y_j. The method evaluates an approximate value for $\overline{q_{i,j}^{n+1}}$ using forward differencing for the spatial derivatives. This can be considered as the ($n + \frac{1}{2}$)th step. The corrector step uses this approximate value with backward spatial differences to obtain a new value for $q_{i,j}^{n+1}$. The steps could be reversed by using backward differences in the predictor step and forward differences in the corrector step. The scheme is stable for CFL numbers up to 1. The predictor step can be considered to be a temporary, approximate, or ad hoc prediction for $q_{i,j}$. The method is second-order accurate in time and space. If the equations are linear, MacCormack's scheme is identical to the Lax–Wendroff scheme (1960). A detailed discussion of accuracy, stability, domain of dependence, nonlinear instability and some numerical examples can be found in MacCormack (1969).

One of the most successful splitting *implicit* techniques is due to Beam and Warming (1976) and Briley and McDonald (1977). Let us consider Eq. B.40. This equation can be expressed as (similar to Eq. B.27)

$$\frac{\partial q}{\partial t} + A\frac{\partial q}{\partial x} + B\frac{\partial q}{\partial y} = 0, \qquad A = \frac{\partial E}{\partial q} \text{ and } B = \frac{\partial F}{\partial q} \tag{B.42}$$

where A and B are Jacobian matrices (similar to Eqs. B.28, 5.87, and 5.88). A second-order temporal difference scheme for Eq. B.42 can be written as

$$q^{n+1} = q^n - \frac{\Delta t}{2}\left\{\left(\frac{\partial E}{\partial x} + \frac{\partial F}{\partial y}\right)^{n+1} + \left(\frac{\partial E}{\partial x} + \frac{\partial F}{\partial y}\right)^n\right\} + 0(\Delta t)^3 \quad (B.43)$$

A direct solution of Eq. B.43 for q^{n+1} is not possible due to the presence of a nonlinear flux vector at the $(n+1)$th time level. Using Taylor expansion for q^n, we obtain

$$E^{n+1} = E^n + A^n(q^{n+1} - q^n) + 0(\Delta t)^2$$
$$F^{n+1} = F^n + B^n(q^{n+1} - q^n) + 0(\Delta t)^2 \quad (B.44)$$

Equation B.43 can be written as

$$\left[I + \frac{\Delta t}{2}\left(\frac{\partial}{\partial x}A^n + \frac{\partial}{\partial y}B^n\right)\right](q^{n+1} - q^n) = -\Delta t\left[\frac{\partial E}{\partial x} + \frac{\partial F}{\partial y}\right]^n + 0(\Delta t^3)$$
$$(B.45)$$

where I is the identity matrix. This is a linear equation for q. Even though this equation is linear in q^n, the resulting algebraic equation is extremely difficult to solve. As mentioned earlier, one of the efficient ways to solve the equation is by splitting or approximate factorization. The factorization suggested by Beam and Warming (1976) is given by

$$\left(I + \frac{\Delta t}{2}\frac{\partial}{\partial x}A^n\right)\left(I + \frac{\Delta t}{2}\frac{\partial}{\partial y}B^n\right)(q^{n+1} - q^n) = -\Delta t\left(\frac{\partial E}{\partial x} + \frac{\partial F}{\partial y}\right)^n + 0(\Delta t)^3$$
$$(B.46)$$

The factorization error is of third order, same as the original scheme. Equation B.46 can now be solved in two steps using the ADI scheme as follows:

$$\left(I + \frac{\Delta t}{2}\frac{\partial}{\partial x}A^n\right)(\Delta q^*) = -\Delta t\left[\frac{\partial E}{\partial x} + \frac{\partial F}{\partial y}\right]^n$$

$$\left(I + \frac{\Delta t}{2}\frac{\partial}{\partial y}B^n\right)\Delta q^n = \Delta q^*, \quad \text{where } q^{n+1} = q^n + \Delta q^n \quad (B.47)$$

The spatial derivatives in the above equation are finite differences that produce a block tridiagonal system. These can be solved by efficient techniques as described earlier. Each of the one-dimensional splits of the matrix result in consistent approximation to the original partial differential equation. The scheme is called a consistently split linearized block implicit scheme (Briley and McDonald, 1977). The method is second-order accurate and unconditionally stable. This method has been widely coded to predict external and internal flows with great success.

The above scheme (Eq. B.47) is a centered (about $n + \frac{1}{2}$) time difference (trapezoidal) formula. Warming and Beam (1978) provided a generalization of the factorization scheme to other finite difference formulations. Their generalization of the scheme is given by

$$\left(I + \frac{\theta\,\Delta t}{1 + \zeta} \frac{\partial}{\partial x} A^n \right) \Delta q^* = -\frac{\Delta t}{1 + \zeta}\left[\frac{\partial E}{\partial x} + \frac{\partial F}{\partial y} \right]^n + \frac{\zeta}{1 + \zeta} \Delta q^{n-1}$$

$$\left(I + \frac{\theta\,\Delta t}{1 + \zeta} \frac{\partial}{\partial y} B^n \right) \Delta q^n = \Delta q^* \qquad (B.48)$$

$\theta = \frac{1}{2}$, $\zeta = 0$ is the trapezoidal formula; $\theta = 1$, $\zeta = 0$ is the Euler implicit method and $\theta = 1$, $\zeta = \frac{1}{2}$ is the three-point backward scheme. Application of this technique to turbomachinery flow field computation is described in Section 5.7.5.

B.6 GRID GENERATION

The numerical techniques described earlier discretize the equations of motion with the proper choice of grid size ($\Delta x, \Delta y, \Delta z$). Inaccuracy or cumbersome coding may result if these grids are not properly chosen from the point of view of numerical technique used and the complexity of the body shape. For example, a body-fitted coordinate system (which follows the body contour) enables efficient coding as well as improved accuracy. There are several textbooks as well as excellent review articles on grid generation techniques. Only a brief outline of the techniques is given below.

One of the most convenient and accurate methods of numerically resolving flows in turbomachinery passages is to transform Eq. 1.18 into a body-fitted coordinate system. This is common to all computational techniques described in Chapter 5. A general transformation is given by

$$\xi = \xi(x, y, z), \qquad \eta = \eta(x, y, z), \qquad \zeta = \zeta(x, y, z)$$

where ξ coordinate follows the blade profile and η is nearly normal to it, and ζ would then be nearly normal to ξ and η. A nearly orthogonal system is sought. Equation 1.18 can be transformed into this new coordinate system, and these are given in Appendix A.

Three types of grids are used in turbomachinery practice, depending on the type of machinery, technique, and the flow regime. For example, space marching and a thin leading edge requires an "H" grid; and a thick leading edge and time-marching solution is better done with a "C" or an "O" grid. In the H grid, the singular periodic corner points (leading and trailing edges) are identified and the periodic boundaries are extended to upstream and downstream. In the C grid, the region is constructed with the trailing edge singular points only. This enables fine resolution of the flow near the leading edge. The H grids also can be placed very close to the leading-edge boundary. In the O grid, the grid contours (ξ = constant line) are closed, with no corner points.

Reviews on the grid generation techniques for external and internal flows are given by Thompson et al. (1985) and Camarero et al. (1986). Numerical grid generation has replaced earlier techniques that were restricted to specific body shapes or numerical techniques. The basic concepts behind numerical grid generation is to (a) establish convenient, strategic distribution of points in the physical domain to ensure an accurate and stable solution and (b) enable uniform distribution of grid points in the computational domain. The grids should be smooth and orthogonal or nearly orthogonal at surfaces. In this procedure, the body shape is selected as one of the coordinates, enabling accurate and easy application of boundary conditions. This will also enable uniformity of grid size in the computational domain, while the corresponding grid size in the physical plane may be nonuniform and unequally spaced. The techniques employed for grid generation can be classified as conformal mapping, differential, and algebraic techniques.

The conformal mapping technique essentially follows the mapping techniques described in Section 3.3.1. An analytical function can be used to transform a complex shape into a simple one. In the techniques described in Section 3.3.1, the transformation was from a complex shape to a circle. In numerical grid generation, however, a rectangular or near-orthogonal grid is generated. For example, the cascade transformation can be used to transform the physical plane into a single profile, which can then be transformed into straight lines using the Schwarz–Christoffel transformation. This is restricted essentially to two-dimensional flows. Details can be found in Ives and Liutermoza (1977).

The differential technique could be parabolic, hyperbolic, or elliptic depending on the type of PDE employed to generate grids. Elliptic grid generation is by far the most common. In this technique, the curvilinear body-fitted coordinates in the physical plane are defined as the solution of the set of elliptic PDEs, one in each coordinate direction. If $\xi(x, y)$ and $\eta(x, y)$ are functions which enable mapping between physical and computational space, the mapping is required to satisfy the Poisson equation:

$$\xi_{xx} + \xi_{yy} = P, \qquad \eta_{xx} + \eta_{yy} = Q \qquad (B.49)$$

where $\xi(x, y)$ and $\eta(x, y)$ specify mapping of physical space to computational space. The mapping is carried out by specifying the grid points in the (x, y) on the boundaries of the physical domain. The interior points are then governed by Eq. B.49. The terms P and Q control the grid spacing, angle, and so on. With proper control of the boundaries, either O, C, or H grids can be generated. One widely used elliptic grid generation code is due to Sorensen (1980). Even though this technique is time-consuming, it provides smooth, nonoverlapping grids with very high flexibility in regard to grid spacing, grid intersection with the boundary, and so on, and therefore it is widely used.

As the name implies, known algebraic expressions are used to transform the body shape in the physical domain into a rectangular shape in the computational domain in the algebraic grid generation technique. In the algebraic grid generation scheme, known functions are used to map the curvilinear coordinate system in the physical space to a convenient (usually rectangular) system in the computational domain. Eiseman (1982) and Smith (1983) provide details of the basic technique for two-dimensional and three-dimensional algebraic grid generation methods. This technique is based on the use of interpolation functions. The approach is to employ algebraic expressions of the physical grid as a function of the computational grid. Many advanced interpolation schemes optimize the grid generation through variational methods. The major advantages of algebraic methods (as compared to differential methods) are the speed, simplicity, and flexibility. This method is fast and accurate, has considerable flexibility, and is also widely used in both two and three dimensions.

In addition to the above techniques, there are more specialized procedures. For example, in the adaptive grid technique, the grids are adapted to the physics of the flow field and modified at each iteration or cycle. The location/spacing of the grid is controlled depending on the gradients of flow variables. For example, in a high-gradient region (e.g., shock wave as it evolves during the computation), the grids may be clustered. It is beyond the scope of this book to cover this aspect. The reader is referred to Thompson et al. (1985) for details. Details on implementation of local grid refinement methods for viscous flows are given by Kallinderis and Baron (see Murthy and Brebbia, 1990).

One of the variations of the adaptive grid technique (where the grid is varied continuously during the solution to capture high-gradient areas) is the solution adaptive grid using unstructured mesh. Such a technique is valid for general geometry. This allows the mesh to be concentrated better in regions of complex flow/geometry (as in cooling passages and in tip clearance, secondary flow, and shock–boundary-layer interaction regions). Dawes (1993) has developed unstructured grid for such turbomachinery applications.

Turbomachinery blade geometries are among the most complex and wide varieties of geometries encountered. This includes large camber, low solidity, blunt and thin leading and trailing edges, spanwise twist, lean, and dihedral. Wide variations of this from one geometry to another make it difficult to develop a unified grid generation program. Fully elliptic grid generators are expensive and cumbersome, with difficulty in controlling orthogonality of the grids. Algebraic grid generators provide better control and boundary points and are inexpensive. One of the most successful grid generation techniques is the combination of the elliptic technique and the algebraic technique, to take advantage of the strength of each of these techniques. Beach (1990) developed a method of interacting these techniques (differential and algebraic) to derive the most desirable grids for turbomachinery. Algebraic grids are maintained near the blade to control orthogonality, and an elliptic technique is used to smooth out the grids at other locations. The following steps are used:

1. Generate the boundary surface grid from the geometry specified.
2. Grid lines are packed from the leading edge to the trailing edge and from the pressure to the suction surfaces using the hyperbolic tangent function and from the hub to the tip using exponential functions.

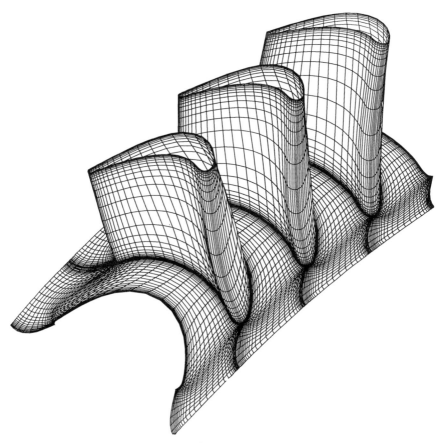

Figure B.6 Typical turbomachinery grid (only surface grids are shown for clarity) (courtesy T. Beach).

3. Cubic splines are used to model the blade surface and to interpolate the coordinates into an axisymmetric through-flow grid.

4. Interactive generation of grids on a blade-to-blade surface is carried out.

5. Several blade-to-blade surfaces are generated, and the radial continuity is maintained by spline interpolation.

6. Steps 1–6 are repeated until desired smooth and orthogonal (near surface) grids are generated.

The grids generated by this procedure for a typical turbomachinery blade row are shown in Fig. B.6.

BOOKS ON
TURBOMACHINERY

Balje, O. E., 1981, *Turbomachines: A Guide to Design, Selection and Theory*, John Wiley & Sons, New York.

Bashta, T. M., Rudnev, S. S., Nekrasov, B. B., 1982, *Hydraulics and Hydraulic Machines*, M. Mashinostronie, Moscow (in Russian).

Bidard, R., and Bonnin, 1979, *Energetique et Turbomachines*, Eyrolles, France.

Bolcs, A. and Suter, P., 1986, *Transsonische Turbomaschinen*, G. Braun, Karlsruhe.

Brennen, C. E., 1995, *Hydrodynamics of Pumps*, Oxford University Press.

Csanady, G. T., 1964, *Theory of Turbomachinery*, McGraw-Hill, New York.

Cumpsty, N. A., 1989, *Compressor Aerodynamics*, Longman Group, New York.

Dixon, S. L., 1984, *Fluid Mechanics and Thermodynamics of Turbomachinery*, Pergamon Press, Elmsford, NY.

Eck, B., 1973, *Fans*, Pergamon Press, Elmsford, NY.

Eckert, B., and Schnell, E., 1980, *Axial und Radial Kompressoren*, 2nd Ed., Springer Verlag.

Gostelow, J. P., 1984, *Cascade Aerodynamics*, Pergamon Press, Elmsford, NY.

Hawthorne, W. R., 1964, *Aerodynamics of Turbines and Compressors*, Princeton University Press, Princeton, NJ.

Holschevnikov, K. V., Emin, O. N., Mitrohin, V. T., 1986, *Theory and design of blade machine*, M. Mashinostronie, Moscow (in Russian).

Horlock, J. H., 1973, *Axial Flow Compressors*, Krieger Publishing Co., Melbourne, FL.

Horlock, J. H., 1973a, *Axial Flow Turbines*, Krieger Publishing Co., Melbourne, FL.

Li, G., Chen, N. X., and Qiang, G., 1980, *Aerothermodynamics of Axial Turbomachinery for Marine Gas Turbines* (in Chinese), Defence Industry Press, Beijing.

Scholz, N., 1965, *Aerodynamik der Schaufelgitter* Verlag G. Braun, Karlsruhe (Translation in AGARDograph No. AG 220, 1977).

Traupel, W., 1977, 1982, *Thermische Turbomaschinen*, Vols. I (1977) and II (1982), Springer, New York.

Turton, R. K., 1984, *Principles of Turbomachinery*, Spon, London.

Vavra, M. H., 1960, *Aerothermodynamics and Flow in Turbomachines*, John Wiley & Sons, New York.

Wallis, R. A., 1983, *Axial Flow Fans and Ducts*, John Wiley & Sons, New York.

Whitfield, A., and Baines, N. C., 1990, *Design of Radial Turbomachines*, Longman Group, New York.

Wilson, D. G., 1984, *The Design of High EfficiencyTurbomachinery and Gas Turbines*, MIT Press, Cambridge, MA.

Wislicenus, G. F., 1965, *Fluid Mechanics of Turbomachinery*, Dover, New York.

REFERENCES

Abdallah, S., and Henderson, R. E., 1987, "Improved Approach to Streamline Curvature Method in Turbomachinery," *J. Fluids Eng.*, Vol. 109.

Abdelhamid, A. N., and Bertrand, J., 1979, "Distinctions Between Two Types of Self Existed Gas Oscillations in Vaneless Radial Diffusors," ASME Paper 79-GT-58.

Abhari, R. S., and Epstein, A. H., 1994, "An Experimental Study of Film Cooling in a Rotating Transonic Turbine," *J. Turbomachinery*, Vol. 116, p. 63.

Abhari, R. S., et al., 1992, "Comparison of Time Resolved Turbine Rotor Blade Heat Measurements and Numerical Solution," *J. Turbomachinery*, Vol. 114, p. 818.

Abuaf, N., and Kercher, D. M., 1992, "Heat Transfer and Turbulence in a Turbulated Blade Cooling Circuit," ASME Paper 92-GT-187.

Adamczyk, J. J., 1985, "Model Equations for Simulating Flows in Multistage Turbomachinery," ASME Paper 85-GT-226.

Adamczyk, J. J., Celestina, M. L., and Greitzer, E. M., 1991, "The Role of Tip Clearance in High Speed Fan Stage," ASME Paper 91-GT-83.

Adamczyk, J. J., et al., 1990, "Simulation of 3D Viscous Flow Within a Multistage Turbine," *J. Turbomachinery*, Vol. 112, p. 370.

Adams, E., 1956, "Anwendung der Karman-Tsien Regel auf ebene kompressible Gitterstromungen im Unterschallberich," DFL Bericht 010, West Germany.

Adkins, G. G., and Smith, L. H., 1982, "Spanwise Mixing in Axial Flow Turbomachines," *J. Eng. Power*, Vol. 104, pp. 97–110.

Adler, D., 1980, "Status of Centrifugal Impeller Internal Aerodynamics, Part I: Inviscid Flow Prediction Methods," *J. Eng. Power*, Vol. 102, pp. 728–737; "Part II: Experiments and Influence of Viscosity," *J. Eng. Power*, Vol. 102, pp. 738–746.

AGARD, 1982, "Centrifugal Compressor Flow Phenomena and Performance," AGARD CP 282.

AGARD, 1987, "Advanced Technology for Aero Gas Turbine Components," AGARD CP 421.

AGARD, 1989, "Blading Design for Axial Flow Turbomachines," AGARD LS 167.

Ainley, D. G., 1948, "Performance of Axial Flow Turbines," *Proc. Inst. Mech. Eng.* (*London*), Vol. 159, p. 230 (1951).

Ainley, D. G., 1963, "An Experiment Single Stage Air Cooled Turbine, Part II," *Proc. Inst. Mech. Eng.*, Vol. 167, p. 351.

Ainley, D. G., and Mathieson, G. C. R., 1951, "A Method of Performance Estimation of Axial Flow Turbines," British ARC R & M 2974 (also see ARC R & M 2891).

Amano, R. S., and Pavelic, V., 1994, "A Study of Rotor Cavities and Heat Transfer in a Cooling Process in a Gas Turbine," ASME Paper 92-GT-358, *J. Turbomachinery*, Vol. 116, p. 333.

Ameri, A. A., Sockol, P. M., and Gorla, R. S. R., 1992, "Navier–Stokes Analysis of Turbomachinery Blade External Heat Transfer," *J. Propulsion and Power*, Vol. 8, No. 2.

Anand, A. K., and Lakshminarayana, B., 1975, "Three Dimensional Turbulent Boundary Layer in a Rotating Helical Channel," *J. Fluids Eng.*, Vol. 97, Series I, No. 2, pp. 197–210 (also see NASA CR 2888, 1977).

Anand, A. K., and Lakshminarayana, B., 1977, "An Experimental and Theoretical Investigation of Three-Dimensional Turbulent Boundary Layer Inside an Axial Flow Inducer," NASA CR 2888.

Anand, A. K., and Lakshminarayana, B., 1978, "An Experimental Study of Three-Dimensional Turbulent Boundary Layer and Turbulence Characteristics Inside a Turbomachinery Rotor Passage," *J. Eng. Power*, Vol. 100, pp. 676–691.

Anderson, D. A., Tannehill, J. C., and Pletcher, R. H., 1984, *Computational Fluid Mechanics and Heat Transfer*, Hemisphere Publishing Co., New York.

Anderson, J. D., 1982, *Modern Compressible Flow with Historical Perspectives*, McGraw-Hill, New York.

Anderson, J. D., 1995, *Computational Fluid Dynamics: The Basics with Applications*, McGraw-Hill, New York.

Anderson, L. R., and Heiser, W. H., 1969, "Systematic Evaluation of Cooled Turbine Efficiency,"ASME 69-GT-63.

Anderson, O. L., 1987, "Calculation of Three-Dimensional Boundary Layers on Rotating Turbine Blades, *J. Fluids Eng.*, Vol. 109, pp. 41–50.

Andrews, and Bradley, 1957, "Heat Transfer to Turbine Blades," AGARD CP 294.

Antaki, J. F., et al., 1993, "In *vivo* Evaluation of the Nimbus Axial Flow Ventricular Assist System," *J. Amer. Soc. for Artificial Internal Organs*, Vol. 39, No. 3, p. 231.

Arndt, W. E., 1984, "Prop Fans Go Full Scale," *Aerospace America*, p. 100.

Arora, S. C., and Abdelmesseh, W., 1985, "Pressure Drop and Heat Transfer Characteristics of Circular and Oblong Low Aspect Ratio Pin Fins," AGARD CP 390.

Arts, T., et al., 1992, "Aerothermal Investigation of Highly Loaded Transonic Linear Turbine Cascade," *J. Turbomachinery*, Vol. 114, p. 147 (also see VKI TN 174).

Ashworth, D. A., et al., 1985, "Unsteady Aerodynamic and Heat Transfer Process in a Transonic Turbine," *J. Eng. Gas Turbines and Power*, Vol. 107, 4, pp. 1022–1030.

Back, et al., 1970, "Effect of Wall Cooling on the Mean Structure of Turbulent Boundary Layer in Low Speed Gas Flow," *J. Heat Mass Transfer*, Vol. 13, pp. 1029–1047.

Baines, N. C., et al., 1986, "The Aerodynamic Development of Highly Loaded Nozzle Guide Vane," ASME Paper 86-GT-229.

Baker, A. J., 1983, *Finite Element Computational Fluid Mechanics*, McGraw-Hill, New York.

Baldwin, B. S., and Lomax, H., 1978, "Thin Layer Approximations and Algebraic Model for Separated Flow," AIAA Paper 78-0257.

Balje, O. E., 1981, *Turbomachines: A Guide to Design, Selection and Theory*, John Wiley & Sons, New York.

Balsa, T. F., and Mellor, G. L., 1972, "The Simulation of Axial Compressor Performance Using an Annulus Wall Boundary Layer Theory," AGARD AG 164, p. 373.

Bammert, K., and Sandstede, H., 1980, "Measurements of the Boundary Layer Development Along a Turbine Blade with Rough Surfaces," ASME Paper 80-GT-40.

Bansod, P., and Rhie, C. M., 1990, "Computation of Flow Through a Centrifugal Impeller with Tip Leakage," AIAA Paper 90-2021.

Bario, F., Barral, L., and Bois, G., 1989, "Air Test Flow Analysis of the Hydrogen Pump of the Vulcan Type," *Pumping Machinery*, ASME FED, Vol. 81, pp. 149–155.

Barnett, M., Hobbs, D. E., and Edwards, D. E., 1990, "Inviscid–Viscous Interaction Analysis of Compressor Cascade Performance," ASME 90-GT-15.

Basson, A. H., 1992, "Numerical Simulation of Steady Three-Dimensional Flows in Axial Turbomachinery Blade Rows," Ph.D. Thesis, Department of Aerospace Engineering, Pennsylvania State University.

Basson, A. H., and Lakshminarayana, B., 1993, "Numerical Simulation of Tip Clearance Effects in Turbomachinery," ASME Paper 93-GT-316 (to be published in *J. Turbomachinery*, July 1995).

Basson, A. H., and Lakshminarayana, B., 1994, "An Artificial Dissipation Formulation for Semi-Implicit Pressure Based Solution Scheme for Inviscid and Viscous Flows," *Int. J. Comp. Fluid Dynamics*, Vol. 2, p. 253.

Batchelor, G. K., 1956, *The Theory of Homogeneous Turbulence*, Cambridge University Press, New York.

Batchelor, G. K., 1967, *Fluid Dynamics*, Cambridge University Press, New York.

Bauer, F., Garabedian, P., and Korn, D., 1977, *Supercritical Wing Sections*, Vols. I, II, and III, Springer-Verlag, New York.

Bauermeister, K. J., and Hubert, G., 1962, "Secondary Flow Losses in Axial Flow Turbine Cascade," Aerodynamische Versuch AVA Report, 62-02, p. 21.

Bayley, F. J., 1977, "Performance and Design of Transpiration Cooled Turbine Blade," AGARD CP 229.

Beach, T. A., 1990, "An Interactive Grid Generation Procedure for Axial and Radial Flow Turbomachinery," AIAA Paper 90-0344.

Beam, R. M., and Warming, R. F., 1976, "An Implicit Finite-Difference Algorithm for Hyperbolic Systems in Conservation Law Form," *J. Comput. Phys.*, Vol. 22, pp. 87–110.

Beam, R. M., and Warming, R. F., 1978, "An Implicit Factored Scheme for the Compressible Navier–Stokes Equations," *AIAA J.*, Vol. 16, No. 4, pp. 393–402.

Behlke, R. F., 1986, "Development of Second-Generation Controlled Diffusion Airfoils for Multistage Compressors," *J. Turbomachinery*, Vol. 108, p. 32.

Behlke, R. F., et al., 1979, "Core Compressor Exit Stage Study II. Final Report," NASA CR 159812, pp. 1–88.

Behlke, R. F., Brooky, J. D., and Canal, E., 1983, "Study of Controlled Diffusion Stator Blading II," NASA CR 167995, pp. 1–128.

Bellows, W. J., and Mayle, R. E., 1986, "Heat Transfer Downstream of a Leading Edge Separation Bubble," *J. Turbomachinery*, Vol. 108 (also ASME Paper 86-GT-59).

Bergeles, G., et al., 1980, "Double Row Discrete Hole Cooling: An Experimental and Numerical Study," *J. Eng. Power*, Vol. 102, p. 498.

Bergeles, G., et al., 1981, "The Prediction of Three Dimensional Discrete Hole Cooling Processes," Part 2, Turbulent Flow," *J. Heat Transfer*, Vol. 103, p. 141.

Bergeron, P., 1940, *Trans. of the ASME*, Vol. 62, p. 162.

Bertin, J. J., and Smith, M. L., 1989, *Aerodynamics for Engineers*, Prentice-Hall, Englewood Cliffs, NJ.

Bhinder, F. S., 1970, "Investigation of Flow in the Nozzleless Spiral Casing of a Radial Inward-Flow Gas Turbine," in *Axial and Radial Turbomachinery*, British Institute of Mechanical Engineer,

Bidard, R., and Bonnin, 1979, *Energetique et Turbomachines*, Eyrolles, France.

Bindon, J. P., 1987, "Pressure and Flow Field Measurements of Axial Flow Turbine Tip Clearance Flow in a Cascade," *Proc. Inst. Mech. Eng.* (*London*), pp. 43–52; also see ASME Paper 87-GT-230 and *J. Turbomachinery*, Vol. 115, pp. 257–263, 1989.

Birch, N. T., 1987, "Navier–Stokes Predictions of Transition, Loss and Heat Transfer in a Turbine Cascade," ASME Paper 87-GT-22.

Blair, M. F., 1983, "Effect or Freestream Turbulence Effect on Turbulent Boundary Layer Heat Transfer and Mean Profile Development, Parts 1 and 2," *J. Heat Transfer*, Vol. 105, p. 33.

Blair, M. F., 1994, "An Experimental Study of Heat Transfer in a Large Scale Turbine Rotor Passage," *J. Turbomachinery*, Vol. 116, p. 1 (also see UTRC Report R91-970057-3).

Bois, G., Geai, P., and Vouillarment, A., 1994, "Experimental Study and Numerical Comparison of Flow Characteristics on an Air Test Model of the Inducer of the L_{H_z} Vulcan Engine," in *Proceedings of the 5th International Symposium on Transport Phenomena of Rotating Machines* (*ISROMAC-5*), Vol. A, Maui, Hawaii, May 8–11, 1994.

Bolcs, A., and Suter, P., 1986, *Transsonische Turbomaschinen*, G. Braun, Karlsruhe.

Boldman, D. R., et al., 1983, "Experimental Evaluation of Shockless Supercritical Airfoils in a Cascade," AIAA Paper 83-0003.

Boonkkamp, T., and Ten, J. H. M., 1988, "Residual Smoothing for Accelerating the ADI Iteration Method for Elliptic Difference Equations," *ZAMM*, Vol. 68, pp. 445–453.

Borisenko, A. I., 1962, *Gazovaya Dynamika Dvigateley*, Gosudarstvennoye, Moscow (Gas Dynamics of Engines, translated into English DDC-AD 609-452 and N65-23315, available from NASA scientific and information facility).

Boussinesq, J., 1877, "Theorie de l'Ecoulement Tourbillant," Memoires Presents par Divers Savants Sciences Mathematique et Physiques, Academie des Sciences, Paris, France, Vol. 23, p. 46.

Boyd, G. L., et al., 1983, "Advanced Gas Turbine Ceramic Development," *20th Automotive Technology Development Conference Coordinator Meeting*, Society of Automotive Engineers, pp. 189–198.

Boyle, M. T., and Hoose, R. V., 1989, "Endwall Heat Transfer in a Vane Cascade Passage and in a Curved Duct," ASME Paper 89-GT-90.

Boyle, R. J., 1991, "Navier–Stokes Analysis of Turbine Blade Heat Transfer," *J. Turbomachinery*, Vol. 113, p. 392.

Bradshaw, P. (ed.), 1970, *Turbulence: Topics in Applied Physics*, Vol. 12, Springer-Verlag, New York.

Bradshaw, P., 1973, "Effects of Streamline Curvature on Tubulent Flow," AGARDograph 169.

Bradshaw, P., 1974, "Prediction of Shear Layers in Turbomachinery," in *Fluid Mechanics, Acoustics and Design of Turbomachinery*, NASA SP 304.

Bradshaw, P., Cebeci, T., and Whitelaw, J. H., 1981, *Engineering Calculation Methods for Turbulent Flow*, Academic Press, New York.

Brandt, A., 1979, "Multigrid Level Adaptive Computations in Fluid Dynamics," AIAA Paper 79-1455.

Brennen, C. E., 1995, *Hydrodynamics of Pumps*, Oxford University Press.

Brideman, M. J., Cherry, D. G., and Pederson, J., 1983, "Low Pressure Turbine Scaled Test Vehicle Performance Report," NASA CR 168290, GE R83AEB143.

Briley, W. R., 1974, "Numerical Method for Predicting Three-Dimensional Steady Viscous Flows in Ducts," *J. Comput. Physics*, Vol. 14, No. 1, pp. 8–28.

Briley, W. R., and McDonald, H., 1977, "Solution of the Multidimensional Compressible Navier–Stokes Equations by a Generalized Implicit Method," *J. Comput. Physics*, Vol. 24, pp. 372–397.

Briley, W. R., and McDonald, H., 1984, "Three Dimensional Viscous Flow with Large Secondary Vorticity," *J. Fluid Mech.*, Vol. 144, pp. 47–77.

Bristow, D. R., 1976, "A New Surface Singularity Method for Multi-element Airfoil Analysis and Design," AIAA Paper 76-20.

Bristow, D. R., 1977, "Recent Improvements in Surface Singularity Method for Flow Field Analysis," AIAA Paper 77-641.

Brown, A., and Saluja, C. L., 1979, "Film Cooling from Three Rows of Holes on Adiabatic, Constant Heat Flux and Isothermal Surfaces in the Presence of Variable Freestream Velocity Gradients and Turbulence Intensity," ASME Paper 79-GT-24.

Bryce, J. D., et al., 1985, "The Design, Performance and Analysis of High Work Capacity Transonic Turbine," ASME Paper 85-GT-15.

Burdsall, E. A., Canal, E., and Lyons, K. A., 1979, "Core Compressor Exit Study, Part 1—Aerodynamic and Mechanical Design," NASA CR 159714, September 1979, "Part 2—Final Report," NASA CR 159812.

Burmeister, L. C., 1983, *Convective Heat Transfer*, John Wiley & Sons, New York.

Bushnell, P., 1988, "Measurement of the Steady Surface Pressure on a Single Rotation Large Scale Advanced Prop Fan Blade at $M = 0.03$ to 0.78," NASA CR 182124.

Butler, K. C., et al., 1992, "Development of Axial Flow Blood Pump LVAS," *Amer. Soc. for Artificial Internal Organs*, Vol. 38, No. 3, p. 296.

By, R., and Lakshminarayana, B., 1991, "Static Pressure Measurements in a Torque Converter Stator," SAE Special Publication, SP 879.

By, R., Kunz, R., and Lakshminarayana, B., 1993, "Navier–Stokes Analysis of the Pump Flow Field of an Automotive Torque Converter," *Proceedings 1993 ASME Pump Symposium* (to be published in *J. Fluids Eng.*, 1995).

Calvert, W. J., 1982, "An Inviscid–Viscous Interaction Treatment to Predict Blade-to-Blade Performance of Axial Compressors with Leading Edge Shocks," ASME 82-GT-135.

Calvert, W. J., et al., 1989, "Performance of a Highly-Loaded HP Compressor," ASME Paper No. 89-GT-24.

Camarero, R., et al., 1986, "Introduction to Grid Generation Systems in Turbomachinery," VKI L. S. 1986-02.

Cambier, L., and Veuillot, J. P., 1985, "Application of a Multi-domain Approach to the Computation of Compressible Flows in Cascades," Proc. Int. Soc. Air Breathing Engines (ISABE), published by AIAA.

Camci, C., 1989, "An Experimental and Numerical Investigation of Near Cooling Holes Heat Fluxes on a Film Cooled Turbine Blade," *J. Turbomachinery*, Vol. 111, p. 63.

Camci, C., and Arts, A., 1986a, "Short Duration Measurements and Numerical Simulation of Heat Transfer Along the Suction Surface of a Film Cooled Gas Turbine Blade," *J. Eng. for Gas Turbines and Power*, Vol. 107, p. 991.

Camci, C., and Arts, A., 1986b, "Experimental Heat Transfer Investigation Around the Film Cooled Leading Edge of a High Pressure Gas Turbine Rotor Blade," *J. Eng. for Gas Turbines and Power*, Vol. 107, p. 1016.

Came, P. M., and Marsh, H., 1974, "Secondary Flow in Cascades: Two Simple Derivations for the Components of Vorticity," *J. Mech. Eng. Sci.*, 16, pp. 391–401.

Cantrell, H. N., and Fowler, J. E., 1959, "The Aerodynamic Design Of Two-Dimensional Turbine Cascades for Incompressible Flow with a High-Speed Computer," *Trans. ASME 81 D*, pp. 349–361.

Caretto, L. S., et al., 1972, "Two Calculation Procedures for Steady Three Dimensional Flows with Recirculation," in *Proc. of the Third International Conference on Numerical Methods in Fluid Dynamics*, Springer-Verlag, New York, pp. 60–68.

Carter, J. E., 1978, "Inverse Boundary-Layer Theory and Comparison with Experiment," NASA TP-1208.

Carter, J. E., Edwards, D. E., and Werle, M. J., 1980, "A New Coordinate Transformation for Turbulent Boundary Layers," NASA CP 2166.

Cebeci, T., and Smith, A. M. O., 1974, *Analysis of Turbulent Boundary Layers*, Academic Press, New York (also see *Appl. Math Mech.*, Vol. 15, 1974).

Cebeci, T., Kaups, K., and Ramsey, J. A., 1977, "A General Method for Calculating Three Dimensional Compressible Laminar and Turbulent Boundary Layers," NASA CR 2777.

Cebeci, T., Khattab, A. A., and Stewartson, K., 1979, "Prediction of Three Dimensional Laminar and Turbulent Boundary Layers on Bodies of Revolution at High Angles of Attack," *Turbulent Shear Flows, II* (L. J. S. Bradbury et al., eds.), Springer-Verlag, New York.

Cetin, M., et al., 1987, "Application of Modified Loss and Deviation Correlations to Transonic Axial Compressors," AGARD-R-745.

Chakravarthy, S. R., and Osher, S., 1983, "Numerical Experiments with the Osher Upwind Scheme for the Euler Equations," *AIAA J.*, Vol. 21, No. 9, p. 1241.

Chan, D. C., and Sheedy, K. P., 1990, "Turbulent Flow Modeling of a Three-Dimensional Turbine," AIAA-90-2124.

Chan, D. C., Sindhir, M. M., and Gosman, A. D., 1987, "Numerical Simulation of Passages with Strong Curvature and Rotation Using a Three-Dimensional Navier Stokes Solver," AIAA Paper 87-1354.

Chan, D. C., et al., 1988, "On the Development of a Reynolds Averaged Navier Stokes Equation Solver for Turbomachinery," in *Proceedings of the 2nd International Symposium on Rotating Machinery* (*ISROMAC-2*), Honolulu, Hawaii.

Chapra, S. C., and Canale, R. P., 1988, *Numerical Methods for Engineers*, McGraw-Hill, New York.

Chapyglin, S. A., 1944, "On Gas Jets," NASA TN 1063.

Chauvin, J., Breugelmans, F., and Janigro, A., 1969, "Supersonic Compressors," in *Supersonic Turbojet Propulsion Systems and Components* (J. Chauvin, ed.), AGAR-Dograph 120, pp. 59–178.

Chauvin, J., Sieverding, C., and Griepentrog, H., 1970, "Flow in Cascade with a Transonic Regime," in *Flow Research on Blading* (L. S. Dzung, ed.), Elsevier, Amsterdam, pp. 151–189.

Chen, N. X., and Dai, L. H., 1987, "A Time Marching Method for Calculating S_2 Stream Surface Viscous Flow in a Single Stage Compressor Rotor," ASME Paper 87-GT-67.

Chen, N. X., and Zhang, F. X., 1987, "A Generalized Numerical Method for Solving Direct, Inverse, Hybrid Problems of Blade Cascade Flow by Using Streamline Coordinate Equation," ASME Paper 87-GT-29 (also see *J. Turbomachinery*, October 1986, 86-GT-189, 86-GT-30).

Chen, S. C., and Eastland, A. H., 1990, "Forced Response of Turbomachinery Blades Due to Passing Wakes," AIAA Paper 90-2353.

Chen, S. C., and Schwab, J. R., 1988, "Three Dimensional Elliptic Grid Generation Technique with Application to Turbomachinery Cascades," NASA TM 101330.

Cherrett, M. A., and Bryce, J. D., 1992, "Unsteady Viscous Flow in a High Speed Core Compressor," *J. Turbomachinery*, Vol. 114, pp. 287–294.

Cherrett, M. A., Bryce, J. D., and Ginder, R. B., 1994, "Unsteady Three-Dimensional Flow in a Single Stage Transonic Fan, Parts 1 and 2," ASME Paper 94-GT-223 and 94-GT-224 (to be published in *J. Turbomachinery*).

Chien, K., 1982, "Prediction of Channel and Boundary Layer Flows with a Low Reynolds Number Turbulence Model," *AIAA J.*, Vol. 20, No. 1, pp. 33.

Chima, R. V., 1985, "Inviscid and Viscous Flows in Cascades with an Explicit Multi-grid Scheme," *AIAA J.*, Vol. 23, No. 10, pp. 1556–1563.

Chima, R. V., 1987, "Explicit Multigrid Algorithm for Quasi Three Dimensional Viscous Flows in Turbomachinery," *J. Propulsion and Power*, Vol. 3, No. 5, pp. 1556–1563.

Chima, R. V., and Strazisar, A. J., 1983, "Comparison of Two and Three-Dimensional Flow Computations with Laser Anemometer Measurements in a Transonic Compressor Rotor," *J. Eng. for Power*, Vol. 105, p. 596.

Chima, R. V., and Yakota, J. W., 1990, "Numerical Analysis of Three-Dimensional Viscous Internal Flows," *AIAA J.*, Vol. 28, No. 5, p. 795.

Chima, R. V., Turkel, E., and Schaffer, S., 1987, "Comparison of Three Explicit Multigrid Methods for the Euler and Navier–Stokes Equations," NASA TM 88878.

Choi, D., and Knight, C. J., 1988, "Computation of 3D Viscous Cascade Flow," AIAA Paper 88-0363.

Choi, D., and Knight, C. J., 1991, "Improved Turbulence Modeling for Gas Turbine Applications," AIAA Paper 91-2339.

Chorin, A. J., 1968, "Numerical Solution of the Navier–Stokes Equations," *Math. Comput.*, Vol. 22, pp. 745–762.

Clifford, R. J., 1985, "Rotating Heat Transfer Investigations on a Multipass Cooling Geometry," AGARD CP 390.

Clifford, R. J., Morris, W. D., and Hasagawa, S. P., 1985, "An Experimental Study of Local and Mean Heat Transfer in a Triangular Section Rotating Duct in the Orthogonal Mode," *J. Eng. Power*, Vol. 107, p. 661 (ASME Paper 84-GT-142).

Coakley, T. J., 1983, "Turbulence Modeling Methods for the Compressible Navier-Stokes Equations," AIAA Paper 83-1693.

Connell, S. D., and Stow, P., 1986, "The Pressure Correction Method," *Computers and Fluids*, Vol. 14, No. 1, pp. 1–10.

Cooper, P., and Bosch, H., 1966, "Three-Dimensional Analysis of Inducer Fluid Flow," NASA CR 54836.

Cooper, P. (ed.), 1989, "Pumping Machinery," ASME, FED, Vol. 81.

Cooper, P. (ed.), 1993, "Pumping Machinery," ASME, FED, Vol. 154.

Copenhaver, W. W., Hah, C., and Puterbuagh, S., 1993, "Three Dimensional Flow Phenomena in a Transonic, High Through Flow, Axial Flow Compressor Stage," *J. Turbomachinery*, Vol. 115, p. 240.

Cousteix, J., 1986, "Three Dimensional Boundary Layers—Introduction to Calculation Methods," AGARD R 471.

Crank, J., and Nicholson, P., 1947, "A Practical Method for Numerical Evaluation of Solutions of Partial Differential Equations of the Heat Conduction Type," *Proc. Cambridge Philos. Soc.*, Vol. 43, pp. 50–67.

Crawford, M. E., and Kays, W. M., 1975, "STAN5-A Program for Numerical Computation of Two-Dimensional Internal/External Boundary Layer Flow," Stanford University, M.E. Report HMT 23.

Crawford, M. E., and Kays, W. M., 1976, "A Program for Numerical Computation of Two-Dimensional Internal and External Boundary Layer Flows," NASA CR 2742.

Crawford, M. E., and Stephens, C., 1988, "TEXTAN Primer," Mechanical Engineering Department, University of Texas, Austin.

Crawford, M. E., Kays, W. M., and Moffat, R. J., 1980, "Full Coverage Film Cooling on Flat, Isothermal Surfaces: A Summary Report on Data and Predictions," NASA CR 3219 (also see *J. Eng. Power*, Vol. 102, p. 1006).

Crook, A. J., et al., 1993, "Numerical Simulation of Compressor Endwall and Casing Treatment Phenomena," *J. Turbomachinery*, Vol. 115, p. 501.

Crouse, J. E., and Sandercock, D. M., 1967, "Blade Element Performance of Two Stage Axial Flow Pump with Tandem Row Inlet Stage," NASA TND 3962.

Csanady, G. T., 1964, *Theory of Turbomachines*, McGraw-Hill, New York.

Cumpsty, N. A., 1989, *Compressor Aerodynamics*, Longman Group, New York.

Cunha, F. J., 1992, "Turbulent Flow and Heat Transfer in Gas Turbine Blade Cooling Passage," ASME Paper 92-GT-239.

Daniels, L. C., 1979, "Film Cooling of Gas Turbine Blades," Ph.D. Thesis, Oxford University (OUEL Report No. 1302/79).

Daniels, L. C., and Shultz, D. L., 1982, *Heat Transfer Rate to Blade Profiles—Theory and Measurement in Transient Facilities*, VKI Lecture Series, Vols. 1 and 2.

Davino, R., and Lakshminarayana, B., 1982, "Characteristics of Mean Velocity in the Annulus Wall Region at the Exit of a Turbomachinery Rotor Passage," *AIAA J.*, Vol. 20, No. 4, pp. 528–535.

Davis, R. L., Hobbs, D. E., and Weingold, H. D., 1988, "Prediction of Compressor Cascade Performance Using a Navier–Stokes Technique," *J. Turbomachinery*, Vol. 110, No. 4, p. 520.

Davis, W. R., and Miller, D. A. J., 1972, "A Discussion of the Matrix Technique Applied to Fluid Flow Problems," *Can. A.S.I. Trans.*, Vol. 5, No. 2, p. 64.

Davis, W. R., and Miller, D. A. J., 1975, "A Comparison of Matrix and Streamline Curvature Methods of Axial Flow Turbomachinery Analysis, from a User's Point of View," *J. Eng. for Power*, Vol. 97, p. 549.

Dawes, W. N., 1987, "A Numerical Analysis of Viscous Flow in a Transonic Compressor Rotor and Comparison with Experiment," *J. Eng. for Gas Turbine and Power*, Vol. 109, p. 33.

Dawes, W. N., 1991, "The Simulation of Three-Dimensional Viscous Flow in Turbomachinery Geometries Using a Solution-Adaptive Unstructured Mesh Methodology," ASME 91-GT-124.

Dawes, W. N., 1993, "The Extension of Solution Adaptive 3D Navier–Stokes Solver Towards Geometries of Arbitrary Complexities," ASME Paper 92-GT-363, *J. Turbomachinery*, Vol. 115, p. 283.

Day, I. J., 1992, *Stall and Surge in Axial Flow Compressors*, VKI Lecture Series 1992-02, January 1992.

Dean, R. C., 1959, "On the Necessity of Unsteady Flow in Fluid Machinery," *J. Basic Eng.*, Vol. 81, p. 24.

Deissler, R. G., 1959, "Turbulent Flows and Heat Transfer," in *High-Speed Aerodynamics and Jet Propulsion*, C. C. Lin, ed., Princeton University Press, Vol. 5, Princeton, NJ, p. 288.

Delaney, R. A., 1983, "Time-Marching Analysis of Steady Transonic Flow in Turbomachinery Cascades Using Hopscotch Method," *J. Eng. for Power*, Vol. 105, p. 272.

Demuren, H. O., 1976, "Aerodynamic Performance and Heat Transfer Characteristics of High Pressure Ratio Transonic Turbines," Sc.D. Thesis, Department of Aero and Astro, M.I.T., Cambridge, MA.

Denton, J. D., 1975, "A Time Marching Method for Two- and Three-Dimensional Blade-to-Blade Flow," Aeronautical Research Council (U.K.) R & M 3775.

Denton, J. D., 1982, "An Improved Time-Marching Method for Turbomachinery Flow Calculation," ASME Paper 82-GT-239.

Denton, J. D., 1993, "Loss Mechanisms in Turbomachines," *J. Turbomachinery*, Vol. 115, p. 621.

Denton, J. D., and Xu, L., 1990, "The Trailing Edge Loss of Transonic Turbine Blades," *J. Turbomachinery*, Vol. 112, No. 2, p. 277.

DeRuyck, J., and Hirsch, C., 1981, "Investigations of Axial Compressor End-Wall Boundary Layer Prediction Method," *J. Eng. Power*, Vol. 103, pp. 20–33.

DeRuyck, J., Hirsch, C., and Kool, P., 1979, "An Axial Compressor End-Wall Boundary Layer Calculation Method," *J. Eng. Power*, Vol. 101, No. 2, pp. 233–249.

Dickmann, H. E., and Weissinger, J., 1955, "Berurag zur Theorie Optimaler Dusen-schrauben (Durtdusen)" *Jahrbuch Det Schiffbautech-Nichen Gesselschaft*, 49, Springer-Verlag, New York, p. 254.

Dixon, S. L., 1974, "Secondary Vorticity in Axial Compressor Blade Rows," NASA SP 304, Vol. 1, p. 173.

Dixon, S. L., 1984, *Fluid Mechanics and Thermodynamics of Turbomachinery*, Pergamon Press, Elmsford, NY.

Dommasch, D. O., Sherby, S. S., and Connolly, T. F., 1967, *Airplane Aerodynamics*, Pitman Publishing, New York.

Doorly, D. J., 1987, "Modeling of Unsteady Flow in a Turbine Rotor Passage," ASME Paper 87-GT-197.

Douglas, J., and Gunn, J. E., 1964, "A General Formulation of Alternating Direction Method, Part 1: Parabolic and Hyperbolic Equations," *Numerische Mathematik*, Vol. 6, pp. 428–453.

Drela, M., 1983, "A New Transformation and Integration Scheme for the Compressible Boundary Layer Equations and Solution of Behavior at Separation," MIT Gas Turbine Lab Report, 172.

Dring, R. P., and Heiser, W. H., 1978, "Turbine Aerodynamics," *The Aerothermodynamics of Aircraft Gas Turbines*, (G. C. Oates, ed.), AFAPL TR 78-52, Chapter 18. (Also published by AIAA in its educational series.)

Dring, R. P., and Joslyn, H. D., 1981, "Measurement of Turbine Rotor Flows," *J. Eng. Power*, Vol. 103, p. 400.

Dring, R. P., and Joslyn, H. D., 1985, "An Assessment of Single- and Multi-stage Compressor Flow Modelling," Final Report for Naval Air Systems Command Contract No. N00014-84-C-0354, AD, B102 101.

Dring, R. P., et al., 1982a, "An Investigation of Axial Compressor Rotor Aerodynamics," *J. Eng. Power*, Vol. 104, p. 84.

Dring, R. P., et al., 1982b, "Turbine Rotor–Strator Interaction," *J. Eng. for Power*, Vol. 104, p. 729.

Dring, R. P., et al., 1986, "Effects of Inlet Turbulence and Rotor/Stator Interactions on the Aerodynamics and Heat Transfer of a Large Rotating Turbine Model," Vol. 1 (NASA CR 4079) and Vols. III and IV (NASA CR 179469).

Dulikravich, G. S., 1988, "Analysis of Artificial Dissipation Models for the Transonic Full Potential Equations," *AIAA J.*, Vol. 26, No. 10, pp. 1238–1245.

Dullenkopf, K., and Mayle, R. E., 1994, "The Effect of Incident Turbulence and Moving Wakes on Laminar Heat Transfer in Gas Turbines," *J. Turbomachinery*, Vol. 116, p. 23.

Dunavant, J. C., and Erwin, J. R., 1956, "Investigations of Related Series of Turbine Blade Profiles in Cascades," NASA TN-3802.

Dunham, J., 1970, "A Review of Cascade Data on Secondary Losses in Turbines," *J. Mech. Eng. Sci.*, Vol. 12, pp. 48–59.

Dunker, R. J., Strinning, P. E., and Weyer, W. B., 1977, "Experimental Study of Flow Field in a Transonic Axial Compressor Rotor by Laser Velocimetry and Comparison with Through Flow Calculations," ASME Paper 77-GT-28.

Dunn, M. G., 1985, "Turbine Heat Flux Measurements: Influence of Slot Injection on Vane Trailing Edge Heat Transfer and Influence of Rotor on Vane Heat Transfer," *J. Eng. Power*, Vol. 107, pp. 76–83.

Dunn, M. G., 1986, "Heat Flux Measurements for the Rotor of a Full Stage Turbine, Part I—Time Averaged Results, Part II—Time Resolved Measurements," *J. Turbomachinery*, Vol. 108, p. 90.

Dunn, M. G., et al., 1994. "Time Averaged Heat Transfer and Pressure Measurments and Comparison with Prediction for a Two Stage Turbine," *J. Turbomachinery*, Vol. 116, p. 14.

Eck, B., 1973, *Fans*, Pergamon Press, Elmsford, NY.

Eckardt, D., 1976, "Detailed Flow Investigations Within a High-Speed Centrifugal Compressor Impeller," *J. Fluids Eng.*, Vol. 98, p. 390.

Eckardt, D., 1980, "Flow Field Analysis of Radial and Backswept Centrifugal Impellers, Part I," in *Performance Prediction of Centrifugal Compressors and Pumps*, S. Gopalakrishna et al., eds., ASME.

Eckert, E. R. G., 1983, "Analysis of Film Cooling and Full Coverage Film Cooling of Gas Turbine Blades," in *Proceedings of the 1983 International Gas Turbine Congress*, Tokyo, Japan, p. 109.

Eckert, E. R. G., and Drake, R. M., 1972, *Analysis of Heat and Mass Transfer*, McGraw-Hill, New York.

Eckert, B., and Schnell, B., 1980, *Axial and Radial Compressors* (in German), 2nd ed., Springer-Verlag, New York.

Eiseman, P. R., 1982, "Coordinate Generation with Precise Controls Over Mesh Properties," *J. Comput. Phys.*, Vol. 47, p. 331.

El-Husayne, H. A., Taslim, M. E., and Kercher, D. M., 1994, "Experimental Heat Transfer Investigation of Stationary and Orthogonally Rotating Asymmetric and Symmetric Heated Smooth and Turbulated Channels," *J. Turbomachinery*, Vol. 116, p. 124.

Engeda, A., and Rautenberg, M., 1987, "Comparisons of Relative Tip Clearance on Centrifugal Impellers," ASME Paper 87-GT-11.

Erickson, L. E., and Billdal, J. T., 1989, "Validation of 3-D Euler and Navier–Stokes Finite Volume Solver for Radial Compressor," AGARD CP-437.

Eroglu, H., and Tabakoff, W., 1989, "LDV Measurements and Investigation of Flow Field through Radial Turbine Guide Vanes," ASME Paper 89-GT-162.

Escudier, M. P., Bornstein, J., and Zehnder, N., 1980, "Observations with LDA Measurements of a Confined Vortex Flow," *J. Fluid Mech.*, Vol. 98, p. 49.

Ewen, J. S., et al., 1973, "Investigations of the Performance of a Small Axial Turbine," ASME Paper 73-GT-3.

Ewing, B. A., et al., 1980, "High Temperature Radial Turbine Demonstration," AIAA Paper 80-0301.

Fagan, J. R., and Fleeter, S., 1991, "LDV Measurements of a Mixed Flow Impeller at Design and Near Stall," ASME Paper 91-GT-310.

Falcao, A. F. D., 1975, "Three-Dimensional Potential Flow Through a Rectilinear Cascade of Blades," *Ing. Arch.*, Vol. 44, p. 27.

Falchetti, F., 1992, "Advanced CFD Simulation and Testing of Blading in the Multistage Environment," AIAA Paper 92-3040.

Fan, S., and Lakshminarayana, B., 1993, "A Simplified Reynolds Stress Model for Unsteady Turbulent Boundary Layers," AIAA Paper 93-0204.

Fan, S., and Lakshminarayana, B., 1994, "Computation and Simulation of Wake Generated Unsteady Pressure and Boundary Layer in Cascades: Parts 1 & 2," ASME Paper 94-GT-140 and 94-GT-141 (to be published in *J. Turbomachinery*).

Fan, S., Lakshminarayana, B., and Barnett, M., 1993, "A Low Reynolds Number $k-\epsilon$ Model for Unsteady Boundary Layers," *AIAA J.*, Vol. 31, p. 1777.

Farrell, C., and Adamczyk, J., 1982, "Full Potential Solution of Transonic Quasi-3D Flow Through a Cascade Using Artificial Compressibility," *J. Eng. Power*, Vol. 104, p. 143.

Favre, J. N., Avellan, F., and Rhyming, I. L., 1987, "Cavitation Performance Improvement by Using a 2D Inverse Method for Hydraulic Runner Design," in *Proceedings of the International Conference on Inverse Design Concepts and Optimization in Engineering Science*, Pennsylvania State University.

Ferziger, J. H., 1981, *Numerical Methods for Engineering Application*, John Wiley & Sons, New York.

Florschuetz, L. W., and Tseng, H. H., 1984, "Effect of Non-uniform Geometries on Flow Distributions and Heat Transfer Characteristics for Arrays of Impingement Cooling," ASME Paper 84-GT-156.

Flot, R., and Papailiou, K., 1975, "Couches Limites et Effects d'Extremities dans les Turbomachines," *Metraflu*, Contract DRME 73/373, Lyon, France.

Fougeres, J. M., and Heider, R., 1994, "Three-Dimensional Navier Stokes Prediction of Heat Transfer with Film Cooling," ASME Paper 94-GT-14.

Fourmaux, A., and LeMeur, A., 1987, "Computation of Unsteady Phenomena in Transonic Turbines and Compressors," ONERA (France), TP No. 1987-131.

Fox, R. W., and McDonald, A. T., 1985, *Introduction to Fluid Mechanics*, John Wiley & Sons, New York.

Freeman, C., and Cumpsty, N., 1989, "A Method for the Prediction of Supersonic Compressor Blade Performance," ASME Paper 89-GT-326.

Fujitani, K., Himerno, R., and Takagi, M., 1988, "Computational Study of Flow Through a Torque Converter," SAE Paper 881746.

Fullagar, K. P. L., 1974, "The Design of Air Cooled Blades," British Aeronautical Research Council Report 35684, HMT 361.

Galmes, J. M., and Lakshminarayana, B., 1984, "Turbulence Modeling for Three Dimensional Turbulent Shear Flows Over Curved Rotating Bodies," *AIAA J.*, Vol. 22, No. 1, pp. 1420–1428.

Garabedian, P., and Korn, D., 1976, "A Systematic Method for Computer Design of Supercritical Airfoils in Cascades," *Commun. Pure Appl. Math.*, Vol. 29, p. 369.

Garg, V. K., and Gaugler, R. E., 1993, "Heat Transfer in Film Cooled Turbine Blades," ASME Paper 93-GT-81.

Garrick, J. E., 1944, "On the Plane Potential Flow Past a Lattice of Arbitrary Airfoils," NACA Report 778.

Gaugler, R. E., 1985, "A Review of Boundary Layer Transition Data for Turbine Applications," ASME Paper 85-GT-83.

Gaugler, R. E., and Russell, L. M., 1980, "Streamline Flow Visualization Study of a Horseshoe Vortex in a Large Scale, Two Dimensional Turbine Stator Cascade," ASME Paper 80-GT-4.

Gearhart, W. S., 1966, "Selection of Propulsor for Submersible System," *J. Aircraft*, Vol. 3, No. 1, p. 84.

Gibson, M. M., 1978, "An Algebraic Stress and Heat Flux Model for Turbulent Shear Flow with Streamline Curvature," *Int. J. Heat Mass Transfer*, Vol. 21, pp. 1609–1617.

Gibson, M. M., and Launder, B., 1976, "On the Calculation of Horizontal, Turbulent, Free Shear Flows Under Gravitational Influence," *J. Heat Transfer*, Vol. 98, p. 81.

Gibson, M. M., and Rodi, W., 1981, "A Reynolds Stress Closure Model of Turbulence Applied to the Calculation of a Highly Curved Mixing Layer," *J. Fluid Mech.*, Vol. 103, p. 161.

Giesing, J. P., 1964, "Extension of the Douglas Neumann Program to Problems of Lifting Infinite Cascades," Douglas Aircraft Company Report LB 31653.

Giles, M. B., 1990a, "Non-reflecting Boundary Conditions for Euler Equation Calculations," *AIAA J.*, Vol. 28, No. 12, p. 2080.

Giles, M. B., 1990b, "Stator/Rotor Interaction in a Transonic Turbine," *J. Propulsion and Power*, Vol. 6, pp. 621–627.

Gillis, J. C., and Johnston, J. P., 1983, "Turbulent Boundary Layer Flow and Structure on a Convex Wall and Its Development on a Flat Wall," *J. Fluid Mech.*, Vol. 135, pp. 123–153.

Gladden, H. J., et al., 1982, "Heat Transfer Results and Operational Characteristics of the NASA Lewis Research Center Hot Section Cascade Facility," ASME 85-GT-82.

Glassman, A. J., 1972, *Turbine Design and Application*, Volumes I, II, and III, NASA SP 290 (reprinted in one volume, 1995).

Goldman, L. J., and Gaugler, R. E., 1980, "Prediction Method for Two Dimensional Aerodynamic Losses of Cooled Turbines Using Integral Boundary Layer Parameters," NASA TP 1623.

Goldman, L. J., and Seaholtz, R. G., 1982, "Laser Anemometer Measurements in Annular Turbine Cascade of Core Turbine Vanes and Comparison with Theory," NASA TP 2018.

Goldstein, R. J., 1971, *Film Cooling Advances in Heat Transfer*, Vol. 7, Academic Press, New York.

Goldstein, R. J., et al., 1985, "Effect of Surface Roughness on Film Cooling Performance," ASME Paper 84-GT-41, *J. Eng. for Power*, Vol. 107.

Gostelow, J. P., 1964, "Potential Flow Through Cascades: A Comparison Between Exact and Approximate Solutions," British Aeronautical Research Council CP 807.

Gostelow, J. P., 1964, "Potential Flow Through Cascades—Extensions to an Exact Theory," British Aeronautical Research Council CP 808.

Gostelow, J. P., 1965, "The Calculation of Incompressible Flow Through Cascades of Highly Cambered Blades," in *Proceedings of a Seminar on Advanced Problems in Turbomachinery, Part II*, VKI, (K. Cassidy and J. Chauvin, eds.), Rhode St. Genese, Belgium, pp. 1–18.

Gostelow, J. P., 1984, *Cascade Aerodynamics*, Pergamon Press, Elmsford, NY.

Goto, A., 1992, "Study of Internal Flows in Mixed Flow Pump Impeller at Various Tip Clearance Using 3D Viscous Flow Calculations," *J. Turbomachinery*, Vol. 114, p. 373.

Goulas, A., and Mealing, B., 1984, "Flow at the Tip of a Forward Curved Centrifugal Fan," ASME Paper 84-GT-222.

Gourlay, R. A., and Morris, J. L., 1972, "Hopscotch Difference Method for Non-Linear Hyperbolic Systems," IBM *J. Res. Dev.*, Vol. 16, No. 6, pp. 349–353.

Govindan, T. R., and Lakshminarayana, B., 1988, "A Space Marching Method for the Computation of Viscous Internal Flows," *Computers and Fluids*, Vol. 16, No. 1, pp. 21–39.

Govindan, T. R., Briley, W. R., and McDonald, H., 1991, "Generalized Three Dimensional Viscous Primary/Secondary Flow Equations," *AIAA J.*, Vol. 29, No. 3, p. 361.

Govindan, T. R., Levy, R., and Shamroth, S. J., 1984, "Computation of the Tip Vortex Generation Process for Ship Propellers," Scientific Research Associates Report R83-920021-F (Hartford, CT) (also see *Proceedings of the 17th Symposium on Naval Hydrodynamics*, The Hague, 1988).

Graziani, R. A., et al., 1980, "An Experimental Study of Endwall and Airfoil Surface Heat Transfer in a Large Scale Turbine Blade Cascade," *J. Eng. Power*, Vol. 102, p. 257.

Greenwood, D. T., 1988, *Principles of Dynamics*, Prentice-Hall, Englewood Cliffs, NJ, p. 49.

Gregory-Smith, D. G., and Cleak, J. G. E., 1992, "Secondary Flow Measurements in a Turbine Cascade with High Inlet Turbulence," *J. Turbomachinery*, Vol. 114, p. 173.

Gregory-Smith, D. G., and Graves, C. P., 1983, "Secondary Flows and Losses in a Turbine Cascade," in *Viscous Effects in Turbomachines*, AGARD CP 351, Paper 17.

Gregory-Smith, D. G., Graves, C. P., and Walsh, J. A., 1988, "Growth of Secondary Losses and Vorticity in a Turbine Cascade," *J. Turbomachinery*, Vol. 110, No. 1, p. 1.

Greitzer, E. M., 1981, "Stability of Pumping System," *J. Fluids Eng.*, Vol. 103, p. 193.

Greitzer, E. M., 1985, "Flow Instabilities in Turbomachines," *Thermodynamics and Fluid Dynamics of Turbomachinery*, Vol. 2 (A. S. Ucer et al., eds.), Martinus Nijhoff, The Hague.

Griffin, L. W., and McConnaughey, H. V., 1989, "Prediction of Aerodynamic Environment and Heat Transfer for Rotor–Stator Configurations," ASME Paper 89-GT-89.

Guderly, K. G., 1962, *Theory of Transonic Flow*, Pergamon Press, Elmsford, NY.

Guidez, J., 1989, "Study of Convective Heat Transfer in Rotating Coolant Channel," *J. Turbomachinery*, Vol. 111, p. 43–50.

Gundy-Burlet, K. L., et al., 1990, "Temporal and Spatial Resolved Flow in a Two-Stage Axial Compressor," ASME Paper 90-GT-299.

Gusakova, E. A., 1960, "Three-Dimensional Flow in Turbine Cascades" (in Russian), Chapter 7, in *Aerodynamics of Blade Cascades* (W. S. Joukowski, ed.), Gosenergoid Inc., Moscow.

Gusakova, E. A., Mikhailova, V. A., and Tyryshkin, V. G., 1960, "Features of the Flow Over the End Parts of Nonshrouded Blades and Their Influence on the Efficiency of a Turbine Stage," *Teploenergetika*, Vol. 17, No. 4.

Haas, J. E., and Kofskey, M. G., 1979, "Effect of Tip Clearance and Configuration on Overall Performance of a 12.77 cm Tip Diameter Axial Flow Turbine," ASME Paper 79-GT-42.

Hadid, A. H., et al., 1988, "Convergence and Accuracy of Pressure Based Finite Difference Schemes for Incompressible Viscous Flow Calculations in a Non-orthogonal Coordinate System," AIAA Paper 88-3529.

Hafez, M., South, J., Murman, E., 1979, "Artificial Compressibility Methods for Numerical Solution of Full Potential Equation," *AIAA J.*, Vol. 17, p. 838.

Hah, C., 1984, "A Navier–Stokes Analysis of Three-Dimensional Turbulent Flows Inside Turbine Blade Rows at Design and Off-Design Conditions," *J. Eng. Power*, Vol. 106, pp. 421–429.

Hah, C., 1986, "A Numerical Modeling of Endwall and Tip Clearance Flow of an Isolated Compressor Rotor," *J. Eng. Power*, Vol. 108, pp. 15–21.

Hah, C., and Krain, H., 1990, "Secondary Flows and Vortex Motion in a High Effiency Backswept Impeller at Design and Off-Design Conditions," *J. Turbomachinery*, Vol. 112, p. 7.

Hah, C., and Lakshminarayana, B., 1980, "The Prediction of Two and Three Dimensional Asymmetric Turbulence Wakes—A Comparison of Performance of Three Turbulence Models for the Effects of Curvature and Rotation," *AIAA J.*, Vol. 18, No. 8, p. 1196.

Hale, A. A., Davis, M. W., and Kneile, K. R., 1994, "Turbine Engine Analysis Compressor Code: TEACC," AIAA Paper 94-0148.

Hall, E. J., 1989, "Simulation of Time Dependent Compressible Viscous Flows Using Central and Upwind Biased Finite Difference Technique," Ph.D. Thesis, Iowa State University.

Hall, E. J., and Delaney, R. A., 1991, "Time-Dependent Aerodynamic Analysis of Ducted and Unducted Prop Fan at Angle of Attack," ASME Paper 91-GT-190.

Hall, E. J., Delaney, R. A., and Pletcher, R. H., 1990, "Simulation of Time Dependent Viscous Flows Using Central and Upwind-Biased Finite Difference Techniques," AIAA Paper 90-3012-CP.

Hall, E. J., Topp, D. A., and Delaney, R. A., 1994, "Aerodynamic and Heat Transfer Analysis of Discrete Site Film Cooled Turbine Airfoils," AIAA paper 94-3070.

Hall, W. S., and Thwaites, B., 1964, "On the Calculation of Cascade Flows," British Aeronautical Research Council CP 806.

Haller, B. R., and Camus, J. J., 1984, "Aerodynamic Loss Penalty Produced by Film Cooling of Transonic Turbine Blades" (ASME Paper 83-GT-77), *J. Eng. Power*, Vol. 106, p. 198.

Han, J. C., Zhang, Y. M., and Lee, C. P., 1994, "Influence of Surface Heating Condition on Local Heat Transfer in a Rotating Square Channel with Smooth Walls and Radial Outward Flow," *J. Turbomachinery*, Vol. 116, p. 149.

Hanjalic, K., and Launder, B. E., 1972, "A Reynolds Stress Model of Turbulence and Its Application to Thin Shear Flows," *J. Fluid Mech.*, Vol. 52, p. 609.

Hannis, J. M., and Smith, M. K. D., 1982, "The Design and Test of Air Cooled Blading for an Industrial Gas Turbine," ASME Paper 82-GT-229.

Hanson, D. B., 1982, "Compressible Lifting Surface Theory for Propeller Performance Calculation," AIAA Paper 82-0200.

Hanson, D. B., 1983, "Compressible Helicoidal Surface Theory for Propeller Aerodynamics and Noise," *AIAA J.*, Vol. 21, p. 881.

Harasgama, S. P., and Burton, C. D., 1992, "Film Cooling Research on Endwall of a Turbine Nozzle Guide Vane in a Short Duration Annular Cascade—Parts 1 and 2," *J. Turbomachinery*, Vol. 114, p. 734.

Harstel, J. E., 1972, "Prediction of Effects of Mass Transfer Cooling on the Blade Row Efficiency of Turbine Aerofoils," AIAA Paper 72-11.

Harvey, W. B., et al., 1982, "Rotor Redesign for a Highly Loaded 1800 ft/sec Tip Speed, Part II, Laser Doppler Velocimeter," NASA CR 167957.

Hathaway, M. D., et al., 1993, "Experimental and Computational Investigation of the NASA Low Speed Centrifugal Compressor Flow Field," *J. Turbomachinery*, Vol. 115, p. 527.

Hauser, C., et al., 1979, "Turbomachinery Technology," NASA CP 2092.

Hawkins, R. C., 1984, "Unducted Fan for Tomorrow's Subsonic Propulsion," *Aerospace America*, p. 52.

Hawthorne, W. R., 1951, "Secondary Circulation in Fluid Flow," *Proc. Roy. Soc. A*, Vol. 206, p. 374.

Hawthorne, W. R., 1955a, "The Growth of Secondary Circulation in Frictionless Flow," *Proceedings Cambridge Philos. Soc.*, Vol. 51, No. 4, p. 737.

Hawthorne, W. R., 1955b, "Some Formulae for the Calculation of Secondary Flows in Cascades," British Aeronautical Research Council Report 17,519.

Hawthorne, W. R., 1964, *Aerodynamics of Turbines and Compressors*, Princeton University Press, Princeton, NJ.

Hawthorne, W. R., 1966, "On the Theory of Shear Flow," MIT Gas Turbine Lab, Report 88.

Hawthorne, W. R., 1974, "Secondary Vorticity in Stratified Compressible Fluids in Rotating Systems," CUED/A-Turbo/TR 63, University of Cambridge, England (also see discussion of paper by B. Lakshminarayana, 1975.)

Hawthorne, W. R., and Horlock J. H., 1962, "Actuator Disc Theory of the Incompressible Flow in Axial Flow Compressors," *Proc. Inst. Mech. Eng.*, Vol. 176, No. 30, p. 789.

Hawthorne, W. R., et al., 1984, "Theory of Blade Design for Large Deflections—Parts I and II," *J. Eng. Power*, Vol. 106, p. 346.

Head, M. R., 1958, "Entrainment in the Turbulent Boundary Layer," British Aeronautical Research Council R & M 3152.

Hearsey, R. M., 1986, "Practical Compressor Aerodynamic Design," in *Advanced Topics in Turbomachinery Technology*, D. Japikse, ed., Concepts ETI, Inc., Norwich, VT.

Heiser, W. H., 1978, "Modern Turbine Technology," in *Turbomachinery Fluid Mechanics*, Iowa State University Lecture Notes.

Hempel, H., and Friedrich, R., 1978, "Profile Loss Characteristics and Heat Transfer for Full Coverage Film Cooled Blading," ASME Paper 78-GT-98.

Herrig, L. J., Emery, J. C., and Erwin, J. R., 1957, "Systematic Two-Dimensional Cascade Tests of NACA 65 Series Compressor Blades at Low Speeds," NACA TN 3916.

Hess, J. L., and Smith, A. M. O., 1966, "Calculation of Potential Flow About Arbitrary Bodies," *Prog. Aero. Sci.*, Vol. 8, pp. 1–138.

Hill, P. G., and Peterson, C. R., 1992, *Mechanics and Thermodynamics of Propulsion*, 2nd ed., Addison-Wesley, Reading, MA.

Hines, A. B., 1982, "Turbulence Modeling," *Aeronaut. J.*, Aug.–Sept., p. 269.

Hinze, J. O., 1975, *Turbulence*, McGraw-Hill, New York.

Hirsch, C., 1974, "End-Wall Boundary Layers in Axial Compressors," *J. Eng. Power*, Vol. 96, pp. 413–426.

Hirsch, C., 1976, "Flow Prediction in Axial Flow Compressors Including End Wall Boundary Layers," ASME Paper 76-GT-72.

Hirsch, C., 1990, *Numerical Computation of Internal and External Flows*, Vols. 1 and 2, John Wiley & Sons, New York.

Hirsch, C., and Dring, R. P., 1987, "Throughflow Models for Mass and Momentum Averaged Variables," ASME Paper 87-GT-52.

Hirsch, C., and Warzee, G., 1976, "A Finite Element Method for Through Flow Calculations in Turbomachines," *J. Fluids Eng.*, Vol. 98, pp. 403–421.

Hirsch, C., and Warzee, G., 1979, "An Integrated Quasi Three-Dimensional Finite Element Calculation Program for Turbomachinery Flows," *J. Eng. Power*, Vol. 101, pp. 141–148.

Hirschel, E. H., and Kordulla, W., 1981, *Shear Flows in Surface Oriented Coordinates*, Notes on Numerical Fluid Mechanics, Vieweg, Germany.

Ho, Y., and Lakshminarayana, B., 1993, "Computation of Unsteady Viscous Flow Through Turbomachinery Blade Row Due to Upstream Rotor Wakes, ASME Paper 93-GT-321 (to be published in *J. Turbomachinery*, Oct. 1995).

Ho, Y., and Lakshminarayana, B., 1994, "Computational Modeling of Three-Dimensional Endwall Flow Through a Turbine Rotor Cascade with Strong Secondary Flows," ASME Paper 94-GT-136 (to be published in *J. Turbomachinery*).

Hobbs, D. E., and Weingold, H. D., 1984, "Development of Controlled Diffusion Airfoils for Multistage Compressor Applications," *J. Eng. for Gas Turbines and Power*, Vol. 106, p. 271.

Hobson, D. E., 1979, "Shock Free Transonic Flow in Turbomachinery Cascade," Cambridge University Report CUED/A Turbo/65 (also as Ph.D. Thesis, Cambridge University).

Hobson, G. V., and Lakshminarayana, B., 1991, "Prediction of Cascade Performance Using an Incompressible Navier–Stokes Technique," *J. Turbomachinery*, Vol. 113, p. 561.

Hodge, R. I., 1958, "A Turbine Nozzle Cascade for Cooling Studies," British Aeronautical Research Council CP 491 and 492.

Hodson, H. P., and Dominy, R. G., 1987, "Three-Dimensional Flow in a Low Pressure Turbine Cascade at Design Condition," *J. Turbomachinery*, Vol. 109, p. 177.

Hodson, H. P., Huntsman, I., and Steele, A. B., 1993, "An Investigation of Boundary Layer Development in a Multistage LP Turbine," ASME Paper 93-GT-310.

Hoffmann, K. A., 1989, *Computational Fluid Dynamics for Engineers*, Engineering Education System (Texas).

Horlock, J. H., 1971, "On Entropy Production in Adiabatic Flow in Turbomachines," ASME *J. Basic Eng.*, Vol. 93, p. 587.

Horlock, J. H., 1973, *Axial Flow Compressors*, Kreiger Publishing Co., Melbourne, FL.

Horlock, J. H., 1973a, *Axial Flow Turbines*, Kreiger Publishing Co., Melbourne, FL.

Horlock, J. H., 1977, *Actuator Disk Theory*, McGraw-Hill, New York.

Horlock, J. H., and Lakshminarayana, B., 1973, "Secondary Flows: Theory, Experiment, and Application in Turbomachinery Aerodynamics," *Ann. Rev. Fluid Mech.*, Vol. 5, pp. 247–280.

Horlock, J. H., and Marsh, H., 1971, "Flow Models for Turbomachines," *J. Mech. Eng. Sci.*, Vol. 13, p. 358.

Horlock, J. H., and Perkins, H. J., 1974, "Annulus Wall Boundary Layers in Turbomchines," AGARDograph No. 185.

Horlock, J. H., et al., 1966, "Wall Stall in Compressor Cascades," *J. Basic Eng.*, Vol. 88, p. 637.

Hosny, W. M., et al., 1985, "Energy Efficient Engine—High Pressure Compressor Component Report," NASA CR 174955, GE Report R85AEB441.

Hourmouziadis, J., Buckl, F., and Bergmann, P., 1987, "Development of the Profile Boundary Layer in a Turbine Ennvironment," *J. Turbomachinery*, Vol. 109, p. 177.

Howard, M. A., Walton, J. H., and Uppington, D. C., 1987, "Computer System for Aerodynamic and Thermal Design of Turbines," Institute of Mechanical Engineers, London, Paper C33/187.

Howell, A. R., 1942, "Present Basis of Axial Flow Compressor Design, Part I, Cascade Theory and Performance," British ARC R & M 2095.

Howell, A. R., 1948, "Note on the Theory of Arbitrary Airfoils in Cascade," *Philos. Mag.*, Vol. 39, pp. 913–927.

Howells, R. W., and Lakshminarayana, B., 1977, "Three Dimensional Potential Flow and Effects of Blade Dihedral in Axial Flow Propeller Pumps," *J. Fluids Eng.*, Vol. 99, p. 167.

Hsu, C. H., Chen, Y. M., and Liu, C. H., 1990, "Preconditioned Upwind Methods to Solve 3D Incompressible Navier Stokes Equations for Viscous Flows," AIAA Paper 90-1496.

Huber, F. W., and Ni, R. H., 1989, "Application of a Multistage 3D Euler Solver to the Design of Turbines for Advanced Propulsion Systems," AIAA Paper 89-2578.

Huber, F. W., et al., 1992, "Design of Advanced Turbopump Drive Turbines for National Launch System Application," AIAA Paper No. 92-3221.

Humble, L. V., Lowdermilk, W. H., and Desmon, L. G., 1951, "Measurement of Average Heat Transfer and Friction Coefficients for Subsonic Flow of Air in Smooth Tubes at High Surface and Fluid Temperatures," NACA Report 1020.

Hunter, I. H., and Cumpsty, N. A., 1982, "Casing Wall Boundary Layer Development Through an Isolated Compressor Rotor," *J. Eng. Power*, Vol. 104, pp. 805–818.

Huntsman, I., and Hodson, H. P., 1994, "An Experimental Assessment of the Aerodynamic Performance of a Low Speed Radial Inflow Turbine," AIAA Paper 94-2932.

Huo, Y., and Wu, W., 1986, "Numerical Solution of Transonic Stream Function Equation on S_1 Stream Surface in Cascade," ASME Paper 86-GT-110.

Hutchings, B., and Iannuzzelli, R., 1987, "Taking the Measure of Fluid Dynamic Software," *Mech. Eng.*, Vol. 109, No. 5, p. 72.

Hylton, L. D., et al., 1981, "Experimental Investigation of Turbine Endwall Heat Transfer," Wright Patterson AFB Report AFWAL-TR-81-2077, Vols. I and II.

Hylton, L. D., et al., 1988, "Effects of Leading Edge and Downstream Film Cooling on Turbine Vane Heat Transfer," NASA CR 182133.

Inoue, M., Kuroumaru, M., and Fukuhara, M., 1986, "Behavior of Tip Leakage Flow Behind an Axial Compressor Rotor," *J. Eng. Power*, Vol. 108, No. 1, pp. 7–14.

Issa, R. I., 1985, "Solution of the Implicitly Discretized Fluid Flow Equations by Operator Splitting," *J. Comput. Physics*, Vol. 62, pp. 40–65.

Ito, S., Eckert, E. R. G., and Goldstein, R. J., 1980, "Aerodynamic Loss in a Gas Turbine Stage with Film Cooling," *J. Eng. Power*, Vol. 102.

Ives, D. C., and Liutermuza, J. F., 1977, "Analysis of Transonic Cascade Flow Using Conformal Mapping and Relaxation Techniques," *AIAA J.*, Vol. 15, p. 647–652.

Iyer, V., and Harris, J., 1989, "Three Dimensional Compressible Boundary Layer Calculations to Fourth Order Accuracy on Wings and Fuselages," AIAA 89-0130.

Jabbari, M. Y., and Goldstein, R. J., 1978a, "Adiabatic Wall Temperature and Heat Transfer Through Two Row Holes," *J. Eng. Power*, Vol. 100, No. 2, pp. 303–307.

Jabbari, M. Y., and Goldstein, R. J., 1978b, "Effect of Mainstream Acceleration on Adiabatic Wall Temperature and Heat Transfer Downstream of Gas Injection," *Heat Transfer 1978*, Vol. 5, Hemisphere Publishing Company, New York, pp. 249–254.

James, P. W., 1987, "Excess Streamwise Vorticity and its Role in Secondary Flow," *Proc. Inst. Mech. Eng.*, Vol. 201, p. 413.

Jameson, A., 1974, "Iterative Solution of Transonic Flow Over Airfoils and Wings Including Flows at Mach 1," *Commun. Pure. Appl. Math*, Vol. 27, pp. 283–309.

Jameson, A., and Baker, T. J., 1984, "Multigrid Solutions of Euler Equations for Aircraft Configurations," AIAA Paper 84-0093.

Jameson, A., and Turkel, E., 1981, "Implicit Schemes and LU Decompositions," *Math. Comput.*, Vol. 37, pp. 385–390.

Jameson, A., Schmidt, W., and Turkel, E., 1981, "Numerical Solutions of the Euler Equations by Finite Volume Methods Using Runga–Kutta Time Stepping Schemes," AIAA Paper 81-1259.

Japikse, D., 1984, *Turbomachinery Diffuser Design Technology*, DTS-1, Concepts ETI, Inc., Norwich, VT.

Jennions, I. K., and Stow, P., 1985, "A Quasi Three-Dimensional Turbomachinery Blade Design System, Parts I and II," *J. Eng. for Gas Turbines and Power*, Vol. 107, pp. 301–316.

Jennions, I. K., and Stow, P., 1986, "Importance of Circumferential Non-Uniformities in a Particle Averaged Quasi Three-Dimensional Turbomachine Design System," *J. Eng. for Gas Turbines and Power*, Vol. 108, pp. 240–245.

Jennions, I. K., and Turner, M. G., 1993, "Three-Dimensonal Navier–Stokes Computations of Transonic Fan Flow Using an Explicit Flow Solver and an Implicit $k\epsilon$ Solver," *J. Turbomachinery*, Vol. 115, p. 249.

Jeracki, R. J., et al., 1979, "Wind Tunnel Performance of Four Energy Efficient Propellers Designed for Mach 0.8 Cruise," SAE Paper No. 790573 (also see NASA TM 82891, 1982).

Johnsen, I. A., and Bullock, R. O. (eds.), 1965, "Aerodynamic Design of Axial Flow Compressors," NASA SP 36.

Johnson, B. V., et al., 1994, "Heat Transfer in Rotating Serpentine Passages with Trips Skewed to the Flow," *J. Turbomachinery*, Vol. 116, p. 113.

Johnson, M. W., and Moore, J., 1980, "The Development of Wake Flow in a Centrifugal Impeller," *J. Eng. Power*, Vol. 102, p. 382.

Johnston, F. T., and Rubbert, P. E., 1975, "Advanced Panel-Type Influence Coefficient Methods Applied to Subsonic Flows," AIAA Paper 75-50.

Johnston, J. P., 1974, "The Effects of Rotation on Boundary Layers in Turbomachine Rotors," NASA SP 304, p. 207.

Johnston, J. P., Halleen, R. M., and Lezius, D. K., 1972, "Effects of Spanwise Rotation on the Structure of Two-Dimensional Fully Developed Turbulent Channel Flow," *J. Fluid Mech.*, Vol. 56, Part 3, pp. 533–557.

Jones, W. P., and Launder, B. E., 1972, "The Prediction of Laminarization with a Two Equation Model of Turbulence," *Int. J. Heat Mass Transfer*, Vol. 15, p. 301.

Jones, W. P., and Launder, B. E., 1973, "The Calculation of Low Reynolds Number Phenomena, with a Two Equation Model of Turbulence," *Int. J. Heat Mass Transfer*, Vol. 16, pp. 1119–1130.

Jorgensen, P. C. E., and Chima, R. V., 1989, "An Unconditionally Stable Runga–Kutta Method for Unsteady Flows," AIAA Paper 89-205.

Joslyn, H. D., et al., 1989, "Multistage Compressor Airfoil Aerodynamics," *J. Propulsion and Power*, Vol. 5, p. 457.

Joukowsky, M. I., 1954, "Calculation of Flow Through a Cascade of Blades and Design of Cascades from a Given Velocity Distribution," (in Russian) *CKTI*, Vol. 27.

Kacker, S. C., and Okapuu, U., 1982, "A Meanline Prediction Method for Axial Flow Turbine Efficiency," *J. Eng. for Power*, Vol. 104, pp. 111–119.

Kamijo, K., and Hirata, K., 1985, "Performance of Small High Speed Cryogenic Pumps," *J. Fluids Eng.*, Vol. 107, p. 197.

Kang, S., Lin, F., and Wang, Z., 1989, "A Method for Calculating Axial Turbomachine End Wall Turbulent Boundary Layers," ASME Paper 89-GT-15.

Kantrowitz, A., 1950, "The Supersonic Axial-Flow Compressor," NACA Report 974.

Karamachetti, K., 1966, *Principles of Ideal-Fluid Aerodynamics*, John Wiley & Sons, New York.

Karimipanath, M. T., and Olsson, 1993, "Calculation of Three Dimensional Boundary Layers on Rotor Blades Using Integral Method," *J. Turbomachinery*, Vol. 115, p. 342.

Katsanis, T., 1966, "Use of Arbitrary Quasi Orthogonals for Calculating Flow Distribution in a Turbomachine," *J. Eng. Power*, Vol. 88.

Katsanis, T., 1968, "Computer Program for Calculating Velocities and Streamlines on a Blade-to-Blade Stream Surface of a Turbomachine," NASA TND 4525 (also see NASA TND 5427, 1969).

Katsanis, T., 1969, "Fortran Program for Calculating Transonic Velocities on a Blade-to-Blade Surface of a Turbomachine," NASA Technical Note D-5427.

Katsanis, T., and McNally, W. D., 1977, "Revised Fortran Program for Calculating Velocities and Streamlines on the Hub-Shroud Mid-Channel Flow Surface of an Axial- or Mixed-Flow Turbomachine, Vols. I and II," NASA TN D 8430 and TN D 8431 (also see NASA TND 2546, 1964).

Katz, J., and Plotkin, A., 1991, *Low Speed Aerodynamics*, McGraw-Hill, New York.

Kays, W. M., and Crawford, M. E., 1993, *Convective Heat and Mass Transfer*, McGraw-Hill, New York.

Keffer, J. F., and Bains, W. D., 1963, "The Round Turbulent Jet in a Cross Wind," *J. Fluid Mech.*, Vol. 15, No. 4, p. 481.

Keller, H. B., 1970, "A New Difference Scheme for Parabolic Problems," in *Numerical Solution of Partial Differential Equations*, Vol. 2 (J. Bramble et al., eds.), Academic Press, New York.

Keller, H. B., 1978, "Numerical Methods in Boundary Layer Theory," *Ann. Rev. Fluid Mech.*, Vol. 10, pp. 417–433.

Keller, H. B., and Cebeci, T., 1972, "Accurate Numerical Methods for Boundary Layer Flows," *AIAA J.*, Vol. 10, pp. 1193–1199.

Kellogg, O. D., 1953, *Foundations of Potential Theory*, Dover, New York.

Kerrebrock, J. L., 1977, *Aircraft Engines and Gas Turbines*, MIT Press, Cambridge, MA.

Kershaw, D. S., 1978, "The Incomplete Conjugate Gradient Modeling for the Iterative Solution of Linear Equations," *J. Comput. Physics*, Vol. 26.

Kim, J., and Simon, T. W., 1988, "Measurement of the Turbulent Transport of Heat and Momentum on a Convexly Curved Boundary Layers," *J. Eng. Power*, Vol. 110, p. 80.

Kiock, R., et al., 1985, "The Boundary Layer Behavior of an Advanced Gas Turbine Rotor Blade Under the Influence of Simulated Film Cooling," AGARD CP 390.

Kirtley, K., 1987, "A Coupled Parabolic Marching Method for the Prediction of Three-Dimensional Viscous Turbomachinery Flows," Ph.D. Thesis, Department of Aerospace Engineering, Pennsylvania State University.

Kirtley, K., and Lakshminarayana, B., 1988, "Computation of Three Dimensional Turbulent Flow Using a Coupled Parabolic Marching Method," *J. Turbomachinery*, Vol. 110, p. 549.

Kirtley, K., Warfield, M., and Lakshminarayana, B., 1986, "A Comparison of Computational Methods for Three Dimensional Turbulent Turbomachinery Flows," AIAA Paper 86-1599 (also see *J. Propulsion*, Vol. 4, No. 3, 1988, p. 20).

Kirtley, K., Beach, T., and Adamczyk, A., 1990, "Numerical Analysis of Secondary Flow in a Two Stage Turbine," AIAA Paper 90-2356.

Klein, A., 1966, "Untersuchungen über die Einfluss der Zuströmgrenzschicht auf die Sekundärströmung in den Beschaufelungen von Axialturbinen," *Forsch. Ing.*, Bd 32, Nr 6.

Kline, S. J., Cantwell, B. J., and Lilley, G. M. (eds.), 1982, *Complex Turbulent Flows*, Vols. I, II, and III, Stanford University.

Knapp, R. T., Daily, J. W., and Hammitt, F. G., 1970, *Cavitation*, McGraw-Hill, New York.

Knight, C. J., and Choi, D., 1989, "Development of a Viscous Cascade Code Based on Scalar Implicit Factorization," *AIAA J.*, Vol. 27, No. 5, p. 581.

Koch, C. C., 1981, "Stalling Pressure Rise Capability of an Axial Compressor Stage," *J. Eng. Power*, Vol. 103, pp. 645–656.

Koch, C. C., Smith, L. H., 1976, "Loss Sources and Magnitudes in Axial Flow Compressor," *J. Engineering Power*, Vol. 98, pp. 411–424.

Kofskey, M. G., and Nusbaum, W. J., 1967, "Performance Evaluation of a Two-Stage Axial Flow Turbine for Two Values of Tip Clearance," NASA TND 4388.

Kohler, H., et al., 1977, "Hot Cascade Test Results of Cooled Turbine Blades and Their Application to Actual Engines," AGARD CP 227.

Kollen, O., and Koschel, W., 1985, "Effect of Film Cooling on the Aerodynamic Performance of a Turbine Cascade," AGARD CP 390.

Kollmann, W. (ed.), 1980, *Prediction Methods for Turbulent Flows*, Hemisphere Publishing Co., New York.

Kooper, F. C., et al., 1981a, "High Pressure Turbine Supersonic Cascade Technology Report," NASA CR 165567.

Kooper, F. C., et al., 1981b, "Experimental Investigation of Endwall Profiling in a Turbine Vane Cascade," *AIAA J.*, Vol. 19, pp. 1033–1044.

Korn, A., 1899, *Textbook on Potential Theory*, Berlin.

Kost, F. H., and Holmes, A. T., 1985, "Aerodynamic Effect of Coolant Ejection in the Rear Part of Transonic Rotor Blades," AGARD CP 390.

Koyama, H., et al., 1979, "Stabilizing and Destabilizing Effects of Coriolis Force on Two Dimensional Laminar and Turbulent Boundary Layers," *J. Eng. Power*, Vol. 101, p. 23.

Krain, H., 1988, "Swirling Impeller Flow," *J. Turbomachinery*, Vol. 110, pp. 122–128.

Kraft, H., 1958, "Development of a Laminar Wing Type Turbine Bucket," *Z. Angew. Math. Phys.*, p. 404.

Krimmerman, Y., and Adler, D., 1978, "The Complete Three Dimensional Flow Field in Turbo Impellers," *J. Mech. Eng. Sci.*, Vol. 20, No. 3, p. 149.

Krishnamoorthy, V., 1982, "Effect of Turbulence on Heat Transfer in a Laminar and Turbulent Boundary Layer Over a Gas Turbine Blade," ASME Paper No. 82-GT-146.

Kruse, F., 1977, "Investigation of Temperature Distribution Near Film Cooled Airfoils," AGARD CP 229.

Kruse, H., 1985, "Effect of Hole Geometry, Wall Curvature, and Pressure Gradient on Film Cooling Downstream of a Single Row," AGARD CP 390.

Kulisa, P., LeBoeuf, F., and Perrin, G., 1992, "Computation of Wall Boundary Layer with Discrete Jet Injections," *J. Turbomachinery*, Vol. 114, p. 756.

Kulisa, P., 1989, "Etude Theorique Des Ecoulements Visqueux en Presence D'injections Parietales en vue du Refroidissement des Turbines," Doctoral Thesis, Ecole Centrale de Lyon, France.

Kumada, M., et al., 1994, "Tip Clearance Effects on Heat Transfer and Leakage Flows on the Shroud Wall Surface in an Axial Flow Turbine," *J. Turbomachinery*, Vol. 116, p. 39.

Kunz, R., and Lakshminarayana, B., 1992a, "Stability of Explicit 3D Navier Stokes Procedures Using $k\epsilon$ and $k\epsilon$/Algebraic Reynolds Stress Model," *J. Comput. Phys.*, Vol. 103, p. 141.

Kunz, R., and Lakshminarayana, B., 1992b, "Navier Stokes Ivestigation of a Transonic Centrifugal Compressor Stage Using an Algebraic Reynolds Stress Model," AIAA Paper 92-3311.

Kunz, R., and Lakshminarayana, B., 1992c, "Explicit Navier–Stokes Computation of Cascade Flow Using the $k\epsilon$ Turbulence Model," *AIAA J.*, Vol. 30, No. 1, pp. 13–22.

Kunz, R., and Lakshminarayana, B., 1992d, "Three-Dimensional Navier Stokes Computation of Turbomachinery Flows Using an Explicit Numerical Procedure and a Coupled $k\epsilon$-Turbulence Model," *J. Turbomachinery*, Vol. 114, p. 627.

Kunz, R., Lakshminarayana, B., and Basson, A., 1993, "Investigation of Tip Clearance Phenomena in an Axial Compressor Cascade Using Euler and Navier–Stokes Procedures," *J. Turbomachinery*, Vol. 115, No. 3, p. 453.

Kuo, C. R., and Hwang, G. J., 1994, "Aspect Ratio Effect on Convective Heat Transfer of Radially Outward Flow in Rectangular Rotating Ducts," *Intl. J. Rotating Machinery*, Vol. 1, No. 1, p. 1.

Kwak, D., et al., 1986, "A Three-Dimensional Incompressible Navier–Stokes Flow Solver Using Primitive Variables," *AIAA J.*, Vol. 24, pp. 390–398.

Kynast, G., 1960, "Anwendung und Vergleich Verschiedener Verfahren Zur Berechnung Der Druckverteilung an Schaufelprofilen inkompressibler Strömung," DFL-Bericht 062.

Lai, Y. G., and So, R. M. C., 1990, "Near Wall Modeling of Turbulent Heat Fluxes," *Int. J. Heat and Mass Transfer*, Vol. 33, p. 1429.

Lakomy, C., 1971, "Theoretical Solution of the Boundary Problems of Subsonic Aerodynamics," *Appl. Sci. Res.*, Vol. 24, pp. 403–421.

Lakomy, C., 1974, "Theoretical Solution of High Subsonic Flow Past Two-Dimensional Cascades of Airfoils," ASME Paper 74-GT-91.

Lakshminarayana, B., 1970, "Methods of Predicting the Tip Clearance Effects in Axial Flow Turbomachines," *J. Basic Eng.*, Vol. 92, pp. 467–482.

Lakshminarayana, B., 1974, "Experimental and Analytical Investigation of flow Distribution in Rocket Pump Inducers," *Fluid Mechanics, Acoustics and Design of Turbomachines*, NASA SP 304, Part II, p. 689.

Lakshminarayana, B., 1975, "The Effects of Inlet Temperature Gradients on Turbomachinery Performance," *J. Eng. Power*, Vol. 97, pp. 64–74.

Lakshminarayana, B., 1978, "On the Shear Pumping Effect in Rocket Pump Inducers," *Pumps; Analysis, Design and Application* (U.S. Library of Congress Card No. 78-50501), Worthington Pump Inc., p. 49–68.

Lakshminarayana, B., 1980, "An Axial Flow Research Compressor Facility Designed for Flow Measurement in Rotor Passages," *J. Fluids Eng.*, Vol. 102, pp. 402–411.

Lakshminarayana, B., 1982, "Fluid Dynamics of Non-cavitating Inducers—A Review," *J. Fluids Eng.*, Vol. 104, pp. 411–427.

Lakshminarayana, B., 1986, "Turbulence Modelling for Complex Shear Flows," *AIAA J.*, Vol. 24, No. 12, pp. 1900–1917.

Lakshminarayana, B., 1991, "An Assessment of Computational Fluid Dynamics Techniques in the Analysis and Design of Turbomachinery," *J. Fluids Eng.*, Vol. 113, p. 315.

Lakshminarayana, B., and Davino, R., 1980, "Mean Velocity and Decay Characteristics of the Wake of an Axial Flow Compressor Guide Vane and Stator Blade," *J. Eng. Power*, Vol. 102, p. 50.

Lakshminarayana, B., and Gorton, C. A., 1977, "Three-Dimensional Flow Field in Rocket Pump Inducers—Part II: 3-D Viscid Flow Analysis and Hot Wire Data on 3-D Mean Flow and Turbulence Inside the Rotor Passage," *J. Fluids Eng.*, Vol. 99, Series I, No. 1, pp. 176–186.

Lakshminarayana, B., and Govindan, T. R., 1981, "Analysis of Turbulent Boundary Layer on Cascade and Rotor Blades of Turbomachinery," *AIAA J.*, Vol. 19, pp. 1333–1341.

Lakshminarayana, B., and Horlock, J. H., 1962, "Tip-Clearance Flow and Losses for an Isolated Compressor Blade," British ARC R & M 3316.

Lakshminarayana, B., and Horlock, J. H., 1963, "Review: Secondary Flows and Losses in Cascades and Axial-Flow Turbomachines," *Int. J. Mech. Sci.*, Vol. 5, pp. 287–307.

Lakshminarayana, B., and Horlock, J. H., 1967a, "Effects of Shear Flows on the Outlet Angle in Axial Compressor Cascades—Methods of Prediction and Correlation with Experiments," *J. Basic Eng.*, Vol. 89, pp. 191–200.

Lakshminarayana, B., and Horlock, J. H., 1967b, "Leakage and Secondary Flows in Compressor Cascades," British ARC, R & M 3483.

Lakshminarayana, B., and Horlock, J. H., 1973, "Generalized Secondary Vorticity Expressions Using Intrinsic Coordinates," *J. Fluid Mech.*, Vol. 59, Part 1, pp. 97–115. (Also see Corrigenda, *J. Fluid Mech.*, Vol. 226, pp. 661–663, 1991.)

Lakshminarayana, B., and Murthy, K. N. S., 1988, "Laser Doppler Velocimeter Measurement of Annulus Wall Boundary Layer Development in a Compressor Rotor," *J. Turbomachinery*, Vol. 110, pp. 377–385.

Lakshminarayana, B., and Popovski, P., 1987, "Three Dimensional Boundary Layer on a Compressor Rotor Blade at Peak Pressure Rise Coefficient,"*J. Turbomachinery*, Vol. 109, pp. 91–98.

Lakshminarayana, B., and Reynolds, B., 1980, "Turbulence Characteristics in the Near Wake of a Compressor Rotor Blade," *AIAA J.*, Vol. 18, pp. 1354–1362.

Lakshminarayana, B., and Sitaram, N., 1984, "Wall Boundary Layer Development Near the Tip Region of an IGV of an Axial Flow Compressor," *J. Eng. Power*, Vol. 106, pp. 337–345.

Lakshminarayana, B., Jabbari, A., and Yamaoka, H., 1972, "Turbulent Boundary Layer on a Rotating Helical Blade," *J. Fluid Mech.*, Vol. 51, p. 545.

Lakshminarayana, B., Pouagare, M., and Davino, R., 1982, "Three-Dimensional Flow Field in the Tip Region of a Compressor Rotor Passage—Parts I and II," *J. Eng. Power*, Vol. 104, No. 4, pp. 760–781.

Lakshminarayana, B., et al., 1983, "Annulus Wall Boundary Layer Development in a Compressor Stage Including the Effects of Tip Clearance," AGARD CP 351.

Lakshminarayana, B., Sitaram, N., and Zhang, J., 1986, "Endwall and Profile Losses in a Low Speed Axial Flow Compressor Rotor," *J. Eng. for Gas Turbines and Power*, Vol. 108, pp. 22–31.

Lakshminarayana, B., Zhang, J., and Murthy, K. N. S., 1990, "An Experimental Study on the Effects of Tip Clearance on Flow Field and Losses in an Axial Flow Compressor Rotor,"*Z. Flugwissenschaften Weltraumforschung*, Vol. 14, pp. 273–281.

Lakshminarayana, B., et al., 1994, "experimental Investigation of the Flow Field in a Multistage Axial Flow Compressor," ASME Paper 94-GT-455.

Lakshminarayana, B., Zaccaria, M., and Marathe, B., 1995, "The Structure of Tip Clearance Flow in an Axial Flow Compressor," *Proceedings of the 10th ISABE Meeting*, Nottingham, England; *J. Turbomachinery*, July 1995.

Lamb, H., 1945, *Hydrodynamics*, 6th ed., Dover, New York, p. 224.

Lane, J. M., 1981, "Cooled Radial Inflow Turbines for Advanced Gas Turbine Engines," ASME Paper 81-GT-213.

Langston, L. S., 1980, "Crossflow in a Turbine Cascade Passage," *J. Eng. Power*, Vol. 102, p. 866.

Langston, L. S., Nice, M. L., and Hooper, R. M., 1977, "Three-Dimensional Flow Within a Turbine Passage," *J. Eng. Power*, Vol. 99, pp. 21–28.

Large, G. D., and Meyer, L. J., 1982, "Cooled Variable Area Radial Turbine," NASA CR 165408.

Launder, B. E., 1971, "An Improved Algebraic Stress Model of Turbulence," Mechanical Engineering Department Report TM/TN/A8, Imperial College, London.

Launder, B. E., 1988, "On the Computation of Convective Heat Transfer in Complex Turbulent Flows," *J. Heat Transfer*, Vol. 110, p. 112.

Launder, B. E., Reece, B. J., and Rodi, W., 1975, "Progress in the Development of a Reynolds Stress Turbulence Closure," *J. Fluid Mech.*, Vol. 68, Part 3, pp. 537–566.

Launder, B. E., Reynolds, W. C., Rodi, W., 1984, *Turbulence Models and Their Applications*, Vol. 2, Eyrolles, Paris.

Law, C. H., and Puterbaugh, S., 1988, "Parametric Blade Study Test Report—Rotor Configuration 6," AFWAL (Wright Aeronautical Laboratory) TR-88-2112.

Law, C. H., and Wadia, A. R., 1993, "Low Aspect Ratio Transonic Rotor—Parts 1 and 2," *J. Turbomachinery*, Vol. 115, pp. 218–226.

Law, C. H., and Wennerstrom, A. J., 1986, "Performance of Two Transonic Axial Flow Compressor Rotors Incorporating Inlet Swirl," ASME Paper 86-GT-31.

Lawaczeck, O. K., 1972, "Calculation of the Flow Properties Up- and Downstream of and Within a Supersonic Turbine Cascade," ASME Paper 72-GT-47.

Lax, P. D., and Wendroff, B., 1960, "Systems of Conservation Laws," *Commun. Pure Appl. Math.*, Vol. 13, pp. 217–237.

Lax, P. D., and Wendroff, B., 1964, "Difference Schemes for Hyperbolic Equations with High Order Accuracy," *Commun. Pure Appl. Math.*, Vol. 17, pp. 381–398.

Leach, K. P., 1983, "Energy EfficientEngine—High Pressure Turbine Component Rig Performance Test," NASA CR 168189, PWA-5594-243.

Leboeuf, F., 1984, "Secondary Flows and End Wall Boundary Layers in Axial Turbomachines," Von Karman Institute LS 84-05.

Leboeuf, F., and Naviere, H., 1985, "Etudes Experimentales et Theoriques des Couches Visqueuses Parietales Dans Un Compresseur Mono-Etage Transsonique," AGARD CP 351.

Leboeuf, F., et al., 1982, "Experimental Study and Theoretical Prediction of Secondary Flows in a Transonic Axial Flow Compressor," ASME Paper 82-GT-14.

Lee, C. P., and Han, J. C., 1983, "Application of Adiabatic Cooling Effectiveness to Non-adiabatic Blade Cooling Design," ASME Paper 83-JPGC-GT-1.

LeGrives, E., 1978, "Mixing Process Induced by the Vorticity Associated with the Penetration of the Jet into a Cross Flow," *J. Eng. Power*, Vol. 100, pp. 465–475.

LeGrives, E., 1986, "Cooling Techniques for Modern Gas Turbines," Chapter 4 in *Topics in Turbomachinery Technology* (D. Japikse, ed.), Concepts ETI, Inc., Norwich, VT.

LeGrives, E., and Nicolas, J. J., 1977, "Method Nouvelle de Calcul de L'Efficacitede Refroidissement Des Aubes de Turbine Par Film D'Air," AGARD CP 229.

LeMeur, A., 1988, "Three Dimensional Unsteady Flow Computation in a Transonic Axial Turbine Stage," ONERA (France), TP No. 1988-2.

Leonard, B. P., 1979, "A Stable and Accurate Convective Modeling Procedure Based on Quadratic Upstream Interpolation," *Computer Methods Appl. Mech. Eng.*, Vol. 19, pp. 59–98.

Leonard, D., 1990, "Subsonic and Transonic Cascade Design," Von Karman Institute, LS 90-8.

Levine, P., 1957, "The Two Dimensional Inflow Conditions for a Supersonic Compressor with Curved Blades," *J. Appl. Mech.*, Vol. 24, pp. 165–169.

Levy, S., 1954, "Effect of Large Temperature Changes (Including Viscous Heating) upon Laminar Boundary Layers with Variable Free Stream Velocity," *J. Aero. Sci.*, Vol. 21, p. 459.

Lewis, R., and Hill, J. M., 1971, "The Influence of Sweep and Dihedral in Turbomachinery Blade Rows," *J. Mech. Eng. Sci.*, Vol. 13, No. 4.

Lewis, R. I., 1982, "A Method for Inverse Aerofoil and Cascade Design by Surface Vorticity," ASME Paper 82-GT-154.

Leylek, J. H., and Wisler, D. C., 1991, "Mixing in Axial Flow Compressors: Conclusions Drawn from 3D Navier–Stokes Analysis and Experiments," *J. Turbomachinery*, Vol. 113, p. 139.

Leylek, J. H., and Zerkle, R. D., 1994, "Discrete Jet Film Cooling: A Comparison of Computational Results with Experiments," *J. Turbomachinery*, Vol. 116, p. 358.

Lichtfuss, H. J., and Starken, H., 1972, "Supersonic Exit Flow of Two Dimensional Cascades," ASME Paper 72-GT-49.

Lichtfuss, H. J., and Starken, H., 1974, "Supersonic Cascade Flow," *Prog. Aeronaut. Sci.*, Vol. 15, pp. 37–149.

Lieblein, S., 1965, "Experimental Flow in Two-Dimensional Cascade," in *Aerodynamic Design of Axial Flow Compressors*, NASA SP 36.

Lieblein, S., and Stockman, N. O., 1972, "Compressibility Correction for Internal Flow Solutions," *J. Aircraft*, Vol. 9, No. 4, pp. 312–313.

Liepmann, H. W., and Roshko, A., 1957, *Elements of Gas Dynamics*, John Wiley & Sons, New York.

Liess, C., 1973, "Film Cooling with Ejection from a Row of Inclined Circular Holes," Von Karman Institute TN 97.

Ligrani, P. M., and Mitchell, S. W., 1992, "Interactions Between Embedded Vortices and Injectant from Film Cooling Holes with Compound Angle Orientations in a Turbulent Boundary Layer," ASME Paper 92-GT-199 (*J. Turbomachinery*, Vol. 116, p. 80, 1994).

Liu, D. K., and Ji, L. J., 1988, "Calculation of Complete Three-Dimensional Flow in a Centrifugal Rotor with Splitter Blades," ASME Paper 88-GT-93.

Liu, H. C., et al., 1979, "An Application of 3D Viscous Flow Analysis to the Design of Low Aspect Ratio Turbine," ASME Paper 79-GT-53.

Liu, J. S., Sockol, P. M., and Prahl, J. M., 1989, "Navier–Stokes Cascade Analysis with a Stiff $k\epsilon$ Turbulence Solver," AIAA Paper 88-0594.

Liu, J. S., Zedan, M., and Buzzola, R., 1991, "A Comparison Between Two Three-Dimensional Codes and Experimental Results for High Work Turbine," AIAA Paper 91-0341.

Liu, X., and Rodi, W., 1992, "Measurement of Unsteady Flow and Heat Transfer in a Linear Turbine Cascade," ASME Paper 92-GT-323 (to be published in *J. Turbomachinery*).

Lokai, V. I., and Limanski, A. S., 1975, "Influence of Rotation on Heat and Mass Transfer in Radial Cooling Passages of Turbine Blades," *Izvestiya Vuz*, Aviatsionnaya Teknika, Vol. 18, No. 3, p. 69.

Lordi, J. A., and Homicz, G. F., 1981, "Linearized Analysis of the Three Dimensional Compressible Flow Through a Rotating Blade Row," *J. Fluid Mech.*, Vol. 103, pp. 413–442.

Louis, J. F., 1977, "Systematic Study of Heat Transfer and Film Cooling Effectiveness," AGARD CP 229.

Lowery and Vachon, 1975, "The Effect of Turbulence and Heat Transfer from Heated Cylinders," *Intl. J. Heat and Mass Transfer*, Vol. 18, p. 1229.

Ludwieg, H., and Tillman, W., 1949, "Untersuchungen uber die Wandschubspannung in Turbulenten Reibungsschichten," *Ing. Arch.*, Vol. 17, p. 288.

Lumley, J. L., 1970, *Stochastic Tools in Turbulence*, Academic Press, New York.

Lumley, J. L., 1980, *Prediction Methods for Turbulent Flows* (W. Kollmann, ed.), Hemisphere Publishing Company, New York.

Luo, J., and Lakshminarayana, B., 1995a, "Navier Stokes Analysis of Turbine Flow Field and Heat Transfer," *J. Propulsion and Power*, Vol. 11, No. 2, p. 221.

Luo, J., and Lakshminarayana, B., 1995b, "Prediction of Strongly Curved Turbulent Duct Flows with Reynolds Stress Model," AIAA Paper 95-2241.

Lyman, F. A., 1993, "On the Conservation of Rothalpy in Turbomachinery," *J. Turbomachinery*, Vol. 115, p. 520.

MacCormack, R. W., 1969, "The Effect of Viscosity in Hypervelocity Impact," AIAA Paper 69-345.

MacCormack, R. W., and Baldwin, B. S., 1975, "A Numerical Method for Solving the Navier–Stokes Equations with Application to Shock–Boundary Layer Interactions," AIAA Paper 75-1.

Mager, A., 1952, "Generalization of Boundary Layer Momentum Integral Equations Including Those of Rotating Systems," NACA TR 1067.

Marathe, B., Lakshminarayana, B., and Dong, Y., 1994, "Experimental and Numerical Investigation of Stator Exit Flow Field of an Automotive Torque Converter," ASME Paper 94-GT-32 (also see ASME 95-GT-231 and 95-GT-232).

Marble, F., 1964, "Three-Dimensional Flow in turbomachines," in *Aerodynamics of Turbines and Compressors*, W. R. Hawthorne, ed., Vol. 10, Princeton University Press, Princeton, NJ.

Marchal, P., and Sieverding, C. H., 1977, "Secondary Flows Within Turbomachinery Bladings," AGARD CP 214, Paper 11.

Marchant, R. D., Howe, D. C., and Williams, M. C., 1989, "Energy EfficientEngine High-Pressure Compressor Performance Report," NASA CR 182219, pp. 1–176.

Marsh, H., 1968, "A Computer Program for the Through Flow Fluid Mechanics in an Arbitrary Turbomachine Using a Matrix Method," British ARC R & M 3509.

Martelli, F., and Boretti, A., 1985, "A Simple Procedure to Compute Losses in Transonic Turbine Cascades," ASME Paper 85-GT-21.

Martensen, E., 1959, "The Calculation of the Pressure Distribution on a Cascade of Thick Airfoils by Means of Fredholm's Integral Equations of the Second Kind," NASA TT F 702, 1971. (Original article in German: Max Planck Institute for Flow Research Report No. 23, 1959. Also see *Arch. Rat. Mech. Anal.*, Vol. 3, pp. 235–270, 1959.)

Martinelli, L., 1987, "Calculation of Viscous Flows with Multigrid Methods," Ph.D. Thesis, MAE Department, Princeton University.

Marvin, J. G., 1983, "Turbulence Modeling for Computational Aerodynamics," *AIAA J.*, Vol. 21, No. 7, p. 941.

Mavriplis, F., 1971, "Aerodynamic Research on High Lift Systems," *Can. Aeronaut. Space J.*, pp. 175–184.

Mayle, R. E., et al., 1976, "Effect of Streamline Curvature on Film Cooling," *J. Heat Transfer*, Vol. 98, p. 240.

Mayle, R. E., 1991, "The Role of Laminar-Turbulent Transition in Gas Turbine Engines," *J. Turbomachinery*, Vol. 113, p. 509.

Mazher and Giddens, 1991, "New Pressure Correction Equation for Incompressible Internal Flows," *AIAA J.*, Vol. 29, No. 3, p. 418.

McCormick, B. W., 1979, *Aerodynamics, Aeronautics and Flight Mechanics*, John Wiley & Sons, New York.

McDonald, P. W., 1971, "The Computation of Transonic Flow Through Two-Dimensional Gas Turbine Cascades," ASME Paper 71-GT-89.

McDonald, P. W., 1981, "A Comparison Between Measured and Computed Flow Fields in a Transonic Compressor Rotor," *J. Eng. for Power*, Vol. 104.

McDonel, J. D., and Eiswerth, J. E., 1977, "Effects of Film Injection on Performance of a Cooled Turbine," AGARD CP 229.

McFarland, E. R., 1982, "Solution of Plane Cascade Flow Using Surface Singularity Method," *J. Eng. for Power*, Vol. 104, No. 3, pp. 668–674.

McFarland, E. R., 1984, "A Rapid Blade-to-Blade Solution for Use in Turbomachinery Design," *J. Eng. Power*, Vol. 106, No. 2, p. 376.

McIntyre, E. A., Jr., 1976, "Design of Transonic Cascade by Conformal Transformation Methods," Courant Institute Math Science Report COO-3077-136, New York.

McNally, W. D., and Sockol, P. M., 1981, "Computational Methods for Internal Flows with Emphasis on Turbomachinery," NASA TM82764 (condensed version in *J. Fluids Eng.*, Vol. 107, 1985).

Mee, D. J., et al., 1992, "An Examination of the Contributions to Loss on a Transonic Turbine Blade in a Cascade," *J. Turbomachinery*, Vol. 114, pp. 155–162 (also see pp. 163–172 in Vol. 114 for boundary layer data by the same authors).

Mehmel, D., 1962, "Clearance Flow in Cascades," *Ing. Arch.*, Vol. 31, p. 294.

Mellor, G. L., and Balsa, T. F., 1972, "The Prediction of Axial Flow Compressor Performance with Emphasis on the Effect of Annulus Wall Boundary Layers," AGARDograph No. 164, p. 367.

Mellor, G. L., and Herrig, H. J., 1973, "A Survey of Mean Turbulent Field Closure Methods," *AIAA J.*, Vol. 11, p. 590.

Mellor, G. L., and Wood, G. M., 1971, "An Axial Compressor Endwall Boundary Layer Theory," *J. Basic Eng.*, Vol. 93D, p. 300.

Merkle, C. M., and Tsai, P. Y. L., 1986, "Application of Runga–Kutta Schemes in Incompressible Flows," AIAA Paper 86-0553.

Metzger, D. E., 1985, "Cooling Techniques for Gas Turbine Airfoils—A Survey," AGARD CP 390.

Metzger, D. E., and Afgan, N. H. (ed.), 1984, *Heat and Mass Transfer in Rotating Machinery*, Hemisphere Publishing Company, New York.

Metzger, D. E., et al., 1981, "Developing Heat Transfer in Rectangular Ducts with Arrays of Short Fins," ASME Paper 81-WA/HT-6.

Metzger, F. B., and Rohrbach, C., 1979, "Aeroacoustic Design of the Prop Fan," AIAA Paper 79-0610.

Milne-Thompson, L. M., 1960, *Theoretical Hydrodynamics*, Macmillan, New York.

Minassian, L. M., 1975, "A Study of Multi-element Cascades and Airfoils," ASME Paper 75-WA/FE3.

Mizuki, S., et al., 1980, "Prediction of Jet and Wake Flow within Centrifigal Impeller Channel," in *Performance Prediction of Centrifugal Compressors and Pumps*, ASME.

Mochizuki, S., et al., 1994, "Heat Transfer in Serpentine Flow Passages with Rotation," *J. Turbomachinery*, Vol. 116, p. 133.

Moeckel, W. E., 1949, "Approximated Method for Predicting Form and Location of Detached Shock Waves Ahead of Plane or Axially Symmetrical Bodies," NACA TN 1921.

Monson, D. J., Seegmiller, H. L., and McConnaughey, P. K., 1989, "Comparison of LDV Measurements and Navier–Stokes Solutions in a Two-Dimensional 180° Turn Around Duct," AIAA Paper 89-0275.

Moore, J. G., 1985, "An Elliptic Procedure for 3D Viscous Flow," in *3D Computation Techniques Applied to Internal Flows in Propulsion Systems*, AGARD LS-140.

Moore, J., and Moore, J. G., 1980, "Three-Dimensional Viscous Flow Calculations for Assessing the Thermodynamic Performance of Centrifugal Compressors," AGARD CP 282.

Moore, J., and Moore, J. G., 1981, "Calculations of Three Dimensional Viscous Flow and Wake Development," *J. Eng. Power*, Vol. 103, pp. 367–372.

Moore, J. and Moore, J. G., 1985, "Performance Evaluation of Linear Turbine Cascade Using Three-Dimensional Viscous Flow Calculation," *J. Eng. for Gas Turbines and Power*, Vol. 107, p. 969–975.

Moore, J. and Ransmayr, A., 1984, "Flow in a Turbine Cascade—Part I: Losses and Leading Edge Effects," *J. Eng. for Gas Turbines and Power*, Vol. 106, No. 2, pp. 400–408.

Moore, J., Moore, J. G., and Timmis, P. H., 1984, "Performance Evaluation of Centrigual Compressor Impellers Using Three-Dimensional Viscous Flow Calculations," *J. Eng. for Gas Turbines and Power*, Vol. 106, p. 475.

Moore, J., LeFur, T., and Moore, J. G., 1990, "Computational Study of 3D Turbulent Air Flow in a Helical Rocket Pump Inducer," AIAA Paper 90-2123.

Moore, R. D., 1982, "Rotor Tip Clearance Effects on Overall Blade Element Performance of Axial Flow Transonic Fan Stage," NASA TP 2049.

Moore, R. D., and Reid, L., 1982, "Performance of Single Stage Axial Flow Transonic Compressor with Rotor and Stator Aspect Ratio of 1.63 and 1.77 Respectively, and with Design Pressure Ratio of 2.05," NASA TP 2001.

Mori, Y., Fukuda, T., and Nakayama, W., 1971, "Convective Heat Transfer in a Rotating Radial Circular Cylinder," *Intl. J. Heat Mass Transfer*, Vol. 14, p. 1807.

Morris, A. L., et al., 1972, "High Loading 1800 ft/sec Tip Speed Transonic Compressor Fan Stage, Phase I: Aerodynamic and Mechanical Design," NASA CR 120907.

Morris, A. W. H., 1977a, "Experimental Evaluation of a Transpiration Cooled Nozzle Guide Vane," AGARD CP 229.

Morris, W. D., 1977b, "Flow and Heat Transfer in Rotating Coolant Passages," AGARD CP 229.

Morris, W. D., and Ayhan, T., 1979, "Observations on the Influence of Rotation on Heat Transfer in the Coolant Channels of Gas Turbine Rotor Blades," *Proc. Inst. Mech. Eng.*, Vol. 193, No. 21, p. 303 (also see AGARD CP 390, 1985).

Moss, R. W., and Oldfield, M. L. G., 1992, "Measurement of the Effect of Free Stream Turbulence Length Scale on Heat Transfer," ASME Paper 92-GT-244.

Moustapha, S. H., et al., 1987, "Influence of Rotor Blade Aerodynamic Loading on the Performance of a Highly Loaded Turbine Stage," *J. Turbomachinery*, Vol. 109, pp. 155–162.

Moustapha, S. H., Carscallen, W. E., and McGreachy, J. D., 1993, "Aerodynamic Performance of a Transonic Low Aspect Ratio Turbine Nozzle," *J. Turbomachinery*, Vol. 113, p. 400.

Muller, U. R., 1982, "Measurement of the Reynolds Stresses and the Mean-Flow Field in a Three Dimensional Pressure Driven Boundary Layer," *J. Fluid Mech.*, Vol. 119, pp. 121–150.

Mulloy, J. M., and Weber, H. G., 1982, "A Radial Inflow Turbine Impeller for Improved Performance," ASME Paper 82-GT-101.

Murthy, K. N. S., and Lakshminarayana, B., 1986, "Laser Doppler Velocimeter Measurement in Tip Region of a Compressor Rotor," *AIAA J.*, Vol. 24, No. 5, pp. 807–814.

Murthy, K. N. S., and Lakshminarayana, B., 1987, "The Hubwall Boundary Layer Development and Losses in an Axial Flow Compressor Rotor Passage," *Z. Flugwissenschaften Weltraumforschung*, Vol. 11, pp. 1–11.

Murthy, T. K. S., and Brebbia, C. A., eds., 1990, *Computational Methods in Viscous Aerodynamics*, Elsevier, New York.

Nagano, Y., and Kim, C., 1988, "A Two Equation Model for Heat Transport in Wall Turbulent Shear Flows," *J. Heat Transfer*, Vol. 110, p. 583.

Nakahashi, K., 1989, "Navier–Stokes Computation of Two- and Three-Dimensional Cascade Flow Fields," *J. Propulsion and Power*, Vol. 5, No. 3, p. 320.

Namba, M., 1974, "Lifting Surface Theory for Rotating Subsonic or Transonic Blade Row," British ARC R & M 3740.

Narasimha, R., 1985, "The Laminar-Turbulent Zone in the Boundary Layer," *Prog. Aerospace Sci.*, Vol. 22, p. 29.

Nash, J. F., and Patel, V. C., 1972, *Three Dimensional Turbulent Boundary Layers*, Scientific and Business Consultants, Inc., Atlanta, GA.

Nealy, D. A., et al., 1984, "Measurement of Heat Transfer Distribution Over the Surfaces of Highly Loaded Nozzle Guide Vane," *J. Eng. Power*, Vol. 106, p. 149 (also see NASA CR 168015, November 1982; and NASA CR 174827, 1985).

Ni, R. H., 1982, "A Multiple Grid Scheme for Solving the Euler Equation," *AIAA J.*, Vol. 20, No. 11, pp. 1565–1571.

Ni, R. H., and Bogoian, J., 1989, "Prediction of 3-D Multi-stage Turbine Flow Field Using a Multiple Grid Euler Solver," AIAA Paper 89-0203.

Ni, R. H., and Sharma, O., 1990, "Using 3D Euler Simulations to Assess Effects of Periodic Unsteady Flow Through Turbines," AIAA Paper 90-2357.

Nirmalan, V., and Hylton, L. D., 1990, "An Experimental Study of Turbine Vane Heat Transfer with Leading Edge and Downstream Film Cooling," *J. Turbomachinery*, Vol. 112, p. 447.

Nojima et al., 1988, "Development of Aerodynamic Design System for Centrifugal Compressors," Mitsubishi Heavy Industries, Technical Review, Vol. 25, No. 1.

Norton, J. M., et al., 1989, "Energy EfficientEngine Hollow Fan Blade Technology, Vol. 1. Shroudless High Aspect Ratio Fan Rig, Part I: Design and Performance Test," NASA CR 182220, pp. 1–186; "Part II: Laser Doppler Velocimeter Data," NASA CR 182221, pp. 1–269.

Novak, O., 1967, "Flow in the Entrance Region of a Supersonic Cascade," *Storjnicky Cusopis* XIX C-2-3 Seite 138.

Novak, R. A., 1976, "Flow Field and Performance Map Computation for Axial Flow Compressors and Turbines," *Modern Prediction Methods for Turbomachine Performance*, AGARD LS-83.

Novak, R. A., and Hearsey, R. M., 1977, "A Nearly Three Dimensional Intrablade Computing System for Turbomachinery, Parts I & II," *J. Fluids Eng.*, Vol. 99, p. 154–166.

Oates, G. C., 1972, "Actuator Disc Theory for Incompressible Highly Rotating Flows," *J. Basic Eng.*, Vol. 94, p. 613.

Oates, G. C., 1984, *Aerothermodynamics of Gas Turbines and Rockets*, AIAA Educational Series, p. 336.

Okapuu, V., 1974, "Some Results from Tests on High Work Axial Gas Generator Turbine," ASME Paper 74-GT-81.

Okurounmu, O., and McCune, J. E., 1974, "Lifting Surface Theory of Axial Compressor Blade Rows," *AIAA J.*, Vol. 12, pp. 1363–1380.

Oldfield, M. L. G., et al., 1981, "Boundary Layer Studies on Highly Loaded Cascades Using Heated Thin Films and a Traversing Probe," *J. Turbomachinery*, Vol. 103, p. 237.

Osher, S., and Solomon, F., 1982, "Upwind Schemes for Hyperbolic Systems of Conservation Laws," *Math. Comput.*, Vol. 38, p. 239.

Oswatitsch, K., 1955, "Uber Die Stromung In einem Uber Schallgitter," *Allg. Warmtechnik*, Vol. 6, pp. 9–11.

Oswatitsch, K., and Rhyming, I., 1957, "Uber den Kompressibilitätseinfluss bei ebenen Schaufelgittern starker umlenkung," DVL (Deutsche Versuchsanstalt Für Luft-und-Raumfahrt, Germany)-Bericht 28.

Özisik, M. N., 1985, *Heat Transfer—A Basic Approach*, McGraw-Hill, New York.

Pal, P., 1965, "Untersuchungen über den Interferenzeinfluss bei Strömungen durch Tandem-Schaufelgitter," *Ing. Arch. 34*, Vol. 3, pp. 173–193.

Pampreen, R. C., 1993, *Compressor Stall and Surge*, Concepts, Inc. (Library of Congress Card No. 92-70348).

Pandya, A., and Lakshminarayana, B., 1983, "Investigation of the TIp-Clearance Flow Inside and at the Exit of a Compressor Rotor Passage—Part I: Mean Velocity Field," and "Part II: Turbulence Properties," *J. Eng. Power*, Vol. 105, pp. 1–17.

Panton, R. L., et al., 1980, "Flight Measurements of a Wing Tip Vortex," *J. Aircraft*, Vol. 17, p. 250.

Patankar, S. V., 1980, *Numerical Heat Transfer and Fluid Flow*, Hemisphere Publishing Co., New York.

Patankar, S. V., 1988, "Recent Developments in Computational Heat Transfer," *J. Heat Transfer*, Vol. 110, p. 1037.

Patankar, S. V., and Spalding, D. B., 1970, *Heat and Mass Transfer in Boundary Layers*, Intertext.

Patankar, S. V. and Spalding, D. B., 1972, "A Calculation Procedure for Heat, Mass, and Momentum Transfer in Three-Dimensional Parabolic Flows," *Int. J. Heat Mass Transfer*, Vol. 15, pp. 1787–1806.

Patel, V. C., Rodi, W., and Scheurer, G., 1985, "A Review and Evaluation of Turbulence Models for Near Wall and Low Reynolds Number Flows," *AIAA J.*, Vol. 23, pp. 1308–1319.

Peaceman, D. W., and Rachford, H. H., 1959, "The Numerical Solution of Parabolic and Elliptic Differential Equations," *J. Soc. Ind. Appl. Math*, Vol. 3, pp. 28–41.

Pederson, D. R., Eckert, E. R. G., and Goldstein, R. J., 1977, "Film Cooling with Large Density Difference Between the Mainstream and the Secondary Fluid," *J. Heat Transfer*, Vol. 99, pp. 620–627.

Peyeret, R., and Taylor, T. D., 1983, *Computational Methods for Fluid Flow*, Springer-Verlag, New York.

Pierzga, M. J., and Wood, J. R., 1985, "Investigations of the Three Dimensional Flow Field Within Transonic Compressor Rotor: Experiment and Analysis," *J. Eng. for Gas Turbines and Power*, Vol. 107, pp. 436–449.

Pietrzyk, J. R., Bogard, D. G., and Crawford, M. E., 1990, "Effects of Density Ratio on the Hydrodynamics of Film Cooling," *J. Turbomachinery*, Vol. 112, pp. 437–443 (also see Vol. 111, pp. 139–145, 1989).

Platzer, M. F., and Carta, F. O. (eds.), 1987, *AGARD Manual on Aeroelasticity of Turbomachines, Vol. 1, Unsteady Turbomachinery Aerodynamics*, AGARD-AG-298.

Pollard, D., and Wordsworth, 1962, "A Comparison of Two Methods for Predicting the Potential Flow Around Arbitrary Airfoils in Cascades," British ARC CP 619.

Popovski, P., and Lakshminarayana, B., 1986, "Laser Anemometer Measurements in a Compressor Rotor Flow Field at Off-Design Conditions," *AIAA J.*, Vol. 24, pp. 1337–1345.

Pouagare, M., and Lakshminarayana, B., 1982, "Development of Secondary Flow and Vorticity in Curved Ducts, Cascades, and Rotors, Including Effects of Viscosity and Rotation," *J. Fluids Eng.*, Vol. 104, pp. 505–512.

Pouagare, M., and Lakshminarayana, B., 1983, "Computation and Turbulence Closure Models for Shear Flows on Rotating Curved Bodies," in *Proceedings of the 4th Turbulent Shear Flow Symposium* (L. J. S. Brabury et al., eds)., Springer-Verlag, New York.

Pouagare, M., and Lakshminarayana, B., 1986, "A Space Marching Method for Viscous Incompressible Internal Flows," *J. Comput. Phys.*, Vol. 64, No. 2, pp. 389–414.

Pouagare, M., Murthy, K. N. S., and Lakshminarayana, B., 1983, "Three-Dimensional Flow Field Inside the Passage of a Low Speed Compressor Rotor," *AIAA J.*, Vol. 21, No. 12, p. 1679.

Pouagare, M., Galmes, J. M., and Lakshminarayana, B., 1985, "An Experimental Study of the Compressor Rotor Blade Boundary Layer," *J. Eng. for Gas Turbines and Power*, Vol. 107, pp. 364–373.

Povinelli, L., 1984, "Assessment of Three-Dimensional Inviscid Codes and Loss Correlations for Turbine Aerodynamics," ASME Paper 84-GT-187.

Prager, W., 1928, "The Pressure Distribution on Bodies in Plane Potential Flow," *Physik. Z.*, Vol. 29, pp. 865–869.

Pratap, V. S., and Spalding, D. B., 1976, "Fluid Flow and Heat Transfer in Three-Dimensional Duct Flows," *Int. J. Heat Mass Transfer*, Vol. 19, pp. 1183–1188.

Prato, J., and Lakshminarayana, B., 1993, "Experimental Investigation of Rotor Wake Structure at Peak Pressure Rise Coefficientand Effects of Loading," *J. Turbomachinery*, Vol. 115, No. 3, p. 487.

Priddy, W. J., and Bayley, F. J., 1988, "Turbulence Measurements in Turbine Blade Passages and Implications for Heat Transfer," *J. Turbomachinery*, Vol. 110, p. 73.

Prust, H. W., Jr., 1972, "Analytical Study of the Effect of Coolant Flow Variables on the Kinetic Energy Output of a Cooled Blade Row," AIAA Paper 72-12.

Prust, H. W., Jr., 1978, "Two Dimensional Cold Air Cascade Study of Film Cooled Turbine Stator Blade," Part IV, NASA TP 1151, Part V, NASA TP 1204.

Pulliam, T. H., and Steger, J. L., 1978, "On Implicit Finite Difference Simulations of Three Dimensional Flow," AIAA Paper 78-10 (published in *AIAA Journal*, Feb. 1980).

Pulliam, T. H., 1986, "EfficientSolution Methods for the Navier–Stokes Equations," in *Numerical Techniques for Viscous Flow Computations in Turbomachinery Bladings*, VKI LS, 1986-02.

Rai, M. M., 1987, "Navier–Stokes Simulations of Rotor/Stator Interaction Using Patched and Overlaid Grids," *J. Propulsion and Power*, Vol. 3, No. 5, pp. 387–396.

Rai, M. M., 1989, "Unsteady Three-Dimensional Navier–Stokes Simulations of Turbine Rotor–Stator Interaction, Parts I and II," *J. Propulsion and Power*, Vol. 5, No. 3, p. 305.

Rai, M. M., and Chakravarthy, S. R., 1986, "An Implicit Form for the Osher Upwind Scheme," *AIAA J.*, Vol. 24, No. 5, p. 735.

Rains, D. A., 1954, "Tip Clearance Flows in Axial Flow Compressors and Pumps," California Institute of Technology, Hydrodynamics and Mechanical Engineering Laboratories, Report No. 5.

Raithby, G. D., 1976, "Skew Upstream Differencing Schemes for Problems Involving Fluid Flow," *Computer Methods Appl. Mech. Eng.*, Vol. 9, pp. 153–164.

Ramshaw, J. D., and Mousseau, V. A., 1990, "Accelerated Artificial Method for Steady State Incompressible Flow Calculations," *Computers and Fluids*, Vol. 18, pp. 361–367.

Raukhman, B. S., 1971, "Flow of an Incompressible Fluid Past a Cascade of Profiles on an Axisymmetric Stream Surface in a Layer of Variable Thickness," Izvestiya Akademii Nauk SSSR, Mekhanika Zhidkosti i Gaza, pp. 83–89.

Ravindranath, A., and Lakshminarayana, B., 1980, "Mean Velocity and Decay Characteristics of the Near and Far Wake of a Moderately Loaded Compressor," *J. Eng. Power*, Vol. 102, No. 3, pp. 535–548.

Ravindranath, A., and Lakshminarayana, B., 1981, "Structure and Decay Characteristics of Turbulence in the Near and Far Wake of a Moderately Loaded Compressor Rotor Blade," *J. Eng. Power*, Vol. 103, No. 1, pp. 131–140.

Rehbach, J., 1960, "Investigation of Clearance Flow in Turbine Cascades," (in German), *Forsch. Ing. Wes.*, Vol. 26, p. 83.

Reiss, W., and Blocker, U., 1987, "Possibilities for On-Line Surge Suppression by Fast Guide Vane Adjustment in Axial Compressors," AGARD CP 421.

Renken, J., 1976, "Calculation of 3D Cascade Flow by Means of a First Order Panel Method," Deutsche Luft- und Raumfahrt (DLR), Report DLR-FB 76-64.

Reynolds, B., and Lakshminarayana, B., 1979, "Characteristics of Lightly Loaded Fan Rotor Blade Wakes," NASA CR-3188, pp. 1–180.

Rhie, C. M., 1985, "A Three Dimensional Passage Flow Analysis Method Aimed at Centrifugal Compressors," *Computers and Fluids*, Vol. 13, No. 4, pp. 443–460.

Rhie, C. M., 1986, "A Pressure Based Navier–Stokes Solver Using the Multigrid Method," AIAA Paper 86-0207.

Rhie, C. M., and Chow, W. L., 1983, "Numerical Study of the Turbulent Flow Past an Airfoil With Trailing Edge Separation," *AIAA J.*, Vol. 21, pp. 1525–1532.

Rhie, C. M., and Stowers, S. T., 1987, "Navier–Stokes Analysis for High Speed Flows Using Pressure Correction Algorithm," AIAA Paper 87-1980.

Rhie, C. M. et al., 1994, "Advanced Transonic Fan Design Procedure based on Navier–Stokes Method," *J. Turbomachinery*, Vol. 116, p. 291.

Rice, I. G., 1979, "Steam Cooled Blading in Combined Reheat Gas Turbine Reheat Turbine Cycle," Part 1 and 2, ASME 79-GPC-GT 2 and 3 (see also 83-GT-85, 83-GT-86).

Riegels, F. W., 1961, "Fortschritte in der Berechnung der Strömung durch Schaufelgitter," *Z. Flugwissenschaften*, 9, pp. 2–15.

Roache, P., 1972, *Computational Fluid Dynamics*, Hermosa Publishers, Albuquerque, New Mexico.

Roberts, W. B., Serovy, G. K., and Sandercock, D. M., 1988, "Design Point Variation of Three-Dimensional Losses and Deviation for Axial Compressor Middle Stages," *J. Turbomachinery*, Vol. 110, p. 426.

Robertson, J. M., 1965, *Hydrodynamics in Theory and Application*, Prentice-Hall, Englewood Cliffs, NJ.

Robinson, C. J., 1992, "Endwall Flows and Blading Design for Axial Flow Compressors," VKI Lecture Series 1992-02.

Rodgers, C., 1980, "Efficiencyof Centrifugal Impellers," AGARD CP 282, Paper 22.

Rodi, W., 1976, "A New Algebraic Relation for Calculating Reynolds Stresses," *ZAMM*, Vol. 56, p. 219.

Rodi, W., 1982, "Examples of Turbulence Models for Incompressible Flows," *AIAA J.*, Vol. 20, p. 872.

Rodi, W., and Scheuerer, G., 1983, "Calculaton of Curved Shear Layers with Two Equation Turbulence Models," *Phys. Fluids*, Vol. 26, pp. 1422–1436.

Rogers, S. E., Chang, J. L. C., and Kwak, D., 1987, "A Diagonal Algorithm for the Method of Pseudo-compressibility," *J. Comput. Phys.*, Vol. 73, pp. 364–379.

Rogers, S. E., Kwak, D., and Kiris, C., 1989, "Numerical Solution of Incompressible Navier–Stokes Equations for Steady State and Time Dependent Problems," AIAA Paper 89-0463 (also see NASA TM 102183).

Rohlik, H. E., 1983, "Current and Future Technology Trends in Radial and Axial Gas Turbines," NASA TM 83414.

Rosen, R., and Facey, J. R., 1987, "Civil Propulsion Technology for the Next Twenty Five Years," *Proceedings of the International Symposium on Air Breathing Engines (ISABE)*, Cincinnati, OH, AIAA.

Rued, K. P., and Wittig, S., 1985, "Freestream Turbulence and Pressure Gradient Effects on Heat Transfer and Boundary Layer Development on Highly Cooled Surfaces," *J. Eng. Gas Turbines and Power*, Vol. 107, pp. 54–59.

Rued, K., and Wittig, S., 1986, "Laminar and Transitional Boundary Layer Structures in Accelerating Flow with Heat Transfer," ASME Paper 86-GT-97.

Sabersky, R. H., and Acosta, A. J., 1964, *Fluid Flow*, Macmillan, New York, p. 294.

Sakurai, T., 1975, "Flow Separation and Performance of Decelerating Channels for Centrifugal Turbomachines," *J. Eng. Power*, Vol. 97, pp. 388–394.

Sanz, J. M., 1983, "Design of Supercritical Cascades with High Solidity," *AIAA J.*, Vol. 21, No. 9, pp. 1289–1293.

Sanz, J. M., 1987, "Lewis Inverse Design Code," NASA TP 2676 (also see *AIAA J.*, Vol. 22, p. 950, 1984).

Schiff, L. B., and Steger, J. L., 1980, "Numerical Simulation of Steady Supersonic Viscous Flow," *AIAA J.*, Vol. 18, No. 12, pp. 1421–1430.

Schilke, P. W., and DeGeorge, C. L., 1983, "Testing and Evaluation of a Water Cooled Gas Turbine Nozzle," ASME Paper 83-GT-240.

Schlichting, H., 1955, "Berechnung der reibungslosen inkompressiblen Strömung für ein vorgegebenes ebenes Schaufelgitter," *VDI Forschungsheft*, 447, pp. 1–35.

Schlichting, H., 1979, *Boundary Layer Theory*, 7th ed., McGraw-Hill, New York.

Schlichting, H., and Truckenbroadt, E., 1979, *Aerodynamics of the Airplane*, McGraw-Hill, New York.

Schmidt, J. F., et al., 1984, "Redesign and Cascade Tests of a Supercritical Controlled Diffusion Stator Blade Section," AIAA Paper 84-1207.

Scholz, N., 1965, *Aerodynamik der Schaufelgitter*, Band I, Verlag G. Braun, Karlsruhe (translated by A. Klein, AGARDograph No. AG 220, 1977).

Schonung, B., and Rodi, W., 1987, "Prediction of Film Cooling by a Row of Holes with a Two Dimensional Boundary Layer Procedure," ASME Paper 87-GT-122.

Schreiber, H. A., 1988, "Experimental Investigations on Shock Losses of Transonic and Supersonic Compressor Cascades," AGARD CP-401, Paper No. 11.

Schreiber, H. A., and Starken, H., 1984, "Experimental Cascade Analysis of a Transonic Compressor Rotor Blade Section," *J. Eng. for Gas Turbines and Power*, Vol. 106, p. 288.

Schreier, S., 1982, *Compressible Flow*, Wiley-Interscience, New York.

Schulz, N., and Gallus, H., 1988, "Experimental Investigation of Three-Dimensional Flow in an Annular Cascade," *J. Turbomachinery*, Vol. 110, No. 4, p. 46.

Schulz, N., Gallus, H., and Lakshminarayana, B., 1990, "Three-Dimensional Separated Flow Field in the Endwall Region of an Annular Compressor Cascade in the Presence of Rotor–Stator Interaction, Parts I & II," *J. Turbomachinery*, Vol. 112, p. 669.

Schwab, J. R., 1982, "Aerodynamic Performance of High Turning Core Turbine Vanes in a Two Dimensional Cascade," AIAA Paper 82-1288, also as NASA TM 82894.

Schweitzer, J. K., and Fairbanks, J. W., 1981, "18:1 Pressure Ratio Axial/Centrifugal Compressor Demonstration Program," AIAA Paper 81-1479.

Scott, J. N., 1985, "Numerical Solution of the Navier–Stokes Equations for 3-D Internal Flows, An Emerging Capability," AGARD LS 140.

Sears, W. R., 1960, *Small Perturbation Theory*, Princeton University Press, Princeton, NJ.

Sehra, A. K., and Kerrebrock, 1981, "Blade-to-Blade Effects on Mean Flow in Transonic Compressors," *AIAA J.*, Vol. 19, pp. 476–483.

Senoo, Y., and Ishida, M., 1986, "Pressure Loss Due to the Tip Clearance of Impeller Blades in Centrifugal and Axial Blowers," *J. Eng. for Gas Turbines and Power*, Vol. 108, pp. 32–37.

Senoo, Y., and Ishida, M., 1987, "Deterioration of Compressor Performance Due to Tip Clearance of Centrifugal Impellers," *J. Turbomachinery*, Vol. 109, p. 55.

Serovy, G. K., 1978, "Axial Flow Compressor Aerodynamics," Chapter 17 in *Aerothermodynamics of Aircraft Gas Turbine Engines*, U.S. Air Force Report AFA APL R78-52 (also published in AIAA Educational Series).

Shapiro, A. H., 1953, *The Dynamics and Thermodynamics of Compressible Fluid Flow*, Vols. 1 & 2, Ronald Press, New York.

Sharma, O. P., and Butler, 1987, "Predictions of Endwall Losses and Secondary Flows in Axial Flow Turbine Cascades," *J. Turbomachinery*, Vol. 109, p. 229.

Sharma, O. P., et al., 1982, "Low Pressure Turbine Subsonic Cascade Technology Report," NASA CR 165592.

Sharma, O. P., et al., 1985, "Three-Dimensional Unsteady Flow in an Axial Turbine," *J. Propulsion and Power*, Vol. 1, p. 29.

Shaw, G. J., and Sivaloganathan, S., 1988, "On the Smoothing Properties of the SIMPLE Pressure-Correction Algorithm," *Int. J. for Numerical Methods in Fluids*, Vol. 8, pp. 441–461.

Sheih, L. F., and Delaney, R. A., 1987, "An Accurate and EfficientEuler Solver for Three Dimensional Turbomachinery Flows," *J. Turbomachinery*, Vol. 109.

Shepherd, D. G., 1956, *Principles of Turbomachinery*, Macmillan, New York.

Sherman, F. S., 1990, *Viscous Flow*, McGraw-Hill, New York.

Sherstyuk, A. N., and Kosmin, V. M., 1966, "Meridional Profiling of Vaneless Diffusers," *Thermal Eng.*, Vol. 13, pp. 64–69.

Sherstyuk, A. N., and Sekolov, A. I., 1969, "The Effect of Slope of Vaneless Diffuser Walls on Characteristics of Mixed Flow Compressor," *Thermal Eng.*, Vol. 16, pp. 116–121.

Shih, T. H., and Lumley, J. L., 1986, "Second Order Closure Modelling of Near Wall Turbulence," *J. Phys. Fluids*, Vol. 29, pp. 971–975.

Shih, T. H., Lumley, J. L., and Janicka, J., 1987, "Second Order Modelling of Variable-Density Mixing Layer," *J. Fluid Mech.*, Vol. 108, p. 93.

Shirakura, M., 1972, "Potential Flow About Arbitrary Thick Blades of Large Camber in Cascade," *Proceedings of the Second International JSME Symposium on Fluidics and Fluid Machines*, Tokyo, Vol. 1, pp. 71–82.

Shyy, W., 1994, *Computational Modeling for Fluid Flow and Interfacial Transport*, Elsevier Science Publishing, New York.

Sieverding, C., 1976, "Etude Experimentale de al Pression de Culot dans des Grilles de Turbine Transsonique," Von Karman Institute, CR 1976-21.

Sieverding, C. H., 1983, "The Influence of Trailing Edge Ejection on the Base Pressure in Transonic Turbine Cascades," *J. Eng. for Power*, Vol. 105, p. 215.

Sieverding, C. H., 1985, "Recent Progress in an Understanding of Basic Aspects of Secondary Flow in Turbine Blade Passages," *J. Eng. for Gas Turbines and Power*, Vol. 107, p. 248.

Sieverding, C. H., and Wilputte, P. H., 1981, "Influence of Mach Number and Endwall Cooling on Secondary Flows in a Straight Nozzle Cascade," *J. Eng. Power*, Vol. 103, p. 257.

Simon, H., and Bohn, D., 1974, "Experimental Investigations of a Recently Developed Supersonic Compressor Stage," ASME Paper 74-GT-116.

Simon, H., Wallman, T., and Monk, T., 1987, "Improvements in Performance Characteristics of Single Stage and Multistage Centrifugal Compressors with Simultaneous Adjustment of Inlet Guide Vanes and Diffuser Vanes," *J. Turbomachinery*, Vol. 109, p. 41.

Simon, T. W., and Moffat, R. J., 1985, "Turbulent Boundary Layer Heat Transfer Experiments: A Separate Effects Study on a Convexly Curved Wall," *J. Heat Transfer*, Vol. 105, No. 4, pp. 835–840.

Simoneau, R. J., and Simon, F. F., 1993, "Progress Towards Understanding and Predicting Heat Transfer in the Turbine Gas Path," *Intl. J. Heat Fluid Flow*, Vol. 14, p. 106.

Simpson, R. L., 1985, "Two Dimensional Separated Flows," AGARDograph 287.

Singh, U. K., 1982, "Computation and Comparison with Measurements of Transonic Flow in an Axial Compressor Stage with Shock Boundary Layer Interactions," *J. Eng. Power*, Vol. 104, pp. 510–515.

Sinha, A., Bogard, D. G., and Crawford, M. E., 1991a, "Film Cooling Effectiveness Downstream of a Single Row of Holes with Variable Density Ratio," *J. Turbomachinery*, Vol. 113, p. 442.

Sinha, A., Bogard, D. G., and Crawford, M. E., 1991b, "Gas Turbine Cooling Flow Field Due to a Second Row of Holes," *J. Turbomachinery*, Vol. 113, p. 450.

Sitaram, N., and Lakshminarayana, B., 1983, "Endwall Flow Characteristics and Overall Performance of an Axial Flow Compressor Stage," NASA CR 3671.

Sjolander, S. A., and Cao, D., 1994, "Measurements of the Flow in an Idealized Tip Gap," ASME Paper 94-GT-74.

Smith, A. G., 1957, "On the Generation of the Streamwise Component of Vorticity for Flows in Rotating Passages," *Aeronautical Quarterly*, Vol. 8, p. 369.

Smith, D. J. L., and Frost, D. H., 1970, "Calculation of the Flow Past Turbomachine Blades," in *Axial and Radial Turbomachinery*, Vol. 2, Institute of Mechanical Engineers (London), paper No. 27, p. 72 (also see NASA SP304, 1974).

Smith, G. D., 1978, *Numerical Solution of Partial Differential Equations: Finite Difference Methods*, Clarendon Press, Oxford.

Smith, L. H., 1955, "Secondary Flow in Axial-Flow Turbomachinery," *Trans. ASME*, Vol. 77, pp. 1065–1076.

Smith, L. H., 1968, "The Radial Equilibrium Equation of Turbomachinery," *J. Eng. Power*, Vol. 88, pp. 1–12.

Smith, L. H., 1970, "Casing Boundary Layers in Multistage Compressors," in *Flow Research on Blading*, (L. S Dzung, ed.), Elsevier Publishing Co., New York.

Smith, L. H., 1987, "Unducted Fan Aerodynamic Design," *J. Turbomachinery*, Vol. 109, pp. 313–324.

Smith, P. D., 1982, "Numerical Computation of Three Dimensional Boundary Layers," *Three Dimensional Turbulent Boundary Layers* (H. H. Fernolz and E. Krause, eds.), Springer-Verlag, New York.

Smith, R. E., 1983, "Three Dimensional Algebraic Grid Generation," AIAA Paper 83-1904.

Snyder, P. H., and Roelke, R. J., 1990, "Design of an Air Cooled Metallic High Temperature Radial Turbine," *J. Propulsion and Energy*, Vol. 6, No. 3, p. 283.

Sobieczky, H., and Dulikravich, G. S., 1982, "A Computation Method for Transonic Turbomachinery Cascades," ASME Paper 82-GT-117.

Soderberg, O. E., 1958, "Secondary Flow and Losses in a Compressor Cascade," MIT Gas Turbine Lab Report, 46.

Sommer, T. P., et al., 1993, "Near Wall Variable Prandtl Number Turbulence Model for Compressible Flow," *AIAA J.*, Vol. 31, p. 27.

Sorensen, R., 1980, "A Computer Program to Generate Two-Dimensional Grids About Airfoils and Other Shapes by the Use of the Poisson Equation," NASA TM 81198.

Sovran, G., and Klomp, E. D., 1967, "Experimentally Determined Optimum Geometry for Rectilinear Diffusers with Retangular, Conical or Annular Cross Section," in Sovran (ed.)," *Fluid Mechanics of Internal Flow*, Elsevier, Amsterdam.

Spalding, D. B., 1972, "A Novel Finite Difference Formulation for Differential Expressions Involving Both First and Second Derivatives," *Int. J. Num. Methods in Eng.*, Vol. 4, pp. 551–559.

Spalding, D. B., 1980, "A Mathematical Modeling of Fluid Dynamics, Heat Transfer and Mass Transfer Processes," Imperial College (London) Report HTS/8011.

Speidel, L., 1954, "Berechnung der Stromungsverluste von Ungestaffelten ebenen Schaufelgittern," *Ing. Arch.*, Vol. 22, pp. 295–322.

Spence, D. A., and Beasley, J. A., 1960, "The Calculation of Lift Slopes, Allowing for Boundary Layer, with Application to RAE 101 & 104 Airfoils," British ARC R & M 3137.

Spencer, E. A., 1945, "Performance of an Axial Flow Pump," *Proc. Inst. Mech. Eng.* (*Lond*), Vol. 143.

Squire, H. B., and Winter, K. G., 1951, "The Secondary Flow in a Cascade of Airfoils in Non-uniform Stream," *J. Aeronaut. Sci.*, Vol. 18, p. 271.

Starken, H., 1971, "Untersuchung der Stromung in ebenen Uberschallverzogerungsgittern," DLR FB, 71-99.

Starken, H., 1978, "Transonic and Supersonic Flows in Cascades," Lecture Notes, ASME/Iowa State University Turbomachinery Institute.

Starken, H., 1993, "Basic Fluid Dynamic Boundary Conditions of Cascade Wind Tunnels," in *Advanced Methods of Cascade Testing*, AGARD AG 328.

Starken, H., Zhong, Y., and Schreiber, H. A., 1985, "Mass Flow Limitation of Supersonic Bladerows Due to Leading Edge Blockage," *Int. J. Turbo and Jet Engines*, Vol. 2, p. 285.

Stauter, R. C., 1993, "Measurement of the Three-Dimensional Tip Region Flow Field in an Axial Compressor," *J. Turbomachinery*, Vol. 115, p. 468.

Stauter, R. C., Dring, R. P., and Carta, F. O., 1991, "Temporally and Spatially Resolved Flow in a Two-Stage Axial Compressor, Part I—Experiment," *J. Turbomachinery*, Vol. 113, pp. 219–226.

Steger, J., 1978, "Implicit Finite Difference Simulation of Flow About Arbitrary Geometries With Application to Airfoils," *AIAA J.*, Vol. 16, No. 7, pp. 679–686.

Stephen, H. E., and Hobbs, D. E., 1979, "Design and Performance Evaluation of Supercritical Air Foils for Axial Flow Compressors," U.S. Navy, NAVAIR Report FT-11455.

Stephens, H. E., 1978, "Application of Supercritical Airfoil Technology for Compressor Cascades: Comparison of Experimental and Theoretical Results," AIAA Paper 78-1138.

Stodola, A., 1924, *Steam And Gas Turbines*, Springer, New York.

Stone, H. L., 1968, "Iterative Solution of Implicit Approximation of Multidimensional Partial Differential Equations," *SIAM J. Num. Anal.*, Vol. 5, pp. 530–555.

Stow, P., 1989, "Blading Design for Multistage HP Compressors," AGARD LS 167.

Strazisar, A., 1985, "Investigation of Flow Phenomena in a Transonic Fan Rotor Using Laser Anemometry," *J. Turbomachinery*, Vol. 107, p. 427 (also see Strazisar et al., NASA TP 2879, 1989).

Stuart-Mitchell, R. W., and Andries, J., 1978, "Heat Transfer Characteristics of the Closed Syphon System," AGARD CP 229.

Subramanian, S. V., 1989, "Three Dimensional Multi-Grid Navier–Stokes Computations for Turbomachinery Applications," AIAA Paper 89-2453 (also see AIAA Paper 89-1818).

Subramanian, S. V., and Bozzola, R., 1987, "Numerical Simulation of Three Dimensional Flow Fields in Turbomachinery Blade Rows Using the Compressible Navier–Stokes Equations," AIAA Paper 87-1314 (also see AIAA Papers 85-1332, 89-2453).

Sullivan, T. J., 1987, "Aerodynamic Performance of a Scale Model, Counter Rotating Unducted Fan," AGARD C.P.421, p. 22-9.

Suryavamshi, N., and Lakshminarayana, B., 1992, "Numerical Prediction of Wakes in Cascades and Compressor Rotors Including the Effects of Mixing—Part I and II," *J. Turbomachinery*, Vol. 114, pp. 607–626.

Suryavamshi, N., et al., 1994, "Experimental Investigation of the Unsteady Flow Field of an Embedded Stage in a Multistage Axial Flow Compressor," in *Proceedings of Unsteady Flows in Aeropropulsion: Recent Advances in Experimental and Computational Methods*, ASME.

Swift, W. L., et al. (ed.), 1983, "Performance Characteristics of Hydraulic Turbines and Pumps," ASME Fed Vol. 6.

Swihart, J. M., 1987, "U.S. Aeronautical R & D Goals, S.S.T. Bridge to the Next Century," *Proceedings of the International Symposium Air Breathing Engines (ISABE)*, Cincinnati, OH, AIAA.

Tabakoff, W., 1984, "A Review of Turbomachinery Performance Deterioration Exposed to Solid Particulate Environment," *J. Fluids Eng.*, Vol. 106, June Issue.

Tabakoff, W., and Hamed, A., 1975, "Theoretical and Experimental Study of Flow Through Turbine Cascade with Coolant Flow Injection," AIAA Paper No. 75-843.

Takagi, T., 1986, "Experiments and Performance Evaluation of a Transonic Axial Flow Turbine with a Variable Nozzle," ASME Paper 86-GT-214.

Tamura, A., and Lakshminarayana, B., 1975, "Three-Dimensional Flow Field Due to a Vortex and Source Spanning an Annulus," *Int. J. Eng. Sci.*, Vol. 13, pp. 549–561.

Tamura, A., and Lakshminarayana, B., 1976, "Assessment of Three-Dimensional Inviscid Effects Using Simple Models," *J. Fluids Eng.*, Vol. 98, pp. 163–172.

Tanaka, H., 1983, "A Survey of Gas Turbine Technology and Research Work in Japan," *Proceedings of the International Gas Turbine Conference*, Tokyo.

Taulbee, D. B., Tran, L., and Dunn, M. G., 1989, "Stagnation Point and Surface Heat Transfer for a Turbine Stage: Prediction and Comparison with Data," *J. Turbomachinery*, Vol. 111, No. 1, pp. 28–35.

Taylor, A. M. K., Whitelaw, J. H., and Yanneskis, M., 1981, "Measurement of Laminar and Turbulent Flow in a Curved Duct with Thin Inlet Boundary Layer," NASA CR 3367.

Tekriwal, P., 1994, "Heat Transfer Predictions in rotating Radial Smooth Channel," ASME Paper 94-GT-196.

Tennekes, H., and Lumley, J. L., 1972, *A First Course in Turbulence*, MIT Press, Cambridge, MA.

Thibaud, F., Drost, A., and Sottas, G., 1989, "Validation of Euler Code for Hydraulic Turbine," AGARD CP-437.

Thompkins, W. T., 1982, "A FORTRAN Program for Calculating Three-Dimensional, Inviscid, Rotational Flows with Shock Waves in Axial Compressor Blade Rows. I. User's Manual," NASA CR 3560.

Thompkins, W. T., and Usab, W. J., Jr., 1982, "A Quasi-Three Dimensional Blade Surface Boundary Layer Analysis of Rotating Blades," *J. Eng. Power*, Vol. 104, p. 439.

Thompson, J. F., 1984, "A Survey of Dynamically Adaptive Grids in the Numerical Solution of PDE," AIAA Paper 84-1606 (also in *App. Num. Math*, Vol. 1, 1985).

Thompson, J. F., Warsi, Z. U. A., and Mastin, C. W., 1985, *Numerical Grid Generation*, North-Holland, Amsterdam.

Thulin, R. D. et al., 1982, "Energy EfficientEngine; High Pressure Turbine Detailed Design Report," Pratt & Whitney, PWA-5594-171, NASA CR 165608.

Thwaites, B., 1960, *Incompressible Aerodynamics*, Oxford University Press, London.

Tillman, E. S., and Jen, H. F., 1984, "Cooling Studies at the Leading Edge of a Film Cooled Airfoil," *J. Eng. Power*, Vol. 106.

Traupel, W., 1945, "Die Berechnung der Strömung durch Schaufelgitter," *Sulzer Techn. Rundschau* Nr. 1, pp. 25–42.

Traupel, W., 1948, "Zur Potentialtheorie des Schaufelgitters," *Sulzer Techn. Rundschau* Nr. 2, pp. 12–30.

Tsakonas, S., 1979, "Propeller Blade Distribution Due to Loading and Thickness Effect," *J. Ship Res.*, Vol. 23, pp. 89–107.

Tsien, H. S., 1939, "Two-Dimensional Subsonic Flow of Compressible Fluids," *J. Aerospace Sci.*, Vol. 6, No. 10.

Turner, A. B., 1971, "Local Heat Transfer Measurements on a Gas Turbine Blade," *J. Mech. Eng. Sci.*, Vol. 13, No. 1.

Turner, A. B., Tarada, F. H. A., and Bayley, F. J., 1985, "Effect of Surface Roughness on Heat Transfer to Gas Turbine Blades," AGARD CP 390.

Turton, R. K., 1984, *Principles of Turbomachinery*, Spon., London.

Tyner, T. M., and Sullivan, J. P., 1990, "Design and Performance of a Small High Speed Axial Compressor," AIAA Paper 90-1911.

Tyson, H. N., 1952, "Three-Dimensional Interference Effect of a Finite Numer of Blades in an Axial Turbomachine," California Institute of Technology Report E19.1.

Ubaldi, M., Zunino, P., and Cattanei, A., 1993, "Relative Flow and Turbulence Measurements Downstream of a Backward Swept Centrifugal Impeller," *J. Turbomachinery*, Vol. 115, pp. 543–551.

Ucer, A. S., Stow, P., and Hirsch, C. (ed.), 1985, *Thermodynamics and Fluid Mechanics of Turbomachinery*, Martinus Nijhoff Publishers, The Hague.

Ucer, A. S., and Shreeve, R. P., 1992, "A Viscous Axisymmetric Throughflow Prediction Model for Multistage Compressors," ASME Paper 92-GT-293.

Vanfossen, G. J., and Simoneau, R. J., 1985, "Preliminary Results of a Study of the Relationship Between Freestream Turbulence and Stagnation Region Heat Transfer," ASME Paper 85-GT-84.

Vanfossen, G. J., and Simoneau, R. J., 1994, "Stagnation Region Heat Transfer: The Influence of Turbulence Parameters, Reynolds Number and Body Shape, NASA TM 106504.

Vanka, S. P., 1986, "Block Implicit Multigrid Calculation of the Two Dimensional Recirculating Flow," *Computer Methods Appl. Mech. Eng.*, Vol. 59, pp. 29–48.

Vatsa, V. N., 1985, "A Three-Dimensional Boundary Layer Analysis Including Heat Transfer and Blade Rotation Effects," in *Proceedings of the Third Symposium on Numerical and Physical Aspects of Aerodynamic Flows* (Ed., T. Cebeci), Long Beach, CA.

Vavra, M. H., 1960, *Aerothermodynamics and Flow in Turbomachines*, John Wiley & Sons, New York.

Veuillot, J. P., and Cambier, L., 1986, "A Sub-domain Approach for the Computation of Inviscid Flows," ONERA TP No. 1984-86.

VKI Lecture Series, 1982, "Film Cooling and Turbine Blade Heat Transfer," VKI LS 82-02, Vols. 1 and 2.

VKI Lecture Series, 1985, "Tip Clearance Effects in Axial Turbomachines," VKI LS 1985, April 15–19 (repeated at Pennsylvania State University, April 14–18, 1986).

VKI Lecture Series, 1986, "Convective Heat Transfer and Film Cooling in Turbomachinery," LS 86-06.

Vogel, T., 1991, "Computation of 3D Viscous Flow and Heat Transfer Prediction for the Application of Film Cooled Gas Turbine Blades," AGARD CP 510, p. 7.1.

Von Der Nuell, W. T., 1964, "The Radial Turbine," in *Aerodynamics of Turbines and Compressors*, (W. R. Hawthorne, ed.), Princeton University Press, Princeton, NJ.

Von Doormal, J. P. and Raithby, G. D., 1984, "Enhancement of the SIMPLE Method for Predicting Incompressible Fluid Flow," *Num. Heat Transfer*, Vol. 67, p. 147.

Von Karman, T., 1941, "Compressibility Effects in Aerodynamics," *J. Aeronaut. Sci.*, Vol. 8, No. 9.

Von Mises, R., 1958, *Mathematical Theory of Compressible Fluid Flow*, Academic Press, New York.

Vuillez, C., and Veuillot, J. P., 1990, "Quasi 3D Viscous Flow Computations in Subsonic and Transonic turbomachinery Blading," AIAA Paper 90-2126.

Wadia, A. R., 1983, "Numerical Calculations of Time Dependent Three-Dimensional Viscous Flow in a Blade Passage with Tip Clearance," AIAA Paper 83-1171.

Wadia, A. R., and Law, C. H., 1993, "Low Aspect Ratio Transonic Rotors Part 2: Influence of Location of Maximum Thickness on Transonic Compressor Performance," *J. Turbomachinery*, Vol. 115, p. 226 (also see Part 1, p. 218).

Wadia, A. R., and Nealy, D. A., 1985, "Development of a Design Model for Airfoil Leading Edge Film Cooling," ASME Paper 85-GT-120.

Wagner, J. H., Dring, R. P., and Joslyn, H. D., 1985, "Inlet Boundary Layer Effects in Axial Compressor Rotor Parts I & II," *J. Eng. for Gas Turbines and Power*, Vol. 107, pp. 374-386.

Wagner, J. H., et al., 1986, "Heat Transfer Experiments with Turbine Airfoil Internal Flow Passages," ASME Paper 86-GT-133 (also see ASME Paper 89-GT-272).

Wagner, J. H., et al., 1991, "Heat Transfer in Rotating Passages with Smooth Walls and Radial Outward Flow," *J. Turbomachinery*, Vol. 113, p. 321.

Wallis, R. A., 1983, *Axial Flow Fans and Ducts*, John Wiley & Sons, New York.

Wang, B., et al., 1986, "Transonic Flow Along Arbitrary Stream Filament of Revolution Resolved by Separate Computations with Shock Fitting," ASME Paper 86-GT-30.

Wang, J. H., et al., 1985a, "Airfoil Heat Transfer Calculation Using a Low Reynolds Number Version of a Two Equation Turbulence Model," ASME Paper 84-GT-261, *J. Eng. for Power*, Vol. 107.

Wang, K. C., 1971, "On the Determination of Zones of Influence and Dependence for Three Dimensional Boundary Layer Equations," *J. Fluid Mech.*, Vol. 48, Part 2, pp. 397-404.

Wang, Q., and Yu, H., 1988, "A Unified Solution Method for the Flow Calculations Along S_1 and S_2 Stream Surfaces Used for the Computer Aided Design of Centrifugal Compressor," ASME Paper 88-GT-237.

Wang, Q., Zhu, G., and Wu, C. H., 1985, "Quasi Three-Dimensional and Fully Three-Dimensional Rotational Flow Calculations in Turbomachines," *J. Eng. for Gas Turbines and Power*, Vol. 107, pp. 277-285.

Wang, Z., et al., 1992, "Experimental Studies on the Mechanism and Control of Secondary Losses in Turbine Cascades," *J. Thermal Sci.*, Vol. 1, No. 3, p. 149.

Warfield, M., and Lakshminarayana, B., 1987a, "Calculation of Three-Dimensional Locally Elliptic Flow with a Zonal Equation Method," *Proceedings of the 8th CFD Conference*, Honolulu, Hawaii, AIAA Paper 87-1141CP.

Warfield, M., and Lakshminarayana, B., 1987b, "Computation of Rotating Turbulent Flow with an Algebraic Reynolds Stress Model," *AIAA J.*, Vol. 25, No. 7, pp. 957-964.

Warfield, M., and Lakshminarayana, B., 1989, "Calculation of Three Dimensional Turbomachinery Rotor Flow Using Pseudo Compressibility and Zonal Techniques," *Z. Flugwiss, Weltraumforsch*, Vol. 13, p. 31.

Warming, R. F., and Beam, R. M., 1978, "On the Construction and Application of Implicit Factored Schemes for Conservation Laws," *SIAM-AMS Proc.*, Vol. 11, pp. 85–119.

Waterman, W. F., 1986, "Tip Clearannce Loss Correlations and Analysis," VKI/PSU Lecture Course on *Tip Clearance Effects in Axial Flow Turbomachines*, PSU, p. 14–18.

Weber, K. F., and Delaney, R. F., 1991, "Viscous Analysis of Three-Dimensional Turbomachinery Flows Using an Implicit Solver," ASME GT-91-205.

Weinig, F. S., 1935, *Die Strömung un die Schaufeln Von Turbomachinen*, J. A. Barth, Leipzig.

Weinig, F. S., 1957, "A New Approach to the Theory of Thin, Slightly Cambered Profiles," *J. Appl. Mech.*, Vol. 99, pp. 177–182.

Weinig, F. S., 1964, "Theory of Two Dimensional Flow Through Cascades," Chapter 1 in *Aerodynamics of Turbines and Compressors*, (W. R. Hawthorne, ed.), Princeton University Press, Princeton, NJ.

Wennerstrom, A., 1986, "Low Aspect Ratio Axial Flow Compressors, Why and What it Means," Society of Automotive Engineers, *Special Publication No.* 683.

White, F., 1979, *Fluid Mechanics*, McGraw-Hill, New York.

White, F., 1991, *Viscous Fluid Flow*, McGraw-Hill, New York.

Whitfield, A., 1990, "A Preliminary Design of Radial Turbines," *J. Turbomachinery*, Vol. 112, No. 1, p. 50.

Whitfield, A., and Baines, N. C., 1990, *Design of Radial Tubomachines*, Longman Group, New York.

Whitney, W. S., Schum, H. J., and Behering, F. P., 1972, "Cold Air Investigation of a Turbine for High Temperature Engine Application," NASA TND 6960.

Wiesner, F. J., 1967, "A Review of Slip Factors for Centrifugal Impellers," *J. Eng. Power*, Vol. 89, pp. 556–572.

Wiggins, J. O., 1986, "The 'Axi-Fuge,' A Novel Compressor," ASME Paper 86-GT-224.

Wilcox, D. C., 1993, *Turbulence Modeling for CFD*, DCW Industries, Inc.

Wilcox, D. C., and Rubesin, M. W., 1980, "Progress in Turbulence Modeling for Complex Flow Fields Including the Effect of Compressibility," NASA TP 1517.

Wilkinson, D. H., 1972, "Calculation of Blade-to-Blade Flow in a Turbomachine by Streamline Curvature," *British ARC R & M* 3704.

Wilson, D. G., 1991, *The Design of High EfficiencyTurbomachinery and Gas Turbines*, MIT Press, Cambridge, MA.

Wisler, D. C., 1985, "Loss Reduction in Axial Flow Compressors Through Low Speed Testing," *J. Eng. for Gas Turbines and Power*, Vol. 107, p. 354 (also see NASA CR 16554, December 1981).

Wisler, D. C., 1989, "Advanced Compressor and Fan Systems," Lecture Notes, General Electric Company (unpublished).

Wislicenus, G. F., 1965, *Fluid Mechanics of Turbomachinery*, Dover, New York.

Wislicenus, G. F., 1986, "Preliminary Design of Turbo Pumps and Related Machinery," NASA Reference Publication (RP) 1170.

Wolf, J., and Moskowitz, S., 1985, "Development of the Transpiration Air Cooled Turbine for High Temperature Dirty Gas Streams," ASME Paper 83-GT-84, 1984, *J. Eng. Power*, Vol. 107.

Wood, J. R., Adam, P. W., and Buggole, A. E., 1983, "NASA Low Speed Centrifugal Compressor for Fundamental Research," NASA TM 83398.

Worth, E., and Plehn, N., 1990, "Application of 3D Viscous Code in the Design of a High Performance Compressor," AIAA Paper 90-1914.

Wu, C. H., 1952, "A General Theory of Three-Dimensional Flow in Subsonic and Supersonic Turbomachine in Radial, Axial and Mixed Flow Types," NACA TN 2604.

Wu, C. H., and Brown, C. A., 1952, "A Theory of Direct and Inverse Problems of Compressible Flow Past Cascades of Arbitrary Airfoils," *J. Aero. Sci.*, Vol. 19, pp. 183–196 (also see NASA TN 2407 and 2455).

Wu, C. H., and Wang, B., 1984, "Matrix Solution of Compressible Flow on S_1 Surface Through a Turbomachine Blade Row with Splitter Vanes or Tandem Blades," *J. Eng. Power*, Vol. 106, (ASME Paper 83-GT-10).

Wu, C. H., and Wu, W., 1954, "Analysis of Tip-Clearance Flow in Turbomachines," Polytechnic Institute of Brooklyn, TR-1.

Wu, W., Wu, C. H., and Yu, D., 1985, "Transonic Cascade Flow Solved by Separate Supersonic and Subsonic Computations with Shock Fitting," *J. Eng. for Gas Turbines and Power*, Vol. 107, pp. 329–336.

Yakota, J. W., 1990, "Diagonally Inverted LU Factored Implicit Scheme for the Three Dimensional Navier–Stokes Equation," *AIAA J.*, Vol. 28, No. 9, p. 1642.

Yamaguchi, S., 1964, "On the Inlet-Flow Field for a Two-Dimensional Supersonic Cascade with Curved Entrance Regions," *Bull. JSME*, Vol. 7, No. 25, pp. 91–95.

Yamamoto, A., 1987, "Production and Development of Secondary Flow and Losses in Two Types of Turbine Cascades—Parts I & II," *J. Turbomachinery*, Vol. 109, pp. 186–200.

Yamamoto, A., 1989, "Endwall Flow/Loss Mechanisms in a Linear Cascade with Blade Tip Clearance," *J. Turbomachinery*, Vol. 111, p. 264.

Yamamoto, A., and Yamasi, R., 1985, "Production and Development of Secondary Flows and Losses Within a Three-Dimensional Turbine Cascade," ASME Paper 85-GT-217.

Yanenko, N. N., 1971, *The Method of Fractional Steps: The Solution of Problems of Mathematical Physics in Several Variables*, Springer-Verlag, New York.

Yaras, M. I., and Sjolander, S. A., 1990, "Development of the Tip Leakage Flow Downstream of a Planar Cascade of Turbine Blades," *J. Turbomachinery*, Vol. 112, p. 609.

Yeoh, C. C., and Young, J. B., 1982, "Non-Equilibrium Streamline Curvature Through Flow Calculations in Wet Steam Turbines," *J. Eng. Power*, Vol. 104, p. 489.

Yokoyama, E., 1961, "Comparative Study of Tip Clearance Effects in Compressors and Turbines," MIT, Gas Turbine Laboratory, Report No. 63.

York, R. E., and Woodard, H. S., 1976, "Supersonic Compressor Cascades—An Analysis Of The Entrance Region Flow Field Containing Detached Shock Waves," *J. Eng. Power*, Vol. 98, pp. 247–257.

Yoshida, T., et al., 1982, "Analysis and Test of Coolant Flow and Heat Transfer of an Advanced Full Coverage Film Cooled Laminated Turbine Vane," ASME Paper 82-GT-131.

Zaccaria, M., and Lakshminarayana, B., 1995a, "Investigation of Three Dimensional Flow Field at the Exit of a Turbine Nozzle," *J. Propulsion and Power*, Vol. 11, p. 55.

Zaccaria, M., and Lakshminarayana, B., 1995b, "Unsteady Flow Due to Nozzle Wake Interaction with the Rotor in an Axial Flow Turbine," Part 1: Rotor Passage Flow Field; Part 2: Rotor Exit Flow-Field; ASME Papers 95-GT-295, -296.

Zannetti, L., and Larocca, F., 1990, "Inverse Design Methods for Internal Flows," Von Karman Institute LS 90-8.

Zhang, J., and Lakshminarayana, B., 1990, "Computation and Turbulence Modeling for Three-Dimensional Turbulent Boundary Layers Including Turbomachinery Rotor Blades," *AIAA J.*, Vol. 28, No. 11, pp. 1861–1969.

Zhao, X., 1986, "Streamfunction Solution of Transonic Flow Along S_2 Streamsurface of Axial Turbomachines," *J. Eng. for Gas Turbines and Power*, Vol. 108.

Zhu, X., and Wang, Z., 1987, "A Discussion of the Mean S_2 Stream Surfaces Applied to Quasi-Three Dimensional Calculation Programs of Turbomachinery," ASME Paper 87-GT-150.

Zierke, W C., Straka, W. A., and Taylor, P. D., 1993, "The High Reynolds Number Flow Through an Axial Flow Pump," PSU ARL Report 93-12 (also see ASME Papers 94-GT-453 and 94-GT-454).

Zimmerman, H., 1990, "Calculation of Two- and Three-Dimensional Flow in a Transonic Turbine Cascade with Particular Regard to the Losses," AIAA Paper 90-1542.

Zukauskas, A., 1972, "Heat Transfer from Tubes in Cross Flow," *Adv. Heat Transfer*, Vol. 8, pp. 93–160.

Zwifel, O., 1945, "The Spacing of Turbomachinery Blading, Especially with Large Angular Deflection," *Brown Boveri Rev.*, Vol. 32, p. 12.

Index